PRINCIPLES OF MASS SPECTROMETRY APPLIED TO BIOMOLECULES

WILEY-INTERSCIENCE SERIES IN MASS SPECTROMETRY

Series Editors:

Dominic M. Desiderio
Departments of Neurology and Biochemistry
University of Tennessee Health Science Center

Nico M. M. Nibbering
Vrije Universiteit Amsterdam, The Netherlands

John R. de Laeter • *Applications of Inorganic Mass Spectrometry*

Michael Kinter and Nicholas E. Sherman • *Protein Sequencing and Identification Using Tandem Mass Spectrometry*

Chhabil Dass, *Principles and Practice of Biological Mass Spectrometry*

Mike S. Lee • *LC/MS Applications in Drug Development*

Jerzy Silberring and Rolf Eckman • *Mass Spectrometry and Hyphenated Techniques in Neuropeptide Research*

J. Wayne Rabalais • *Principles and Applications of Ion Scattering Spectrometry: Surface Chemical and Structural Analysis*

Mahmoud Hamdan and Pier Giorgio Righetti • *Proteomics Today: Protein Assessment and Biomarkers Using Mass Spectrometry, 2D Electrophoresis, and Microarray Technology*

Igor A. Kaltashov and Stephen J. Eyles • *Mass Spectrometry in Biophysics: Conformation and Dynamics of Biomolecules*

Isabella Dalle-Donne, Andrea Scaloni, and D. Allan Butterfield • *Redox Proteomics: From Protein Modifications to Cellular Dysfunction and Disease*

Julia Laskin and Chava Lifshitz • *Principles of Mass Spectrometry Applied to Biomolecules*

PRINCIPLES OF MASS SPECTROMETRY APPLIED TO BIOMOLECULES

Edited by

JULIA LASKIN, PhD
Pacific Northwest National Laboratory
Richland, Washington

CHAVA LIFSHITZ, PhD
The Hebrew University
Jerusalem, Israel

WILEY-INTERSCIENCE
A JOHN WILEY & SONS, INC., PUBLICATION

Published by John Wiley & Sons, Inc., Hoboken, New Jersey
Published simultaneously in Canada

Library of Congress Cataloging-in-Publication Data:

Principles of mass spectrometry applied to biomolecules/edited by Julia Laskin, Chava Lifshitz.
p. cm.
Includes bibliographical references and index.

ISBN-13 978-0-471-72184-0 (cloth)
ISBN-10 0-471-72184-0 (cloth)

1. Mass spectrometry. 2. Biomolecules–Analysis. I. Laskin, Julia, 1967-II. Lifshitz, Chava.
QP519.9.M3P77 2006
543′.65–dc22 2006043900

10 9 8 7 6 5 4 3 2 1

This book is dedicated to the memory of Chava Lifshitz—one of the pioneers of the field of gas-phase ion chemistry and fundamental mass spectrometry—a great scientist, an excellent mentor, and a good friend.

—Julia Laskin

CONTENTS

CONTRIBUTORS

Richard L. Beardsley, Department of Chemistry, Box 210041, University of Arizona, 1306 East University Avenue, Tucson, AZ 85721-0041

Kathrin Breuker, Institute of Organic Chemistry and Center for Molecular Biosciences Innsbruck (CMBI), University of Innsbruck, Innrain 52a, A-6020 Innsbruck, Austria

Guilong Cheng, Department of Chemistry, Box 210041, University of Arizona, 1306 East University Avenue, Tucson, AZ 85721-0041

R. Graham Cooks, Department of Chemistry, Purdue University, 560 Oval Drive, West Lafayette, IN 47907-2038

Robert C. Dunbar, Chemistry Department, Case Western Reserve University, Cleveland, OH 44106

You-Jun-Fu, Department of Physics, Washington State University, 2710 University Drive, Richland, WA 99352; W. R. Wiley Environmental Molecular Sciences Laboratory and Chemical Sciences Division, Pacific Northwest National Laboratory, MS K8-88, P.O. Box 999, Richland, WA 99352

R. Benny Gerber, Department of Chemistry, University of California, Irvine, CA 92697; Department of Physical Chemistry and the Fritz Haber Research Center, The Hebrew University, Jerusalem 91904, Israel

Markus Gerhards, Heinrich-Heine Universität Düsseldorf, Institut für Physikalische Chemie I, Universitätstrasse 26.33.O2, 40225 Düsseldorf, Germany

Bogdan Gologan, Department of Chemistry, Purdue University, 560 Oval Drive, West Lafayette, IN 47907-2038

M. Kirk Green, McMaster Regional Centre for Mass Spectrometry, Department of Chemistry, McMaster University, Hamilton, Canada

William L. Hase, Department of Chemistry and Biochemistry, Texas Tech University, Lubbock, TX 79409-1061

Kristin A. Herrmann, Department of Chemistry, Box 210041, University of Arizona, 1306 East University Avenue, Tucson, AZ 85721-0041

Amy E. Hilderbrand, Department of Chemistry, Box 210041, University of Arizona, 1306 East University Avenue, Tucson, AZ 85721-0041

Alan C. Hopkinson, Centre for Research in Mass Spectrometry and the Department of Chemistry, York University, 4700 Keele Street, Toronto, Ontario, Canada M3J 1P3

Julia Laskin, Fundamental Sciences Division, Pacific Northwest National Laboratory, P.O. Box 999 K8-88, Richland, WA 99352

Carlito B. Lebrilla, Department of Chemistry, University of California, Davis, CA 95616

Chava Lifshitz, Department of Physical Chemistry and The Farkas Center for Light Induced Processes, The Hebrew University of Jerusalem, Jerusalem 91904, Israel

Scott A. McLuckey, Department of Chemistry, Purdue University, 560 Oval Drive, West Lafayette, IN 47907-2084

Samy O. Meroueh, Department of Chemistry and Biochemistry, University of Notre Dame, Notre Dame, IN 46556-5670

Asif Rahaman, Department of Chemistry and Biochemistry, Texas Tech University, Lubbock, TX 79409-1061

Carol V. Robinson, The University Chemical Laboratory, University of Cambridge, Lensfield Road, Cambridge CB2 1EW, United Kingdom

Dorit Shemesh, Department of Physical Chemistry and The Fritz Haber Research Center, The Hebrew University, Jerusalem 91904, Israel

K. W. Michael Siu, Centre for Research in Mass Spectrometry and the Department of Chemistry, York University, 4700 Keele Street, Toronto, Ontario, Canada M3J 1P3

Frank Sobott, Structural Genomics Consortium, University of Oxford, Botnar Research Centre, Oxford OX3 7LD, United Kingdom

Kihyung Song, Department of Chemistry, Korea National University of Education, Chongwon, Chungbuk 363-791, Korea

Jiangping Wang, Department of Chemistry, Wayne State University, Detroit, MI 48202

Lai-Sheng Wang, Department of Physics, Washington State University, 2710 University Drive, Richland, WA 99352; W. R. Wiley Environmental Molecular Sciences Laboratory and Chemical Sciences Division, Pacific Northwest National Laboratory, MS K8-88, P.O. Box 999, Richland, WA 99352

Ping Wang, Department of Chemistry, The University of Akron, Akron, OH 44325

Xue-Bin Wang, Department of Physics, Washington State University, 2710 University Drive, Richland, WA 99352; W. R. Wiley Environmental Molecular Sciences Laboratory and Chemical Sciences Division, Pacific Northwest National Laboratory, MS K8-88, P.O. Box 999, Richland, WA 99352

Chrys Wesdemiotis, Department of Chemistry, The University of Akron, Akron, OH 44325

Justin M. Wiseman, Department of Chemistry, Purdue University, 560 Oval Drive, West Lafayette, IN 47907-2038

Vicki H. Wysocki, Department of Chemistry, Box 210041, University of Arizona, 1306 East University Avenue, Tucson, AZ 85721-0041

Xin Yang, Department of Physics, Washington State University, 2710 University Drive, Richland, WA 99352; W. R. Wiley Environmental Molecular Sciences Laboratory and Chemical Sciences Division, Pacific Northwest National Laboratory, MS K8-88, P.O. Box 999, Richland, WA 99352

Qingfen Zhang, Department of Chemistry, Box 210041, University of Arizona, 1306 East University Avenue, Tucson, AZ 85721-0041

Roman Zubarev, Laboratory for Biological and Medical Mass Spectrometry Uppsala University, Box 583, Uppsala S-751 23, Sweden

PREFACE

The introduction of biological molecules into the gas phase by matrix-assisted laser desorption/ionization (MALDI) and electrospray ionization (ESI) has led to a revolution in biological mass spectrometry. The analytical aspects are a success story. Molecular weights can be determined with a high precision, peptide sequencing is now done with great success, and even higher-order structures of peptides and proteins can be accessed using mass spectrometry. Exceptionally high sensitivity, high mass resolution, and inherent speed are the key factors that positioned mass spectrometry at the forefront of analytical techniques for identification and characterization of biomolecules.

This success is based largely on the principles of mass spectrometry that have been developed since the mid-1970s for small organic molecules. However, studies of biomolecules in the gas phase have also revealed a number of challenges associated with the flexibility and the size of these species. For example, it was difficult to achieve efficient fragmentation of large molecules using traditional mass spectrometric approaches. Understanding of fundamental limitations of the existing ion activation techniques resulted in development of novel analytical approaches for studying fragmentation of large molecules in the gas phase. Improved identification of biomolecules in real-world applications is facilitated by understanding of their fragmentation mechanisms and the effect of the primary and the secondary structure on the observed fragmentation patterns.

Because of the large size, conformational flexibility, and the ability of biomolecules to hold multiple charges, studies of biomolecular gas-phase ion chemistry have opened a number of new and exciting areas of research. Multiply charged biomolecules are excellent targets for studying ion–ion chemistry and processes following capture of low-energy electrons. Various approaches are being developed to gain phenomenological understanding of the formation and fragmentation of hydrogen-rich radical cations, molecular radical cations, and radical anions of peptides and proteins. Development of new approaches for studying thermochemistry of gas-phase biomolecules and their dissociation energetics is at the forefront of the field. Vibrational spectroscopy of biomolecular ions is another area of research that is currently undergoing an explosive growth. In parallel, new high-resolution spectroscopic techniques have been successfully applied to larger systems, providing feedback to mass spectrometric studies. Reactivity of mass-selected biomolecules with solid targets has a potential for preparation of novel surfaces relevant for a variety of applications in biology and biotechnology.

In addition, there are several basic aspects related to the physics of the various problems that have remained unanswered. For example, the question of ergodicity and/or statistical versus nonstatistical behavior in the breakup of biomolecules has been raised in connection with several methods, including electron capture dissociation (ECD) or photodissociation. The old questions that were raised many years ago concerning organic molecules are again at the forefront—do gas phase biomolecules undergo intramolecular vibrational redistribution (IVR) prior to dissociation? Are all vibrational modes involved in IVR? Is there site selectivity and charge-directed reactivity? The mere fact that a large protein fragments on the short timescale of mass spectrometry, which is an absolute necessity in terms of analysis and sequencing, is somewhat surprising in view of our previous knowledge of dissociation of relatively small organic molecules in the gas phase and its description using statistical theories [Rice–Ramsperger–Kassel–Marcus/quasiequilibrium theory (RRKM/QET) and the like].

This book is a collection of reviews on fundamental aspects underlying mass spectrometry of biomolecules. The various selected topics have been arranged in three parts: (1) structures and dynamics of gas-phase biomolecules; (2) activation, dissociation, and reactivity; and (3) thermochemistry and energetics.

Fundamental mass spectrometry has always been strongly linked to a variety of gas-phase spectroscopic techniques, which provide unique insights on the structure and dynamics of ions and molecules in the gas phase. High-resolution UV and IR spectroscopy discussed in Chapter 1 allows study of the structure and dynamics of individual conformers of neutral biomolecules, exploring the effect of the solvent on the intrinsic properties of these molecules, and molecular recognition by examining the behavior of gas-phase clusters of biomolecules. Chapter 2 gives an example of high-resolution photodetachment phoelectron spectroscopy studies of electron transfer in iron–sulfur (Fe–S) clusters. In particular, this technique is used to explore the effect of solvents and protein environment on the electronic properties of the cubane-type [4Fe–4S] cluster—the most common agent for electron transfer and storage in metalloproteins.

Ion–molecule reactions and H/D (hydrogen/deuterium) exchange studies have traditionally been used in mass spectrometry for structure determinations. Chapter 3 gives an overview of the application of these techniques to studies of structures and conformations of gas-phase biomolecules. While spectroscopic techniques are currently limited to relatively small systems, mass spectrometry has been used to investigate quaternary structures of large protein complexes. Experimental approaches utilized in such studies are summarized in Chapter 4. Protein structures and folding in the gas phase is discussed in Chapter 5. Understanding protein dynamics in the absence of solvent—the driving force and the timescale of protein folding in the gas phase—is important for separating the effect of solvent from the effect of the intrinsic properties of proteins on their dynamics in solution.

The dynamics of the intramolecular vibrational energy redistribution (IVR) in gas-phase biomolecules is discussed in Chapters 6 and 7. Classical trajectory simulations using semiempirical PM3 potential energy surfaces described in Chapter 6 are instrumental for understanding ultra fast dynamics following photoionization of biomolecules and the validity of statistical theories of dissociation of these large floppy molecules. Studies of gas-phase ion chemistry of peptides and proteins revealed a variety of very interesting phenomena, some of which (e.g., electron capture dissociation and photodissociation) were described as nonergodic processes that circumvent IVR. The pros and cons of IVR and ergodic behavior in biomolecules based on the available experimental findings are discussed in Chapter 7.

Gas-phase fragmentation of protonated peptides is an important prerequisite for peptide and protein identification using tandem mass spectrometry (MS/MS). Understanding mechanistic aspects of peptide fragmentation as a function of peptide sequence and conformation summarized in Chapter 8 plays a central role in the interpretation of MS/MS spectra and refining strategies for database searching. Most mass spectrometric studies utilize closed-shell biomolecules (protonated or cationized on metals) generated using soft ionization techniques. Formation and dissociation of peptide radical cations described in Chapter 9 is a new rapidly growing field in gas-phase ion chemistry of biomolecules. These ions are formed by gas-phase fragmentation of complexes of the corresponding neutral peptide with transition metals and various organic ligands.

Collisional activation and multiphoton excitation are conventionally used for identification of biomolecules in a variety of mass spectrometric applications. Current status of multiphoton excitation, spectroscopy, and photodissociation of gas-phase biomolecules is summarized in Chapter 10. Chapter 11 presents classical trajectory simulations of the energy transfer in collisions of ions with atomic neutrals and surfaces. The phenomena observed following ion–surface collisions and the instrumentation involved in such studies are presented in Chapter 12 with particular emphasis on soft landing of biological molecules on a variety of surfaces. Soft landing can be utilized for a very specific modification of surfaces using a beam of mass-selected ions of any size and composition or for separating and preparing biomolecules on substrates in pure form for subsequent analysis.

Another method of ion activation in biological mass spectrometry relies on capture of low-energy electrons by multiply charged ions. Electron capture dissociation (ECD), discussed in Chapter 13, opens up a variety of unique dissociation pathways and provides information on the structure of the ion that is complementary to collisional or multiphoton excitation. Chapter 14 presents the fundamental principles of ion–ion chemistry of biomolecules. Ion–ion reactions provide a means of manipulating charge states of multiply charged peptides and proteins. Charge reduction by reactions of multiply charged biomolecules with singly charged ions of opposite polarity has developed as a powerful tool for structural elucidation of peptides and proteins.

Mass spectrometry has been widely utilized for thermochemical determinations. However, studying thermochemistry and dissociation energetics of peptides and proteins is challenging because most of the well-developed experimental approaches that have been successfully employed in the studies of small and medium-size ions are simply not applicable to the fragmentation of large molecules. Chapter 15 presents an overview of mass spectrometric approaches that have been utilized for thermochemical determinations of biomolecules and discusses the current status and limitations of these techniques, focusing on determination of proton affinities and alkali metal affinities of biomolecules. Chapter 16 describes the experimental approaches developed for studying the energetics and entropy effects in peptide and protein dissociation reactions.

Finally, we would like to acknowledge the authors of the chapters, who have invested a considerable amount of time and effort and prepared high-quality reviews for this book. Special thanks go to Jean Futrell for his generous help on various stages of this project and insightful feedback on the contents of several chapters. We are also thankful to many other colleagues who provided their comments and suggestions on the contents of this book.

Julia Laskin and Chava Lifshitz

PART I

STRUCTURES AND DYNAMICS OF GAS-PHASE BIOMOLECULES

1

SPECTROSCOPY OF NEUTRAL PEPTIDES IN THE GAS PHASE: STRUCTURE, REACTIVITY, MICROSOLVATION, MOLECULAR RECOGNITION

MARKUS GERHARDS

Heinrich-Heine Universität Düsseldorf
Institut für Physikalische Chemie I
Düsseldorf, Germany

Principles of Mass Spectrometry Applied to Biomolecules, edited by Julia Laskin and Chava Lifshitz

1.1. INTRODUCTION AND HISTORICAL BACKGROUND

As reported in previous chapters of the book, it has been a great challenge to transfer large molecules in the gas phase without dissociation. The investigations focus on a pure mass spectrometric analysis, but no spectroscopic information on the analyzed species is available. To obtain more information on the energy of different electronic states as well as the structure and dynamical changes of the investigated isolated species, the pure mass spectrometry has to be combined with different spectroscopic techniques. The motivation is strongly triggered by the following questions:

(1) What are the driving forces for protein folding or aggregation of peptides?
(2) How does solvation change the secondary structure of peptides, and how can this process be influenced, i.e. in our investigations can we perform experiments on mass-selected peptides and can we add, for instance, one water molecule after the other in order to determine how the structures will change?

By answering these questions on a molecular level, we may contribute to explanations of how structures and dynamics of peptides can be understood or predicted. The main focus of this chapter is a review on the most important combined spectroscopic and mass spectrometric analyses. This chapter focuses only on neutral amino acids and peptides; the spectroscopic investigation of ionic species is another rapidly growing field and will not be discussed here.

As mentioned in other chapters, large charged molecules can be transferred into the gas phase by applying MALDI (Karas and Hillenkamp 1988), ESI (Fenn et al. 1989), or LILBID (laser-induced liquid beam ion description) (Kleinekofort et al. 1996) and other sources. Neutral molecules can be transferred by heating sources, but in the case of pure amino acids or peptides, the molecules can easily fragment by elimination of CO_2. Different sources for transferring neutral species into the gas phase are discussed in this chapter. A major breakthrough was the introduction of laser desorption sources (see Section 1.2) in combination with supersonic cooling and laser ionization (of the neutral desorbed species). The combination of this pure mass spectrometry on selected neutral species (which are ionized for detection as cations) with spectroscopic techniques was triggered by the pioneering work of Levy and coworkers (Cable et al. 1987, 1988a,b; Rizzo et al. 1985, 1986b). Starting from the analysis of amino acids by a combination of laser desorption and fluorescence spectroscopy or resonant multiphoton ionization, the Levy group increases the size of the investigated species up to tripeptides (Cable et al. 1987, 1988a,b). The spectroscopic results yield information on the vibrations of the S_1 state, especially in the low-frequency region up to several hundred wavenumbers. The amide I or amide II region as well as NH stretching modes could not be investigated. Although the work of Levy's group lead to phantastic spectroscopic results, the main drawback was that spectra could not be clearly interpreted: (1) it could not be excluded that the spectra result from an overlay of different isomers, and (2) the computer power available in the late 1980s made it impossible to get any reliable prediction of vibrational spectra of different isomers. Furthermore, the structures of S_1 states can still not be predicted with an accuracy available for S_0 states. Even S_0 state calculations on relative energies and vibrations of tripeptides with hundreds of possible isomers are a challenge with respect to available computer resources. Additionally, anharmonicities must be accounted for, especially for low-frequency vibrations. All these problems made it nearly impossible to get a reliable interpretation of the spectra obtained by Levy's group. With the development of new spectroscopic techniques (the double-resonance methods), different isomers could experimentally be distinguished and vibrational frequencies in the amide I or II region as well as the NH stretching modes could be recorded. The relevant techniques will be described in Section 1.2. Additionally, the rapid increase of computer power since the 1990s has made it possible to get reliable predictions of the structure of isolated large molecules in the gas phase. As a consequence of the technical improvements, several investigations on amino acid model systems, amino acids and peptides began in the late 1990s and have become a rapidly growing field of scientific research, which will be described in detail in Sections 1.3–1.6.

Although this chapter focuses mainly on different laser spectroscopic methods, new developments within the field of microwave spectroscopy are also discussed, yielding very high resolution spectra and thus precise geometric information.

In the following sections different techniques are reviewed and then their applications starting with selected model systems up to the peptides are discussed. An outlook on remaining issues is given at the end of the chapter.

1.2. EXPERIMENTAL SETUPS AND METHODS

1.2.1. Laser Spectroscopic Methods and Microwave Spectroscopy

Several spectroscopic techniques have been developed in order to analyze the electronic ground and excited states of isolated biomolecules in the gas phase. Both rotational resolution and vibrational spectra yield information on the structure of the investigated systems. The fluorescence techniques offer some insight, especially in the "low frequency" region of the S_0 and S_1 states of the investigated amino acids and peptides (see Sections 1.4–1.6). In dispersed fluorescence (DF) spectroscopy (see Figure 1.1a) the excitation laser frequency is fixed and the fluorescence light is dispersed, yielding information on the S_0 state. In the case of the laser-induced fluorescence (LIF) technique, the excitation laser is scanned and the integral of the complete fluorescence light is detected (Figure 1.1b). This method gives information on the (low) vibrational frequencies of the electronically excited state. Another method used to obtain frequencies of the S_1 state is the resonant two-photon ionization (R2PI) method (Figure 1.1c). As in LIF, the excitation laser is scanned, but now a second photon from the same laser (one-color R2PI) or another UV laser (two-color R2PI) is used to ionize the investigated species. Ions are observed efficiently only when the first laser is in resonance with a vibrational (rotational) level of the S_1 state. If this vibronic level lives long enough with respect to the pulse duration of the laser, the absorption of a second laser photon is enhanced, leading to

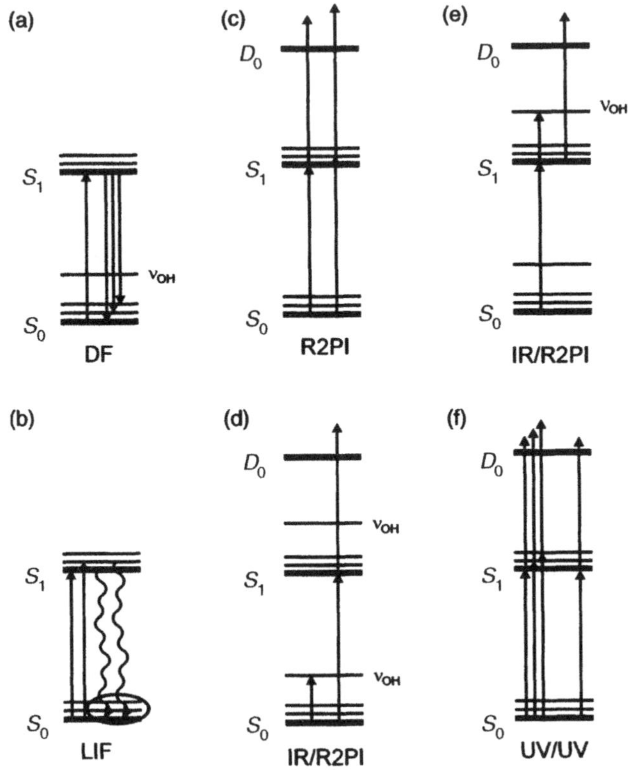

FIGURE 1.1. Different laser spectroscopic techniques applied to isolated amino acids or peptides: (a) dispersed fluorescence, analyzing vibrations in the S_0 state; (b) laser-induced fluorescence to analyze vibrations in the S_1 state; (c) resonant 2-photon ionization applied for investigating vibrations of the excited state; (d, e) infrared/R2PI technique to investigate IR spectra for the electronic ground and excited state, respectively; (f) UV/UV hole burning to analyze different isomers appearing in one R2PI spectrum (for further details, see text).

the production of ions. It should be mentioned that this chapter focuses mainly on investigations with nanosecond (ns) laser systems.

A major advantage of the R2PI method compared with the LIF technique is the possibility of combining spectroscopy with mass spectrometry; thus, ions produced by the R2PI process can be mass-selected and one thus obtains direct information of the investigated species (see Figure 1.2 and Section 1.2). In a molecular-beam experiment containing mainly the monomer, usually other species are present, such as clusters of the monomer or clusters with water, which can often not be removed completely from the gas lines. Thus R2PI spectroscopy is a mass- and isomer-selective method.

The very important window of the NH stretching modes as well as the amide I and II modes has been opened by the development of combined IR/UV techniques. The combination of IR spectroscopy with both fluorescence (IR/LIF) and the R2PI technique (IR/R2PI; see Figure 1.1d) are used. In both methods the UV laser photon is fixed to one electronic transition that belongs to a selected isomer (originating

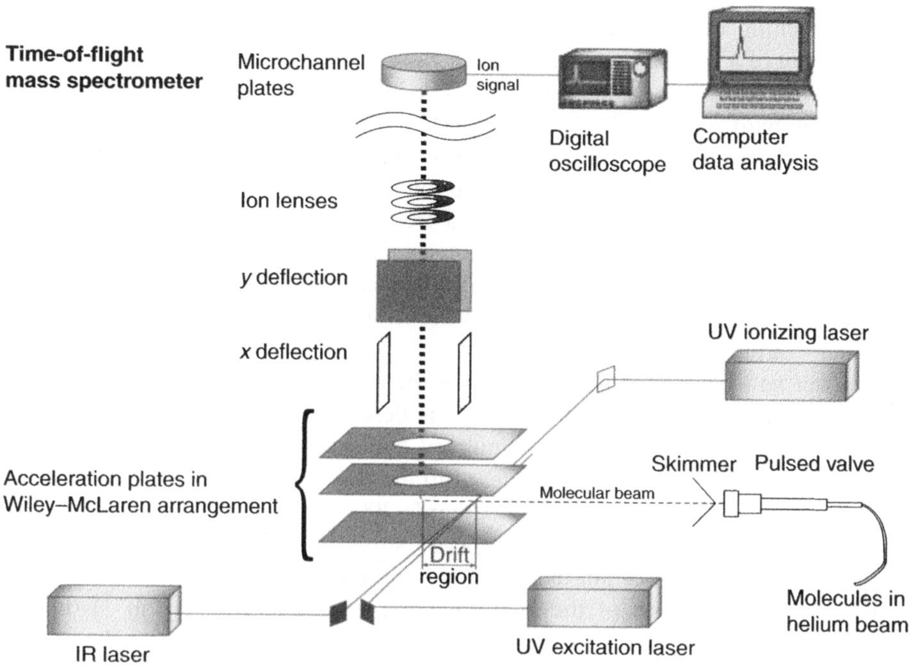

FIGURE 1.2. Experimental setup to analyze the investigated species by R2PI, IR/R2PI, or UV/UV hole-burning spectroscopy. The ions produced by two UV photons of a neutral species are mass-analyzed in a linear time-of-flight spectrometer. The peptides are introduced via a coexpansion with a rare gas (He or Ar) in a pulse valve. This valve can be heated, or a laser desorption (ablation) source can be located in front of the valve (not shown).

from its vibrational ground state in the S_0 state). By scanning an IR laser, different vibrational modes in the S_0 state can be excited. If the IR laser is in resonance with a vibrational level of the S_0 state, the vibrationless ground state is depopulated. By firing the UV laser after the IR laser, the efficiency of the UV excitation is reduced, since the number of molecules in the vibrationless ground state is reduced by the resonant IR excitation. A reduction in UV laser excitation efficiency leads to a decrease of the fluorescence signal (in the case of the IR/LIF technique) or to a decrease of the R2PI signal (in the case of the IR/R2PI method). Thus both methods indirectly yield an IR spectrum of a selected isomer in the S_0 state by recording the intensity of the LIF or R2PI signal as a function of the chosen IR wavelength. Like the R2PI technique, the IR/R2PI method is also mass- and isomer-selective. This method is also state-selective with respect to the different vibrational levels of the S_0 state. Historically, the first IR/R2PI spectrum was recorded by Page et al. (1988); a plethora of publications have followed in this field (see Sections 1.3–1.6), beginning with the investigations of Brutschy (Riehn et al. 1992), Mikami (Tanabe et al. 1993), and Zwier (1996) and their coworkers. In these publications also the abbreviations IR/UV double resonance, IR hole burning, and RIDIR spectroscopy are chosen instead of IR/R2PI. The authors' group published the first IR/R2PI spectrum in the

C=O stretching region (Gerhards et al. 2002). This became possible due to the development of a new laser system in the nanosecond regime that produces narrow-bandwidth IR light (better than 0.1 cm^{-1}) with high energy (~1 mJ now from 4.7 to 10 μm) (Gerhards et al. 2002; Gerhards 2004).

The region of C=O stretching vibrations is important since the amide I and amide II vibrations are significant for the description of peptide structures. Instead of the laser described earlier (Gerhards 2004), a free-electron laser (Oepts et al. 1994) is used in different applications on peptides (see Sections 1.3–1.5). This laser system is very powerful (typically ~50 mJ in one macropulse) and covers a region of ~40–2200 cm^{-1} but has the drawback of a relative low spectral resolution (~15 cm^{-1} around 6 μm).

Another new developed laser system for the region of ~6 μm (which would be suitable for the investigations of peptides) is described by Kuyanov et al. (2004). Here the IR light is produced by stimulated backward Raman scattering in solid *para*-hydrogen at 4 K pumped by a near-infrared OPO/OPA system. In contrast to the generation of IR light by DFM (Gerhards 2004), the bandwidth is larger (0.4 instead of 0.1 cm^{-1}) and the output energy strongly depends on the frequency, ranging from 1.7 mJ at 4.4 μm to 120 μJ at 8 μm (Kuyanov et al. 2004).

IR light in the region of the NH stretching vibrations (~3450 cm^{-1}) or OH stretching modes (~3650 cm^{-1}, important for the investigation of hydrated clusters) is usually generated by difference frequency mixing, an OPO/OPA process [see, e.g., Huisken et al. (1993)] or by a combination of DFM and OPA [e.g., see Unterberg et al. (2000)]. Details on the laser are partly given in the references on applications of IR/LIF and IR/R2PI spectroscopy.

Finally, it should be mentioned that the IR/R2PI (IR/LIF) method can also be used to determine vibrational transitions in the electronically excited state (Figure 1.1e). This method has been introduced by Ebata et al. (1996). By applying the IR/R2PI technique for the S_1 state, this state is excited by one UV photon and then the depopulation of the S_1 state via a subsequent IR excitation is detected by the decrease of the R2PI signal caused by the ionization with a second UV photon. (In the case of the IR/LIF method for the S_1 state, no second UV photon is necessary. Here the decrease of the fluorescence caused by the IR excitation is determined.)

A further double-resonance technique used for the analysis of peptides is the UV/UV hole-burning method (see Figure 1.1f) first applied by Lippert and Colson (1989) to the phenol(H_2O) cluster. In contrast to the IR/R2PI technique, not an IR laser but a UV laser is scanned, while a second UV laser (fired after the first scanning UV laser) is fixed to one wavelength, such as the electronic origin of one isomer. This method is used to determine whether different electronic transitions belong to the same isomer, i.e. when the first (scanning) UV laser is in resonance with an electronic transition that belongs to the isomer excited by the second laser, either the fluorescence or ion signal caused by the second UV laser decreases, since excitations of the first laser already removed parts of the molecules in the beam. In contrast to the IR/UV technique, both laser photons of the UV/UV method lead to a fluorescence or ion signal. Thus the signals resulting from first and second laser have to be separated. The lasers are usually fired within 100–300 ns so that laser with foci of ~1–2 mm can still spatially overlap. For instance: If both the first and second laser create ions by a two photon absorption, either (1) the resolution of the spectrometer must be good enough to separate the ions produced by the two lasers or (2) the ions are separated by the use of fast high-voltage switches that accelerate the ions produced by the first laser into the opposite direction.

The double-resonance techniques IR/R2PI and UV/UV hole burning lead to a selection of isomers and identification of the different species by their IR spectra that can be fully recorded in the range of all characteristic vibrational transitions. Another technique that provides additional information on the dynamical behavior of a flexible molecule is (infrared–population transfer spectroscopy (IR-PTS) (Dian et al. 2002b) as well as the hole-filling method, both introduced by Zwier and coworkers (Dian et al. 2002b, 2004b) (see Figure 1.3). By applying this method, the

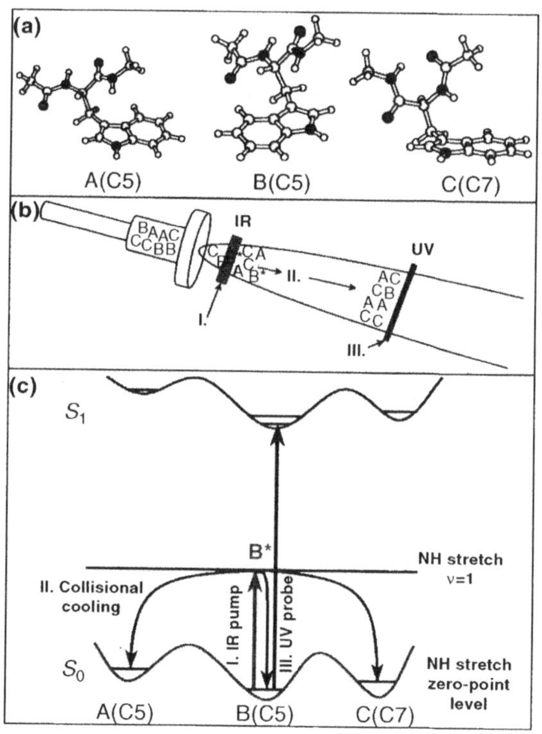

FIGURE 1.3. (a) Three isomers of NATMA (Ac–Trp–NHMe) (see Section 1.4 for further details); (b, c) schemes for IR population transfer (IR/PT) and hole-filling spectroscopy [21]. By exciting one isomer selectively with an IR photon, the two other isomers can be populated. After a further collisional cooling within the expanding beam, the cold molecules of B are analyzed via a UV probe laser. Because of a loss of population in B after IR excitation, the fluorescence quantum yield caused by the UV probe laser is reduced (see Figure 1.14). By scanning the UV laser, one can determine the distribution with respect to all isomers (hole-filling spectroscopy). [Figure taken from Dian et al. (2002b).]

molecules in a molecular beam are excited twice. First the molecules are excited directly behind the nozzle of a pulse valve, a region where collisional cooling still takes place. If different conformers of a peptide molecule are in the beam, IR excitation of one conformer (via, e.g. a NH stretching mode) may lead to formation of another conformer. By further expansion in the jet, the remaining collisional cooling can freeze out the new population induced by the IR excitation. The population of different species can be recorded via the LIF spectroscopy (or, in principle, also via the R2PI method). By comparing the LIF spectra obtained with or without the first IR excitation, the change of the population is determined. The difference between the IR-PTS and IR hole-filling methods is that in the former procedure, the IR laser is scanned and the UV laser is fixed to a resonant transition of one selected conformer; thus, the efficiency of forming a selected isomer with respect to the IR excitation is determined. In the hole-filling method the UV laser is scanned whereas the IR laser is fixed to one vibrational transition. Here the quantum yield of forming different isomers after one selected IR excitation is obtained. These new methods have been successfully applied to different species, including protected amino acids (Dian et al. 2002b, 2004a,b); see Section 1.4.

All techniques mentioned so far need an aromatic chromophore. The infrared/resonance energy transfer (IR/RET) method developed by Desfrancois and coworkers (Lucas et al. 2004) offers the possibility of recording IR spectra of species without such a chromophore. This technique is an extension of the RET method, where a neutral molecule (with a large total dipole moment of approximately >2D) collides with a Xe atom that has been excited to a high-lying Rydberg state. The collision of the excited Xe atom with the neutral species induces a resonant electronic energy transfer leading to an anionic species. The produced dipole-bound anions or even quadrupole-bound anions (DBAs or QBAs) have an excess electron in a very diffuse orbital and should retain the structure of the neutral parent. Since almost no internal energy is added, this very soft ionization process occurs without fragmentation even for weakly bound clusters. Because of the influence on the dipole moment of the formerly neutral molecule on ionization, this technique is mass- and structure-sensitive. If the investigated molecule is resonantly excited by an IR photon prior to the RET process, the stored energy can break weak intermolecular bonds (in clusters) or lead to an autodetachment of the DBA on ionization by RET. Thus the RET with IR excitation of the neutral species is lower compared to the RET obtained for a molecule in its vibrational ground state, resulting in a depletion of the anion signal. This method describes a very useful supplement to other techniques that require aromatic chromophores. With this method the spectra of the water dimer and the formamide water complex (Lucas et al. 2004) have been recorded and have shown very good agreement to earlier gas-phase and matrix investigations. In a more recent publication the same group examined formamide and its dimer in order to test the capability for monomers and strongly bound cluster with binding energies higher than the excess electron-binding energy (EBE) of the DBA (Lucas et al. 2005).

A classical method applied to determine structural parameter is microwave spectroscopy. A significant process with respect to efficiency and analysis of supersonically cooled molecules has been obtained by the development of molecular beam–Fourier transform microwave spectroscopy (MB-FTMW) (Balle and Flygare 1981; Legon 1983; Andresen et al. 1990; Harmony et al. 1995; Grabow et al. 1996; Storm et al. 1996; Suenram et al. 1999). Here the molecular beam is expanded in a Fabry–Perot cavity and polarized by a short microwave pulse along the cavity axis. The coherent radiation emitted by the molecules at their rotational transition frequencies is digitized and Fourier-transformed to obtain a spectrum in the frequency domain. The method becomes of high interest since it can also be coupled with laser ablation sources (Suenram et al. 1989; Lessari et al. 2003). Like IR/RET, microwave spectroscopy requires a dipole moment of the investigated species but no aromatic chromophore. It has the best resolution (in the kHz region) achievable, but no mass resolution can be obtained directly.

Another method, which is a kind of optical analogous to the MB-FTMW method, is rotational coherence spectroscopy (RCS). It is a time-dependent high-resolution method based on quantum beats that arise from the coherent excitation of different rotational levels within a vibronic state. The quantum beats consist of a superposition of rotational levels, and their frequencies are integer multiples of the rotational constants or a combination thereof. Pioneering experiments have been performed by Baskin, Felker, and Zewail (Baskin et al. 1986; Felker et al. 1986; Felker 1992). A linearly polarized picosecond pump pulse excites molecules that have a transition dipole moment parallel to the laser polarization at time zero. Due to the rotation of the molecule at different rotational velocities, the alignment is lost and recurs after characteristic time periods that are related to the rotational constants. These time periods and the correlating rotational constants can be probed by a second time-delayed pulse. In this way one obtains the rotational constants and therefore information on the structure. With respect to the topic of this chapter, the method has been used only for an amino acid model system (Connell et al. 1990), although it may also be applicable to larger systems [see, e.g., Weichert et al. (2001) and Riehn (2002)].

All methods described above use a molecular-beam technique. An alternative approach to obtain ultracold spectra is the use of helium droplets to investigate the vibrational spectra of amino acids. In studies by Vilesov, Toennies, and coworkers amino acids such as Trp, Tyr, (Lindinger et al. 1999; Toennies and Vilesov 2004) the protected amino acid Ac–Trp–NH_2 (NATA; see Section 1.4) and the amino acid model system tryptamine (see Section 1.3) (Lindinger et al. 1999; Toennies and Vilesov 2004) are embedded in a beam of superfluid ^4He. Two detection schemes for the resonant UV excitation are discussed: (1) laser-induced fluorescence (LIF) is recorded or (2) the depletion of the He droplet size is analyzed; i.e., the evaporation of He atoms after ionization of the embedded amino acid can be detected in a mass spectrometer as a function of the excitation wavelength of a UV laser. This technique in combination with IR excitation in the region of the OH stretching vibrations has also been used to analyze glycine (Huisken et al. 1999) and its dimer (Chocholousova et al. 2002). In the case of glycine all three conformers observed for this amino acid by microwave spectroscopy (see text below) have been identified by their vibrational transitions. For the glycine dimer it has been concluded in comparison with high-level ab initio calculations that a stacked structure is formed (Chocholousova et al. 2002). At the moment the extension of the He droplet techniques to larger peptides is still a part of future work. [For further techniques including IR/IR double-resonance methods, which have not yet been applied to amino acids, see, e.g., Douberly et al. (2005)]. It is interesting to note that isomers observed in He cluster beams can differ from those observed in supersonic jet experiments. This can result either from the very low temperatures (0.38 K) achieved in a He droplet or directly from He aggregation (see discussion below).

Finally it should be mentioned that for some amino acids it was possible to obtain FTIR spectra just by applying a rapid heating source (Linder et al. 2005); specifically, by applying this technique the decomposition could be delayed for seconds offering a small time window to detect the unfragmented acid.

1.2.2. Some Experimental Setups: Mass Spectrometry, Double-Resonance Spectroscopy, and Sources

A principal arrangement of an apparatus with pulsed jet expansion used to analyze (bio)molecules (by spectroscopy and mass spectrometry) in the gas phase is shown in Figure 1.2. This setup (which is also used in the author's group) contains a differentially pumped arrangement with at least two, usually three, chambers to transfer the investigated species from the source chamber to the mass spectrometer. The mass spectrometer is usually a time-of-flight (TOF) mass spectrometer. Two configurations are used: (1) a linear arrangement (LTOF; shown in Figure 1.2) or (2) a reflectron (RETOF) arrangement (Mamyrin et al. 1973; Kühlewind et al. 1984; Bergmann et al. 1989; Boesl et al. 1992). In a reflectron, ions are first decelerated and than focused on a detector in order to correct for different velocities of species with the same mass. LTOF and RETOF can be arranged in such a way that the mass spectrometer is oriented either perpendicular to or in line of the molecular-beam direction. In a perpendicular arrangement (see Figure 1.2) nearly no velocity component of the investigated species resulting from the molecular-beam expansion need to be taken into account.

In the case of a fluorescence spectrometer usually only one vacuum chamber is used containing the source and the optics to focus the fluorescence light on a detector. For resolution of the fluorescence spectrum, a monochromator is placed between the chamber and the detector; this can be either a photomultiplier or a charge-coupled device (CCD) camera. By using a CCD camera, a complete spectrum can be recorded for every laser shot.

For very high-resolution experiments (rotationally resolved electronic spectroscopy) a doubly skimmed continuous-beam apparatus is used that requires approximately three or four differentially pumped vacuum chambers. The fluorescence obtained after excitation with a very high resolution continuous-wave (CW) ring dye laser system is collected on a photomultiplier combined with a photon counter (Majewski and Meerts 1984; Pratt 1998; Schmitt et al. 2005).

The most important part of all apparatus to produce a sufficient amount of non dissociated amino acid and peptides is the source. The sources discussed in this chapter produce neutral species only. Mainly two different types are used to bring neutral amino acids and peptides into the gas phase. If protected (also called "modified" or "capped") amino acids are used (see text below), a special heating source can be used for larger peptides [see, e.g., Fricke et al. (2004) and Section 1.5]. In these sources the distance between a small stainless-steel chamber containing the substance and the pulse valve is as short as possible. Furthermore, the stainless-steel chamber and connectors can be protected by a glass surface, and the heating should be extremely homogenous. In this case temperatures below 200°C can be used to yield sufficient amounts of larger peptides in the gas phase. In another heating arrangement the substance is not in the high-pressure region behind the valve but is placed in front of the pulse valve (Snoek et al. 2000), i.e. the substance is connected with the vacuum of the source chamber. This method has also been used for fragile unprotected amino acids [see, e.g., Section 1.4.1 (Snoek et al. 2000)]. As a result of the surrounding vacuum, temperatures below 150°C can be used to bring enough molecules into the gas phase.

Similar to the heating sources, the laser desorption technique is now well established to transfer a sufficient amount of peptides into the gas phase. Comparison between thermal heating and laser desorption techniques reveals that the heating techniques are simpler and often lead to a more stable signal. With the use of laser depletion techniques (IR/UV of UV/UV), the non-background-free methods need a very stable baseline of ions produced by a UV laser set to a fixed frequency (see methods discussed above). Furthermore, the amount of substance needed is usually lower in the case of a heating source. Modern developments of laser desorption sources focus on this problem. If there is no way to transfer a peptide into the gas phase by heating without dissociation, the desorption is a very gentle method and the only alternative. Different groups described their desorption sources (Grotemeyer et al. 1986; Tembreull and Lubman 1986, 1987a,b; Cable et al. 1988a,b; Meijer et al. 1990; Cohen et al. 2000; Piuzzi et al. 2000; Hünig et al. 2003); in principle, the investigated substance is mixed with a matrix substrate. In most cases graphite (Meijer et al. 1990; Cohen et al. 2000; Piuzzi et al. 2000; Hünig et al. 2003) is used, and in modern sources it is pressed together with a (small) amount of the substance. In the pioneering spectroscopic work of Levy's group a dye has been used as matrix (Cable et al. 1988a,b). In a desorption process the matrix absorbs the energy of a short laser pulse (e.g., a 10 ns pulse of a Nd:Yag laser at 1064 nm with 1 mJ energy), thus, the matrix is heated up extremely fast and the embedded molecules of interest are evaporated into the gas phase. In a subsequent step the evaporated molecules are cooled in a molecular beam, generated by expanding a rare gas by a pulsed valve. The source is located directly in front of the nozzle. It can be a rotating rod or pellet coated by the matrix (or the pure substance) or only a fixed disk. The complete process of shock heating and cooling is so efficient that the peptides do not fragment. Some groups also describe desorption sources by using pure substances (without matrix) (Tembreull and Lubman 1986, 1987a,b; Grotemeyer et al. 1996). Although these sources may also be described as ablation sources (see text below), the main idea of all desorption arrangements for neutral species is the decoupling of the process of ionization from bringing a neutral peptide into the gas phase and cooling it. This general idea was the key to the ability to transfer large fragile and "cold" molecules with sufficient amounts into the gas phase, making it possible to apply different spectroscopic techniques to the neutral species described in Section 1.2.1.

Closely related to the desorption source is the laser ablation source (Bondybey and English 1981; Dietz et al. 1981) using a pure substance as target instead of a matrix including the substance of interest. After laser ablation small damages remain on the surface, i.e. substantial amounts of substance are removed. Most of the laser experiments presented in this chapter are related to experiments using repetition rates of 10 or 20 Hz; hence the desorption or ablation sources are especially constructed for "lower" repetition rates. Ablation sources are now also reported for higher rates in the kHz region (Smits et al. 2003). This is very useful since the fs (femtosecond) laser systems operate at kHz rates. Another interesting application is the combination of laser ablation sources with microwave spectroscopy (Suenram et al. 1989; Lessari et al. 2003) (see Section 1.2.1). Of course, the use of ablation methods usually requires a large amount of substance, which can limit the applications for larger peptides.

1.3. SPECTROSCOPY ON SELECTED AMINO ACID MODEL SYSTEMS

To calculate the potential energy surface (PES) of amino acids and peptides, the application of modern double-resonance techniques began with model systems. These systems, like *N*-phenylformamide (Manea et al. 1997; Dickinson et al. 1999; Fedorov and Cable 2000; Robertson 2000; Mons et al. 2001; Robertson and Simons 2001), 2-phenylacetamide (Robertson et al. 2001), and *N*-benzylformamide (Robertson et al. 2000) (see Figure 1.4), contain one H—N—C=O bond to get a model of an amide bond in a peptide. It has to be pointed out that these molecules give no realistic description of peptide binding since the connectivity in such molecules is not identical to the one observed for a real amino acid or peptide; i.e. in a peptide α-C atoms are located at each end of the amide group and pure NH_2 groups are located at the end of a peptide. Although the model character of the investigated amides was quite limited, the investigations give a first insight into the structural assignments of isolated amids. Furthermore, hydration shells have been extensively investigated on the basis of the model systems. The aromatic chromophore has been chosen in all model systems, since it allows application of the IR/R2PI technique.

In the case of *N*-phenylformamide a cis and trans conformer with respect to the H—N—C=O bond can be observed; the cis conformer is slightly more stable than the trans arrangement (Figure 1.4a) (Manea et al. 1997; Dickinson et al. 1999). For *N*-benzylformamide (Robertson 2000), the trans conformer becomes much more stable (Figure 1.4d), similar to the situation in a real peptide (see discussion below). The introduction of a CH_2 group leads to increased flexibility (benzyl instead of phenyl) and also allows a better description of an amino acid backbone. In the case of 2-phenylacetamide (Robertson et al. 2001) (see Figure 1.4f), the "sidechain" is

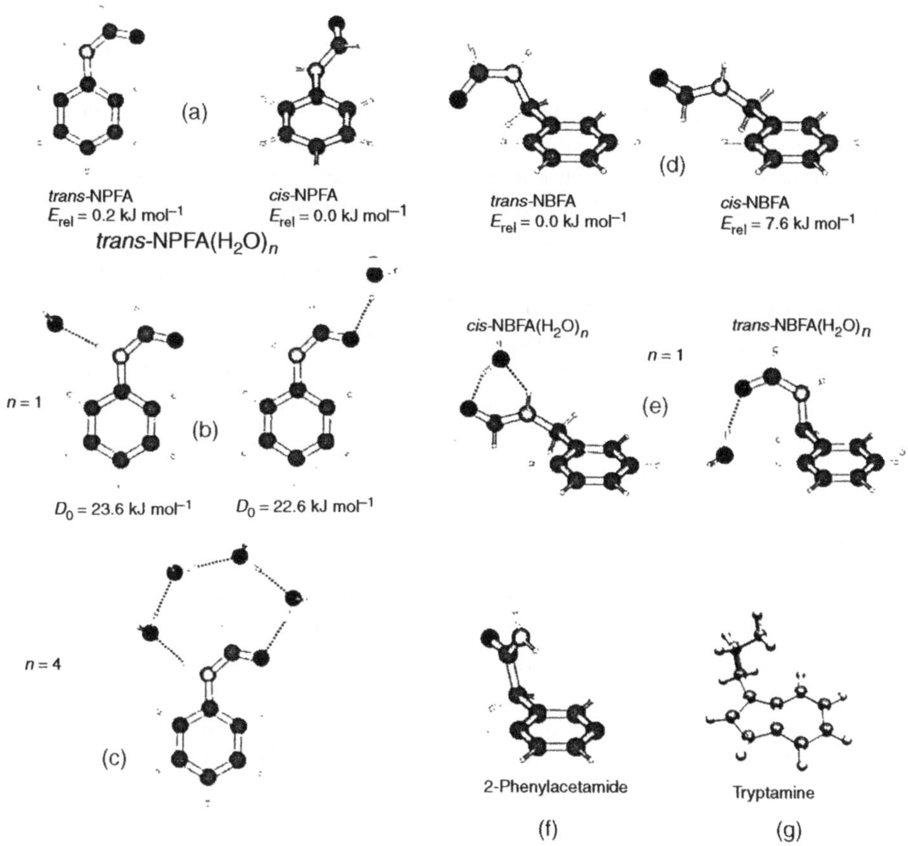

FIGURE 1.4. Different amino acid model systems: (a) *N*-phenylformamide and its cluster with (b) one and (c) three water molecules; (d) *N*-benzylformamide and its cluster with (e) one water molecule; (f) 2-phenylacetamide and (g) tryptamine. [Figure partly taken from Robertson and Simons (2001).]

attached to the carbonyl function, yielding a primary amide group that contains a NH_2 group.

The model systems selected have been good candidates for investigation of the stepwise hydration. In the case of *N*-phenylformamide the binding energy resulting from a proton donor bond of one water molecule to the C=O group or a proton acceptor bond to the N–H group are nearly identical (Mons et al. 2001) (Figure 1.4b). In contrast to the monomer the trans conformer of the *N*-phenylformamide moiety is more stable after hydration, and this conformation is conserved in all clusters with water (Dickinson et al. 1999; Fedorov and Cable 2000). By stimulated emission pumping, it was possible to investigate the shuttling of water from one binding site to the other (Clarkson et al. 2005a). In the case of the cluster containing two water molecules water dimers are connected to either the C=O or the N–H function

(Dickinson et al. 1999; Fedorov and Cable 2000). The isomer containing a N—H donor group is additionally stabilized by an O—H$\cdots\pi$ interaction. In the case of the cluster with three water molecules a cyclic arrangement between a donor bond (C=O$\cdots$H) and an acceptor (N—H$\cdots$O) bond is obtained (Figure 1.4c). This geometric arrangement resembles that of a proton wire structure (Tanner et al. 2003; Smedarchina et al. 2000), which may lead to a formation of zwitterionic structures in amino acids. In contrast to *N*-phenylformamide/water clusters, the cis conformation is preferred in the *N*-benzylformamide moiety of its hydrated clusters. Similar to the *N*-phenylformamide clusters, different isomers are obtained for mono- and dihydrated *N*-benzylformamide (see Figure 1.4e) (Robertson et al. 2000). An interesting structural element is the donor–acceptor arrangement, where one water molecule is hydrogen-bonded simultaneously to the N—H group and the C=O group. The experience obtained from analyzing these small model systems can be applied for investigation of hydrated amino acids and peptides (see discussion below).

Several other molecules can be considered as amino acid analogs [see, e.g., the early work by Martinez et al. (1991, 1993). Tryptamine (see Figure 1.4g) has been investigated by several groups and is selected for review here. This molecule contains the indole chromophore similar to the amino acid tryptophan (see Section 1.4).

The first molecular-beam experiments on neutral tryptamine in the gas phase were performed by Levy and coworkers (Park et al. 1986). By applying laser-induced fluorescence excitation (LIF) and resonant two-photon spectroscopy (R2PI) six bands in the electronic S_1–S_0 spectrum corresponding to six different conformers have been observed. The geometries of five of the six conformers could be determined by recording high-resolution fluorescence excitation spectra leading to the assignment of three gauge and two eclipsed structures concerning the position of the amino group with respect to the indole ring. It has also been shown that one transition in the electronic spectrum consists of two very close-lying bands that could (in the earlier study) not be discriminated by rotational analysis (Philips and Levy 1986, 1988). Furthermore, the lowest excited electronic state could be determined as ^1L_b state, as in tryptophan. Subsequent experiments with deuterated tryptamine confirmed the results and yielded a more precise determination of the position of the amino group (Wu and Levy 1989). Further structure analyses were done with the aid of rotational coherence (Connell et al. 1990) and microwave spectroscopy (Caminati 2004). From the microwave spectra in combination with ab initio calculations, Caminati and coworkers assigned two conformers (Caminati 2004). Zwier and coworkers investigated tryptamine with UV/UV hole burning and IR/R2PI spectroscopy and were able to determine in combination with density functional theory calculations seven structures including an interpretation of the two close-lying bands that could not be distinguished before (Carney and Zwier 2000). The same group also measured the energy barriers of the conformational isomerization of tryptamine by applying stimulated emission pumping–hole-filling spectroscopy (SEP-HFS) and stimulated emission pumping–population transfer spectroscopy (SEP-PTS) (Dian et al. 2004a; Clarkson et al. 2005b,c). (In these techniques either the UV dump laser of a SEP pump/dump process or the third laser that analyzes the conformers formed after the SEP process are scanned.)

Pratt and coworkers detected additional subbands in the rotationally resolved spectra of two bands and postulated a hindered motion that may account for the appearance of these subbands (Nguyen et al. 2005).

Investigations on microsolvated clusters of tryptamine have been performed to a much lesser amount than for the monomer. Park et al. (1986) observed that the number of possible conformers collapses to one conformer on cluster formation with methanol. Sipior and Sulkes (1988) yielded the same result for the clusters with methanol, ethanol and the water dimer by applying LIF spectroscopy. An interesting deviation of this behaviour has been observed for the cluster with dioxane (Peteanu and Levy 1988), which can be only a hydrogen acceptor. Clusters of tryptamine with up to three water molecules have been investigated by recording R2PI and IR/R2PI spectra (Carney et al. 2001). The first water molecule is attached to the lone pair of the amino group, and the clusters with two and three water molecules form a bridge to the indolic NH group. Connell et al. (1990) deduced a bridged monohydrated cluster using rotational coherence spectroscopy. This result has been confirmed by Schmitt et al. (2005), who recorded the rotationally resolved laser-induced fluorescence spectrum of six monomer conformers and the monohydrated cluster of tryptamine and observed, in combination with ab initio calculations, an additionally stabilizing interaction of water with one of the aromatic C—H bonds. This leads to a situation where the energy difference to all other possible clusters of tryptamine with water is much higher than between the different monomers.

The very similar melantonin (*N*-acetyl-5-methoxytryptamine) and its cluster with water has also been investigated via IR/UV spectroscopy (Florio et al. 2002; Florio and Zwier 2003). Five conformers for the monomer—three structures with trans and two with cis configuration of the amide group—four monohydrated, and two dihydrated conformers have been detected and assigned.

1.4. DOUBLE-RESONANCE AND MICROWAVE SPECTROSCOPY ON AMINO ACIDS

The three natural aromatic amino acids are phenylalanine (Phe), tyrosine (Tyr), and tryptophan (Trp) (see Figure 1.5). Because of their aromatic chromophores (benzene in Phe, phenol in Tyr, indole in Trp), they can be investigated by the mass- and isomer-selective double-resonance techniques (UV/UV and IR/UV). In this section applications particularly to the aromatic amino acids phenylalanine (Phe) and tryptophan (Trp) are summarized, as these compounds have been the most extensively investigated.

1.4.1. Phenylalanine

The first aromatic amino acid investigated with IR/UV double-resonance spectroscopy was Phe (Snoek et al. 2000). From LIF spectra and through saturation measurements performed earlier by Levy's group (Martinez et al. 1992), at least five isomers have been postulated. It is a very nice textbook example how this

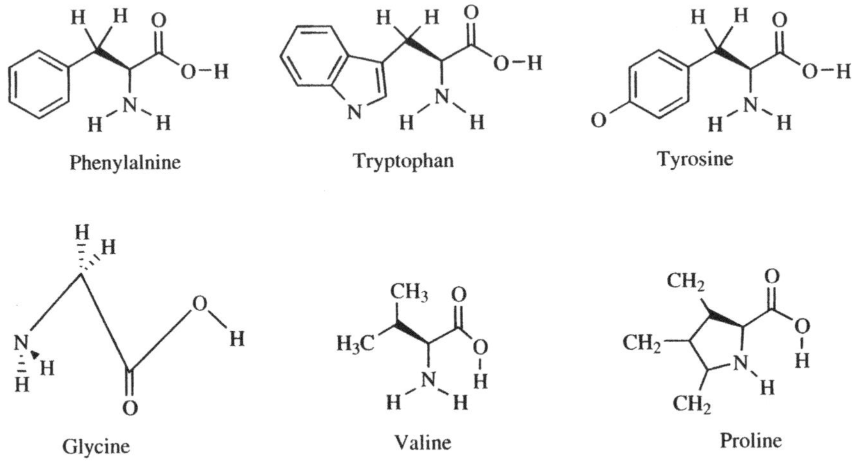

FIGURE 1.5. Six natural amino acids containing the three aromatic ones (Phe, Trp, Tyr). The structure shown for glycine is the most stable one in the gas phase.

assumption can be proved by UV/UV double resonance spectroscopy (see Figure 1.6b). Here six isomers have different $S_1 \leftarrow S_0$ excitation energies. For five isomers IR/R2PI spectra (see Figure 1.6c) can be recorded, giving a hint on the strucutral arrangements. The calculated structures of Phe isomers (see Figure 1.6a)

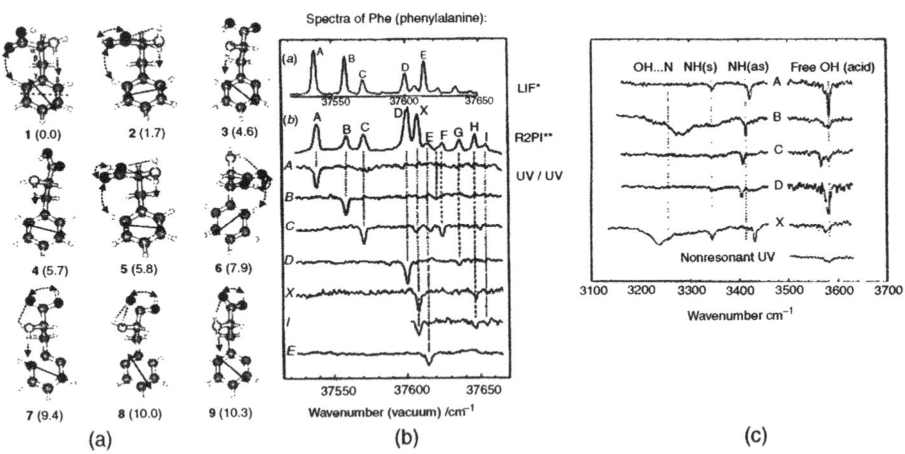

FIGURE 1.6. (a) Nine most stable structures of Phe calculated at the MP2 level [6-311 G(*d,p*) basis set] (Snoek et al. 2000). The positions of the $S_1 \leftarrow S_0$ transition moments as obtained from CIS calculations are also shown by arrows. (b) LIF (Martinez et al. 1992), R2PI and UV/UV hole-burning (Snoek et al. 2000) spectrum of Phe. The hole-burning spectrum clearly indicates that at least six different isomers (A,B,C,D,X,E) belong to the R2PI spectrum and (c) IR/R2PI spectra of isomers A,B,C,D,X. A correlation of the experimentally observed species to the nine possible structures is discussed in the text. [Parts (a) and (b) taken from Robertson and Simons (2001); part (c) taken from Snoek et al. (2000).]

are dominated by intramolecular hydrogen-bonds between the NH_2 and COOH groups. The variety of different isomers results partly from the fact that both functional groups can act as hydrogen donor and hydrogen acceptor . In this first study the six isomers have been correlated with the six most stable structures of Phe (see Figure 1.6).

In further studies Kim and coworkers investigated the isomers of Phe by analyzing their ionization potentials. To obtain these values, the second laser of the R2PI method (see Figure 1.1b) is scanned and the ion current is recorded as function of the frequency of the second ionizing laser (Lee et al. 2002a,b, 2003). No sharp onset of the ion current spectra is obtained, indicating a strong geometry change, but according to the onset positions, the spectra can be divided into two groups; four isomers have onsets of the ion currents at ~8.8 eV, two isomers have onsets above 9 eV. This difference should be a result of the relative positions between the phenyl ring and the "backbone" of Phe; i.e. the ionization potential depends on the interaction of the ionized phenol chromophore with a NH_2 group that can undergo a hydrogen bond to the phenyl ring. If no significant π interaction exists, the ionization potential is expected to be lower than in the case of a strong π interaction. The results obtained from these studies on ionization potentials fit very well with the one obtained by Snoek et al. (2000), except one isomer (E), which has to be reassigned.

In a subsequent study Simons and coworkers again investigated the six Phe isomers by performing a rotational band contour analysis of the R2PI spectra (Lee et al. 2004). It turns out that the assignment of four conformers (1 = X, 2 = D, 3 = B, 6 = C; see abbreviations in Figure 1.6) is in agreement with the assignment given on the basis of the IR/R2PI spectra (Snoek et al. 2000). A reassignment has been made with respect to isomer A, which is now correlated with structure 7, and finally isomer E has been reassigned to structure 9 (see Figure 1.6a). It is interesting to note that the most stable isomers are found experimentally and have also been clearly identified by different spectroscopic methods, but there are also missing isomers (especially 4 and 5) that are lower in energy than isomers 7 and 9.

Another aspect discussed by Lee et al. (2004) is the S_1 lifetime of the isomers obtained from pump-probe delayed ionization. In contrast to all isomers with S_1 state lifetimes at about 80 ns (isomers 2,3,6,7) or 120 ns (isomer 9), the main isomer, isomer 1, has a short lifetime of only 20 ns. This conformer is stabilized by a number of intramolecular hydrogen bonds (see Figure 1.6a) that allow for the possibility of efficient nonradiative decay pathways.

In another strategy not the pure amino acids but their derivatives, protected at the *N* and *C* termini, are investigated; specifically, the NH_2 group is acetylated and the OH group of the carboxyl rest is replaced by either an ester or an amide function.

While the investigation of pure Phe yields an excellent overview on the PES resulting from the interaction of acid and base end groups, analysis of protected amino acids (and peptides) now offers good models for the description of an inner part of a large peptide molecule, since the interactions can be formed only between functional groups of the "backbone."

The structure of a peptide is characterized by a Ramachandran plot (Ramachandran et al. 1966) containing the angles ϕ and ψ of the backbone of

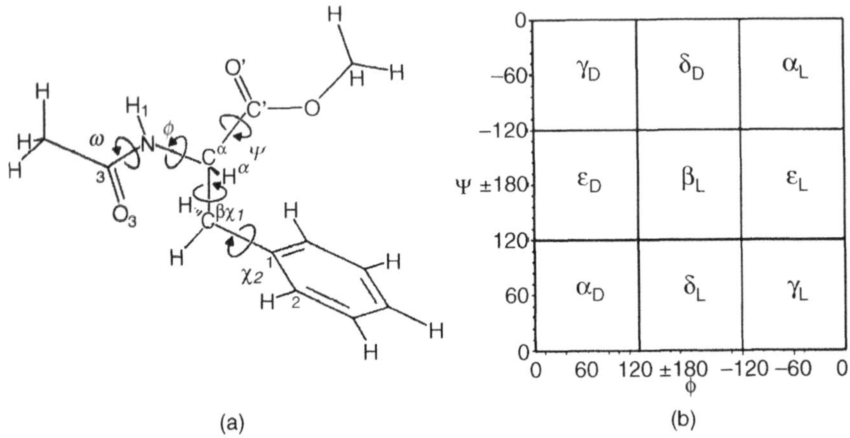

FIGURE 1.7. (a) Structure of Ac–Phe–OMe; the five most important angles (Φ, ψ, χ_1, χ_2, and ω) are displayed that describe backbone (Φ, ψ) and sidechain (χ_1, χ_2) arrangements as well as the cis/trans isomerism (ω). In Ac–Phe–OMe the *N* terminus of Phe is acetylated and the *C* terminus contains an ester function. (b) Ramachandran plot exhibiting the nine different backbone conformations with respect to the angles Φ and ψ. [Figure partly taken from Gerhards and Unterberg (2002).]

each amino acid. This nomenclature can also be applied to protected amino acids (see Figure 1.7). An appropriate example is the definition of the angles in Ac–Phe–OMe (IUPAC 1974; Schulz and Schirmer 1979; Gerhards and Unterberg 2002): $\phi = <(C', C^{\alpha}, N, C_3)$ and $\psi = <(O_1, C', C^{\alpha}, N)$ with $-180° < \phi, \psi \leq 180°$; see Figure 1.7. In their theoretical work on different partially protected amino acids, Perczel et al. (1991) pointed out that each of angles ϕ and ψ can have three different minimum-energy positions leading to a total number of at most nine conformational "backbone" orientations. To describe the nine possible arrangements, the abbreviations α_D, α_L, β_L, γ_L, γ_D, ε_D, ε_L, δ_L, and δ_D are used (Perczel et al. 1991). The isomers $\alpha_D(\phi \approx 0\text{–}120°, \psi \approx 0\text{–}120°)$ and $\alpha_L(\phi \approx -120\text{–}0°, \psi \approx -120\text{–}0°)$ describe helical structures, $\beta_L(\phi \approx -120\text{–}120°, \psi \approx -120\text{–}120°)$ describes a β-sheet-related conformation, $\gamma_L(\phi \approx -120\text{–}0°, \psi \approx 0\text{–}120°)$, and $\gamma_D(\phi \approx 0\text{–}120°, \psi \approx -120\text{–}0°)$ describe the inverse and normal γ turns, and $\varepsilon_D(\Phi \approx 0\text{–}120°, \psi \approx -120\text{–}120°)$ and $\varepsilon_L(\Phi \approx -120\text{–}0°, \psi \approx -120\text{–}120°)$ represent the inverse and normal polyproline II structures. The corresponding angles of the δ structures are $\delta_L(\Phi \approx -120\text{–}120°, \psi \approx 0\text{–}120°)$ and $\delta_D(\Phi \approx -120\text{–}120°, \psi \approx -120\text{–}0°)$. The torsional angles $\chi_1 = <(N, C^{\alpha}, C^{\beta}, C_1)$ and $\chi_2 = <(C^{\alpha}, C^{\beta}, C_1, C_2)$ are defined to characterize the structure of the sidechain of Ac–Phe–OMe. The angles χ_1 and χ_2 describe rotations about the C^{α}–C^{β} and C^{β}–C_1 axes, respectively. In the case of the angle χ_1, three positions of the phenyl group turn out to be minima. These positions are assigned as $g+$ (gauche $+$, $\chi_1 \sim 60°$), a (anti, $\chi_1 \sim 180°$), and $g-$ (gauche $-$, χ_1 about $-60°$) (IUPAC 1974; Gelin and Karplus 1979). Another important parameter describing the relative position of the NH_1 group with respect to the C_3O_3 group is the angle ω. This angle defines a trans or cis conformation with respect to the amide group. Usually the trans structures of the amino acids turn out to be more stable than the corresponding cis arrangements. With respect to the angles ϕ, ψ, and χ_1 27 different conformers (nine backbone arrangements multiplied by three different positions of χ_1) can be minima on the potential energy surface of Ac–Phe–OMe, but only one structure has been obtained in the R2PI spectra (Gerhards and Unterberg 2002; Gerhards et al. 2002). IR/R2PI spectra have been recorded in the region of the NH stretching modes and in the region of the amide I/II vibrations (C=O stretching modes, NH bending vibrations) (Gerhards and Unterberg 2002; Gerhards et al. 2002). By comparing the experimentally observed vibrational frequencies with the values obtained from ab initio and DFT calculations it can be concluded that Ac–Phe–OMe forms a β_L structure (see Figure 1.8a) with a sidechain in either anti or gauche + position; that is, either a β_L(a) or a $\beta_L(g+)$ structure is formed. Further investigations up to the fingerprint region clearly indicate that a $\beta_L(g+)$ structure is formed for the monomer (Fricke et al. 2006a).

The Ac–Phe–NHMe molecule is another example of a protected amino acid on the basis of Phe. With the choice of the NHMe protection group, it contains two amide groups (O=C–N–H) and is therefore a dipeptide model. In the R2PI

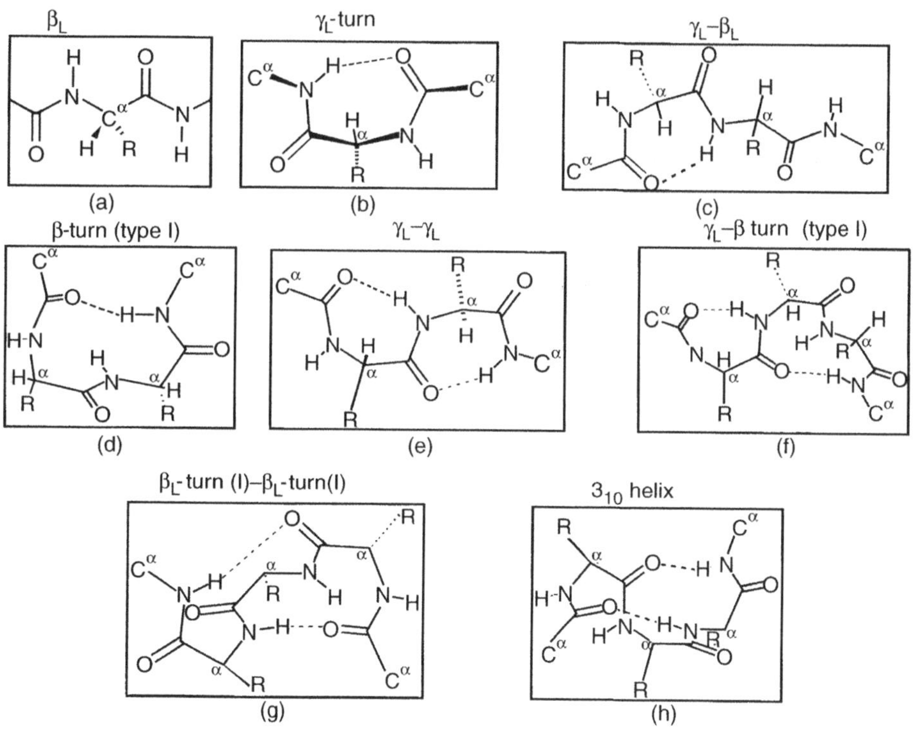

FIGURE 1.8. Different binding motifs describing backbone arrangements in peptides: (a) β-sheet-related structure (β_L, C5 arrangement) (b) γ-turn (γ_L, C7 structure), (c) combined γ_L–β_L structure, (d) β-turn (type I) (C10 structure), (e) γ_L–γ_L arrangement, (f) combined γ_L–β-turn (type I) structure, (g) successive β-turn (type I) arrangement, and (h) 3_{10} helical structure [identical to a combined β-turn (type III) – β-turn (type I) arrangement].

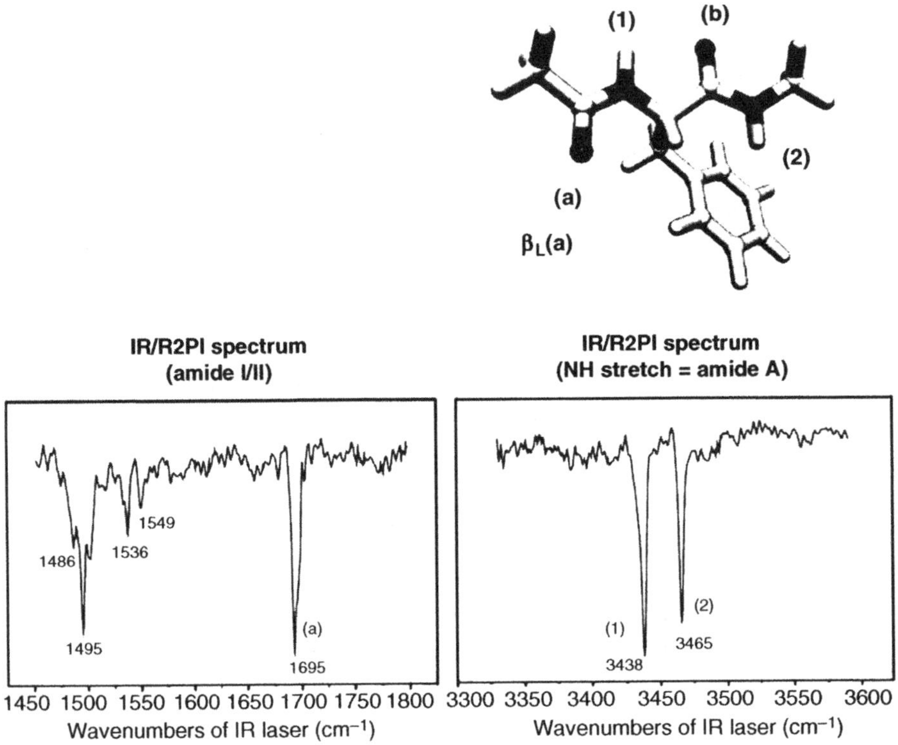

FIGURE 1.9. Structure and IR/R2PI spectrum of the dipeptide model Ac–Phe–NHMe (Gerhards et al. 2004). The spectrum is shown in regions of both amide I/II transitions (1450–$1800\,cm^{-1}$) and N–H stretching (amide A) vibrations. All vibrational transitions in the NH stretching region are significant above $3400\,cm^{-1}$, indicating a β-sheet-related structure. [Figure partly taken from Gerhards et al. (2004).]

spectrum of Ac–Phe–NHMe, five isomers are obtained (Gerhards et al. 2004). In agreement with ab initio and DFT calculations, the most stable structure is again a β_L(a) arrangement (see Figures 1.8a and 1.9), but the second and third most stable ones contain intramolecular hydrogen bonds and are γ_L-turn arrangements (see Figure 1.8b, in the case of an ester protection group these types of structure are not possible). It should be mentioned that the γ-turn structures are also called C7 structures and that because of the interaction of the H atom at the NH–CH_3 group and the O atom of the acetylated carboxyl group, both atoms are in 1,7 position to each other (see Figure 1.8b). The β-sheet-related structures are called C5 structures because of the possible weak interaction between the H and O atoms of the parallel orientated NH and carbonyl groups which are in 1,5 position (see Figures 1.7a and 1.8a).

As an example, the IR/R2PI spectra of the β_L(a) isomer of Ac–Phe–NHMe is shown in Figure 1.9. It is very typical in a stretched (β-sheet-related) arrangement for all NH stretching frequencies to be above $3400\,cm^{-1}$, since the corresponding

NH groups are free and not hydrogen-bonded. It is a special feature of this isomer of Ac–Phe–NHMe that only one vibration in the C=O stretching region can be observed. In agreement with the calculations, the predicted intensity of the second C=O stretching mode is much lower. In the case of the γ_L-turn structures the frequency of one hydrogen-bonded NH group is located below 3400 cm^{-1}, whereas the remaining NH stretching mode is above 3400 cm^{-1} (Gerhards and Unterberg 2002; Gerhards et al. 2004). Furthermore, transitions that belong to hydrogen-bonded NH groups are usually broader than those observed for free NH stretching modes. In general it can be stated that information on the principal structure of the investigated species can be derived from the position, intensity, and width of a vibrational transition. If, in addition, the vibrational transitions have been recorded in a wide spectral range (fingerprint, amide I, II, NH stretching modes, overtones), much information is available for spectral interpretation. In similar systems (e.g., Ac–Trp–NHMe) β-sheet-related structures and γ-turn structures are also observed.

A third example of a protected amino acid based on Phe is Ac–Phe–NH_2. In this case the NH_2 group is acetylated and the pure amide of the acid is used. This model might be less efficient for describing the inner structure of a peptide (which contains no primary NH_2 groups), but it offers an additional NH stretching frequency that can be useful for structural interpretations. Analysis of Ac–Phe–NH_2 by applying the IR/R2PI yields three conformers (Chin et al. 2005d). As in Ac–Phe–NHMe (Gerhards et al. 2004), the most stable one turns out to be a β-sheet-related structure whereas the two other structures have γ_L-turn binding motifs. It is very interesting to note that the conformer containing a γ_L(gauche+) arrangement shows a very low fluorescence quantum yield (Chin et al. 2005d). This may result from a N–H bond to the π-system leading to fast radiationless decay pathways (see discussion below).

The γ-turn structure type has also been observed in Z–Pro–NHMe (with Z = C_6H_5–CH_2–O; for Pro, see Figure 1.5). This system is based on the amino acid prolin, and since this species has no chromophore in the sidechain, an aromatic chromophore has been introduced in the protecting group (Compagnon et al. 2005).

The γ-turn structures are important elements for describing secondary structure binding motifs. Further secondary structure types include helices, β sheets, and β turns (see Figures 1.8 and 1.10). β-Turn and helix models are presented below, but even with protected amino acids, the most simple β-sheet structures can be discussed here. A β-sheet model is an aggregate of at least two (protected) amino acids or peptides in which the amino acids have the same orientation as the amino acids in a large peptide (see Figure 1.10). In order to discuss β-sheet model systems, the use of protected instead of unprotected amino acids has the advantage that the most polar groups are now located at the "backbone" of the molecule and not at the end. In unprotected amino acids [see, e.g., the six most stable isomers of Phe in Figure 6 of Snoek et al. (2000)], the COOH acid and NH_2 amino group tend to form intermolecular hydrogen bonds, whereas molecules like Ac–Phe–OMe or Ac–Phe–NHMe (or Ac–Phe–NH_2) prefer stretched β-sheet-related structures, which are ideal candidates for dimer formation. These dimers could be suitable examples for β-sheet models leading to the question of whether driving forces exist for isolated molecules in the gas phase (without any environment) to form a more complex secondary structure-binding motif.

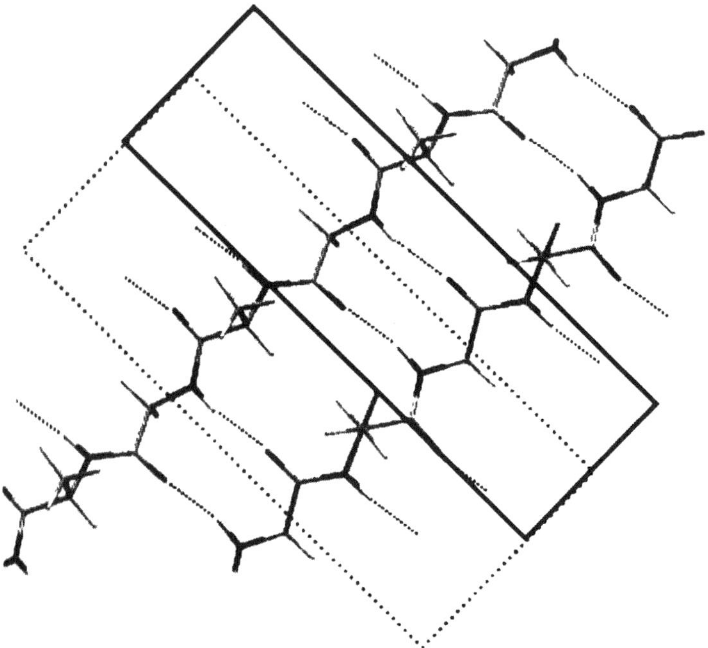

FIGURE 1.10. Part of a silk protein containing an antiparallel β-sheet secondary structure. The solid rectangle represents a small β-sheet model with a small distance between the neighboring N–H···O=C bonds, whereas the rectangle with the dotted lines contains a model with a large N–H···O=C distance. Both structures can be taken as subpattern of a large β-sheet. By looking at the whole area covered by the two rectangles a unit cell of a complete antiparallel β sheet is described; thus, by repeating this unit (containing both subpatterns) a complete β sheet can be generated (Gerhards et al. 2006a).

Indeed, in the case of Ac–Phe–OMe and Ac–Phe–NHMe analysis of IR/R2PI spectra in combination with ab initio calculations show that the dimers are β-sheet model systems (see Figure 1.11). For the dipeptide model Ac–Phe–NHMe there are two different β-sheet model systems: an outer-bound arrangement with the NH groups of the *N* termini (NHMe groups) hydrogen-bonded two the CO group of the *C* termini (acetyl groups) (Figure 1.11a) and an inner-bound structure with the NH groups of the amino acid hydrogen-bonded to the CO groups of the amide group (Figure 1.11b). By comparing the IR/R2PI spectra with the vibrational spectra predicted for the different structures, it could be deduced that the outer bound dimer is formed (Gerhards et al. 2004). In the case of Ac–Phe–OMe only an inner-bound β-sheet arrangement can be formed and has been experimentally determined (Gerhards and Unterberg 2002; Gerhards et al. 2002). It is significant that more than one inner- or outer-bound dimer can be formed; three possible orientations of the sidechain (*g*+, *g*−, *a*) for each monomer yield nine possible inner- or outer-bound dimer structures. The number of NH and CO stretching frequencies observed in the IR/R2PI spectra indicates that symmetric structures are observed, namely, structures in which the two monomers

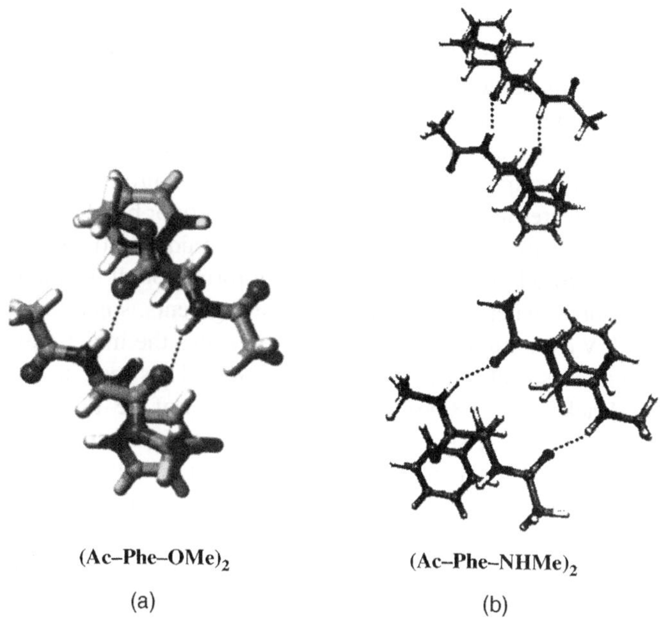

FIGURE 1.11. β-Sheet model systems for the two subpatterns described in Figure 1.10: (a) for the dimer of Ac–Phe–OMe an inner-bound arrangement has been observed (Gerhards et al. 2002; Gerhards and Unterberg 2002); (b) in the case of Ac–Phe–NHMe both inner- and outer-bound structures are possible but only the outer-bound arrangement has been observed experimentally (Gerhards et al. 2004). [Figure partly taken from Gerhards et al. (2004).]

have the same sidechain orientation. Thus only three possible structures remain for the inner- or outer-bound dimers. In the case of Ac–Phe–NHMe only a $\beta_L(g+)$-$\beta_L(g+)$ arrangement fits to the experimentally observed frequencies and intensity pattern both in the C=O stretching and NH stretching regions (Gerhards et al. 2004). For (Ac–Phe–OMe)$_2$ both a $\beta_L(g+)$-$\beta_L(g+)$ and a $\beta_L(a)$-$\beta_L(a)$ arrangement fits to the experimentally observed spectra (Gerhards and Unterberg 2002; Gerhards et al. 2002). Again investigations of the fingerprint region may indicate which structure is observed. Since much information on the backbone can be derived from IR spectroscopy, the very detailed analysis of sidechain orientation may be an interesting task for rotationally resolved high-resolution measurements, which should now be possible for systems of the size of a (protected) amino acid or its dimer.

For a systematic analysis for building larger β-sheet models saying it more general, to build up larger secondary structure elements, the size of the system will be increased by adding one and more amino acids in a systematic way. This leads to the formation of peptide structures and will be discussed in the next section.

In order to investigate the process of microsolvation, clusters of Phe with water are of interest. With respect to the pure amino acid Phe, Kim and coworkers analyzed the cluster with one water molecule by applying the R2PI method (Lee et al. 2002a,b). The structure was identified by a conformer-selective depletion of monomer signals as a result of hydration. It is concluded that the cluster can be derived from monomer A (see Figure 1.6; A = 7), which does not change much in geometry due to complexation with one water molecule. In agreement with ab initio calculations [MP2 single-point energies on DFT-optimized geometries, 6-31 + G(*d*) basis set], the most stable structure contains a water molecule bridging the carboxyl group. Thus the OH group of the carboxyl groups acts as proton donor to the water molecule, whereas water itself undergoes a hydrogen donor bond to the C=O group of COOH. It might be useful to apply the IR/R2PI technique to the Phe/water cluster system since this method yields direct structural information. It is important that this technique also works, when the excited clusters fragments, since the IR excitation occurs prior to UV excitation; i.e. the IR photon excites the intact cluster.

The effect of microsolvation an a protected amino acid has been investigated by the author's group (Fricke et al. 2006b) analyzing the cluster of Ac–Phe–OMe with up to three water molecules. These structures have been analyzed by applying the IR/R2PI method in the region of the NH and OH stretching region as well as in the region of the amide I and II vibrations. In contrast to unprotected amino acids, water cannot be aggregated to the very polar COOH and NH_2 end groups, but aggregation of the backbone can be discussed here, as it is extremely important in a large extended peptide. Two methods of aggregating one water molecule are shown in panels (a) and (b) of Figure 1.12. There is competition between the NH (proton donor) and CO (proton acceptor) group of the amide bond. In structure (a) a bridged arrangement between NH and one CO group is formed, whereas in (b) one water molecule is attached as a proton donor to the CO of the amide group. It is interesting to note that

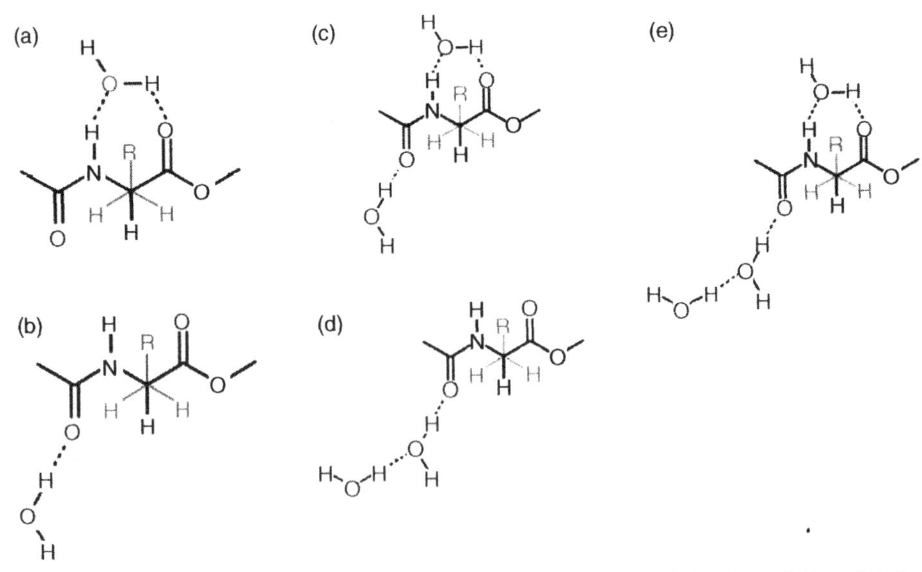

FIGURE 1.12. Experimentally observed structures of the cluster of Ac–Phe–OMe with (a,b) one, (c,d) two, and (e) three water molecules (Fricke et al. 2006b). The competing effect of water acting as proton donor or acceptor is clearly indicated.

both structures are obtained experimentally (Fricke et al. 2006b). For the cluster with two water molecules, a structure has in fact been obtained where one water molecule is located on each side of Ac–Phe–OMe (see Figure 1.12c). This structure is an overlay of the two structures obtained for the cluster with one water molecule. But additionally a second isomer has been observed where a water dimer undergoes a hydrogen donor bond to the amide CO group (Figure 1.12d) (Fricke et al. 2006b). Finally, only one isomer has been obtained for the cluster with three water molecules, consisting of an aggregation with one molecule acting as bridge between the NH and CO group at one side and with a water dimer hydrogen-bonded to the CO group at the other side. With respect to the pure backbone the first solvation "shell" is complete by the addition of three water molecules. The very interesting result is that in the case of Ac–Phe–OMe the first solvation shell has virtually no influence on the backbone; i.e. the principal orientation of a β-sheet-related stretched structure remains. It will be interesting to see how this changes if the solvation of larger peptides is investigated step by step (i.e., water molecule by water molecule). There should be very nice examples of how molecular beam experiments start with a stepwise solvation and how this compares to solution conditions.

1.4.2. Tryptophan

The most efficient amino acid with respect to absorbance and ion efficiency is Trp. By applying the laser desorption technique a R2PI and a LIF spectrum was obtained in the 1980s by Levy and coworkers (Rizzo et al. 1985, 1986a,b; Philips et al. 1988). From power dependence measurements it can be assumed that six isomers contribute to the R2PI signal; i.e. transitions in the R2PI spectrum with different intensities (arising from different transition moments for the $S_1 \leftarrow S_0$ transitions) belong to the same conformer, if power saturation leads to nearly the same intensities of the peaks (Rizzo et al. 1986b). Peaks belonging to different conformers do not have the same intensities because of the different populations in the S_0 state. A further proof for the existence of six isomers was deduced from DF spectra recorded via the transitions of the S_1 state. The DF spectra are not identical, and in particular a broad redshifted fluorescence observed for isomer A (the conformer with the lowest excitation energy) leads to the assumption that a zwitterionic form may be formed in the electronically excited state (Rizzo et al. 1986a). Another explanation results from a strong interaction of ^1L_a and ^1L_b electronically excited states in the case of conformer A. According to time-resolved fluorescence decay measurements, conformer A turns out to be the species with the shortest lifetime of the electronically excited state (Engh et al. 1986; Philips et al. 1988). Later experiments show that LIF and DF spectra have a nearly mirror-image appearance (Snoek et al. 2001), and broadening in the DF spectrum may result from an unresolved vibronic structure of Trp that is enhanced (due to large Franck–Condon factors) in the DF spectrum.

The early work by the Levy group gave very significant but generally indirect hints on different conformers of Trp. The application of double resonance techniques made it possible to test these assumptions. Indeed, both UV/UV

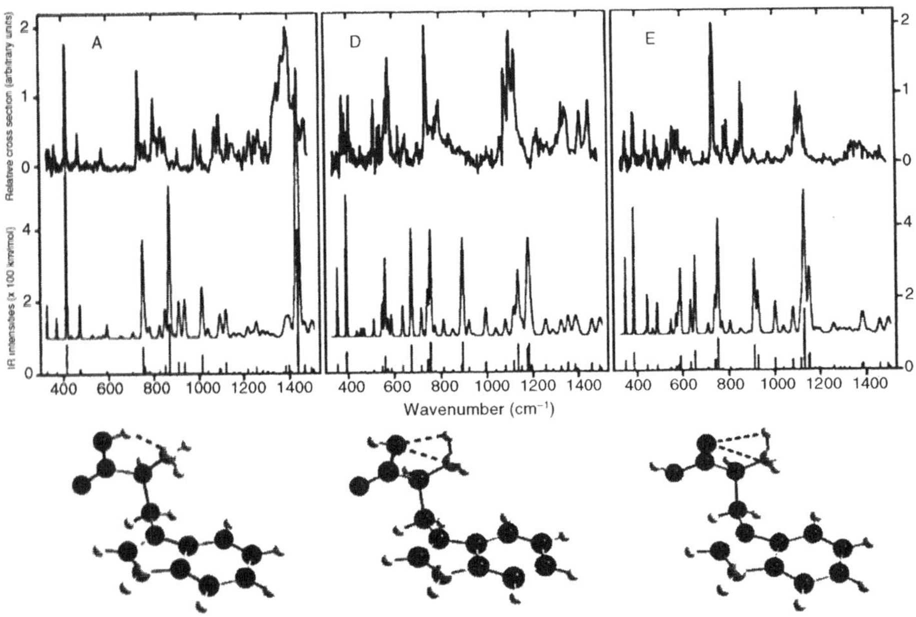

FIGURE 1.13. IR/R2PI spectra of three isomers of Trp in the fingerprint and amide II region from 330 to $1500\,cm^{-1}$ (Bakker et al. 2003). The experimental spectra (upper traces) are compared to the calculated spectra (lower trace) obtained at DFT [B3LYP/6-31 + G(*d*)] level. These spectra are given both as stick spectrum and as convolution with the spectral profile of the free-electron laser (Oepts et al. 1994). The corresponding structures of isomers A, D, and E are given at the bottom (for further details, see text). [Figure taken from Bakker et al. (2003).]

hole-burning (Piuzzi et al. 2000) and IR/R2PI (Snoek et al. 2001) spectroscopy confirmed the existence of six isomers of Trp. Investigations in the region of the NH stretching modes lead to an assignment of at least five isomers. The most stable structure (A) (see Figure 1.13a) contains an intramolecular hydrogen bond between the carboxyl OH group and the NH_2 group with the hydrogen atoms of the NH_2 group showing in the direction of the aromatic indole ring of the sidechain (Snoek et al. 2001). In the second most stable structure (E) the NH_2 group acts as proton donor to the carboxylic C=O group. The assignments of structures A and E have been further confirmed by measurements in the IR fingerprint region from 330 to $1500\,cm^{-1}$ (Bakker et al. 2003) (see Figure 1.13). A third isomer (D) contains a hydrogen bond between a (donor) NH_2 group and the carboxylic OH group acting as proton acceptor. Although this assignment is still not completely confirmed by IR spectroscopy, the agreement between calculated and experimental spectra in both the fingerprint and NH stretching regions supports the assignment. It is interesting to note that investigations of Trp in helium droplets (Lindinger et al. 2001) lead to a complete suppression of structure A, which turns out to be the most stable one in molecular-beam experiments and in all calculations performed on this system up to the MP2 level, including triple-ζ basis sets (Snoek et al. 2001). A possible explanation for this experimental result may be aggregation of He to the aromatic chromophore (indole ring) leading to destabilization of structure A. This result would show that the superfluid He medium can have a significant influence on the most stable structural arrangements.

Starting from bare Trp, the microsolvation process has been investigated by adding water molecules. In two studies water clusters have been investigated by the combination of IR/UV spectroscopy and DFT (ab initio) calculations (Snoek et al. 2002; Çarçabal et al. 2004). The investigations performed in the amide A region (Snoek et al. 2002) have been extended to the fingerprint region using a free-electron laser (Çarçabal et al. 2004). A significant problem in obtaining a correct assignment results from the fragmentation of higher clusters to the monohydrated cluster. The final conclusion is that the experimentally obtained IR spectrum results mainly from a triply hydrated cluster. For the Trp/water cluster, it is important to note that the Trp conformer observed within the cluster is not identical with the one obtained for the bare molecule; i.e. water is not simply attached to Trp monomer but also rearranges Trp to a conformation that is optimal for formation of a solvation shell. Furthermore, it could be deduced that in clusters with three or fewer water molecules, no zwitterion is formed. In an early paper the additions of other molecules (methanol, ethanol, chloroform, benzene, and acetone) were investigated by LIF, fluorescence lifetime, and dispersed fluorescence spectroscopy (Teh et al. 1989). From electronic spectroscopy, it could be derived that the solvent molecule is attached to the α-amine (i.e., to the NH group of the backbone) and not to the NH group of the indole moiety.

Similar to Phe, several derivatives of Trp have been investigated by various methods. In early papers by the Levy group (Park et al. 1986; Tubergen et al. 1990), the following derivatives have been analyzed by LIF, R2PI, and dispersed fluorescence spectroscopy: 3-indole acetic acid, 3-indole propionic acid, tryptamine, *N*-acetyltryptophan ethyl ester (NATE), different tryptophan amides, *N*-acetyltryptophan, *N*-acetyltryptophan amide (NATA), and *N*-acetyltryptophan–dimethylamide (NATDMA). In the latter one Trp is acetylated at the base group and amidated at the acid side (by using either the pure amide or a dimethylated amid). The idea behind these investigations was to find out why similar species show complete different electronic spectra; specifically, some spectra exhibit a pronounced low-frequency region and a broad redshifted fluorescence whereas other conformers of the same species exhibit sharp-resonance fluorescence spectra and no pronounced low-frequency transitions. It was assumed that the possibility of forming an intramolecular hydrogen bond between the NH and a carbonyl group can be responsible for the different electronic spectra. Also, the relative positions of backbone and sidechain (indole ring) are important for the spectroscopic results. It is important that indole possess two close-lying electronic states (^1L_a and ^1L_b). The spectroscopic results are a consequence of the perturbation in the electronic structure induced by the choice of derivatives and the conformation of the investigated species. This aspect has also been addressed by the Zwier group, who published a series of papers on NATA, especially *N*-acetyltryptophan–methyl amide (NATMA,

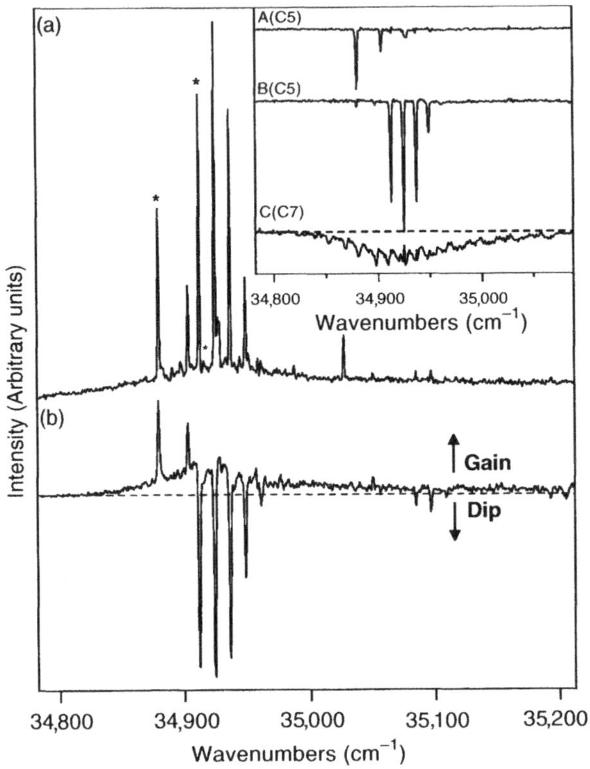

FIGURE 1.14. (a) LIF spectrum of NATMA (Ac–Trp–NHMe) analyzed by UV/UV hole-burning spectroscopy (inset). Thus the contribution of the three NATMA isomers (see Figure 1.3) is determined; (b) hole-filling spectrum obtained by exciting one NH stretching mode of isomer B. By scanning the UV laser, one can observe a decrease of the fluorescence resulting from isomer B and an increase of the fluorescence resulting from isomers A and C (Dian et al. 2002b). With respect to C, only a broad transition appears, indicating very efficient relaxation pathways in the electronically excited state (Dian et al. 2003) (for further details, see text). [Figure taken from Dian et al. (2002b).]

Ac–Trp–NHMe) (Dian et al. 2002a,b; 2003, 2004a,b,c; Evans et al. 2004). Similar to Ac–Phe–NHMe (see discussion above), NATMA (Ac–Trp–NHMe) represents a dipeptide model. By applying UV/UV hole-burning spectroscopy, the electronically excited states of three different conformers of NATMA have been determined. In contrast to the two other conformers, a third one shows a broad electronic transition (see hole-burning spectrum in the inset of Figure 1.14 and structures in Figure 1.3) (Dian et al. 2002a). By further applying IR/UV double-resonance spectroscopy, the vibrations are obtained and compared with calculated spectra (for details concerning the strategy for calculation of spectra, see Section 1.7). It turns out that the species with the broad electronic transition forms a γ-turn (C5) arrangement. The unusual behavior of this γ-turn structure is explained by a possible coupling of the $^1\pi\sigma^*$ state with the ^1L_b state. This coupling may result from the position and strength of the dipole moment as well as the orientation of the sidechain with respect to the backbone (with a NH group that might be directed on the π system). A direct hint to a coupling of the $^1\pi\sigma^*$ state with the ^1L_b state can be obtained from IR/R2PI spectra of the S_1 state. In the spectra belonging to the γ-turn isomer with a broad spectral UV absorption, the indole stretching vibration is not obtained, indicating that the $^1\pi\sigma^*$ state is dissociative along the NH stretching coordinate (Dian et al. 2003). The result shows very clearly that photophysics can be strongly conformation-dependent and that it cannot be induced only by a solvent environment.

NATMA has not only been of interest with respect to the investigation of different electronic states but is also used to introduce the IR-PTS and IR hole-filling techniques [see Figure 1.14 (Dian et al. 2002b, 2004c)]. In the case of NATMA the population could be redistributed very efficiently between different isomers, but no real vibrational selectivity and only a moderate conformer selectivity have been observed. In other words, by exciting one isomer, such as the most stable one (C5, β-sheet-related), the other two isomers are formed, but never an isomer that does not belong to the most three stable ones. The distribution has been simulated by applying forcefield potentials (Amber, OPLS) in combination with RRKM calculations (Evans et al. 2004). The experimental results are simulated as a function of excess energy and cooling rate. An important factor is the influence of the cooling process to the redistribution; that is, after IR excitation, the collisional cooling with helium can also lead to redistributions. By applying different backing pressures as well as different x/D distances (where $x =$ distance to nozzle, $D =$ nozzle hole diameter) the cooling conditions are varied. For a special x/D value, pressure and fixed IR and UV wavelengths, IR/UV spectra are recorded for different time delays between IR and UV lasers. The x/D value turns out to be extremely sensitive, since cooling occurred within a few hundred nanoseconds down to <200 ns (Dian et al. 2004b).

To obtain a clearer insight into the important factors to obtain vibrationally induced population transfer between different isomers, investigations have also been extended to melantonin, which offers a larger signal than NATMA and possesses only one NH group in the backbone (Dian et al. 2004b). It turns out that the transfer can be changed by the cooling conditions; furthermore, population redistribution after IR excitation can be observed between different trans isomers, but no cis structure is observed. This results from the barrier height of the trans–cis isomerization, which cannot be reached by absorption of one IR photon; thus the chosen method is suitable for determination of barrier heights (Dian et al. 2004b). The promising investigations with IR-PT and hole-filling spectroscopy should be continued to answer questions about mode selectivity in (chemical) processes. As mentioned above, many parameters have to be taken into account (e.g., selected IR wavelength, cooling rates, number of isomers, relative energies of the isomers, PES). Furthermore, investigations of the S_1 state may be taken into account requiring a description more complicated than that for the S_0 state resulting from perturbations induced by different electronically excited states.

Another reason to investigate (protected) Trp is the analysis of secondary-structure-binding motifs. According to R2PI and IR/R2PI spectroscopy, and in

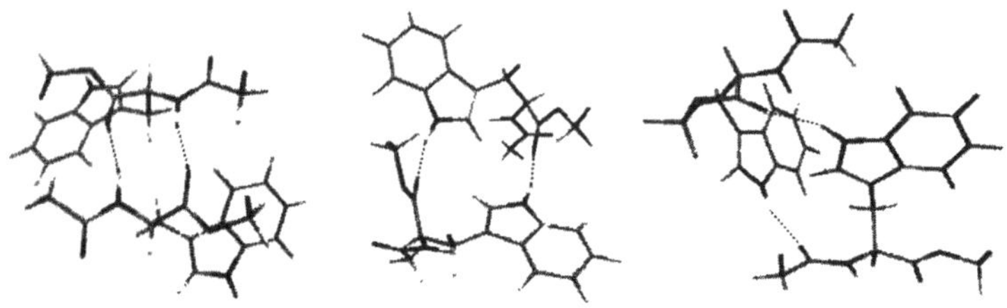

(a) (b) (c)

FIGURE 1.15. Different structures of the Ac–Trp–OMe dimer (Gerlach et al. 2005); (a) β-sheet model system with N—H···O=C bonds between the backbones; (b) dimer with hydrogen bonds between the indolic NH groups and the C=O groups of the ester functions. It is called a symmetric dimer since the two hydrogen bonds are of the same type. In contrast to this arrangement structure (c) is an asymmetric dimer where one indolic NH group is hydrogen-bonded to the ester C=O whereas the other indolic NH undergoes a hydrogen bond to the amide C=O group. [Figure taken from Gerlach et al. (2005).]

agreement with ab initio calculations, it turns out that the Trp monomer has an β-sheet-related structure (Gerlach et al. 2005). This result is similar to the one obtained for the protected amino acids Ac–Phe–OMe and Ac–Phe–NHMe (Gerhards and Unterberg 2002; Gerhards et al. 2002, 2004), which both have β-sheet-related structures as the most stable ones. As in the case of these Phe-based protected amino acids (Ac–Phe–OMe and Ac–Phe–NHMe) it would be possible for the Ac–Trp–OMe dimer to form a β-sheet model system, but the main difference between the amino acid Phe and Trp results from the existence of a polar group in the sidechain of Trp; specifically, the NH group of indole can undergo intermolecular hydrogen bonds to the CO groups of the second Trp moiety (see Figure 1.15b,c). Indeed, analysis of the $(Ac–Trp–OMe)_2$ dimer exhibits a structure in which one of the indolic NH groups is hydrogen-bonded via the ester C=O group whereas the other one is hydrogen-bonded to the acetylic CO group (Gerlach et al. 2005). This structure is an asymmetric arrangement (see Figure 1.15c) whereas in a sysmmetric arrangement (Figure 1.15b), both indolic NH groups would be hydrogen-bonded to either the ester CO groups or the acetylic CO groups. A β-sheet model system (Figure 1.15a) with hydrogen bonds between the NH groups of the backbone to the ester CO groups (which are orientated parallel to the backbone NH groups) cannot be observed since the indolic NH groups are more acidic. This remarkable difference between Phe and Trp indicates the strong influence of the sidechain with respect to structural arrangements. This influence of the sidechain can be further complicated when hydrogen bonds of the NH groups to the π system of the sidechain are possible. Investigations of different protected amino acids show that the interplay between sidechains and backbones is important for the structure of an isolated peptide or an aggregate. In particular, sidechains with polar groups (like NH or OH) might be avoided if a β-sheet model should be observed. An interesting question is whether water added, for example, to the dimer of Ac–Trp–OMe leads to a rearrangement of the dimer to a β-sheet model with solvated NH groups of the indolic sidechain or whether the free functional groups of the backbone (NH or CO) are solvated, conserving the structure of the unsolvated dimer. Experiments seem to indicate that the first water molecule is inserted in the NH···O=C hydrogen-bonds of the dimer. This may indicate that with a larger amount of solvent molecules, a β-sheet arrangement of the Ac–Trp–OMe dimer could be obtained.

1.4.3. Applications of Microwave Spectroscopy

Free amino acids have been investigated intensively with the aid of microwave spectroscopy in the gas phase. Glycine, the simplest amino acid, has been the subject of many investigations. Starting with the early work of Brown et al. (1978) as well as Suenram and Lovas (1978), the microwave spectrum of glycine has been revisited several times for a complete structural analysis of three possible conformers of glycine (Berulis et al. 1985; Bludsky et al. 2000; Brauer et al. 2004; Brown et al. 1978; Godfrey and Brown 1995; Godfrey et al. 1996; Guélin and Cernicharo 1989; Hollis et al. 1980; Lovas et al. 1995; McGlone et al. 1999; Miller et al. 2005; Senent et al. 2005; Snyder et al. 1983; Suenram and Lovas 1978, 1980; Suenram et al. 1989). In the most stable structure the NH_2 group undergoes a bifurcated intramolecular hydrogen bond to the C=O group (isomer I). The same is observed for alanine (Blanco et al. 2004; Brauer et al. 2004; Császár 1996; Godfrey 1993; Godfrey et al. 1996; McGlone et al. 1999).

Histamine has been investigated by Vogelsanger et al. (1991) by recording the rotational spectrum in a free-expansion jet spectrometer. Four conformers have been observed (Vogelsanger et al. 1991). Two different conformers of alanine have been observed by Godfrey et al. (1996) with the use of a Stark-modulated free-expansion jet spectrometer. The development of the abovementioned ablation technique (Suenram et al. 1999; Lessari et al. 2003) opened the window to other amino acids that could not be investigated in the gas phase because of decomposition on thermal heating or if the heating were less effective. Alonso and coworkers recorded the jet-cooled rotational spectra of proline, valine (Lessari et al. 2002, 2004), and alanine (Blanco et al. 2004) by applying laser ablation molecular-beam Fourier transform microwave spectroscopy (MB-FTMW). In the case of proline (Allen et al. 2004; Czinski and Császár 2003), two conformers have been assigned that have a hydrogen bond between the carboxylic OH group and the lone pair of the nitrogen. This is in contrast to the observed most stable conformers of glycine and alanine with the abovementioned bifurcated intramolecular hydrogen bond between the NH_2 and C=O group. For valine two conformers have been determined with the two different types of hydrogen bonding as described before.

Several derivatives and analogs of amino acids have been investigated by recording rotational spectra since the mid-1990s. Tubergen and coworkers recorded the rotational spectra of seven isotopomers of alaninamid (Lavrich et al. 1999). They deduced a structure with a hydrogen bond between the amide and amine with the amide group acting as donor. In a later work Lavrich and Tubergen investigated the cluster of alaninamid with water, which is an interesting example of a hydrogen bonded amino acid derivative (without aromatic chromophore) (Lavrich and

Tubergen 2000). Here water acts both as hydrogen donor to the C=O group and as hydrogen acceptor to the amide NH_2 group, forming a cyclic hydrogen-bonded structure like the one observed for isomer of Ac–Phe–OMe$(H_2O)_{1-3}$ (see Figure 1.12a,c,e).

Also the dipeptide model *N*-acetyl-alanine N′-methylamide, and its double $^{15}N(^{15}N_2)$ isotopomer, have been successfully examined by means of MB-FTMW spectroscopy. For this larger system a folded γ-type structure (Lavrich et al. 2003b) has been observed, similar to the one obtained for other dipeptide models like AcPheNHMe or AcTrpNHMe (see discussion above). Furthermore derivatives of valine, glycine, and proline have been studied. The investigation of valinamide yielded no unambiguous agreement between theoretical and experimental results, but it has been deduced from ab initio calculations that the three most stable conformers have intramolecular hydrogen bonds (Lavrich et al. 2002). For *N*-acetylglycine only one conformer has been observed, with an intramolecular hydrogen bond between the carbonyl group of the acid and the amide proton (Lovas et al. 2004). Two conformers of each molecule have been detected and assigned in the experimental studies on 4(*S*)- and 4(*R*)-hydroxyproline (Lessari et al. 2005). Again, each conformer exhibits an intramolecular hydrogen bond, either between the lone pair of the nitrogen atom in the ring and the carboxylic OH group or between NH and C=O.

Since cylcopeptides are of increasing interest, the investigation of diketopiperazine has offered an interesting insight for conformational properties of this class of structure (Bettens et al. 2000). Only one isomer has been observed that adopts a structure other than in solid phase as obtained from X-ray crystallography.

Furthermore model systems for peptides (Lavrich et al. 2003a; Ohashi et al. 2004) and the analogs of tyrosine and tryptophan, tyramine (Melandri and Maris 2004) and tryptamine (see Section 1.3), have been studied by means of microwave spectroscopy.

Finally, it should be mentioned that all isolated amino acids investigated in the gas phase up to now are not zwitterions. These structures result from an aggregation with water leading to the question of how many water molecules are necessary to form a zwitterion. Except for theoretical approaches [see, e.g., Ding and Krogh-Jespersen (1992), Jensen and Gordon (1995), Gutowski et al. (2000), Kassab et al. (2000), Chaudhari et al. (2004, 2005a,b)], Bowen as well as Johnson and coworkers [Diken et al. 2004, 2005; Xu et al. 2003] analyzed clusters of glycine/water anions by applying mass spectrometry, photoelectron spectroscopy (Xu et al. 2003), and IR photodissociation techniques (Diken et al. 2004, 2005). Although these systems are anions, it is argued that the role of the excess electron is subordinate to that of hydration. Thus it can be estimated that the results obtained from the anions are at least a lower limit for the number of water molecules needed to form a zwitterion in the case of the neutral cluster. Bowen and coworkers derive from the onset of the formation of intact cluster ions at $n = 5$ (five water molecules) that the zwitterion structure is formed. In contrast to these results, Johnson and coworkers observed the formation of glycine $H_2O)^-_{1,2}$ clusters. Future IR investigations on efficiently cooled glycine/water anions will give a hint as to which structure is formed for the different cluster sizes. The first IR photodissociation spectrum recorded for a "warm" glycine$(H_2O)_6^-$ cluster is not unambiguous (Diken et al. 2005).

In addition to the clusters of glycine with water, Bowen and coworkers investigated the anionic clusters of Trp and Phe with water yielding a solvation of four water molecules until the zwitterion is formed (Xu et al. 2003).

1.5. SPECTROSCOPIC ANALYSIS OF PEPTIDE STRUCTURES

Starting with the amino acids, the investigation of peptides is motivated by the incentive to obtain information on the driving forces of protein folding or protein aggregation. In gas-phase experiments the isolated peptides can be investigated without an environment. Thus the influence of the environment can be estimated. The dependence on the formation of different secondary structures with respect to a sequence of amino acids can also be determined. Even for protected amino acids different secondary-structure-binding motifs (γ-turn structures, β-sheet-related arrangements, and β-sheet model systems) have been discussed; see Section 1.4. A further aggregation of amino acids would give insight into the selectivity and efficiency of intermolecular peptide bonds. Finally, the successive addition of water (process of microsolvation) makes it possible to analyze the influence to the structure of peptides due to solvation.

By going from amino acids to peptides, three basic types are investigated: (1) unprotected peptides with a free acid COOH group as well as a NH_2 group, (2) (partially) protected peptides with the COOH group acetylated and a NH_2 group that is either free or methylated, and finally (3) cyclopeptides in which the ends of a peptide (COOH and NH_2) are condensed. The spectroscopic investigations performed on these different structures are discussed in the following paragraph.

As mentioned in the introduction, the first work on peptides was been performed by Levy and coworkers by applying R2PI and LIF spectroscopy on the dipeptides Trp–Gly, Gly–Trp, Phe–Trp, Trp–Phe, and Trp–Trp as well as the tripeptides Trp–Gly–Gly and Gly–Gly–Trp (Cable et al. 1987, 1988a,b). More than 10 years later de Vries and coworkers continued this work by investigating a series of dipeptides (Gly–Tyr, Ala–Tyr, Tyr–Ala) as well as the tripeptide Phe–Gly–Gly (Cohen et al. 2000). In this work UV/UV hole-burning spectroscopy has also been applied in order to find out how many conformers of the different species contribute to the R2PI spectrum. Furthermore a tentative assignment of possible structures is given by comparing spectral features of the R2PI spectrum with calculations performed for the corresponding species in its S_0 state. Similar investigations (by applying UV/UV double-resonance spectroscopy) have been performed by Kleinermanns and coworkers (Hünig et al. 2003) on the dipeptides Gly–Trp, Trp–Ser, and Pro–Trp (see Figure 1.16).

All these applications offer the possibility of determining the number of conformers, but in contrast to UV/UV double-resonance spectroscopy, application of IR/UV double-resonance methods makes it possible to get information on the structure in the S_0 state. Thus, similar to the analyses performed on the amino acids,

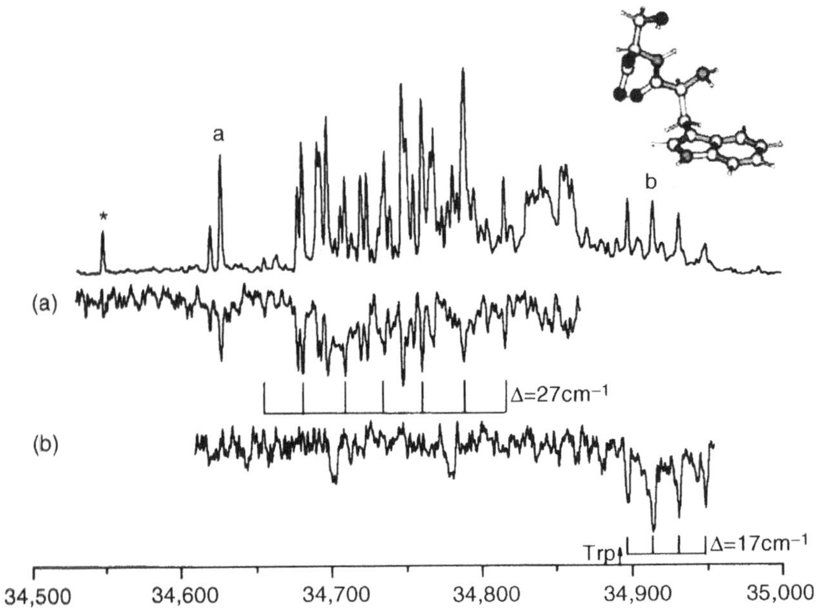

FIGURE 1.16. R2PI and UV/UV hole-burning spectra of the dipeptide Trp–Ser indicating that two isomers are present (Hünig et al. 2003). Furthermore, prominent low-frequency progressions are observed. The structure shown in this figure is not necessarily correlated with one of the spectra since no direct structural information is obtained from the UV spectra. (For further details, see text.) [Figure taken from Hünig et al. (2003).]

structural assignments can be derived by comparing experimentally observed vibrational frequencies with calculated values and values obtained for smaller subunits; thus, if a dipeptide is analyzed, the information obtained for the amino acid building blocks gives further information on the assignment of the dipeptide. The author's group first applied the IR/R2PI technique to a peptide, the protected dipeptide Ac–Val–Phe–OMe (Unterberg et al. 2003) (see Figure 1.17). The spectra have been recorded in the region of the NH and the amide I/II region. The use of our new high-resolution IR system in the region of the amide I/II region enables us to resolve close-lying C=O stretching frequencies. From the information obtained for Ac–Phe–OMe and in agreement with HF calculations on the dipeptide, a β-sheet-related structure could be deduced for this species. Interestingly, only one isomer has been obtained for the dipeptide, indicating that even in the gas phase conformational preselections are performed. This is partly an effect of the protection groups that exclude isomers that form intermolecular hydrogen-bonded structures between the acid and base end groups.

Starting in 2004, a series of papers have been published on (protected) peptides by applying the IR/R2PI technique. Mons and coworkers investigated different dipeptides (tripeptide models) in order to find out models for various secondary structure elements (Chin et al. 2004, 2005b,c). The conformers of Ac–Pro–Phe–NH_2

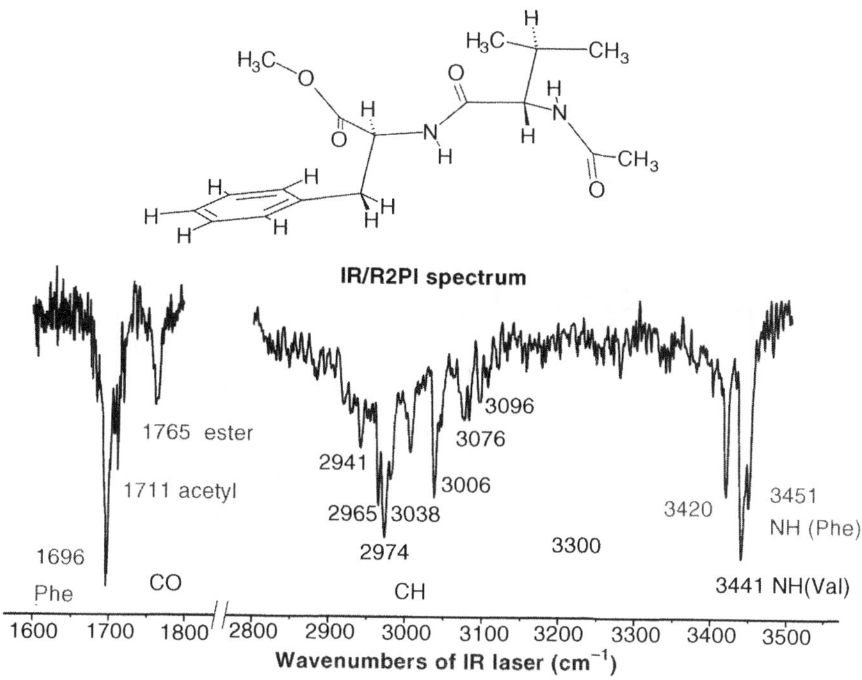

FIGURE 1.17. IR/R2PI spectrum of Ac–Val–Phe–OMe in the amide I (C=O), CH stretching and amide A (NH stretching) region (Unterberg et al. 2003). The structure shown in this figure results from an analysis of the structure-sensitive IR spectrum in combination with ab initio calculations. The NH transitions above 3400 cm^{-1} (with one overtone at 3420 cm^{-1}) indicates a β-sheet-related arrangement. For further details, see the article by Unterberg et al. (2003). [Figure taken from Unterberg et al. (2003).]

and Ac–Phe–Pro–NH_2 (Chin et al. 2004) as well as Ac–Phe–Gly–NH_2 and Ac–Gly–Phe–NH_2 (Chin et al. 2005c) (see Figure 1.18) have been investigated. Furthermore, the series of tripeptide models Ac–X–Phe–NH_2 (X = Gly,Ala,Val) (Chin et al. 2005b) are analyzed by IR/R2PI spectroscopy in combination with ab initio and DFT calculations. In the case of Ac–Pro–Phe–NH_2 a sequence of two γ_L-turn structures turns out to be the most stable arrangement (see Figure 1.8e), whereas in the case of Ac–Phe–Pro–NH_2 a $\beta_L(g-)$-γ_L see similar structure in Figure 1.8c arrangement is most stable and a less intense conformer forms a β-turn structure (Figure 1.8d; these structures are also called C10 structures). Although the introduction of Pro in a sequence of a peptide may favor the formation of β-turn structures, it turns out to be not a very prominent component. There is always a balance between the strength of intermolecular hydrogen bonds (which are stronger in γ-turn structures and relatively weak in β-turns, indicated by a small redshift of the hydrogen-bonded NH group) and the strain in the peptide backbone (which is lower in the case of the β-turn structure). Additionally, NH$\cdots\pi$ interactions can come into play so that a special structure cannot be predicted a priori by simply selecting a special sequence of amino acids.

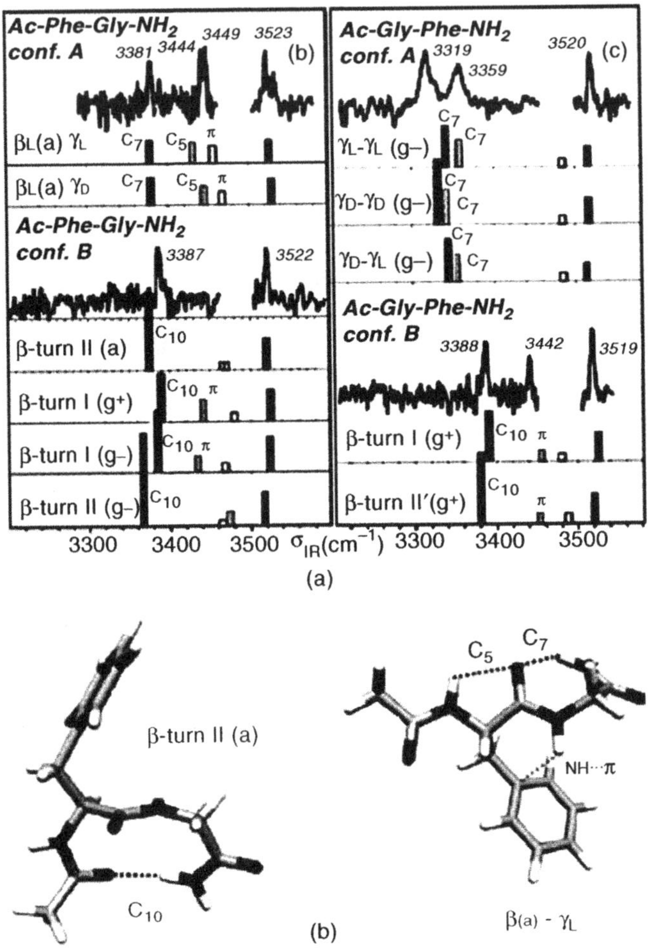

FIGURE 1.18. (a) IR/R2PI spectra in the amide A (NH stretching) region of different Ac–Phe–Gly–NH_2 and Ac–Gly–Phe–NH_2 isomers. The most prominent isomers A of the tripeptide models Ac–Phe–Gly–NH_2 and Ac–Gly–Phe–NH_2 possess a β_L–γ_L arrangement (cf. similar structure in Figure 1.8c) and a γ_L–γ_L (cf. Figure 1.8e) arrangement, respectively. A second component B represent different β-turn arrangements (types II and I) for the two tripeptide models. The calculated scaled vibrational frequencies are obtained from DFT calculations [B3LYP/6-31 + G(*d*)]. For further details, see Chin et al. (2005).] [Figure taken from Chin et al. (2005c).] (b) The most stable structures of Ac–Phe–Gly–NH_2. The β_L(a)–γ_L arrangement includes not only C5 and C7 interaction but also a N–H$\cdots\pi$ interaction between backbone and sidechain stabilizing the binding motif. [Figure taken from Chin et al. (2005c).]

The competing effects of the different interactions lead to two different conformers of the tripeptide models Ac–Phe–Gly–NH_2 and Ac–Gly–Phe–NH_2. The most prominent isomer of Ac–Phe–Gly–NH_2 shows a β_L(a)-γ_L structure and a weaker isomer forms a β-turn. In the case of Ac–Gly–Phe–NH_2 a sequence of two γ-turns (see Figure 1.18) and a β-turn structure is formed (Chin et al. 2005c). For

Ac–Phe–Gly–NH_2 a type II β-turn is preferred whereas for Ac–Gly–Phe–NH_2 a type I turn is most probable. The different types of β-turns differ with respect to the orientation of the backbone [characterized by two sets of angles ψ and ϕ (Venkatachalan 1986; Hutchinson and Thornton 1994)]. Similar structural arrangements have also been observed for other dipeptides in the series Ac–X–Phe–NH_2 (where X = Ala, Val); specifically, two isomers with either γ-turn or β-turn binding motifs have been identified (Chin et al. 2005b).

The work on the dipeptides was further extended to protected tripeptides that serve as tetrapeptide models. In more recent publications by the Mons group the systems Ac–Phe–Gly–Gly–NH_2 (Chin et al. 2005a) and Ac–Ala–Phe–Ala–NH_2 and the peptides with the two other sequences Ac–Ala–Ala–Phe–NH_2 and Ac–Phe–Ala–Ala–NH_2 (Chin et al. 2005e) were investigated. Ac–Phe–Gly–Gly–NH_2 forms a structure with successive β-turns (see Figure 1.8g); specifically two β-turns are formed in a sequence, without building a helical arrangement. Ac–Phe–Ala–Ala–NH_2 contains a combination of a β-sheet-related structure and two γ turns [$\beta_L(a)$-γ_L-γ_L] and Ac–Ala–Ala–Phe–NH_2 forms a structure containing a β-turn and a γ_L-structure binding motif (Chin et al. 2005e) (see Figure 1.8f). The third isomer, Ac–Ala–Phe–Ala–NH_2, serves as the first model system to describe a 3_{10} helical arrangement, where two successive β turns (type III and approximately type I) are observed, indicating the beginning of a helical arrangement consisting of three amino acid residues in one winding (Chin et al. 2005e) (Figure 1.8h).

In the author's group we continued the work on dipeptides with the tripeptide model Ac–Val–Tyr(Me)–NHMe (Fricke et al. 2004). The chromophore is tyrosine, containing an OMe group instead of an OH group in order to avoid any intramolecular hydrogen bonds to the OH group of the side chain. Only one isomer has been obtained for this species, and the structure could be interpreted as a β_L–β_L arrangement (a fully β-sheet-related structure) (Fricke et al. 2004) (see Figure 1.19). Since spectral assignments that could be obtained from β-turn arrangements are very similar, the complete spectrum of the tripeptide model up to 1000 cm^{-1} in the fingerprint region has been recorded [see Figure 1.19 (Fricke and Gerhards 2005)]. It can be stated that Ac–Val–Tyr(Me)–NHMe is the spectroscopically best investigated molecule of this size; the region of the NH stretching modes *and* the region of the amide I/II region up to the fingerprint at 1000 cm^{-1} have been recorded with our high-resolution (!) laser system; especially in the region from 1000 to 1800 cm^{-1}, the very close-lying vibrational transitions are resolved. Nevertheless, the patterns observed for two completely different structural arrangements described here are (accidentally) extremely identical. It is a good hint that spectral assignments are occasionally ambiguous; although very extensive spectral information has been obtained, it also indicates (see, e.g., the interpretation on Ac–Phe–OMe in Section 1.4) that one can achieve the most reliable assignment only if spectral information in the complete spectral region from fingerprint to the NH stretching region is available.

Finally, the tetrapeptide model Ac–Leu–Val–Tyr(Me)-NHMe has been investigated, indicating two structures that contain either a set of three successive γ turns or a combination of a β turn and a γ turn (Fricke et al. 2006c). The reason for choosing this special variant of tetrapeptide model results from the idea of mimicking a model for

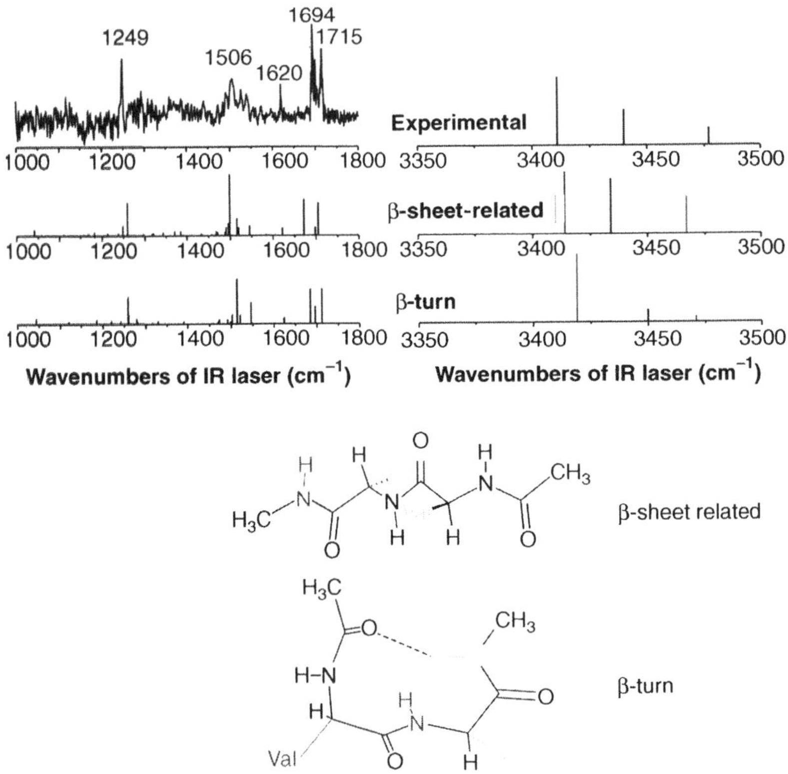

FIGURE 1.19. IR/R2PI spectrum of the tripeptide model Ac–Val–Tyr(Me)-NHMe in the amide A (NH stretching) and amide I/II region [see also Fricke et al. (2004)] as well as in the fingerprint region up to $1000\,cm^{-1}$. By comparing the experimental spectrum with the calculated ones [DFT level, B3LYP functional, 6-31 + G(*d*) basis set], no unambiguous assignment between a β-sheet-related structure and β-turn structure is possible, indicating the difficulties that may arise in some cases despite the availability of a significant amount of spectral information.

the KLVFF (Lys–Leu–Val–Phe–Phe), peptide sequence (in our model Tyr instead of Phe), which is important for the discussion of Alzheimer disease (see Section 1.6).

It can be stated that most work has been done up to now on the protected peptides to ensure a systematic investigation of secondary structure elements and to get an insight into the driving forces and mechanisms of peptide folding with respect to isolated molecules. With respect to the unprotected peptides, essentially two papers on the systems Trp–Gly, Gly–Trp, and Trp–Gly–Gly have been published by applying IR/R2PI spectroscopy first in the region of the NH stretching vibrations (Hünig and Kleinermanns 2004) and later also in the fingerprint region by using a free-electron laser (Bakker et al. 2005). In the case of Trp–Gly at least four conformers have been observed, but IR spectra could be obtained for only two of them. The most prominent isomer is nearly unfolded but contains a N–H···NH_2 interaction between the NH group of the Trp residue and the NH_2 group of the free acid. In the case of Gly–Trp two conformers are observed, and both conformers show (as in Trp–Gly) a N–H···NH_2 interaction, but one conformer is more stretched (unfolded) whereas the other one contains a strong OH···O=C hydrogen bond between the COOH group of glycine and the CO group of Trp. For the tripeptide Trp–Gly–Gly two isomers are observed. One isomer shows a strong OH···O=C hydrogen bond between the COOH group of the first and second glycine residues. Furthermore, a hydrogen bond between the NH of the indole group and the C=O group of COOH appears. Except a N–H···NH_2 interaction the second isomer remains nearly unfolded (stretched). The structural assignments are not completely unambiguous on the basis of the experimentally observed IR data, but because of the additional free COOH and NH_2 functions, more structures have to be calculated for the free peptides than for the protected systems. These groups can undergo several hydrogen bonds [see, e.g., the simple Phe molecule (Snoek et al. 2000)], leading to a more complex analysis of the structural assignment.

The third class of important peptides are the cyclopeptides in which *N* and *C* termini of one peptide are condensed. These molecules are of interest with respect to various biological functions, such as for ion transport mechanisms (Duax et al. 1996; Kubik et al. 2002). The first mass-selected spectroscopic investigations on isolated peptides were performed by Weinkauf and coworkers on cyclic Trp–Gly by applying R2PI and UV/UV double-resonance spectroscopy (Wiedemann et al. 2004). The first IR/R2PI study was performed by de Vries and coworkers on cyclo(Phe–Ser) (Abo-Riziq et al. 2005). Five different isomers have been observed (also by applying UV/UV double-resonance spectroscopy). The IR spectra of the isomers as well as the corresponding structures and their calculated vibrational frequencies are shown in Figure 1.20. The excellent agreement between calculated and experimental values shows that identification of different isomers can be very straightforward.

With respect to the microsolvation of peptides, only one paper has been published, including an aggregate of a peptide with water, the report on Ac–Val–Tyr(Me)–NHMe$(H_2O)_1$ (Fricke et al. 2004). From the IR/R2PI spectrum it has been concluded that the water molecule acts as hydrogen acceptor and is hydrogen-bonded to the NHMe protection group. Investigation of larger clusters (with more water molecules) as well as analyses of different peptide/water clusters will be an important topic for future experiments. It should be mentioned that observation of larger clusters of peptides with several water molecules by mass spectrometry is not problematic, but strong fragmentation of these clusters after UV excitation makes it difficult to obtain an unambiguous correlation of observed IR/R2PI spectrum and cluster size; thus it will be absolutely necessary to perform two-color R2PI spectra (see Figures 1.1 and 1.2), which may lead to a significant reduction of cluster fragmentation.

A significant amount of work is in progress to obtain a further extension of the different types of chosen peptides (protected or unprotected peptides, cyclopeptides). The field of analyzing large peptides in the gas phase is rapidly growing, and some groups have already succeeded in investigating larger systems. With respect to the unprotected species, the delta sleep-inducing nona peptide (DISP) has been investigated by using a free-electron laser in the region 600–2200 cm^{-1}; but only with these experiments in the fingerprint and amide I/II regions, it was impossible to

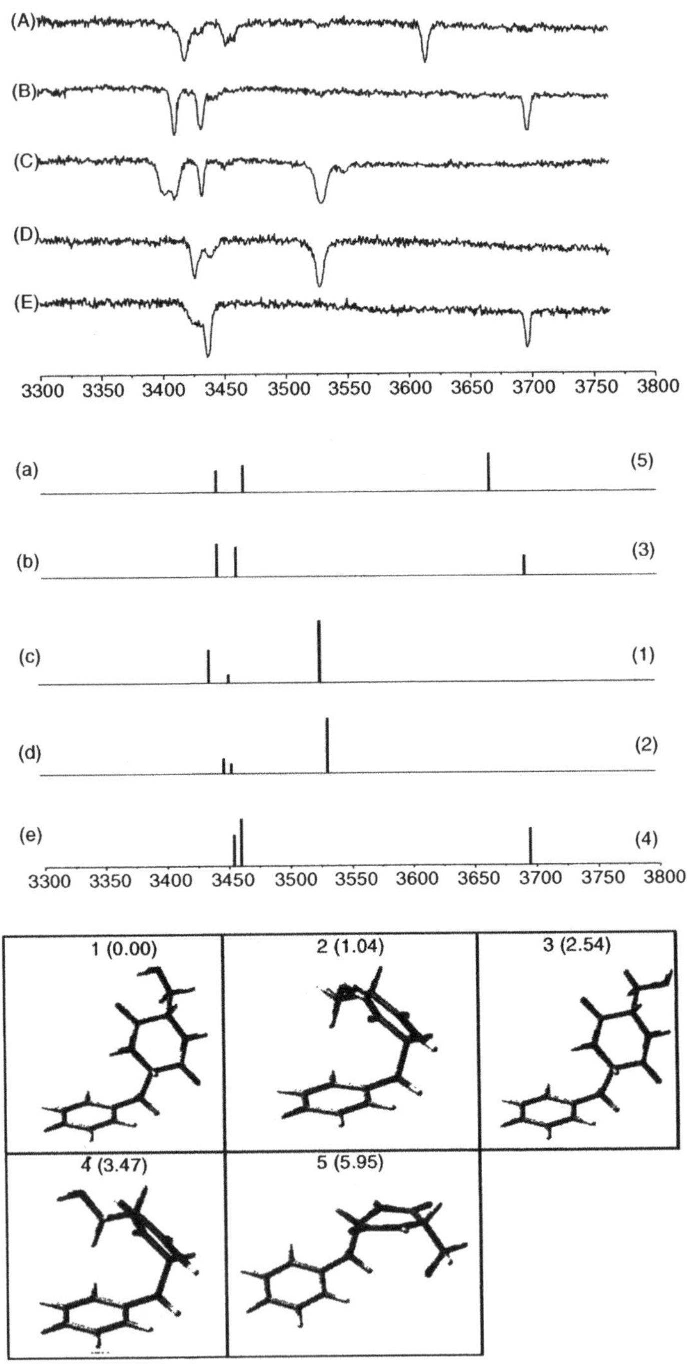

FIGURE 1.20. IR/R2PI spectra of the cyclopeptide Phe–Ser for all five experimentally observed isomers. The vibrational frequencies are calculated at the DFT [B3LYP/6-31G(*d,p*)] level. The calculated frequencies are in excellent agreement with the experimental values leading to an unambiguous assignment (Abo-Riziq et al. 2005) to the five structures given at the bottom. [Figure taken from Abo-Riziq et al. (2005).]

distinguish between folded and stretched conformations (Bakker et al. 2005). However, the experiment is a first step, indicating the possibility of investigating larger systems. Similarly, the cyclic gramicidin S, including 10 amino acid residues, has been investigated by applying IR/R2PI in the region of the NH stretching modes. Here different isomers have been observed (de Vries 2005). Finally, a large β-sheet model system formed by an aggregation of two tripeptide models (dimer of Ac–Val–Tyr–NHMe) has been investigated in the region of the NH stretching and in the region of C=O stretching vibrations using our new high-resolution laser system (Gerhards et al. 2006a). The dimer is the first β-sheet model that contains both the "inner"- and "outer"-bound binding motifs (see Figure 1.10); thus it serves as a unit cell model from which a large β sheet can be formed by simply multiplying the β-sheet unit.

All strategies and results have in common that the combined IR/UV spectroscopy is a very powerful tool for larger systems, as well; although R2PI spectra might be broad, the corresponding IR/R2PI spectra exhibit well-resolved spectral transitions. Furthermore, the width is an indication of the kind of transition, since narrow peaks indicate a free NH or CO group, whereas broader transitions arise from hydrogen-bonded structures, but these tendencies do not significantly depend on the size of the system. The question as to why IR spectroscopy is much less sensitive than UV spectroscopy to the broadening of resonances with respect to the size of the system is not easy to answer. It may result partly from the situation that in a large molecule some of the significant vibrational transitions are still localized. It can also be argued that using a UV laser only a small fragment of molecules are selected and that (for some systems) no significant number of different vibrational frequencies exists within this selection.

To obtain reliable results from combined IR/UV investigations, a clear strategy has to be performed:

(1) The approach must be systematic, starting from small systems and proceeding to larger ones. As described in this review, this technique has proved very successful during the last year.
(2) A broad spectral region within the IR has to be covered, which is supported by the introduction of new laser sources within the nanosecond regime.
(3) If a system still cannot unambiguously be assigned with respect to a very small number of alternative structures, high-resolution (rotationally resolved) spectroscopy may be applied, which (at the moment) is more limited by the size of the system than the IR/UV methods.
(4) In addition to all measurements, extensive calculations are required.

Usually first a selection of isomers should be performed by application of forcefield programs and molecular dynamics methods. The most stable conformers are then calculated at the ab initio and DFT levels and, if possible, also at the MP2 level of theory (see Section 1.7). For future work, inclusion of anharmonic couplings might also become more important, especially with respect to the spectra in the mid-IR region below 1000 cm^{-1}, which belongs in the "real" fingerprint region.

1.6. MOLECULAR RECOGNITION

The process of molecular recognition is extremely important for drug design; i.e. molecules that have a pharmaceutical effect should aggregate selectively and efficiently to a "receptor." Following the idea of producing pharmaceutically relevant substances, one must use molecules that fit exactly to a binding position. The process of blocking (binding) special positions is one strategy to inhibit the effect of proteins that are responsible for diseases; for instance, the development to larger β-sheet aggregates causes BSE/Creutzfeldt–Jakob and other prion-related diseases. β-Amyloids are discussed as a primary cause of Alzheimer disease (Mestel 1996; Prusiner 1996; Riesner et al. 1996; Lansbury 1996; Carrell and Lomas 1997; Wisniewski et al. 1998; Pitschke et al. 1998). The KLVFF sequence is a fragment in the central region of Aβ, which can be inhibited by blocking this sequence (Kirsten and Schrader 1997; Rzepecki et al. 2004a,b; Rzepecki and Schrader 2005; Schrader and Kirsten 1996; Kirsten and Schrader 1997; Tjernberg et al. 1996). All proteins with pathogenic effects responsible for BSE and Creutzfeldt–Jakob and Alzheimer diseases have prominent β-sheet structure regions. Using the strategy to develop template molecules that can aggregate to relevant β-sheet sequences, the class of aminopyrazole molecules has been tested (Kirsten and Schrader 1997; Rzepecki et al. 2004b; Schrader and Kirsten 1996). The aminopyrazoles can form two tautomers; thus they are able to aggregate to a peptide by forming two or three hydrogen bonds (see Figure 1.21). The efficiency and selectivity of the templates with respect to different peptide backbones require a detailed analysis of the molecular interactions, especially the hydrogen bond strength between template and peptide. To determine how these molecules bind with respect to the number of intermolecular hydrogen bonds and their relative bond strengths, the author's group undertook a systematic investigation of different isolated peptide/template clusters.

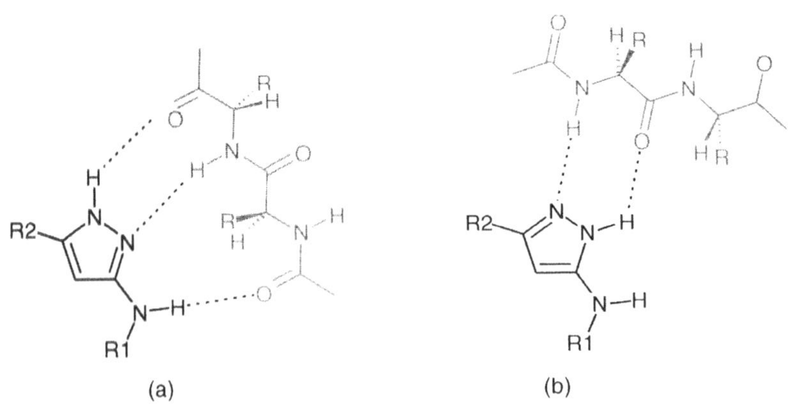

FIGURE 1.21. Different possibilities of aminopyrazole derivatives to aggregate to a peptide backbone. Two different tautomers can be formed, and only the one shown in (a) is able to undergo three intermolecular hydrogen bonds (binding sites). The number and strengths of the bonds are responsible for the selectivity and efficiency of the aggregation and is influenced by the choice of the R1 and R2 substitutions.

The initial investigations were performed on the most simple model systems, clusters of pyrazole or (methyl)aminopyrazole with the protected amino acid Ac–Phe–OMe (Unterberg et al. 2002). The resulting cluster forms two very stable hydrogen bonds indicated by strong redshifts of the NH stretching frequencies (compared to the monomers) in the IR/R2PI spectra. Further investigations are extended to clusters of dipeptide (Ac–Val–Phe–OMe) and tripeptide models [Ac–Val–Tyr(Me)–NHMe] with different aminopyrazols (with changes of the substitution at R1 and R2 positions; see Figure 1.21) (Gerhards et al. 2006b). These studies clearly reveal a strong competition between intermolecular hydrogen bonds within the peptide and intramolecular hydrogen bonds to the aminopyrazoles depending on the acidity of the aminopyrazole (Gerhards et al. 2006b).

Another very interesting class of peptides are the carbopeptoides. Here sugar molecules are incorporated in a peptide strand and are used to mimic secondary structure elements. This concept is also used in peptidomimetics (drug design). The carbopeptoids may work as a rigid template to influence the conformation of a peptide backbone. The first molecular-beam experiments on this type of molecule were performed by the Simons group (Jokusch and Simons 2005).

The objective of all these investigations, in cooperation with organic and pharmaceutical chemists, is to contribute to the development of efficient drugs to understand their actions on a molecular level, leading to new strategies as to how a most appropriate molecule may be designed. It is an interesting side aspect that a pharmaceutically relevant molecule should not exceed a molecular weight of 500 amu, since otherwise it cannot be resorbed by the organs (e.g., stomach, intestine). This limitation to "smaller" molecules may enable a very detailed analysis of these systems by the extremely specific double-resonance methods.

1.7. CALCULATIONS AND ASSIGNMENT OF VIBRATIONAL FREQUENCIES

One major problem in determining the correct structure formed in molecular-beam experiments is the large variety of possible isomers on the PES of an amino acid or peptide. As mentioned above, even for a protected amino acid 27 minima must be considered if only the two backbone angles ψ and ϕ and the angle χ_1 characterizing the sidechain are taken into account. If additional angles such as ω (characterizing the cis/trans position) or χ_2 (characterizing the sidechain; see Figure 1.7) and if more than one set of ψ and ϕ angles have to be considered, the number of possible isomers increases very rapidly. If larger peptides or clusters of peptides with solvent molecules (especially water) are examined, the number of isomers increases further. To identify the most stable isomers, it is necessary to scan the PES first on the basis of a low-cost method; thus, forcefield in combination with molecular dynamic calculations are performed by using different potentials such as Amber (Cornell et al. 1995; Kollman et al. 1997), OPLS-AA (Jorgensen et al. 1996), or CFF (Maple et al. 1998; CFF 2000). The first two potentials are used, for example, by Wales' and Zwier's group (Evans et al. 2004) whereas CFF has been chosen in our group (Gerhards et al. 2004). CFF is a class II forcefield that is parameterized on the basis of ab initio data and may thus be useful for predicting vibrational frequencies. The following procedure of selecting different conformers was chosen in the author's group. In a first step, the conformational landscape of the system of interest was scanned by a modified quenched-dynamics (Stillinger and Weber 1983; Kratochvil et al. 2000) technique. Beginning with an arbitrary starting geometry, a short-time molecular dynamics run (typically ~2 ps with 1 fs timestep) was performed at a certain temperature T. The final geometry was fully optimized and stored for later analysis. This minimized structure was reheated to the temperature T (i.e., a new set of impulse coordinates was generated) and another short-time dynamics run was performed. This procedure was executed until no new minima were found and was repeated with different simulation temperatures. All minima were sorted with respect to their relative energy.

After selecting a set of the most stable conformers obtained from the combined forcefield and molecular dynamic calculations, ab initio and DFT calculations are performed for these species in order to determine reliable vibrational frequencies (discussed below) and relative energies. Different levels of ab initio calculations have been used, starting with Hartree–Fock (HF) and density functional methods (DFT) up to Møller–Plesset (MP2) calculations. For larger systems fully optimized DFT calculations performed for a series of possible isomers are time-consuming; thus only MP2 single-point calculations can be performed on a DFT-optimized geometry. It should be pointed out that in the case of energetically very close-lying isomers a full optimization almost at the MP2 level is necessary, which, of course, also requires large basis sets (at least triple-ζ level). The RI-MP2 method can then be a helpful tool (Feyereisen and Fitzgerald 1993; Weigend et al. 1998). However, the fully optimized MP2 calculations for large systems (di- and tripeptides) are still very time-consuming.

The following summary results from all publications on the different types of protected amino acids and peptides mentioned above (Gerhards et al. 2002, 2004; Fricke et al. 2004, 2006; Snoek et al. 2000, 2001; Gerhards, Unterberg 2002; Chin et al. 2005a–e; Compagnon et al. 2005; Bakker et al. 2003; Dian et al. 2002, 2003; Unterberg et al. 2003; Huenig, Kleinermanns 2004; Bakker et al. 2005; Abo-Riziq et al. 2005). Vibrational frequencies give a direct hint of geometric arrangements. In an unprotected amino acid the free OH stretching vibration is located around 3580 cm^{-1} and shows a strong redshift to 3220–3280 cm^{-1} if it is involved in hydrogen bonding. The antisymmetric NH stretching vibration has been observed between 3400 and 3430 cm^{-1} and the symmetric NH stretching vibration between 3340 and 3370 cm^{-1}. By going to a protected amino acid with a free NH_2 group in the O=C–NH_2 amide unit, the antisymmetric NH stretching vibration is usually observed at ~3510–3550 cm^{-1}. An unperturbed (i.e., non-hydrogen-bonded) symmetric NH stretching frequency of a NH_2 group is located at ~3425 cm^{-1}. In the case of protected amino acids and peptide containing an NHMe group instead of a NH_2 group, the NH stretching frequency is at ~3470 cm^{-1} if the NH group is not involved in a hydrogen bond. The NH stretching modes that belong to a free NH group inside a peptide backbone are usually located at ~3400–3470 cm^{-1}. The exact frequency strongly depends on the chosen amino acid and on the secondary structure element where the acid is implemented; for instance, if a free NH group is part of a β-sheet-related or a γ-turn structure, its value is slightly different. However, for these fine tunings more datasets are required from further investigation performed on dipeptide and tripeptide model systems. Especially in the region at $3400\,cm^{-1}$, frequencies are found that belong to weakly hydrogen-bonded NH modes that appear, for example, in β-turn structures. Interactions of the NH group with the π system leads only to a small redshift of the observed frequency. Frequencies that are significantly below $3400\,cm^{-1}$ definitely belong to a hydrogen-bonded NH group. The redshift strongly depends on the strength of the hydrogen-bond; i.e. the very significant value of a hydrogen-bonded stretching mode can range from $\sim$3260 to $3390\,cm^{-1}$. Usually these hydrogen-bonded NH groups are involved in γ-turn arrangements. If a NH_2 group is involved in a hydrogen-bonded structure, the frequencies of not only the bonded NH group but also the asymmetric NH stretching vibration decrease. The latter frequency is now in the lower part of the 3510–$3550\,cm^{-1}$ range mentioned above. The most constant vibrational frequency belongs to the free NH group of the indole sidechain in Trp. Independently of the type of peptide or amino acid taken into account, the frequency differs by only a few wavenumbers at $\sim$3520–$3525\,cm^{-1}$.

In peptides the characterization of NH and CO groups is most important, where the CO stretching frequencies are responsible for the amide I region at $\sim$1600–$1800\,cm^{-1}$. If the CO stretching frequency belongs to a carboxyl group of a free acid, it is located at $\sim 1790\,cm^{-1}$; in the case of an ester the frequencies are $\sim 1760\,cm^{-1}$. If the CO group belongs to an amide group in the backbone of a peptide or to an amide group of the protection group at the end of the peptide, its frequency is $\sim 1700\,cm^{-1}$, ranging from 1650 to $1730\,cm^{-1}$. The differences between the C=O stretching vibrations can become very small (a few wavenumbers), but the pattern obtained for similar binding motifs is very significant and enables one to distinguish between different backbone conformers. The same holds for the location of the hydrogen-bonded C=O groups. The redshift of the corresponding stretching modes due to the hydrogen bonding is small (sometimes only $10\,cm^{-1}$) but significant.

In order to resolve the specific pattern of very close-lying CO stretching vibrations, the use of a high-resolution laser system is absolutely necessary; otherwise, only one broad nonsignificant transition covering all C=O stretching frequencies (amide I vibrations) would be observed. Finally it should be mentioned that in the region from 1000 to $\sim 1400\,cm^{-1}$ the upper part of the fingerprint region is located, covering especially C—H bending vibrations. In the range from $\sim$1400 to $1600\,cm^{-1}$ the amide II region has its window covering the NH bending modes.

In all publications on peptide models mentioned above the vibrational frequencies from the upper fingerprint region ($>1000\,cm^{-1}$) up to the region of the NH and OH stretching modes (in the case of free COOH groups or hydrated cluster) are calculated on the basis of a normal-mode analysis. Different levels have been chosen starting from HF up to DFT (usually with the B3LYP functional) and the MP2 method (usually only as single-point calculations on an optimized structure). Because of very good error compensations (between dipole strengths and correlation effects), it turns out that HF calculations using a small basis set [3-21G(*d*)] yield very useful results if only the scaled vibrational frequencies are taken into account; thus, spectral interpretation may be achieved by this simple method, but the energetic order of different isomers can be completely wrong. As much as possible, the higher-level methods are strongly required, but a large basis set, especially in the case of MP2 calculations, is necessary. Especially for the dipeptides the choice of DFT calculations using the B3LYP functional and the double-ζ 6-31 + G(d) basis set (which partly contains polarization and diffuse functions) becomes a kind of standard for frequency calculations. Another ansatz has been tested by using the frequencies directly obtained from forcefield calculations (Jansen et al. 2006). Here CFF has been chosen, and except for the description of C=O groups in hydrogen bonds and N–H groups in π bonds, some promising results have been obtained. The systematic use of lower-level methods in order to obtain frequency predictions is strongly required because of the rapid progress in investigating the frequencies of larger peptides experimentally.

Especially for the interpretation of the fingerprint region below 1000 cm^{-1}, the harmonic approximation are often no longer useful. Thus anharmonic couplings between different vibrational modes should be calculated [see, e.g., Brauer et al. (2005)]. If it were possible to reliably predict these vibrations, an enormous number of vibrational frequencies would yield a very significant interpretation of the correct conformer.

1.8. SUMMARY AND OUTLOOK

The applications presented in this chapter show that the variety of different experimental techniques enables us to analyze (1) structures in different electronic states and (2) the dynamical behavior (with respect to both electronical and vibrational excitation). The methods give a detailed insight on a molecular level in the building blocks of peptides. Experimentally high-resolution spectroscopy and isomer-selective double-resonance techniques in combination with mass spectrometry are chosen and combined with a theoretical analysis starting from forcefield up to high-level ab initio calculations. Compared to the status described in earlier reviews [Pratt, (1998), Robertson and Simons (2001), Simons (2003), Zwier (2001); see also special editions of *Mol. Phys.* **20**, 2005; *Phys. Chem. Chem. Phys.* **6**, 2004; *Eur. Phys. J. D*, 2002), very rapid development has been taken place in the last year. Starting with amino acids and dipeptides, the sizes are systematically increased in order to analyze the process of peptide folding, as well as the efficiency and selectivity of peptide aggregation.

Except for the neutral species presented in this chapter, another (future) field of research is the investigation of protonated (or more generally, charged) species with highly selective (double-resonance) methods. This has not been reviewed in this chapter, but one example showing that large protonated peptides can be investigated by IR spectroscopy in the gas phase is the analysis of cytochrom *c* by using a free-electron laser in combination with an ICR cell (Oomens et al. 2005). Here the fragmentation due to resonant IR excitations is recorded. Although no spectral resolution with respect to each amino acid can be excpected for this large protein, the

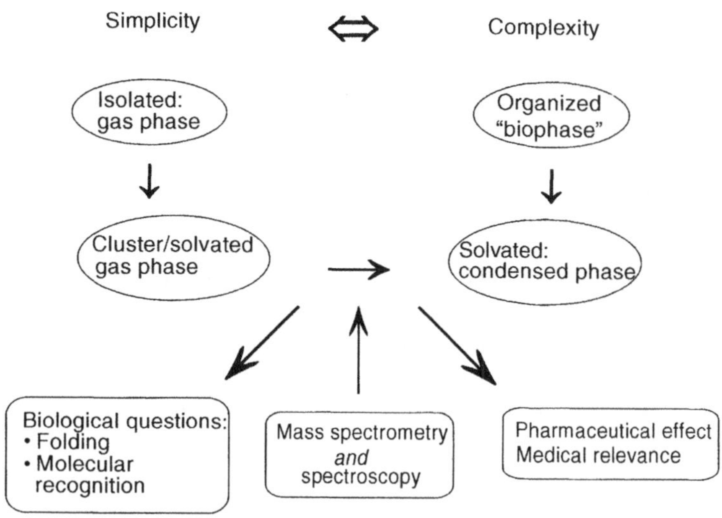

FIGURE 1.22. Scheme supplemented to the one shown in Simons (2003). It shows the two different directions starting either from the isolated molecule in the gas phase or the complex biophase in order to describe the structure and functionality of a peptide. One tool used to analyze biological relevant structures and processes on a molecular level very precisely is a combination of spectroscopic and mass spectrometric studies. [Figure partly redrawn from Simons (2003).]

possibilities of examining large isolated peptides (proteins) by IR spectroscopy is shown.

A nice overview showing the direction of different approaches to analyze structure and functionality of biomolecules is given in Figure 1.22. One extreme is the biophase from which larger isolated (solvated) aggregates are extracted and analyzed in the condensed phase. These systems can be investigated with respect to their biological function but cannot be analyzed precisely on a molecular level. The other extreme is the isolated (small) molecule in the gas phase, which can be described in full detail (rotational constants, vibrational frequencies) leading to a relative precise determination of its structure and dynamics. These systems can be clustered to form microsolvation shells or peptide aggregates to simulate extended secondary structures. On one hand, the difference between isolated species and the influence of an environment (analyzed in the condensed phase) is determined. On the other hand, the stepwise influence of an environment can be simulated (by adding, e.g., water, other peptides, or templates, molecule by molecule). While the sizes of peptides and their solvation shells investigated in the gas phase increase, the sizes of systems analyzed in solution can be reduced to get a clear indication of the influence of *one* special sequence (e.g., KLVFF) in a protein. We are approaching a closer understanding from both sides – gas phase and condensed phase – in order to understand on a molecular level the mechanisms responsible for the action of molecular recognition and the formation of secondary (and perhaps in the future also tertiary) structures.

REFERENCES

Abo-Riziq AG, Crews B, Bushnell JE, Callahan MP, de Vries MS (2005): Conformational analysis of cyclo(Phe-Ser) by UV-UV and IR-UV double resonance spectroscopy and ab initio calculations. *Mol. Phys.* **103**:1491–1495.

Allen WD, Czinki E, Csaszar AG (2004): Molecular structure of proline. *Chem. Eur. J.* **10**:4512–4517.

Andresen U, Dreizler H, Grabov J-U, Stahl W (1990): An automatic molecular beam microwave Fourier transform spectrometer. *Rev. Sci. Instrum.* **61**:3694–3699.

Bakker JM, Aleese LM, Meijer G, von Helden G (2003): Fingerprint IR spectroscopy to probe amino acid conformations in the gas phase. *Phys. Rev. Lett.* **91**:203003/1–203003/4.

Bakker JM, Plützer C, Hünig I, Häber T, Compagnon I, von Helden G, Meijer G, Kleinermanns K (2005): Folding structures of isolated peptides as revealed by gas-phase mid-infrared spectroscopy. *Chem. Phys. Chem.* **6**:120–128.

Balle TJ, Flygare WH (1981): Fabry-Perot cavity pulsed Fourier transform microwave spectrometer with a pulsed nozzle particle source. *Rev. Sci. Instrum.* **52**:33–45.

Baskin JS, Felker PM, Zewail AH (1986): Doppler-free time-resolved polarization spectroscopy of large molecules: Measurement of excited state rotational constants. *J. Chem. Phys.* **84**:4708–4710.

Bergmann T, Martin TP, Schaber H (1989): High-resolution time-of-flight mass-spectrometer. *Rev. Sci. Instrum.* **60**:792–793.

Berulis II, Winnewisser G, Krasnov VV, Sorochenko RL (1985): A search for interstellar glycine. *Sov. Astron. Lett.* **11**:251–253.

Bettens FL, Bettens RPA, Brown RD, Godfrey PD (2000): The microwave spectrum, structure, and ring-puckering of the cyclic dipeptide diketopiperazine. *J. Am. Chem. Soc.* **122**:5856–5860.

Blanco S, Lesarri A, López JC, Alonso JL (2004): The gas-phase structure of alanine. *J. Am. Chem. Soc.* **126**:11675–11683.

Bludsky O, Chocholousova J, Vacek J, Huisken F, Hobza P (2000): Anharmonic treatment of the lowest-energy conformers of glycine: A theoretical study. *J. Chem. Phys.* **113**: 4629–4635.

Boesl U, Weinkauf R, Schlag EW (1992): Reflectron time-of-flight mass-spectrometry and laser excitation for the analysis of neutrals, ionized molecules and secondary fragments. *Int. J. Mass Spectrom Ion Proc.* **112**:121–166.

Bondybey VE, English JH (1981): Laser-induced fluorescence of metal-clusters produced by laser vaporization—gas-phase spectrum of pb-2. *J. Chem. Phys.* **74**:6978–6979.

Brauer B, Chaban GM, Gerber RB (2004): Spectroscopically-improved, semi-empirical potentials for biological molecules: Calculations for glycine, alanine and proline. *Phys. Chem. Chem. Phys.* **6**:2543–2556.

Brauer B, Gerber RB, Kabelac M, Hobza P, Bakker JM, Riziq AGA, de Vries MS (2005): Vibrational spectroscopy of the G···C base pair: Experiment, harmonic and anharmonic calculations, and the nature of the anharmonic couplings. *J. Phys. Chem. A.* **109**: 6974–6984.

Brown RD, Godfrey PD, Storey JWV, Bassez MP (1978): Microwave-spectrum and conformation of glycine. *J. Chem. Soc. Chem. Commun.* **1**:547–548.

Cable JR, Tubergen MJ, Levy DH (1987): Laser desorption molecular beam spectroscopy: The electronic spectra of tryptophan peptides in the gas phase. *J. Am. Chem. Soc.* **109**:6198–6199.

Cable JR, Tubergen MJ, Levy DH (1988a): Electronic spectroscopy of small tryptophan peptides in supersonic molecular beams. *J. Am. Chem. Soc.* **110**:7349–7355.

Cable JR, Tubergen MJ, Levy DH (1988b): The electronic spectra of small peptides in the gas phase. *Faraday Discuss. Chem. Soc.* **86**:143–152.

Caminati W (2004): The rotational spectra of conformers of biomolecules: Tryptamine. *Phys. Chem. Chem. Phys.* **6**:2806–2809.

Çarçabal P, Kroemer RT, Snoek LC, Simons JP, Bakker JM, Compagnon I, Meijer G, von Helden G (2004): Hydrated complexes of tryptophan: Ion dip infrared spectroscopy in the "molecular fingerprint" region, 100–2000 cm-1. *Phys. Chem. Chem. Phys.* **6**:4546–4552.

Carney JR, Zwier TS (2000): The infrared and ultraviolet spectra of individual conformational isomers of biomolecules: tryptamine. *J. Phys. Chem. A.* **104**:8677–8688.

Carney JR, Dian BC, Florio GM, Zwier TS (2001): The role of water bridges in directing the conformational preferences of 3-indole-propionic acid and tryptamine. *J. Am. Chem. Soc.* **123**:5596–5597.

Carrell RW, Lomas DA (1997): Conformational disease. *Lancet* **350**:134–138.

CFF Consistent Force Field and DISCOVER Molecular Simulation Program, Version 2000. Molecular Simulations Inc., San Diego.

Chaudhari A, Sahu PK, Lee SL (2004): Microsolvation effect on alanine: A many-body interaction study using ab initio method. *J. Mol. Struct. (Theochem)* **683**:115–119.

Chaudhari A, Lee SL (2005a): Computational study of glycine–(water)3 complex by density functional method. *Chem. Phys.* **310**:281–285.

Chaudhari A, Lee SL (2005b): Density functional theory study of contribution of many-body energies to binding energy for alanine–(water)(4) complex. *Int. J. Quantum Chem.* **102**:174–177.

Chin W, Mons M, Dognon J-P, Piuzzi P, Tardivel B, Dimicoli I (2004): Competition between local conformational preferences and secondary structures in gas-phase model tripeptides as revealed by laser spectroscopy and theoretical chemistry. *Phys. Chem. Chem. Phys.* **6**:2700–2709.

Chin W, Compagnon I, Dognon J-P, Canuel C, Piuzzi F, Dimicoli I, von Helden G, Meijer G, Mons M (2005a): Spectroscopic evidence for gas-phase formation of successive b-turns in a three-residue peptide chain. *J. Am. Chem. Soc.* **127**:1388–1389.

Chin W, Dognon J-P, Canuel C, Piuzzi F, Dimicoli I, Mons M, Compagnon I, von Helden G, Meijer G (2005b): Secondary structures of short peptide chains in the gas phase: Double resonance spectroscopy of protected dipeptides. *J. Chem. Phys.* **122**:054317/1–054317/8.

Chin W, Dognon J-P, Piuzzi F, Tardivel B, Dimicoli I, Mons M (2005c): Intrinsic folding of small peptide chains: Spectroscopic evidence for the formation of β-turns in the gas phase. *J. Am. Chem. Soc.* **127**:707–712.

Chin W, Mons M, Dognon J-P, Mirasol R, Chass G, Dimicoli I, Piuzzi F, Butz P, Tardivel B, Compagnon I, von Helden G, Meijer G (2005d): The gas-phase dipeptide analogue acetyl-phenylalanyl-amide: A model for the study of side chain/backbone interactions in proteins. *J. Phys. Chem. A* **109**:5281–5288.

Chin W, Piuzzi F, Dognon J-P, Dimicoli I, Tardivel B, Mons M (2005e): Gas phase formation of a 310-helix in a three-residue peptide chain: Role of side chain-backbone interactions as evidenced by IR-UV double resonance experiments. *J. Am. Chem. Soc.* **127**:11900–11901.

Chocholousova J, Vacek J, Huisken F, Werhahn O, Hobza P (2002): Stacked structure of the glycine dimer is more stable than the cyclic planar geometry with two O–H···O hydrogen bonds: Concerted action of empirical, high-level nonempirical ab initio, and experimental studies. *J. Phys. Chem.* **A106**:11540–11549.

Clarkson JR, Baquero E, Shubert VA, Myshakin EM, Jordan KD, Zwier TS (2005a): Laser-initiated shuttling of a water molecule between H-bonding sites. *Science* **307**:1443–1446.

Clarkson JR, Baquero E, Zwier TS (2005b): Direct measurement of the energy thresholds to conformational isomerization. II. 3-Indole-propionic acid and its water-containing complex. *J. Chem. Phys.* **122**:214312/1–214312/8.

Clarkson JR, Dian BC, Moriggi L, DeFusco A, McCarthy V, Jordan KD, Zwier TS (2005c): Direct measurement of the energy thresholds to conformational isomerization in trypatamine: experiment and theory. *J. Chem. Phys.* **122**:214311/1–214311/15.

Cohen R, Brauer B, Nir E, Grace L, de Vries MS (2000): Resonance-enhanced multiphoton ionization spectroscopy of dipeptides. *J. Phys. Chem. A.* **104**:6351–6355.

Compagnon I, Oomens J, Bakker J, Meijer G, von Helden G (2005): Vibrational spectroscopy of a anon-aromatic amino acid-based model peptide: Identification of the γ-turn motif of the peptide backbone. *Phys. Chem. Chem. Phys.* **7**:13–15.

Connell LL, Corcoran TC, Joireman PW, Felker PM (1990): Conformational analysis of jet-cooled tryptophan analogs by rotational coherence spectroscopy. *Chem. Phys. Lett.* **166**:510–516.

Cornell WD, Cieplak P, Bayly CI, Gould IR, Merz KM, Ferguson DM, Spellmeyer DC, Fox T, Caldwell JW, Kollman PA (1995): A 2nd generation force-field for the simulation of proteins, nucleic-acids, and organic-molecules. *J. Am. Chem. Soc.* **117**:5179–5197.

Császár A (1996): Conformers of gaseous a-alanine. *J. Phys. Chem.* **100**:3541–3551.

Czinski E, Császár AG (2003): Conformers of gaseous proline. *Chem. Eur. J.* **9**:1008–1019.

de Vries M (2005): Private communication.

Dian BC, Longarte A, Mercier S, Evans DA, Wales DJ, Zwier TS (2002a): The infrared and ultraviolet spectra of single conformations of methyl-capped dipeptides: N-Acetyl tryptophan amide and N-acetyl tryptophan methyl amide. *J. Chem. Phys.* **117**:10688–10702.

Dian BC, Longarte A, Zwier TS (2002b): Conformational dynamics in a dipeptide after single-mode vibrational excitation. *Science* **296**:2369–2373.

Dian BC, Longarte A, Zwier TS (2003): Hydride stretch infrared spectra in the excited states of indole and its derivatives: Direct evidence for the 1ps* state. *J. Chem. Phys.* **118**: 2969–2706.

Dian BC, Clarkson JR, Zwier TS (2004a): Direct measurement of energy thresholds to conformational isomerization in tryptamine. *Science* **303**:1169–1173.

Dian BC, Florio GM, Clarkson JR, Longarte A, Zwier TS (2004b): Infrared-induced conformational isomerization and vibrational relaxation dynamics in melatonin and 5-methoxy-N-acetyl tryptophan methyl amide. *J. Chem. Phys.* **120**:9033–9046.

Dian BC, Longarte A, Winter PR, Zwier TS (2004c): The dynamics of conformational isomerization in flexible biomolecules. I. Hole-filling spectroscopy of N-acetyl tryptophan methyl amide and N-acetyl tryptophan amide. *J. Chem. Phys.* **120**:133–147.

Dickinson JA, Hockridge MA, Robertson EG, Simons JP (1999): Molecular and supramolecular structures of N-phenyl formamide and its hydrated clusters. *J. Phys. Chem. A* **103**:6938–6949.

Dietz TG, Duncan MA, Powers DE, Smalley RE (1981): Laser production of supersonic metal cluster beams. *J. Chem. Phys.* **74**:6511–6512.

Diken EG, Hammer NI, Johnson MA (2004): Preparation and photoelectron spectrum of the glycine molecular anion: Assignment to a dipole-bound electron species with a high-dipole moment, non-zwitterionic form of the neutral core. *J. Chem. Phys.* **120**: 9899–9902.

Diken EG, Headrick JM, Johnson MA (2005): Photoelectron spectroscopy of the [glycine $\cdot(H_2O)_{1,2}](-)$ clusters: Sequential hydration shifts and observation of isomers. *J. Chem. Phys.* Article 224317.

Ding Y, Krogh-Jespersen K (1992): The glycine zwitterion does not exist in the gas phase: Results from a detailed ab initio electronic structure study. *Chem. Phys. Lett.* **199**:261–266.

Douberly GE, Merritt JM, Miller RE (2005): IR-IR double resonance spectroscopy in helium nanodroplets: Photo-induced isomerization. *Phys. Chem. Chem. Phys.* **7**:463–468.

Duax WL, Griffin JF, Langs DA, Smith GD, Grochulski P, Pletnev V, Ivanov V (1996): Molecular structure and mechanisms of action of cyclic and linear ion transport antibiotics. *Biopolymers* **40**:141–155.

Ebata T, Mizuochi T, Watanabe T, Mikami N (1996): OH stretching vibratios of phenol-$(H_2O)_1$ and phenol-$(H_2O)_3$ in the S_1 state. *J. Phys. Chem.* **100**:546–550

Engh RA, Chen LX-Q, Fleming GR (1986): Conformational dynamics of tryptophan: A proposal for the origin of the non-exponential fluorescence decay. *Chem. Phys. Lett.* **126**:365–372.

Evans DA, Wales DJ, Dian BC, Zwier TS (2004): The dynamics of conformational isomerization in flexible biomolecules. II. Simulating isomerizations in a supersonic free jet with master equation dynamics. *J. Chem. Phys.* **120**:148–157.

Fedorov AV, Cable JR (2000): Spectroscopy of hydrogen-bonded formanilide clusters in a supersonic jet: Solvation of a model trans amide. *J. Phys. Chem. A* **104**:4943–4952.

Felker PM, Baskin JS, Zewail AH (1986): Rephasing of collisionless molecular rotational coherence in large molecules. *J. Phys. Chem.* **90**:724–728.

Felker PM (1992): Rotational coherence spectroscopy: Studies of the geometries of large gas-phase species by picosecond time-domain methods. *J. Phys. Chem.* **96**:7844–7857.

Fenn JB, Mann M, Meng CK, Wong SF, Whitehouse CM (1989): Electrospray ionization for mass-spectrometry of large biomolecules. *Science* **246**:64–71.

Feyereisen M, Fitzgerald G (1993): Use of approximate integrals in ab initio theory. An application in MP2 energy calculations. *Chem. Phys. Lett.* **208**:359–363.

Florio GM, Christie RA, Jordan KD, Zwier TS (2002): Conformational preferences of jet-cooled melatonin: Probing trans- and cis-amide regions of the potential energy surface. *J. Am. Chem. Soc.* **124**:10236–10247.

Florio GM, Zwier TS (2003): Solvation of a flexible biomolecule in the gas phase: the ultraviolet and infrared spectroscopy of melatonin-water clusters. *J. Phys. Chem. A.* **107**:974–983.

Fricke H, Gerlach A, Unterberg C, Rzepecki P, Schrader T, Gerhards M (2004): Structure of the tripeptide model Ac-Val-Tyr(OMe)-NHMe and its cluster with water investigated by IR/UV double resonance spectroscopy. *Phys. Chem. Chem. Phys.* **6**:4636–4641.

Fricke H, Gerhards M (2005): Unpublished results

Fricke H, Gerlach A, Gerhards M (2006a): Structure of a β-sheet model system in the gas phase: Analysis of the fingerprint region up to 10 μm. *Phys. Chem. Chem. Phys.* **8**: 1660–1662.

Fricke H, Unterberg C, Gerlach A, Gerhards M (2006b): IR/UV double resonance spectroscopy and theoretical investigations applied to the clusters of the solvated protected amino acid Ac-Phe-OMe$(H_2O)_{1\text{-}3}$ in the gas phase, to be submitted.

Fricke H, Schrader T., Gerhards M (2006c): The tetrapeptide model Ac-Leu-Val-Tyr-NHMe in the gas phase: Discussion of different secondary structures, to be submitted.

Gelin BR, Karplus M (1979): Side-chain torsional potentials—effect of dipeptide, protein, and solvent environment. *Biochemistry* **18**:1256–1268.

Gerhards M, Unterberg C (2002): Structures of the protected amino acid Ac-Phe-OMe and its dimer: A ß-sheet model system in the gas phase. *Phys. Chem. Chem. Phys.* **4**:1760–1765.

Gerhards M, Unterberg C, Gerlach A (2002): Structure of a β-sheet model system in the gas phase: analysis of the C=O stretching vibrations. *Phys. Chem. Chem. Phys.* **4**:5563–5565.

Gerhards M (2004): High energy and narrow bandwidth mid IR nanosecond laser system. *Opt. Commun.* **241**:493–497.

Gerhards M, Unterberg C, Gerlach A, Jansen A (2004): β-Sheet model systems in the gas phase: Structures and vibrations of Ac–Phe–NHMe and its dimer $(\text{Ac–Phe–NHMe})_2$. *Phys. Chem. Chem. Phys.* (special issue on bioactive molecules in the gas phase) **6**:2682–2690.

Gerhards M, Fricke H, Schrader T (2006a): Investigation of secondary structure elements by IR/UV double resonance spectroscopy : Analysis of an isolated β-sheet model system, to be submitted.

Gerhards M, Unterberg C, Fricke H, Gerlach A, Schrader T (2006b): Clusters of protected dipeptides with pyrazole derivatives: The efficiency of forming a β-sheet model system in the gas phase. to be submitted.

Gerlach A, Unterberg C, Fricke H, Gerhards M (2005): Structures of Ac-Trp-OMe and its dimer $(\text{Ac-Trp-OMe})_2$ in the gas phase: Influence of a polar group in the side chain. *Mol. Phys.* (J. Simons special issue) **103**:1521–1529.

Godfrey PD, Firth S, Hatherley LD, Brown RD, Pierlot AP (1993): Millimeter-wave spectroscopy of biomolecules: Alanine. *J. Am. Chem. Soc.* **115**:9687–9691.

Godfrey PD, Brown RD (1995): Shape of glycine. *J. Am. Chem. Soc.* **117**:2019–2023.

Godfrey PD, Brown RD, Rodgers FM (1996): The missing conformers of glycine and alanine: Relaxation in seeded supersonic jets. *J. Mol. Struct.* **376**:65–81.

Grabow JU, Stahl W, Dreizler H (1996): A multioctave coaxially oriented beam-resonator arrangement Fourier-transform microwave spectrometer. *Rev. Sci. Instrum.* **67**:4072–4084.

Grotemeyer J, Boesl U, Walter K, Schlag EW (1986): A general soft ionization method for mass spectrometry: Resonanace-enhanced multi-photon ionization of biomolecules. *Org. Mass Spectrom.* **21**:645–653.

Guélin M, Cernicharo J (1989): In Winnewisser G, Armstrong JT (eds), *The Physics and Chemistry of InterstellarMolecular Clouds*, Springer, Berlin.

Gutowski M, Skurski P, Simons J (2000): Dipole-bound anions of glycine based on the zwitterion and neutral structures. *J. Am. Chem. Soc.* **122**:10159–10162.

Harmony MD, Beran KA, Angst DM, Ratzlaff KL (1995): A compact hot-nozzle Fourier-transform microwave spectrometer. *Rev. Sci. Instrum.* **66**:5196–5202.

Hollis JM, Snyder LE, Suenram RD, Lovas FJ (1980): A search for the lowest-energy conformer of interstellar glycine. *Astrophys. J.* **241**:1001–1006.

Huisken F, Kulcke A, Voelkel D, Laush C, Lisy JM (1993): New infrared injection-seeded optical parametric oscillator with high-energy and narrow bandwidth output. *Appl. Phys. Lett.* **62**:805–807.

Huisken F, Werhahn O, Ivanov AY, Krasnokutski SA (1999): The O-H stretching vibrations of glycine trapped in rare gas matrices and helium clusters. *J. Chem. Phys.* **111**: 2978–2984.

Hünig I, Seefeld KA, Kleinermanns K (2003): REMPI and UV-UV double resonance spectroscopy of tryptophan ethylester and the dipeptides tryptophan-serine, glycine-tryptophan and proline-tryptophane. *Chem. Phys. Lett.* **369**:173–179.

Hünig I, Kleinermanns K (2004): Conformers of the peptides glycine-tryptophan, tryptophan-glycine and tryptophan-glycine-glycine as revealed by double resonance laser spectroscopy. *Phys. Chem. Chem. Phys.* **6**:2650–2658.

Hutchinson EG, Thornton JM (1994): A revised set of potentials for beta-turn formation in proteins. *Protein Sci.* **3**:2207–2216.

IUPAC (1974): IUB Commission on Biochemical Nomenclature, *Pure Appl. Chem.* **40**: 291.

Jansen A, Fricke H, Gerhards M (2006): Structure and vibrations of amino acids and small peptides: A detailed analysis at the force field level of theory. *J. Phys. Chem. A* to be submitted

Jensen JH, Gordon MS (1995): On the number of water molecules necessary to stabilize the glycine zwitterion. *J. Am. Chem. Soc.* **117**:8159–8170.

Jokusch R, Simons JP (2005): Private communication.

Jorgensen WL, Maxwell DS, TiradoRives J (1996): Development and testing of the OPLS all-atom force field on conformational energetics and properties of organic liquids. *J. Am. Chem. Soc.* **118**:11225–11236.

Karas M, Hillenkamp F (1988): Laser desorption ionization of proteins with molecular masses exceeding 10000 daltons. *Anal. Chem.* **60**:2299–2301.

Kassab E, Langlet J, Evleth E, Akacem Y (2000): Theoretical study of solvent effect on intramolecular proton transfer of glycine. *J. Mol. Struct. (Theochem)* **531**:267–282.

Kirsten C, Schrader T (1997): Intermolecular b-sheet stabilization with aminopyrazoles. *J. Am. Chem. Soc.* **119**:12061–12068.

Kleinekofort W, Pfenninger A, Plomer T, Griesinger C, Brutschy B (1996): Observation of noncovalent complexes using laser-induced liquid beam ionization desorption. *Int. J. Mass Spectrom. Ion Proc.* **156**:195–202.

Kollman P, Dixon R, Cornell W, Fox T, Chipot C, Pohorille A (1997): *Comput. Sim. Biomol. Syst.* **3**:83.

Kratochvil M, Šponer J, Hobza P (2000): Global minimum of the adenine···thymine base pair corresponds neither to Watson-Crick nor to Hoogsteen structures. Molecular dynamic/quenching/AMBER and ab initio beyond Hartree-Fock studies. *J. Am. Chem. Soc.* **122**:3496–3499.

Kubik S, Kirchner R, Nolting D, Seidel J (2002): A molecular oyster: A neutral anion receptor containing two subunits with a remarkable sulfate affinity in aqueous solution. *J. Am. Chem. Soc.* **124**:12752–12760.

Kühlewind H, Neusser HJ, Schlag EW (1984): Production and unimolecular decay-rate of internal energy selected molecular-ions in a laser reflectron time-of-flight mass-spectrometer. *J. Phys. Chem.* **88**:6104–6106.

Kuyanov KE, Momose T, Vilesov AF (2004): Solid hydrogen Raman shifter for the mid-infrared range (4.4-8 μm). *Appl. Opt.* **43**:6023–6029.

Lansbury PT (1996): A reductionist view of Alzheimer's disease. *Acc. Chem. Res.* **29**: 317–321.

Lavrich RJ, Farrar JO, Tubergen MJ (1999): Heavy-atom structure of alaninamide from rotational spectroscopy. *J. Phys. Chem. A* **103**:4659–4663.

Lavrich RJ, Tubergen MJ (2000): Conformation and hydrogen bonding in the alaninamide-water van der Waals complex. *J. Am. Chem. Soc.* **122**:2938–2943.

Lavrich RJ, Torok CR, Tubergen MJ (2002): Effect of the bulky side chain on the backbone structure of the amino acid derivative valinamide. *J. Phys. Chem. A* **106**:8013–8018.

Lavrich RJ, Hight Walker AR, Plusquellic DF, Kleiner I, Suenram RD, Hougen JT, Fraser GT (2003a): Conformational analysis of the jet-cooled peptide mimetic ethylacetamidoacetate from torsion-rotation spectra. *J. Chem. Phys.* **119**:5497–5504.

Lavrich RJ, Plusquellic DF, Suenram RD, Fraser GT, Hight Walker AR, Tubergen MJ (2003b): Experimental studies of peptide bonds: Identification of the C7eq conformation of the alanine dipeptide analog N-acetyl-alanine N′-methylamide from torsion-rotation interactions. *J. Chem. Phys.* **118**:1253–1265.

Lee KT, Sung J, Lee KJ, Kim SK, Park YD (2002a): Resonant two-photon ionization study of jet-cooled amino acid: L-phenylalanine and its monohydrated complex. *J. Chem. Phys.* **116**:8251–8254.

Lee KT, Sung J, Lee KJ, Park YD, Kim SK (2002b): Conformation-dependent ionization energies of L-phenylalanine. *Angew. Chem. Int. Ed.* **41**:4114–4117.

Lee KT, Sung JH, Lee KJ, Kim SK, Park YD (2003): Conformation-dependent ionization of L-phenylalanine: Structures and energetics of cationic conformers. *Chem. Phys. Lett.* **368**:262–268.

Lee Y, Jung J, Kim B, Butz P, Snoek LC, Kroemer RT, Simons JP (2004): Alanyl side chain folding in phenylalanine: Conformational assignments through ultraviolet rotational band contour analysis. *J. Phys. Chem. A* **108**:69–73.

Legon AC (1983): Pulsed-nozzle, Fourier-transform microwave spectroscopy of weakly bound dimers. *Annu. Rev. Phys. Chem.* **34**:275–300.

Lesarri A, Mata S, Cocinero EJ, Blanco S, López JC, Alonso JL (2002): The structure of neutral proline. *Angew. Chem. Int. Ed.* **41**:4673–4676.

Lesarri A, Mata S, López JC, Alonso JL (2003): A laser-ablation molecular-beam Fourier-transform microwave spectrometer: The rotational spectrum of organic solids. *Rev. Sci. Instrum.* **74**:4799–4804.

Lesarri A, Cocinero EJ, López JC, Alonso JL (2004): The shape of neutral valine. *Angew. Chem. Int. Ed.* **43**:605–610.

Lesarri A, Cocinero EJ, López JC, Alonso JL (2005): Shape of 4(S)- and 4(R)-hydroxyproline in gas phase. *J. Am. Chem. Soc.* **127**:2572–2579.

Linder R, Nispel M, Häber T, Kleinermanns K (2005): Gas-phase FT-IR-spectra of neutral amino acids. *Chem. Phys. Lett.* **409**:260–264.

Lindinger A, Toennies JP, Vilesov AF (1999): High resolution vibronic spectra of the amino acids tryptophan and tyrosine in 0.38 K cold helium droplets. *J. Chem. Phys.* **110**:1429–1436.

Lindinger A, Lugovoj E, Toennies JP, Vilesov AF (2001): Splitting of the zero phonon lines of indole, 3-methyl indole, tryptamine and N-acetyl tryptophan amide in helium droplets. *Z. Phys. Chem.* (Munich) **215**:401–416.

Lipert RJ, Colson SD (1989): Persistent spectral hole burning of molecular clusters in a supersonic jet. *J. Phys. Chem.* **93**:3894–3896.

Lovas FJ, Kawashima Y, Grabow J-U, Suenram RD, Fraser GT, Hirota E (1995): Microwave spectra, hyperfine structure, and electric dipole moments for conformers I and II of glycine. *Astrophys. J.* **455**:L201-L204.

Lovas FJ, Lavrich RJ, Plusquellic DF (2004): Methy torsional state analysis of the jet-cooled microwave spectrum of N-acetylglycine. *J. Mol. Spectrosc.* **228**:251–257.

Lucas B, Lecomte F, Reimann B, Barth HD, Gregoire G, Bouteiller Y, Schermann JP, Desfrancois C (2004): A new infrared spectroscopy technique for structural studies of mass-selected neutral polar complexes without chromophore. *Phys. Chem. Chem. Phys.* **6**:2600–2604.

Lucas B, Gregoire G, Lecomte F, Reimann B, Schermann JP, Desfrancois C (2005): Infrared spectroscopy of mass-selected neutral molecular systems without chromophore: The formamide monomer and dimer. *Mol. Phys.* **103**:1497–1503.

Majewski W, Meerts WL (1984): Near-UV spectra with fully resolved rotational structure of naphthalene and perdeuterated naphthalene. *J. Mol. Spectrosc.* **104**:271–281.

Mamyrin BA, Karataev VI, Shmikk DV, Zagulin VA (1973): Mass reflectron. New nonmagnetic time-of-flight high-resolution mass spectrometer. *Zh. Eksp. Teor. Fiz.* **64**:82–89.

Manea VP, Wilson KJ, Cable JR (1997): Conformations and relative stabilities of the cis and trans isomers in a series of isolated N.-phenylamides. *J. Am. Chem. Soc.* **119**: 2033–2039.

Maple JR, Hwang M-J, Jalkanen KJ, Stockfisch TP, Hagler AT (1998): Derivation of class II force fields. V. Quantum force field for amides, peptides, and related compounds. *J. Comput. Chem.* **19**:430–458.

Martinez SJ III, Alfano JC, Levy DH (1991): The electronic spectroscopy of tyrosine and phenylalanine analogs in a supersonic jet: Acidic analogs. *J. Mol. Spectrosc.* **145**: 100–111.

Martinez SJ III, Alfano JC, Levy DH (1992): The electronic spectroscopy of the amino acids tyrosine and phenylalanine in a supersonic jet. *J. Mol. Spectrosc.* **156**:421–430.

Martinez SJ III, Alfano JC, Levy DH (1993): The electronic spectroscopy of tyrosine and phenylalanine analogs in a supersonic jet: Basic analogs. *J. Mol. Spectrosc.* **158**:82–92.

McGlone SJ, Godfrey PD (1995): Rotational spectrum of a neurohormone: β-Alanine. *J. Am. Chem. Soc.* **117**:1043–1048.

McGlone SJ, Elmes PS, Brown RD, Godfrey PD (1999): Molecular structure of a conformer of glycine by microwave spectroscopy. *J. Mol. Struct.* **486**:225–238.

Meijer G, de Vries MS, Hunziker HE, Wendt HR (1990): Laser desorption jet-cooling of organic molecules. *Appl. Phys. B* 395–403.

Melandri S, Maris A (2004): Intramolecular hydrogen bonds and conformational properties of biogenic amines: A free-jet microwave study of tyramine. *Phys. Chem. Chem. Phys.* **6**:2863–2866.

Mestel R (1996): Putting prions to the test. *Science* **273**:184–189.

Miller TF III, Clary DC, Meijer AJHM (2005): Collision-induced conformational changes in glycine. *J. Chem. Phys.* **122**:244323/1–244323/13.

Mons M, Dimicoli I, Tardivel B, Piuzzi F, Robertson EG, Simons JP (2001): Energetics of the gas phase hydrates of trans-formamilidine: A microscopic approach to the hydration sites of the peptide bond. *J. Phys. Chem. A* **105**:969–973.

Nguyen TV, Korter TM, Pratt DW (2005): Tryptamine in the gas phase. A high resolution laser study of the structural and dynamic properties of its ground and electronically excited states. *Mol. Phys.* **103**:1603–1613.

Oepts D, van der Meer AFG, van Amersfoort PW (1994): The free-electron-laser user facility FELIX. *Infrared Phys. Technol.* **36**:297–308.

Ohashi N, Hougen JT, Suenram RD, Lovas FJ, Kawashima Y, Fujitake M, Pyka J (2004): Analysis and fit of the Fourier-transform microwave spectrum of the two-top molecule N-methylacetamide. *J. Mol. Spectrosc.* **227**: 28–42.

Oomens J, Polfer N, Moore DT, van der Meer L, Marshall AG, Eyler JR, Meijer G, von Helden G (2005): Charge-state resolved mid-infrared spectroscopy of a gas-phase protein. *Phys. Chem. Chem. Phys.* **7**:1345–1348.

Page RH, Shen YR, Lee YT (1988): Local modes of benzene and benzene dimer, studied by infrared-ultraviolet double-resonance in a supersonic beam. *J. Chem. Phys.* **88**:4621–4636.

Park YD, Rizzo TR, Peteanu LA, Levy DH (1986): Electronic spectroscopy of tryptophan analogs in supersonic jets: 3-Indole acetic acid, 3-indole propionic acid, tryptamine, and N-acetyl tryptophan ethyl ester. *J. Chem. Phys.* **84**:6539–6549.

Perczel A, Ángyán JG, Kajtár M, Viviani W, Rivail JL, Marcoccia JF, Csizmadia IG (1991): Peptide models. 1. Topology of selected peptide conformational potential energy surfaces (glycine and alanine derivatives). *J. Am. Chem. Soc.* **113**:6256–6265.

Peteanu LA, Levy DH (1988): Spectroscopy of complexes of tryptamine and 3-indolepropionic acid with various solvents. *J. Phys. Chem.* **92**:6554–6561.

Philips LA, Levy DH (1986): Determination of the transition moment and the geometry of tryptamine by rotationally resolved electronic spectroscopy. *J. Phys. Chem.* **90**:4921–4923.

Philips LA, Levy DH (1988): Rotationally resolved electronic spectroscopy of tryptamine conformers in a supersonic jet. *J. Chem. Phys.* **89**:85–90.

Philips LA, Webb SP, Martinez SJ III, Fleming GR, Levy DH (1988): Time-resolved spectroscopy of tryptophan conformers in a supersonic jet. *J. Am. Chem. Soc.* **110**:1352–1355.

Pitschke M, Prior R, Haupt M, Riesner D (1998): Detection of single amyloid beta-protein aggregates in the cerebrospinal fluid of Alzheimer's patients by fluorescence correlation spectroscopy. *Nat. Med.* **4**:832–834.

Piuzzi F, Dimicoli I, Mons M, Tardivel B, Zhao Q (2000): A simple laser vaporization source for thermally fragile molecules coupled to a supersonic expansion: Application to the spectroscopy of tryptophan. *Chem. Phys. Lett.* **320**:282–288.

Pratt DW (1998): High resolution spectroscopy in the gas-phase: Even large molecules have well-defined shapes. *Annu. Rev. Phys. Chem.* **49**:481–530.

Prusiner SB (1996): Human prion diseases and neurodegeneration. Prions Prions Prions. *Curr. Top. Microbiol. Immunol.* **207**:1–17.

Ramachandran GN, Sasisekharan V, Ramakrishnan C (1966): Molecular structures of polyglycine II. *Biochim. Biophys. Acta.* **112**:168–170.

Riehn C, Lahmann C, Wassermann B, Brutschy B (1992): IR depletion spectroscopy—a method for characterizing a microsolvation environment. *Chem. Phys. Lett.* **197**:443–450.

Riehn C (2002): High-resolution pump-probe rotational coherencespectroscopy—rotational constants and structure of ground and electronically excited states of large molecular systems. *Chem. Phys.* **283**:297–329.

Riesner D, Kellings K, Post K, Wille H, Serban H, Baldwin M, Groth D (1996): Disruption of prion rods generates 10-nm spherical particles having a high alpha-helical content and lacking scrapie infectivity. *J. Virol.* **70**:1714–1722.

Rizzo TR, Park YD, Peteanu L, Levy DH (1985): Electronic spectrum of the amino acid tryptophan cooled in a supersonic molecular beam. *J. Chem. Phys.* **83**:4819–4820.

Rizzo TR, Park YD, Levy DH (1986a): Dispersed fluorescence of jet-cooled tryptophan: excited state conformers and intramolecular exciplex formation. *J. Chem. Phys.* **85**: 6945–6951.

Rizzo TR, Park YD, Peteanu L, Levy DH (1986b): The electronic spectrum of the amino acid tryptophan in the gas phase. *J. Chem. Phys.* **84**:2534–2541.

Robertson, EG (2000): IR-UV ion-dip spectroscopy of N-phenyl formamide, and its hydrated clusters. *Chem. Phys. Lett.* **325**:299–307.

Robertson EG, Hockridge MR, Jelfs PD, Simons JP (2000): IR-UV ion-dip spectroscopy of N-benzylformamide clusters: Stepwise hydration of a model peptide. *J. Phys. Chem. A* **104**:11714–11724.

Robertson EG, Hockridge MR, Jelfs PD, Simons JP (2001): IR-UV ion-depletion and fluorescence spectroscopy of 2-phenylacetamide clusters: Hydration of a primary amide. *Phys. Chem. Chem. Phys.* **3**:786–795.

Robertson EG, Simons JP (2001): Getting into shape: conformational and supramolecular landscapes in small biomolecules and their hydrated clusters. *Phys. Chem. Chem. Phys.* **3**:1–18.

Rzepecki P, Gallmeier H, Geib N, Cernovska K, König B, Schrader T (2004a): New heterocyclic β-sheet ligands with peptidic recognition elements. *J. Org. Chem.* **69**:5168–5178.

Rzepecki P, Nagel-Stegers L, Feuerstein S, Linne U, Molt O, Zadmard R, Aschermann K, Wehner M, Schrader T, Riesner D (2004b): Prevention of Alzheimer's disease associated Ab aggregation by rationally designed nonpeptidic b-sheet ligands. *J. Biol. Chem.* **279**:47497–47505.

Rzepecki P, Schrader T (2005): β-sheet ligands in action: KLVFF recognition by aminopyrazole hybrid receptors in water. *J. Am. Chem. Soc.* **127**:3016–3025.

Schmitt M, Böhm M, Ratzer C, Vu C, Kalkman I, Meerts WL (2005): Structural selection by microsolvation: Conformational locking of tryptamine. *J. Am. Chem. Soc.* **127**: 10356–10364.

Schrader T, Kirsten C (1996): Intermolecular stabilization of b-sheet conformation in dipeptides. *Chem. Commun.* **17**:2089–2090.

Schulz GE, Schirmer RH (1979): *Principles of Protein Structure*, Springer-Verlag, New York.

Senent ML, Villa M, Dominguez-Gomez R, Fernandez-Clavero A (2005): Ab initio study of the far infrared spectrum of glycine. *Int. J. Quantum. Chem.* **104**:551–561.

Simons JP (2003): Getting into shape? The interaction of molecules of biological interest with molecules of water: From simplicity to complexity. *C. R. Chimie* **6**:17–31.

Sipior J, Sulkes M (1988): Spectroscopy of tryptophan derivatives in supersonic expansions—addition of solvent molecules. *J. Chem. Phys.* **88**:6146–6156.

Smedarchina Z, Siebrandt W, Fernández-Ramos A, Gorb L, Leszczynski J (2000): A direct-dynamics study of proton transfer through water bridges in guanine and 7-azaindole. *J. Chem. Phys.* **112**:566–573.

Smits M, de Lange CA, Ullrich S, Schultz T, Schmitt M, Underwood JG, Shaffer JP, Rayner DM, Stolow A (2003): Stable kilohertz rate molecular beam laser ablation sources. *Rev. Sci. Instrum.* **74**:4812–4817.

Snoek LC, Robertson EG, Kroemer RT, Simons JP (2000): Conformational landscapes in amino acids: Infrared and ultraviolet ion-dip spectroscopy of phenylalanine in the gas phase. *Chem. Phys. Lett.* **321**:49–56.

Snoek LC, Kroemer RT, Hockridge MR, Simons JP (2001): Conformational landscapes of aromatic amino acids in the gas phase: Infrared and ultraviolet ion-dip spectroscopy of tryptophan. *Phys. Chem. Chem. Phys.* **3**:1819–1826.

Snoek LC, Kroemer RT, Simons JP (2002): A spectroscopic and computational exploration of tryptophan-water cluster structures in the gas phase. *Phys. Chem. Chem. Phys.* **4**: 2130–2139.

Snyder IE, Hollis JM, Suenram RD, Lovas FJ, Brown IW, Buhl D (1983): An extensive galactic search for conformer-II glycine. *Astrophys. J.* **268**:123–128.

Softley T, Haward B (eds) (2005), *Molecular Physics*—a special issue in honor of Prof. John P. Simons (*Mol. Phys.* **103**), Taylor & Francis.

Stillinger FH, Weber TA (1983): Inherent structure in water. *J. Phys. Chem.* **87**:2833–2840.

Storm V, Dreizler H, Consalvo D, Grabov JU, Merke I (1996): A newly designed molecular beam Fourier transform microwave spectrometer in the range 1–4 GHz. *Rev. Sci. Instrum.* **67**:2714-2719.

Suenram RD, Lovas FJ (1978): Millimeter wave spectrum of glycine. *J. Mol. Spectrosc.* **72**:372–382.

Suenram RD, Lovas FJ (1980): Millimeter wave spectrum of glycine—a new conformer. *J. Am. Chem. Soc.* **102**:7180–7184.

Suenram RD, Lovas FJ, Matsumura K (1989): Laboratory measurement of the 101–000 transition and electric dipole moment of SiC_2. *Astrophys. J.* **342**:L103-L105.

Suenram SD, Grabow JU, Zuban A, Leonov I (1999): A portable, pulsed-molecular-beam, Fourier-transform microwave spectrometer designed for chemical analysis. *Rev. Sci. Intrum.* **70**:2127–2135.

Tanabe S, Ebata T, Fujii M, Mikami N (1993): OH stretching vibrations of phenol-$(H_2O)_N(N=1-3)$ complexes observed by IR-UV double-resonance spectroscopy. *Chem. Phys. Lett.* **215**:347–352.

Tanner C, Manca C, Leutwyler S (2003): Probing the threshold to H atom transfer along a hydrogen-bonded ammonia wire. *Science* **302**:1736–1739.

Teh CK, Sipior J, Sulkes M (1989): Spectroscopy of tryptophan in supersonic expansions: addition of solvent molecules. *J. Phys. Chem.* **93**:5393–5400.

Tembreull R, Lubman DM (1986): Pulsed laser desorption with resonant two-photon ionization detection in supersonic beam mass spectrometry. *Anal. Chem.* **58**:1299–1303.

Tembreull R, Lubman DM (1987a): Pulsed laser desorption of biological molecules in supersonic beam mass spectrometry with resonant two-photon ionization detection. *Anal. Chem.* **59**:1082–1088.

Tembreull R, Lubman DM (1987b): Resonant two-photon ionization of small peptides using pulsed laser desorption in supersonic beam mass spectrometry. *Anal. Chem.* **59**:1003–1006.

Tjernberg LO, Näslund J, Lindqvist F, Karlström A, Thyberg J, Terenius L, Nordstedt C (1996): Arrest of beta-amyloid fibril formation by a pentapeptide ligand. *J. Biol. Chem.* **271**:8545–8548.

Toennies JP, Vilesov AF (2004): Superfluid helium droplets: A uniquely cold nanomatrix or molecules and molecular complexes. *Angew. Chem. Int. Ed.* **43**:2622–2648.

Tubergen MJ, Cable JR, Levy DH (1990): Substituent effects on the electronic spectroscopy of tryptophan derivatives in jet expansions. *J. Chem. Phys.* **92**:51–60.

Unterberg C, Jansen A, Gerhards M (2000): Ultraviolet/infrared-double resonance spectroscopy and ab initio calculations on the indole(+) and indole(H2O)(1)(+) cations. *J. Chem. Phys.* **113**:7945–7954.

Unterberg C, Gerlach A, Schrader T, Gerhards M (2002): Clusters of a protected amino acid with pyrazole derivatives: b-Sheet model systems in the gas phase. *Eur. Phys. J. D* **20**:543–550.

Unterberg C, Gerlach A, Schrader T, Gerhards M (2003): Structure of the protected dipeptide Ac-Val-Phe-OMe in the gas phase: Towards a b-sheet model system. *J. Chem. Phys.* **118**:8296–8300.

Venkatachalan CM (1986): *Biopolymers* **6**:1425–1436.

Vogelsanger B, Godfrey PD, Brown RD (1991): Rotational spectra of biomolecules–histamine. *J. Am. Chem. Soc.* **113**:7864–7869.

Weichert A, Riehn C, Barth HD, Lembach G, Zimmermann M, Brutschy B, Podenas D (2001): Implementation of a high-resolution two-color spectrometer for rotational coherence spectroscopy in the picosecond domain. *Rev. Sci. Instrum.* **72**:2697–2708.

Weigend F, Haser M, Patzelt H, Ahlrichs R (1998): RI-MP2: optimized auxiliary basis sets and demonstration of efficiency. *Chem. Phys. Lett.* **294**:143–452.

Wiedemann S, Metsala A, Nolting D, Weinkauf R (2004): The dipeptide cyclic(glycyltryptophanyl) in the gas phase: A concerted action of density functional calculations, S0-S1 two-photon ionization, spectral UV/UV hole burning and laser photoelectron spectroscopy. *Phys. Chem. Chem. Phys.* **6**:2641–2649.

Wisniewski T, Aucouturier P, Soto C, Frangione B (1998): The prionoses and other conformational disorders. *Amyloid. Int. J. Exp. Clin. Invest.* **5**:212–224.

Wu YR, Levy DH (1989): Determination of the geometry of deuterated tryptamine by rotationally resolved electronic spectroscopy. *J. Chem. Phys.* **91**:5278–5284.

Xu SJ, Nilles M, Bowen KH (2003): Zwitterion formation in hydrated amino acid, dipole bound anions: How many water molecules are required? *J. Chem. Phys.* **119**:10696–10701.

Zwier TS (1996): The spectroscopy of solvation in hydrogen-bonded aromatic clusters. *Annu. Rev. Phys. Chem.* **47**:205–241.

Zwier TS (2001): Laser spectroscopy of jet-cooled biomolecules and their water-containing clusters: water bridges and molecular conformation. *J. Phys. Chem. A* **105**:8827–8839.

2

PROBING THE ELECTRONIC STRUCTURE OF Fe–S CLUSTERS: UBIQUITOUS ELECTRON TRANSFER CENTERS IN METALLOPROTEINS USING ANION PHOTOELECTRON SPECTROSCOPY IN THE GAS PHASE

XIN YANG, XUE-BIN WANG, YOU-JUN FU, AND LAI-SHEN WANG*

Department of Physics
Washington State University
Richland, WA
and
W. R. Wiley Environmental Molecular Sciences Laboratory and Chemical Sciences Division Pacific Northwest National Laboratory
Richland, WA

*To whom correspondence should be addressed. (ls.wang@pnl.gov).

Principles of Mass Spectrometry Applied to Biomolecules, edited by Julia Laskin and Chava Lifshitz

2.1. INTRODUCTION

Iron–sulfur clusters are found in all forms of life, which constitute the active sites of a growing list of proteins in such essential life-sustaining processes as respiration, nitrogen fixation, and photosynthesis (Spiro 1982). The most prototypical and ubiquitous Fe–S cluster is the cubane-type [4Fe–4S] cluster, which, in addition to its catalytic and regulatory roles, appears to be nature's favorite agent for electron transfer and storage, such as in ferredoxins (Fds), high-potential iron proteins (HiPIPs), and the integral machineries of hydrogenases and nitrogenases (Bernnert et al. 1997; Einsle et al. 2002; Peters et al. 1998). In proteins, the cubane [4Fe–4S] unit is usually coordinated by the amino acid cysteine (Figure 2.1). The [4Fe–4S] core functions as electron transfer agent usually between the following oxidation states: $[4Fe–4S]^{1+} \leftrightarrow [4Fe–4S]^{2+} \leftrightarrow [4Fe–4S]^{3+}$. A fourth state, the all-ferrous

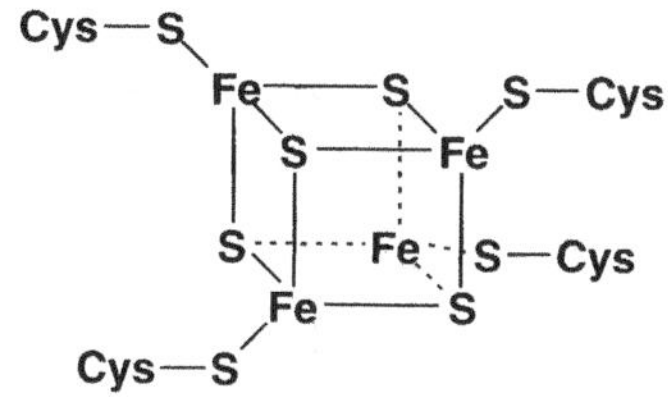

FIGURE 2.1. Schematic structure of the [4Fe–4S] active site in proteins. [Reprinted with permission from Wang XB, Niu S, Yang X, Ibrahim SK, Pickett CJ, Ichiye T, Wang LS, *J. Am. Chem. Soc.* **125**:14072–14081, 2003 (Wang et al. 2003). Copyright (2003) American Chemical Society.]

species $[4Fe\text{–}4S]^0$, was also detected in the iron protein of nitrogenase (Musgrave et al. 1998; Yoo et al. 1999).

It is important to understand the intrinsic electronic structure of the Fe–S clusters and modifications by their surroundings in order to understand the properties and functionalities of iron–sulfur proteins. Various spectroscopic techniques and theoretical methods have been used to study the magnetic and electronic structures of the Fe–S clusters in both synthetic analogs and proteins (Bernnert et al. 1997; Holm et al. 1996; Glaser et al. 2000, 2001). Extensive theoretical work using broken-symmetry density functional theory have shown that the $[4Fe\text{–}4S]^{2+}$ core can be viewed as a two-layer system, where two high-spin Fe in each [2Fe–2S] sublayer are coupled ferromagnetically and the two [2Fe–2S] sublayers are coupled antiferromagnetically to give a low-spin state (Aizman and Case 1982; Noodleman and Baerends 1984; Noodleman et al. 1995). The theoretical results are consistent with experimental observations from Mössbauer and EPR spectroscopy.

The reduction potential is one of the most important functional characteristics for an electron transfer protein. For the [4Fe–4S] core, the redox couple in Fds involves $[Fe_4S_4(S_{Cys})_4]^{2-}/[Fe_4S_4(S_{Cys})_4]^{3-}$ with reduction potentials ranging from −645 to 0 mV. The redox couple in HiPIP involves $[Fe_4S_4(S_{Cys})_4]^{1-}/[Fe_4S_4(S_{Cys})_4]^{2-}$ with reduction potentials ranging from 50 to 450 mV; the $[Fe_4S_4(S_{Cys})_4]^{2-}/[Fe_4S_4(S_{Cys})_4]^{3-}$ redox couple in HiPIP was estimated to lie near −1000 mV (Stephens et al. 1996). Since the free energy of a reduction reaction can be divided into the intrinsic free energy of the prosthetic group, the [4Fe–4S] core, and the extrinsic free energy due to the protein surrounding the solvent at the redox site ($\Delta G^\circ = \Delta G_{int} + \Delta G_{env} + \Delta G_{int/env}$) (Ichiye 1999), identifying the intrinsic and environmental determinants that lead to this large variation in reduction potentials is crucial for understanding the functions of the Fe–S proteins. Major environmental factors contributing to the reduction potentials of Fe–S proteins have been suggested to include the H bonding to the cysteine and bridging sulfide ligands, dipole interactions of the Fe–S cluster with the solvent and the protein sidechain/backbone, and electrostatic interactions (Adman et al. 1975; Backes et al. 1991; Pickett and Ryder 1994). For instance, the larger number of NH···S H bonds to the redox site in the Fds than in the HiPIPs, as well as the more solvent-exposed redox sites in the Fds, was suggested to be responsible for the differences in reduction potentials.

Alteration of the reduction potentials of the $[4Fe–4S]^{2+}/[4Fe–4S]^{1+}$ couple have been reported by changing the cluster environment (Chen et al. 1999; Zhou et al. 2000). Theoretical calculations of the reduction potentials of the [4Fe–4S] cluster were also carried out in which environmental effects were evaluated using a continuum dielectric model (Mouesca et al. 1994; Torres et al. 2003). However, the environmental effects on the electronic structure and redox properties of the [4Fe–4S] core remain uncertain.

Investigation of Fe–S proteins in the gas phase had been a formidable challenge. The underlying difficulty was how to produce proteins, which exist in the solutions intact to the gas phase. It has become possible only after development of soft ionization techniques such as electrospray ionization (ESI) and matrix-assisted laser desorption/ionization (MALDI) (Fenn 2003; Tanaka 2003; Karas and Hillenkamp 1988). Since then, gas-phase study of Fe–S proteins and other biological molecules, largely using mass spectrometry, has been actively pursued. Many techniques based on mass spectrometry have been developed to study varieties of issues such as protein–protein, protein–ligand interactions (Wells et al. 2003; Daneshfar et al. 2004), peptide conformation change (Kaleta and Jarrold 2003), and even kinetics of enzyme catalysis (Li et al. 2003). Surface-induced dissociation and collision-activated dissociation have been applied to obtain the energetic and structural information of biomolecules (Laskin et al. 2003, 2004; Smith et al. 1990; Loo et al. 1993). Structure, thermochemistry, and reactivity of small Fe–S clusters were investigated in the gas phase using selected ion-beam techniques (Koszinowski et al. 2002, 2004; Kretzschmar et al. 2003).

Dominant peaks of certain compositions in the mass spectrum (magic number) often indicate very special structures and stability of these clusters, such as C_{60} (Kroto et al. 1985), metallocarbohedrenes (Metal–Car) (Ti_8C_{12}) (Guo et al. 1992a,b), and more recently octamer of serine (Cooks et al. 2001). The proposed or hypothesized structures need to be proved theoretically and experimentally. For example, the reactions of Metal–Car with polar molecules were carried out to support the proposed Metal–Car structures (Guo et al. 1993; Deng et al. 1996). Anion photoelectron spectroscopy, as described in detail below, has been applied to probe the electronic and structural information for C_{60} since the "soccer ball" structure was proposed (Yang et al. 1987; Gunnarsson et al. 1995; Wang et al. 1999b, 2005), as well as for Ti_8C_{12} (Li et al. 1997). Combining with theoretical computations, photoelectron spectra often provide key complementary information regarding the geometric structures of the investigated clusters, and further illustrate their electronic structures.

Photoelectron spectroscopy (PES) is a powerful experimental technique to probe the electronic structure of matter (Rabalais 1977; Eland 1984). It has been widely used to study singly charged anions, providing information about electron affinities and low-lying electronic states for a vast number of neutral species and clusters. The observed features represent transitions from the electronic ground state of the anion to the ground and excited states of the corresponding final molecule (neutral or the anion with one less charge). Unlike optical absorption, which is subject to stringent selection rules, PES of an anion can access optically "dark" electronic states of the final molecule. Gas-phase PES is an ideal experimental technique used to study the electronic structure and chemical bonding of the Fe–S clusters without perturbation of the solvents, crystal field or the protein environment, yielding the intrinsic properties of the Fe–S clusters and providing the basis for elucidating the complex cluster–protein interactions. In addition, photodetachment involving removal of an electron from a molecule ($AB^{n-} \rightarrow AB^{(n-1)} + e^-$), is an oxidation process. The measured adiabatic electron detachment energy (ADE) reflects the energy difference between the oxidized and reduced species in the gas phase, providing the intrinsic reduction potential (Wang and Wang 2000c), $\Delta G_{\text{int}} \approx \Delta E_{\text{int}} = -\text{ADE} = -(\text{VDE} + \lambda_{\text{oxd}})$, where VDE represents the vertical electron detachment energy and λ_{oxd}, the oxidation reorganization energy. We have developed an experimental technique, which couples an electrospray ionization source (ESI) with time-of-flight mass spectrometer (TOFMS) and a magnetic bottle photoelectron spectrometer (Wang et. al. 1999a). ESI is a versatile technique, allowing ionic species, especially multiply charged anions in solution samples, to be transported into the gas phase. Our research has shown that the new ESI-MS-PES technique is ideal for investigating electronic properties of multiply charged anions in the gas phase, as well as anionic metal complexes commonly present in solution (Wang and Wang 1999a, 2000a,b; Wang et al. 2001, 2002).

Using this technique, we have performed systematic studies on a series of Fe–S complex anions as the analogs of the active centers of different Fe–S proteins (Wang et al. 2003, Yang et al. 2002, 2003a,b,c Niu et al. 2003, 2004; Fu et al. 2004, Zhai et al. 2004). In this chapter we focus on our comprehensive investigation of [4Fe–4S] cubane complexes. We have studied the intrinsic electronic structure of a series of free cubane complexes $[Fe_4S_4L_4]^{2-}$ (L = SH,SC_2H_5,Cl,Br,I) and the Se-substituted $[Fe_4Se_4L_4]^{2-}$ (L = Cl,S_2H_5) species (Wang et al. 2003). The influences of both the terminal ligands and the bridging ligands (substitution of S by Se) on the electronic structure of the cubane core are investigated. The PES spectral features confirm the low-spin two-layer model for the $[4Fe–4S]^{2+}$ core and its "inverted-level scheme." The measured ADEs provide the intrinsic oxidation energies of the $[Fe_4S_4L_4]^{2-}$ complexes. We further investigate the influence of the terminal ligands on the cubane core in mixed-ligand systems, $[Fe_4S_4L_{4-x}L'_x]^{2-}$ ($x = 0$–4), formed by ligand substitution reactions Fu et al. 2004). We discovered that all the mixed-ligand species gave similar PES spectral features, whereas their electron binding energies are very sensitive to the terminal ligand substitution. A linear relationship between the binding energy and the substitution number was shown for each series, suggesting that each ligand contributes independently to the total binding energy. We observed surprisingly symmetric fissions in cubane dianions, $[Fe_4S_4L_4]^{2-} \rightarrow 2[Fe_2S_2L_2]^-$ (L = Cl, Br, SC_2H_5) in collision-induced dissociation (CID) experiments (Yang et al. 2002; Yang et al., 2003c; Niu et al. 2004). PES spectra show that the parent and the fission fragments have similar electronic structures and confirm the unusual spin couplings in the $[Fe_4S_4L_4]^{2-}$ clusters, which contain two valent-delocalized and ferromagnetically coupled Fe_2S_2 subunits. A series of singly charged fragments $[Fe_4S_4L_n]^-$ (L = Cl,Br,I; $n = 0$–4) were also observed in CID experiments. Our PES data clearly reveal a behavior of sequential oxidation of the cubane over five formal oxidation states: $[4Fe–4S]^{-} \rightarrow [4Fe–4S]^{0} \rightarrow [4Fe–4S]^{+} \rightarrow [4Fe–4S]^{2+} \rightarrow [4Fe–4S]^{3+}$, demonstrating the robustness and electron–storage capability of the [4Fe–4S] cubane (Zhai et al. 2004).

This chapter is organized as follows. In Section 2.2, we describe the experimental approach. In Sections 2.3–2.6, we review our investigation on the [4Fe–4S] cubane complexes starting from the $[Fe_4S_4L_4]^{2-}$ dianions to their CID fragments. Results and discussion are presented for each system including the relevant literature. Conclusions are given in Section 2.7.

2.2. EXPERIMENTAL TECHNIQUES

The details of our apparatus and its performance have been given elsewhere (Wang et al. 1999a). Figure 2.2 shows a schematic view of our experimental apparatus, which consists of an electrospray ionization source, a modified Wiley–McLaren TOF mass spectrometer, and a magnetic bottle TOF photoelectron analyzer.

2.2.1. Electrospray Ionization Source

The sample solutions are prepared by dissolving $(t\text{-}Bu_4N)_2[Fe_4S_4L_4]^{2-}$ in O_2-free acetonitrile at $\sim 10^{-3}$ M concentration. A quartz syringe with a stainless-steel needle is used for the electrospray. The sample solution is sprayed into the ambient atmosphere through a fused-silica needle (0.01 mm in diameter), which is connected to the end of the syringe needle. The stainless-steel syringe needle is biased at −2.2 kV for the formation of negatively charged liquid droplets. A microprocessor-controlled syringe pump (World Precision Instrument SP100i) is used to deliver the solution. The highly charged liquid droplets are fed into a desolvation capillary heated to ~80°C. The charged liquid droplets are broken down and desolvated in the capillary, transferring the ionic species present in the solution sample into the gas phase.

2.2.2. Ion Trap Time-of-Flight Mass Spectrometry

Ions emerging from the desolvation capillary are focused and transmitted using a RF-only quadrupole ion guide to a quadrupole ion trap (R. M. Jordan Company, Grass Valley, CA). The trap has 3-mm-diameter holes on each of the two endcaps, allowing ions to enter and exit the trap. It operates at a background pressure of about 10^{-4} Torr, so that ions can be confined to the center of the trap through collisions with the background gas. The ions are accumulated in the trap for 0.1 s before being pulsed out into the extraction zone of a TOF mass spectrometer.

Our mass spectrometer uses a modified Wiley–McLaren arrangement. The major modification involves an addition of a short field-free region between the two acceleration stages of the original Wiley–Mclaren design, which allows a high mass resolution to be achieved even for a large volume of ion package. The anions are

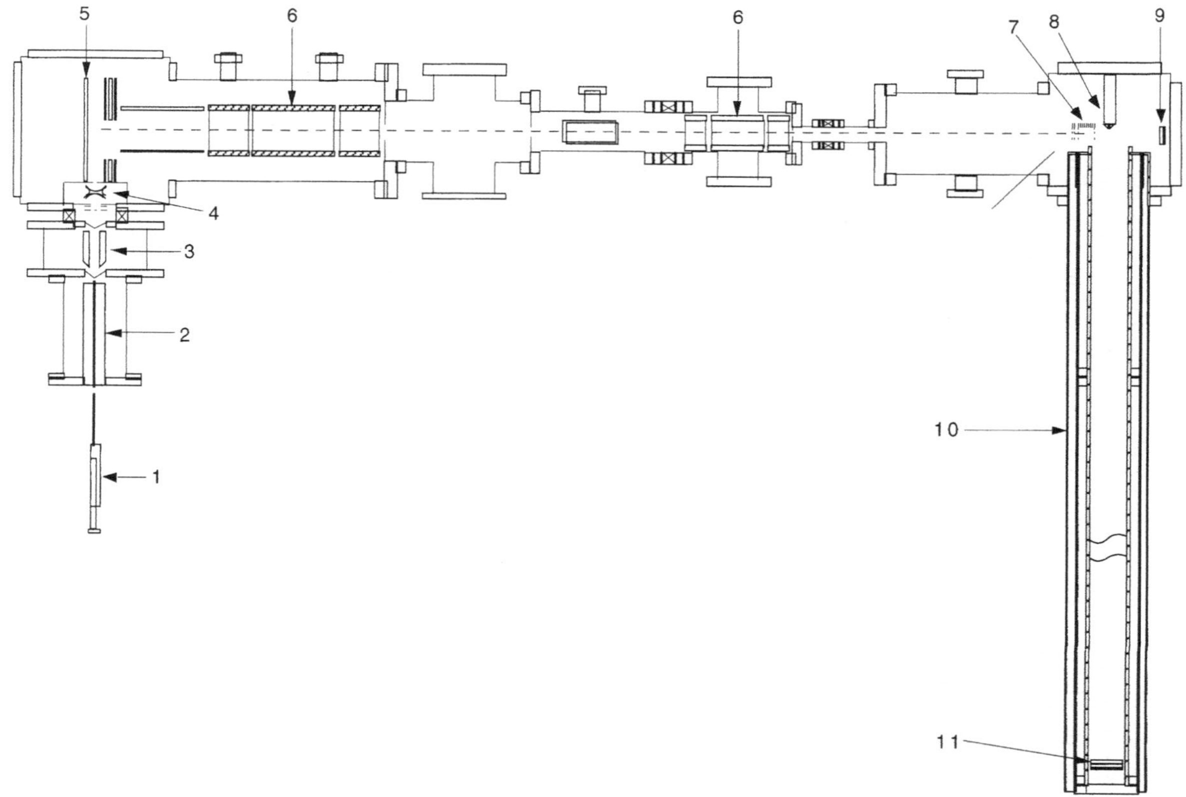

FIGURE 2.2. Schematic view of the electrospray–photodetachment apparatus: (1) syringe; (2) heated desolvation capillary; (3) radiofrequency quadrupole ionguide; (4) quadrupole iontrap; (5) time-of-flight mass spectrometer extraction stack; (6) Einzel lens assembly; (7) three-grid mass–gate and momentum decelerator assembly; (8) permanent magnet (NdFeB); (9) 40-mm dual-microchannel plate inline ion detector; (10) 4-m time-of-flight tube with a low-field solenoid and double-layer μ-metal shielding; (11) Φ18-mm Z-stack microchannel plate photoelectron detector. [Reprinted with permission from Wang LS, Ding CF, Wang XB, Barlow SE, *Rev. Sci. Instrum.* **70**:1957–1966, 1999 (Wang et al. 1999a). Copyright (1999) American Institute of Physics.]

extracted perpendicularly to the TOF axis with a high-voltage pulse (−1.25 kV, 100 μs). The anion beam is deflected, focused, and collimated along the 2.5-m-long field-free flight tube to a set of inline microchannel plate (MCP) detectors in the electron detachment chamber.

For PES experiments, only the anions of interest are allowed to enter the electron detachment zone. The target anions are selected using a mass gate and decelerated by a momentum decelerator. The middle electrode of the three-grid mass gate is biased at −1300 V. The mass gate high voltage is pulsed to ground potential at a predefined time, which ensures that only the anions of interest pass unaffected into the electron detachment zone while stopping all other ions. Exiting the mass gate, the mass-selected anions enter the momentum decelerator. A +3000 V high-voltage pulse is applied across the deceleration region consisting of a series of 11 electrodes. The pulsewidth is determined by the mass intensity of the anions and the amount of deceleration desired. This deceleration step is crucial to improve the PES resolution due to the minimization of the Doppler broadening caused by the anion-beam velocity.

2.2.3. Time-of-Flight Photoelectron Spectroscopy (PES)

Our photoelectron analyzer is a 4π solid-angle magnetic bottle time-of-flight PES spectrometer, and located at the end of the TOF mass spectrometer (Figure 2.2). The high magnetic field is generated by a permanent magnet mounted on a translation stage so that it can be adjusted to find the position for the best resolution or to minimize the photoelectron noise at high photon energies. The total field-free electron flight tube is 4 m long with low-field solenoid and μ-metal shielding. The electron detector is a fast *z* stack consisting of a set of three MCPs.

Photodetachment are studied at five photon energies: 532 nm (2.331 eV), 355 nm (3.496 eV), and 266 nm (4.661 eV) from a Nd:YAG laser, and 193 nm (6.424 eV) and 157 nm (7.866 eV) from an excimer laser. All experiments are performed at 20 Hz repetition rate with the ion beam switched off at alternating laser shots for background subtraction, which is critical for high-photon-energy experiments (>4.661 eV), due to background noise. Photoelectrons are collected at nearly 100% efficiency. Photoelectron time-of-flight spectra are collected and then converted to kinetic energy spectra, calibrated by the known spectra of I^- and O^-. The electron binding energy spectra presented are obtained by subtracting the kinetic energy spectra from the detachment photon energies. The energy resolution ($\Delta E/E$) is about 2%, specifically, ~10 meV for 0.5-eV electrons, as measured from the spectrum of I^- at 355 nm.

2.3. THE INTRINSIC ELECTRONIC STRUCTURE OF THE CUBANE [4Fe–4S] CLUSTER

We conducted a systematic and extensive study of a series of free cubane complexes, $[Fe_4S_4L_4]^{2-}$ (L = SH,SC_2H_5,Cl,Br,I) and the Se-substituted $[Fe_4Se_4L_4]^{2-}$ (L = Cl,

SC_2H_5) species using photoelectron spectroscopy and density functional theory (DFT) calculations (Wang et al. 2003). Similar spectral features were observed for all the cubane complexes, revealing their similar electronic structure and the robustness of the [4Fe–4S] core as a modular unit. The spectral features confirmed the low-spin two-layer model for the $[Fe_4S_4]^{2+}$ core and its "inverted-level scheme" molecular orbital diagrams. The measured adiabatic detachment energies (ADEs) provided the intrinsic oxidation energies of the $[Fe_4S_4L_4]^{2-}$ complexes. The influences of both the terminal ligands and the bridging ligands (substitution of S by Se) on the electronic structure of the cubane core were investigated.

2.3.1. Photoelectron Spectra of $[Fe_4S_4L_4]^{2-}$ and $[Fe_4Se_4L_4]^{2-}$ Complexes

The 193-nm spectra of $[Fe_4S_4L_4]^{2-}$ (L = SC_2H_5,SH,Cl,Br,I) are shown in Figure 2.3. The spectral patterns for all species show certain similarities. A weak but well-defined threshold feature *X* was observed in the spectra of all the samples. A second band *A*, well separated from the threshold band, was not well resolved in the spectrum of $[Fe_4S_4(SC_2H_5)_4]^{2-}$ (Figure 2.3a), but became a well-defined peak in the spectra of $[Fe_4S_4(SH)_4]^{2-}$ and the halide complexes. The relative intensities of the bands *X* and *A*, their bandwidths, as well as the *X*–*A* energy gap (~0.75 eV), are almost identical in all the spectra. High-binding-energy features in the spectra exhibit differences due to the different terminal ligands. A prominent feature emerges in the spectrum of $[Fe_4S_4Cl_4]^{2-}$ at ~4.4 eV. It shifts to lower binding energies in the spectra of the bromide and iodide complexes and is easily recognized as the halogen ligand feature (Figure 2.3c–e). The spectra of the SC_2H_5 and the SH complexes are nearly identical, except that the broad feature at the very-high-binding-energy side (~5 eV) is more intense in the spectrum of SC_2H_5 (Figure 2.3a), reflecting the contribution from the C_2H_5 groups. In addition to the prominent halide features, the major difference between the spectra of the halide complexes and those of the SC_2H_5 and SH complexes is in the spectral range above the A band. The overall spectral intensities in these ranges are higher in the spectra of the SC_2H_5 and SH complexes, reflecting the contribution from the terminal ligand S. Despite the overall similarities of spectral patterns among the five complexes, their electron binding energies were observed to depend on the type of ligands, increasing in the direction $SC_2H_5- > SH- > Cl- > Br- > I$.

Substitution of the bridging S by Se has little effect on the PES spectra, namely, the electronic structure of the cubane core. Both the spectral patterns and the electron binding energies of the two $[Fe_4Se_4L_4]^{2-}$ species are almost identical to those of their $[Fe_4S_4L_4]^{2-}$ counterparts except for the slight change of the relative intensities of bands *X* and *A* and the energy gap between them. The *A* band is relatively stronger in the spectra of the Se-bridged complexes. The *X*–*A* energy gaps of $[Fe_4Se_4(SC_2H_5)_4]^{2-}$ and $[Fe_4Se_4Cl_4]^{2-}$ are the same (~0.69 eV), which is slightly smaller than those observed for the $[Fe_4S_4L_4]^{2-}$ species (~0.75 eV).

The ADE and vertical detachment energy (VDE) of the threshold peak (*X*) in the spectrum of each complex are listed in Table 2.1. Because of the lack of vibrational resolution, the ADEs were determined by drawing a straight line along the leading

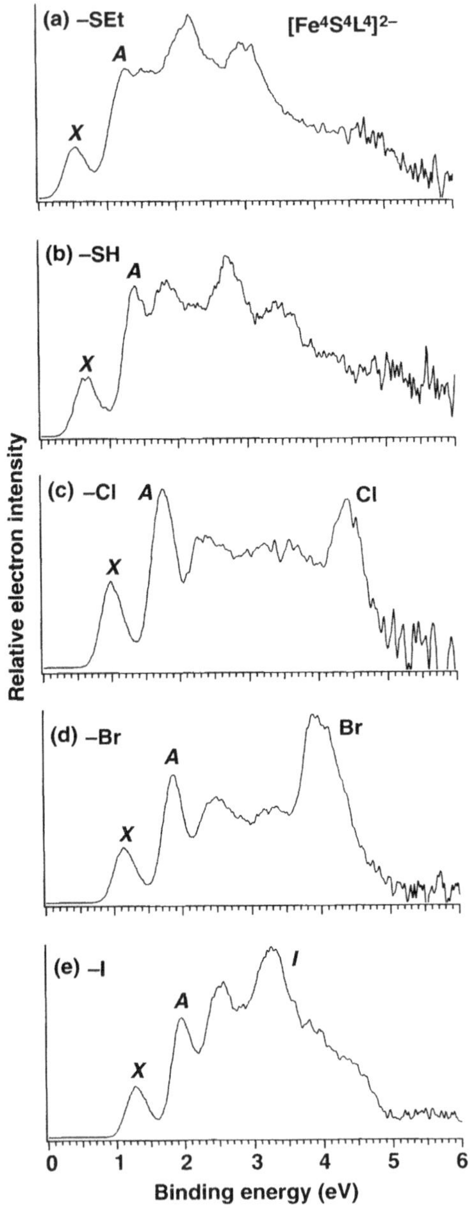

FIGURE 2.3. Photoelectron spectra of (a) $[Fe_4S_4(SC_2H_5)_4]^{2-}$, (b) $[Fe_4S_4(SH)_4]^{2-}$, (c) $[Fe_4S_4Cl_4]^{2-}$, (d) $[Fe_4S_4Br_4]^{2-}$, (e) $[Fe_4S_4I_4]^{2-}$ at 193 nm (6.424 eV). [Reprinted with permission from Wang XB, Niu S, Yang X, Ibrahim SK, Pickett CJ, Ichiye T, Wang LS, *J. Am. Chem. Soc.* **125**:14072–14081, 2003 (Wang et al. 2003). Copyright (2003) American Chemical Society.]

TABLE 2.1. Experimental Adiabatic (ADE), Vertical (VDE) Detachment Energies, Repulsive Coulomb Barriers (RCBs), and Oxidation Reorganization Energies (λ_{oxd}) for $[Fe_4X_4L_4]^{2-}$ (X = S, Se; L = SC_2H_5,SH,Cl,Br,I)

	ADE		VDE			
	Exp.[a]	Calc.[b]	Exp.[c]	Calc.[b]	RCB	λ_{oxd}
$[Fe_4S_4(SC_2H_5)_4]^{2-}$	0.29	0.16	0.52	0.47	1.6	0.23
$[Fe_4S_4(SH)_4]^{2-}$	0.39	0.41	0.63	0.80	1.5	0.24
$[Fe_4S_4Cl_4]^{2-}$	0.76	0.69	1.00	1.00	2.0	0.24
$[Fe_4S_4Br_4]^{2-}$	0.90	—	1.13	—	1.9	0.23
$[Fe_4S_4I_4]^{2-}$	1.06	—	1.28	—	1.8	0.22
$[Fe_4Se_4(SC_2H_5)_4]^{2-}$	0.28	0.16	0.51	0.39	1.6	0.23
$[Fe_4Se_4Cl_4]^{2-}$	0.72	—	0.94	—	2.0	0.22

[a]The estimated uncertainty for the ADEs is ±0.10 eV.

[b]Theoretical ADEs and VDEs for several complexes at B3LYP/6-31++G** level are shown for comparison. All energies are in eV.

[c]The estimated uncertainty for the VDEs is ±0.06 eV.

Source: Reprinted with permission from Wang XB, Niu S, Yang X, Ibrahim SK, Pickett CJ, Ichiye T, Wang LS, *J. Am. Chem. Soc.* **125**:14072–14081, 2003 (Wang et al. 2003). Copyright (2003) American Chemical Society.

edge of the threshold band and then adding a constant to the intersection with the binding energy axis to account for the instrumental resolution at the given energy range. This procedure was rather approximate, but consistent data were obtained from the spectra taken at different photon energies. The VDE was measured straightforwardly from the peak maximum.

2.3.2. Photon-Energy-Dependent Studies and the Repulsive Coulomb Barriers (RCBs)

PES spectra of $[Fe_4S_4L_4]^{2-}$ (L = SC_2H_5,SH) and $[Fe_4Se_4(SC_2H_5)_4]^{2-}$ were taken at all five photon energies, whereas only 355, 266, 193, and 157 nm spectra were taken for the four complexes with halide ligands because of their higher electron binding energies. In the lower-photon-energy spectra, high-binding-energy features observed in higher-photon-energy spectra disappeared as a direct consequence of the repulsive Coulomb barrier (RCB) in multiply charged anions (Wang and Wang 2000a,b). From these data the barrier height for each dianion can be estimated, as discussed below.

One unique property of multiply charged anions is the existence of intramolecular Coulomb repulsion between the excess charges. When an electron is removed from a multiply charged anion (AB^{n-}), the two photoproducts $[AB^{(n-1)-} + e^-]$ are both negatively charged. Superposition of the long-range Coulomb repulsion between the outgoing electron and the remaining anion and the short-range electron binding produces an effective potential barrier for the outgoing electron (Wang et al. 1998a,b, 1999c; Wang and Wang 2000a,b). If the detachment photon energy is below the top of the RCB, no electron detachment will occur even if the photon energy is above the asymptotic electron binding energy. In this case, detachment can take place only through electron tunneling, which depends exponentially on the energy difference between the photon energy and the RCB top and become negligible if the photon energy is far below the barrier top. When the photon energy is close to the RCB top, the detachment signal may be reduced. In the tunneling regime, the appearance of the PES peak tends to shift to the lower-binding-energy side, due to a convolution of Franck–Condon factors and tunneling probabilities, which depend on the electron kinetic energies exponentially. The intramolecular Coulomb repulsion and the resulting RCB have profound effects on the chemical and physical properties of multiply charged anions. We have shown that the Coulomb repulsion is equal in magnitude to the RCB if the detached electron corresponds to the negative charge carrier or is localized on the charge carrier group (Wang et al. 1998b; 2000a). In general, the RCB decreases with increasing physical sizes of the anions.

The RCB effects on the PES data were seen most clearly in the photon-energy-dependent PES spectra, where the high-binding-energy features observed at high photon energies were severely cut off in the low-photon-energy spectra. On the basis of the spectral cutoff, the magnitude of the RCB could be estimated by subtracting the binding energies at the cutoff point from the photon energies. Here we use the spectra of $[Fe_4Se_4Cl_4]^{2-}$ as an example (Figure 2.4). The *B* band was strong in the 266 nm spectrum (Figure 2.4b), but disappeared completely in the 355 nm spectrum (Figure 2.4a), indicating that the 355-nm photon (3.496 eV) lies below and the 266 nm photon (4.661 eV) lies above the top of the RCB corresponding to this detachment channel. Thus the RCB must be larger than ~1.1 eV (355 nm $h\nu$—VDE of *B* state, i.e., 3.5 − 2.2 eV) and less than ~2.5 eV (266 nm $h\nu$—VDE of *B* state, 4.661 − 2.2 eV). The strong *X* feature in the 355 nm spectrum also suggested the RCB is less than 2.5 eV (355 nm $h\nu$—VDE of the *X* state, 3.5 − 1.0 eV). The relatively weak *A* band and its apparent shift to lower binding energy indicated that the higher-binding-energy part of this band was cut off by the RCB. On the basis of the relative intensities between the *A* and *X* bands, we estimated that the cutoff point was around 1.5 eV in the 355 nm spectrum, implying a RCB of ~2.0 eV (3.5 − 1.5 eV). This value is in the range bracketed above and is consistent with the cutoff in the 193 and 157 nm spectra (Figure 2.4c,d). Therefore, we concluded that the RCB of $[Fe_4Se_4Cl_4]^{2-}$ should be around ~2.0 eV. Similarly, we estimated the RCBs for all the dianions from the photon-energy-dependent PES spectra, as given in Table 2.1. We noticed that the RCBs of the three halide complexes decrease from $[Fe_4S_4Cl_4]^{2-}$ to $[Fe_4S_4I_4]^{2-}$, due to the increasing physical size, namely, the increasing Fe–halide bond lengths. The RCBs of the two $[Fe_4Se_4L_4]^{2-}$ species are identical to those of their S counterparts.

2.3.3. Theoretical Results on the Cubane [4Fe–4S] Cluster

The broken-symmetry DFT method (Parr and Yang 1989), specifically with the Becke three-parameter hybrid exchange functional (Becke 1993) and the Lee–Yang–Parr correlation functional (B3LYP) (Lee et al. 1988) using two different basis sets,

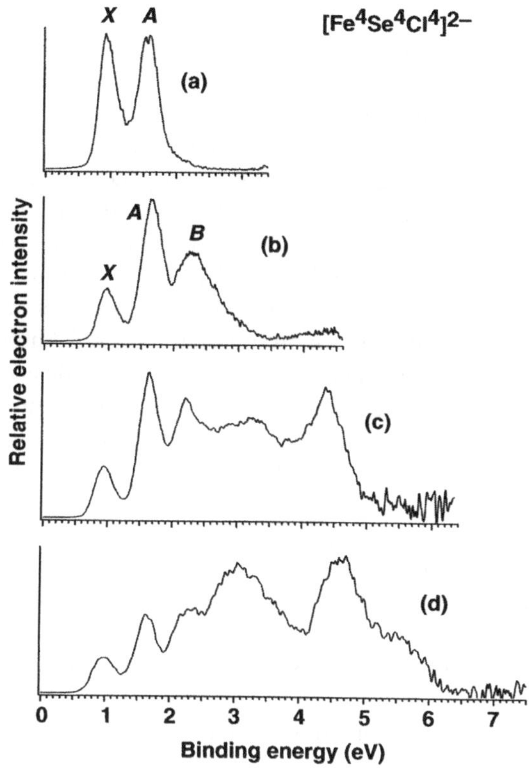

FIGURE 2.4. Photoelectron spectra of $[Fe_4Se_4Cl_4]^{2-}$ at (a) 355 nm (3.496 eV), (b) 266 nm (4.661 eV), (c) 193 nm (6.424 eV), (d) 157 nm (7.866 eV). [Reprinted with permission from Wang XB, Niu S, Yang X, Ibrahim SK, Pickett CJ, Ichiye T, Wang LS, *J. Am. Chem. Soc.* **125**:14072–14081, 2003 (Wang et al. 2003). Copyright (2003) American Chemical Society.]

6-31G** and 6-31(++)G** (Rassolov et al. 1998; Francl et al. 1982), was utilized for the geometry optimizations, electronic structure, and energy calculations of the $[Fe_4S_4L_4]^{3-}$/$[Fe_4S_4L_4]^{2-}$/$[Fe_4S_4L_4]^{1-}$ (L = SCH_3,SH,Cl) and $[Fe_4Se_4(SCH_3)_4]^{2-}$/$[Fe_4Se_4(SCH_3)_4]^{1-}$ redox couples. We used the simpler $-SCH_3$ ligand in all the calculations, instead of the more complex $-SC_2H_5$ ligand, which does not significantly change the electronic properties of the complexes. No symmetry constraints were imposed during geometry optimizations, and each structure was confirmed to be a ground-state structure by several separate calculations on different possible configuration states. This procedure is necessary because the electronic structure is very sensitive to the iron–sulfur cluster structure. The calculated energies were refined at the B3LYP/6-31$(++)_S$G**//B3LYP/6-31G** level, where *sp*-type diffuse functions were added to the 6–31G** basis set of the sulfur and chlorine atoms, which significantly improved the accuracy of the calculated oxidation potentials of the iron–sulfur redox couples. The ADE of $[Fe_4S_4L_4]^{2-}$ was calculated as the total energy difference between the ground states of $[Fe_4S_4L_4]^{2-}$ and $[Fe_4S_4L_4]^{-}$; the VDE was calculated as the energy difference between the ground state of $[Fe_4S_4L_4]^{2-}$ and the energy of $[Fe_4S_4L_4]^-$ at the geometry of $[Fe_4S_4L_4]^{2-}$. All calculations were performed using the NWChem program package (PNNL 2003). The molecular orbital visualizations were performed using the extensible computational chemistry environment (ECCE) application software (Black et al. 2000).

Generally, the cubane [4Fe–4S] redox site can be regarded as the coupling of two [2Fe–2S] redox layers according to broken-symmetry DFT calculations (Aizman and Case 1982; Noodleman and Baerends 1984; Noodleman et al. 1995). One possibility for the coupling of the two layers is a high-spin ferromagnetically coupled state, in which the spins of the $[Fe_4S_4(SR)_4]^{2-}$ site are all aligned in a parallel manner in the two redox layers with 20 Fe $3d$ α electrons $(d_{Fe-Fe})^{\alpha}$ and 2 Fe $3d$ β electrons $(d_{Fe-Fe})^{\beta}$, resulting in a high spin state ($S = \frac{18}{2}$). Another possibility is a low-spin state, in which the symmetry is broken because while each redox layer of the $[Fe_4S_4(SR)_4]^{2-}$ site couples ferromagnetically ($S = \frac{9}{2}$), the spins of the two redox layers couple antiferromagnetically, giving rise to the low-spin state ($S = 0$). As show schematically in Figure 2.5 for three oxidation levels of the [4Fe–4S] cubane, our DFT calculations showed that for $[Fe_4S_4(SCH_3)_4]^{3-/2-/1-}$ the low-spin state is favored by ~0.7 eV relative to the high-spin state, which is in good agreement with the previous broken-symmetry DFT calculations (Noodleman and Baerends 1984; Noodleman et al. 1995) and experimental observations from Mössbauer and EPR spectroscopy (Spiro 1982).

Our molecular orbital (MO) analysis showed that the spin-coupled interaction may split the MOs of the four individual iron sites into the Fe($3d$) majority-spin orbitals and Fe($3d$) minority-spin orbitals of the spin-coupled broken-symmetry

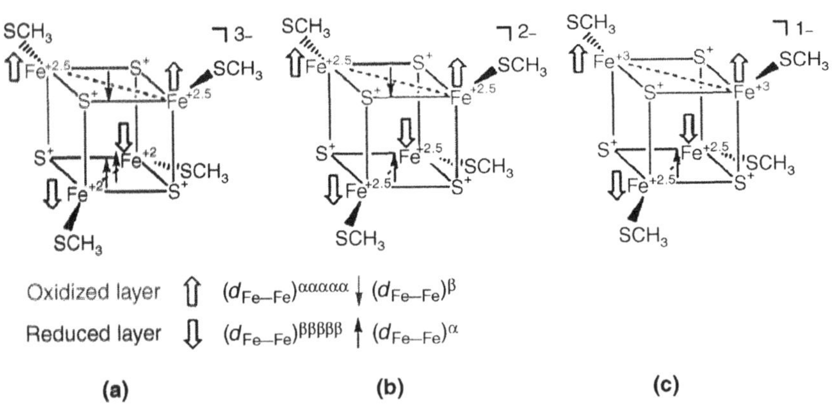

FIGURE 2.5. Schematic models of the spin couplings between the two redox sublayers of [4Fe-4S] cubane complexes at three different oxidation states: (a) 3^-, (b) 2^-, and (c) 1^-. In each sublayer, the two high-spin Fe centers couple ferromagnetically, while those in the two sublayers couple antiferromagnetically to give a low spin state. The 10 majority spins (represented by the large hallow arrows) are stabilized relative to the minority spin (represented by the small arrows), which is delocalized between the two Fe centers in each sublayer. [Reprinted with permission from Wang XB, Niu S, Yang X, Ibrahim SK, Pickett CJ, Ichiye T, Wang LS, *J. Am. Chem. Soc.* **125**:14072–14081, 2003 (Wang et al. 2003). Copyright (2003) American Chemical Society.]

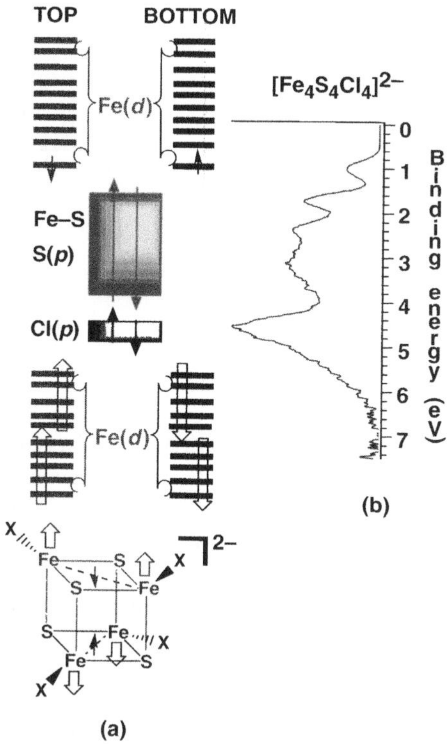

FIGURE 2.6. Schematic molecular orbital diagram showing the "inverted level scheme" and the spin coupling for the $[Fe_4S_4Cl_4]^{2-}$ cubane complex (a) in comparison with its photoelectron spectrum (b). [Reprinted with permission from Wang XB, Niu S, Yang X, Ibrahim SK, Pickett CJ, Ichiye T, Wang LS, *J. Am. Chem. Soc.* **125**:14072–14081, 2003 (Wang et al. 2003). Copyright (2003) American Chemical Society.]

state (i.e., $[Fe_4]^{10+}$), which interacted with the MOs of the terminal S(3*p*) and bridged S*(3*p*), to generate the higher-lying minority spin orbitals and a set of lower-lying Fe(3*d*) majority-spin orbitals (Figure 2.6). According to the broken-symmetry DFT calculations, the Fe(3*d*) majority-spin states stabilized by ~5–6 eV relative to the minority-spin levels. The mainly ligand MOs are energetically situated in between the minority and majority-spin levels of Fe, giving rise to the "inverted-level scheme" (Aizman and Case 1982; Noodleman and Baerends 1984; Noodleman et al. 1995), where the ligand levels are higher in energy than the Fe 3*d* levels (the majority-spin levels), as shown schematically in Figure 2.6a. The single valence-delocalized minority spin in each sublayer of $[Fe_2S_2]$ occupies the highest occupied molecular orbital (HOMO) of $[Fe_4S_4]^{2+}$, and this electron is transferred in an oxidation reaction or photodetachment of the $[Fe_4S_4L_4]^{2-}$ complexes.

Figure 2.7 illustrates the HOMO and the lowest unoccupied molecular orbital (LUMO) from our calculations and the schematic MO diagrams. The HOMO of $[Fe_4S_4(SCH_3)_4]^{2-}$ (oxidized MO in Figure 2.7b) exhibits a terminal Fe–S anti bonding character (σ^*_{Fe-S}), with a strong bonding interaction between the two Fe on

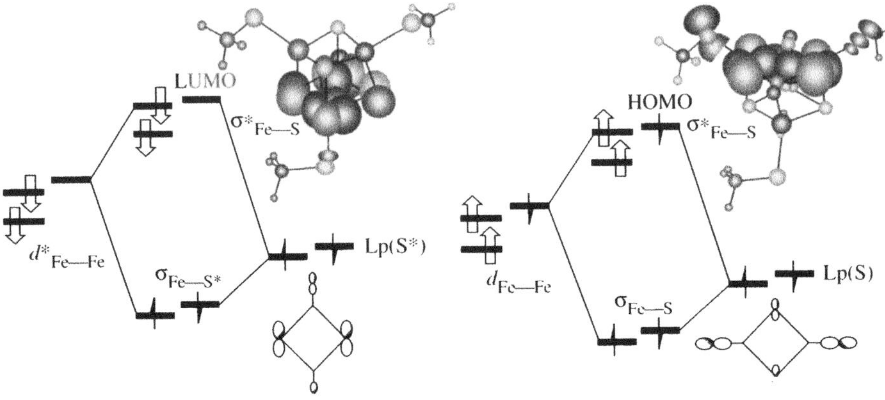

(a) (b)

FIGURE 2.7. Molecular orbital pictures of (a) LUMO and (b) HOMO of $[Fe_4S_4(SCH_3)_4]^{2-}$, and schematic molecular orbital correlation diagrams. [Reprinted with permission from Wang XB, Niu S, Yang X, Ibrahim SK, Pickett CJ, Ichiye T, Wang LS, *J. Am. Chem. Soc.* **125**:14072–14081, 2003 (Wang et al. 2003). Copyright (2003) American Chemical Society.]

one sublayer. The LUMO (reduced MO in Figure 2.7a) shows a bridging Fe–S* antibonding character ($\sigma^*_{\text{Fe–S*}}$), with a strong antibonding interaction between the two Fe on one sublayer. When $[Fe_4S_4(SR)_4]^{2-}$ is oxidized to $[Fe_4S_4(SR)_4]^{1-}$, the process involves removal of one electron from the HOMO $\sigma^*_{\text{Fe–S}}$. The sublayer on which this oxidation process occurs is called the *oxidized layer.* When $[Fe_4S_4(SR)_4]^{2-}$ is reduced to $[Fe_4S_4(SR)_4]^{3-}$, one electron is added to the LUMO $\sigma^*_{\text{Fe–S*}}$ on the *reduced layer,* as shown schematically in Figures 2.5 and 2.7.

The B3LYP/6-31G** optimized geometries of $[Fe_4S_4(SCH_3)_4]^{3-/2-/1-}$ were calculated and compared with the X-ray crystal structures of $[Fe_4S_4(SPh)_4]^{3-/2-/1-}$ (Carney et al. 1988; Excoffon et al. 1991; O'Sullivan and Millar 1985). It is shown that from $[Fe_4S_4(SCH_3)_4]^{2-}$ to $[Fe_4S_4(SCH_3)_4]^{3-}$ both the Fe_{red}–Fe_{red} and the Fe_{red}–S^*_{red} distances in the reduced layer increase. From $[Fe_4S_4(SCH_3)_4]^{2-}$ to $[Fe_4S_4(SCH_3)_4]^{1-}$, the Fe_{oxd}–Fe_{oxd} distance in the oxidized layer increases, but the Fe_{oxd}–S_{oxd} bond lengths decrease. All these bond length changes are consistent with the HOMO and LUMO analysis presented above. We note that from $[Fe_4S_4(SR)_4]^{2-}$ to $[Fe_4S_4(SR)_4]^{3-}$ or $[Fe_4S_4(SR)_4]^{1-}$, the Fe–S* bond lengths within the reduced or oxidized layers tend to increase. The geometries of the reduced and oxidized sites determine these intrinsic electronic structures and pin down the locations of the redox electrons. The ADEs and VDEs of $[Fe_4S_4L_4]^{2-}$ ($L = SC_2H_5$,SH,Cl) and $[Fe_4Se_4(SC_2H_5)_4]^{2-}$ were calculated and compared with the experimental data (Table 2.1).

2.3.4. PES Spectra and Electronic Structures

The PES features shown in Figure 2.3 represent transitions from the ground state of the $[Fe_4S_4L_4]^{2-}$ dianions to the ground and excited states of the corresponding singly charged anions. Within the single-particle approximation, the PES features can be viewed as removing electrons from the occupied MOs of the parent anions. Therefore, unlike other various experimental methods based on electronic transitions from occupied MOs to empty or partially occupied MOs, PES provides a direct map of the occupied MOs.

The most striking feature in all the PES spectra shown in Figures 2.3 is the weak threshold peak *X* regardless of the ligand type, suggesting that it should have the same origin in all the species. This feature, corresponding to removal of the most loosely bound electron, implies that the HOMOs of all the $[Fe_4S_4L_4]^{2-}$ complexes are the same. In our previous PES study of the [1Fe] Fe–S complex, we observed a similar threshold feature present in all the ferrous Fe(II) complexes, but not in the ferric Fe(III) complexes (Yang et al. 2003a). This band was assigned to removal of the Fe 3*d* minority-spin electron: $Fe^{II} \rightarrow Fe^{III}$. Similar observations in the PES spectra of $[Fe_4S_4L_4]^{2-}$ and $[Fe_4Se_4L_4]^{2-}$ dianions suggest that the HOMOs of the [4Fe–4S] or [4Fe–4Se] complexes may have the same character as those of the [1Fe] ferrous complexes because the [4Fe] complexes all contain two ferrous centers formally. This observation is consistent with the MO levels from broken-symmetry DFT calculations, which showed that the HOMOs of the $[Fe_4S_4]^{2+}$ core contain two degenerate Fe 3*d* minority-spin levels each from one $[Fe_2S_2]$ sublayer. The *X* band corresponds to ionization from the minority-spin levels, as schematically shown in Figure 2.5b for the $[Fe_4S_4Cl_4]^{2-}$ complex.

The second photoelectron band *A* also shows some similarity in all the complexes (Figures 2.3). Even the *A*–*X* energy gap (0.75 eV) is identical for all the $[Fe_4S_4L_4]^{2-}$ complexes except for $[Fe_4S_4I_4]^{2-}$, which has a slightly smaller *A*–*X* gap (0.68 eV). The *A*–*X* energy gap in the two Se–cubane complexes is also identical (0.69 eV), with a magnitude slightly smaller than that of their S counterparts. However, the relative intensity of the *A* band increased in the spectra of the two $[Fe_4Se_4L_4]^{2-}$ species. This dependence of band *A* on the bridge ligands (S or Se) is consistent with the electronic structure of the [4Fe–4S] cubane from the broken-symmetry DFT calculations (Figure 2.6a). The valence electronic configuration of each [2Fe–2S] sublayer is $[(\sigma_{Fe-S})^{\alpha\beta}(\sigma_{Fe-S^*})^{\alpha\beta}(\sigma^*_{Fe-S})^{\alpha}(\sigma^*_{Fe-S^*})^{\alpha}(\sigma^*_{Fe-S})^{\beta}(\sigma^*_{Fe-S^*})]$. The HOMO-1 $[(\sigma^*_{Fe-S*})^{\alpha}]$, which corresponds to the *A* band, is a σ antibonding orbital between Fe and the bridge ligand with mainly ligand characters. The stronger σ donor Se* in the [4Fe-4Se] complex increases the energy level of this orbital and thus decreases the *X*–*A* energy gap. Furthermore, the ionization cross section is expected to be higher for the Se-dominated MO, consistent with the enhanced intensity of the *A* band in the Se–cubane complexes. The $[Fe_4S_4I_4]^{2-}$ complex is special because of the large and soft terminal ligand I^-. Its electron distribution is more diffused, influencing the HOMO-1 orbital and reducing the *A*–*X* energy separation in $[Fe_4S_4I_4]^{2-}$.

Therefore, the current PES data directly show the metal character of the HOMO and the bridge ligand character of the HOMO-1 in the [4Fe] cubane complexes. The higher-binding-energy features up to 5 eV in Figure 2.3 change with the different terminal ligands. This part of the spectrum is due to the ionization from bonding and antibonding MOs of Fe–S and Fe–S*, S lone pairs, and other ligand-based MOs (Figure 2.6). The halide features were easily recognizable in the four halide-coordinated cubanes (Figures 2.3). The Fe 3*d* majority spin electrons possess too high binding energies to be clearly observed at 193 nm. The higher-binding-energy tails in the 157 nm data (e.g., see Figure 2.4) may contain detachment from the Fe 3*d* majority-spin electrons, as shown schematically in Figure 2.6.

The substantial energy gap between the threshold feature *X* and the second ionization band *A* indicates that HOMO-1 is well separated from HOMO. HOMO-1 in the Fe(II) (d^6) complex would be HOMO in the Fe(III) (d^5) case. Thus the wide energy gap of *X*–*A* bands suggests stability of the high-spin d^5 electron configuration in Fe(III) complexes and provides direct electronic structural basis for the fact that the $[4Fe–4S]^{2+}$ cubane core, which contains two Fe(II) centers, is used as a reducing agent in HiPIPs or other Fe–S proteins.

The current PES data also provide direct experimental confirmation for the inverted-level scheme and the broken-symmetry DFT description of the electronic structure and spin couplings of the cubane [4Fe–4S] core (Figure 2.6). The similarity among the PES data of all the complexes confirms the robustness of the $[4Fe–4S]^{2+}$ core as a modular unit and demonstrates that PES is capable of probing its intrinsic electronic structure.

2.3.5. Ligand Effects on the Electron Binding Energies and Redox Potentials

A one-electron oxidation reaction, aside from solvation effects, is similar to electron detachment in the gas-phase. Therefore, the gas phase ADEs should be inherently related to oxidation potentials, except that the solvation effects are absent in the electron detachment in vacuum. As discussed above, the threshold feature *X* in the PES spectrum of $[Fe_4S_4L_4]^{2-}$ complex corresponds to removing a minority spin 3*d* electron from the HOMO of each species. This detachment process represents an oxidation of the [4Fe–4S] core: $[4Fe–4S]^{2+} \rightarrow [4Fe–4S]^{3+}$. The ADE of the *X* band thus represents the gas-phase oxidation potential of the $[Fe_4S_4L_4]^{2-}$ complexes. Consequently, the width of the *X* feature directly reflects the geometry changes after one electron is transferred, and hence is related to the intrinsic reorganization energy (λ_{oxd}) on oxidization of $[Fe_4S_4L_4]^{2-}$ (Wang and Wang 2000c; Sigfridson et al. 2001a,b). The VDE and ADE differences, which characterize λ_{oxd}, of all the complexes are listed in Table 2.1. The λ_{oxd} value is identical for all the seven cubane complexes within our experimental uncertainties, again implying that the $[4Fe–4S]^{2+}$ core is identical regardless of the ligand type or the Se substitution and further confirming its robustness as a modular unit. Our measured λ_{oxd} value (0.23 eV) for the cubane complexes is in good agreement with a previous calculation [18.3 kJ/mol (0.19 eV)] (Sigfridson et al. 2001a,b).

As shown in Table 2.1 and Figures 2.3, the ADEs of the cubane complexes are influenced largely by the terminal ligand but not sensitive to the bridging S* or Se* ligand. The strong electron donor ligands, $–SC_2H_5$ and –SH, yield much lower ADEs than the electron withdrawing halide ligands, suggesting that the HOMO of $[Fe_4S_4L_4]^{2-}$ has considerable contribution from the ligand orbitals, in addition to the Fe 3*d* character. As shown in Figure 2.7, the HOMO of $[Fe_4S_4(SCH_3)_4]^{2-}$ arises from the interaction between the high-lying occupied minority-spin Fe(3*d*) orbitals and the terminal ligand lone pairs, Lp(S). Raising the energy of Lp(S) causes the energy gap between the Lp and the Fe $d_{\text{Fe–Fe}}$ set to diminish and enhances the interaction between the ligand and Fe. This would destabilize the $\sigma^*_{\text{Fe–S}}$ orbital, resulting in a decrease of the oxidation energy. Thus, the terminal thiolate ligands, which are strong σ donors, decrease the ADE and VDE of $[Fe_4S_4(SCH_3)_4]^{2-}$ and make it easy to be oxidized.

However, substitution of the bridging S* by Se* in $[Fe_4Se*_4(SR)_4]^{2-}$ has little effect on the ADE and VDE, compared to the S counterparts. This is consistent with the nature of the $\sigma^*_{\text{Fe–S}}$ orbital, which has little contribution from the bridging S* (Figure 2.7).

On the other hand, the LUMO of $[Fe_4S_4(SCH_3)_4]^{2-}$ arises from the interaction between the minority-spin Fe(3*d*) orbitals and the bridging ligand lone pairs, Lp(S*) (Figure 2.7). The bridging Se* substitution in $[Fe_4Se*_4(SR)_4]^{2-}$ leads to an increase in energy of the $\sigma^*_{\text{Fe–Se}*}$ orbital (LUMO) and a decrease in energy of the $\sigma_{\text{Fe–Se}*}$ orbital. Thus the redox potential of the $[Fe_4S_4L_4]^{3-}/[Fe_4S_4L_4]^{2-}$ couple depends mainly on the electron donor tendency of the bridging ligands.

2.4. TERMINAL LIGAND INFLUENCE ON THE [4Fe–4S] CUBANE CORE IN MIXED-LIGAND SYSTEMS

The ligand environmental effect on the electronic structure and redox properties of cubane core was further probed in several series of ligand-substituted analog complexes: $[Fe_4S_4Cl_{4-x}(CN)_x]^{2-}$, $[Fe_4S_4Cl_{4-x}(SCN)_x]^{2-}$, $[Fe_4S_4Cl_{4-x}(OAc)_x]^{2-}$ (OAc: acetate), $[Fe_4S_4(SC_2H_5)_{4-x}(OPr)_x]^{2-}$ (Opr: propionate), and $[Fe_4S_4(SC_2H_5)_{4-x}Cl_x]^{2-}$ ($x = 0$–4) (Fu et al. 2004). The mixed-ligand cubane complexes were prepared by the ligand substitution reaction in solution. Using the ESI-MS-PES technique, we were able to select species from an equilibrium solution of mixed-cubane complexes, $[Fe_4S_4L_{4-x}L'_x]^{2-}$, and study their electronic structures in the gas phase in a ligand-specific fashion. This technique also avoided the influence of the solvation effects so that the changes of the intrinsic electronic structure due to sequential substitution and asymmetric terminal ligand coordination were studied systematically.

PES spectral features for all the ligand-substituted complexes are similar, suggesting that the mixed-ligand coordination does not perturb the electronic structure of the cubane core significantly. The terminal ligands, however, have profound effects on the electron binding energies of the cubane and induce significant shifts of the PES spectra, increasing in the order $SC_2H_5^- \rightarrow Cl^- \rightarrow OAc^-/OPr^- \rightarrow CN^- \rightarrow SCN^-$. A linear relationship between the electron binding energies and the substitution number x was observed for each series, indicating that each ligand contributes independently and additively to the total binding energy. Our study reveals the electrostatic nature of the interaction between the [4Fe–4S] cubane core and its coordination environment and provides further evidence for the electronic and structural stability of the cubane core and its robustness as a structural and functional unit in Fe–S proteins.

2.4.1. Mass Spectrometric Detection of Ligand Substitution Reaction in Solution

Ligand substitution reactions at the terminal positions of the [4Fe–4S] cubane were reported to take place readily (Cleland et al. 1983; Wong et al. 1978; Que et al. 1974a,b; Depamphilis et al. 1974; Ohno et al. 1991). However, using conventional purification and characterization methods, it is difficult to obtain pure samples of each of the mixed-ligand products from the equilibrium mixture. Here we used ESI to transfer mixed-ligand [4Fe–4S] complexes from the following solution reactions to the gas phase for size-selected PES studies,

$$[Fe_4S_4Cl_4]^{2-} + xCN^- \rightarrow [Fe_4S_4Cl_{4-x}(CN)_x]^{2-} + xCl^- \quad (2.1)$$

$$[Fe_4S_4Cl_4]^{2-} + xSCN^- \rightarrow [Fe_4S_4Cl_{4-x}(SCN)_x]^{2-} + xSCN^- \quad (2.2)$$

$$[Fe_4S_4Cl_4]^{2-} + xOAc^- \rightarrow [Fe_4S_4Cl_{4-x}(OAc)_x]^{2-} + xCl^- \quad (2.3)$$

$$[Fe_4S_4(SEt)_4]^{2-} + xOPr^- \rightarrow [Fe_4S_4(SEt)_{4-x}(OPr)_x]^{2-} + xSEt^- \quad (2.4)$$

$$[Fe_4S_4(SEt)_4]^{2-} + [Fe_4S_4Cl_4]^{2-} \rightarrow [Fe_4S_4(SEt)_{4-x}Cl_x]^{2-} + [Fe_4S_4(SEt)_xCl_{4-x}]^{2-} \quad (2.5)$$

It should be noted that the different products ($x = 0$–4) in these reactions coexist in equilibrium. Figure 2.8 shows the ESI mass spectrum of reaction (2.2) taken with a mixture of $[Fe_4S_4Cl_4]^{2-}$ and SCN^- in 1 : 1 Cl^-/SCN^- molar ratio. Five $[Fe_4S_4Cl_{4-x}(SCN)_x]^{2-}$ species were observed, ranging from the parent ($x = 0$) to the completely substituted species ($x = 4$). Obvious differences of the isotopic pattern among the five groups of peaks were observed, corresponding to the different numbers of Cl^- in each species. Varying the molar ratio of $[Fe_4S_4Cl_4]^{2-}$ and SCN^- in the initial solution changed the relative intensities among the five species. During

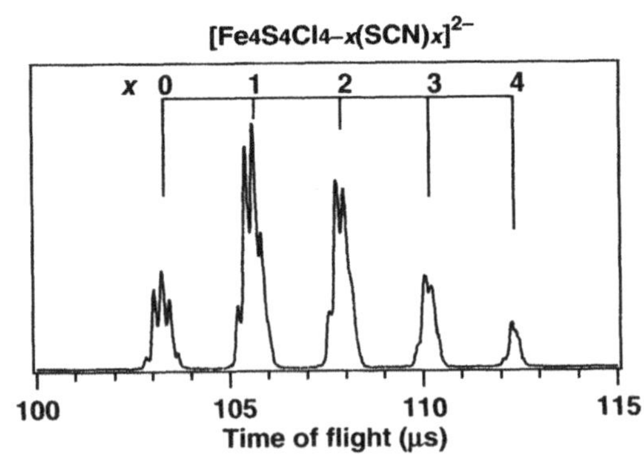

FIGURE 2.8. Electrospray mass spectrum of $[Fe_4S_4Cl_{4-x}(SCN)_x]^{2-}$ from reactions of $(t\text{-}Bu_4N)_2[Fe_4S_4Cl_4]$ and $(t\text{-}Bu_4N)SCN$ solutions (1 : 1 Cl^-/SCN^- molar ratio) in O_2-free acetonitrile. [Reprinted with permission from Fu YJ, Yang X, Wang XB, Wang LS, *Inorg. Chem.* **43**:3647–3655, 2004 (Fu et al. 2004). Copyright (2004) American Chemical Society.]

PES experiments, the mass intensity of a given x was optimized by changing the relative concentrations of the initial reactants.

2.4.2. Photoelectron Spectra of $[Fe_4S_4L_{4-x}L'_x]^{2-}$ ($x = 0$–4)

The PES spectra of $[Fe_4S_4Cl_{4-x}(CN)_x]^{2-}$, $[Fe_4S_4Cl_{4-x}(SCN)_x]^{2-}$, $[Fe_4S_4Cl_{4-x}(OAc_x]^{2-}$, $[Fe_4S_4(SEt)_{4-x}Cl_x]^{2-}$, and $[Fe_4S_4(SEt)_{4-x}(OPr)_x]^{2-}$ were taken at various photon energies. Overall, the spectra of the mixed-ligand complexes are similar to those of the precursor complexes $[Fe_4S_4(SEt)_4]^{2-}$ and $[Fe_4S_4Cl_4]^{2-}$. All the spectra exhibit a weak but well-defined threshold feature X in the lower-binding-energy range, followed by an intense and well-defined band (A) and continuous spectral transitions at high binding energies. In each series, the electron binding energies of $[Fe_4S_4L_{4-x}L'_x]^{2-}$ change systematically with the substitution number. Significant spectral cutoff was observed in all the PES spectra due to the repulsive Coulomb barrier (RCB).

For example, the PES spectra of $[Fe_4S_4Cl_{4-x}(CN)_x]^{2-}$ ($x = 0$-4) are shown in Figure 2.9 at 157, 193, and 266 nm. All the spectra have similar spectral features, which rigidly shift to higher binding energies with increasing numbers of the CN^- ligand. In the 157 nm spectra (Figure 2.9a), the broad and intense feature in the high-binding-energy range due to the Cl^- terminal ligands was clearly observed. The intensity of the Cl^- feature decreased with increasing substitution, accompanied by the appearance of a new feature due to CN^- at even higher binding energies. At $x = 4$, the band due to Cl^- disappeared completely and the feature from CN^- was dominant. At 193 nm (Figure 2.9b), the CN^- features disappeared as a result of the

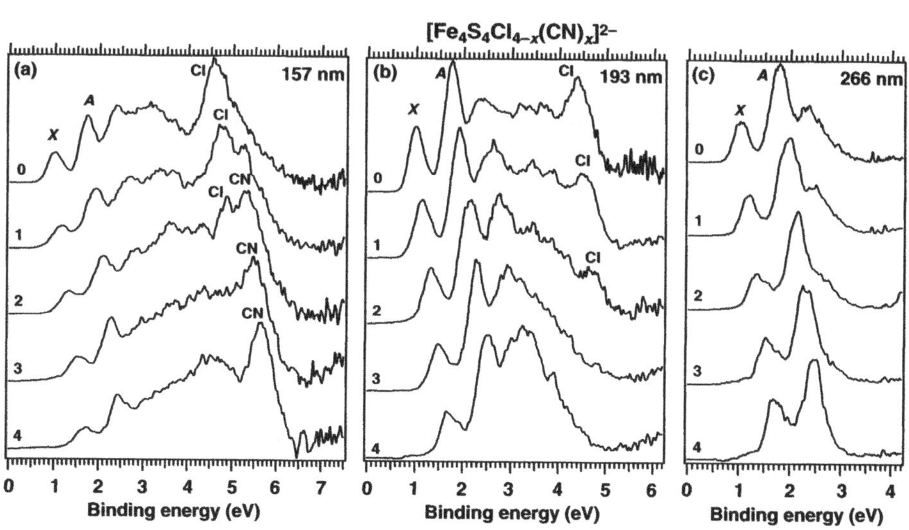

FIGURE 2.9. Photoelectron spectra of ligand substitution series $[Fe_4S_4Cl_{4-x}(CN)_x]^{2-}$ ($x = 0$-4) at (a) 157 nm, (b) 193 nm, and (c) 266 nm. [Reprinted with permission from Fu YJ, Yang X, Wang XB, Wang LS, *Inorg. Chem.* **43**:3647–3655, 2004 (Fu et al. 2004). Copyright (2004) American Chemical Society.]

TABLE 2.2. Adiabatic (ADE) and Vertical (VDE) Electron Binding Energies for *X* and *A* Bands, Separation between *X* and *A* Bands [$\Delta_{(A-X)}$], and Bandwidth of *X* Band (λ_{oxd}) from the Photoelectron Spectra of $[Fe_4S_4L_{4-x}L'_x]^{2-}$ (All Energies in eV)

	ADE	VDE(*X*)	VDE(*A*)	$\Delta_{(A-X)}$	λ_{oxd}
$[Fe_4S_4Cl_4]^{2-}$	0.80 (8)	1.01 (6)	1.78 (6)	0.77	0.21
$[Fe_4S_4Cl_3(CN)]^{2-}$	0.97 (8)	1.17 (6)	2.00 (6)	0.83	0.20
$[Fe_4S_4Cl_2(CN)_2]^{2-}$	1.09 (10)	1.36 (8)	2.15 (4)	0.79	0.27
$[Fe_4S_4Cl(CN)_3]^{2-}$	1.27 (8)	1.53 (6)	2.30 (6)	0.77	0.26
$[Fe_4S_4(CN)_4]^{2-}$	1.47 (11)	1.69 (6)	2.46 (6)	0.77	0.22
$[Fe_4S_4Cl_3(SCN)]^{2-}$	1.06 (8)	1.30 (5)	2.07 (7)	0.77	0.24
$[Fe_4S_4Cl_2(SCN)_2]^{2-}$	1.36 (9)	1.61 (5)	2.31 (5)	0.70	0.25
$[Fe_4S_4Cl(SCN)_3]^{2-}$	1.61 (10)	1.88 (10)	2.55 (6)	0.67	0.27
$[Fe_4S_4(SCN)_4]^{2-}$	1.84 (10)	2.10 (8)	2.76 (5)	0.66	0.26
$[Fe_4S_4Cl_3(OAc)]^{2-}$	0.93 (9)	1.19 (7)	1.92 (8)	0.73	0.26
$[Fe_4S_4Cl_2(OAc)_2]^{2-}$	1.04 (9)	1.29 (6)	2.10 (8)	0.81	0.25
$[Fe_4S_4Cl(OAc)_3]^{2-}$	1.11 (12)	1.48 (10)	2.22 (6)	0.74	0.37
$[Fe_4S_4(OAc)_4]^{2-}$	1.18 (12)	1.72 (8)	2.40 (10)	0.68	0.54
$[Fe_4S_4(SEt)_4]^{2-}$	0.29 (8)	0.52 (6)	1.20 (6)	0.68	0.23
$[Fe_4S_4(SEt)_3Cl]^{2-}$	0.41 (6)	0.60 (8)	1.33 (6)	0.73	0.19
$[Fe_4S_4(SEt)_2Cl_2]^{2-}$	0.52 (8)	0.71 (6)	1.43 (5)	0.72	0.19
$[Fe_4S_4(SEt)Cl_3]^{2-}$	0.62 (8)	0.82 (6)	1.56 (6)	0.74	0.20
$[Fe_4S_4(SEt)_3(OPr)]^{2-}$	0.57 (8)	0.75 (6)	1.47 (8)	0.72	0.18
$[Fe_4S_4(SEt)_2(OPr)_2]^{2-}$	0.78 (10)	1.03 (8)	1.75 (6)	0.72	0.25
$[Fe_4S_4(SEt)(OPr)_3]^{2-}$	1.01 (10)	1.32 (6)	2.08 (8)	0.76	0.31
$[Fe_4S_4(OPr)_4]^{2-}$	1.22 (12)	1.56 (12)	2.34 (8)	0.78	0.34

Source: Reprinted with permission from Fu YJ, Yang X, Wang XB, Wang LS, *Inorg. Chem.* **43**:3647–3655, 2004 (Fu et al. 2004). Copyright (2004) American Chemical Society.

RCB, whereas the *X* and *A* bands were better resolved. At 266 nm, all the higher binding energy features were cut off by the RCB and only the two lowest-binding-energy bands (*X* and *A*) were observed intact.

The ADE and VDE of the threshold band (*X*) and the VDE of the second detachment band (*A*) for all the species are given in Table 2.2. The *X* bandwidth was also given in Table 2.2 as λ_{oxd} calculated by [VDE(*X*) – ADE(*X*)]. In general, the *X* bandwidth did not vary significantly for the different species, except for the carboxylate-coordinated systems, in particular, $[Fe_4S_4(OAc)_4]^{2-}$ and $[Fe_4S_4(OPr)_4]^{2-}$.

As shown in Section 2.3.5, the values of the intrinsic reorganization energy (λ_{oxd}) can be obtained from the difference between VDE and ADE of the *X* band, as given in Table 2.2. The λ_{oxd} values are almost identical for all the species, ranging from 0.20 to 0.25 eV, except for the two fully carboxylate-coordinated complexes. These values are also similar to the homoligand complexes, $[Fe_4S_4L_4]^{2-}$ (L = –SH, –SEt, –Cl, –Br, –I). The nearly constant reorganization energy indicates the stability of the cubane core with respect to the terminal ligands. It is particularly surprising that even the mixed-ligand species did not give rise to a substantially broader *X* band.

Only $[Fe_4S_4(OAc)_4]^{2-}$ and $[Fe_4S_4(OPr)_4]^{2-}$ gave a much broader *X* band, suggesting a much larger geometry change on removal of an electron from the HOMO of the two fully carboxylate-coordinated cubanes. Preliminary theoretical calculations indicated that structural isomers may exist with the carboxylate coordinated cubanes, which could be an alternative explanation for the broad *X* band in these species.

2.4.3. Influence of Terminal Ligands on Electronic Structure and Redox Properties of the Mixed-Ligand Cubane Complexes

Our PES study on cubane complexes with four identical ligands $[Fe_4S_4L_4]^{2-}$ (L = –SH, –SEt, –Cl, –Br, –I), as well as the Se-substituted, has confirmed the two-layer "inverted-level scheme," based on the broken-symmetry DFT calculations for the electronic structure of the cubanes (Wang et al. 2003). In the mixed-ligand complexes, $[Fe_4S_4L_{4-x}L'_x]^{2-}$, because of the asymmetry induced by the two different ligands, L and L′, it was not clear whether the two-layer inverted-level scheme would still be applicable. The asymmetry of the ligand environment makes the two sublayers nonequivalent and can cause an energetic shift of the two sublayers. This shift should be reflected in the width of the *X* band, which is due to removal of the two minority electrons.

The PES spectra of all the mixed-ligand cubane exhibited spectral features nearly identical to those of the parents except for the systematic increase in binding energies and the additional ligand-induced bands at higher binding energies. In particular, the spectral features in the lower-binding-energy side were almost identical in all the cubane complexes. Even the separation between the *X* and *A* bands was nearly identical in all the spectra, as shown in Table 2.2. The fact that the *X* bandwidth is also nearly identical suggested that the asymmetry of the coordination environment in the mixed-ligand cubane was not significant enough to change the relative energies of the two layers. These observations suggested that, while the electron binding energies (oxidation potentials) of the cubane complexes are very sensitive to the terminal ligands, the electronic structures entailed in the two-layer "inverted-level scheme" are not sensitive to the terminal ligand environment, providing further evidence for the robustness of the cubane as a structural and functional unit.

Although the terminal ligands do not seem to change the electronic structures of the cubane significantly, they do have dramatic effects on the electron binding energies of the cubane, which are related to the intrinsic redox properties of the cubane complexes. As shown in Section 2.3.5, the electron binding energies of the cubane are strongly dependent on the terminal ligands and are related to the electron-donating withdrawing capability of the terminal ligands. The HOMO of the $[Fe_4S_4L_4]^{2-}$ complex consists of Fe–Fe bonding and Fe–L antibonding interactions. Strong electron donors, such as SEt^- and SH^-, destabilize the HOMO, resulting in much lower electron binding energies, whereas the halogen ligands are strong electron-withdrawing ligands and their cubane complexes all have rather high electron binding energies. The same trend was observed in the current study; both

CN^- and SCN^- are strong electron-withdrawing ligands, and they significantly increase the electron binding energies of the cubane complexes.

2.4.4. Observation of Linear Relations of Binding Energies versus x in $[Fe_4S_4L_{4-x}L'_x]^{2-}$

An interesting observation in this study is the linear relationship between the electron binding energies and the substitution number (x) in the mixed-ligand cubanes, as shown in Figure 2.10, where the ADE and the VDE of both the X and A bands are plotted against the substitution number x. We found that the binding energy of the mixed-ligand complex, $[Fe_4S_4L_{4-x}L'_x]^{2-}$, can be expressed as

$$BE = BE_1 + \gamma x \tag{2.6}$$

where BE_1 is the binding energy of $[Fe_4S_4L_4]^{2-}$ ($x = 0$) and γ is the slope.

The linear relationship suggests that each of the four terminal ligands contribute to the total binding energy of the cubane complex independently and additively. Therefore, the BE can be written as

$$BE = BE_0 + (4 - x)\delta_1 + x\delta_2 \tag{2.7}$$

where BE_0 is the electron binding energy of the bare cubane core $[Fe_4S_4]^{2+}$; δ_1 and δ_2 are contributions to the binding energy from ligands L and L′, respectively. Equation (2.7) can be rearranged as follows:

$$BE = BE_0 + 4\delta_1 + x(\delta_2 - \delta_1) \tag{2.8}$$

Comparing Eqs. (2.6) and (2.8), we see that the slope γ equals $(\delta_2 - \delta_1)$, that is, the difference of the contributions to the binding energy by ligands L′ and L.

We obtained the following slopes from Figure 2.10: $\delta_{CN^-} - \delta_{Cl^-} = 0.16\,eV$; $\delta_{SCN^-} - \delta_{Cl^-} = 0.26\,eV$; $\delta_{Cl^-} - \delta_{SEt^-} = 0.13\,eV$; $\delta_{OAc^-} - \delta_{Cl^-} = 0.15\,eV$; $\delta_{OPr^-} - \delta_{SEt^-} = 0.27\,eV$ for both the VDE(X) and VDE(A) curves. The same slope was found for the ADE curves of the CN^-/Cl^-, SCN^-/Cl^-, and SEt^-/Cl^- systems as the VDE curves. But the slopes of the ADE curves for the OAc^-/Cl^- and OPr^-/SEt^- systems were different from those of the VDE curves, $\delta_{OAc^-} - \delta_{Cl^-} = 0.10\,eV$; $\delta_{OPr^-} - \delta_{SEt^-} = 0.22\,eV$, as can be seen clearly from Figure 2.10d,e. This was caused by the broadening of the X band in the carboxylate-coordinated complexes.

Thus, if the binding energies of two cubane complexes, $[Fe_4S_4L_4]^{2-}$ and $[Fe_4S_4L'_4]^{2-}$, are known, one can predict the binding energies of the L/L′ mixed-ligand complexes, $[Fe_4S_4L_{4-x}L'_x]^{2-}$ because the slope can be calculated: $\delta_2 - \delta_1 = (BE_2 - BE_1)/4$. For example, from the known VDEs for $[Fe_4S_4Cl_4]^{2-}$ (1.00 eV) and $[Fe_4S_4I_4]^{2-}$ (1.66 eV), we can predict the VDEs for the three mixed-ligand complexes, $[Fe_4S_4Cl_3I]^{2-}$, $[Fe_4S_4Cl_2I_2]^{2-}$, and $[Fe_4S_4ClI_3]^{2-}$, to be 1.16, 1.33, and 1.49 eV, respectively. These predicted values are indeed in good agreement with our experimental measurements (not shown).

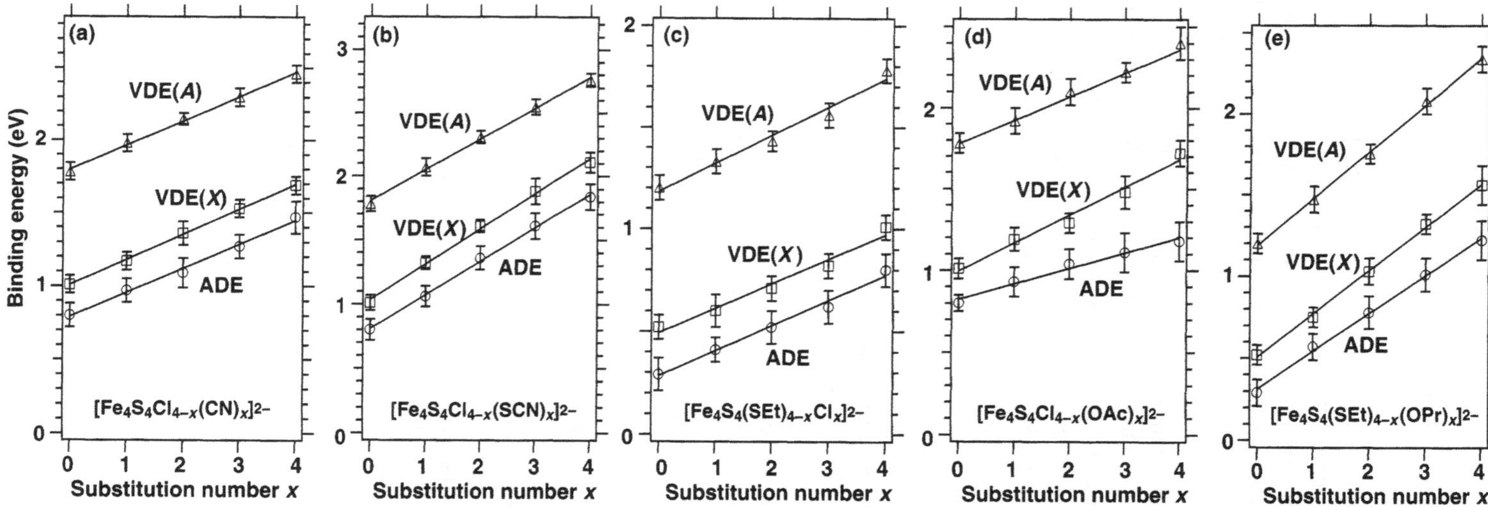

FIGURE 2.10. Linear relationships between binding energies [ADE, VDE(X), VDE(A)] and substitution number x: (a) $[Fe_4S_4Cl_{4-x}(CN)_x]^{2-}$; (b) $[Fe_4S_4Cl_{4-x}(SCN)_x]^{2-}$; (c) $[Fe_4S_4(SEt)_{4-x}Cl_x]^{2-}$; (d) $[Fe_4S_4Cl_{4-x}(OAc_x]^{2-}$; (e) $[Fe_4S_4(SEt)_{4-x}(OPr)_x]^{2-}$. The solid lines are drawn from linear fittings. [Reprinted with permission from Fu YJ, Yang X, Wang XB, Wang LS, *Inorg. Chem.* **43**:3647–3655, 2004 (Fu et al. 2004). Copyright (2004) American Chemical Society.]

2.4.5. Comparison with Redox Potentials of Mixed-Ligand Cubanes in Solution

Redox potentials are known for many cubane complexes with four identical ligands (Que et al. 1974a,b; DePamphilis et al. 1974; Ohno et al. 1991; Zhou et al. 1996, 1997; Ciurli et al. 1990). Redox potentials for some mixed-ligand complexes are also known (Johnson and Holm 1978). For example, ligand substitution reactions between $[Fe_4S_4(SCH_2Ph)_4]^{2-}$ and CH_3COCl and $(CH_3CO)_2O$ were studied to give two series of mixed-ligand substitution species, $[Fe_4S_4(SCH_2Ph)_{4-x}Cl_x]^{2-}$ and $[Fe_4S_4(SCH_2Ph)_{4-x}(OAc)_x]^{2-}$, with $x = 1$–4 (Johnson and Holm 1978). It was found that each substitution by Cl^- or OAc^- induced a positive reduction potential shift by about 100 meV; thus, a linear relationship between the reduction potentials and x was observed in the two mixed-ligand complexes. This observation is very similar to our observation in the gas phase for the electron binding energies (oxidation potentials) of the mixed-ligand cubane complexes. The solution results concerned a reduction of a 2-cubane to a 3-cubane, which involved addition of an electron to the LUMO of the 2-cubane complexes, whereas the gas-phase data were related to the oxidation of 2-cubanes to 1-, which involved removal of an electron from the HOMO. However, as our PES results have shown, all the molecular orbitals of the cubane complexes rigidly shift with the terminal ligands. We expect the LUMO of the cubane to shift accordingly. Thus, similar linear behaviors between our gas-phase data and the solution-phase redox potentials suggest that our gas-phase data can be reliably used to extrapolate to behaviors in solution for the mixed-ligand complexes. These results give the intrinsic redox potentials of the cubane complexes and provide an experimental basis for partitioning of the intrinsic and extrinsic factors to the redox potentials in solution.

The independent contribution of the ligand toward the redox potentials of transition metal complexes, namely, the "ligand additivity" model, has been well documented in inorganic chemistry, albeit mostly for mononuclear redox species (Treichel et al. 1972; Pickett and Pletcher 1975; Sarapu and Fenske 1975; Lever 1990). The "ligand additivity" toward the redox potential of the [4Fe–4S] complexes further confirms the robustness of the [4Fe–4S] cubane as a structural and functional unit. It shows that the terminal ligands act as perturbations on the electronic structure of the cubane.

2.5. OBSERVATION OF SYMMETRIC FISSION OF DOUBLY CHARGED CUBANE COMPLEXES: $[Fe_4S_4L_4]^{2-}$ (L = CL, BR, SET)

The conversion of the Fe–S cluster was studied for the first time in the gas phase by collision-induced dissociation (CID) and PES experiments (Yang et al. 2002, 2003c; Niu et al. 2004). The CID experiments were conducted by applying a negative DC voltage (~ -2 V) to the skimmer after the desolvation capillary (Figure 2.2). The dominant CID channel for [4Fe–4S] cubane dianions is the symmetric fission, $[Fe_4S_4L_4]^{2-} \rightarrow 2[Fe_2S_2L_2]^-$. The PES data indicate that both the parent and the daughter anions have similar electronic structures. Both Coulomb repulsion and the antiferromagnetic coupling of two layers in the cubane dianion play important roles in the fission process. The observation of this symmetric fission and the similar electronic structure for the parent and the daughter anions provides direct evidence of the unique layered structure of the $[Fe_4S_4L_4]^{2-}$ cluster and its antiferromagnetic coupling. Our gas-phase experiments suggest that solution phase conversions between $[Fe_4S_4]$ and $[Fe_2S_2]$ assemblies in proteins may also involve related fission chemistry with reactive $[Fe_2S_2L_2]^-$ intermediates.

2.5.1. Collision-Induced Dissociation (CID)

The mass-selected CID experiments were performed on a commercial LCQ (Finnigan, San Jose, CA) electrospray/quadrupole ion trap mass spectrometer to obtain high-resolution mass spectra. The dianions of interest, $[Fe_4S_4X_4]^{2-}$ (X = Cl,Br,SEt), were first isolated in the trap by ejecting all other anions. After isolation, an excitation AC voltage was applied to the endcaps to drive collisions of the isolated anions with the background gas (10^{-4} Torr N_2), facilitating the CID process. The Mathieu parameter q_z value for resonance excitation was 0.25. The ion excitation time for CID was 30 ms. The amplitude of the excitation AC voltage used for CID was optimized in each experiment. It was ramped up as relative collision energy (CE) from 0% to 100%, which corresponded to 0–2.5 V zero-to-peak resonant excitation potential as calibrated by the manufacturer. The contents of the ion trap were then analyzed to detect the CID products.

Mass spectra of isolated $[Fe_4S_4L_4]^{2-}$ (L = Cl,Br,SEt) and their CID products are presented in Figure 2.11. For $[Fe_4S_4Cl_4]^{2-}$ and $[Fe_4S_4Br_4]^{2-}$, the symmetric fission was the only CID channel, forming $[Fe_2S_2X_2]^-$ (X = Cl,Br). The fission products $[Fe_2S_2X_2]^-$, which have the same mass-to-charge ratios as those of the parent dianions, can be identified easily by the isotope patterns. The singly mass peak isolated for $[Fe_4S_4Cl_4]^{2-}$ (Figure 2.11a) was due mainly to the isotopmer $[Fe_4S_4{}^{35}Cl_3{}^{37}Cl]^{2-}$ ($m/z = 247$). It split into two dominating isotopic combinations: $[Fe_2S_2{}^{35}Cl_2]^-$ ($m/z = 246$) and $[Fe_2S_2{}^{35}Cl^{37}Cl]^-$ ($m/z = 248$) in the CID products. The parent dianion peak ($m/z = 247$) was almost invisible in the CID mass spectrum, suggesting that the fission channel is very efficient. The CID behavior of $[Fe_4S_4Br_4]^{2-}$ (Figure 2.11b) was identical to that of $[Fe_4S_4Cl_4]^{2-}$. An isolated isotopic peak was $[Fe_4S_4{}^{79}Br_2{}^{81}Br_2]^{2-}$ ($m/z = 336$), from which three fission isotopmers were observed: $[Fe_2S_2{}^{79}Br_2]^-$ ($m/z = 334$), $[Fe_2S_2{}^{79}Br^{81}Br]^-$ ($m/z = 336$), and $[Fe_2S_2{}^{81}Br_2]^-$ ($m/z = 338$). The observed $[Fe_4S_4Cl_3O_2]^-$ and $[Fe_4S_4Br_3O_2]^-$ mass signals were due to ligand exchange reactions with residual background O_2 during the CID processes. It appeared that this reactive channel was very efficient for $[Fe_4S_4Br_4]^{2-}$, because the product, $[Fe_4S_4Br_3O_2]^-$, dominated the CID spectra.

The CID spectra of $[Fe_4S_4(SEt)_4]^{2-}$ are more complicated, as shown in Figure 2.11c. We isolated the major isotopmer of $[Fe_4S_4(SEt)_4]^{2-}$ at $m/z = 298$. The CID product at $m/z = 596$ was from a new channel: the collision-induced electron detachment of $[Fe_4S_4(SEt)_4]^{2-}$. The two weak peaks emerging from the sides of the

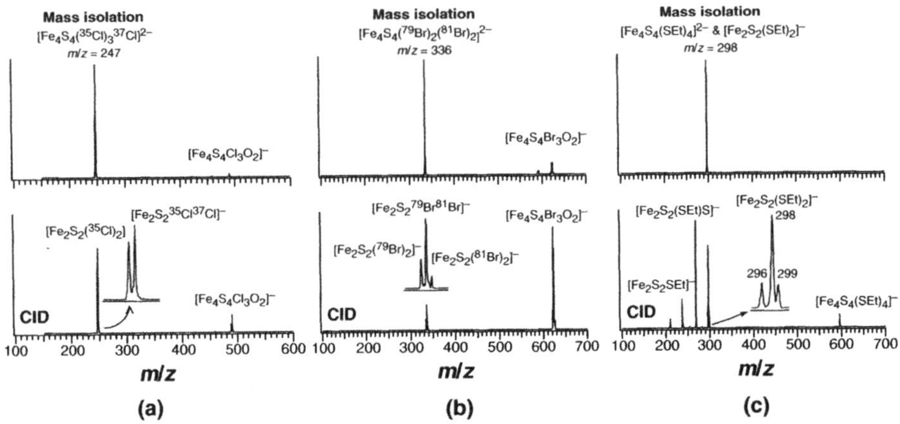

FIGURE 2.11. Collision-induced dissociation (CID) of mass-selected anions of (a) $[Fe_4S_4Cl_4]^{2-}$, (b) $[Fe_4S_4Br_4]^{2-}$, and (c) $[Fe_4S_4(SEt)_4]^{2-}$. Ligand exchange reaction with residual background O_2 was observed during the CID experiments for $[Fe_4S_4Cl_4]^{2-}$ and $[Fe_4S_4Br_4]^{2-}$. All CID fragments were confirmed by their appropriate isotope patterns as shown. [Reprinted with permission from Yang X, Wang XB, Niu S, Pickett CJ, Ichiye T, Wang LS, *Phys. Rev. Lett.* **89**:163401-1–163401-4, 2002 (Yang et al. 2002). Copyright (2002) American Physical Society.]

parent dianions at $m/z = 296$, 299 were due to isotopmers of the symmetric fission product $[Fe_2S_2(SEt)_2]^-$. Their intensities relative to the original $m/z = 298$ peak indicate a quite complete dissociation for the parent dianions. Three small fragments were observed and identified as $[Fe_2S_2(SEt)S]^-$, $[Fe_2S_2(SEt)]^-$, and $[FeS(SEt)_2]^-$. We should point out that, unlike the cases for $[Fe_4S_4Cl_4]^{2-}$ and $[Fe_4S_4Br_4]^{2-}$, for which a single isotopmer of the doubly charged anion could be isolated, the m/z 298 mass peak isolated for $[Fe_4S_4(SEt)_4]^{2-}$ has identical m/z as the fission product, $[Fe_2S_2(SEt)_2]^-$. However, the isotope pattern in the CID spectra, the observation of the electron loss channel in the CID, as well as our PES data, showed that the $[Fe_2S_2(SEt)_2]^-$ ion was less than 10% in the isolated $m/z = 298$ peak in the CID experiment of $[Fe_4S_4(SEt)_4]^{2-}$.

2.5.2. Comparison between PES Spectra of Parent and Daughter Anions

PES spectra of $[Fe_2S_2L_2]^-$ (L = Cl,Br,SEt) are shown in Figure 2.12b and compared with the spectra of parent dianions (Figure 2.12a). The daughter anions have much higher electron binding energies because they are now singly charged and lack the intracluster Coulomb repulsion present in the parent dianions. But the spectral features are very similar in each case, suggesting that the daughter anions and parent dianions have similar electronic structures.

The electronic structure for $[Fe_4S_4L_4]^{2-}$ cubane dainions has been well studied both experimentally and theoretically and are presented in Section 3 (also in Figure 2.13a). On the fission of $[Fe_4S_4L_4]^{2-}$, two possible $[Fe_2S_2L_2]^-$ daughter anions may be produced: either a high-spin product ($S = \frac{9}{2}$) or a low-spin product

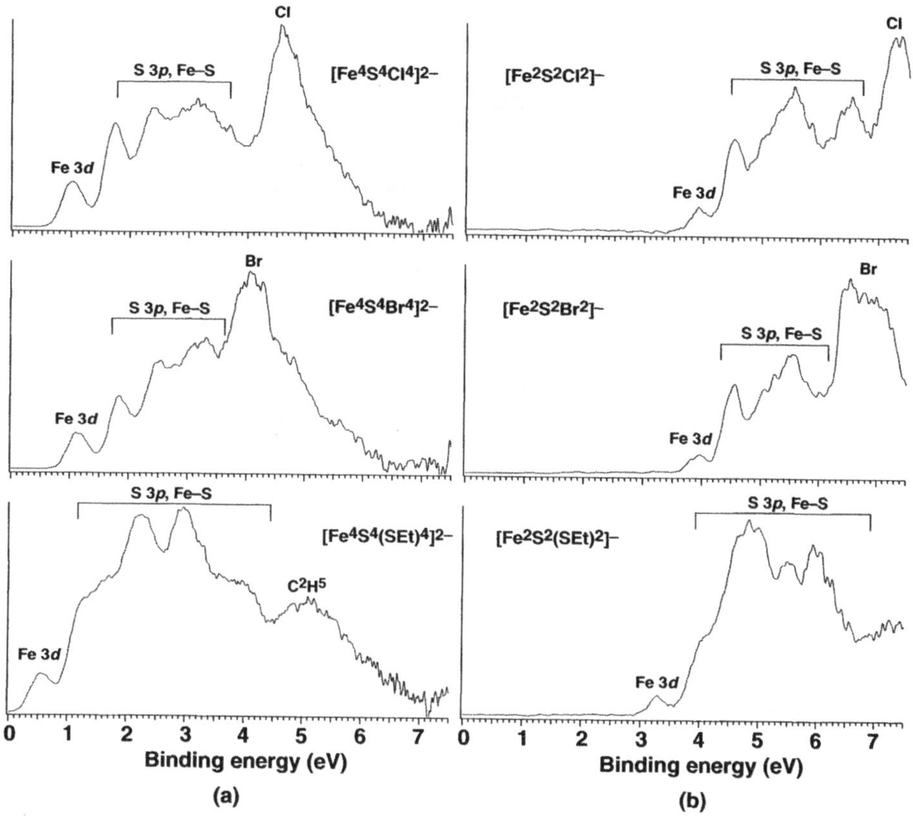

FIGURE 2.12. Photoelectron spectra at 157 nm of (a) $[Fe_4S_4L_4]^{2-}$ (L = Cl, Br, SEt) and (b) their fission daughter ions, $[Fe_4S_4L_4]^-$. [Reprinted with permission from Yang X, Wang XB, Niu S, Pickett CJ, Ichiye T, Wang LS, *Phys. Rev. Lett.* **89**:163401-1–163401-4, 2002 (Yang et al. 2002). Copyright (2002) American Physical Society.]

($S = \frac{1}{2}$). Previous experimental and theoretical studies of the reduced $[Fe_2S_2R_4]^{3-}$ species showed that the $[2Fe\text{–}2S]^{1+}$ core contains a valence-localized Fe^{3+} site and a valence-localized Fe^{2+} site with a net spin $S = \frac{1}{2}$ (Noodleman and Baerends 1984; Noodleman et al. 1995; Mouesca et al. 1994; Torres et al. 2003). We also performed DFT calculation on one of the daughter ion $[Fe_2S_2Cl_2]^-$ (Niu et al. 2004). The calculation agree with previous studies and show that the low-spin state of $[Fe_2S_2Cl_2]^-$ is more stable than the high-spin state by 0.58 eV at the B3LYP/6-31(++)s** level. Analyses of the molecular orbitals and the spin densities show that the majority and minority spin Fe 3*d* orbitals of the high-spin state are entirely delocalized over both Fe centers, while those of the low-spin state are localized on the individual Fe sites with antiferromagnetic coupling (Figure 2.13b). The energy scheme for $[Fe_2S_2Cl_2]^-$ is very similar to that of parent dianions, giving similar spectral features in the PES spectra.

Further confirmation of the DFT results is made by the PES measurements. The B3LYP/6-31(++)$_S$G** calculations show that the calculated ADE and relaxation

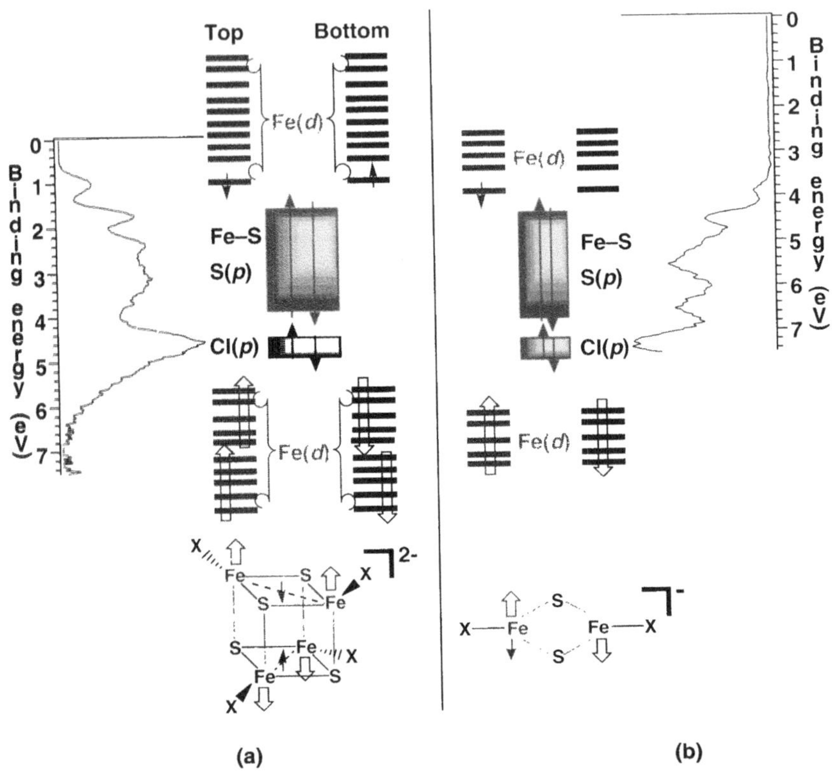

FIGURE 2.13. Schematic inverted energy schemes and comparison with experimental PES data for (a) $[Fe_4S_4Cl_4]^{2-}$ and (b) $[Fe_2S_2Cl_2]^{-}$. The thick arrows represent Fe d^5 electron configurations, and the small arrows represent a single electron spin. [Reprinted with permission from Yang X, Wang XB, Niu S, Pickett CJ, Ichiye T, Wang LS, *Phys. Rev. Lett.* **89**:163401-1–163401-4, 2002 (Yang et al. 2002). Copyright (2002) American Physical Society.]

energy (λ_{oxd}) of the low-spin $[Fe_2S_2Cl_2]^{-}$ are 3.97 and 0.15 eV, respectively, in very good agreement with the experimental values of 3.80 and 0.13 eV. On the other hand, a high-spin $[Fe_2S_2Cl_2]^{-}$ would have significant geometry distortions on photodetachment of the delocalized electron, resulting in a larger ADE of 4.09 and a much larger λ_{oxd} of 0.64 eV, inconsistent with the PES measurements. Although a more stable high-spin species of $[Fe_2S_2Cl_2]^{-}$ with $S = \frac{7}{2}$ is lower by 0.23 eV in energy than that of the $S = \frac{9}{2}$ state (through an α spin flip), the larger ADE of 4.22 eV and the larger λ_{oxd} of 0.27 eV of the $S = \frac{7}{2}$ state also disagree with the PES values.

2.5.3. Mechanism of Symmetric Fission: Intracluster Coulomb Repulsion and Antiferromagnetic Coupling

The symmetric fission observed in the CID of $[Fe_4S_4L_4]^{2-}$ (X = Cl,Br,SEt) was totally unexpected because it involved the breaking of four strong Fe–S bonds. Even more surprisingly, for $[Fe_4S_4Cl_4]^{2-}$ and $[Fe_4S_4Br_4]^{2-}$, the symmetric fission was the only CID channel. Why is this possible among so many other possible fragmentation channels, such as electron detachment or ligand elimination? First, the intracluster coulomb repulsion due to the two excess charges in $[Fe_4S_4L_4]^{2-}$ must play a key role in the fission process, analogous to that in atomic nuclei(Bohr and Wheeler 1939) or multiply ionized metal clusters (Näher et al. 1997; Bréchignac et al. 1990). Intramolecular Coulomb repulsion is unique to multiply charged systems, due to the presence of more than one excess charges in these systems. The importance of the Coulomb repulsion for the fission process is demonstrated by the CID of the singly charged $[Fe_4S_4L_4]^-$ anions. The singly charged parent anions have similar structure and bonding as the $[Fe_4S_4L_4]^{2-}$ dianions, but no symmetric fission was observed in their CID, due to the absence of the strong intracluster Coulomb repulsion in these singly charged species. As reported above, the magnitude of the intracluster Coulomb repulsion can be obtained by estimating the repulsive Coulomb barrier (RCB) of the multiply charged anions from their photoelectron spectra. The RCB for removing an electron in $[Fe_4S_4L_4]^{2-}$ (X = Cl,Br,SEt) was estimated to be ~2 eV (Wang et al. 2003). The same amount of Coulomb repulsion is available for the symmetric fission channel.

Further insight into the symmetric fission mechanisms was provided by considering the electronic structures of the $[Fe_4S_4L_4]^{2-}$ species. Broken-symmetry DFT calculations showed that the $[Fe_4S_4]^{2+}$ core basically contains two valent-delocalized, ferromagnetically coupled $[Fe_2S_2]$ sublayers, which in turn are antiferromagnetically coupled to give the low-spin state. Thus the $[Fe_4S_4L_4]^{2-}$ clusters can be viewed as two ferromagnets aligned oppositely but held together by the four strong Fe–S bonds. On symmetric fission, the $[Fe_2S_2L_2]^-$ daughter ions also exhibit antiferromagnetic coupling. Hence, there is not only a strong intracluster Coulomb repulsion but also a strong magnetic repulsion in the $[Fe_4S_4L_4]^{2-}$ clusters. Both effects are critical for the symmetric fission.

To further confirm the interpretation above, we carried out full geometry optimizations and energy calculations on the $[Fe_4S_4Cl_4]^{2-}$ fission along both the high-spin and low-spin fission pathways at the B3LYP/6-31G** and B3LYP/6-31(++)$_S$G**//B3LYP/6-31G** levels (Niu et al. 2004). As shown in Figure 2.14, there are two possible symmetric fission channels with either low-spin or high-spin $[Fe_2S_2Cl_2]^-$ daughter ions. We found that the symmetric fission along the high-spin fission channel to generate two high-spin daughter anions is endothermic by 1.26 eV with a very high barrier of 2.65 eV while the reaction along the low-spin fission channel in the gas phase was found to be slightly endothermic by 0.09 eV with a relatively low barrier of 1.51 eV. Apparently the low-spin fission channel is a favorable pathway both thermodynamically and kinetically. The average Fe–S bond energy in the $[Fe_4S_4]^{2+}$ cubane core was estimated to be ~2 eV. Thus, about 8 eV energy would be needed for the symmetric fission, which involves the breaking of four Fe–S bonds. The 2 eV Coulomb energy and the 1.2 eV (1.26 - 0.09 eV) energy due to the antiferromagnetic coupling account for almost half of the energy required for the symmetric fission. The remaining endothermicity is recovered from the stronger Fe–S bonds in the $[Fe_2S_2]$ daughter anions relative to that in the parent. The

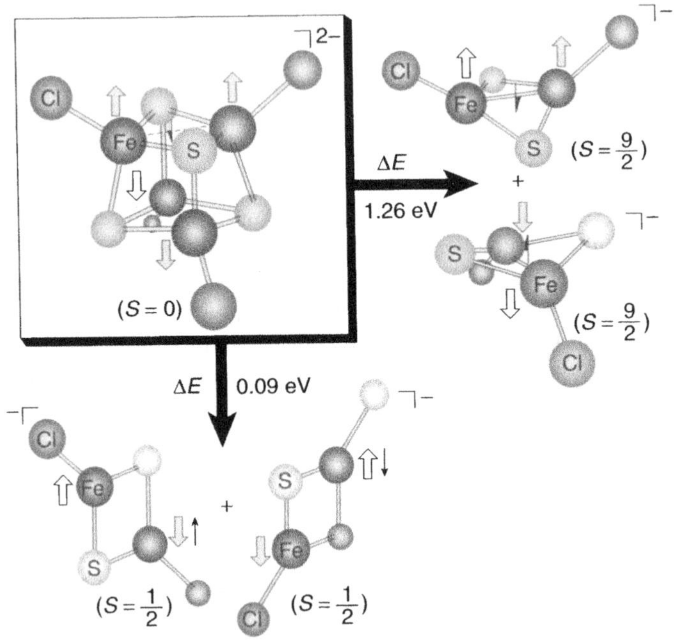

FIGURE 2.14. Calculated energetics for two symmetric fission channels of $[Fe_4S_4Cl_4]^{2-}$. [Reprinted with permission from Yang X, Wang XB, Niu S, Pickett CJ, Ichiye T, Wang LS, *Phys. Rev. Lett.* **89**:163401-1–163401-4, 2002 (Yang et al. 2002). Copyright (2002) American Physical Society.]

near-thermoneutrality for the symmetric fission of $[Fe_4S_4Cl_4]^{2-}$ to two antiferromagnetic $[Fe_2S_2Cl_2]^{-}$ is consistent with our observation that the electron loss channel for the two halogen-ligated complexes is not competitive with the fission channel because of their relative high electron binding energies (0.76 eV for $[Fe_4S_4Cl_4]^{2-}$, and 0.90 eV for $[Fe_4S_4Br_4]^{2-}$; see Table 2.1). Thus, both the intracluster Coulomb repulsion and the antiferromagnetic coupling are important for the observed symmetric fission in the $[Fe_4S_4X_4]^{2-}$ clusters. More details about the kinetic behavior of the fission reaction, such as the transition states and possible fission intermediates along the two fission channels, were presented in our previous publication about fission mechanism (Niu et al. 2004).

2.5.4. Implications for Conversions between [4Fe–4S] and [2Fe–2S] Clusters in Proteins

Iron–sulfur clusters have a remarkable ability for conversion and interconversion in both the free and protein-bound conditions. The most common transformation is the interconversion between the cubane $[Fe_4S_4]^{2+}$ and the cuboidal $[Fe_3S_4]^{+}$ observed in the *Desulfovibrio gigas* ferredoxin II and the enzyme aconitase (Moura et al. 1982; Kent et al. 1982). In the chemical synthesis, $[Fe_4S_4]^{2+}$ can be obtained through the following reactions: $2[Fe_2S_2(SR)_4]^{2-} \leftrightarrow 2[Fe_2S_2(SR)_4]^{3-} \rightarrow [Fe_4S_4(SR)_4]^{2-}$

(Bernert et al. 1997). The first example of a direct $[Fe_4S_4]^{2+} \rightarrow [Fe_2S_2]^{2+}$ conversion in protein was reported in 1984 when the oxidized Fe protein from *Azotobacter vinelandii* nitrogenase exposed to a chelator (α,α'-dipyridyl) in the presence of MgATP (Anderson and Howard 1984). In 1997, an almost quantitative conversion of $[Fe_4S_4]^{2+}$ to $[Fe_2S_2]^{2+}$ was observed on exposure of the FNR (fumarate nitrate reduction) protein of *Escherichia coli* to dioxygen (Khoroshilova et al. 1997). This protein is a transcriptional activator that controls numerous genes required for the synthesis of components of the anaerobic respiratory pathways of *E. coli*. When isolated aerobically, FNR is inactive and occurs as a 30-kD monomer. The active protein is dimeric containing one $[Fe_4S_4]^{2+}$ cluster per subunit (Khoroshilova et al. 1995). On exposure to dioxygen, these $[Fe_4S_4]^{2+}$ clusters are readily converted to $[Fe_2S_2]^{2+}$ in high yield. The $[Fe_2S_2]^{2+}$ cluster form of FNR is much more stable to oxygen, but was unable to sustain biological activity. The $[Fe_2S_2]^{2+}$ cluster can be largely reconverted to the $[Fe_4S_4]^{2+}$ form on reduction with dithionite in vitro. The same cluster conversion also occurs in vivo on exposure to O_2 (Popescu et al. 1998). These investigations demonstrate that the $[Fe_4S_4]^{2+} \leftrightarrow [Fe_2S_2]^{2+}$ conversion has important biological implications, but the reaction mechanisms remain unknown.

The symmetric fission channel can be viewed as a $[Fe_4S_4]$–$[Fe_2S_2]$ conversion reaction in the gas phase. It is interesting to compare it with a similar observation in the oxidized FNR protein in solution. The gas-phase symmetric fission channel is a nonredox reaction; the products remain the same oxidation state $[Fe_2S_2]^{+}$ as in the parent dianions. Because the fission product is exactly half that of the parent cluster, one parent dianion $[Fe_4S_4L_4]^{2-}$ should generate two $[Fe_2S_2L_2]^{-}$ fragments. But in the solution reaction, two Fe^{2+} and two S^{2-} per cluster are lost during the conversion from the 4Fe to the 2Fe cluster. In the history of Fe–S protein chemistry, the $[Fe_3S_4]^{+}$ cluster is the primary product of oxidation and one Fe^{2+} is obviously set free first. If this also holds for the 4Fe $\rightarrow$ 2Fe conversion, the $[Fe_3S_4]^{+}$ cluster would be an important transient intermediate in the degradation pathway. To confirm this mechanism, numerous attempts were made to detect the very easily observable EPR signals ($g = 2.01$) of the $[Fe_3S_4]^{+}$ clusters by rapid freezing during or after oxidation of anaerobically isolated FNR by air or ferricyanide, but never more than $\sim$5% of the clusters originally present were in the 3Fe state. Our observation of the 4Fe cluster symmetric fission in the gas phase may provide a new angle for the interpretation of the 4Fe $\rightarrow$ 2Fe conversion in FNR protein; intracluster Coulomb repulsion and the antiferromagnetic coupling may also play an important role in this case and the two Fe and two S may depart the 4Fe cluster simultaneously, analogous to the gas-phase symmetric fission process.

2.6. SEQUENTIAL OXIDATION OF THE [4Fe–4S] CUBANE CLUSTER: FROM $[4Fe–4S]^{-}$ to $[4Fe–4S]^{3+}$

Series of partially coordinated [4Fe–4S] cubane clusters $Fe_4S_n^-$ ($n = 4$–6) and $Fe_4S_4L_n^-$ ($L = Cl,Br,I$; $n = 1$–4) were produced by laser vaporization and ESI-CID experiments (Zhai et al. 2004). The electronic structure for each fragment was probed by photoelectron spectroscopy. Low binding energy features derived from minority-spin Fe 3*d* electrons were clearly distinguished from S-derived bands. We showed that the electronic structure of the simplest $Fe_4S_4^-$ cubane cluster can be described by the two-layer spin coupling model previously developed for the [4Fe] cubane analogs. The photoelectron data revealed that each extra S atom in $Fe_4S_5^-$ and $Fe_4S_6^-$ removes two minority-spin Fe 3*d* electrons from the [4Fe–4S] cubane core, and each halogen ligand removes one Fe 3*d* electron from the cubane core in the $Fe_4S_4L_n^-$ complexes, clearly revealing a behavior of sequential oxidation of the cubane over five formal oxidation states: $[4Fe\text{–}4S]^- \rightarrow [4Fe\text{–}4S]^0 \rightarrow [4Fe\text{–}4S]^+ \rightarrow [4Fe\text{–}4S]^{2+} \rightarrow [4Fe\text{–}4S]^{3+}$. This work shows the electron-storage capability of the [4Fe–4S] cubane, contributes to understanding of its electronic structure, and further demonstrates the robustness of the cubane as a structural unit and electron transfer center.

2.6.1. Production of Bare and Partially Coordinated [4Fe–4S] Cubane Clusters Using Laser Vaporization and CID

The experiments on bare iron–sulfur clusters $Fe_4S_n^-$ ($n = 4$–6) were carried out using another photoelectron spectroscopy apparatus in our laboratory with a laser vaporization cluster source (Wang et al. 1995). Briefly, a mixed Fe : S target (10 : 1 molar ratio) was laser-vaporized (10–20 mJ from the second harmonic of a Nd:YAG laser) in the presence of a helium carrier gas (10 atm stagnation pressure). The intense carrier gas pulse mixes with and cools the laser-induced plasma. Various $Fe_mS_n^-$ clusters were produced, and the cluster/He mixture underwent a supersonic expansion and entered the extraction area of the TOF mass spectrometer. The clusters of interest were mass-selected for the PES study.

Partially coordinated clusters $Fe_4S_4L_n^-$ (L = Cl,Br,I; $n = 1$–4) were produced by performing CID on the singly charged $[Fe_4S_4L_4]^-$ complexes, which were produced by collision-induced electron detachment from the corresponding $[Fe_4S_4L_4]^{2-}$ parent dianions. Subsequent CID required for Fe–L bond scission from the singly charged $Fe_4S_4L_3^-$ fragment to produce the smaller $Fe_4S_4L_n^-$ singly charged fragments, namely, $Fe_4S_4L_3^- \rightarrow Fe_4S_4L_2^- + L$. For L = Cl, we could not observe this bond scission process, but for L = I, this process can go all the way down to the bare $Fe_4S_4^-$. Bare $Fe_4S_5^-$ and $Fe_4S_6^-$ clusters were also produced from CID of $[Fe_4S_4(SEt)_4]^{2-}$.

2.6.2. PES of $Fe_4S_n^-$ (n = 4–6)

In contrast to the intensive investigations on the [4Fe–4S] active sites in proteins and the cubane [4Fe–4S] cores in analog complexes, little is known about the bare $Fe_4S_4^-$ cluster (Nakajima et al. 1997; Yu et al. 1993; Koszinowski et al. 2004). An interesting question concerns its ground-state structure. Does it also possess a cubane-type structure? We were able to generate this cluster from both our laser vaporization and electrospray sources. The ion formation process is completely different in the two ion sources. In laser vaporization, clusters are formed through aggregation of atoms, and in general the lowest-energy structures are produced. On the other hand, the $Fe_4S_4^-$ clusters were formed by successive loss of I ligands from $Fe_4S_4I_4^-$ through CID in the electrospray source. It was expected the cubane structure to be maintained.

Photoelectron spectra of $Fe_4S_n^-$ ($n = 4$–6) produced from laser vaporization are shown in Figures 2.15 and 2.16. In Figure 2.15a, the 355 nm spectrum of $Fe_4S_4^-$ revealed three bands (*X*, *A*, and *B*). The *X* band with a vertical detachment energy (VDE) of 2.37 eV was relatively sharp. The well-defined onset of feature *X* allows a fairly accurate ADE of 2.30 ± 0.02 eV to be obtained, which represents the electron affinity of the corresponding neutral Fe_4S_4 species. At 266 nm (Figure 2.15b), the *B* band was better defined, and another intense band *C* was revealed at 3.5 eV. The 193 nm spectrum (Figure 2.15c) showed the overall PES pattern of $Fe_4S_4^-$: three

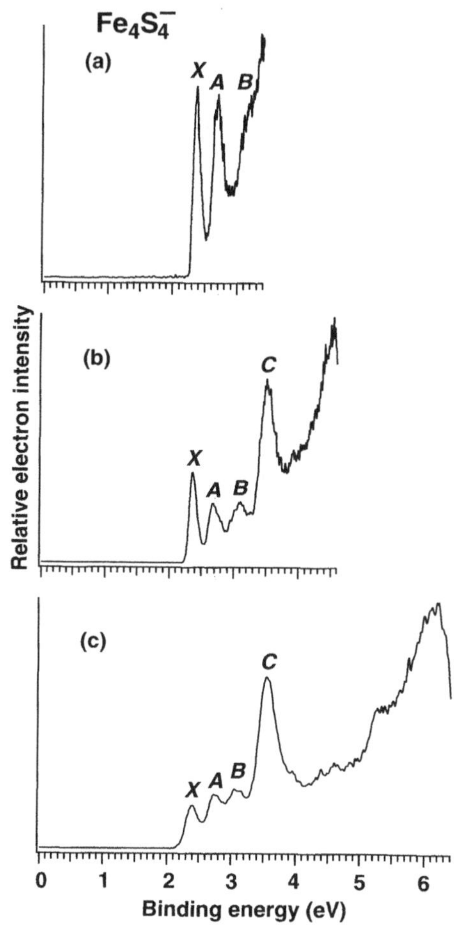

FIGURE 2.15. Photoelectron spectra of $Fe_4S_4^-$ at (a) 355 nm, (b) 266 nm, and (c) 193 nm. [Reprinted with permission from Zhai HJ, Yang X, Fu YJ, Wang XB, Wang LS, *J. Am. Chem. Soc.* **126**:8413–8420, 2004 (Zhai et al. 2004). Copyright (2004) American Chemical Society.]

weak low-binding-energy bands (*X*, *A*, and *B*) followed by an intense and broader *C* band. The higher-binding-energy part of the 193 nm spectrum appeared to be continuous, indicative of the high density of electronic states.

The spectra of $Fe_4S_5^-$ and $Fe_4S_6^-$ are shown in Figure 2.16 at two photon energies. For $Fe_4S_5^-$, three well-defined bands (*X*, *A*, and *B*) were observed at 266 nm (Figure 2.16a); the *A* band overlapped with the more intense *B* band. The well-resolved *X* band yielded a VDE of 3.56 eV and an ADE of 3.48 eV for the detachment transition from the ground state of $Fe_4S_5^-$ to that of Fe_4S_5. The ADE of $Fe_4S_5^-$ increased significantly compared to that of $Fe_4S_4^-$. The higher-binding-energy part of the $Fe_4S_5^-$ spectrum as revealed at 193 nm (Figure 2.16d) also appeared to be

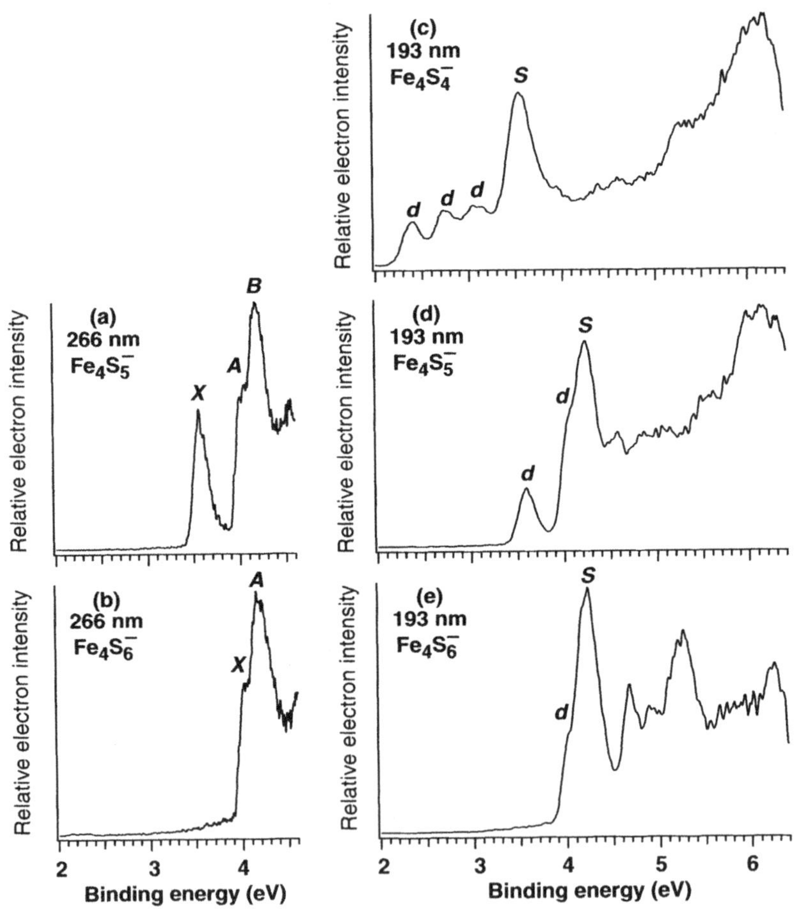

FIGURE 2.16. Photoelectron spectra of $Fe_4S_5^-$ and $Fe_4S_6^-$ at 266 and 193 nm. The 193-nm spectrum of $Fe_4S_4^-$ is compared with those of $Fe_4S_5^-$ and $Fe_4S_6^-$. The *d* labels indicate features from detachment of Fe 3*d* electrons, and *S* denotes features derived from S 3*p*-based molecular orbitals. [Reprinted with permission from Zhai HJ, Yang X, Fu YJ, Wang XB, Wang LS, *J. Am. Chem. Soc.* **126**:8413–8420, 2004 (Zhai et al. 2004). Copyright (2004) American Chemical Society.]

continuous. For $Fe_4S_6^-$, the 266 nm spectrum (Figure 2.16b) revealed two partially overlapped bands (*X* and *A*). An ADE of 3.94 eV was estimated from the sharp onset of band *X*, which also defined the electron affinity of the Fe_4S_6 neutral cluster. The 193 nm spectrum revealed more well-defined features beyond 4.5 eV (Figure 2.16e).

The 193 m spectrum of $Fe_4S_4^-$ is compared with those of $Fe_4S_5^-$ and $Fe_4S_6^-$ in Figure 2.16. The most remarkable feature in the three spectra is the intense band labeled *S*, meaning from S 3*p*-derived molecular orbitals (MOs), as will be discussed later. The *S* band is similar in the three spectra and has the same VDE in the cases of $Fe_4S_5^-$ and $Fe_4S_6^-$. This band becomes the demarcation line in the spectra. Features to the left of the *S* band at lower binding energies are weaker in intensity and diminish in number. These weak features are labeled *d*, meaning minority-spin Fe 3*d*-derived bands. To the right of the *S* band at higher binding energies all spectra became very complicated, and only that of $Fe_4S_6^-$ exhibited resolved features. The binding energies of the spectra increase significantly, in particular from Fe_4S^- to $Fe_4S_5^-$. The observed VDEs for the *d* and *S* features and the ground-state ADEs are given in Table 2.3.

Figure 2.17 compares the PES spectra of $Fe_4S_n^-$ ($n = 4$–6) produced from laser vaporization and electrospray sources. They are essentially identical, except that the spectrum from the CID product was broader; this was due to the fact that the CID products were relatively hot with high internal energies, which could not be effectively cooled during the ion transport and trapping. On the other hand, in the laser vaporization source the supersonic expansion provided moderate cooling and

TABLE 2.3. Measured Adiabatic and Vertical Detachment Energies (eV) of Low-Binding-Energy Features from Photoelectron Spectra of $Fe_4S_n^-$ ($n = 4$–6) and $Fe_4S_4L_n^-$ (L = Cl, Br, I; $n = 1$–4)

			VDE[a]			
Species	Cubane Core	ADE[a,b]		*d* band		*S* band
$Fe_4S_4^-$	$[Fe_4S_4]^-$	2.30 (2)	2.37 (2)	2.70 (2)	3.09 (2)	3.51 (2)
$Fe_4S_5^-$	$[Fe_4S_4]^+$	3.48 (2)	3.56 (2)	4.05 (3)	—	4.20 (2)
$Fe_4S_6^-$	$[Fe_4S_4]^{3+}$	3.94 (2)	4.05 (3)	—	—	4.20 (2)
$Fe_4S_4Cl_4^-$	$[Fe_4S_4]^{3+}$	4.62 (5)	4.85 (4)	—	—	5.56 (4)
$Fe_4S_4Cl_3^-$	$[Fe_4S_4]^{2+}$	4.23 (5)	4.42 (4)	4.74 (6)	—	5.20 (4)
$Fe_4S_4Br_4^-$	$[Fe_4S_4]^{3+}$	4.56 (5)	4.83 (4)	—	—	5.46 (4)
$Fe_4S_4Br_3^-$	$[Fe_4S_4]^{2+}$	4.23 (5)	4.41 (4)	4.72 (6)	—	5.15 (4)
$Fe_4S_4Br_2^-$	$[Fe_4S_4]^+$	3.76 (5)	4.08 (6)	—	—	4.70 (6)
$Fe_4S_4I_4^-$	$[Fe_4S_4]^{3+}$	4.47 (5)	4.70 (4)	—	—	5.24 (4)
$Fe_4S_4I_3^-$	$[Fe_4S_4]^{2+}$	4.20 (5)	4.36 (4)	4.65 (6)	—	5.01 (4)
$Fe_4S_4I_2^-$	$[Fe_4S_4]^+$	3.76 (5)	4.08 (4)	—	—	4.63 (4)
$Fe_4S_4I^-$	$[Fe_4S_4]^0$	3.03 (10)	3.23 (6)	3.62 (6)	—	4.36 (6)

[a]Numbers in parentheses represent experimental uncertainties in the last digits.
[b]Also represent the adiabatic electron affinities of the corresponding neutral species.

Source: Reprinted with permission from Zhai HJ, Yang X, Fu YJ, Wang XB, Wang LS, *J. Am. Chem. Soc.* **126**:8413–8420, 2004 (Zhai et al. 2004). Copyright (2004) American Chemical Society.

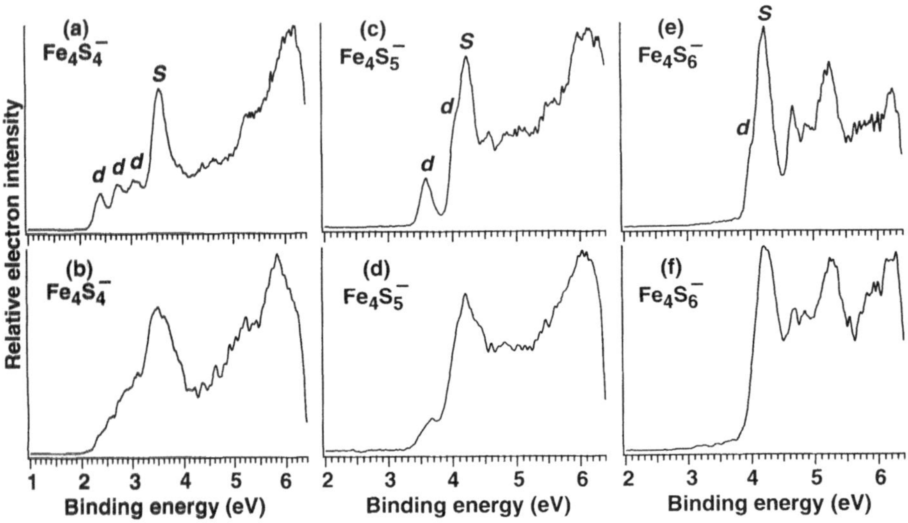

FIGURE 2.17. Comparison of photoelectron spectra of $Fe_4S_4^-$, $Fe_4S_5^-$, and $Fe_4S_6^-$ produced from two different ion sources. *Top*—from laser vaporization of a Fe/S mixed target; *bottom*—from collision-induced dissociation of doubly charged anions from an electrospray source. [Reprinted with permission from Zhai HJ, Yang X, Fu YJ, Wang XB, Wang LS, *J. Am. Chem. Soc.* **126**:8413–8420, 2004 (Zhai et al. 2004). Copyright (2004) American Chemical Society.]

in general produces cluster anions with vibrational temperatures slightly below room temperature. Thus, the identity of the PES spectra of $Fe_4S_n^-$ from two totally different formation processes indicated that the bare $Fe_4S_n^-$ cluster indeed possesses a cubane-like structure, probably not too different from that in the $Fe_4S_4L_4^-$ complexes.

2.6.3. PES of $Fe_4S_4Cl_n^-$ ($n = 3,4$), $Fe_4S_4Br_n^-$ ($n = 2$–4), and $Fe_4S_4I_n^-$ ($n = 0$–4)

Under the CID conditions in our electrospray source, we were able to observe only the singly charged $Fe_4S_4Cl_4^-$ and its CID product by losing one Cl atom, $Fe_4S_4Cl_3^-$. The PES spectra of these two complexes at 193 and 157 nm are shown in Figure 2.18. The spectra of $Fe_4S_4Cl_4^-$ displayed three features, labeled *d* and *S*, as well as an intense band at 6.2 eV (Figure 2.18c). The *d* band had a relatively weak intensity and yielded a rather high ADE (4.62 eV) for $Fe_4S_4Cl_4^-$. The spectra of $Fe_4S_4Cl_3^-$ shifted to lower binding energies with an intense band labeled *S*, which is similar to that in the spectra of $Fe_4S_4Cl_4^-$. There are two weaker features at the lower-binding-energy side, labeled *d*. The separation between the first *d* band and the *S* band are almost identical to that in the spectra of $Fe_4S_4Cl_4^-$. With the exception of the extra *d* band in $Fe_4S_4Cl_3^-$, the overall PES spectral patterns of the two $Fe_4S_4Cl_n^-$ complexes are similar, and both also show some similarity to the spectra

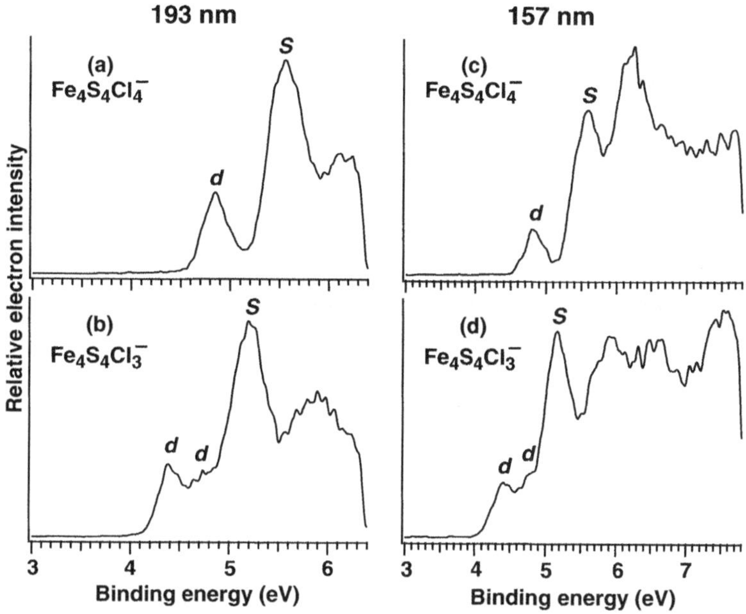

FIGURE 2.18. Photoelectron spectra of $Fe_4S_4Cl_n^-$ ($n = 3,4$) at 193 and 157 nm. See Figure 2.16 cation for labels *d* and *S*. [Reprinted with permission from Zhai HJ, Yang X, Fu YJ, Wang XB, Wang LS, *J. Am. Chem. Soc.* **126**:8413–8420, 2004 (Zhai et al. 2004). Copyright (2004) American Chemical Society.]

of the doubly charged $[Fe_4S_4Cl_4]^{2-}$ complex as shown in Figure 2.3. The ADEs for the first *d* band and the VDEs for all the *d* and *S* bands are given in Table 2.3.

For the Br-ligated complexes, two CID products ($Fe_4S_4Br_3^-$ and $Fe_4S_4Br_2^-$) were observed under our experimental conditions. The spectra of $Fe_4S_4Br_4^-$ and $Fe_4S_4Br_3^-$ (not shown) are nearly identical to those of the corresponding Cl-ligated complexes. The spectra of $Fe_4S_4Br_2^-$ shifted further to lower binding energies. The *S* band is similar to that in the spectra of $Fe_4S_4Br_3^-$, but the lower-binding-energy part of the $Fe_4S_4Br_2^-$ spectra seemed to be more complicated with unresolved bands, although only one *d* band was labeled.

For the I-ligated cubane, the loss of all the ligands down to the bare $Fe_4S_4^-$ core was observed in the CID owing to the relatively weaker Fe–I bond. This series of complexes gave us the most systematic and complete dataset, as shown in Figure 2.19. As *n* decreases, we observed that the spectra systematically shift to lower binding energies. For $n = 4,3,2$, the I-ligated complexes gave rise to spectra similar to those of the corresponding Br-ligated complexes. For $n = 1,0$, more *d* bands were observed at the lower-binding-energy side. Two were discernible in the spectra of $Fe_4S_4I^-$ and three in those of the bare $Fe_4S_4^-$. Although the higher-binding-energy side of the $Fe_4S_4^-$ spectra showed some difference, the spectra of $Fe_4S_4^-$ follow the general trend of the $Fe_4S_4I_n^-$ series, suggesting that the cubane core survived the CID processes and maintained a structure similar to that in the $Fe_4S_4I_n^-$

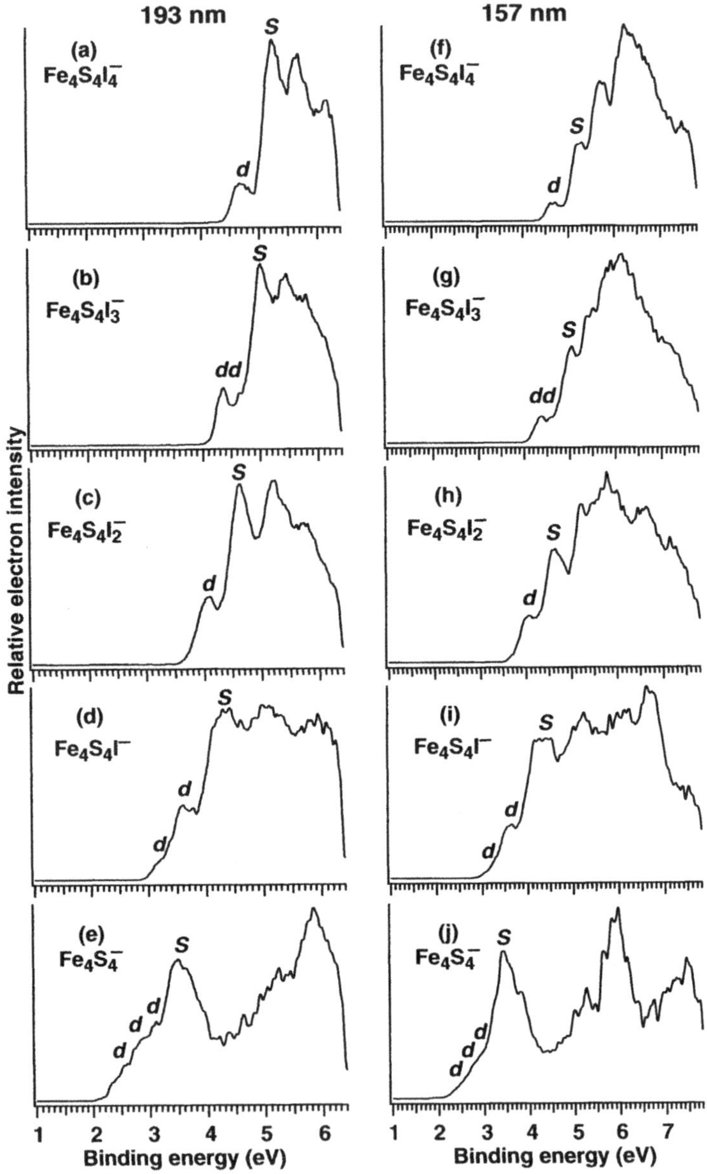

FIGURE 2.19. Photoelectron spectra of $Fe_4S_4I_n^-$ ($n = 0$-4) at 193 and 157 nm. See Figure 2.16 cation for labels *d* and *S*. [Reprinted with permission from Zhai HJ, Yang X, Fu YJ, Wang XB, Wang LS, *J. Am. Chem. Soc.* **126**:8413–8420, 2004 (Zhai et al. 2004). Copyright (2004) American Chemical Society.]

complexes. The ADEs and VDEs for the *d* and *S* bands of $Fe_4S_4I_n^-$ are also given in Table 2.3.

2.6.4. Electronic Structures of $Fe_4S_4L_4^-$ and $Fe_4S_6^-$ with $[Fe_4S_4]^{3+}$ Cubane Core

In $[Fe_4S_4L_4]^{2-}$, there are two energetically equivalent minority spins, one in each [2Fe–2S] sublayer (Figure 2.6), whereas in the singly charged $Fe_4S_4L_4^-$ complexes, there is only one minority-spin electron. The detachment of this electron gave rise to the weak *d* band in the PES spectra of $Fe_4S_4L_4^-$, as summarized in Figure 2.20a–c for L = Cl,Br,I. The second PES band (*S*) corresponds to detachment primarily from S-based MOs. We noted that the PES pattern of the $Fe_4S_4L_4^-$ singly charged species are very similar to those of the corresponding doubly charged $[Fe_4S_4L_4]^{2-}$ complexes (Figure 2.3), except that the binding energy of the singly charged species is much higher because of the absence of the intramolecular Coulomb repulsion present in the doubly charged anions. The relative intensity of the first band is weaker in the PES spectra of the singly charged species than that observed for the doubly charged anions because this feature corresponds to only one minority-spin electron in the singly charged complexes. Therefore, our PES data of the singly charged $Fe_4S_4L_4^-$ complexes with a $[4Fe–4S]^{3+}$ oxidation state is consistent with the inverted-level scheme shown in Figure 2.6. All these complexes have a spin of $\frac{1}{2}$, due to the presence of the single minority-spin electron.

The PES spectrum of $Fe_4S_6^-$ is compared to those of $Fe_4S_4L_4^-$ in Figure 2.20. As discussed above, this cluster is expected to have a cubane core with a $[4Fe–4S]^{3+}$ oxidation state. Although the spacing of the first two features in the PES spectrum of $Fe_4S_6^-$ (Figure 2.20d) is much smaller, their relative intensities are similar to those in the spectra of $Fe_4S_4L_4^-$. Thus, the weak low-binding-energy feature (*d*), which appeared as a shoulder, should correspond to the single minority-spin electron in the $[4Fe–4S]^{3+}$ cubane core, and the intense band (*S*) should correspond to S-based MOs. Thus, the electronic structure of $Fe_4S_6^-$ can also be described by the inverted-level scheme with a spin of $\frac{1}{2}$, similar to the $Fe_4S_4L_4^-$ complexes. The two extra S atoms most likely coordinate each to two Fe atoms in the two sublayers of the cubane core. The smaller *d–S* band spacing and the different spectral pattern in the higher-binding-energy part in the spectrum of $Fe_4S_6^-$ are due to the difference between the S ligand and the halogen ligands. We noted that the *d–S* band spacing decreases slightly from $Fe_4S_4Cl_4^-$ to $Fe_4S_4I_4^-$, as the terminal ligand becomes less electron-withdrawing from Cl to I. Sulfur as a divalent terminal ligand should be even less electron-withdrawing relative to I, and the smaller *d–S* spacing in $Fe_4S_6^-$ is consistent with this trend.

2.6.5. Electronic Structure of $Fe_4S_4L_3^-$ (L = Cl,Br,I) with $[Fe_4S_4]^{2+}$ Cubane Core and Partial Coordination Effects on the Cubane

The partially coordinated $Fe_4S_4L_3^-$ complexes contain a cubane core with a $[4Fe–4S]^{2+}$ oxidation state, which is similar to that in the $[Fe_4S_4L_4]^{2-}$ doubly charged complexes. However, in $Fe_4S_4L_3^-$ the two sublayers are no longer equivalent, due to

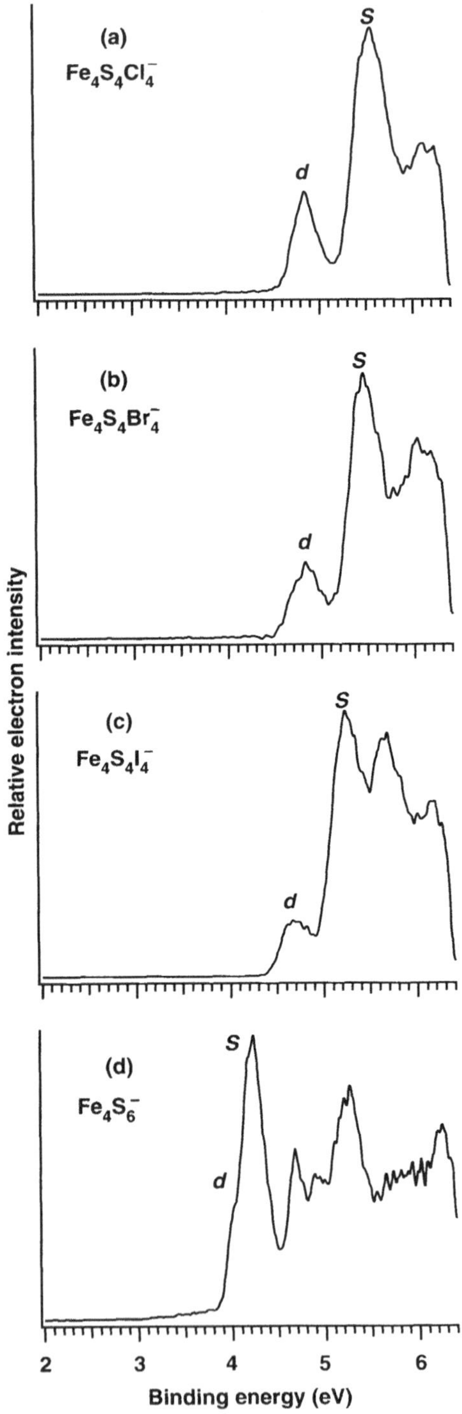

FIGURE 2.20. Comparison of photoelectron spectra of all species with a $[Fe_4S_4]^{3+}$ core. [Reprinted with permission from Zhai HJ, Yang X, Fu YJ, Wang XB, Wang LS, *J. Am. Chem. Soc.* **126**:8413–8420, 2004 (Zhai et al. 2004). Copyright (2004) American Chemical Society.]

the absence of one ligand. This asymmetry should induce a splitting in the minority-spin levels, which are no longer equivalent energetically. The minority-spin level (the HOMO of $[Fe_4S_4L_4]^{2-}$) involves Fe–Fe bonding interactions and Fe–L antibonding interactions within each [2Fe–2S] sublayer (Wang et al. 2003). The minority-spin level in the sublayer with only one ligand should be energetically stabilized and the one in the sublayer with two ligands should remain the same as in $[Fe_4S_4L_4]^{2-}$. A splitting of the minority-spin levels was indeed evident in the PES spectra of the three $Fe_4S_4L_3^-$ complexes, as summarized in Figure 2.21. More interestingly, the separation between the first *d* band and the *S* band are nearly identical in the spectra of $Fe_4S_4L_3^-$ and $[Fe_4S_4L_4]^{2-}$, in complete agreement with the inverted-level scheme, suggesting that the first *d* band came from detachment of the minority-spin electron in the [2Fe–2S] sublayer with two ligands and the second *d*

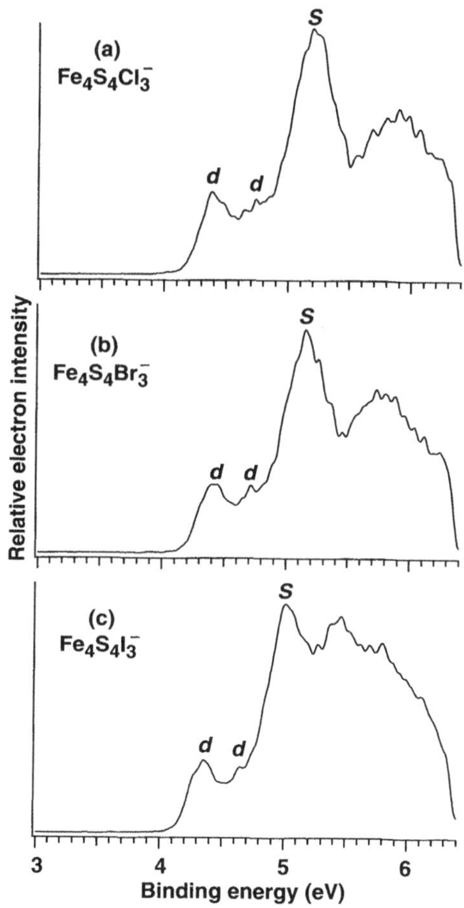

FIGURE 2.21. Comparison of photoelectron spectra of all species with a $[Fe_4S_4]^{2+}$ core. [Reprinted with permission from Zhai HJ, Yang X, Fu YJ, Wang XB, Wang LS, *J. Am. Chem. Soc.* **126**:8413–8420, 2004 (Zhai et al. 2004). Copyright (2004) American Chemical Society.]

band derived from detachment of the minority electron in the sublayer with only one ligand.

2.6.6. Electronic Structures of $Fe_4S_4L_2^-$ (L = Br,I) and $Fe_4S_5^-$ with $[Fe_4S_4]^+$ Cubane Core

These three species should each contain a $[4Fe–4S]^+$ core with three minority-spin electrons. The spin coupling in these species is expected to be complicated and depends on how the two sublayers are divided, either as two [2Fe–2S–L] layers, each coordinated with one ligand, or one [2Fe–2S–2L] layer and one [2Fe–2S] layer without any terminal ligand. In either way, the two sublayers cannot be equivalent because of the odd number of the minority-spin electrons, which should be distributed in the two sublayers as 1 : 2 ratio (giving a spin $\frac{1}{2}$ state). Although only one *d* band is labeled in Figure 2.22a,b, the low-binding-energy part of the

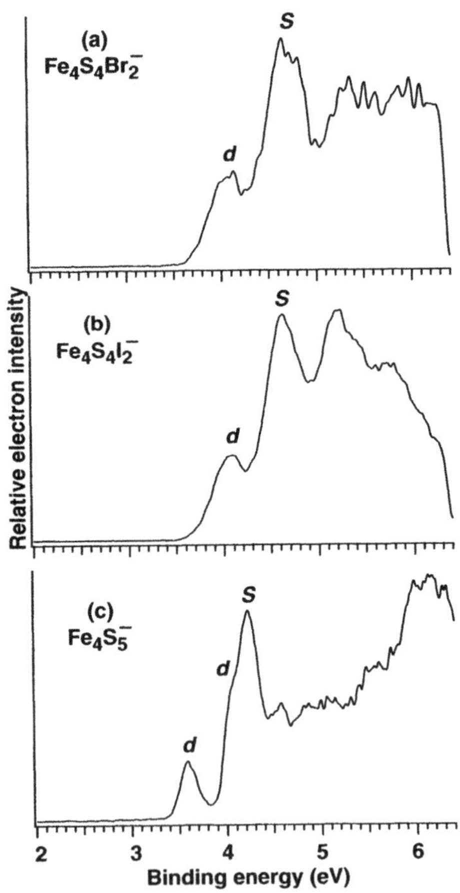

FIGURE 2.22. Comparison of photoelectron spectra of all species with a $[Fe_4S_4]^+$ core. [Reprinted with permission from Zhai HJ, Yang X, Fu YJ, Wang XB, Wang LS, *J. Am. Chem. Soc.* **126**:8413–8420, 2004 (Zhai et al. 2004). Copyright (2004) American Chemical Society.]

PES spectra of $Fe_4S_4Br_2^-$ and $Fe_4S_4I_2^-$ were complicated and were not well resolved, suggesting perhaps a combination of all the abovementioned possibilities.

The PES spectrum of $Fe_4S_5^-$ is compared to those of $Fe_4S_4L_2^-$ in Figure 2.22. The low-binding-energy part of the $Fe_4S_5^-$ spectrum was much better resolved with two well-defined *d* bands. The extra S should coordinate to two Fe atoms in the [4Fe–4S]$^+$ core, defining one sublayer. The three minority spins may distribute with two minority spins in the layer that is not coordinated by the fifth S atom, giving rise to a spin $\frac{1}{2}$ state for $Fe_4S_5^-$. The first *d* band in the PES spectrum of $Fe_4S_5^-$ then came from detachment of the highest occupied minority-spin electron. Depending on the magnitude of the splitting between the two sublayers, either a spin 0 or a spin $\frac{2}{2}$ state can result for the ground state of neutral Fe_4S_5.

2.6.7. Electronic Structures of $Fe_4S_4I^-$ and $Fe_4S_4^-$

$Fe_4S_4I^-$ should contain a [4Fe–4S]^0 cubane core with four minority spins and all Fe are in the ferrous state. Thus each sublayer of the cubane core should have two minority-spin electrons, resulting in a spin 0 state for $Fe_4S_4I^-$. However, the two sublayers are inequivalent, which can result in a splitting of the minority-spin levels in the two sublayers, similar to the case in the $Fe_4S_4L_3^-$ complexes. This splitting was evident in the PES spectra of $Fe_4S_4I^-$ (Figure 2.19d). If the two sublayers were equivalent, two *d* bands would be expected with similar intensities. The low-binding-energy part of the PES spectra of $Fe_4S_4I^-$ was complicated, indicating more transitions were congested in this part of the spectrum and giving direct evidence for the splitting in the minority-spin levels due to the asymmetry of the coordination environment in $Fe_4S_4I^-$. The all-ferrous [4Fe–4S]^0 center of the Fe protein was found in *Azotobacter vinelandii*. One Fe site was shown to be unique presumably due to environmental or geometric asymmetries in the protein. The $Fe_4S_4I^-$ complex may be considered the simplest model system for the all-ferrous center because the single I ligand generates a unique Fe site naturally.

There should be five minority spins in the bare $Fe_4S_4^-$ cluster. They should fill three levels in one sublayer and two in the other sublayer, according to the inverted level scheme of Figure 2.6. This would result in a spin $\frac{1}{2}$ state for $Fe_4S_4^-$. Detachment from the highest occupied minority-spin level gave rise to the *X* band in the PES spectra (Figure 2.15) and resulted in a spin 0 state for the ground state of neutral Fe_4S_4. Detachment from the other occupied minority-spin levels would lead to either a spin 0 or spin $\frac{2}{2}$ excited state for neutral Fe_4S_4. If the splitting between these spin states is small relative to our spectral resolution (~30 meV in the relevant spectral range of Figure 2.15), only one PES band would be resulted. This indeed appeared to be the case since only three well-resolved *d* bands (X, A, B) were observed in the PES spectra of $Fe_4S_4^-$. Thus, our PES data suggested that the inverted-level scheme devised for the cubane core is also applicable to describe the electronic structure of the bare $Fe_4S_4^-$ and Fe_4S_4 clusters.

2.6.8. Electron Storage and Sequential Oxidation of the [4Fe–4S] Cubane Cluster

Iron has two common oxidation states: Fe^{2+} and Fe^{3+}, which are cycled in redox reactions involving Fe. The strong spin polarization stabilizes the d^5 majority spins and destabilizes the single minority spin in the d^6 electron configuration of Fe^{2+}, making the Fe^{2+}/Fe^{3+} redox couple one of the most favorite in chemistry and biochemistry. The redox capability of all Fe–S clusters and proteins rely on this redox couple. In the cubane, the four Fe centers can store up to four minority-spin electrons in the all-ferrous $[4Fe\text{–}4S]^0$ oxidation states. In principle, all of these four electrons are available for electron transfer reactions, leading to the all-ferric $[4Fe\text{–}4S]^{4+}$ oxidation state and giving the cubane cluster an extraordinary capacity for electron storage.

It is noteworthy that the bare Fe_4S_4 cluster possesses a cubane-type structure and its electronic structure can be described by the inverted-level scheme. This proves the stability of the cubane structural feature and provides further support for its robustness as a modular functional unit in analog complexes and proteins. The fact that the bare $Fe_4S_4^-$ cluster possesses a cubane-type structure enables us to access the wide range of oxidation states of the cubane core in the gas phase. A density functional study investigated all five oxidation states of the analog complex $[Fe_4S_4(SCH_3)_4]^{n-}$ ($n = 0$–4) (Torres et al. 2003). However, only the $n = 1$ and $n = 2$ species would be accessible in the gas phase, because the species with $n > 2$ would not be stable as gaseous species because of the strong intramolecular Coulomb repulsion. The current investigation takes advantage of the variable terminal ligands to access a wide range of oxidation states for the cubane core all in the form of singly charged anions.

The $Fe_4S_4^-$ cluster in fact has five minority spins, even though this oxidation state is not accessible in either analog complexes or proteins. In $Fe_4S_5^-$, two minority-spin electrons are transferred from the cubane core to the extra S, resulting in a $[4Fe\text{–}4S]^+$ oxidation state with three minority spins. In $Fe_4S_6^-$, a $[4Fe\text{–}4S]^{3+}$ oxidation is achieved. Thus, from $Fe_4S_4^-$ to $Fe_4S_6^-$ a sequential oxidation of the cubane core is observed and the electron binding energies of the clusters also increase with the number of extra S ligands (Figure 2.16). Although the partially coordinated halogen complexes ($Fe_4S_4L_n^-$) were produced from the fully coordinated $Fe_4S_4L_4^-$, the series of species can also be viewed as a sequential oxidation of the cubane core from the bare $Fe_4S_4^-$ cluster because the core oxidation state increases by one with each additional halogen ligand in $Fe_4S_4L_n^-$. In the case of $Fe_4S_4I_n^-$, a full range of oxidation states of the cubane core is accessed from $[4Fe\text{–}4S]^- \rightarrow [4Fe\text{–}4S]^0 \rightarrow [4Fe\text{–}4S]^+ \rightarrow [4Fe\text{–}4S]^{2+} \rightarrow [4Fe\text{–}4S]^{3+}$ with $n = 0,1,2,3,4$, respectively. The number of minority-spin electrons decreases from $5 \rightarrow 4 \rightarrow 3 \rightarrow 2 \rightarrow 1$ along the same sequence.

Figure 2.23 displays the ADEs of the threshold d band in the four series of $Fe_4S_4L_n^-$ species with respect to the formal oxidation states of the [4Fe–4S] cubane core. The ADEs represent the gas-phase oxidation potential of the corresponding complexes and increase in each series with the oxidation states of the cubane core.

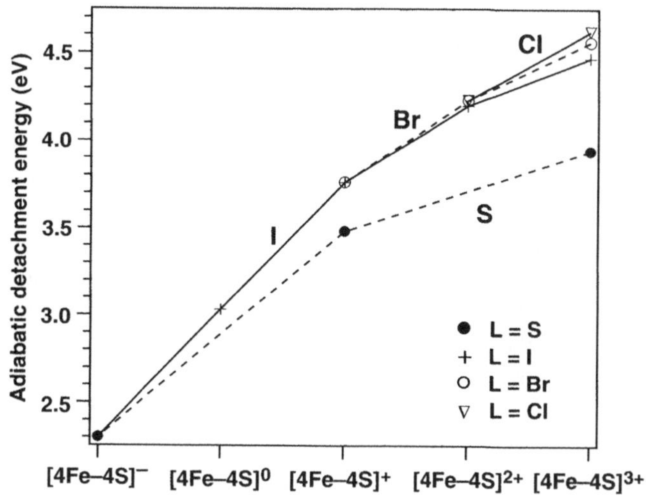

FIGURE 2.23. Adiabatic detachment energies versus the oxidation states of the [4Fe-4S] cubane core. [Reprinted with permission from Zhai HJ, Yang X, Fu YJ, Wang XB, Wang LS, *J. Am. Chem. Soc.* **126**:8413–8420, 2004 (Zhai et al. 2004). Copyright (2004) American Chemical Society.]

For the $[4Fe–4S]^{3+}$ core, the ADEs are extremely high, suggesting that it is much more difficult to oxidize the cubane core to the all-ferric form $[4Fe–4S]^{4+}$. The terminal ligands also have significant influences on the ADEs of the cubane complexes with the order $S < I \sim Br \sim Cl$. We observed that the extra S leads to a 1.18 eV increase of ADE in $Fe_4S_5^-$ relative to that in $Fe_4S_4^-$. This is almost twice as large an increase as the effect of two iodine atoms. This observation is consistent with the divalent nature of the extra S as terminal ligands in $Fe_4S_5^-$.

2.7. CONCLUSIONS

This review summarizes our most recent effort to probe the electronic structures of Fe–S clusters in the gas phase using photoelectron spectroscopy. We demonstrated that our ESI-TOFMS-PES technique is very powerful for transporting ionic and redox species from solution to the gas phase and investigating their electronic structures without solvent perturbation. Several conclusions can be drawn from these studies.

The intrinsic electronic structure of a series of [4Fe–4S] cubane complexes, $[Fe_4S_4L_4]^{2-}$ ($L = SH, SC_2H_5, Cl, Br, I$) and the Se-substituted species $[Fe_4Se_4L_4]^{2-}$ ($L = Cl, SC_2H_5$) were studied using photoelectron spectroscopy and broken-symmetry density functional calculations. All the cubane complexes exhibit similar spectral features, showing the robustness of the [4Fe–4S] cluster as a modular unit. The spectral features confirm the low-spin two-layer model for the $[4Fe–4S]^{2+}$ core and its "inverted-level scheme" molecular orbital diagrams. We found the ADEs, which provide the intrinsic oxidation potentials of the $[Fe_4S_4L_4]^{2-}$ complexes, to be very sensitive to the terminal ligands, but independent of the substitution of the bridging inorganic S by Se. The DFT calculations revealed that the HOMO of $[Fe_4S_4L_4]^{2-}$ is derived from the interaction of the Fe 3*d* minority-spin orbitals and the terminal ligand lone-pair electrons and that the HOMO energy depends on the electron donor property of the terminal ligands, consistent with the experimental observation of the variation of the ADEs with the terminal ligands.

The terminal ligand effect on the electronic structure and redox properties of the cubane core were further studied in a series of mixed-ligand complexes, $[Fe_4S_4L_{4-x}L'_x]^{2-}$ ($x = 0$–4), formed by ligand substitution reactions. The PES spectra showed that the asymmetric coordination environment has no major influence on the electronic structure of the cubane. However, significant and systematic changes in electron binding energies were observed with each substitution of the terminal ligand and the electron binding energies increase in the order $SEt^- \rightarrow Cl^- \rightarrow OAc^- \sim OPr^- \rightarrow CN^- \rightarrow SCN^-$, consistent with the increase of the electron-withdrawing capability of each ligand type. A linear relationship was observed for each mixed-ligand system between the electron binding energies and the substitution number, suggesting that the contribution of each ligand toward the electron binding energy of the cubane, is independent and additive. The linear relationship reveals the electrostatic nature of the interaction between the cubane and the terminal ligands and validates the approach to partition the extrinsic contributions to the cubane redox potentials due to different environmental factors.

Conversion of the Fe–S clusters was studied in the gas phase with aid of CID experiments. We observed symmetric fission for the [4Fe–4S] cubane dianions, $[Fe_4S_4L_4]^{2-} \rightarrow [Fe_2S_2L_2]^2$ (L = Cl,Br,SEt). The DFT calculations show that the fission along low-spin channel is favored both thermodynamically and kinetically. Both Coulomb repulsion and antiferromagnetic coupling of the cubane dianions play important roles in the symmetric fission. The PES data reveal that the electronic structures are quite similar for the fission product and parent dianion, providing direct evidence of the spin couplings and the inverted-energy schemes within these cube-like clusters. Our gas phase observation on the symmetric fission may provide a new angle for the interpretation about the interconversion of $[Fe_4S_4]$ to $[Fe_2S_2]$ in proteins.

Series of partially coordinated [4Fe–4S] cubane clusters produced by laser vaporization and CID provide ideal model system to study the electronic structures of cubane core at different oxidation states. Five oxidation states of the cubane core, $[4Fe–4S]^- \rightarrow [4Fe–4S]^0 \rightarrow [4Fe–4S]^+ \rightarrow [4Fe–4S]^{2+} \rightarrow [4Fe–4S]^{3+}$, were accessed by varying the terminal ligand numbers. Spectral features due to the detachment of the minority-spin Fe 3*d* electrons were observed at the lowest binding energies and were readily recognized. Experimental evidence of the bare $Fe_4S_4^-$ cluster possessing the cubane-type structure proves the stability of the cubane structural unit. The behavior of sequential oxidation further confirms the electron storage capability of the cubane core and its robustness as nature's favorite electron transfer center.

ACKNOWLEDGMENT

We thank Dr. Shuqiang Niu and Prof. Toshiko Ichiye for their insightful theoretical effort on the electronic structure of Fe–S clusters in collaboration with our experimental work. We also thank Prof. R. H. Holm for providing us initially the $Fe_4S_4L_4^{2-}$ and $Fe_4Se_4L_4^{2-}$ samples and Prof. C. J. Pickett for providing us additional [4Fe–4S] cubane samples. We thank Dr. Hua-Jin Zhai, who performed the experiment on the bare Fe–S clusters. Support by the National Institutes of Health (GM-63555) is gratefully acknowledged. The experimental work and some of the calculations (using the molecular science computing facility) were performed at the W. R. Wiley Environmental Molecular Sciences Laboratory, a national scientific user facility sponsored by the DOE Office of Biological and Environmental Research and located at Pacific Northwest National Laboratory, which is operated for DOE by Battelle.

REFERENCES

Adman ET, Watenpaugh KD, Jensen LH (1975): NH—S hydrogen bonds in Peptococcus aerogenes ferredoxin, Clostridium pasteurianum rubredoxin, and Chromatium high potential iron protein. *Proc. Natl. Acad. Sci. USA* **72**:4854–4858.

Aizman A, Case DA (1982): Electronic structure calculations on active site models for 4-Fe, 4-S iron-sulfur proteins. *J. Am. Chem. Soc.* **104**:3269–3279.

Anderson GL, Howard JB (1984): Reactions with the oxidized iron protein of Azotobacter vinelandii nitrogenase: Formation of a 2Fe center. *Biochemistry* **23**:2118–2122.

Backes G, Mino Y, Loehr TM, Meyer TE, Cusanovich MA, Sweeney WV, Adman ET, Sanders-Loehr J (1991): The environment of Fe_4S_4 clusters in ferredoxins and high-potential iron proteins. New information from x-ray crystallography and resonance Raman spectroscopy. *J. Am. Chem. Soc.* **113**:2055–2064.

Becke AD (1993): Density-functional thermochemistry III. The role of exact exchange. *J. Chem. Phys.* **98**:5648–5652.

Bernert H, Holm RH, Münck E (1997): Iron-sulfur clusters: Nature's modular, multipurpose structures. *Science* **277**:653–659.

Black G, Chase J, Didier B, Feller D, Gracio D, Jones D, Jurrus E, Keller T, Matsumoto S, Mendoza E, Nordwall D, Olander M, Palmer B, Peden N, Schuchardt K, Stephan E, Taylor H, Thomas G, Vorpagel E, Windus T (2000): *Ecce, A Problem Solving Environment for Computational Chemistry, Software Version 1.5*; Pacific Northwest National Laboratory, Richland, WA 99352-0999, USA.

Bohr N, Wheeler JA (1939): The mechanism of nuclear fission. *Phys. Rev.* **56**:426–450.

Bréchignac C, Cahuzac FC, Frutos M (1990): Asymmetric fission of $Na^{++}{}_n$ around the critical size of stability. *Phys. Rev. Lett.* **64**:2893–2896.

Carney MJ, Papaefthymiou GC, Whitener MA, Spartalian K, Frankel RB, Holm RH (1988): Alternative spin states in synthetic analogs of biological clusters: spin-quartet ground states and structures of tetrakis(benzenethiolato)tetrasulfidotetraferrate(3-) and tetrakis(benzenethiolato)tetraselenidotetraferrate(3-) as their tetramethylammonium salts. *Inorg. Chem.* **27**:346–352.

Chen K, Tilley GJ, Sridhar V, Prasad GS, Stout CD, Armstrong FA, Burgess BK (1999): Alteration of the reduction potential of the $[4Fe–4S]^{2+/+}$ cluster of *Azotobacter vinelandii* ferredoxin I. *J. Biol. Chem.* **274**:36479–36487.

Ciurli S, Carrie M, Weigel JA, Carney MJ, Stack TDP, Papaefthymiou GC, Holm RH (1990): Subsite-differentiated analogs of native iron sulfide $[4Fe–4S]^{2+}$ clusters: Preparation of clusters with five- and six-coordinate subsites and modulation of redox potentials and charge distributions. *J. Am. Chem. Soc.* **112**:2654–2664.

Cleland WE, Holtman DA, Sabat M, Ibers JA, DeFotis GC, Averill BA (1983): Effects of phenoxide ligation on iron-sulfur clusters. Preparation and properties of $[Fe_4S_4(OAr)_4]^{2-}$ ions and the structure of bis(tetraethylammonium) tetrakis(phenolatosulfidoferrate) ($[(C_2H_5)_4N]_2[Fe_4S_4(OC_6H_5)_4]$). *J. Am. Chem. Soc.* **105**:6021–6031.

Cooks RG, Zhang D, Koch KJ (2001): Chiroselective self-directed octamerization of serine: Implications for homochirogenesis. *Anal. Chem.* **73**:3646–3655.

Daneshfar R, Kitova EN, Klassen JS (2004): Determination of protein-ligand association thermochemistry using variable-temperature nanoelectrospary mass spectrometry. *J. Am. Chem. Soc.* **126**:4786–4787.

DePamphilis BV, Averill BA, Herskovitz T, Que L, Holm RH (1974): Synthetic analogs of the active sites of iron-sulfur proteins. VI. Spectral and redox characteristics of the tetranuclear clusters $[Fe_4S_4(SR)_4]^{2-}$. *J. Am. Chem. Soc.* **96**:4159–4167.

Deng HT, Kerns KP, Castleman Jr AW (1996): Studies of Met-Car adducts: $Ti_8C_{12}^+$ $(M)_n$ (M = halogens, π-bonding molecules, and polar molecules. *J. Am. Chem. Soc.* **118**: 446–450.

Einsle O, Tezcan FA, Andrade SLA, Schmid B. Yoshida M, Howard JB, Rees DC (2002): Nitrogenase MoFe-protein at 1.16 Å resolution: A central ligand in the FeMo-cofactor. *Science* **297**:1696–1700.

Eland JHD (1984) *Photoelectron Spectroscopy*, 2nd ed., Butterworth, London.

Excoffon P, Laugier J, Lamotte B (1991): Influence of a phase transition in the solid state on the structure of a synthetic iron-sulfur cubane. *Inorg. Chem.* **30**:3075–3081.

Fenn JB (2003): Electrospary wings for molecular elephants (Nobel lecture). *Angew. Chem. Int. Ed.* **42**:3871–3894.

Francl MM, Pietro WJ, Hehre WJ, Binkley JS, Gordon MS, DeFrees DJ, Pople JA (1982): Self-consistent molecular orbital methods. XXIII. A polarization-type basis set for second-row elements. *J. Chem. Phys.* **77**:3654–3665.

Fu YJ, Yang X, Wang XB, Wang LS (2004): Terminal ligand influence on the electronic structure and intrinsic redox properties of the $[Fe_4S_4]^{2+}$ cubane clusters. *Inorg. Chem.* **43**:3647–3655.

Glaser T, Hedman B, Hodgson KO, Solomon EI (2000): Ligand K-edge X-ray absorption spectroscopy: A direct probe of ligand-metal covalency. *Acc. Chem. Res.* **33**:859–868.

Glaser T, Bertini I, Moura JJG, Hedman B, Hodgson KO, Solomon EI (2001): Protein effects on the electronic structure of the $[Fe_4S_4]^{2+}$ Cluster in ferredoxin and HiPIP. *J. Am. Chem. Soc.* **123**:4859–4860.

Gunnarsson O, Handschuh H, Bechthold PS, Kessler B, Ganteför G, Eberhardt W (1995): Photoemission spectra of C_{60}^-: Electron-phonon coupling, Jahn-Teller effect, and superconductivity in the fullerides. *Phys. Rev. Lett.* **74**:1875–1878.

Guo BC, Kerns KP, Castleman AW (1992a): $Ti_8C_{12}^+$ - Metallo-carbohedrenes—a new class of molecular clusters. *Science* **255**:1411–1413.

Guo BC, Wei S, Purnell J, Buzza S, Castleman AW (1992b): Metallo-carbohedrenes [$M_8C_{12}^+$ (M = V, Zr, Hf, and Ti)]—a class of stable molecular cluster ions. *Science* **256**: 515–516.

Guo BC, Kerns KP, Castleman Jr AW (1993): Reactivities of $Ti_8C_{12}^+$ at thermal energies. *J. Am. Chem. Soc.* **115**:7415–7418.

High Performance Computational Chemistry Group, Pacific Northwest National Laboratory (PNNL) (2003): *NWChem, A Computational Chemistry Package for Parallel Computers, Version 4.5*; Pacific Northwest National Laboratory, Richland, WA 99352, USA.

Holm RH, Kennepohl P, Solomon EI (1996): Structural and functional aspects of metal sites in biology. *Chem. Rev.* **96**:2239–2314.

Ichiye T (1999). Computational studies of redox potentials of electron transfer proteins. In Pratt LR, Hummer G (eds), *Simulation and Theory of Electrostatic Interactions in Solution*, American Institute of Physics, pp. 431–452.

Johnson RW, Holm RH (1978): Reaction chemistry of the iron-sulfur protein site analogs $[Fe_4S_4(SR)_4]^{2-}$. Sequential thiolate ligand substitution reactions with electrophiles. *J. Am. Chem. Soc.* **100**:5338–5344.

Kaleta DT, Jarrold MF (2003): Helix-turn-helix motifs in unsolvated peptides. *J. Am. Chem. Soc.* **125**:7186–7187.

Karas M, Hillenkamp F (1988): Laser desorption ionization of proteins with molecular masses exceeding 10,000 daltons. *Anal. Chem.* **60**:2299–2301.

Kent TA, Dreyer JL, Kennedy MC, Huynh BH, Emptage MH, Bennert H, Münck E (1982): Mössbauer studies of beef heart aconitase: Evidence for facile interconversions of iron-sulfur clusters. *Proc. Natl. Acad. Sci. USA* **79**:1096–1100.

Khoroshilova N, Bennert H, Kiley PJ (1995): Association of a polynuclear iron-sulfur center with a mutant FNR protein enhances DNA binding. *Proc. Natl. Acad. Sci. USA* **92**:2499–2503.

Khoroshilova N, Popescu CV, Münck E, Bernert H, Kiley PJ (1997): Iron-sulfur cluster disassembly in the FNR protein of *Escherichia coli* by O2: [4Fe–4S] to [2Fe–2S] conversion with loss of biological activity. *Proc. Natl. Acad. Sci. USA* **94**:6087–6092.

Koszinowski K, Schröder D, Schwarz H, Liyanage R, Armentrount PB (2002): Thermochemistry of small cationic iron-sulfur clusters. *J. Chem. Phys.* **117**:10039–10056.

Koszinowski K, Schröder D, Schwarz H (2004): Formation and reactivity of gaseous iron-sulfur clusters. *Eur. J. Inorg. Chem.* 44–50.

Kretzschmar I, Schröder D, Schwarz H, Armentrout PB (2003): Structure, thermochemistry, and reactivity of MS_n^+ cations (M = V, Mo; $n = 1$–3) in the gas phase. *Int. J. Mass. Spectrom.* **228**:439–456.

Kroto HW, Heath JR, O'Brien SC, Curl RF, Smalley RE (1985): C_{60}: Buckminsterfullerene. *Nature* **318**:162–163.

Laskin J, Bailey TH, Futrell JH (2003): Shattering of peptide ions on self-assembled monolayer surfaces. *J. Am. Chem. Soc.* **125**:1625–1632.

Laskin J, Beck KM, Hache JJ, Futrell JH (2004): Surface-induced dissociation of ions produced by matrix-assisted laser desorption ionization in a Fourier transform ion cyclotron resonance mass spectrometer. *Anal. Chem.* **76**:351–356.

Lee C, Yang W, Parr RG (1988): Development of the Colle-Salvetti correlation-energy formula into a functional of the electron density. *Phys. Rev. B* **37**:785–789.

Lever ABP (1990): Electrochemical parametrization of metal complex redox potentials, using the ruthenium(III)/ruthenium(II) couple to generate a ligand electrochemical series. *Inorg. Chem.* **29**:1271–1285.

Li S, Wu H, Wang LS (1997): Probing the electronic structure of metallocarbohedrenes: M_8C_{12} (M = Ti, V, Cr, Zr, and Nb). *J. Am. Chem. Soc.* **119**:7417–7422.

Li Z, Sau AK, Shen S, Whitehouse C, Baasov T, Anderson KS (2003): A snapshot of enzyme catalysis using electrospray ionization mass spectrometry. *J. Am. Chem. Soc.* **125**:9938–9939.

Loo JA, Edmonds CG, Smith RD (1993): Tandem mass-spectrometry of very large molecules. 2. Dissociation of multiply charged proline-containing proteins from electrospray ionization. *Anal. Chem.* **65**:425–438.

Mouesca JM, Chen JL, Noodleman L, Bashford D, Case DA (1994): Density functional/Poisson-Boltzmann calculations of redox potentials for iron-sulfur clusters. *J. Am. Chem. Soc.* **116**:11898–11914.

Moura JJ, Moura I, Kent TA, Lipscomb JD, Huynh BH, LeGall J, Xavier AV, Munck E (1982): Interconversions of [3Fe-3S] and [4Fe–4S] clusters. Mossbauer and electron paramagnetic resonance studies of Desulfovibrio gigas ferredoxin II. *J. Biol. Chem.* **257**:6259–6267.

Musgrave KB, Angove HC, Burgess BK, Hedman B, Hodgson KO (1998): All-ferrous titanium(III) citrate reduced Fe protein of nitrogenase: An XAS study of electronic and metrical structure. *J. Am. Chem. Soc.* **120**:5325–5326.

Näher U, Bjørnholm S, Frauendorf S, Garcias F, Guet C (1997): Fission of metal clusters. *Phys. Rep.* **285**:245–320.

Nakajima A, Hayase T, Hayakawa F, Kaya K (1997): Study on iron–sulfur cluster in gas phase: Electronic structure and reactivity. *Chem. Phys. Lett.* **280**:381–389.

Niu S, Wang XB, Nichols JA, Wang LS, Ichiye T (2003): Combined quantum chemistry and photoelectron spectroscopy study of the electronic structure and reduction potentials of rubredoxin redox site analogues. *J. Phys. Chem. A* **107**:2898–2907.

Niu S, Wang XB, Yang X, Wang LS, Ichiye T (2004): Mechanistic insight into symmetric fission of [4Fe–4S] analogue complexes and implications for cluster conversions in iron-sulfur proteins. *J. Phys. Chem. A* **108**:6750–6757.

Noodleman L, Baerends EJ (1984): Electronic structure, magnetic properties, ESR, and optical spectra for 2-iron ferredoxin models by LCAO-X.alpha. Valence bond theory. *J. Am. Chem. Soc.* **106**:2316–2327.

Noodleman L, Peng CY, Case DA, Mouesca JM (1995): Orbital interactions, electron delocalization and spin coupling in iron-sulfur clusters. *Coord. Chem. Rev.* **144**:199–244.

Ohno R, Ueyama N, Nakamura A (1991): Influence of the distal para substituent through NH—S hydrogen bonds on the positive shift of the reduction potentials of $[Fe_4S_4(\text{Z-cys-Gly-NHC}_6H_4\text{-p-X})_4]^{2-}$ (X = H, OMe, F, Cl, CN) complexes. *Inorg. Chem.* **30**:4887–4891.

O'Sullivan T, Millar MM (1985): Synthesis and study of an analog for the $[Fe_4S_4]^3+$ center of oxidized high potential iron-sulfur proteins. *J. Am. Chem. Soc.* **107**:4096–4097.

Parr RG, Yang W (1989) *Density-Functional Theory of Atoms and Molecules*, Oxford Univ. Press, Oxford.

Peters JW, Lanzilotta WN, Lemon BJ, Seefeldt LC (1998): X-ray crystal structure of the Fe-only hydrogenase (CpI) from *Clostridium pasteurianum* to 1.8 angstrom resolution. *Science* **282**:1853–1858.

Pickett CJ, Pletcher DJ (1975): Electrochemical measurements as a probe for the investigation of the structure of metal carbonyls. *J. Organomet. Chem.* **102**:327–333.

Pickett CJ, Ryder KS (1994): Bioinorganic reaction centres on electrodes. Modified electrodes possessing amino acid, peptide and ferredoxin-type groups on a poly(pyrrole) backbone. *J. Chem. Soc. Dalton* 2181–2187.

Popescu CV, Bates DM, Bennert H, Münck E, Kiley PJ (1998): Mössbauer spectroscopy as a tool for the study of activation/inactivation of the transcription regulator FNR in whole cells of *Escherichia coli. Proc. Natl. Acad. Sci. USA* **95**:13431–13435.

Que L, Bobrik MA, Ibers JA, Holm RH (1974a): Synthetic analogs of the active sites of iron-sulfur proteins. VII. Ligand substitution reactions of the tetranuclear clusters $[Fe_4S_4(SR)_4]^{2-}$ and the structure of bis(tetramethylammonium) [tetra-.mu.-sulfide-tetra-kis(benzenethiolato)tetrairon. *J. Am. Chem. Soc.* **96**:4168–4178.

Que L, Anglin JR, Bobrik MA, Davison A, Holm RH (1974b): Synthetic analogs of the active sites of iron-sulfur proteins. IX. Formation and some electronic and reactivity properties of iron sulfide (Fe_4S_4) glycyl-L-cysteinylglycyl oligopeptide complexes obtained by ligand substitution reactions. *J. Am. Chem. Soc.* **96**:6042–6048.

Rabalais JW (1977) *Principles of Ultraviolet Photoelectron Spectroscopy*, Wiley, New York.

Rassolov VA, Pople JA, Ratner MA, Windus TL (1998): 6-31G* basis set for atoms K through Zn. *J. Chem. Phys.* **109**:1223–1229.

Sarapu A, Fenske RF (1975): Transition metal-isocyanide bond. Approximate molecular orbital study. *Inorg. Chem.* **14**:247–253.

Sigfridsson E, Olsson MHM, Ryde U (2001a): Inner-sphere reorganization energy of iron-sulfur clusters studied with theoretical methods. *Inorg. Chem.* **40**:2509–2519.

Sigfridsson E, Olsson MHM, Ryde U (2001b): A comparison of the inner-sphere reorganization energies of cytochromes, iron-sulfur clusters, and blue copper proteins. *J. Phys. Chem. B* **105**:5546–5552.

Smith RD, Loo JA, Barinaga CJ, Edmonds CG, Udseth HR (1990): Collisional activation and collision-activated dissociation of large multiply charged polypeptides and proteins produced by electrospray ionization. *J. Am. Soc. Mass Spectrom.* **1**:53–65.

Spiro TG (1982): *Iron-Sufur Proteins*, Wiley-Interscience, New York.

Stephens PJ, Jollie DR, Warshel A (1996): Protein control of redox potentials of iron-sulfur proteins. *Chem. Rev.* **96**:2491–2514.

Tanaka K (2003): The origin of macromolecule ionization by laser irradiation (Nobel lecture). *Angew. Chem. Int. Ed.* **42**:3861–3870.

Torres RA, Lovell T, Noodleman L, Case DA (2003): Density functional and reduction potential calculations of Fe_4S_4 clusters. *J. Am. Chem. Soc.* **125**:1923–1936.

Treichel PM, Durren GE, Mueh HJ (1972): Manganese(I) isocyanide complexes. *J. Organomet. Chem.* **44**:339–352.

Wang LS, Cheng HS, Fan J (1995): Photoelectron spectroscopy of size-selected transition metal clusters: Fe, $n = 3$–24. *J. Chem. Phys.* **102**:9480–9493.

Wang LS, Ding CF, Wang XB, Nicholas JB (1998a): Probing the potential barriers and intramolecular electrostatic interactions in free doubly charged anions. *Phys. Rev. Lett.* **81**:2667–2670.

Wang XB, Ding CF, Wang LS (1998b): Photodetachment spectroscopy of a doubly charged anions: Direct observation of the repulsive Coulomb barrier. *Phys. Rev. Lett.* **81**:3351–3354.

Wang LS, Ding CF, Wang XB, Barlow SE (1999a): Photodetachment photoelectron spectroscopy of multiply charged anions using electrospray ionization. *Rev. Sci. Instrum.* **70**:1957–1966.

Wang XB, Ding CF, Wang LS (1999b): High resolution photoelectron spectroscopy of C_{60}^-. *J. Chem. Phys.* **110**:8217–8220.

Wang XB, Ding CF, Wang LS (1999c): Electron tunneling through the repulsive Coulomb barrier in photodetachment of multiply charged anions. *Chem. Phys. Lett.* **307**:391–396.

Wang XB, Wang LS (1999d): Observation of negative electron-binding energy in a molecule. *Nature* **400**:245–248.

Wang XB, Wang LS (1999e): Photodetachment of free hexahalogenometallate doubly charged anions in the gas phase: $[ML_6]^{2-}$, (M = Re, Os, Ir, Pt; L = Cl and Br). *J. Chem. Phys.* **111**:4497–4509.

Wang XB, Nicholas JB, Wang LS (2000a): Intramolecular Coulomb repulsion and anisotropies of the repulsive Coulomb barrier in multiply charged anions. *J. Chem. Phys.* **113**:653–661.

Wang LS, Wang XB (2000b): Probing free multiply charged anions using photodetachment photoelectron spectroscopy. *J. Phys. Chem. A* **104**:1978–1990.

Wang XB, Wang LS (2000c): Probing the electronic structure of redox species and direct determination of intrinsic reorganization energies of electron transfer reactions. *J. Chem. Phys.* **112**:6959–6962.

Wang XB, Wang LS, Brown R, Schwerdtfeger P, Schröder D, Schwarz H (2001): The electronic structure of $CuCl_2$ and $CuBr_2$ from anion photoelectron spectroscopy and ab initio calculations. *J. Chem. Phys.* **114**:7388–7395.

Wang XB, Inscore FE, Yang X, Cooney JJA, Enemark JH, Wang LS (2002): Probing the electronic structure of $[MoOS_4]^-$ centers using anionic photoelectron spectroscopy. *J. Am. Chem. Soc.* **124**:10182–10191.

Wang XB, Niu S, Yang X, Ibrahim SK, Pickett CJ, Ichiye T, Wang LS (2003): Probing the intrinsic electrinic structure of the cubane [4Fe-4S] cluster: Nature's favorite cluster for electron transfer and storage. *J. Am. Chem. Soc.* **125**:14072–14081.

Wang XB, Woo HK, Wang LS (2005): Vibrational cooling in qa cold ion trap: Vibrationally resolved photoelectron spectroscopy of cold C_{60}^- anions. *J. Chem. Phys.* **123**:051106-1–051106-4.

Wells JM, Chrisman PA, McLuckey SA (2003): Formation and characterization of protein-protein complexes in vacuo. *J. Am. Chem. Soc.* **125**:7238–7249.

Wong GB, Bobrik MA, Holm RH (1978): Inorganic derivatives of iron sulfide thiolate dimers and tetramers: synthesis and properties of the halide series $[Fe_2S_2X_4]^{2-}$ and $[Fe_4S_4X_4]^{2-}$ (X = chlorine, bromine, iodine). *Inorg. Chem.* **17**:578–584.

Yang SH, Pettiette CL, Conceicao J, Cheshnovsky O, Smalley RE (1987): UPS of Buckminsterfullerene and other large clusters of carbon. *Chem. Phys. Lett.* **139**:233–237.

Yang X, Wang XB, Niu S, Pickett CJ, Ichiye T, Wang LS (2002): Coulomb- and antiferromagnetic-induced fission in doubly charged cubelike Fe–S clusters. *Phys. Rev. Lett.* **89**:163401-1–163401-4.

Yang X, Wang XB, Fu YJ, Wang LS (2003a): On the electronic structure of [1Fe] Fe–S complexes from anionic photoelectron spectroscopy. *J. Phys. Chem. A* **107**:1703–1709.

Yang X, Razavet M, Wang XB, Pickett CJ, Wang LS (2003b): Probing the electronic structure of the di-iron subsite of [Fe]-hydrogenase: A photoelectron spectroscopic study of Fe(I)-Fe(I) model complexes. *J. Phys. Chem. A* **107**:4612–4618.

Yang X, Wang XB, Wang LS (2003c): Collision-induced symmetric fission of doubly-charged cubelike $[Fe_4S_4X_4]^{2-}$ clusters. *Int. J. Mass Spectrom.* **228**:797–805.

Yoo SJ, Angove HC, Burgess BK, Hendrich MP, Münck E (1999): Mössbauer and integer-spin EPR studies and spin-coupling analysis of the $[4Fe\text{–}4S]^0$ cluster of the Fe protein from *Azotobacter vinelandii* nitrogenase. *J. Am. Chem. Soc.* **121**:2534–2545.

Yu Z, Zhang N, Wu X, Gao Z, Zhu Q, Kong F (1993): The production and photodissociation of iron–sulfur cluster ions. *J. Chem. Phys.* **99**:1765–1770.

Zhai HJ, Yang X, Fu YJ, Wang XB, Wang LS (2004): Sequential oxidation of the cubane [4Fe–4S] cluster from $[4Fe\text{–}4S]^-$ to $[4Fe\text{–}4S]^{3+}$ in $Fe_4S_4L_n^-$ complexes. *J. Am. Chem. Soc.* **126**:8413–8420.

Zhou J, Hu Z, Munck E, Holm RH (1996): The cuboidal Fe_3S_4 cluster: Synthesis, stability, and geometric and electronic structures in a non-protein environment. *J. Am. Chem. Soc.* **118**:1966–1980.

Zhou C, Holm RH (1997): Comparative isotropic shifts, redox potentials, and ligand binding propensities of [1:3] site-differentiated cubane-type $[Fe_4Q_4]^{2+}$ clusters (Q = S, Se). *Inorg. Chem.* **36**:4066–4077.

Zhou J. Raebiger JW, Crawford CA, Holm RH (1997): Metal ion incorporation reactions of the cluster $[Fe_3S_4(LS_3)]^{3-}$, containing the cuboidal $[Fe_3S_4]^0$ core. *J. Am. Chem. Soc.* **119**:6242–6250.

Zhou C, Raebiger JW, Segal BM, Holm RH (2000): The influence of net charge on the redox potentials of Fe_4S_4 cubane-type clusters in aprotic solvents. *Inorg. Chim. Acta.* **300–302**: 892–902.

3

ION–MOLECULE REACTIONS AND H/D EXCHANGE FOR STRUCTURAL CHARACTERIZATION OF BIOMOLECULES

M. KIRK GREEN
McMaster Regional Centre for Mass Spectrometry
Department of Chemistry
McMaster University, Hamilton, Canada

CARLITO B. LEBRILLA
Department of Chemistry
University of California, Davis, CA

Principles of Mass Spectrometry Applied to Biomolecules, edited by Julia Laskin and Chava Lifshitz

3.1. INTRODUCTION

Ion–molecule reactions have long been used to probe the gas-phase structure of ions. In the case of biomolecules, the information may be related to primary structure, such as counting labile hydrogens or basic groups, or of a secondary or tertiary nature, such as ionization states of particular groups, molecular conformation, or sites of interaction of complexes. It is clear that the latter sort of information obtained for the gas-phase structure is often different from that for the structure in aqueous solution, where the solvent stabilizes charged groups and hydrophobic interactions play an important role. This can be a disadvantage if the data are interpreted without keeping this fact in view; on the other hand, characterizing and understanding these differences can provide fascinating insight into solvent effects on structure. It should also be kept in mind that the structure of a biomolecule in solution may not be the most biologically relevant information: the "operating environment" of, say, a membrane protein, is quite different from that encountered in a neutral, aqueous solution.

3.2. METHODS

3.2.1. Proton Transfer Reactions

The thermodynamics and methodology of proton transfer reactions are covered thoroughly in Chapter 15, on thermochemistry studies of biomolecules. However, it is useful to briefly review the subject here. The most important thermodynamic parameters of a gas-phase proton transfer reaction are the free energy and the enthalpy of the following reaction:

$$B + H^+ \rightarrow BH^+ \tag{3.1}$$

By convention, the negative of the free energy ($-\Delta G$) of the reaction is called the gas-phase basicity (GB) and the negative of the enthalpy ($-\Delta H$) is the proton affinity (PA). Because the reaction is always exergonic and exothermic, GB and PA are both always positive numbers. Similarly, the gas-phase acidity (GPA) (also abbreviated GA in the literature), ΔG for the reaction

$$AH \rightarrow A^- + H^+ \tag{3.2}$$

is always positive.

For small organic bases, it has been customary to provide the PA, because this parameter is less dependent on temperature. For large biomolecules, containing a large number of base sites and extensive intramolecular interactions, obtaining the PA often requires large approximations regarding intramolecular interactions. In this situation, it is advisable to provide GB rather than PA values. The most direct method for obtaining GB and PA is to perform the proton transfer experiment between the protonated unknown base B and a base R_b of known GB. By using several reference bases, the GB of the compound can be determined:

$$BH^+ + R_b \rightleftharpoons R_bH^+ + B \tag{3.3}$$

This "equilibrium" method has been used extensively used to assign GB and PA values for numerous small organic compounds (Aue and Bowers 1979; Bartmess and McIver 1979). The reaction is performed in the presence of gaseous R_b and B. The equilibrium constant is determined from the expression

$$K = \frac{[R_bH^+][B]}{[BH^+][R_b]}$$

where $[R_bH^+]$ and $[BH^+]$ are the relative intensities of the protonated reference and base, respectively, and $[R_b]$ and [B] are the respective partial pressures. The GB is determined from the expression

$$\Delta G = -RT \ln K$$

Unfortunately, most molecules of biological interest have no appreciable vapor pressure. To deal with this problem, two approaches are commonly used. A simple approach is to produce the protonated species of unknown basicity and to react it with a background pressure of a volatile reference base. This method is commonly called "bracketing."

$$BH^+ + R_b \rightarrow R_bH^+ + B \tag{3.4}$$

Simply, the bracketing method involves monitoring the disappearance of BH^+ as a function of the appearance of R_bH^+. If the intensity of R_bH^+ grows to an appreciable extent during the reaction time, then $GB(R_b) > GB(B)$. In a variation of this, the neutral unknown has been generated in the presence of protonated reference base by a MALDI technique (Gorman and Amster 1993). Using a series of reference bases places B on a relative scale. Measuring the kinetics and comparing efficiencies provides a more accurate method than a qualitative assessment of "appreciable extent" for determining where endergonic and exergonic transitions occur in a series of references bases (Bohme et al. 1980; Büker and Grützmacher 1991). Problems can arise when the basic site in either R_b or B is hindered. For example, a hindered base can lower the efficiency of the reaction and cause a higher GB assignment [Sunner et al. 1989; Meot-Ner (Mautner) and Seck 1991; Wu and Lebrilla 1995].

An alternative method introduced by Cooks and Kruger (1977) relies on the formation of the mixed dimer $(B\cdots H\cdots R_b^+)$. The reaction monitored is the disproportionation reaction shown. The method is commonly referred to as the "kinetic" method:

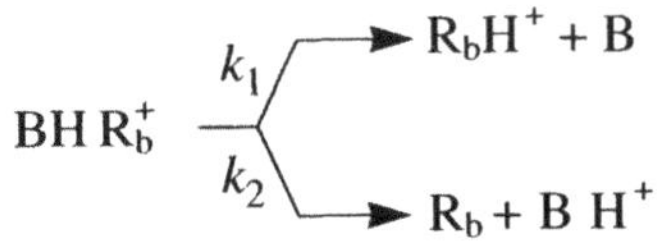

The kinetic method employs the following equation to obtain proton affinities:

$$\ln\frac{k_1}{k_2} = \ln\frac{Q_1^*}{Q_2^*} + \frac{\Delta PA}{RT}$$

where k_1 and k_2 are rate constants for the competing reactions whose ratio is obtained from the relative ion abundance. Q_1^* and Q_2^* are partition functions for the activated complexes. The method works best if the reference (R) and the base (B) are chemically and structurally similar and with the assumption that reverse barriers are close to zero so that

$$\ln\frac{Q_1^*}{Q_2^*} \cong 0$$

and this equation becomes

$$\ln\frac{k_1}{k_2} \cong \frac{\Delta PA}{RT}$$

Thus, from the relative intensity, the relative proton affinity is obtained. In this method, the equation is applied primarily to compounds with a single basic site and very weak or absent intramolecular interactions. In the event that the proton transfer reaction is accompanied by a negligible free-energy barrier, the term $R \ln (Q_1^*/Q_2^*)$ is equal to ΔS and $\Delta S^{\#}$

$$\Delta S = \Delta S^{\#} = R\ln\frac{Q_1^*}{Q_2^*}$$

where ΔS is the difference between the entropy of protonation of the reference base and the unknown basic and $\Delta S^{\#}$ is the entropy difference between activated complexes (Majumdar et al. 1992; Wu and Lebrilla 1995). Thus if intramolecular interactions occur, then ΔS is nonzero and the ratio of the rates is related instead to the difference in GB as shown by the following equation:

$$\ln\frac{k_1}{k_2} \cong \frac{\mathrm{GB}(1) - \mathrm{GB}(2)}{RT}$$

Discussions on the restrictions and the limitations of the kinetic method are also given in a paper by Bliznyuk et al. (1993).

Of course, the conditions presented above are seldom met for molecules of interest; reference bases with close structural/chemical similarities may not be available, intramolecular interactions may be important, and basic sites may be hindered. Additionally, the appropriate value of the temperature, particularly in the kinetic method, is not always clear.

The "extended kinetic method" (Wu et al. 1994) addresses these problems in the kinetic method by varying the collision energy for the dissociation: this changes the effective temperature and allows the extraction of both GB and PA. The "thermokinetic method" (Bouchoux et al. 1996; Bouchoux and Salpin 2003) addresses these problems by writing the following expression for the reaction efficiency k_{exp}/k_{coll} observed in a series of bracketing experiments:

$$\frac{k_{exp}}{k_{coll}} = \frac{1}{1 + \exp\left[\frac{\Delta_i G^\circ + \Delta G_a^\circ}{RT}\right]}$$

where $\Delta_i G^\circ$ is difference between GB(B) and GB(R_i) and ΔG_a° is an empirical correction. In practice the equation is recast as

$$\frac{k_{exp}}{k_{coll}} = \frac{a}{1 + \exp[b(c - \mathrm{GB}(\mathrm{R}_i)]}$$

with $\mathrm{GB}(\mathrm{B}) = c - 1/b$.

The parameters are extracted from a fit to a plot of k_{exp}/k_{coll} versus GB(R_i). GB values obtained by this approach have generally been in good agreement with equilibrium values obtained for small, volatile species.

3.2.2. Hydrogen/Deuterium Exchange

Hydrogen/deuterium (H/D) exchange is an energy-neutral reaction if isotope effects are ignored. Thus, with appropriate control of experimental conditions, these reactions can probe the reaction surface and barrier without the energy differences between reactants and products complicating the picture. Furthermore, it can generally be assumed that the reaction itself has no permanent effect on the structure of the substrate, the product of a single exchange is merely an isotopomer of the reactant, and is otherwise identical in every way. One advantage of H/D exchange over a proton transfer is that, whereas proton transfer reaction generally involves only a single site, H/D reactions can potentially probe several sites in a molecule, if the multiple exchanges observed can be correlated to the different available reactive sites. One drawback is that exchange reactions give only indirect information on thermochemical properties such as GB or PA.

The primary data obtained from H/D exchange experiments is the number of exchangeable hydrogens and in many cases, the rates associated with these exchanges. Understanding of these data may yield information about the location of the ionization site(s), locations of reactive groups, relative proton affinities, and intramolecular hydrogen bonding and conformation. In a typical H/D exchange experiment, an ionized hydrogen-containing substrate is exposed to an atmosphere of the deuterating reagent and the extent and/or rate of the exchange is monitored by following the shift in m/z of the substrate ion signal. The reverse experiment (deuterated substrate and protonating reagent) is seldom used because of the possibility of backward exchange of the substrate during introduction and ionization. Although most of the gas-phase H/D exchange reactions reported in the literature deal with exchange reactions of protonated substrates, reactions involving species charged with other cations (Reyzer and Brodbelt 2000; Solouki et al. 2001; Jurchen et al. 2003; Cox et al. 2004) as well as deprotonated anions (Freitas et al. 1998; Robinson et al. 1998; Hofstadler et al. 2000; Freitas and Marshall 2001) have been studied.

3.2.2.1. Mechanism of H/D Exchange. The generally accepted simplified mechanism (Brauman 1979; Lias 1984) for H/D exchange between a protonated substrate (S) and a deuterating reagent (RD) in the gas phase consists of three steps: (1) the initial formation of a loose hydrogen-bonded complex; (2) complete or partial transfer of the proton to the reagent, which results in isotope scrambling; and finally (3) dissociation of the complex to yield either the original or the exchanged substrate species (Scheme 3.1).

For exchange to be observed, the energy made available by complex formation must be sufficient to overcome the barrier to internal proton transfer. This barrier will depend mainly on the proton affinity difference ($\Delta PA = PA_{substrate} - PA_{reagent}$) between the two unprotonated species.

This mechanism for H/D exchange is supported by the observation that for a large range of ionized substrates and deuterating reagents, there is an approximately inverse correlation between ΔPA and the observed rate of exchange (Hunt and Sethi 1980; Ausloos and Lias 1981), to the point where once ΔPA exceeds a limit of ~80 kJ/mol, no exchange is observed, presumably because the energy made

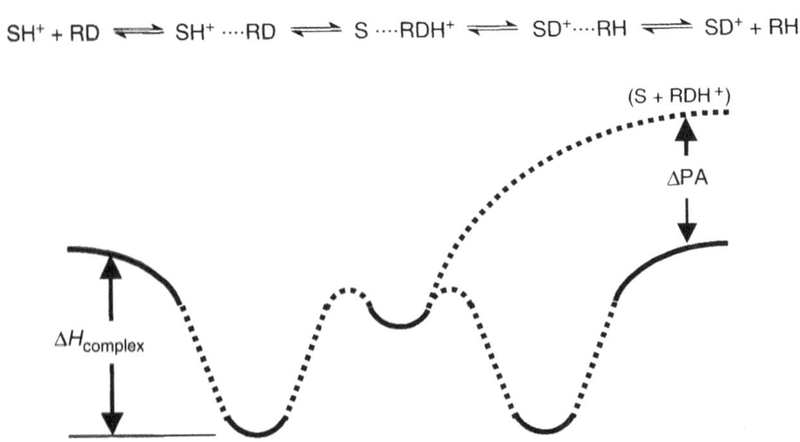

SCHEME 3.1

available by the exothermicity of complex formation is insufficient to overcome the barrier to endothermic proton transfer within the complex. It should be noted that this mechanism assumes that a complex lifetime that is long compared to the timescale for proton transfer. This is probably a good assumption for complexes with hydrogen bonding. For example, it has been estimated (Henchman et al. 1991) that, for the proton-bound dimers of H_2O and NH_3, complex lifetimes at room temperature are 1–2 ns, as compared to a "proton jump" time of <100 ps. Indeed, collisionally stabilized [diglycineH^+–ND_3] complexes have been directly observed in flow tube H/D exchange studies [37].

However, it has become clear for many compounds, particularly biological molecules such as amino acids, proteins, and peptides, the 80 kJ/mol limit for reactivity no longer applies (Cheng and Fenselau 1992; Winger et al. 1992b; Gard et al. 1993; Campbell et al. 1994; Hemling et al. 1994; Reyzer and Brodbelt 2000): significant reactivity may be observed even with ΔPA as high as 200 kJ/mol. This high reactivity is not restricted to amino acids and peptides; it has been found that aliphatic diamines (Green and Lebrilla 1998) and polyamines (Reyzer and Brodbelt 2000) show high reactivity as well. It appears that abnormally high reactivity is observed whenever the substrate contains a potential bridging site where the deuterating reagent may insert and form multiple hydrogen bonds. This can be attributed to a lowering of the barrier to proton transfer within the complex (Gard et al. 1993; Campbell et al. 1994; Gur et al. 1995; Green and Lebrilla 1997). For the reactions of CH_3OD with simple amino acids, Lebrilla and coworkers (Gard et al. 1993) proposed a complex in which the alcohol forms a bridge between the protonated amino group and the carboxyl carbonyl (Scheme 3.2, which shows protonated glycine).

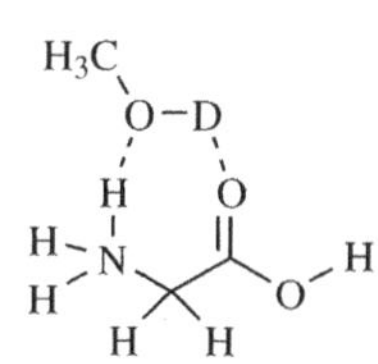

SCHEME 3.2

In this scenario, following formation of the bridged complex, proton transfer to the deuterating agent is accompanied by a simultaneous transfer of the deuteron to the substrate, in effect obviating the need for the formation of a protonated deuterating agent. On the basis of their experimental results with a number of deuterating reagents and a selection of peptides and extensive semiempirical modeling, Campbell et al. (1994, 1995) expanded on this idea and proposed several mechanisms for H/D exchange that depended on ΔPA and structural features of the substrate:

For reagents of relatively low PA, such as D_2O and CD_3OD interacting with an *N* terminus and a carbonyl group, they proposed that exchange proceeds via the appropriately named "relay mechanism," in which a bridged complex similar to that discussed above is formed.

Exchange of a carboxyl terminus with low-PA reagents was proposed to proceed via a similar mechanism, the "flipflop" mechanism (Scheme 3.3).

SCHEME 3.3

ND_3, with a higher PA, is capable of forming a protonated ammonium ion if solvated by additional hydrogen bonding to the peptide. This process is termed the "onium mechanism" (Scheme 3.4).

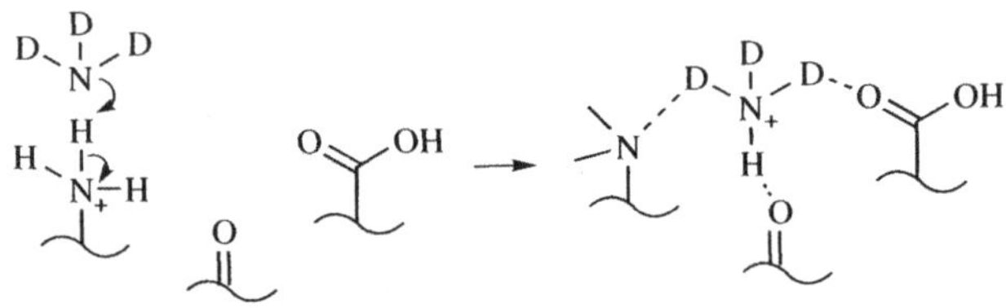

SCHEME 3.4

Exchange of ND_3 with amide protons was proposed to proceed via tautomerization of the amide, with the tautomerized amide stabilizing the protonated ND_3, the "tautomer mechanism" (Scheme 3.5).

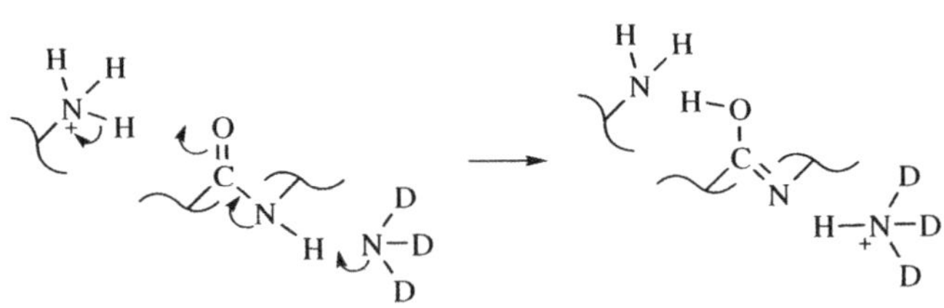

SCHEME 3.5

Exchange of a carboxyl terminus with low-PA reagents was proposed to proceed via formation of a saltbridge that stabilizes the protonated reagent (Scheme 3.6).

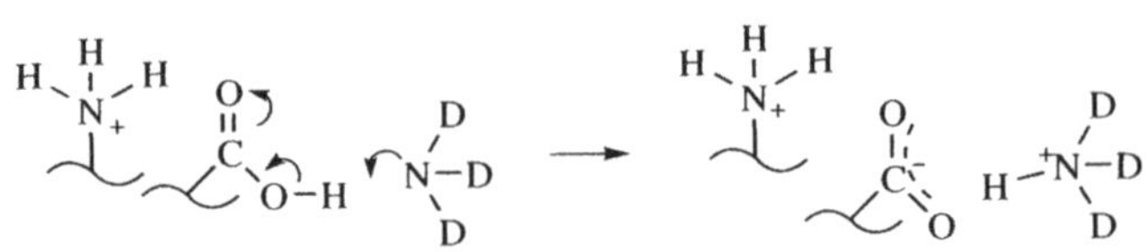

SCHEME 3.6

Thus the propensity of a particular site (or arrangement of sites) to undergo exchange and the mechanism by which this exchange proceeds will be highly dependent not only on ΔPA but also on the relative arrangement of potentially exchangeable groups within the substrate.

3.2.2.2. Kinetic Analysis. While the rates of H/D exchange reactions have most often been measured in Fourier transform ion cyclotron resonance (FTICR) instruments (Campbell et al. 1995; Cassady and Carr 1996; Green and Lebrilla 1997; Freitas and Marshall 2001; Rozman et al. 2003), a number of linear/quadrupole ion trap (Reyzer and Brodbelt 2000; Evans et al. 2003; Mao and Douglas 2003) and flow tube (Kogan et al. 2002; Geller and Lifshitz 2003; Mazurek et al. 2005) studies have been reported. Each instrument has advantages and drawbacks: FTICR offers high resolution and the consequent ability to isolate a particular isotopic peak. The ability to carry out reactions over a wide pressure range (10^{-5}–10^{-9} Torr) and timescale (seconds to hours) gives access to a wide range of reactivity. Ion traps can be operated at relatively high pressures, and are relatively inexpensive instruments. Flow tubes operate on a relatively short timescale, but can operate at very high pressures, allowing the observation of collisionally stabilized species, and allow rigorous control of experimental parameters such as reagent pressure.

In an experiment, one observes the decay of the parent peak at mass M and the growth (and possibly the eventual decay) of peaks at masses $M + n$ corresponding to the products containing n deuterium atoms. To simplify the analysis, it may be desirable to isolate a single peak of the natural isotope cluster prior to reaction; however, this is not necessary, since the exchange information can be deconvoluted from experimental data for the isotope cluster fairly readily. Also, although it is obviously preferable to have sufficient resolution to follow the evolution of the individual isotopic peaks, much can still be learned from monitoring the shift of the average m/z of the isotope cluster or of m/z data points on the envelope (Geller and Lifshitz 2004).

One approach (Campbell et al. 1994; Gard et al. 1994) to analyzing the kinetics is to treat the system as one of successive exchanges with apparent rate constants k_n for each exchange, as illustrated for the deuterating reagent CH_3OD:

$$D_0 + CH_3OD \underset{k_{-1}}{\overset{k_1}{\rightleftharpoons}} D_1 + CH_3OH \tag{3.5}$$

$$D_1 + CH_3OD \underset{k_{-2}}{\overset{k_2}{\rightleftharpoons}} D_2 + CH_3OH \tag{3.6}$$

$$D_2 + CH_3OD \underset{k_{-3}}{\overset{k_3}{\rightleftharpoons}} D_3 + CH_3OH \tag{3.7}$$

$$D_3 + CH_3OD \underset{k_{-4}}{\overset{k_4}{\rightleftharpoons}} D_4 + CH_3OH \tag{3.8}$$

$$\vdots$$

where D_n represents the ion containing n deuteriums. The k_{-n} values are included to account for any protonated impurity in the deuterating reagent, but are assumed to be equal to the corresponding k_n values. The rate expressions are a series of coupled differential equations:

$$-\frac{d[D_0]}{dt} = k_1[D_0][CH_3OD] - k_{-1}[D_1][CH_3OH] \tag{3.9}$$

$$\frac{d[D_1]}{dt} = k_1[D_0][CH_3OD] - k_{-1}[D_1][CH_3OH] - k_2[D_1][CH_3OD] + k_{-2}[D_2][CH_3OH] \tag{3.10}$$

$$\frac{d[D_2]}{dt} = k_2[D_1][CH_3OD] - k_{-2}[D_2][CH_3OH] - k_3[D_2][CH_3OD] + k_{-3}[D_3][CH_3OH] \tag{3.11}$$

$$\frac{d[D_3]}{dt} = k_3[D_2][CH_3OD] - k_{-3}[D_3][CH_3OH] - k_4[D_3][CH_3OD] + k_{-4}[D_4][CH_3OH] \tag{3.12}$$

(etc.)

This series of equations can be solved numerically, and the rate constants can be extracted from the experimental data in an iterative fashion. The isotopic purity of the deuterating reagent can be either assumed or treated as an additional parameter to be extracted from the data.

Alternatively, the system may be treated as one of x independent sites (Green et al. 1995a,b), each following a simple first-order rate law

$$S_j(H) + CH_3OD \underset{k_{-j}}{\overset{k_j}{\rightleftharpoons}} S_j(D) + CH_3OH \tag{3.13}$$

where S_j corresponds to site j and $k_j = k_{-j}$. In this case, the observed populations of the D_n species are related to the populations of the individual sites. For example, the D_1 species observed in a system of two independent sites, a and b, is actually the sum of two distinct species, D_aH_b and H_aD_b. Again, the rate constants may be determined iteratively from the data, but in this case, they represent the rates of exchange at individual sites.

Which treatment is appropriate depends on the objective of a study and the mechanistic details of the system involved. If the object is simply to assess relative reactivities of a series of systems in a semiqualitative manner, the choice is probably not too important. If more insight is desired, more consideration must be given; assuming serial reactions seems unreasonable, particularly in larger, multiply charged ions with many well-separated potential sites of exchange. On the other hand, while in solution-phase H/D exchange studies exchange sites are (rightfully) assumed to be independent (Englander et al. 1997), one must be careful about assuming that all sites are completely independent in gas-phase work; in the example of protonated glycine discussed above, the two exchange sites are close together and the relay mechanism implies interaction. A further consideration when considering kinetic treatments is that neither of the models presented above may be strictly appropriate; for example, one can envision models where exchange occurs at a number of independent sites and the exchanged deuteriums then migrate from these sites—a combination of independent and serial exchange. Furthermore, many exchange studies use reagents that are capable of multiple exchanges; ND_3, one of the most popular reagents because of its high reactivity, has been shown (Campbell et al. 1994) to exchange more than one of its deuteriums in a single encounter with a substrate ion. A rigorous kinetic analysis would take this into account.

Nonetheless, the ability of an independent site treatment to provide data on the relative reactivities of different sites within an ionized biomolecule has proved useful. As mentioned above, the independent site treatment of the data for glycineH^+ yields results consistent with expectations for three identical (amine) sites and one unique (carboxyl) site. Furthermore, using the independent site treatment for a series of glycine-containing dipeptides (He et al. 2001) gave rate constant values that could be readily interpreted in terms of the number and type of sites expected to be available for exchange. The success of this treatment with simple systems gives one some confidence in applying it to more complex systems. For example, site-specific rate constants have been used to validate an ion–zwitterion salt bridge structure (Strittmatter and Williams 2000) for the proton-bound arginine dimer in the gas phase (Reuben et al. 2003). In the reaction with D_2O a single fast exchange and 14 slow exchanges were observed. The most reasonable interpretation of this observation is that the terminal carboxyl of one of the arginines is deprotonated in the gas phase and provides the highly reactive site. As mentioned above, the independent site treatment of the data for glycineH^+ yields results consistent with expectations for three identical (amine) sites and one unique (carboxyl) site. Furthermore, using the independent site treatment for a series of glycine-containing dipeptides (He et al. 2001) gave rate constant values that could be readily interpreted in terms of the number and type of sites available for exchange. The success of this treatment with simple systems gives one some confidence in applying it to more complex systems.

Kinetic analysis for small systems is fairly straightforward, and a simple, dedicated program has been described (He and Marshall 2000) that is available from the authors. However, as the number of exchanges increases above 10–15, calculations become computationally intense to the point where they are impractical. A treatment that extracts independent rate constant distributions from plots of deuterium incorporation versus reaction time using a "maximum entropy" method (MEM) has been reported (Freitas and Marshall 1999) and applied to a peptide. This program is available from the authors. However, the implementation described requires that the isotopic purity of the deuterating regent be known, and the method yields probability distributions for the rate constants rather than precise numbers. More recently a fast algorithim has been described (Reuben et al. 2003) that avoids numerical solution of differential equations and minimizes the Kullback–Leibler information divergence (mutual entropy) between experimental and fit data. The latter aspect minimizes the chances of the algorithm failing to find a global minimum in the fit. However, the computational resources required still become prohibitive for large numbers of exchanges. The authors subsequently dealt with this problem in the case of the serine octamer (33 exchanging hydrogens) (Mazurek et al. 2005) by grouping rate constants together according to different structural models, and using the quality of the fits to the exchange data to distinguish between the structural models (all-zwitterionic or all-neutral) for different ion populations.

3.3. GAS-PHASE PROTON TRANSFER REACTIONS

3.3.1. Gas-Phase Basicities and Protonation Sites of Singly Protonated Amino Acids and Peptides

Gas-phase basicity and proton affinity values for the amino acids have been the subject of numerous studies (Aue and Bowers 1979; Meot-Ner et al. 1979; Locke and McIver 1983; Bojesen 1987; Isa et al. 1990; Li and Harrison 1993; Hunter and Lias 1998) using equilibrium, bracketing, and kinetic methods. The compilation of Hunter and Lias (1998) is generally accepted, although more recent studies indicate that some corrections (Kinser et al. 2002; Bouchoux et al. 2004; Schroeder et al. 2004) may be in order. These corrections arise from the significant protonation entropies consequent to formation of cyclic species when the proton bridges the terminus and a basic sidechain. An examination and reassessment of the data and a discussion of structure and solvation effects are contained in a review by Meot-Ner (2003). The tabulated values for a large subset of compounds suitable as reference bases are considered to be accurate within 4 kJ/mol (Meot-Ner 2003). An early comparison of GB and gas-phase acidity (GPA) for the amino acids with those for carboxylic acids and amines showed that the site of protonation was the more basic amino group rather than the carboxyl group (Aue and Bowers 1979).

A number of studies have examined GB and PA trends in the glycine oligomers (Wu and Fenselau 1992; Wu and Lebrilla 1993; Zhang et al. 1993; Bouchoux and Salpin 2003). All workers found a relatively large increase (~25 kJ/mol) in basicity on going from glycine to diglycine, with progressively smaller increases observed for higher members of the series. This can be interpreted as stabilization of the protonated species by intramolecular hydrogen bonding in the oligomers; the large increase in GB on going from glycine to diglycine can be attributed to the interaction between the amide carbonyl and the protonated amine terminus in diglycine (Green and Lebrilla 1997). This trend is mirrored in the results for alanine and valine oligomers (Wu and Lebrilla 1995). In this work, proton transfer kinetics were exceedingly slow, and this was attributed to a gas-phase conformation in which the nonpolar sidechains hindered access of the base to the protonation site.

Cassady and co-workers studied proton transfer in a series of small peptides (Zhang et al. 1993; McKiernan et al. 1994; Cassady et al. 1995; Carr and Cassady

1996; Ewing et al. 1996), and these results were reexamined by Bouchoux (Bouchoux and Salpin 2003). The more basic the residue next to *N*-terminal glycine, the greater the GB of the peptide, consistent with stabilization of the *N*-terminal protonation site by the amide carbonyl. It should be noted that the proton transfer reactivity of a number of these peptides suggests that several containing glycine and either proline or histidine appear to exhibit two values of GB (Carr and Cassady 1996; Ewing et al. 1996; Bouchoux and Salpin 2003); that is, under the experimental conditions used, there are at least two species of protonated peptide present. Whether this reflects different conformations or different protonation sites is not clear.

3.3.2. Multiply Protonated Peptides and Proteins

When multiple charges are present on a peptide or protein, the increase in GB with size due to intramolecular hydrogen bonding is countered by Coulombic repulsion between charges (Williams 1996; Green and Lebrilla 1997). In fact, it is this dependence of GB on charge state that is the key factor in determining the charge states observed in ESI spectra of multiply charged species (Schnier et al. 1995b)—the maximum charge state observed represents a balance between the GB of the ESI solvent and the GB of the species being observed. An early study of proton transfer reactivity in the cytochrome *c*–dimethyl amine system showed high reactivity for the higher charge states (+14, +15) and very low reactivity for the lower (+9) charge states. The decrease in reactivity with decreasing charge state was attributed to (1) fewer potential reactions sites being available in low charge state ions and (2) a general increase in site basicity due to smaller Coulombic repulsion in the lower charge states. Subsequently, it was noted that charge states of cytochrome *c* deprotonate to some degree even when only water is available as a base (Winger et al. 1992a). Measured rate constants for the deprotonation of ubiquitin by amine bases approached, and in some cases equaled, the collision rate (Cassady et al. 1994). The kinetic plots for the reactions of the basic amines with the +12 charge state showed nonlinearity, implying at least two rate constants and the existence of more than one species. In a follow-up study (Cassady and Car 1996), CID spectra of the two species were obtained by performing CID on the unreacted +12 charge state and comparing this with the CID spectrum of the +12 charge state after the more reactive species had been depleted by proton transfer; the spectra were significantly different. Proton transfer reactivities have been used to infer the existence of more than one distinct species of the same charge state in other studies (Gross et al. 1996; Ewing and Cassady 1999).

The correlation between structure and proton transfer reactivity was explored with the $[M+4]^{4+}$ ions of a series of synthetic peptides, K_4G_8, $(K_2G_4)_2$, and $(KG_2)_4$ (Zhang et al. 1998). K_4G_8 was much more reactive than the other two peptides, with a GB that was significantly higher. These peptides would be expected to protonate on the lysine residues, and the greater reactivity of K_4G_8 was attributed to the greater Coulombic repulsion in this peptide owing to the proximity of the lysines. Molecular modeling indicated that K_4G_8 has a compact structure, while the other two peptides were extended to minimize the Coulombic energy. In a later study (Pallante and Cassady 2002), the same group examined GB trends in a series of doubly charged ions that started with bradykinin and successively removed residues from the *C* terminus. They found that GB_{app} (obtained by bracketing, and uncorrected) of the singly charged ions decreased from 940 kJ/mol for bradykinin to 820 kJ/mol for the truncated tripeptide. Decreases for each residue varied from 40 to 0 kJ/mol, and could be explained by the basicities of the individual residues, Coulombic repulsion, and hydrogen bonding.

Gross and Williams (1995) determined (using the bracketing method) the GBs for the singly and doubly protonated ion of gramicidin *S*. From these values, they determined the Coulombic energy of the doubly charged ion, and combining this with a proton–proton distance obtained from molecular modeling (9.5 Å), they were able to estimate a value of 1.2 for the dielectric polarizability for gramicidin *s*. A further study (Gross and Williams 1996) on doubly charged gramicidin *s* ions showed that protonated gramicidin *S* has a slightly lower GB (by 4.2 kcal/mol) than does gramicidin *s* cationized by an alkali metal ion. This difference was attributed to the greater charge separation between the proton and an alkali metal ion than between two protons. Using the value of dielectric polarizability derived in their earlier work, the authors were able to calculate a proton–metal distance of 11.5 Å, consistent with location of the metal ion on the exterior surface of the peptide. A similar approach to cytochrome *c* yielded a value for the dielectric polarizability of ~2.0 (Schnier et al. 1995a).

A number of studies have compared the reactivity of disulfide-intact and disulfide-reduced species (Ogorzalek Loo and Smith 1994; Gross et al. 1996; Ewing and Cassady 1999; Wang and Cassady 1999). Both the Smith and the Williams groups found that disulfide-intact hen eggwhite lysozyme is more reactive to proton transfer than is the disulfide-reduced form (for the same charge state), and this was attributed to the more compact structure of the disulfide-intact species, with consequent greater Coulombic repulsion and lower GB. Furthermore, Williams et al. were able to compare their measured GB values for disulfide-intact and disulfide-reduced species with those calculated for "native" (the X-ray crystal structure) and "denatured" (fully extended chain) structures, respectively, and showed good agreement. Similarly, for charge states of the same form that showed more than one reactive species, the presence of intermediate, partially folded structures was inferred.

However, it appears that in smaller peptides, reduction of disulfide bridges may not result in consistent changes in reactivity (Wang and Cassady 1999). Molecular modeling results suggest that the important factor is the degree of compactness of the peptide; in cases where disulfide cleavage results in a more compact structure, reduction leads to lower reactivity. Coulombic effects do not appear to play a major role in these peptides.

Lebrilla and coworkers examined chiral specificity in proton transfer using (2*R*)- and (2*S*)-2-butylamine and various charge states of cytochrome *c* (Gong et al. 1999). The *R* isomer of the base was found to be more reactive toward proton transfer than the *S* isomer for the +9, +8, and +7 charge states of cytochrome *c* by nearly an order of magnitude. Furthermore, in some cases such as the reaction of +8 and +7 states, the kinetics indicate the existence of more than one reacting species.

3.4. HYDROGEN/DEUTERIUM EXCHANGE

3.4.1. Exchangeable Hydrogens

Knowing the number of exchangeable hydrogens in a molecule (or a fragment ion) can be a useful aid in structure determination. In an early report, D_2O was used as the reagent gas in chemical ionization to count the number of exchangeable hydrogens in small molecules (Hunt et al. 1972). Complete exchange was observed for molecules such as adenosine, but some propensity for "nonexchangeable" hydrogens to exchange was noted as well. In contrast, Freiser et al. (1975) found this latter process to be fairly efficient in an ion cyclotron resonance instrument with a much lower reagent gas pressure. Later studies have shown high levels of exchange (nearly complete exchange for ions containing up to 25 exchangeable hydrogens) of ESI-produced ions when ND_3 was used as the curtain gas (Hemling et al. 1994) in an ESI source, or when it was leaked into a hexapole ion reservoir (Hofstadler et al. 2000). However, it should be cautioned that isotopic scrambling, and thus the potential for artificially high exchangeable hydrogen counts, has been observed in a hexapole ion reservoir as well (Reed and Kass 2001). Mao and Douglas (2003) reported complete exchange of bradykinin^{2+} ions in 80 s in a linear ion trap.

There is a risk of obtaining an incorrect count of exchangeable hydrogens; too high an energy input from the ionization method itself, or from trapping or transport, can potentially result in the exchange of normally unreactive hydrogens. Although such extra exchanges may be useful in some other applications, they are obviously undesirable for structure elucidation. A more common concern is that at low reagent pressures and short reaction times, incomplete exchange can result in a low count. This becomes more important as the size of the ion increases and accessibility of the exchangeable sites becomes an issue. Because a low, positive ΔPA facilitates exchange, ND_3 is usually the preferred exchange reagent.

3.4.2. Conformations of Peptides and Proteins

Initial studies on gas-phase H/D exchange of proteins showed evidence of a dependence of reactivity on gas-phase conformation (Suckau et al. 1992, 1993; Winger et al. 1992b). Winger et al. (1992b) compared reactivity of native and disulfide-reduced forms of proteins and found higher exchange reactivity with the native forms of the same charge state. Suckau et al. (1993) reported a higher reactivity for the disulfide-reduced form of RNase. In solution-phase exchange, the more compact native forms are less reactive, due to lesser accessibility of the solvent; in the gas phase, when the same charge states are to be compared, accessibility effects may be countered by the lower PA (or GB) to be expected as a result of Coulombic effects in a more compact conformation. In one of these studies

(Suckau et al. 1993), evidence was also reported for the existence of more than one conformation for the same charge state of an ionized protein.

Conformational effects are perhaps easier to resolve in peptides, which are more readily understood and accessible to theoretical calculations. It has already been pointed out that structural effects play a strong role in determining reactivty of a particular site. For example, in protonated glycine, the amine hydrogens exchange very slowly relative to the carboxylic hydrogen (Gard et al. 1993); however, in diglycine, the amine Hs exchange much more quickly than does the carboxylic hydrogen (Green and Lebrilla 1997). As might be expected, higher charge states are generally more reactive (Gross and Williams 1995; Green and Lebrilla 1998; Freitas and Marshall 1999) [with some glaring exceptions (Freitas and Marshall 1999)]; higher charge states have lower PA (or GB), and in principle, more reactive sites.

H/D exchange has been used to address the issue of gas-phase zwitterionic structures for peptides. Although the amino acids exist as zwitterions in solution, none of the simple amino acids or their ions has a zwitterionic structure in the gas phase. However, it has been suggested that a zwitterionic structure should be stable for some peptides ions (Strittmatter and Williams 2000). The observation that exchange behavior for singly protonated bradykinin is almost identical to that for doubly protonated *O*-methyl bradykinin was interpreted as evidence for a zwitterionic structure for the former (Freitas and Marshall 1999): The doubly protonated *O*-methyl peptide, which cannot form a zwitterion, would have both arginine residues protonated, whereas singly protonated bradykinin would only have both arginine residues protonated if it were in the zwitterionic form. More compelling evidence was obtained for the proton-bound dimer of arginine (Reuben et al. 2003; Lifshitz 2004). The site-specific rate constants obtained indicated one fast exchanging site and 14 slow ones, implying a single reactive single carboxyl hydrogen (with four amino and 10 guanidino hydrogens).

Cassady and coworkers have carried out a number of studies in which they examined both proton transfer and H/D exchange reactivity (CD_3OD) (Zhang et al. 1998; Ewing et al. 1999; Wang and Cassady 1999) of selected peptides. Reactivity trends were similar for both reactions; for example, in their study of the multiply protonated $K_4G_8^{4+}$ peptides (Zhang et al. 1998), they found that for H/D exchange (K_4G_8) was significantly more reactive than $(KG_2)_4$ and $(K_2G_4)_2$. It was postulated that this occurred for a similar reason—Coulombic forces between the protons on the K residues lowered the GB of the peptide ion. Modeling indicated that the geometry for formation of the hydrogen-bonded intermediates required in the relay mechanism was more favorable in the case of K_4G_8. This may have also contributed to the higher reactivity. On the other hand, when Beauchamp et al. (Campbell et al. 1995) examined the reactivity of D_2O with Gly_nH^+, $n = 1$–5, they found maximum exchange reactivity for $n = 2,3$. The increase in reactivity on going from $n = 1$ to $n = 2,3$, despite an increase in GB, can be attributed to a more favorable geometry for intermediate formation. The decrease in reactivity for the larger oligomers is due less to their higher GB and more to the inability of D_2O to disrupt the internal hydrogen bonding and effectively form the intermediate required for exchange.

One interesting facet of gas-phase H/D exchange is the choice of reagents available. As discussed above, PA of the reagent plays a major role in determining the rate and mechanism of an exchange, and the behavior of a given substrate varies dramatically with the reagent chosen, in terms of the rate and often extent of exchange. D_2O is relatively slow to react, and is expected to react only with "surface" sites of peptides and proteins (Campbell et al. 1994; Wyttenbach and Bowers 1999); ND_3 reacts more rapidly, and the solvated ammonium ion may have some ability to penetrate the interior of a gas-phase ion. CH_3OD offers intermediate reactivity, and in principle, more simple kinetics because of its single exchangeable deuterium.

A number of studies of exchange reactivity have reported coexisting populations of the same charge state with different reactivities (Zhang et al. 1996; Schaaff et al. 1999; Witt et al. 2002; Mao and Douglas 2003; Takats et al. 2003; Lifshitz 2004; Mazurek et al. 2005; Sawyer et al. 2005); this is usually interpreted as evidence of the existence of different gas-phase conformers, although charge site isomers are also possible. In particular, Sawyer et al. (2005). performed ion mobility measurements on bradykinin 1–5$^+$ in parallel and found extended and compact forms in approximately the same relative abundance as the slow- and fast-exchanging populations, respectively. Furthermore, they were able to rationalize the observed reactivities in terms of their proposed structures: the more compact forms were postulated to be more reactive because the proposed exchange site was more exposed. In the case of the proton-bound dimer of arginine (Lifshitz 2004), the two populations observed were proposed to correspond to the zwitterionic (fast-exchanging) and nonzwitterionic (slow) forms based on kinetic analysis. For the protonated serine octamer (Mazurek et al. 2005), the kinetics were interpreted as supporting a zwitterionic serine structure for the less reactive population, and a nonionic serine structure for the more reactive one. It seems that simply looking at relative reactivity does not allow the distinction between structural features such as extended/compact or zwitterionic/nonzwitterionic; rather, the distinction must be made on a case-by-case basis, with consideration of structural and reactive subtleties.

More recent studies have demonstrated the utility of H/D exchange for separation of gas-phase conformers (Mazurek et al. 2004; Wysocki et al. 2005); in cases in which two (or more) conformers show drastically different rates of exchange, they become separated on the *m/z* axis as the exchange proceeds. This allows for selective CID of the different conformers for structural elucidation.

The effects of noncovalent complex formation are varied; complexation of amino acids with sugars (Green et al. 1995b) and glycine oligomers with crown ethers (Lee et al. 1998) significantly decreases reactivity, which is attributed to hydrogen bonding of the partner with the reactive site. On the other hand, Reid et al. (1998) found that a series of amino acids and small peptides increased in reactivity on dimer formation, and Heck et al. (1998) found that complexation of vancomycins with peptides resulted in either decreases or increases in reactivity, depending on the structure of the vancomycin. Again, broad generalizations are difficult to make.

The effects of metal complexation (usually Na^+) on exchange reactivity have been examined as well (Kaltashov et al. 1997; Freitas and Marshall 1999; Solouki et al. 2001; Jurchen et al. 2003; Cox et al. 2004). There is general agreement that metal ion complexation drastically affects the gas-phase conformation of a peptide; the peptides tend to adopt more compact, ion solvated structures. This might be expected to result in lower reactivity, and indeed, this is sometimes observed (Kaltashov et al. 1997; Solouki et al. 2001; Jurchen et al. 2003). However, enhancements have been noted as well (Freitas and Marshall 1999; Jurchen et al. 2003; Cox et al. 2004; Jurchen et al. 2004). Increases in reactivity have been postulated to be due to stabilization of carboxylate formed by proton transfer to the exchange reagent when carboxyl hydrogens are involved (Jurchen et al. 2003). In support of this, it was noted that species in which all the carboxyl hydrogens plus the ionizing proton are replaced by Na, reactivities are very low.

A major objective of gas-phase H/D exchange of proteins has been directed toward conformational studies. Solution-phase H/D exchange has long found application in determining structural and conformational details of proteins and complexes (Englander et al. 1997; Kaltashov 2005), and mass spectrometry has played an important role in this. Labile hydrogens that are tied up by hydrogen bonding and/or less accessible to solvent because of conformation or noncovalent binding to a partner exchange more slowly than are those that are exposed to solvent. A comparison of gas- and solution-phase conformations would provide insight into the nature of the folding process, particularly with respect to the role of the solvent. The degree to which gas- and solution-phase structures correspond is not always clear; as a generalization, it seems that binding interactions that are due mainly to hydrophobic forces in solution are lost in the gas phase whereas those effects due to nonpolar stacking and hydrogen bonding are retained (Robinson et al. 1996; Barran et al. 2005). Intuitively, one would expect that, as the size of a molecule increases and intramolecular interactions play an increasingly important role, the gas- and solution-phase structures would more closely resemble each other.

Gas-phase H/D exchange of proteins was first reported by Smith and coworkers (Winger et al. 1992b), who reacted electrosprayed proteins with D_2O. They observed qualitative differences in the exchange reactivity for the multiply protonated native and reduced forms of bovine proinsulin and α-lactalbumin. However, as in their proton transfer study (Winger et al. 1992a), the results were contrary to what was expected, based on solution experience; the more compact, native forms of the proteins showed a higher exchange reactivity, whereas these forms generally show less exchange in solution. Again, this difference was interpreted as a case of Coulombic effects dominating over steric effects. As discussed above, a lower PA difference between substrate and reagent is expected to enhance reactivity.

Studies on the reaction of protein ions with D_2O by McLafferty and coworkers (Suckau et al. 1993; Wood et al. 1995) found reactivity trends more in line with the solution-phase behavior—the reduced form of RNase exchanged 4 times as many hydrogens as did the native form. A number of different conformers of cytochrome *c* ions were described, based on the extent of H/D exchange observed. However, the conformers found in the two different studies were generally different, suggesting a dependence on experimental conditions. Although the conformers were stable, the authors were able to interconvert them to some degree; unfolding could be induced by infrared heating, and folding could be induced by deprotonation. Although the conformers observed had a strong dependence on the conditions in the solution from which they were electrosprayed, it was not clear how closely the gas- and solution-phase conformations were related. In later studies (Guan et al. 1996; McLafferty et al. 1998), CID of H/D exchanged cytochrome *c* showed high levels of exchange at the termini of the protein, contrary to the solution-phase results, in which exchange at the termini is relatively slow; this suggests a poor correspondence between gas- and solution-phase structures. Valentine and Clemmer (2002) combined ion mobility and H/D exchange to study the +5 (compact) and +9 (elongated) charge states of cytochrome *c* over a wide temperature range. They found that at lower temperatures the elongated conformer exchanged more hydrogens, but that this was reversed at higher temperatures (although the degree of exchange was greater for both). This was rationalized with a model that assumed intermolecular exchange at protonated sites, followed by intramolecular exchange; it was postulated that at higher temperatures, the elongated conformer exchanged fewer hydrogens because some labile hydrogens were too remote from protonation sites.

A more recent study was able to extract out rate constant conformation for the exchange of ND_3 with the +10 (compact) and +12 (elongated) charge states of cytochrome *c* (Geller and Lifshitz 2004). Although both conformers has several common rate constants, the compact forms showed 10 very slow exchanges that the elongated did not, and similarly, the elongated form had 12 fast exchanges that the compact form did not. The authors proposed that the fast exchanges in the elongated form were a consequence of localized unfolding.

3.4.3. H/D Exchange of Nucleotides

Although nucleosides and nucleotides have not been the subject of nearly as much attention as peptides and proteins, a number of studies have been reported (Freitas et al. 1998; Robinson et al. 1998; Felix et al. 1999; Hofstadler et al. 2000; Freitas and Marshall 2001; Green-Church et al. 2001; Crestoni and Fornarini 2003; Chipuk and Brodbelt, 2005). Both cationic (Felix et al. 1999; Green-Church et al. 2001; Crestoni and Fornarini 2003) and anionic (Freitas et al. 1998; Robinson et al. 1998; Freitas and Marshall 2001; Crestoni and Fornarini 2003; Chipuk and Brodbelt 2005) species have been investigated. For the cationic species, many of the observations for peptides have been reported: the inverse dependence of exchange reactivity on ΔPA, the importance of hydrogen bonded intermediates and the tendency for complete replacement of acidic protons by sodium to strongly inhibit exchange. The anions also show reactivities dependent on ΔPA [(or ΔGPA): D_2S (GPA = 1444 kJ/mol (Mallard 2003)] shows much higher reactivity than does D_2O (GPA = 1605 kJ/mol) (Freitas et al. 1998), and hydrogen bonding in the intermediates plays a key role. Evidence indicates that initial exchange in the anions is at the phosphate, and that exchange at other sites depends on their accessibility to the phosphate (Robinson et al. 1998; Felix et al. 1999; Crestoni and Fornarini 2003). Reports of exchange reactions of oligonucleotides are rare (Hemling et al. 1994; Hofstadler et al. 2000; Chipuk and Brodbelt 2005): Hofstadler et al. (2000) reported that CID of partially exchanged nucleotides showed that greatest exchange occurred at the termini.

3.5. HYDROGEN IODIDE ATTACHMENT

Stephenson and McLuckey (1997) first reported that HI attaches to the basic sites of peptides and proteins. With few exceptions, they found that the number of HI attachments equalled the number of unprotonated basic sites—arginine, lysine, histidine and the *N* terminus, in proteins as large as β-lactoglobulin, namely, charge state + maximum number of HI attachments = sum of basic residues + 1 (assuming an unmodified N terminus). Later papers (Schaaff et al. 1999, 2000) examined the kinetics of attachment; the kinetics were analyzed in a manner similar to that for reactions (3.5)–(3.8) above. As in the case of proton transfer and H/D exchange, it was noted that the kinetics for attachment in some cases indicated the existence of multiple conformations of the peptides. When DI was used, there was a competition between H/D exchange and attachment. It should be mentioned that these studies were carried out in a quadrupole ion trap; attempts to observe HI attachment in an FTICR instrument were unsuccessful (Ewing and Cassady 1999).

3.6. CONCLUSION

Ion–molecule reactions, in particular H/D exchange, have proved their usefulness as probes of gas-phase ion structures, providing insights into both primary and secondary aspects of the gas-phase structure of biomolecules. However, it should be cautioned that these methods cannot be applied blindly and are most useful when applied in combinations with other methods of study.

As computational methods for structural calculations and kinetics analysis become more sophisticated, the potential for ion–molecule reactions to provide more detailed information on gas-phase structure grows. The relation between gas and solution-phase structures of biomolecules and complexes requires further investigation, in terms of both assessing the relevance of gas-phase structural information and understanding better the effects of the solvent environment.

REFERENCES

Aue DH, Bowers MT (1979): Stabilities of positive ions from equilibrium gas-phase basicity measurements. In Bowers MT (ed), *Gas Phase Ion Chemistry*, Academic Press, New York, Vol. 2, pp. 1–51.

Ausloos P, Lias SG (1981): Thermoneutral isotope exchange reactions of cations in the gas phase. *J. Am. Chem. Soc.* **103**:3641–3647.

Barran PE, Polfer NC, Campopiano DJ, Clarke DJ, Langridge-Smith PRR, Langley RJ, Govan JRW, Maxwell A, Dorin JR, Millard RP, Bowers MT (2005): Is it biologically relevant to measure the structures of small peptides in the gas-phase? *Int. J. Mass Spectrom.* **240**: 274–284.

Bartmess JE, McIver RT, Jr. (1979): The gas-phase acidity scale. In Bowers MT (ed), *Gas Phase Ion Chemistry*, Academic Press New York, Vol. 2, pp. 88–121.

Bliznyuk AA, Schaefer HF, III, Amster IJ (1993); Proton affinities of lysine and histidine: A theoretical consideration of the discrepancy between experimental results from the kinetic and bracketing methods. *J. Am. Chem. Soc.* **115**:5149–5154.

Bohme DK, MacKay GI, Schiff HI (1980): Determination of proton affinities from the kinetics of proton transfer reactions. VII. The proton affinities of O_2, H_2, Kr, N_2, Xe, CO_2, CH_4, N_2O, and CO. *J. Chem. Phys.* **73**:4976–4986.

Bojesen G (1987): The order of proton affinities of the 20 common l-.Alpha.-amino acids. *J. Am. Chem. Soc.* **109**(18):5557–5558.

Bouchoux G, Buisson D-A, Colas C, Sablier M (2004): Protonation thermochemistry of a-amino acids bearing a basic residue. *Eur. J. Mass Spectrom.* **10**(6):977–992.

Bouchoux G, Salpin J-Y (2003): Gas-phase basicity of glycine, alanine, proline, serine, lysine, histidine and some of their peptides by the thermokinetic method. *Eur. J. Mass Spectrom.* **9**(4):391–402.

Bouchoux G, Salpin JY, Leblanc D (1996): A relationship between the kinetics and thermochemistry of proton transfer reactions in the gas phase. *Int. J. Mass Spectrom. Ion Proc.* **153**:37–48.

Brauman JL (1979): In Ausloos P (ed), *Kinetics of Ion-Molecule Reactions*, Plenum Press, New York, pp. 153–164.

Büker HH, Grützmacher HF (1991): A fourier transform-ion cyclotron resonance study of steric effects on proton transfer reactions of polyalkyl benzenes. *Int. J. Mass Spectrom. Ion Proc.* **109**:95–104.

Campbell S, Rodgers MT, Marzluff EM, Beauchamp JL (1994): Structural and energetic constraints on gas-phase hydrogen/deuterium exchange reactions of protonated peptides with D_2O, CD_3OD, CD_3CO_2D, and ND_3. *J. Am. Chem. Soc.* **116**:9765–9766.

Campbell S, Rodgers MT, Marzluff EM, Beauchamp JL (1995): Deuterium exchange reactions as a probe of biomolecule structure. Fundamental studies of gas phase h/d exchange reactions of protonated glycine oligomers with D_2O, CD_3OD, CD_3CO_2D. *J. Am. Chem. Soc.* **117**:12840–12854.

Carr SR, Cassady CJ (1996): Gas-phase basicities of histidine and lysine and their selected di- and tripeptides. *J. Am. Soc. Mass Spectrom.* **7**(12):1203–1210.

Cassady CJ, Cam SR, Zhang K, Chung-Phillips A (1995): Experimental and ab initio studies on protonations of alanine and small peptides of alanine and glycine. *J. Org. Chem.* **60**:1704–1712.

Cassady CJ, Carr SR (1996): Elucidation of isomeric structures for ubiquitin [m+12h]12+ ions produced by electrospray ionization mass spectrometry. *J. Mass Spectrom.* **31**:247–254.

Cassady CJ, Wronka J, Kruppa GH, Laukien FH (1994): Deprotonation reaction of multiply protonated ubiquitin ions. *Rapid Commun. Mass Spectrom.* **8**:394–400.

Cheng X, Fenselau C (1992): Hydrogen/deuterium exchange of mass-selected peptide ions with nd3 in a tandem sector mass spectrometer. *Int. J. Mass Spectrom. Ion Proc.* **122**:109–119.

Chipuk JE, Brodbelt JS (2005): Investigating h/d exchange of oligonucleotides using a quadrupole ion trap, *Proc. 53rd ASMS Conf. Mass Spectrometry*, San Antonio, TX.

Cooks RG, Kruger TL (1977): Intrinsic basicity determination using metastable ions. *J. Am. Chem. Soc.* **99**:1279–1281.

Cox HA, Julian RR, Lee S-W, Beauchamp JL (2004): Gas-phase H/D exchange of sodiated glycine oligomers with nd3: Exchange kinetics do not reflect parent ion structures. *J. Am. Chem. Soc.* **126**:6485–6490.

Crestoni ME, Fornarini S (2003): Gas-phase hydrogen/deuterium exchange of adenine nucleotides. *J. Mass Spectrom.* **38**:854–861.

Englander SW, Mayne L, Bai Y, Sosnick TR (1997): Hydrogen exchange: The modern legacy of Linderstrom-Lang. *Protein Sci.* **6**(5):1101–1109.

Evans SE, Lueck N, Marzluff EM (2003): Gas phase hydrogen/deuterium exchange of proteins in an ion trap mass spectrometer. *Int. J. Mass Spectrom.* **222**:175–187.

Ewing NP, Cassady CJ (1999): Effects of cysteic acid groups on the gas-phase reactivity and dissociation of $[M + 4H]^{4+}$ ions from insulin chain B. *J. Am. Soc. Mass Spectrom.* **10**:928–940.

Ewing NP, Zhang X, Cassady CJ (1996): Determination of the gas-phase basicities of proline and its di- and tripeptides with glycine: The enhanced basicity of prolylproline. *J. Mass Spectrom.* **31**:1345–1350.

Felix T, Reyzer M, Brodbelt J (1999): Hydrogen/deuterium exchange of nucleoside analogs in a quadrupole ion trap mass spectrometer. *Int. J. Mass Spectrom.* **190–191**: 161.

Freiser BS, Woodin RL, Beauchamp JL (1975): Sequential deuterium exchange reactions of protonated benzenes with D_2O in the gas phase by ion cyclotron resonance spectroscopy. *J. Am. Chem. Soc.* **97**:6893–6894.

Freitas MA, Marshall AG (1999): Rate and extent of gas-phase hydrogen/deuterium exchange of bradykinins: Evidence for peptide zwitterions in the gas phase. *Int. J. Mass Spectrom.* **182–183**:221–231.

Freitas MA, Marshall AG (2001): Gas phase RNA and DNA ions 2. Conformational dependence of the gas-phase H/D exchange of nucleotide-5′-monophosphates. *J. Am. Soc. Mass Spectrom.* **12**(7):780–785.

Freitas MA, Shi SD-H, Hendrickson CL, Marshall AG (1998): Gas-phase RNA and DNA ions. 1. H/D exchange of the $[M - H]^-$ anions of nucleoside 5′-monophosphates (GMP, dGMP, AMP, dAMP, CMP, dCMP, UMP, dUMP), ribose 5-monophosphate, and 2-deoxyribose 5-monophosphate with D_2O and D_2S. *J. Am. Chem. Soc.* **120**:10187–10193.

Gard E, Green MK, Bregar J, Lebrilla CB (1993): Site specificity in the H/D exchange reactions of gas-phase protonated amino acids with CH_3OD. *Org. Mass Spectrom.* **28**:1632.

Gard E, Green MK, Bregar J, Lebrilla CB (1994): Gas-phase hydrogen/deuterium exchange as a molecular probe for the interaction of methanol and protonated peptides. *J. Am. Soc. Mass Spectrom.* **5**(7):623–631.

Geller O, Lifshitz C (2003): Hydrogen/deuterium exchange kinetics of cytochrome c: An electrospray ionization fast flow experiment. *Israel J. Chem.* **43**(3–4):347–352.

Geller O, Lifshitz C (2004): Applying a new algorithm to H/D exchange of multiply protonated cytochrome c. *Int. J. Mass Spectrom.* **233**:125–129.

Gong S, Camara E, He F, Green MK, Lebrilla CB (1999): Chiral recognition and the deprotonation reaction of gas-phase cytochrome c ions. *Int. J. Mass Spectrom.* **185–187**:401–412.

Gorman GS, Amster IJ (1993): Gas-phase basicity measurements of dipeptides that contain valine. *J. Am. Chem. Soc.* **115**:5729–5735.

Green-Church KB, Limbach PA, Freitas MA, Marshall AG (2001): Gas-phase hydrogen/deuterium exchange of positively charged mononucleotides by use of Fourier-transform ion cyclotron resonance mass spectrometry. *J. Am. Soc. Mass Spectrom.* **12**(3):268–277.

Green MK, Gard E, Bregar J, Lebrilla CB (1995a): H-D exchange kinetics of alcohols and protonated peptides: Effects of structure and proton affinity. *J. Mass Spectrom.* **30**:1103–1110.

Green MK, Lebrilla CB (1997): Ion-molecule reactions as probes of gas-phase structures of peptides and proteins. *Mass Spectrom. Rev.* **16**:53–71.

Green MK, Lebrilla CB (1998): The role of proton-bridged intermediates in promoting hydrogen-deuterium exchange in gas-phase protonated diamines, peptides and proteins. *Rapid Commun. Mass Spectrom.* **14**(9):15–26.

Green MK, Penn SG, Lebrilla CB (1995b): The complexation of protonated peptides with saccarides in the gas phase decreases the rates of H/D exchange reactions. *J. Am. Soc. Mass Spectrom.* **6**:1247–1251.

Gross DS, Schnier PD, Rodriguez-Cruz SE, C. K. Fagerquist, Williams ER (1996): Conformations and folding of lysozyme ions in vacuo. *Proc. Natl. Acad. Sci. USA* **93**:3143–3148.

Gross DS, Williams ER (1995): Experimental measurement of coulomb energy and intrinsic dielectric polarizability of a multiply protonated peptide ion using electrospray ionization ftms. *J. Am. Chem. Soc.* **117**:883–890.

Gross DS, Williams ER (1996): Structure of gramicidin S $[M + H + X]^{2+}$ ions ($X = Li$, Na, K) probed by proton transfer reactions. *J. Am. Chem. Soc.* **118**:202–204.

Guan Z, Wood TD, Kelleher NL, Aaserud DJ, McLafferty FW (1996): Dominant forces in gas-phase conformation of protein ions, *Proc. 44th ASMS Conf. Mass Spectrometry and Allied Topics*, Portland.

Gur EH, de Koning LJ, Nibbering NMM (1995): The bimolecular hydrogen–deuterium exchange behavior of protonated alkyl dipeptides in the gas phase. *J. Am. Soc. Mass Spectrom.* **6**(6):466–477.

He F, Marshall AG (2000): Weighted quasi-newton and variable-order, variable-step adams algorithm for determining site-specific reaction rate constants. *J. Phys. Chem. A* **104**:562–567.

He F, Marshall AG, Freitas MA (2001): Assignment of gas-phase dipeptide amide hydrogen exchange rate constants by site-specific substitution: Glygly. *J. Phys. Chem. B* **105**:2244–2249.

Heck AJR, Jorgensen TJ, O'Sullivan M, Raumer Mv, Derrick PJ (1998): Gas-phase non-covalent interactions between vancomycin-group antibiotics and bacterial cell-wall precursor peptides probed by hydrogen/deuterium exchange. *J. Am. Soc. Mass Spectrom.* **9**:1255–1266.

Hemling ME, Conboy JJ, Bean MF, Mentzer M, Carr SA (1994): Gas phase H/D exchange in esi-ms as a practical tool for structure elucidation. *J. Am. Soc. Mass Spectrom.* **5**:434–442.

Henchman MJ, Smith D, Adams N, G. (1991): Proton motion within proton-bound dimers. A kinetic mechanism for isotope exchange reactions. *Int. J. Mass Spectrom. Ion Proc.* **109**:105.

Hofstadler SA, Sannes-Lowery KA, Griffey RH (2000): Enhanced gas-phase hydrogen-deuterium exchange of oligonucleotide and protein ions stored in an external multipole ion reservoir. *J. Mass Spectrom.* **35**(1):62–70.

Hunt DF, McEwen CN, Upham RA (1972): Determination of active hydrogen in organic compounds by chemical ionization mass spectrometry. *Anal. Chem.* **44**:1292–1294.

Hunt DF, Sethi SK (1980): Gas-phase ion/molcule isotope-exchange reactions: Methodology for cpimtomg judrpgem atoms in specific organic structural environments by chemical ionization mass spectrometry. *J. Am. Chem. Soc.* **102**:6953–6963.

Hunter EP, Lias SG (1998): Evaluated gas phase basicities and proton affinities of molecules: An update. *J. Phys. Chem. Ref. Data* **27**(3):413–656.

Isa K, Omote T, Amaya M (1990): Proton affinities of amino acids. *Org. Mass Spectrom.* **25**:620–628.

Jurchen JC, Cooper RE, Williams ER (2003): The role of acidic residues and of sodium ion adduction on the gas-phase h/d exchange of peptides and peptide dimers. *J. Am. Soc. Mass Spectrom.* **14**:1477–1487.

Jurchen JC, Garcia DE, Williams ER (2004): Further studies on the origins of asymmetric charge partitioning in protein homodimers. *J. Am. Soc. Mass Spectrom.* **15**:1408–1415.

Kaltashov IA (2005): Probing protein dynamics and function under native and mildly denaturing conditions with hydrogen exchange and mass spectrometry. *Int. J. Mass Spectrom.* **240**(3):249–259.

Kaltashov IA, Doroshenko VM, Cotter RJ (1997): Gas phase hydrogen/deuterium exchange reactions of peptide ions in a quadrupole ion trap mass spectrometer. *Proteins Struct. Funct. Genet.* **28**:53–58.

Kinser RD, Nicol G, Ridge DP (2002): The proton affinity and entropy of protonation of lysinamide. The effects of intramolecular proton solvation. *J. Phys. Chem. A* **106**: 9925–9929.

Kogan A, Ustyuzhanin P, Reuben BG, Lifshitz C (2002): Hydrogen/deuterium exchange of monomers and dimers of leucine enkephalin. *Int. J. Mass Spectrom.* **213**:1–4.

Lee SW, Lee HN, Kim HS, Beauchamp JL (1998): Selective binding of crown ethers to protonated peptides can be used to probe mechanisms of H/D exchange and collision-induced dissociation reactions in the gas phase. *J. Am. Chem. Soc.* **120**:5800–5805.

Li XP, Harrison AG (1993): A kinetic approach to the proton affinities of amine bases. *Org. Mass Spectrom.* **28**:366–173.

Lias SG (1984): Thermoneutral isotope exchange reactions in proton-bound complexes of water with organic molecules: Correlations with energetics of formation of the corresponding association ions. *J. Phys. Chem.* **88**:4401–4407.

Lifshitz C (2004): A review of gas-phase H/D exchange experiments: The protonated arginine dimer and bradykinin nonapeptide systems. *Int. J. Mass Spectrom.* **234**:63–70.

Locke MJ, McIver RT, Jr. (1983): Effect of solvation on the acid/base properties of glycine. *J. Am. Chem. Soc.* **105**:4226–4232.

Majumdar TK, Clairet F, Tabet J-C, Cooks RG (1992): Epimer distinction and structural effects on gas-phase acidities of alcohols measured using the kinetic method. *J. Am. Chem. Soc.* **114**:2897–2903.

Mallard WG. (2003): *NIST Chemistry Webbook*, from http://webbook.nist.gov/chemistry/.

Mao D, Douglas DJ (2003): H/d exchange of gas phase bradykinin ions in a linear quadrupole ion trap. *J. Am. Soc. Mass Spectrom.* **14**:85–94.

Mazurek U, Geller O, Lifshitz C, McFarland MA, Marshall AG, Reuben BG (2005): Protonated serine octamer cluster: Structure elucidation by gas-phase h/d exchange reactions. *J. Phys. Chem. A* **109**:2107–2112.

Mazurek U, McFarland MA, Marshall AG, Lifshitz C (2004): Isolation of isomers based on hydrogen/deuterium exchange in the gas phase. *Eur. J. Mass Spectrom.* **10**(6):755–758.

McKiernan JW, Beltrame CEA, Cassady CJ (1994): Gas-phase basicities of serine and dipeptides of serine and glycine. *J. Am. Soc. Mass Spectrom.* **5**(8):718–723.

McLafferty FW, Guan Z, Haupts U, Wood TD, Kelleher NL (1998): Gaseous conformational structures of cytochrome c. *J. Am. Chem. Soc.* **120**:4732–4740.

Meot-Ner (Mautner) M, Seck LW (1991): Proton affinity ladders from variable-temperature equilibrium measurements. 1. A reevaluation of the upper proton affinity range. *J. Am. Chem. Soc.* **114**:4448–4460.

Meot-Ner M (2003): The proton affinity scale, and effects of ion structure and solvation. *Int. J. Mass Spectrom.* **227**(3):525.

Meot-Ner M, Hunter EP, Field FH (1979): Ion thermochemistry of low-volatility compounds in the gas phase. 1. Intrinsic basicities of α-amino acids. *J. Am. Chem. Soc.* **101**:686–689.

Ogorzalek Loo RR, Smith RD (1994): Investigation of the gas-phase structure of electrosprayed proteins using ion–molecule reactions. *J. Am. Soc. Mass Spectrom.* **5**(4):207–220.

Pallante GA, Cassady CJ (2002): Effects of peptide chain length on the gas-phase proton transfer properties of doubly-protonated ions from bradykinin and its n-terminal fragment peptides. *Int. J. Mass Spectrom.* **219**:115–131.

Reed DR, Kass SR (2001): Hydrogen-deuterium exchange at non-labile sites: A new reaction facet with broad implications for structural and dynamic determinations. *J. Am. Soc. Mass Spectrom.* **12**:1163–1168.

Reid GE, O'Hair RAJ, Styles ML, McFadyen WD, Simpson RJ (1998): Gas phase ion–molecule reactions in a modified ion trap: H/D exchange of non-covalent complexes and coordinatively unsaturated platinum complexes. *Rapid Commun. Mass Spectrom.* **12**:1701–1708.

Reuben BG, Ritov Ya, Geller O, McFarland MA, Marshall AG, Lifshitz C (2003): Applying a new algorithm for obtaining site specific rate constants for H/D exchange of the gas phase proton-bound arginine dimer. *Chem. Phys. Lett.* **380**:88–94.

Reyzer ML, Brodbelt JS (2000): Gas-phase H/D exchange reactions of polyamine complexes: $(M + H)^+$, $(M + \text{alkali metal}^+)$, and $(M + 2H)^{2+}$: *J. Am. Soc. Mass Spectrom.* **11**:711–721.

Robinson CV, Chung EW, Kragelund BB, Knudsen J, Aplin RT, Poulsen FM, Dobson C (1996): Probing the nature of noncovalent interactions by mass spectrometry. A study of protein-coa ligand binding and assembly. *J. Am. Chem. Soc.* **118**:8646–8653.

Robinson JM, Greig MJ, Griffey RH, Mohan V, Laude DA (1998): Hydrogen/deuterium exchange of nucleotides in the gas phase. *Anal. Chem.* **70**:3566–3571.

Rozman M, Kazazic S, Klasinc L, Srzic D (2003): Kinetics of gas-phase hydrogen/deuterium exchange and gas-phase structure of protonated phenylalanine, proline, tyrosine and tryptophan. *Rapid Commun. Mass Spectrom.* **17**(24):2769–2772.

Sawyer HA, Marini JT, Stone EG, Ruotolo BT, Gillig KJ, Russell DH (2005): The structure of gas-phase bradykinin fragment 1–5 (RPPGF) ions: An ion mobility spectrometry and

H/D exchange ion-molecule reaction chemistry study. *J. Am. Soc. Mass Spectrom.* **16**: 893–905.

Schaaff TG, James L. Stephenson J, McLuckey SA (1999): The reactivity of gaseous ions of bradykinin and its analogues with hydro- and deuteroiodic acid. *J. Am. Chem. Soc.* **121**:8907–8919.

Schaaff TG, Jr. JLS, McLuckey SA (2000): Reactions of angiotensin ions with hydroiodic acid. *Int. J. Mass Spectrom.* **202**:299–313.

Schnier PD, Gross DS, Williams ER (1995a): Electrostatic forces and dielectric polarizability of multiply protonated gas-phase cytochrome c ions probed by ion/molcule chemistry. *J. Am. Chem. Soc.* **117**:6747–6757.

Schnier PD, Gross DS, Williams ER (1995b): On the maximum charge state and proton transfer reactivity of peptide and protein ions formed by electrospray ionization. *J. Am. Soc. Mass Spectrom.* **6**(11):1086–1097.

Schroeder OE, Andriole EJ, Carver KL, Colyer KE, Poutsma JC (2004): Proton affinity of lysine homologues from the extended kinetic method. *J. Phys. Chem. A* **108**:326–332.

Solouki T, Fort RC, Alomary A, Fattahi A (2001): Gas phase hydrogen/deuterium exchange reactions of a model peptide: FT-ICR and computational analyses of metal induced conformational mutations. *J. Am. Soc. Mass Spectrom.* **12**:1272–1285.

Stephenson J, McLuckey SA (1997): Counting basic sites in oligopeptides via gas-phase ion chemistry. *Anal. Chem.* **69**:281–285.

Strittmatter EF, Williams ER (2000): Structures of protonated arginine dimer and bradykinin investigated by density functional theory: Further support for stable gas-phase salt bridges. *J. Phys. Chem. A* **104**:6069–6076.

Suckau D, Shi Y, Beu SC, Senko MW, Quinn JP, Wampler FM, McLafferty FW (1993): Coexisting stable conformations of gaseous protein ions. *Proc. Natl. Acad. Sci. USA* **90**:790–793.

Suckau D, Shi Y, Quinn JP, Senko MW, Zhang M-Y, McLafferty FW (1992): Conformations of gaseous multiply-charged protein ions by H/D exchange in ESI/FTMS. *Proc. 40th ASMS Conf. Mass Spectrometry and Allied Topics*, Washington, DC, May 31–June 5, 1992.

Sunner J, Hirao K, Kebarle P (1989): Hydride-transfer reactions: Temperature dependence of rate constants for isopropyl(1+) + isobutane = propane + tert-butyl(1+); clusters of isopropyl(1+) and tert-butyl(1+) with propane and isobutane. *J. Phys. Chem.* **93**:4010–4016.

Takats Z, Nanita SC, Schlosser G, Vekey K, Cooks RG (2003): Atmospheric pressure gas-phase H/D exchange of serine octamers. *Anal. Chem.* **75**(22):6147–6154.

Valentine SJ, Clemmer DE (2002): Temperature-dependent H/D exchange of compact and elongated cytochrome c ions in the gas phase. *J. Am. Soc. Mass Spectrom.* **13**(5):506–517.

Wang J, Cassady CJ (1999): Effects of disulfide linkages on gas-phase reactions of small multiply charged peptide ions. *Int. J. Mass Spectrom.* **182/183**:233–241.

Williams ER (1996): Proton transfer reactivity of large multiply charged ions. *J. Mass Spectrom.* **31**:831–842.

Winger BE, Light-Wahl KJ, Richard DS (1992a): Gas-phase proton transfer reactions involving multiply charged cytochrome c ions and water under thermal conditions. *J. Am. Soc. Mass Spectrom.* **3**(6):624–630.

Winger BE, Light-Wahl KJ, Rockwood AL, Smith RD (1992b): Probing qualitative conformation differences of multiply protonated gas-phase proteins via h/d isotopic exchange with d20. *J. Am. Chem. Soc.* **114**:5897–5898.

Witt M, Fuchser J, Baykut G (2002): In-source h/d exchange and ion-molecule reactions using matrix assisted laser desorption/ionization fourier transform ion cyclotron resonance mass spectrometry with pulsed collision and reaction gases. *J. Am. Soc. Mass Spectrom.* **13**(4):308–317.

Wood TD, Chorush RA, Wampler FMI, Little DP, O'Connor PB, McLafferty FW (1995): Gas-phase folding and unfolding of cytochrome c cations. *Proc. Natl. Acad. Sci. USA* **92**:2451–2454.

Wu J, Lebrilla CB (1993): Gas-phase basicities of Gly_n (n=1–5): *J. Am. Chem. Soc.* **115**: 3270–3275.

Wu J, Lebrilla CB (1995): Intrinsic basicity of oligomeric petides that contain glycine, alanine and valine-the effects of the alkyl side chain on proton transfer reactions. *J. Am. Soc. Mass Spectrom.* **6**:91–101.

Wu Z, Fenselau C (1992): Proton affinities of polyglycines assessed by using the kinetic method. *J. Am. Soc. Mass Spectrom.* **3**:863–866.

Wu Z, Fenselau C, Cooks RG (1994): Gas-phase basicities and proton affinities of lysine and histidine measured from the dissociation of proton-bound dimers. *Rapid Commun. Mass Spectrom.* **8**:777–780.

Wysocki VH, Gamange C, Huang Y, Zhang Q, Tan G, Ji L, Herrmann K (2005): Mechanisms of peptide fragmentation. *Proc. 53rd ASMS Conf. Mass Spectrometry*, San Antonio, TX.

Wyttenbach T, Bowers MT (1999): Gas phase conformations of biological molecules: The hydrogen/deuterium exchange mechanism. *J. Am. Soc. Mass Spectrom.* **10**:9–14.

Zhang K, Zimmerman DM, Chung-Phillips A, Cassady CJ (1993): Experimental and ab initio studies of the gas-phase basicities of polyglycines. *J. Am. Chem. Soc.* **115**:10812–10822.

Zhang X, Ewing NP, Cassady CJ (1998): Effects of basic site proximity on deprotonation and hydrogen/deuterium exchange reactions for model dodecapeptide ions containing lysine and glycine. *Int. J. Mass Spectrom.* **175**:159–171.

Zhang Z, Li W, Guan S, Marshall AG (1996): Higher-order structure of gas-phase peptides and proteins from gas-phase exchange experiments. *Proc. 44th ASMS Conf. Mass Spectrometry and Allied Topics*, Portland.

4

UNDERSTANDING PROTEIN INTERACTIONS AND THEIR REPRESENTATION IN THE GAS PHASE OF THE MASS SPECTROMETER

FRANK SOBOTT

Structural Genomics Consortium
University of Oxford
Botnar Research Centre
Oxford, United Kingdom

CAROL V. ROBINSON

The University Chemical Laboratory
University of Cambridge
Cambridge, United Kingdom

Principles of Mass Spectrometry Applied to Biomolecules, edited by Julia Laskin and Chava Lifshitz

4.1. PROTEIN STRUCTURE AND INTERACTIONS IN MASS SPECTROMETRY

It may sometimes come as a surprise to the uninitiated that a method such as mass spectrometry, which is used to weigh ions in the gas phase, can shed light on the structure of biomolecules and how they interact in solution. While the mass is one of the fundamental properties of a molecular entity, often used to identify a known peptide fragment or protein, its knowledge will in most cases also be sufficient to define the stoichiometry of a complex that consists of known components. Key to obtaining structural information are, however, tandem MS techniques (also called MS/MS or MS^n), a quasimultidimensional approach where selected ions are submitted to dissociating conditions and the resulting product ion spectra reveal characteristic fragments or subunits of the species under investigation. Such methods are now widespread in the burgeoning field of proteomics for the identification and quantization of proteins in complex mixtures.

Tandem MS can also be used to detect interacting proteins, in conjunction with tandem affinity purification methods (Gavin et al. 2002; Ho et al. 2002). While this strategy appears well suited for the analysis of the primary structure of proteins (amino acid sequence and posttranslational modifications), more specific information on the secondary and tertiary structures (the fold conformation, interaction interfaces, etc.) can be gleaned by sampling accessible residues with H/D exchange or by introducing covalent amino acid modifications on the surface, approaches that are very amenable to analysis by mass spectrometry (Kaltashov and Eyles 2002).

The majority of research on the structure and interactions of proteins by mass spectrometry is currently done on the peptide level; samples are either digested beforehand ("bottom-up" proteomics) or fragmented within the mass spectrometer ("top-down" proteomics), an approach for which both MALDI and ESI-MS are being employed in conjunction with various separation methods (Reinders et al. 2004). For MALDI-MS the sample is mixed with an excess of matrix material and deposited on a metal surface. On irradiation with short UV or IR laser pulses, ions are desorbed from this target in vacuo. In more recent variants of this technique the solid, or sometimes liquid, target is at atmospheric or intermediate pressure. MALDI produces predominantly singly charged ions and is often combined with specific protease digestion to yield a mass fingerprint for identification of proteins in databases.

Electrospray ionization (ESI), on the other hand, allows direct analysis of dilute aqueous solutions, and as such it is largely the method of choice for the investigation of higher-order structure of biomolecules (Fenn et al. 1989). Liquid sample flowing through a capillary is electrostatically dispersed (electrosprayed) at its tip by means of a high voltage. Thus droplets with an excess of positive or negative charges are formed (depending on the polarity of the voltage) at atmospheric pressure. Subsequent evaporation of solvent and charge-driven breakup of the droplets eventually produces desolvated analyte ions. In this way, series of multiply charged positive or negative ions are formed via protonation or deprotonation of basic or acidic groups, respectively. The maximum number of charges in the series is related to the number of accessible sites in the solution conformation of the protein (Šamalikova et al. 2004; Kaltashov and Mohimen 2005).

We focus here on an emerging method that is being employed to investigate quaternary structure, namely, subunit stoichiometry and transient noncovalent associations, by preserving complexes intact in the mass spectrometer, measuring their mass, and using high-energy collisions to effect their dissociation. This approach has been used successfully to examine the stoichiometry of protein complexes and to monitor changes in response to different solution conditions, including temperature, pH, and concentration of ligands, cofactors and other proteins (Loo 2000; Sobott and Robinson 2002b; Van den Heuvel and Heck 2004; Heck and Van den Heuvel 2004). The vast majority of these studies takes advantage of a miniaturized version of ESI, referred to as *nanoflow* (nano-ESI), where a liquid flow of a few nL/min (nanoliters per minute) is sprayed from metal-coated glass or fused-silica capillaries (Wilm and Mann 1996), or more recently from multiplexed chip-based devices (Keetch et al. 2003).

Structural proteomics increasingly relies on mass spectrometry as a key enabling tool, owing to its specific advantages—notably speed and sensitivity. It is essentially capable of detecting single ions, thus allowing rapid identification of a multitude of different species in parallel. Mass spectrometry can detect analytes in complex conformational or association equilibria next to each other, rather than obtaining an averaged picture or characterizing one favorable species within the mixture only (McCammon et al. 2004). Apart from cooperative effects (Rogniaux et al. 2001), relative binding strengths can also be determined (McCammon et al. 2002). The speed of analysis, with sampling frequencies on the order of several kHz routinely achieved in QqTOF instruments, enables dynamic reactions to be monitored in real time, adding an exciting new dimension to the investigation of transient associations in macromolecular complexes (Sobott et al. 2002b).

4.2. ELECTROSPRAY MASS SPECTROMETRY OF BIOMOLECULES

4.2.1. Native-Like Conditions in Mass Spectrometry

Native-like mass spectrometry requires that noncovalent interactions be preserved from the solution in which they are formed throughout the ionization/desorption process and into the vacuum of the mass spectrometer. At the same time the solvent environment, which constitutes in essence also a noncovalent complex between solvent molecules and the analyte, needs to be removed gently to reveal the embedded ion. The delicate balance between sufficient desolvation of macromolecular ions and the preservation of noncovalent forces poses a formidable challenge for any technique that can achieve phase transfer of the analyte from solution into the gas phase. Currently the most suitable approach for gentle ionization of biomacromolecules from aqueous, buffered solutions is nano-ESI, which is described in detail in Section 4.2.2. Once the ions are formed, it is possible to maintain them intact under appropriate conditions in vacuo where they are essentially isolated, such that a snapshot of the ion distribution in solution can be obtained on the detector of a mass spectrometer (see Section 4.3).

Prerequisite to native-like mass spectrometry is a solution environment that resembles that of the biological system under investigation, such that biomacromolecules adopt their native conformation and association states. When ESI is carried out from acidified solutions containing organic solvent, however, as is common practice for most proteomics applications, including intact protein mass determination, a series of highly charged macromolecular ions is generated in the ion source. Both the organic solvent (acetonitrile or methanol) and the acidic component (usually formic acid) act to partially unfold the protein, thus destroying the native fold and any noncovalent association the proteins might have. This usually results in higher signal intensities, due to the higher availability of protons for analyte charging and higher conductivity of the solution, as well as sharper peaks since attached solvent molecules are more easily removed because of the higher vapor pressure of the organic component. Information on tertiary and quaternary protein structures is, however, lost under these conditions.

Figure 4.1 shows a comparison of the 85-kDa human heat shock protein Hsp90, which performs essential roles as a molecular chaperone and in cell signaling and is

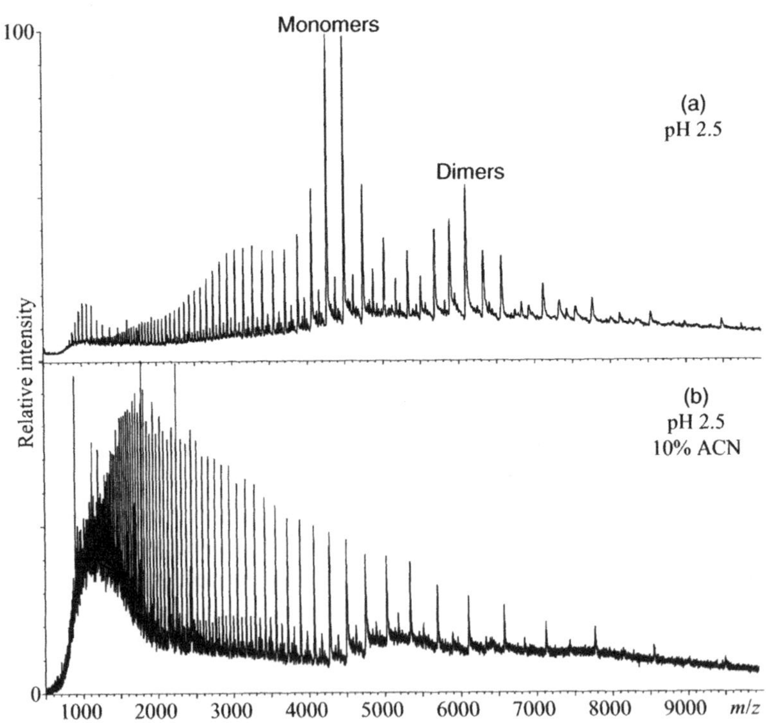

FIGURE 4.1. Nano-ESI MS of Hsp90 sprayed from different nonnative solution environments; (a) acidic solution (25 μM protein in 100 mM aqueous ammonium acetate, adjusted with acetic acid to pH 2.5); (b) acidic solution with organic cosolvent (20 μM protein in 100 mM aqueous ammonium acetate, adjusted with acetic acid to pH 2.5, with 10% acetonitrile).

implicated in cancer development, as sprayed from an acidified aqueous solution without the addition of organic cosolvent [panel (a)] and with an additional 10% acetonitrile in the solvent [panel (b)]. Even at pH 2.5 [panel (a)] there is still some residual native structure left, as evidenced by the dimer signal around m/z 6000. The peaks around m/z 4500 correspond to lowly charged native-like monomer, while the series extends further to lower m/z (higher charge states) including more fully denatured protein around m/z 1000. At acidic pH, the polypeptide chain unfolds partially in solution and is then capable of picking up many more positive charges than in its fully folded state (Grandori 2003). Under conditions that mimic the native environment of the protein, on the other hand (see Figure 4.2), much more compact charge state distributions are observed that correspond to monomeric next to dimeric species. It should be noted here that in spite of the well-defined composition of the noncovalent assemblies, the corresponding signal appears broader than expected purely from the performance of the instrument in this mass range, as is, for example, apparent with calibration compounds (not shown). We will expand on this peak broadening later.

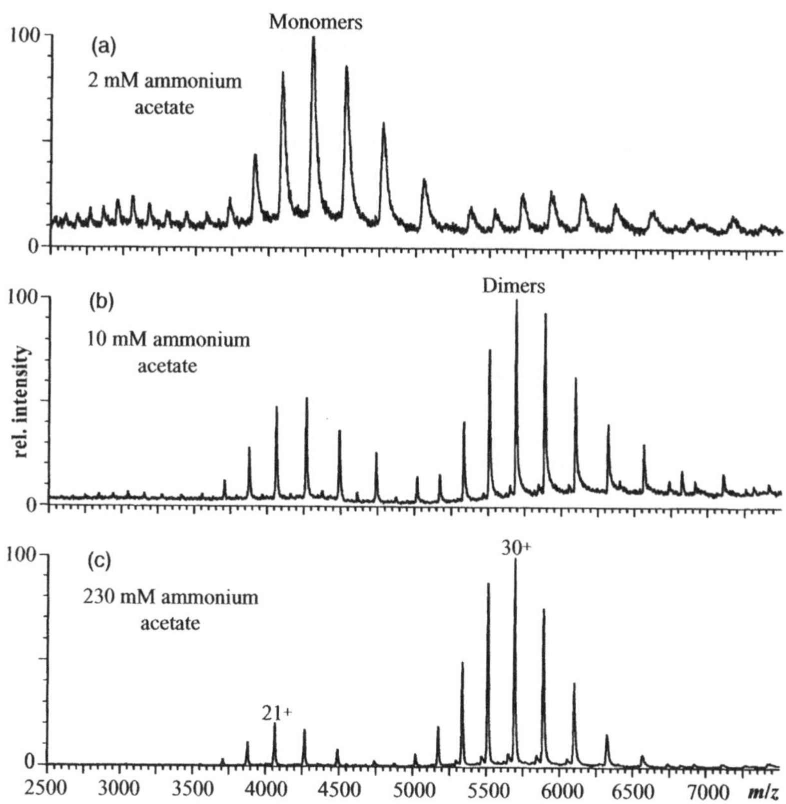

FIGURE 4.2. Human Hsp90 monomer (measured mass 85, 335 Da) and dimer (170,737 Da) in equilibrium at different ammonium acetate concentrations in aqueous solution (pH 7). Other experimental parameters remain unchanged (nano-ESI, protein concentration 25 μM, Waters LCT o-TOF, capillary voltage 2200 V, cone voltage 200 V).

How closely a sample solution mimics the native environment also depends on its ionic strength and temperature. The presence of additional electrolytes influences the polarity of the solution and can also have an effect on the conformation and association/dissociation equilibria of the proteins. There is usually an optimum ammonium acetate (or hydrogen carbonate) concentration range that may differ for various noncovalent complexes. Figure 4.2 illustrates how the monomer–dimer equilibrium of the Hsp90 protein depends on the concentration of ammonium acetate buffer (pH 7) in solution. Little or no dimer is visible at 2 mM buffer [panel (a)], while the dimer dominates at 10 mM [panel (b)] and even more so at 100 mM (not shown) and 230 mM [panel (c)], which represents the optimum condition for this protein complex [for another example, see Benkestock et al. (2004)].

Because of its chemical properties (ionic radii, neutral pH), ammonium acetate closely resembles potassium chloride, which is a salt commonly added to protein solutions. Nano-ESI mass spectrometry is amenable to volatile buffer concentrations of up to several moles per liter and a wide range of pH as long as volatile acids and bases are used. Ammonium acetate buffers up to 9 M have been employed in the investigation of DNA-binding protein complexes (Gupta et al. 2004). The amount of buffer that is added to aqueous solutions also determines the buffer capacity of the solution. It is easily forgotten that the ESI source constitutes an electrochemical cell (Van Berkel et al. 2001), with the spray emitter in positive mode as the anode where oxidation occurs, predominantly oxygen from water molecules, causing the acidity of the solution to increase over time. Figure 4.3 illustrates this effect for a zinc-binding protein, the glucocorticoid receptor DNA-binding domain (GRDBD), which during a 20-min-long continuous spray loses Zn^{2+} ions, as apparent by its apo form, and begins to unfold causing higher charge states to appear at low m/z. This effect is most likely due to a pH drop in the capillary during the (continuous spray, ongoing spray) process. At the same time higher charge states that correspond to a less tightly folded structure appear as well, indicating that the zinc ion stabilizes the more compact conformation of the protein.

Proteins usually undergo conformational changes above a certain temperature. With a variation in solution temperature, a shift in charge states corresponding to more highly charged partially unfolded molecules is observed for hen eggwhite lysozyme using a capillary cooler/heater to vary the solution temperature between 15 and 80°C (Figure 4.4) (Benesch et al. 2003). From plotting the average charge state against the temperature, a clear unfolding transition is observed and a melting point of 43°C determined, in good agreement with an unfolding transition of 44°C as measured by intrinsic tryptophan fluorescence in the same solution (data not shown).

4.2.2. Nano-ESI

Biological mass spectrometry has benefited tremendously from the introduction of a miniaturized version of electrospray (nano-ESI), which uses spray emitters with tip diameters of only a few micrometers to disperse the liquid into very fine sprays. Because of the resulting smaller size of the initial droplets, the sensitivity is

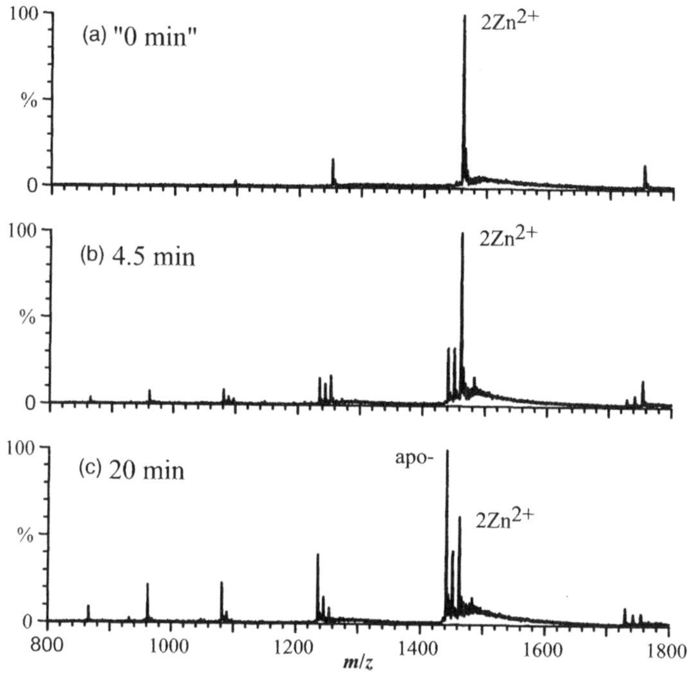

FIGURE 4.3. Changes in the nano-ESI mass spectrum of Zn-binding GRDBD (25 μM in 20 mM aqueous ammonium acetate pH 7) during a 20-min continuous spray. The *apo* form appears after 4.5 min and dominates the spectrum after 20 min.

improved by at least an order of magnitude and desolvation conditions are also gentler than in pneumatically assisted ESI (Wilm and Mann 1996; Juraschek et al. 1999; Karas et al. 2000; Schmidt et al. 2003; El-Faramawy et al. 2005). It also facilitates the use of purely aqueous, native-like solution environments free of organic solvents that are otherwise much more difficult to spray at conventional flow rates using large emitters. Typical samples for nano-ESI contain protein at low micromolar concentration in aqueous ammonium acetate (usually 5–1000 mM) or some other volatile buffer. Then 1–5 μL of this solution are loaded into metal-coated borosilicate glass (Figure 4.5) or quartz capillaries, in online nano-ESI more commonly fused-silica capillaries, and sprayed at flow rates of a few nanolitres per minute.

Key for the advantage of nano-ESI over conventional electrospray is that the initial droplets that are emitted from the Taylor cone are smaller, with diameters on the order of a few hundred nanometers. Basically nano-ESI enters the electrospray process of solvent evaporation and droplet breakup due to Rayleigh fission events (Kebarle and Tang 1993; Fenn 1993), as illustrated in Figure 4.6 (Manil et al. 2003), at a later stage. With the typical protein concentration of 1 μM it can be easily estimated that the nanodroplets will contain at most one protein molecule or complex, rather than several of them, and also fewer additional (counter)ions or

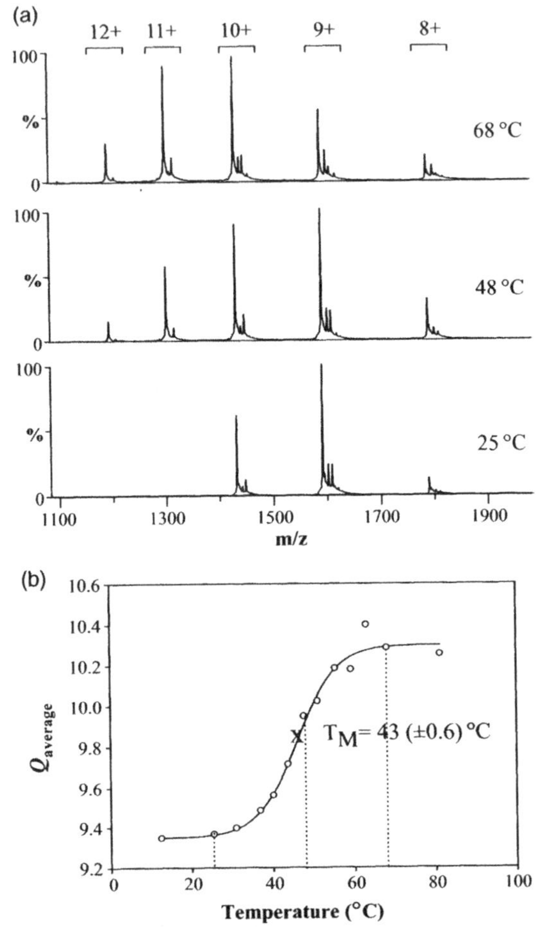

FIGURE 4.4. Nano-ESI mass spectra of hen eggwhite lysozyme measured at different solution temperatures prior to electrospray: (a) room temperature (lower panel), 48°C (middle panel) and 68°C (upper panel); (b) average charge state ($Q_{average}$) versus temperature. An unfolding transition is observed and a melting point of 43.0 (±0.6)°C determined. [Reprinted with permission from Benesch et al. (2003). Copyright (2003) American Chemical Society.]

surface-active molecules, thus minimizing losses due to ion pairing. In addition, fewer solvent molecules need to be removed during the "drying" phase of the ions' journey from solution into vacuum, reducing the need for ion activation and resulting in less energetic conditions in the source (see schematic of a typical nano-ESI ion source in Figure 4.7).

The actual mechanism by which the desolvated gaseous ions are ultimately formed from the highly charged nanodroplets is still under some dispute, but it is thought to be different for small ions compared to larger native-like proteins and

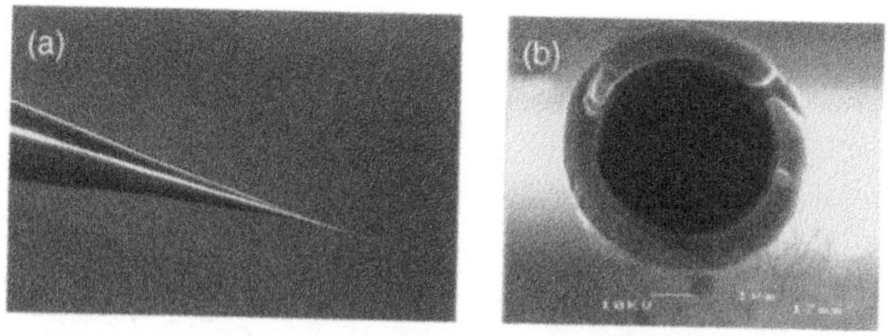

FIGURE 4.5. Typical nano-ESI emitter: (a) side view of an emitter tip that was drawn from an 1-mm-o.d. borosilicate glass tube using a micropipette puller (Flaming/Brown P-97, Sutter Instruments, Novato, CA) and gold-coated using a sputter coater (Harvard Apparatus, Holliston, MA); (b) EM image of the capillary tip after it was clipped under a stereomicroscope to result in a tip inner diameter of 1–5 μm.

complexes (Cole 2000; Heck and Van den Heuvel 2004). While the ion evaporation mechanism is assumed to be dominant in the production of small organic and inorganic ions, it is quite unlikely that large proteins and complexes may evaporate from a droplet. Alternatively, the charge residue mechanism is now widely believed to be dominant in the production of macroions of globular proteins. In this model, the water molecules in the smallest droplets containing a native protein evaporate completely and the protein is charged by the residual charges (e.g., ammonium ions in positive spray mode), which have accumulated on the droplet during the ESI process (see schematic of the ion desolvation process in Figure 4.7). These excess charges end up on surface-exposed basic sites of the protein.

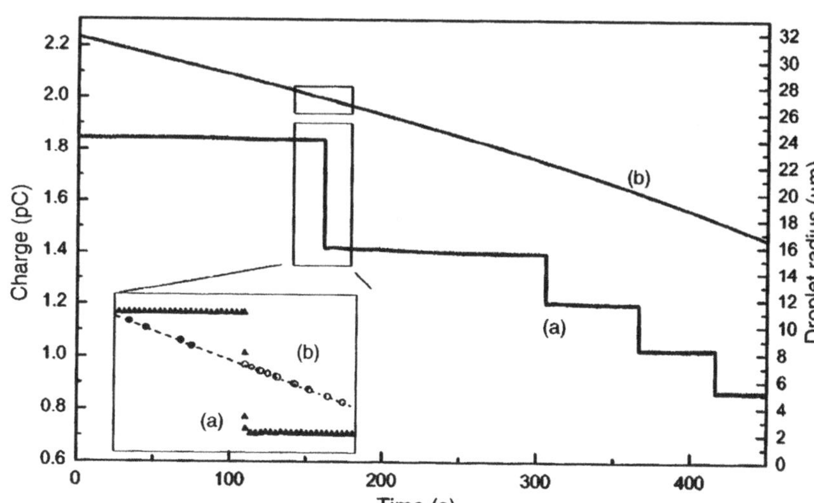

FIGURE 4.6. Time dependence of the charge (a) and the radius (b) of an evaporating ethylene glycol droplet. The inlet shows the first Coulomb instability in more detail. [Reprinted from Manil et al. (2003) with permission from Elsevier.]

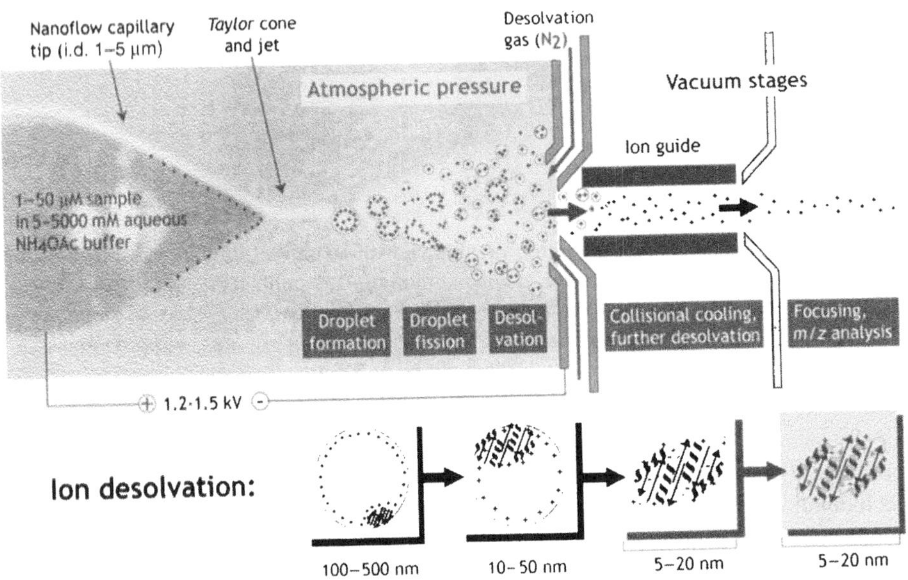

FIGURE 4.7. Schematic view of the nanoelectrospray process. *Upper part*: depiction of the key elements of a typical nanoflow source including the various declustering/desolvation stages. *Lower part*: sketch of the desolvation process that a protein dimer experiences during transition from solution to gas phase.

4.3. ION TRANSMISSION AND ANALYSIS OF NONCOVALENT COMPLEXES

4.3.1. General Aspects of High-m/z Ions

Unlike denatured protein ions, which adopt a partly unfolded, elongated conformation and can therefore accommodate a large number of charges (low m/z, Figure 4.1b), native-like protein ions and noncovalent complexes retain a more globular shape and can carry only as many charges as the critical surface charge density (Rayleigh limit) allows. While the volume of these ions is roughly proportional to the mass, the maximum charge scales only with the surface area. Under native-like conditions large proteins or complexes will therefore appear in comparably low charge states, giving rise to signal at high m/z. For large macromolecular particles, such as the MS2 bacteriophage composed of 180 identical copies of capsid protein, m/z values of over 20 000 have been recorded (Tito et al. 2000). The correlation between the size (mass) of a noncovalent complex and its average charge states was first examined by Ken Standing et al. In a more recent study (Heck and Van den Heuvel 2004) the charge was found to scale roughly with the square root of the mass of the complex, as would be expected for the critical charge density on the surface of near-spherical particles (Figure 4.8). The analysis of large multisubunit complexes therefore requires instrumentation with an extended m/z range.

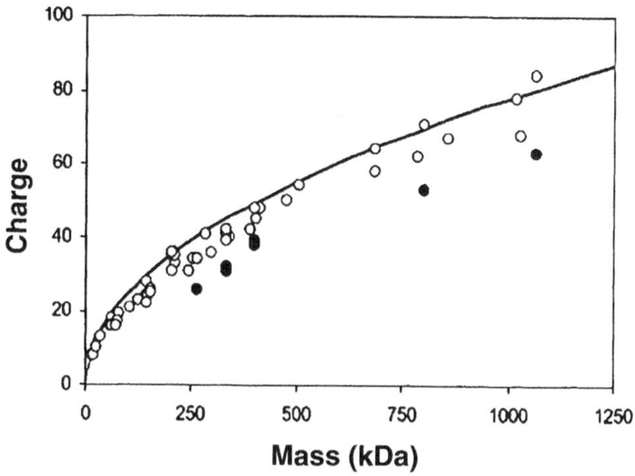

FIGURE 4.8. Average charge of noncovalent protein complexes compared with the maximum charge predicted by the Rayleigh instability limit (solid line). All proteins were sprayed from 50 mM ammonium acetate at neutral pH. Open circles denote positive and filled circles, negative mode. [Reprinted from Heck and Van den Heuvel (2004) with permission from John Wiley & Sons, Inc.]

Another important aspect for work with high-mass ions is that internal energy buildup in collisions with background gas (e.g., nitrogen or argon) becomes less efficient, since the number of degrees of freedom increases more rapidly than does the collision cross section of a particle of increasing mass (volume) (Jellen et al. 2002). Ion activation decreases correspondingly in both the interface of the instrument (collisional cooling/focusing; see Section 4.3.2) and in the collision cell (CID; see Section 4.3.3) at the same voltage and pressure conditions. Since high-mass ions also have a higher density of states and energy can dissipate much more rapidly among their degrees of freedom, it becomes much more difficult to dissociate high-mass ions via gas-phase collisions.

4.3.2. Ion Transfer and Internal Energy of Biomolecules in the Gas Phase

Electrospray ions are generated at atmospheric pressure and introduced into the vacuum chamber of the mass spectrometer in a flowing gas stream (jet), usually through a set of orifices or skimmers (Figure 4.7). As the gas and the still partly solvated ions expand adiabatically through the orifice into vacuum, they are internally cooled and a molecular beam forms (Fenn 2000; Thomson 1997). While the gas in the free jet is accelerated to velocities of several hundred meters per second, even very large ions experience relatively little velocity slip from the supersonic neutral gas velocity. While these speeds may not present a problem for small ions, larger ions can acquire energies in the electronvolt range (Chernushevich and Thomson 2004), making them initially quite "hot" with respect to their translational energy (i.e., fast). The internal degrees of freedom will, however, be rapidly cooled by solvent evaporation and by internal conversion during the adiabatic expansion into vacuum. If ions and solvent vapor cool down uncontrollably, on the other hand, they can cluster and form unspecific aggregates. (One of the key features of John Fenn's first successful ESI mass spectrometer was the ability to control this process.) The internal energy of the ions therefore needs to be carefully controlled, particularly if noncovalent interactions are to be preserved.

In most electrospray instruments the amount of collisional cooling or heating is determined by one or several acceleration voltages (nozzle/skimmer voltage or sample cone/extractor offset). Sometimes additional heating of the ion source is used, but rarely for analysis of noncovalent complexes. The protein molecules undergo collisions with background gas atoms by which they are activated and subsequently lose any remaining, attached solvent molecules (see Figure 4.7). The amount of activation depends on the kinetic energy of the ions, which is determined by the accelerating electric field, as well as the collision number with background gas atoms according to the pressures in the different stages of the instrument. Consequently the association state of a protein appears to depend to a large extent on the choice of voltages and pressures in the ion source (Figure 4.9), and care needs to be taken to optimize these parameters for each system under investigation.

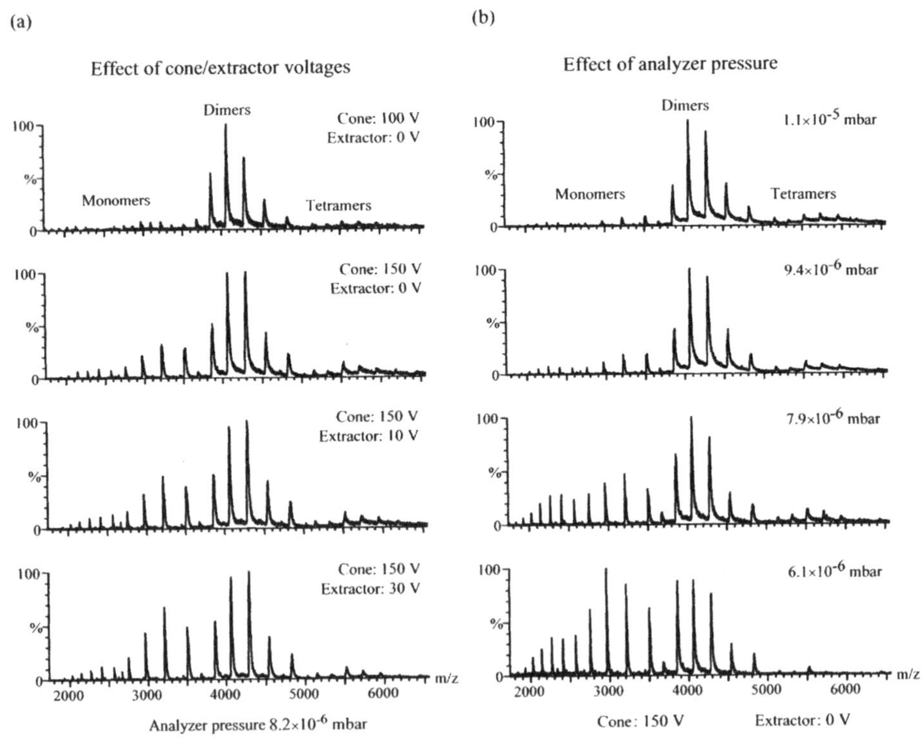

FIGURE 4.9. Effect of various instrumental parameters on the appearance of monomer and dimer signal of biotin synthase sprayed from 20 mM ammonium acetate pH 7 in nano-ESI MS (Waters LCT o-TOF).

Several reports have previously highlighted the important observation that transmission of high-mass ions requires pressures in the first vacuum stages of the mass spectrometer to be increased by reducing the pumping speed or adding a collision gas (Schmidt et al. 2001; Tahallah et al. 2001; Sobott et al. 2002a; Lascoux et al. 2005). In particular, the pressure dependence in the analyzer stage (Figure 4.9b), which is indicative of the pressures in the ion guide (see also Figure 4.11), shows that transmission of intact dimer improves at higher background gas levels, indicating that the noncovalent complex benefits from an increased number of collisions. If these collisions become too energetic, in-source CID occurs and noncovalent interactions are disrupted (Figure 4.9a).

Although the underlying processes are not fully understood, it is clear from work by Igor Chernushevich and Bruce Thomson that a combination of collisional focusing and cooling of the ions is effective in maintaining and transmitting intact complexes through the vacuum interface (Chernushevich and Thomson 2004). Figure 4.10 shows the simulation of ion trajectories for myoglobin [panel (a)] and the 20*S* proteasome [panel (b)] at base pressure as well as increased pressure [panel (c)]. While the off-axis movement and axial speed of myoglobin ions are efficiently dampened at normal operating pressure, much larger noncovalent assemblies such as the 20*S* proteasome complex (a 28-subunit assembly) would not be focused through the pinhole exit of the collisional cooling hexapole. At roughly 4 times higher pressure [panel (c)]; however, the damping effect of the collision gas is restored.

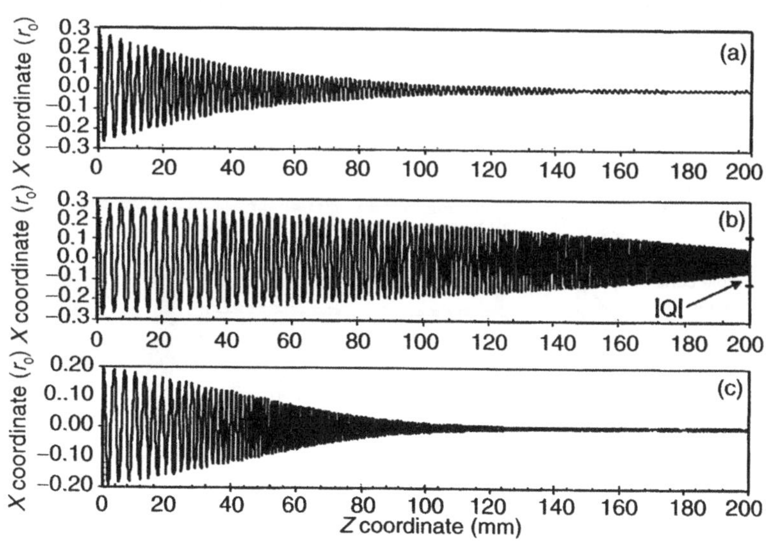

FIGURE 4.10. Simulations of ion trajectories in a collisional cooling quadrupole (the *Z* coordinate coincides with the direction of the ion path): (a) myoglobin ion (16 951Da, collision cross section $\sigma = 2\ 520\ \text{Å}^2$) at 8 mTorr background pressure; (b) proteasome 20*S* ion (692 kDa, $\sigma = 19{,}400\ \text{Å}^2$) at 8 mTorr; (c) proteasome 20*S* ion at 30 mTorr. [Reprinted with permission from Chernushevich and Thomson (2004). Copyright (2004) American Chemical Society.]

A similar collisional cooling effect can be obtained by introducing a flow-restricting sleeve around the front part of the first quadrupole. Additionally, Chernushevich and Thomson introduced some other alternative methods to improve collisional cooling in a QqTOF instrument (Q-Star XL, Sciex, Toronto/Canada) such as trapping of the ions in the quadrupole (Chernushevich and Thomson 2004).

4.3.3. MS/MS of Noncovalent Complexes

At present the mass spectrometric analysis of biomacromolecules is largely the domain of time-of-flight (TOF) mass spectrometry, due to its superior speed and sensitivity as well as virtually unlimited mass range. Combining a quadrupole mass filter with an orthogonal TOF analyzer, in QqTOF-type mass spectrometers, provides supplementary structural information for ions that are isolated in the quadrupole (Q), dissociated in a collision cell (q), and the product ion spectrum analyzed in the TOF analyzer. Conventional tandem mass spectrometers have limitations for studying large noncovalent complexes since the quadrupole mass range is usually limited to m/z 3000–4000, which allows isolation of ions up to a molecular mass of approximately 50–60 kDa. Since noncovalent complexes of MDa sizes give rise to charge states well above the 4000 m/z range, standard instrumentation is unable to isolate ions from these complexes. More recently several QqTOF instruments have been constructed that overcome this limitation with a custom-built quadrupole operating up to more than m/z 30,000 (Sobott et al. 2002a; Chernushevich and Thomson 2004). Figure 4.11 depicts the layout of such a

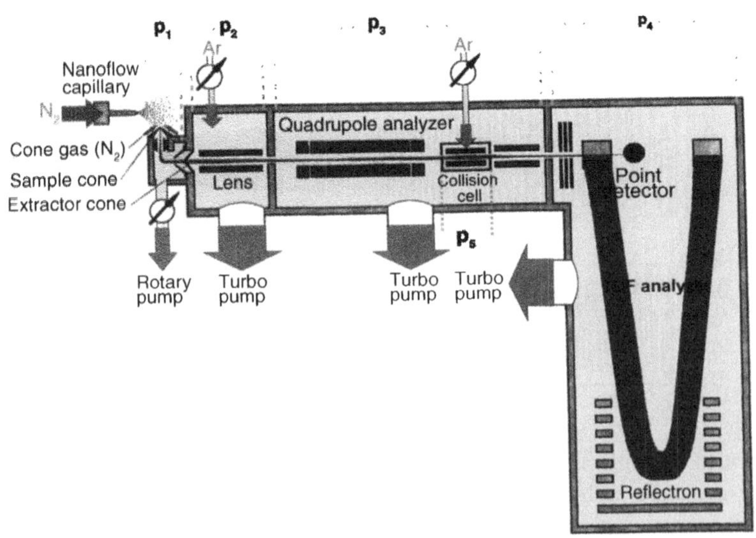

FIGURE 4.11. Layout of a QqTOF instrument used for the investigation of noncovalent complexes, indicating the different pressure regimes for high-mass operation (based on a Waters Q-TOF2 instrument, with Z-spray nanoflow source).

TABLE 4.1.

Pressure	Standard Conditions (mbar)	Conditions for Noncovalent Complexes (mbar)
p_2	10^{-4}	10^{-2}–10^{-3}
p_3	10^{-6}	10^{-4}–10^{-5}
p_4	10^{-7}	10^{-6}–10^{-7}
p_5	$\leq 10^{-3}$	10^{-2}–10^{-3}

high-mass QqTOF instrument, which is based on a Waters Q-TOF2 (Manchester, UK).

The rationale behind the design of these instruments is to obtain a wider range for pressure manipulation as well as an extended m/z range for MS and MS/MS operation. Modifications were introduced to allow operation at higher pressure, notably a *Speedivalve* in the roughing pump line (p_1; see Figure 4.11), an argon bleed valve into the hexapole ion guide (collisional cooling stage, p_2), and a shortened leak capillary into the collision cell (p_5). Table 4.1 gives an overview over the various pressures in the different stages of the instrument under standard conditions, as well as the range of values for noncovalent complexes.

Quadrupoles can transmit ions only to a certain upper limit that depends on the RF amplitude and frequency as well as the inscribed diameter of the rod assembly. The highest m/z (M_{max}) that can be transmitted is given by

$$M_{\text{max}} = \frac{7 \times 10^6 V_m}{f^2 r_0^2}$$

with $V_m \cos(2\pi ft)$ as the RF voltage applied between adjacent rods ($2V_m$ is the peak-to-peak amplitude, f the frequency, and t the time) and r_0 as the inner radius between rods in meters. For a standard Waters Q-TOF, which operates at a frequency of 832 kHz, the mass range of that first stage is limited to m/z 4190. To increase the upper transmission limit, the frequency is lowered to about 300 kHz, which corresponds to a high-mass limit of about 32,000 m/z, while the maximum RF amplitude and rod geometry remain unchanged (Sobott et al. 2002a). As the quadrupole is used only as a mass filter to isolate ions for tandem MS and the product ion spectrum is acquired using the TOF analyzer, the overall mass resolution (which is determined by the TOF) is not compromised significantly by the change in operating frequency of the quadrupole. It should also be noted that in nonresolving (RF-only) mode, ions with more than 5 times the highest set m/z (i.e., around 150,000), although somewhat attenuated, can still easily pass through the quadrupole mass filter. Presently, tandem MS emerges as a key tool for the identification and dissection of large, heterogeneous protein complexes, and its ability to elucidate their exact composition makes it a promising novel approach to structural studies of high-mass biomolecules.

4.4. SOME EXAMPLES

4.4.1. Small-Heat-Shock Proteins and α-Crystallin

The small-heat-shock protein (sHSP) family is both large and diverse, with members found in virtually all organisms. The sHSPs are induced under a variety of stresses and act to increase cellular tolerance to these conditions. This is evidenced by their ability to bind nonnative proteins under different stress conditions, thereby preventing their irreversible aggregation and consequent damage to the cell.

Figure 4.12 shows the nano-ESI QqTOF spectrum of HSP18.1 from *Pisum sativum* [garden pea, (Sobott and Robinson 2002)]. Transformation of the charge state series from an m/z to a mass scale shows three major components with molecular masses around 215 kDa (inset), which correspond to different dodecameric forms of the protein. Isolation of one charge state for each individual peak in the quadrupole mass analyzer of the QqTOF and subsequent collision-induced dissociation reveals the different monomeric components in the product ion spectra. While dodecamer **1** consists exclusively of full-length protein monomers (A) with a molecular mass of 17,974Da (amino acid sequence 1-157), **2** contains additional truncated subunits without the first five (D) or six (B) amino acids, and **3** comprises two species with one C (amino acid sequence 14–157) or two B subunits. Thus the composition of **3** has been found to be either $A_{10}B_2$ or $A_{11}C$, and **2** to be $A_{11}B_1$ or $A_{11}D_1$, while **1** corresponds to the homododecamer A_{12}. This example illustrates how QqTOF-MS can be used to elucidate the heterogeneity of protein complexes and identify modifications in individual subunits.

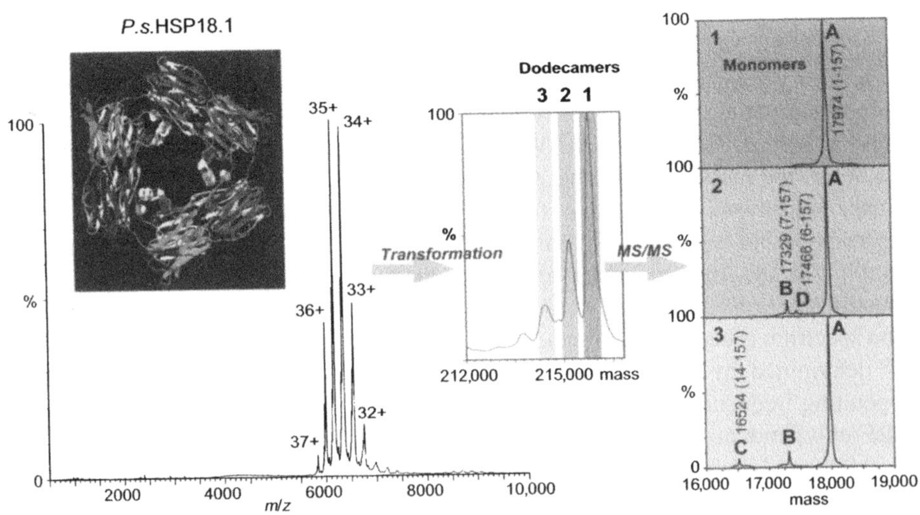

FIGURE 4.12. Nano-ESI MS and MS/MS spectra of *p.s.* HSP18.1 dodecamers acquired with a high-mass QqTOF instrument. The inset represents an X-ray structure of *p.s.* HSP18.1. [Reproduced with permission from Sobott et al. (2002b).]

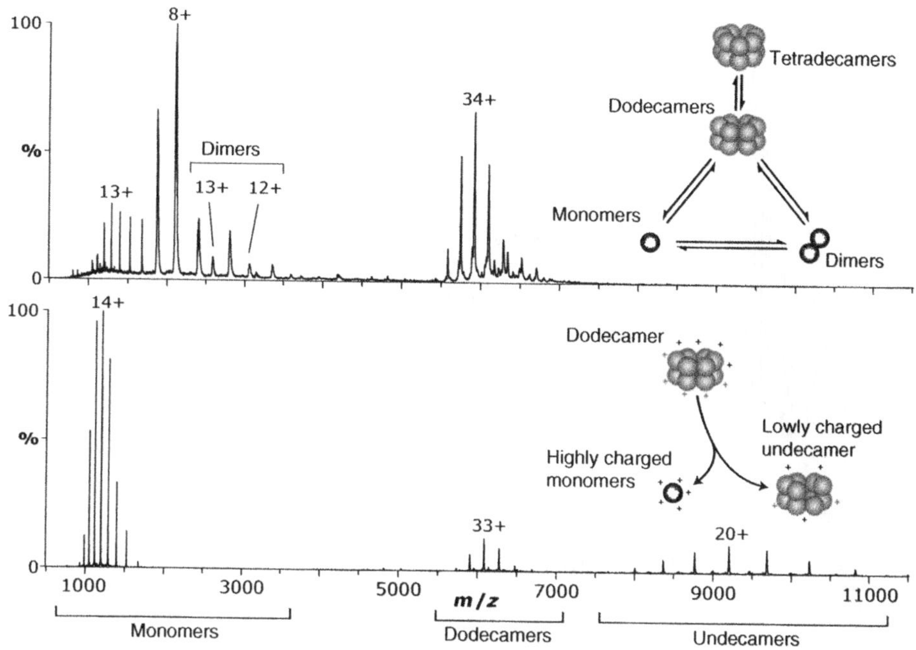

FIGURE 4.13. Solution-phase versus gas-phase activation of *Triticum aestirum* HSP16.9. Starting from intact dodecamers, different species are produced by either solution heating (upper panel) or gas-phase CID (lower panel) and depicted schematically. [Reprinted with permission from Benesch et al. (2003). Copyright (2003) American Chemical Society.]

A comparison between solution-phase and gas-phase activation of *Triticum aestivum* (wheat) HSP16.9 is shown in Figure 4.13 (Benesch et al. 2003). Heating in solution prior to electrospray (upper panel, spectrum at 74°C), using an approach similar to that in Figure 4.4, causes the dodecamers to dissociate into monomers and dimers and also the formation of 14mers. This allows the delineation of solution-phase equilibria (inset, upper panel) that shift away from the dodecameric state as the temperature is increased, revealing an association/dissociation dynamics that could be essential for the chaperone to perform its function. Activation in the gas phase by CID (lower panel, spectrum at 24°C) results in the dodecamers dissociating into highly charged monomers and lowly charged 11mers. Such highly asymmetric gas-phase dissociation is a common feature for CID of noncovalent multisubunit complexes (Jurchen and Williams 2003), and it resembles somewhat the asymmetric fission of electrospray droplets. This charge-driven gas-phase process does not mimic an equilibrium in solution, but is a rather useful tool to determine the composition and stoichiometries of multisubunit complexes.

The small heat shock protein HSP16.5 from *methanococcus janaschii* exists as a discrete 24mer in solution. Tandem-MS of the 24mer peak (46+) at a collision cell argon pressure of 3.5–4.0 × 10^{-2} mbar shows three different dissociation pathways according to collision energy (Figure 4.14) (Benesch et al. 2006). At 60 V asymmetric dissociation into highly charged monomers and 23mer residual

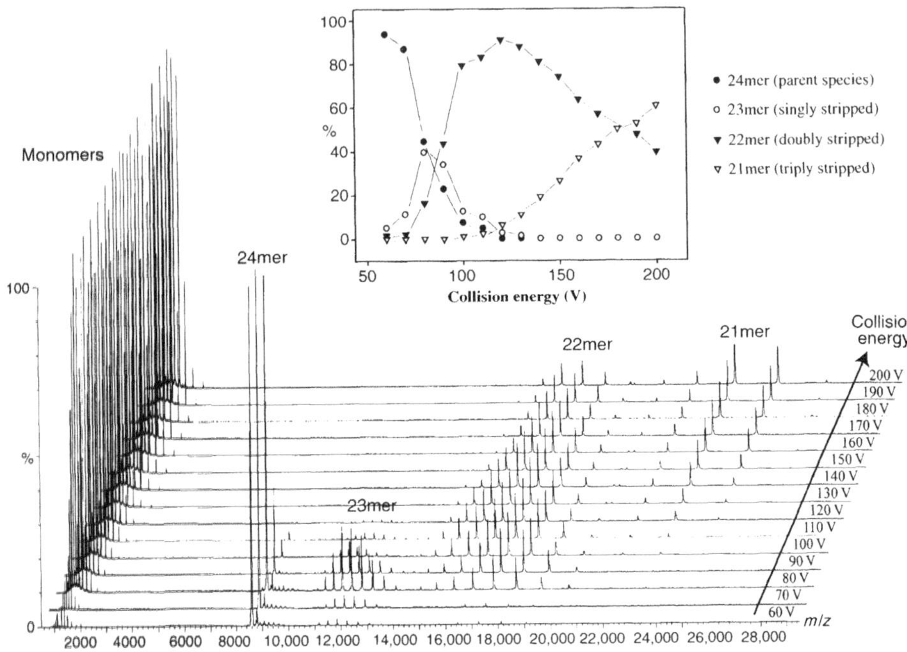

FIGURE 4.14. MS/MS spectra at increasing collision energy of a charge state of the *m.j.* HSP16.5 24mer shows multiple fragmentation of the complex in the collision cell.

complexes sets in. Above 80 V doubly stripped 22mer complexes, and above 120 V triply stripped 21mer complexes appear. At the same time, the monomer charge state distribution shifts to higher m/z. The inset shows relative parent and product ion abundances at different collision energies. These data illustrate the various dissociation products of collisionally activated ions of noncovalent complexes, and that they can undergo repeated fragmentation events provided that sufficient collision energy is available to the ions. Such an approach can also be very useful for the structural elucidation of complex posttranslational modifications, such as glycosylation patterns.

A similar approach can be used to unravel the complex and heterodisperse αB-crystallin assembly, which forms part of the sHSP family. Although systemically expressed, it is found primarily in the eye lens, where it associates with the closely related αA-crystallin into large heterooligomers, which have been shown to display molecular chaperone activity in vitro and to arrest the aggregation of the β- and γ-crystallins in the lens (Aquilina et al. 2004). When analyzed by mass spectrometry, αB-crystallin samples show a very complex, poorly resolved MS spectrum formed by a superposition of peak series of different complex sizes (not shown). By using a tandem MS approach, however, a composite peak at m/z ~10,000 has been selected for CID (Figure 4.15), and the resulting complex product ion spectrum of singly, doubly, and triply stripped complexes illustrates how different species and charge states overlap (Figure 4.15). By summing up over the intensities of all charge states

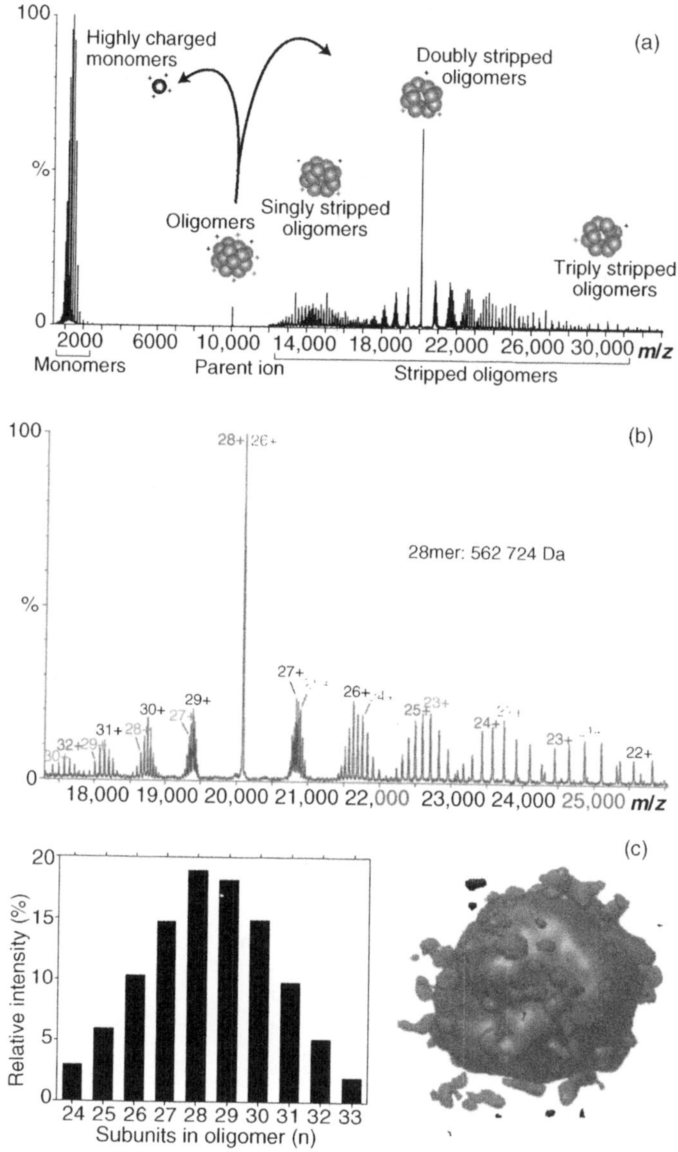

FIGURE 4.15. Nano-ESI tandem MS of heterodisperse bovine αB-crystallin: (a) MS/MS spectra of a composite oligomer peak near m/z 10,000 show highly charged monomers and also singly, doubly, and triply stripped oligomers (summation over a broad range of collision energies up to 200 V); (b) enlarged view of the section with doubly stripped oligomers. Charge states for the 26mer (in light gray) and 28mer (in dark gray) have been assigned. At m/z 20 090 all species with as many charges as subunits, $(n)^{n+}$, coincide in the spectrum. On either side of this peak overlapping charge state series can be distinguished for the different oligomer sizes which are present in the sample. (c) All peak intensities within a charge state series have been summed up to show the oligomer size distribution (n) of the precursor complexes. On the right side a cryo-EM structure of heterodisperse α-crystallin is shown.

of one oligomeric precursor, the size distribution for this αB-crystallin sample has been obtained (Figure 4.15). This size distribution has been found to change with increasing phosphorylation of the protein, demonstrating the effect of aging on lenticular tissue (Aquilina et al. 2004).

4.4.2. The Molecular Chaperonin GroEL

The chaperonins are a structurally conserved class of molecular chaperones that assist, in an ATP-dependent manner, in the efficient folding of a subset of newly synthesized and stress-denatured polypeptide chains. They are found in bacteria, archaea, and eukarya and form double-ring toroidal structures with seven to nine subunits of ~60 kDa per ring. Each ring encloses a central cavity for the binding of a nonnative protein, and cooperates with cofactors of the Cpn10 or HSP10 family (GroES in *E. coli*). We investigated the supramolecular structure of the chaperonin GroEL from *E. coli* from a near-native environment with the high-mass Q-TOF instrument described above (Sobott and Robinson 2004). Figure 4.16 shows a nano-ESI spectrum obtained from aqueous solution buffered to pH 7 using ammonium acetate. Around m/z 12,000 a group of well-resolved peaks is visible to which we assign charge states from +65 to +72 corresponding to a molecular mass of 804, 700 (±100) Da. Comparison with the mass for *E. coli* GroEL subunits as calculated from the amino acid sequence (57,198Da) shows good agreement with 14 times the monomer mass (800,770 Da). It is apparent that the dominant species in this spectrum represents an intact 14mer of GroEL, in accordance with the native double-barrel-shaped structure of the chaperonin (see inset in Figure 4.16). Virtually no dissociation or fragmentation products are visible at lower m/z under these

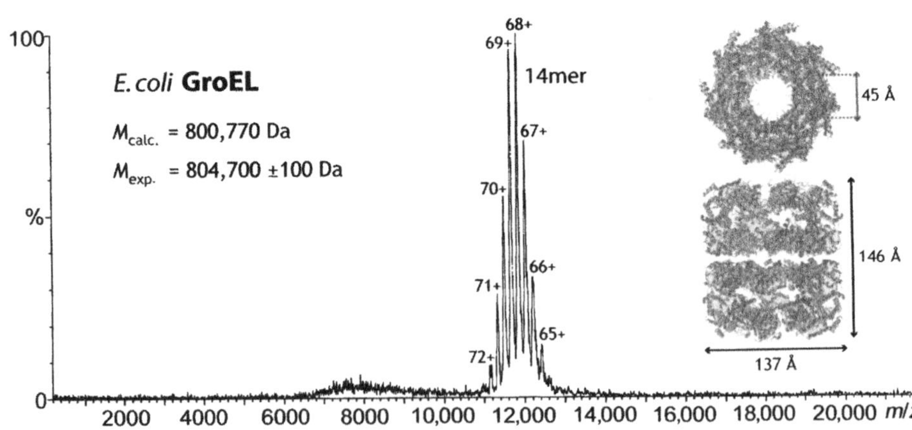

FIGURE 4.16. Nano-ESI spectrum of intact *E. coli* GroEL 14mer (Waters high-mass Q-TOF2; sample cone 150 V, extractor cone 50 V, vacuum pressures $p_2 = 1.1 \times 10^{-2}$ mbar, $p_3 = 8.3 \times 10^{-5}$ mbar and $p_4 = 7.1 \times 10^{-7}$ mbar, without gas in the collision cell. Inset: model of *E. coli* GroEL 14mer X-ray structure. [Reprinted from Sobott and Robinson (2004) with permission from Elsevier.]

conditions. The experimentally determined mass is, in this case, about 4000 Da higher than the calculated value, which amounts to 0.5% of the total mass. The observation of a mass higher than expected from the sum of the components is a common feature in ESI mass spectra of noncovalent complexes. This "noncovalent mass shift" is thought to be caused by molecules and ions from solution that remain attached to the complex in the gas phase—a hypothesis that is investigated further below.

Isolation of a single charge state of the GroEL 14mer at ~12,500 m/z in the quadrupole analyzer is demonstrated in Figure 4.17a (Sobott et al. 2005). For collision-induced dissociation (CID), ions are accelerated with voltage offsets up to 200 V into the gas-filled collision cell, where some of their translational energy is converted into internal energy in a multicollision process. In a similar manner to

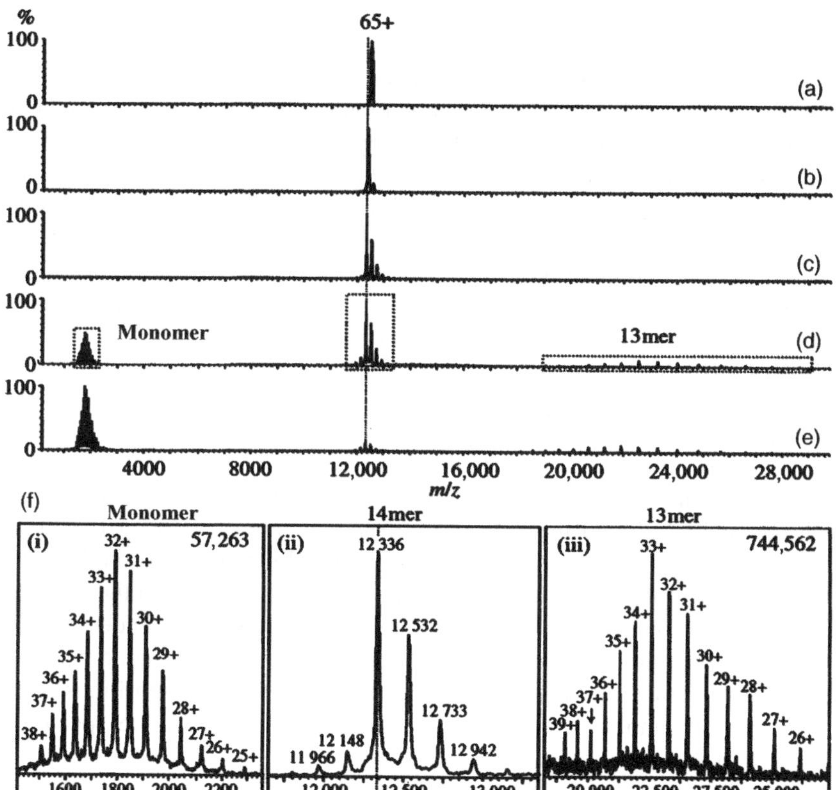

FIGURE 4.17. Tandem MS of the isolated 65+ charge state of GroEL 14mer at m/z 12,500 with 3×10^{-2} mbar argon in the collision cell. Panels (a)–(e): the ion acceleration voltage (collision energy) into the collision cell was varied between 4 V and the maximum value, 200 V. Three insets show enlarged sections of spectrum *d* for the monomer [f(i)], 14mer [f(ii)], and 13mer [f(iii)] signal as well as the mass calculated for the monomer and 13mer from the charge state series (upper right corner). [Reproduced from Sobott et al. (2005).]

desolvation and collisional cooling processes described above, heavier ions require more collisions than lighter ones to build up sufficient internal energy for dissociation to take place and for the resultant product ions to be collisionally cooled. Below 2.5×10^{-2}mbar argon pressure in the collision cell, no CID of GroEL 14mer occurs because of an insufficient number of collisions (data not shown). At 3.0×10^{-2}mbar and with an ion acceleration of up to 150 V (Figure 4.17b,c), no CID products are detected, but the isolated 65+ parent ion peak is narrower and at m/z lower than that in Figure 4.17a. At this intermediate voltage the complex is "heated" to the extent that attached ions and molecules are stripped without dissociating the protein complex. The noncovalent mass shift is significantly reduced to a value representing less than 0.0035% of the total mass. The observed peak coincides with the m/z calculated from the sum of the subunits for the +65 charge state, demonstrating the utility of this approach for accurate mass determination of noncovalent assemblies.

Under certain conditions of pressure and collision cell voltage additional peaks appear both higher and lower than the isolated peak, which coincide with charge states of the intact 14mer. Up to three positive, and interestingly also two negative, charges are lost in energetic collisions with argon gas [see Figure 4.17f(ii)]. While a charge state reduction is readily explained by loss of protons or other attached cations (presumably ammonium ions from the buffer or sodium impurities), the appearance of higher charge states is somewhat unexpected. A gain in charge by attachment of positive ions is energetically very unfavorable and not feasible in the gas phase; we therefore propose that loss of negative charge arises through the detachment of a counterion, presumably acetate from the buffer. This is not as unlikely as it might first appear, since the anion interacts with only a fraction of the overall positive charge of the complex (~+65), while remaining charges are effectively shielded by other protein subunits. If the collision cell voltage exceeds 180 V (Figure 4.17d,e), dissociation products appear in the spectra at low and very high m/z. Two product ion distributions are apparent, corresponding to highly charged GroEL monomers, centred on the +32 charge state, and stripped 13mer complexes centred around the +33 charge state [Figure 4.17f(i),(iii)].While each subunit carries on average 4.6 charges in the intact 14mer complex, on dissociation approximately half the total number of charges are concentrated on the monomeric form. Since most solution-derived ions and molecules are stripped under these high-energy conditions, the masses measured for the highly charged monomer and the residual 13mer formed in tandem MS can delineate subunit compositions of large, heterogeneous assemblies even if their charge states in the TOF-MS spectrum are not fully resolved.

An ion acceleration of 180 V into the collision cell (as in Figure 4.17d) corresponds to a kinetic energy of ~11.7 keV for the +65 charge state of the GroEL 14mer, a figure that may at first seem to be surprisingly high. Following the laws of physics for a collision between a heavy and a light object (conservation of momentum and total energy), however, only a small fraction of the translational energy is actually converted into internal energy of a high-mass ion. This small amount of ion activation in a single collision illustrates the requirement for high gas pressures in the collision cell—just as for collisional cooling in the ion guide. For a rough estimate of collision numbers, we consider a simple hard-sphere collision model and assume the GroEL 14mer to be a sphere with 16 nm diameter, corresponding to a collision cross section of approximately $2 \times 10^{-16}\,m^2$. It can be readily calculated that these ions experience tens of thousands of collisions during their passage through the argon-filled collision cell (length 0.185 m, $p_5 = 3 \times 10^{-2}$mbar), thereby building up high levels of internal energy. Such excited gas-phase protein complexes show two different dissociation pathways: (1) loss of attached molecules and ions at intermediate ion activation and (2) loss of an intact covalent subunit from the noncovalent complex at high activation. The stripping of small particles at intermediate tandem MS collision energies indicates that molecules and ions from the solution were previously attached to the complex. While the amount of additional mass which is found with gas-phase ions of large complexes depends on the desolvation conditions used, specifically, ion acceleration and gas pressures, a noncovalent mass shift persists even under optimum conditions for the preservation of near-native structures in the mass spectrometer.

Why should the dissociation pathway observed in CID differ from that in solution? The absence of bulk water is the key, since in solution binding is in competition with interactions with water molecules. Without surrounding water, there is no hydrophobic effect to drive the burial of hydrophobic surfaces, which arises from the entropy benefit of releasing ordered water molecules at a surface. Since this is a major driving force in protein folding, the native structures of proteins are likely to be more thermodynamically unstable in the gas phase compared to their solution counterparts. Dissociation, however, requires the breaking of electrostatic interactions within the complex, which are now relatively stronger than in solution since the alternative stabilization provided by hydrogen bonding to water is lost. Collisions must provide the necessary energy for breaking these interactions, by activation in the collision cell. It is intriguing to note that CID products from isolated ribosome ions lead to a bias toward the release of only proteins that were not in contact with the rRNA molecules; this is consistent with the higher proportion of charged electrostatic interactions expected in protein-RNA contacts, compared to protein–protein contacts (Hanson et al. 2003). CID will also favor dissociation of subunits with the smallest interface areas with other subunits, making it a means of probing the geometry of large complexes.

4.4.3. Mass Spectrometry Studies of the Ribosome

The ribosome is the macromolecular machinery responsible for manufacturing protein in the cell, with a mass of 2.5 MDa, and consists to about two-thirds of RNA and one-third of protein. Two subunits, the 50*S* and the 30*S*, make up the intact 70*S* particle. Ribosomes represent macromolecular structures that are asymmetric, often heterogeneous, and contain dynamic regions—properties that pose considerable challenges for modern-day structural biology techniques. Yet it is possible, under appropriate conditions, to investigate intact ribosomes in a mass spectrometer (Rostom et al. 2000; Ilag et al. 2005) (Figure 4.18). An elongated and highly flexible

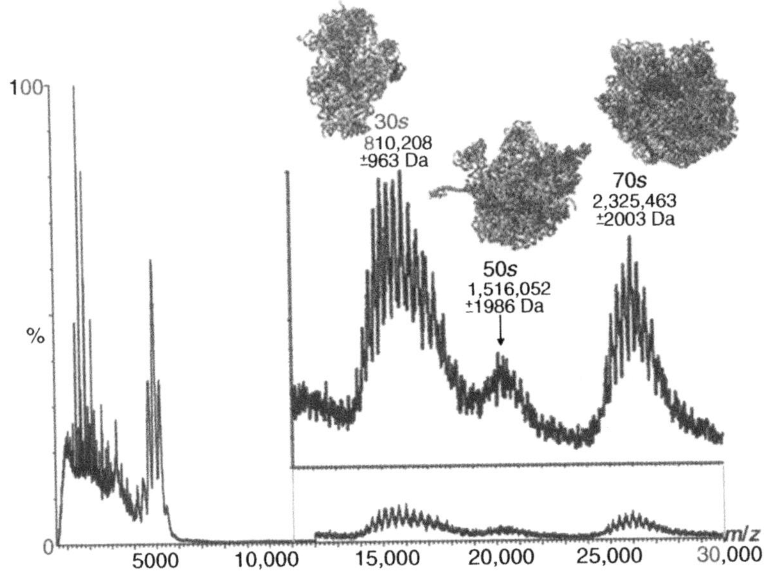

FIGURE 4.18. Intact ribosome from *Thermus thermophilus* as measured by nano-ESI MS in a high-mass QqTOF instrument. Apart from intact 70*S* particles, the two major subunits 30*S* and 50*S* are also being detected, next to another protein complex around *m/z* 5000 (96 kDa, most likely a component of the stalk complex).

region known as the "stalk" protrudes from the 50*S* subunit. This consists of at least one of a pair of L7/L12 dimers (L7 is the *N*-acetylated form of L12). These proteins are unique as they are the only ones present in multiple copies in the prokaryotic ribosome. Tandem MS of the intact 50S ribosomal subunit reveals, exclusively while attached to ribosomes, a phosphorylation of L12, the protein located in multiple copies at the tip of the stalk complex (Figure 4.19).

The studies of the ribosome show that by applying tandem MS approaches to these MDa particles, we can obtain precise information regarding the stoichiometry of associated complexes and identify a posttranslational modification on a protein within the complex.

4.5. CHALLENGES AND FUTURE DIRECTIONS

The development of electrospray by the group of John Fenn and others has led to the fact that very large complexes of biomacromolecules can now be investigated by mass spectrometry. In his words, "It has made it possible to let these elephants fly." Some challenges remain however. Cellular expression levels of many complexes are low, and crucial interactions may be fleeting. Noncovalent assemblies are often composed of multiple components, and the complexes may reorganize rapidly. Tandem affinity purification techniques are currently being developed that are

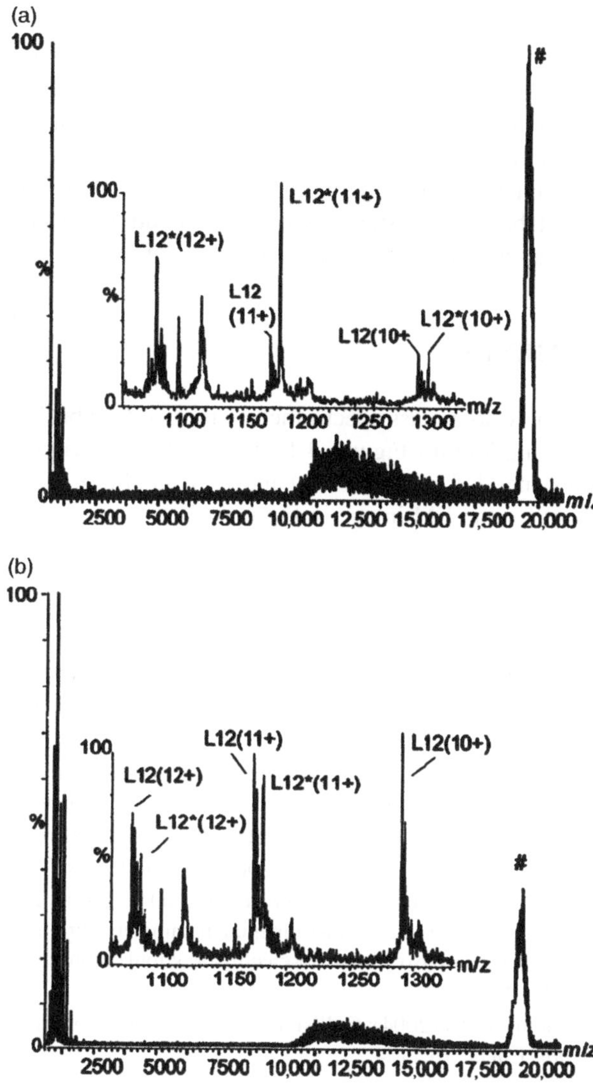

FIGURE 4.19. Tandem MS of the 50*S* subunit releases L12 protein and a phosphorylated form of L12: (a) tandem MS of the 50*S* subunit (precursor ion marked with #) releases L12 and a modified form, denoted L12*; (b) after treatment with a phosphatase in the presence of a kinase inhibitor, peaks assigned to unmodified L12 are more intense indicating that the modification is indeed a phosphorylation. [Reproduced with permission from Ilag et al. (2005). Copyright (2005) National Academy of Sciences, USA.]

amenable to mass spectrometric analysis and allow meeting some of these challenges.

In summary, the current focus on interaction proteomics as well as the ever-increasing size of biomolecular complexes now amenable to analysis by mass spectrometry have driven significant methodological advances in recent years. These together with the exciting possibilities afforded by monitoring real-time reactions that could be applied to processes such as virus assembly, subunit exchange and complex assembly more generally place mass spectrometry at the forefront of technologies for defining dynamic reactions in macromolecular complexes. The ability to define the stoichiometry and topological arrangement of subunits as well as to distinguish both stable and transient associations in complexes ensures that mass spectrometry will play an increasing role in structural proteomics.

ACKNOWLEDGMENTS

We are grateful to the whole mass spectrometry group at Cambridge University, in particular to Zhongping Yao for Figures 4.1 and 4.2, Helena Hernández for Figures 4.3 and 4.9, Justin Benesch for Figures 4.14 and 4.15, and Mark Paine (Queen Mary University, London) for the EM image in Figure 4.5.

REFERENCES

Aquilina JA, Benesch JLP, Ding LL, Yaron O, Horwitz J, Robinson CV (2004): Phosphorylation of αB-crystallin alters chaperone function through loss of dimeric substructure. *J. Biol. Chem.* **279**:28675.

Benesch JLP, Sobott F, Robinson CV (2003): Thermal dissociation of multimeric protein complexes by using nanoelectrospray mass spectrometry. *Anal. Chem.* **75**:2208.

Benesch JLP, Aquilina JA, Ruotolo BT, Sobott F, Robinson CV (2006): The dissociation of large multimeric protein complexes in the gas phase. *Chemistry & Biology.*

Benkestock K, Sundqvist G, Edlund PO, Roeraade J (2004): Influence of droplet size, capillary–cone distance and selected instrumental parameters for the analysis of noncovalent protein–ligand complexes by nano-electrospray ionization mass spectrometry. *J. Mass Spectrom.* **39**:1059.

Chernushevich IV, Thomson BA (2004): Collisional cooling of large ions in electrospray mass spectrometry. *Anal. Chem.* **76**:1754.

Cole RB (2000): Some tenets pertaining to electrospray ionization mass spectrometry. *J. Mass Spectrom.* **35**:763.

El-Faramawy A, Siu KWM, Thomson BA (2005): Efficiency of nano-electrospray ionization. *J. Am. Soc. Mass Spectrom.* **16**:1702.

Fenn JB, Mann M, Meng CK, Wong SF, Whitehouse CM (1989): Electrospray ionization for mass spectrometry of large biomolecules. *Science* **246**:64.

Fenn JB (1993): Ion formation from charged droplets: Roles of geometry, energy, and time. *J. Am. Soc. Mass Spectrom.* **4**:524.

Fenn JB (2000): Mass spectrometric implications of high-pressure ion sources. *Int. J. Mass Spectrom.* **200**:459.

Gavin AC, Bosche M, Krause R, Grandi P, Marzioch M, Bauer A, Schultz J, Rick JM, Michon AM, Cruciat CM, Remor M, Hofert C, Schelder M, Brajenovic M, Ruffner H, Merino A,

Klein K, Hudak M, Dickson D, Rudi T, Gnau V, Bauch A, Bastuck S, Huhse B, Leutwein C, Heurtier MA, Copley RR, Edelmann A, Querfurth E, Rybin V, Drewes G, Raida M, Bouwmeester T, Bork P, Seraphin B, Kuster B, Neubauer G, Superti-Furga G (2002): Functional organization of the yeast proteome by systematic analysis of protein complexes. *Nature* **415**:141.

Grandori R (2003): Electrospray-ionization mass spectrometry for protein conformational studies. *Curr. Org. Chem.* **7**:1589.

Gupta R, Hamdan SM, Dixon NE, Sheil MM, Beck JL (2004): Application of electrospray ionization mass spectrometry to study the hydrophobic interaction between the ε and θ and subunits of DNA polymerase III. *Protein Sci.* **13**:2878.

Hanson CL, Fucini P, Ilag LL, Nierhaus KH, Robinson CV (2003): Dissociation of intact *Escherichia coli* ribosomes in a mass spectrometer. *J. Biol. Chem.* **278**:1259.

Heck AJR, Van den Heuvel RHH (2004): Investigation of intact protein complexes by mass spectrometry. *Mass Spectrom. Rev.* **23**:368.

Ho Y, Gruhler A, Heilbut A, Bader GD, Moore L, Adams SL, Millar A, Taylor P, Bennett K, Boutilier K, Yang L, Wolting C, Donaldson I, Schandorff S, Shewnarane J, Vo M, Taggart J, Goudreault M, Muskat B, Alfarano C, Dewar D, Lin Z, Michalickova K, Willems AR, Sassi H, Nielsen PA, Rasmussen KJ, Andersen JR, Johansen LE, Hansen LH, Jespersen H, Podtelejnikov A, Nielsen E, Crawford J, Poulsen V, Sorensen BD, Matthiesen J, Hendrickson RC, Gleeson F, Pawson T, Moran MF, Durocher D, Mann M, Hogue CWV, Figeys D, Tyers M (2002): Systematic identification of protein complexes in saccharomyces cerevisiae by mass spectrometry. *Nature* **415**:180.

Ilag LL, Videler H, McKay AR, Sobott F, Fucini P, Nierhaus KH, Robinson CV (2005): Heptameric $(L12)_6$/L10 rather than canonical pentameric complexes are found by tandem MS of intact ribosomes from thermophilic bacteria. *Proc. Natl. Acad. Sci. USA* **102**:8192.

Jellen EE, Chappell AM, Ryzhov V (2002): Effects of size of noncovalent complexes on their stability during collision-induced dissociation. *Rapid Commun. Mass Spectrom.* **16**:1799.

Juraschek R, Dülcks T, Karas M (1999): Nanoelectrospray—more than just a minimized-flow electrospray ionization source. *J. Am. Soc. Mass Spectrom.* **10**:300.

Jurchen JC, Williams ER (2003): Origin of asymmetric charge partitioning in the dissociation of gas-phase protein homodimers. *J. Am. Chem. Soc.* **125**:2817.

Kaltashov IA, Eyles SJ (2002): Crossing the phase boundary to study protein dynamics and function: combination of amide hydrogen exchange in solution and ion fragmentation in the gas phase. *J. Mass Spectrom.* **37**:557.

Kaltashov IA, Mohimen A (2005): Estimates of protein surface areas in solution by electrospray ionization mass spectrometry. *Anal. Chem.* **77**:5370.

Karas M, Bahr U, Dülcks T (2000): Nano-electrospray ionization mass spectrometry: Addressing analytical problems beyond routine. *Fres. J. Anal. Chem.* **366**:669.

Kebarle P, Tang L (1993): From ions in solution to ions in the gas phase. *Anal. Chem.* **65**: 972.

Keetch CA, Hernández H, Sterling A, Baumert M, Allen MH, Robinson CV (2003): Use of a microchip device coupled with mass spectrometry for ligand screening of a multi-protein target. *Anal. Chem.* **75**:4937.

Lascoux D, Cravello L, Lemaire D, Forest E (2005): Increased pressure at the electrospray interface dramatically raises sensitivity in analysis of denaturated proteins. *Rapid Commun. Mass Spectrom.* **19**:1758.

Loo JA (2000): Electrospray ionization mass spectrometry: A technology for studying noncovalent macromolecular complexes. *Int. J. Mass Spectrom.* **200**:175.

Manil B, Ntamack GE, Lebius H, Huber BA, Duft D, Leisner T, Chandezon F, Guet C (2003): Charge emission and decay dynamics of highly charged clusters and micro-droplets. *Nucl. Instrum. Meth. Phys. Res. B* **205**:684.

McCammon MG, Scott DJ, Keetch CA, Greene LH, Purkey HE, Petrassi HM, Kelly JW, Robinson CV (2002): Screening transthyretin amyloid fibril inhibitors: Characterization of novel multiprotein, multiligand complexes by mass spectrometry. *Structure* **10**:851.

McCammon MG, Hernández H, Sobott F, Robinson CV (2004): Tandem mass spectrometry defines the stoichiometry and quaternary structural arrangement of tryptophan molecules in the multiprotein complex TRAP. *J. Am. Chem. Soc.* **126**:5950.

Reinders J, Lewandrowski U, Moebius J, Wagner Y, Sickmann A (2004): Challenges in mass spectrometry-based proteomics. *Proteomics* **4**:3686.

Rogniaux H, Sanglier S, Strupat K, Azza S, Roitel O, Ball V, Tritsch D, Branlant G, van Dorsselaer A (2001): MS as a novel approach to probe cooperativity in multimeric enzymatic systems. *Anal. Biochem.* **291**:48.

Rostom AA, Fucini P, Benjamin DR, Juenemann R, Nierhaus KH, Hartl FU, Dobson CM, Robinson CV (2000): Detection and selective dissociation of intact ribosomes in a mass spectrometer. *Proc. Natl. Acad. Sci. USA* **97**:5185.

Šamalikova M, Matečko I, Müller N, Grandori R (2004): Interpreting conformational effects in protein nano-ESI-MS spectra. *Anal. Bioanal. Chem.* **378**:1112.

Schmidt A, Bahr U, Karas M (2001): Influence of pressure in the first pumping stage on analyte desolvation and fragmentation in nano-ESI MS. *Anal. Chem.* **73**:6040.

Schmidt A, Karas M, Dülcks T (2003): Effect of different solution flow rates on analyte ion signals in nano-ESI MS, or: When does ESI turn into nano-ESI? *J. Am. Soc. Mass Spectrom.* **14**:492.

Sobott F, Hernández H, McCammon MG, Tito MA, Robinson CV (2002a): A tandem mass spectrometer for improved transmission and analysis of large macromolecular assemblies. *Anal. Chem.* **74**:1402.

Sobott F, Benesch JLP, Vierling E, Robinson CV (2002b): Subunit exchange of multimeric protein complexes. Real-time monitoring of subunit exchange between small heat shock proteins using electrospray-mass spectrometry. *J. Biol. Chem.* **277**:38921.

Sobott F, Robinson CV (2002): Protein complexes gain momentum. *Curr. Opin. Struct. Biol.* **12**: 729.

Sobott F, Robinson CV (2004): Characterising electrosprayed biomolecules using tandem-MS—the noncovalent GroEL chaperonin assembly. *Int. J. Mass Spectrom.* **236**:25.

Sobott F, McCammon MG, Hernández H, Robinson CV (2005): The flight of macromolecular complexes in a mass spectrometer. *Phil. Trans. Roy. Soc A* **363**:379.

Tahallah N, Pinkse M, Maier CS, Heck AJR (2001): The effect of the source pressure on the abundance of ions of noncovalent protein assemblies in an electrospray ionization orthogonal time-of-flight instrument. *Rapid Commun. Mass Spectrom.* **15**:596.

Thomson BA (1997): Declustering and fragmentation of protein ions from an electrospray ion source. *J. Am. Soc. Mass Spectrom.* **8**:1053.

Tito MA, Tars K, Valegard K, Hajdu J, Robinson CV (2000): Electrospray time-of-flight mass spectrometry of the intact MS2 virus capsid. *J. Am. Chem. Soc.* **122**:3550.

Van Berkel GJ, Asano KG, Schnier PD (2001): Electrochemical processes in a wire-in-a-capillary bulk-loaded, nano-electrospray emitter. *J. Am. Soc. Mass Spectrom.* **12**:853.

Van den Heuvel RHH, Heck AJR (2004): Native protein mass spectrometry: From intact oligomers to functional machineries. *Curr. Opin. Chem. Biol.* **8**:519.

Wilm M, Mann M (1996): Analytical properties of the nanoelectrospray ion source. *Anal. Chem.* **68**:1.

5

PROTEIN STRUCTURE AND FOLDING IN THE GAS PHASE: UBIQUITIN AND CYTOCHROME *c*

KATHRIN BREUKER

Institute of Organic Chemistry and Center for Molecular Biosciences Innsbruck (CMBI)
University of Innsbruck
Innsbruck, Austria

Principles of Mass Spectrometry Applied to Biomolecules, edited by Julia Laskin and Chava Lifshitz

5.1. INTRODUCTION

The fundamental aspects of protein structure and folding in solution have been studied for decades (Cooper 1999; Dinner et al. 2000; Dobson et al. 1998; Franks 2002), yet numerous questions are still unanswered, most of which relate to the effect of a protein's environment on structure and stability. In solution-phase experiments, however, it is very difficult, if not impossible, to separate the contribution of solvent from intrinsic protein stability. An altogether different approach in addressing the effect of hydration is to completely remove any solvent and study the structure and energetics of the protein by itself. This raises a number of new questions:

What do protein structures look like in the complete absence of solvent, and do single minimum energy gas-phase structures exist?

Can proteins in the gas phase fold and unfold, and what are the energetics of gas-phase folding processes?

If folding is possible, what is the major driving force, and on what timescale does it occur?

Most importantly, how can we piece together the solution and gas-phase data to get a clearer picture of the protein folding process?

This chapter discusses gas-phase data on ubiquitin and cytochrome *c* in view of these questions.

A practical motivation for understanding gaseous protein structures is their analysis by top-down mass spectrometry (Ge et al. 2002; Horn et al. 2000; Kelleher et al. 1999; Reid and McLuckey 2002; Strader et al. 2004; Sze et al. 2002, 2003; Zhai et al. 2005). In the top-down approach, the molecular weight measurement of gaseous protein ions from electrospray ionization (ESI) (Fenn et al. 1989; Smith et al. 1990) in a mass spectrometer is directly followed by protein backbone dissociation and fragment ion mass analysis. Especially for larger proteins, however, the extent of sequence information in top-down experiments can be considerably limited by the formation of highly stable gaseous ion conformations that withstand the dissociation step. Preventing the formation of such structures, or finding ways to manipulate them, would greatly increase the range of proteins that can be studied by top-down mass spectrometry.

5.2. PROTEIN FOLDING—STRUCTURE AND ENERGETICS

A biologically active protein usually shows extensive average conformational homogeneity, and is said to have a "native" or "folded" structure. Consequently, any other structure, whether compact or not, can be considered "unfolded" (Cooper 1999), and protein "folding" in solution commonly refers to structural transitions involving the native structure. For proteins in the gas phase, a reference to biological activity obviously is seldom adequate, although evidence for enzymatic activity in the absence of solvent has been reported (He et al. 1999). As will be shown later, structural transitions in gaseous protein ions can indeed be observed, but these do not necessarily involve native structures. For lack of alternative expressions, terms such as "folding" and "partially folded" will nevertheless be used here to also describe structural transitions of non-native structures.

Native protein structures are stabilized by van der Waals forces, electrostatic interactions such as hydrogen bonds and saltbridges, and hydrophobic interactions (Dill 1990). Although these interactions are understood qualitatively, their relative contributions to the stability of a native protein structure are still unclear, and it is not yet possible to reliably predict the native structure for a given protein sequence. This comes, in large part, from the vast complexity of protein–water interactions (Cooper 1999; Franks 2002; Klotz 1993), and one could be tempted to say that the "protein folding problem" is in fact a "protein hydration problem." In an attempt to reduce the complexity of interactions, researchers have begun to study peptide and protein structures in the complete absence of solvent (Breuker 2004; Breuker et al. 2002; Hoaglund-Hyzer et al. 1999; Jarrold 2000; McLafferty et al. 1998; Oh et al. 2002), and even follow their structural changes on the incremental addition of water molecules (Jarrold 1999).

The most obvious difference between hydrated and gaseous proteins is that for the latter, there is no hydrophobic effect (Kauzmann 1959; Southall et al. 2002). As will be shown later, regions stabilized by hydrophobic bonding in native cytochrome *c* are the first to unfold after water removal (Breuker and McLafferty 2005). The loss of hydrophobic bonding can also reduce the stability of α-helical structures, which were suggested to form in a hydrophobic collapse constrained by the requirement that polar groups be either exposed to solvent or form hydrogen bonds (Yang and Honig 1995a). However, peptide hydrogen bonds are generally strengthened in the gas phase (Sheu et al. 2003). Helix stability in the gas phase also depends critically on the charge location, with a *C*-terminal positive charge stabilizing the helix via charge–helix macrodipole interactions (Hudgins and Jarrold 1999; Hudgins et al. 1998; Kohtani et al. 2004). Whether or not β sheets (Yang and Honig 1995b) are elements of stable gaseous protein ion structures is unclear at this point. Molecular modeling could, in principle, tell us more about gaseous protein structures, but a major problem lies in assigning the charge sites of the gaseous ions.

Gaseous as well as solvated proteins spontaneously fold if the free energy of folding ΔG is favorable. Both enthalpic (ΔH) and entropic (ΔS) contributions of all covalent bonds and noncovalent interactions contribute to the free energy of folding ($\Delta G = \Delta H - T\Delta S$), and in solution, the noncovalent interactions include interactions with solvent water. If the solution and gas-phase structures were the same, the effect of hydration on protein stability could easily be calculated from the respective folding free energies. However, as will be shown below, there is no evidence for a native solution structure being stable in the gas phase. In solution (Cooper et al. 2001) as well as in the gas phase (Breuker et al. 2002; Iavarone and Parks 2005), the enthalpic and entropic contributions often act in a compensatory fashion, resulting in relatively small ΔG values and marginal stability of the protein structure. The principal opposing force to protein folding is entropic, simply because the number of possible conformations of a polypeptide chain is much larger in the unfolded state (Dill 1990). Entropy is also the major driving force for the dissociation of proteins and protein–ligand complexes in the gas phase (Laskin and Futrell 2003).

5.3. PROTEINS

The small proteins ubiquitin and cytochrome *c* have been studied extensively in the gas phase and in solution. Their main features relating to structure and folding in solution are summarized in Sections 5.3.1 and 5.3.2.

5.3.1. Ubiquitin

Mammalian ubiquitin is a 76-residue protein with an extensive intramolecular hydrogen-bond network in the native state (Cornilescu et al. 1998; Vijay-Kumar et al. 1987) (Figure 5.1); all except 18 residues (L8, T9, T12, E16, S20, D39, A46, G47, Q49, D52, G53, K63, T66, L71, L73, R74, G75, G76) are involved in hydrogen bonding. The native structure is also stabilized by hydrophobic interactions, with 17 residues (I3, V5, I13, L15, V17, I23, V26, I30, I36, L43, F45, L50, L56, Y59, I61, L67, L69) having more than four hydrophobic contacts with other residues each (Figure 5.1, black circles). The 18 residues without hydrogen bonding are located on the protein surface, whereas the 17 residues with more than four hydrophobic contacts constitute the protein core. The 12 basic (K6, K11, K27, K29, K33, R42, K48, R54, K63, H68, R72, R74) and 11 acidic (E16, E18, D21, E24, D32, E34, D39, E51, D52, D58, E64) residues are surface-exposed and provide additional stabilization through salt link interactions (Loladze et al. 1999). Ubiquitin is no different from most other single-domain proteins, with apparent two-state folding kinetics (Krantz and Sosnick 2000) in that its *N*- and *C*-terminal secondary structure elements are in contact in the native state (Krishna and Englander 2005). Measured two-state folding kinetics do not, however, exclude the possibility of intermediate states, but indicate that the first barrier toward the native state is rate-limiting. The secondary structure of ubiquitin is 16% α-helical and 30% β-sheet, with a single α helix and a mixed (parallel and antiparallel) β sheet consisting of five strands (Cornilescu et al. 1998).

Unfolding of the native ubiquitin structure can be achieved by increasing (Ibarra-Molero et al. 1999a; Wintrode et al. 1994) or decreasing (Ibarra-Molero et al. 1999b) the solution temperature, decreasing the pH (Wintrode et al. 1994), adding guanidinium hydrochloride (Ibarra-Molero et al. 1999a,b; Jourdan and Searle 2001), or organic solvents such as methanol (Brutscher et al. 1997; Jourdan and Searle 2001). In particular, ubiquitin assumes a partially folded "*A* state" in 60% methanol/40% water at pH 2 with a largely conserved native structure in the *N*-terminal half and a nonnative extended helix structure in the *C*-terminal half (Brutscher et al. 1997). On the other hand, stabilization of the native structure has been observed in aqueous salt solutions, an effect that was interpreted in terms of anion binding and

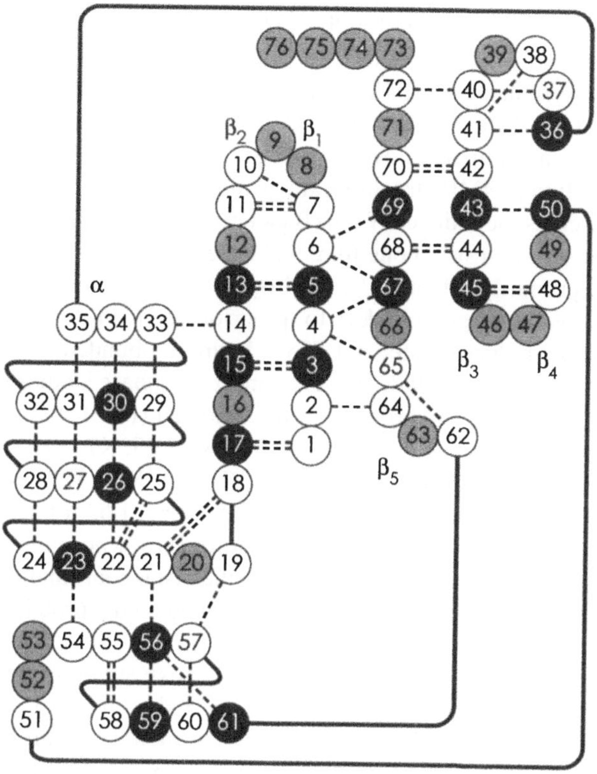

FIGURE 5.1. Schematics of the hydrogen bonding network in native human ubiquitin [conformer 1 from PDB (Berman et al. 2000) entry 1D3Z, reference (Cornilescu et al. 1998)]: hydrogen bonds, dashed lines; residues with more than four hydrophobic contacts, black circles; residues without hydrogen bonding, gray circles. Solid lines connect sequence neighbors for which a side-by-side representation was not possible.

long-range electrostatic interactions (Makhatadze et al. 1998). Destabilizing factors can act in a cumulative fashion; the thermal stability of native ubiquitin decreases when the solution pH decreases from 4 to 2, with unfolding transition temperatures of ~90°C at pH 4 and ~60°C at pH 2 (Wintrode et al. 1994). Solution NMR measurements in the temperature range from 5 to 65°C identified regions of particular thermal lability and stability in native ubiquitin at pH 6.5, and found that the end of β-sheet β_1/β_5 at hydrogen bond Q2/E64 is the most thermolabile, whereas the hydrogen bond I3/L15 between β_1 and β_2 is slightly stabilized at elevated temperatures (Cordier and Grzesiek 2002).

5.3.2. Cytochrome *c*

Eukaryotic cytochromes *c* are single-electron-transfer proteins with highly conserved amino acid sequences and very similar three-dimensional structures in their native states (Banci et al. 1999b), even though their thermodynamic stabilities can vary substantially (Filosa and English 2000; Travaglini-Allocatelli et al. 2004). Most gas-phase studies were carried out with equine and bovine cytochrome *c*, whose amino acid sequences (104 residues) differ only at positions 47, 60, and 89 (equine—T47, K60, T89; bovine—S47, G60, G89). A heme group is covalently bound to the protein chain via thioether linkages at C14 and C17. For native equine cytochrome *c*, solution NMR structures are available for the oxidized (Fe^{III}) (Banci et al. 1997) as well as the reduced (Fe^{II}) (Banci et al. 1999a) state. The Fe^{III} and Fe^{II} forms have a very similar backbone fold, and differ mostly in the reorientation of a few sidechains, primarily affecting the hydrogen-bonding network of the heme. In particular, N52 is hydrogen-bonded to proprionate 7 in the oxidized but to proprionate 6 in the reduced state, and T78 is hydrogen-bonded to M80 only in the reduced state (Banci et al. 1999a). However, most gas-phase studies discussed later in this chapter employed cytochrome *c* in the oxidized (Fe^{III}) state.

In native equine (Fe^{III})cytochrome *c*, the 24 basic (K5, K7, K8, K13, H18, K22, K25, H26, K27, H33, R38, K39, K53, K55, K60, K72, K73, K79, K86, K87, K88, R91, K99, K100) and 12 acidic (D2, E4, E21, D50, E61, E62, E66, E69, E90, E92, D93, E104) residues are surface exposed, except for H18, whose imidazole nitrogen coordinates the heme iron on the proximal side; the distal coordination partner is the M80 sulfur. Whereas the basic residues are distributed on the protein surface almost evenly, the acidic residues pool on the side opposite to the exposed heme edge region, resulting in a large dipole moment of the native structure that is thought to assist the protein in orienting itself toward its redox partners (Koppenol and Margoliash 1982). The heme noncovalent bonding includes hydrophobic contacts with 18 residues (F10, K13, T28, G29, P30, L32, L35, F46, Y48, L64, Y67, L68, K79, I81, F82, I85, L94, L98), hydrogen bonding of the heme proprionates with 4 residues (T49, N52, W59, T78), and coordinative bonding with H18 and M80. Native equine (Fe^{III})cytochrome *c* has five α helices (34% helix content) but no β-sheet structure (Banci et al. 1997).

Unfolding of the native cytochrome *c* structure can be achieved by adding organic solvents to the protein solution (Kamatari et al. 1996; Kaminsky and Davison 1969), increasing the temperature, adding denaturants such as urea or guanidinium hydrochloride (Hagihara et al. 1994), decreasing the solution pH (Goto et al. 1993; Kamatari et al. 1996), or increasing it (Hoang et al. 2002). In hydrogen exchange experiments designed to probe intermediate equilibrium structures, five cooperative folding–unfolding units ("foldons") in equine cytochrome *c* have been identified (Krishna et al. 2003; Maity et al. 2004). Their free energies of unfolding, relative to the native state, lie between 21 and 54 kJ/mol, with the crossed terminal helices (residues 1–18 and 86–104) and the 40–57 segment constituting the most and least stable regions, respectively (Maity et al. 2004).

5.4. DESORPTION/IONIZATION OF PROTEINS

Until the late 1980s, it was not possible to study gaseous proteins because of their negligible volatility. Thermal desorption requires temperatures at which a protein degrades before it can enter the gas phase, and more recent desorption methods such as fast-atom bombardment (FAB) (Barber et al. 1981; Rinehart et al. 1981; Surman and Vickerman 1981) or field desorption (Beckey and Schulten 1975) also turned out to be unsuitable for the intact desorption of larger proteins. The situation changed dramatically with the introduction of the "soft" desorption/ionization methods, matrix-assisted laser desorption/ionization (MALDI) (Karas et al. 1987, 1989; Karas and Hillenkamp 1988) and electrospray ionization (ESI) (Fenn et al. 1989). Although MALDI is capable of transferring very large protein ions into the gas phase, the matrix crystal is quite an artificial environment for a protein, unless, perhaps, frozen water is used as a matrix (Berkenkamp et al. 1996). To be able to at least control protein conformation in the condensed phase, most studies on gaseous protein ion conformations use ESI, which directly transfers proteins from solution into the gas phase. Impressive examples include the desorption/ionization by ESI of ribosomes (Rostom et al. 2000) and other large biomolecular complexes (Sanglier et al. 2003; van Duijn et al. 2005), and even viruses (Bothner and Siuzdak 2004). All data discussed below come from ESI experiments, and most of them are on gaseous protein ions carrying a positive net charge.

5.5. METHODS AND INSTRUMENTATION FOR PROBING GASEOUS PROTEIN STRUCTURE AND ENERGETICS

In the condensed phase, detailed structural information can be obtained from X-ray crystallography or nuclear magnetic resonance (NMR) experiments. However, proteins in the gas phase are not amenable to X-ray analysis, and gas-phase NMR spectroscopy is still limited to rather small molecules of sufficient volatility (Jameson 1991). Structural information on gaseous protein ions is thus obtained by other methods, most of which have been developed or adapted for the structural analysis of protein ions only very recently (as of 2006).

5.5.1. Ion Mobility Measurements

The topic of ion mobility spectrometry for the determination of collision cross sections of protein ions has been excellently reviewed (Clemmer and Jarrold 1997; Hoaglund-Hyzer et al. 1999). Briefly, what is measured in an ion mobility instrument is the time it takes ions to travel through a drift tube filled with buffer gas in the presence of an electric field. Compact and extended structures of protein ions can be separated according to their mobilities—the extended conformers have longer drift times than do the compact conformers because the latter experience fewer decelerating collisions with the buffer gas. It should be noted that when protein ions enter the drift tube, part of their kinetic energy may be converted into internal energy by inelastic collisions. This can cause conformational changes, which is why very low injection energies were used in some experiments. Moreover, it is believed that the conformers observed with high injection energies reflect the most stable gas-phase structures because these ions have unfolded and then refolded on entering the drift tube, thereby overcoming possible barriers in the folding energy landscape (Shelimov and Jarrold 1996).

5.5.2. Gas-Phase H/D Exchange

Gas-phase H/D exchange of protein ions can be performed with a variety of deuterated reagents, including D_2O, CH_3OD, and ND_3. Five different mechanisms for the exchange reaction of protonated glycine oligomers with different reagents have been suggested (Campbell et al. 1995), and experimental data in combination with ab initio calculations have provided convincing evidence that exchange with D_2O proceeds via a "relay" mechanism (Wyttenbach and Bowers 1998). However, Beauchamp pointed out that "It is difficult to draw any general conclusions about the behavior of ND_3 and D_2O as exchange reagents," and that the H/D kinetics may not directly reflect ion structures (Cox et al. 2004). Even so, H/D exchange can separate different conformers of gaseous protein ions (Freitas et al. 1999; McLafferty et al. 1998; Suckau et al. 1993; Winger et al. 1992; Wood et al. 1995). For H/D exchange experiments in a Fourier transform mass spectrometer (FTMS, base pressure $<10^{-9}$ mbar), the ions of interest are first isolated by application of radiofrequency (RF) waveforms. The exchange reagent is then introduced via a leak valve, kept at constant pressure ($\sim 10^{-7}$ mbar) during the exchange period, and pumped away prior to ion detection. Exchange rates are obtained by variation of the exchange time, typically between a few seconds and 2 h (Freitas et al. 1999; McLafferty et al. 1998; Oh et al. 2002; Wood et al. 1995). For H/D exchange in an ion mobility instrument, the helium buffer gas (partial pressure $\sim$2 mbar) is doped with exchange reagent; typical ion residence times in the drift tube are between $\sim$1.5 and 30 ms. Exchange rates are obtained by varying the partial pressure of the deuterating gas up to $\sim$0.7 mbar (Valentine and Clemmer 1997, 2002).

5.5.3. Electron Capture Dissociation

Electron capture dissociation (ECD) (Zubarev et al. 1998, 1999, 2000) is the topic Chapter 13, and only the aspects relating to its capability of conformational probing are briefly outlined here. ECD cleaves the protein backbone N—C_α bond to yield complementary $c/z^\bullet$ fragment ion pairs ($\sim$90%); in an alternative pathway $a^\bullet$ and y fragment ions ($\sim$10%) are formed. It was postulated that the energy available from charge recombination ($\sim$6 eV) produces localized excitation far in excess of that required for backbone dissociation, and that reaction occurs before this energy is randomized over the thousands of vibrational degrees of freedom of the protein ion (Breuker et al. 2004). For proteins as large as ubiquitin and cytochrome c, the excess energy (<3 eV) was insufficient to cause thermal unfolding of the gaseous ions, so that ECD effects only backbone cleavage but causes no significant conformational change (Breuker et al. 2002). Thus the ECD spectrum of $(M + nH)^{n+}$ protein ions will show $c/z^\bullet$ (or $a^\bullet/y$) fragment ions only if they are not held together by noncovalent bonding; the unseparated fragments will instead appear as reduced molecular ions, $(M + nH)^{(n-1)+\bullet}$ (Breuker et al. 2002; Horn et al. 2001; Zubarev et al. 2000). In a typical ECD experiment for probing the higher-order protein structure, ions formed by ESI are transferred through differentially pumped vacuum stages into the trapped ion cell ($<10^{-9}$ mbar) of a Fourier transform mass spectrometer (FTMS) where the ion population is allowed to thermally equilibrate for ~40 s with the cell and vacuum chamber via blackbody infrared radiation (Price et al. 1996) prior to irradiation with low energy (<0.5 eV) electrons (Breuker et al. 2002; Horn et al. 2001).

A related phenomenon is native electron capture dissociation (NECD) (Breuker and McLafferty 2003), which gives detailed structural information on protein ions in the transition from solution to gas phase (Breuker and McLafferty 2005); see Section 5.6 (of this chapter) for details.

5.5.4. Infrared Photodissociation Spectroscopy

Unlike conventional infrared spectroscopy, infrared photodissociation spectroscopy (IRPDS) is a type of "action spectroscopy," meaning that the effect of infrared photon absorption rather than the absorption itself is monitored. This was realized for gaseous ubiquitin ions stored in a FTMS instrument by first breaking the protein backbone with ECD, and then monitoring the dissociation of the reduced molecular ions, $(M + nH)^{(n-1)+\bullet}$, into $c/z^{\bullet}$ fragments for different wavelengths emitted from an optical parametric oscillator (OPO) laser (Oh et al. 2002). In another FTMS-based approach, the low-energy potassium detachment dissociation reaction, $(M + nH + K)^{(n+1)+} \rightarrow (M + nH)^{n+} + K^{+}$, served to obtain infrared spectra of gaseous cytochrome *c* ions using IR irradiation from a free-electron laser (FEL) (Oomens et al. 2005). Yet another experimental strategy uses triflic acid (CF_3SO_3H) in the electrospray solution, which forms abundant protein adducts $(M + nH + CF_3SO_3H)^{n+}$. Protein IR spectra are obtained by monitoring the dissociation reaction $(M + nH + CF_3SO_3H)^{n+} \rightarrow (M + nH)^{n+} + CF_3SO_3H$ on IR radiation emitted from an OPO laser (Lin et al. 2005).

5.6. FROM SOLUTION TO GAS PHASE

Shortly after the introduction of ESI of large biomolecules (Fenn et al. 1989), it was realized that an important factor in determining the protein charge distribution in a positive-ion m/z spectrum is the protein conformation in solution. ESI of solutions in which the protein has a compact conformation ("native ESI") gives narrow distributions and low average charge values, whereas ESI of solutions in which the protein is unfolded gives broad distributions centered around high positive charge values (Chowdhury et al. 1990; Loo et al. 1990; Mirza et al. 1993; Kaltashov and Eyles 2002). For example, ESI of aqueous ubiquitin solutions with 3% methanol produces mostly +7 molecular ions in the pH range 2.5–8.5 (Konermann and Douglas 1998a); increasing the methanol content to 60% at pH 2 (Babu et al. 2001) or heating an aqueous pH 2.8 solution to ~90°C (Mirza et al. 1993) gives broad charge distributions centered around +10. This correlation between solution structure and ESI charge state distributions can be used to study protein folding in solution. For instance, by employing a continuous-flow mixing apparatus directly coupled to an ESI source, the kinetics of ubiquitin folding was monitored with millisecond time resolution (Wilson and Konermann 2003).

However, the charge state distributions in the negative-ion ESI spectra of ubiquitin and cytochrome *c* did not show a clear correlation with solution conformation (Konermann and Douglas 1998b). The positive- and negative-ion ESI spectra of a solution of native cytochrome *c* (97% H_2O/3% MeOH, pH 6.4) showed narrow distributions centered around +7.5 and −5.5, respectively. A solution of acid-unfolded cytochrome *c* (97% H_2O/3% MeOH, pH 2.4) showed a broad distribution centered around +16 in the positive-ion ESI spectrum, whereas its negative-ion spectrum was essentially the same as for the pH 6.4 solution. Similar results were obtained with ubiquitin, for which both acid and base unfolding was reflected in the positive-ion but not the negative-ion ESI spectra (Konermann and Douglas 1998b). A possible explanation for this is that positive- and negative-ion charge state distributions monitor different structural aspects of proteins in solution (Konermann and Douglas 1998b), although the underlying mechanism is still unclear. It should be noted that in the abovementioned study, the positive-ion ESI spectra of cytochrome *c* showing a correlation with solution structure were obtained under "right-way-around" conditions, that is, positive-ion ESI of pH 2.4 and 6.4 solutions in which cytochrome *c* (p*I* > 10) carries a positive net charge. In contrast, no correlation was found for the spectra obtained under "wrong-way-around" conditions, that is, negative-ion ESI of the same pH 2.4 and 6.4 solutions (Konermann and Douglas 1998b). McLuckey and coworkers showed that the ion yield in "wrong-way-around" ESI (Kelly et al. 1992) is at least one order of magnitude lower than in "right-way-around" ESI (Pan et al. 2004), indicating the possibility that different ionization mechanisms dominate "right-way-around" and "wrong-way-round" ESI. Thus an alternative explanation for the observed difference in charge state response to solution conformation in the positive- and negative-ion ESI spectra of cytochrome *c* could be that only "right-way-around," but not "wrong-way-around," ESI reflects the protein conformation in solution. The ubiquitin data are, however, not suitable for testing this hypothesis. The ESI spectra of folded ubiquitin were obtained at pH 7.2 (Konermann and Douglas 1998b), close enough to its isoelectric point of 6.7 (Wilkinson et al. 1980) that neither "right-way-around" nor "wrong-way-around" conditions clearly apply.

Although changes in the protein conformation in solution can alter the charge state distributions in the ESI mass spectra, the absolute charge values in solution and gas phase can be quite different. For example, ubiquitin (p*I* 6.7) (Wilkinson et al. 1980) in an aqueous pH ~6.7 solution has a net charge of zero, but electrospraying this solution in positive-ion mode gives mostly +6 and +7 molecular ions. Moreover, solution additives with no discernible effect on solution conformation can substantially alter the molecular ion charge state distributions in ESI. Replacing ammonium acetate (2 mM) by triethylammonium acetate in aqueous ubiquitin solutions (pH ~6.5) decreases the average charge values from +6 to +4 (Verkerk et al. 2003), and the addition of 20 mM 1,5-diazabicyclo[4.3.0]non-5-ene to aqueous lysozyme or myoglobin solutions while maintaining pH 7 decreases the average charge values from +10 to +3.5 without changing the near-UV CD spectra of the protein solutions (Catalina et al. 2005). These data have been rationalized on the basis of the charge residue model (CRM) (Dole et al. 1968) for protein ion generation in ESI (Verkerk et al. 2003; Catalina et al. 2005). In the CRM, it is assumed that the initially formed droplets on evaporation of the solvent become electrically unstable and break down into smaller drops until droplets containing only one macromolecule per drop result, and that further evaporation of the solvent results in the formation of electrically charged intact gas-phase macromolecules (Dole et al. 1968). In this scenario, electrosprayed biomolecules remain in their solution environment until the very last stages of desolvation, and their solution structure is likely to be maintained at least until then. Furthermore, Mirza and Chait (1997) concluded in a study on the effect of capillary temperature on charge state distributions that thermal denaturation of proteins within the electrospray droplets is an unlikely process. The ion evaporation model (IEM) instead proposes direct emission of ions from the charged droplet surface (Iribarne and Thomson 1976). Although accumulating evidence suggests that CRM is the dominating process in electrospray ionization of proteins (de la Mora 2000; Felitsyn et al. 2002; Iavarone and Williams 2003; Kaltashov and Mohimen 2005; Nesatyy and Suter 2004; Pan et al. 2004; Verkerk and Kebarle 2005; Verkerk et al. 2003), a discussion of the ionization mechanisms in ESI (Constantopoulos et al. 2000; de la Mora et al. 2000; Kebarle 2000; Kebarle and Peschke 2000; Rohner et al. 2004) is beyond the scope of this chapter.

ESI produces gas-phase ions directly from solution, and the experimental conditions can be adjusted such as to minimize collisional or thermal activation of the proteins during the phase transition. However, it cannot be expected that complete dehydration leave a given solution conformation entirely unaffected (Breuker 2004). Site-specific information on the structural changes of a native protein during transfer from solution to gas phase comes from native electron capture dissociation (NECD) (Breuker and McLafferty 2003, 2005). In NECD, a noncovalently bound cytochrome *c* homodimer ion from ESI of aqueous (Fe^{III})cytochrome *c* solutions (~75 μM, pH 3.5–6.5) enters the Fourier transform mass spectrometer (FTMS) via a heated capillary for desolvation, where one of its monomers partially unfolds. This causes proton transfer from the compact monomer to the newly exposed sites of the partially unfolded monomer, and induces a substantial charge asymmetry (Jurchen and Williams 2003; Breuker 2006). In turn, this prompts intermolecular transfer of two electrons to the heme of the partially unfolded monomer, one reducing the heme iron and the other causing protein backbone cleavage (NECD) next to residues in contact with the heme (Breuker and McLafferty 2003, 2005). Thus the NECD fragment ions from a given backbone cleavage site indicate intact noncovalent bonding between the residue next to this cleavage site and the heme, whereas the "missing" cleavages identify regions where native structure is lost on transfer into the gas phase.

In native horse heart (Fe^{III})cytochrome *c* (Banci et al. 1997), 25 out of the total 104 residues are in contact with the heme (F10, K13, H18, T28, G29, P30, L32, L35,

T40, F46, Y48, T49, N52, W59, L64, Y67, L68, T78, K79, M80, I81, F82, I85, L94, L98), so that any loss of native structure can be monitored by NECD with nearly 25% sequence coverage. NECD of equine (Fe^{III})cytochrome *c* identified two different dimer structures A and B (Breuker and McLafferty 2005), with the NECD spectrum of structure B corresponding to the asymmetric dimer unit of horse heart (Fe^{III})cytochrome *c* crystals at low ionic strength (Sanishvili et al. 1995). The solutions containing dimers A or B were prepared in the same way, except that A was stored overnight and B was stored for 3 months at 4°C (Breuker and McLafferty 2005). It was found that more protein–heme interactions in the unfolding monomer are retained for dimer B than A, which was rationalized on the basis of structural stabilization via the native hydrogen bonding network (Breuker and McLafferty 2005). However, both dimers were shown to initially unfold in the same manner by separation of the terminal helices and unfolding of the 18–34 Ω loop; these regions are generally stabilized by hydrophobic bonding in solution (10 hydrophobic bonds between the terminal helices and 3 hydrophobic helix–heme contacts; 5 hydrophobic bonds between the helices and the 18–34 Ω-loop and 4 hydrophobic loop–heme contacts). In contrast, the most stable regions of the native structure in the gas phase are those involved in hydrogen bonding (Breuker and McLafferty 2005).

An analysis of the temperature dependence of NECD product branching ratios revealed the order of stability, and the reverse order of unfolding, for the unfolding monomer in the two dimer structures (A—K79/M80 > Y48/T49 > F82 = L68 > I85 > K13;B—Y48/T49 = W59 = K79/M80 > F46 = N52 = F82 > T40 > L68 = I85 > L35 > K13). This stability order of native horse heart (Fe^{III})cytochrome *c* in a gas-phase environment is essentially the reverse of the unfolding in solution determined by Englander and coworkers (K13 > L68 = L35 > W59 > K79/M80 = F82 = I85 > T40 = F46 = Y48/T49 = N52) (Krishna et al. 2003; Maity et al. 2004). Dehydration of the native protein structure, plus stabilization by the compact monomer, almost inverts the stabilities of intramolecular noncovalent interactions in native cytochrome *c*, demonstrating that the native structure is unstable in the gas phase. The conformational rearrangements on transfer into the gas phase, as well as the increased electrostatic interactions in a gaseous environment, also result in a substantial charge redistribution within the cytochrome *c* ions (Breuker 2006).

Nozzle–skimmer dissociation experiments with +6 and +7 ubiquitin ions from ESI of aqueous pH 5 solutions indicated that the destabilization of the native structure by loss of hydrophobic bonding can be transiently balanced by the stabilization of native hydrogen bonds and saltbridges. This native-like, compact gas-phase structure can exist long enough to resist collisional dissociation in the nozzle–skimmer region, but ultimately unfolds during transfer into the FTMS cell (Zhai et al. 2005).

Eventually, the gaseous protein ions from ESI will rearrange to form stable gas-phase structures. So the question is, what do these look like, and is there a single low-energy gas phase structure, corresponding to the unique native state in solution? In the next section, these issues will be addressed on the basis of ion mobility, ECD, H/D exchange, and IRPDS data on gaseous ubiquitin and cytochrome *c* ions.

5.7. STABLE GAS-PHASE STRUCTURES

5.7.1. Ubiquitin

As all ESI ions originate from solution, and any conformational rearrangements toward more stable gas-phase structures occur on a finite timescale, how do we know whether a given gaseous protein ion structure observed experimentally is actually stable, or merely a transient or intermediate structure? This question can be addressed by storing protein ions from ESI in an ion trap, and monitoring any structural changes that may occur over extended periods of time.

Clemmer and coworkers studied the structures of gaseous +6, +7, and +8 ubiquitin ions from ESI of 49% water/49% methanol/2% acetic acid solutions over ~10 ms–30 s storage times in a quadrupole ion trap (~7×10^{-4} mbar helium, 27°C) that was interfaced with a drift tube for collision cross-sectional measurements (Myung et al. 2002). The experiments were carried out under conditions that caused rapid thermal equilibration and minimal collisional activation of the ions after ESI. At short trapping times (20 ms), the drift time distribution was dominated by peaks assigned to compact structures for the +6 and +7 ions, and partially compact structures for the +8 ions. Increasing the storage time increased the proportion of more extended structures (Figure 5.2), with depletion rate constants of 0.2, 49, and 240 s^{-1} for the compact +6, +7, and partially compact +8 ions, respectively. The threshold times for unfolding were 100, 30, and 25 ms for the +6, +7, and +8 ions, respectively; the possible reasoning behind these induction

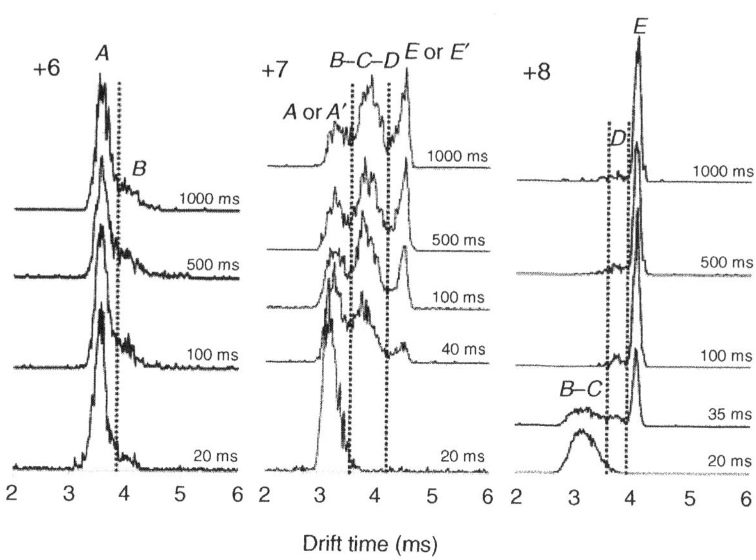

FIGURE 5.2. Unfolding of gaseous ubiquitin ions in a Paul trap. Drift time distributions for the +6, +7, and +8 ions of ubiquitin were recorded after the indicated trapping time. [Reprinted with permission from Myung et al. (2002). Copyright (2002) American Chemical Society.]

periods was attributed to ion cooling below the ambient temperature and conformational freezing as a result of solvent evaporation during the ESI process (Myung et al. 2002). Conformational freezing could also contribute to the transient stability of the native-like, compact +6 and +7 ubiquitin ions observed in nozzle–skimmer dissociation experiments (Zhai et al. 2005). The increased unfolding rates and the decreased threshold times with increasing charge values are consistent with an increase in ion Coulomb energy.

Although an ion's internal energy may initially increase on entering a quadrupole trap as a result of energetic collisions, thermal equilibration by IR radiation and collisional energy exchange with the bath gas should occur on a relatively short timescale (<1 s) (Goeringer and McLuckey 1998; Konn et al. 2005; McLuckey et al. 2000). Yet unfolding of the +7 ubiquitin ions continued during the 30 s storage time (Myung et al. 2002), indicating that the initially observed compact structures are not the most stable conformations in the gas phase. However, the (partially) resolved features in the drift time distributions also show that the conformational rearrangements on desolvation can proceed via discrete intermediate structures.

The collision cross sections of gaseous ubiquitin ions $(M + nH)^{n+}$ recorded immediately after ESI in various ion mobility measurements on different instruments (Figure 5.3) (Valentine et al. 1997; Hoaglund et al. 1998; Li et al. 1999; Purves et al. 2000, 2001; Wyttenbach et al. 2001) show that the charge value n is not generally indicative of a single specific conformation in the gas phase. Instead, multiple conformations are found for each charge value except for +15, with the +5, +6, +7, and +8 ions exhibiting the largest differences in collision cross section of up to about 600 Å^2. For these lower charge states, the data points in Figure 5.3

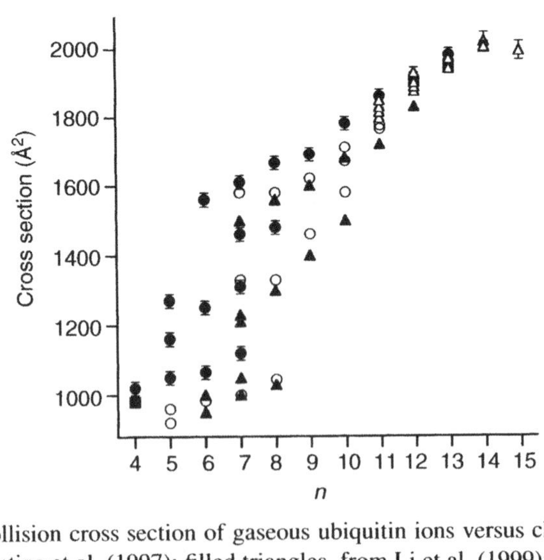

FIGURE 5.3. Collision cross section of gaseous ubiquitin ions versus charge state n: filled circles, from Valentine et al. (1997); filled triangles, from Li et al. (1999); open circles, from Purves et al. (2000); open triangles, from Purves et al. (2001); squares, from Hoaglund et al. (1998).

correspond to reproducible maxima in the drift time distributions with otherwise unresolved features (Li et al. 1999; Valentine et al. 1997). Moreover, peak widths are usually larger than expected for a single conformer, indicating either the presence of multiple substructures or the interconversion of conformers as they travel through the drift tube (Valentine et al. 1997). The ubiquitin ion collision cross sections were found to also depend on the solvent composition used for ESI. The +6 to +11 ions electrosprayed from "pseudonative" solutions (89% water/9% acetonitrile/2% acetic acid) all favored more compact conformers than those from "denaturing" solutions (49% water/49% acetonitrile/2% acetic acid), which can be attributed to either differences in their initial solution structure, or differences in the desolvation/ionization process (Li et al. 1999).

As a general trend, ubiquitin ion collision cross sections increase with increasing charge value and plateau around 2000 $Å^2$ for $n > 12$, although different conformers can still be resolved for the +13 and +14 ions (Purves et al. 2001). The collision cross sections of the +11, +12, and +13 ions from ESI of 49% water/49% acetonitrile/2% acetic acid solutions did not change when increasing the temperature of the metal capillary (through which the ions enter the instrument) from 25 to 132°C (Li et al. 1999), consistent with extended gas-phase structures stabilized by electrostatic repulsion. On the other hand, for ions +6 to +10, the fraction of extended conformers increases with increasing temperature of the capillary (25–132°C) (Li et al. 1999) or ion injection energy (385 vs. 760 eV) (Valentine et al. 1997), similar to the unfolding at extended ion storage times (Myung et al. 2002). Thus, adding internal energy in the desolvation region appears to merely accelerate the desolvation-induced unfolding of the initially compact ions, as also observed with NECD of cytochrome *c* (Breuker and McLafferty 2005).

Does this mean that the initially compact protein ions from ESI simply unfold after desolvation, and that the stable protein conformers in the gas phase have extended structures? To address this question, we also have to consider the possibility that conformational rearrangements in the gas phase may occur on extended timescales. The longest possible ion storage times can be realized in FTMS instruments (Marshall et al. 1998, 2002), in which the structures of desolvated protein ions can be studied by H/D exchange (Campbell et al. 1995; Freitas et al. 1999; Heck et al. 1998; McLafferty et al. 1998; Suckau et al. 1993; Winger et al. 1992; Wood et al. 1995), ECD (Breuker et al. 2002; Horn et al. 2001; Oh et al. 2002), and IRPDS (Oh et al. 2002, 2005; Oomens et al. 2005; Lin et al. 2005).

Gas-phase H/D exchange of ubiquitin +7 ions with D_2O (2.7×10^{-7} mbar) in the trapped ion cell of a FTMS instrument for 3600 s showed that increasing the capillary temperature or the ion accumulation time in an external linear octopole trap increases the proportion of faster exchanging conformers (Freitas et al. 1999). Assuming that H/D exchange rates in the gas phase are correlated with the extent of conformer elongation, these findings are in qualitative agreement with the ion mobility data presented above (Li et al. 1999). However, a possible refolding into more stable gas-phase structures at the later stages of the 3600 s D_2O exposure time may not be reflected in the isotopic profile because the exchange reaction is essentially irreversible in these experiments.

In analogy to H/D exchange in solution (Englander et al. 1997; Engen and Smith 2001; Englander and Krishna 2001), it could be expected that the rate of hydrogen exchange in gaseous protein ions, or the extent of deuterium incorporation at a given exchange time, is smaller for compact than for open conformers. For the gaseous, ubiquitin ions whose collision cross sections generally increase with increasing charge value n (Figure 5.3), this would suggest an increase in H/D exchange with increasing charge value. Surprisingly, for a fixed capillary temperature, external ion accumulation time, and exchange time, the extent of deuterium incorporation into the gaseous +7 to +13 ubiquitin ions by H/D exchange with D_2O or CH_3OD actually decreases with increasing charge value (Freitas et al. 1999; Oh et al. 2002). This puzzling observation has been rationalized by intramolecular charge solvation of protonated residues to protect exchangeable hydrogens, with every added proton protecting additional sites (McLafferty et al. 1998; Oh et al. 2002). Moreover, it is conceivable that the conformational flexibility of the gaseous protein ions affects H/D exchange. In this picture, the more highly charged ions have restricted conformational flexibility and relatively rigid structures as a result of Coulombic repulsion, thus reducing the probability of meeting the high structural demands of the exchange reaction with D_2O or CH_3OD. With decreasing charge and Coulombic repulsion, flexibility increases and conformational substates that allow for H/D exchange become more accessible. This hypothesis is generally supported by the ion mobility data of gaseous ubiquitin ions, which show larger variations in collision cross section for the +5 to +8 ions than for the higher charge states (+9 to +13) (Valentine et al. 1997).

H/D exchange of gaseous ubiquitin ions with ND_3 in a fast flow tube apparatus with continuous collisional activation instead shows the expected correlation of increased exchange rates and maximum number of exchanged hydrogen with increasing charge state (Geller and Lifshitz 2005). A possible explanation for the different exchange behavior is that different exchange reagents were used: D_2O and CH_3OD in the FTMS experiments and ND_3 in the fast flow tube study (Geller and Lifshitz 2005).

Williams and coworkers (Robinson and Williams 2005) studied ubiquitin ions in an FTMS instrument that was interfaced with a FAIMS (high-field asymmetric waveform ion mobility specrometry) device (Guevremont 2004). The ubiquitin conformers were first separated on the basis of their ion mobility values in the FAIMS region, and then subjected to H/D exchange with D_2O (20–40 s) in the FTMS cell. It was concluded that the extent of H/D exchange does not show a significant correlation with collision cross section. For example, FAIMS separated two conformers for the +12 ubiquitin ions that differed in collision cross section by <1%, yet the difference in H/D exchange was substantial, 6 versus 25 hydrogens. On the other hand, only one conformer was observed with H/D exchange of the +9 ubiquitin ions, whereas FAIMS found two conformers that differed in collision cross section by 10% (Robinson and Williams 2005). This suggests that H/D exchange and ion mobility measurements probe different structural aspects of the gaseous protein ions. Whereas collision cross sections depend largely on the overall tertiary structure, H/D exchange appears to be sensitive to secondary structure and/or conformational flexibility.

In most ESI spectra of acidified ubiquitin solutions that contain a substantial proportion (~50%) of organic solvent, such as methanol, the highest charge state observed was +13 (Schnier et al. 1995), although sometimes low-abundant +14 and +15 ions were also detected (Purves et al. 2001; Samalikova and Grandori 2005). H/D exchange with D_2O and CH_3OD (Cassady and Carr 1996; Freitas et al. 1999; Oh et al. 2002), proton transfer (Zhang and Cassady 1996), and ion mobility (Hoaglund et al. 1998; Li et al. 1999; Valentine et al. 1997) data all indicate a single conformation for the +13 ions. In contrast, two conformers for the +13 ion were independently found in H/D exchange experiments using ND_3 in a fast-flow (>1 L/min) tube apparatus (Geller and Lifshitz 2005), and in studies using a FAIMS instrument in which ions are transported by a flowing stream of gas (~0.5 L/min) (Purves et al. 2001). This suggests that protein ion conformation is affected differently in a flow of gas and at static pressure, but the reason for such behavior is unclear at this point. Nevertheless, it is generally agreed that the +13 ubiquitin ions have extended conformations in the gas phase.

While the isotopic profiles from H/D exchange of the +8 to +13 ubiquitin ions indicate one or more individual conformational states for each charge state, those of the +5 to +7 ions are largely unresolved (Freitas et al. 1999), suggesting that the latter ions exist in a multitude of possibly interconverting structures. This agrees well with the ion mobility data that also show unresolved features for the lower charge states, but neither ion mobility data nor H/D exchange profiles have yet provided site-specific structural information on the gaseous ubiquitin conformers. This eventually came from ECD studies on the ubiquitin +6 to +13 ions in a FTMS instrument, in which the +6 ions were from ESI of 99% water/1% NH_4OH solutions, the +7 and +8 ions from 99% water/1% acetic acid solutions, and higher charge states from acidified methanol/water solutions (Breuker et al. 2002). The ions were transferred into the FTMS cell (~10^{-9} mbar) through quadrupole ion guides, trapped with a nitrogen gas pulse (~1.3×10^{-6} mbar peak pressure), and individual charge states were isolated by ejection of all other ions employing SWIFT waveforms. Following isolation and prior to irradiation with low-energy electrons, the ion population was allowed to thermally equilibrate with the ion cell and vacuum chamber walls (25–175°C) for 40 s via blackbody infrared irradiation (Jockusch et al. 1997; Price et al. 1996).

As outlined in Section 5.5, ECD gives separated $c/z^\bullet$ or $a^\bullet/y$ backbone cleavage products only from regions without tertiary noncovalent bonding, whereas the fragment ions that are still joined by noncovalent bonding appear in the spectrum as reduced molecular ions, $(M + nH)^{(n-1)+\bullet}$. Consistent with both acid and thermal unfolding in solution, the yield of separated ECD products at 25°C increased with increasing charge state n of the ubiquitin $(M + nH)^{n+}$ precursor ions, and for a given charge state n with increasing temperature (Breuker et al. 2002). The yield of separated ECD products at 25°C also correlated well with collision cross sections (Li et al. 1999) of ubiquitin ions electrosprayed under similar conditions (Breuker et al. 2002).

The ECD spectrum of ubiquitin +13 ions from ESI of 49% water/49% methanol/ 2% acetic acid solutions showed separated products from cleavage at 55 out of 75 interresidue locations (Breuker et al. 2002), and was essentially unchanged when heating the ion population from 25 to 125°C (Breuker et al. 2004), indicating an extended structure of high thermal stability. Moreover, the ECD cleavages generally peaked at sites three amino acids towards the *N* terminus from the protonated basic residues, consistent with charge solvation in an extended α-helical structure and stabilization via charge–dipole interactions (Breuker et al. 2004). An extremely high thermal stability of α-helical structures in the gas phase was also observed in ion mobility experiments for alanine-based peptides with protonation on the *C*-terminal lysine (Kohtani et al. 2004).

ECD of the +6 and +7 ubiquitin ions from "native" solutions (99% water/1% acetic acid) both gave separated products from cleavage in the *N*-terminal region (residues 1–10), which would not be possible if the native structure had been retained; freeing the β_1 strand (residues 1-7) in the native structure requires its separation from the β_2 and β_5 strands as well as the hydrophobic core (Figure 5.1). Yet these +6 and +7 ubiquitin ions do have tertiary structure even after 40 s storage at 25°C and 10^{-9} mbar, as indicated by the lack of separated ECD cleavage products in other regions (Breuker et al. 2002; Oh et al. 2002). Whether these structures are thermodynamically stable was tested for the +7 ubiquitin ions in a separate experiment. Exposure of the thermally equilibrated (25°C) +7 ion population to IR-laser irradiation (10.6 μm) resulted in their partial unfolding within ~0.1 s, as evidenced by the appearance of new cleavage products and increased fragment ion abundances in the "0.07 s" spectrum of Figure 5.4. With longer delay times between the laser pulse and the ECD event for conformational probing, ion cooling by IR emission caused refolding, as indicated by the continuous decrease in fragment ion abundances with increasing delay time (Figure 5.4). An analysis of the site-specific refolding rates showed that refolding into nearly the original structure occurred within 2 s (Breuker et al. 2002), which is strong evidence for the existence of compact stable conformers of ubiquitin ions in the gas phase. However, longer-term cooling caused folding at some sites even beyond the compactness of the original structure (Breuker et al. 2002). These laser unfolding experiments, as well as the ion mobility and H/D exchange data discussed above, indicate that the folding energy landscape in the gas phase is more shallow and rugged than that in solution.

Can refolding of the gaseous $(M + nH)^{n+}$ ubiquitin ions also be effected by decreasing their net charge, similar to pH-induced refolding reactions in solution (Briggs and Roder 1992)? Clemmer and coworkers showed that exposure of the +6 to +13 ubiquitin ions to proton transfer reagents in the source region of the ion mobility apparatus gave charge-depleted molecular ions with charge values as low as +4 (Valentine et al. 1997). Charge depletion was more extensive with proton transfer reagents of higher gas-phase basicity, consistent with an increased exoergicity of the proton transfer reactions. However, there was no significant change in the total fraction of extended conformers after proton transfer with any of the reagents used. This somewhat surprising result was thought to result from differences in gas-phase acidity of the different conformer types and a higher proton transfer reactivity of the more compact ions (Valentine et al. 1997). An alternative explanation would be the substantial exoergicity of the proton transfer reactions of up to 160 kJ/mol for the

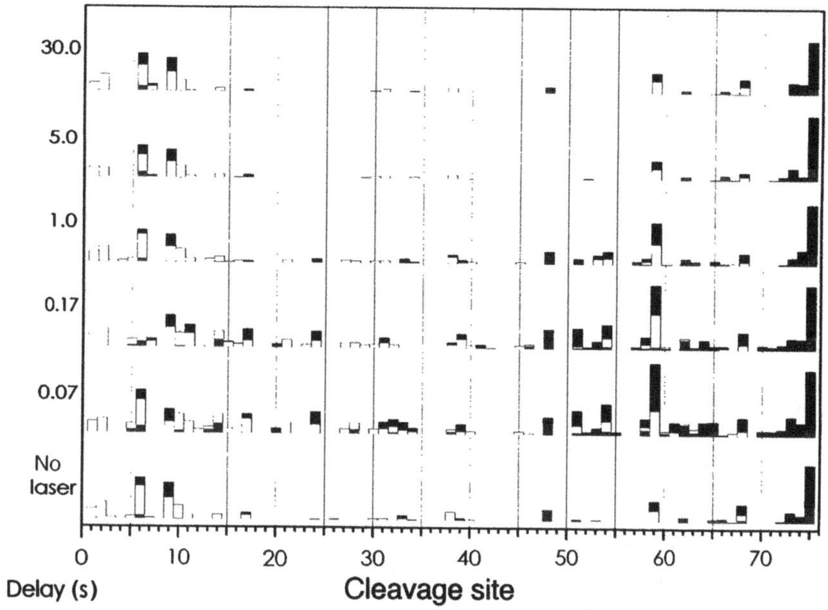

FIGURE 5.4. IR laser unfolding/refolding of gaseous +7 ubiquitin ions monitored by ECD. Black bars represent *c* ions, white bars $z^•$ ions, and gray bars $a^•$ and *y* ions. The delay time between the laser pulse (250 ms) for unfolding and the ECD event (1.2 s) for conformational probing is indicated on the left side of the graph. [Reprinted with permission from Breuker et al. (2002). Copyright (2002) American Chemical Society.]

reaction between $(M + 13H)^{13+}$ ubiquitin ions (apparent gas-phase acidity: 858 kJ/mol) (Zhang and Cassady 1996) and 7-methyl-1,5,7-triazabicyclo[4.4.0]dec-5-ene (MTBD, gas-phase basicity: 1018 kJ/mol), which could increase the protein ion's internal energy and prevent refolding on the timescale of the experiment.

Thermodynamic information on the folding/unfolding reactions of the gaseous +6 to +9 ubiquitin ions was obtained in ECD experiments in the temperature range 25–175°C (Breuker et al. 2002). Equilibrium constants for unfolding were derived from the ratio of separated (unfolded ions) to unseparated (folded ions) ECD products, from which unfolding enthalpies and entropies were determined in a van't Hoff analysis (Breuker et al. 2002). The ΔH and ΔS values, as well as the free energies for unfolding, generally decreased with increasing charge state, consistent with increased Coulombic repulsion. A similar effect of increased protonation on ΔH was observed in solution, where the enthalpy of ubiquitin unfolding decreased monotonically with decreasing pH (Figure 5.5) (Wintrode et al. 1994). The ΔH values for the gaseous +7 ubiquitin ions below and above 100°C were very similar to those of the +6 and +8 ions, respectively, revealing a three-state unfolding mechanism (Breuker et al. 2002), in contrast to the apparent two-state process in solution (Krantz and Sosnick 2000; Wintrode et al. 1994). However, solution measurements cannot differentiate between contributions of individual protein charge states at a given pH, such that a recorded parameter value is merely an average.

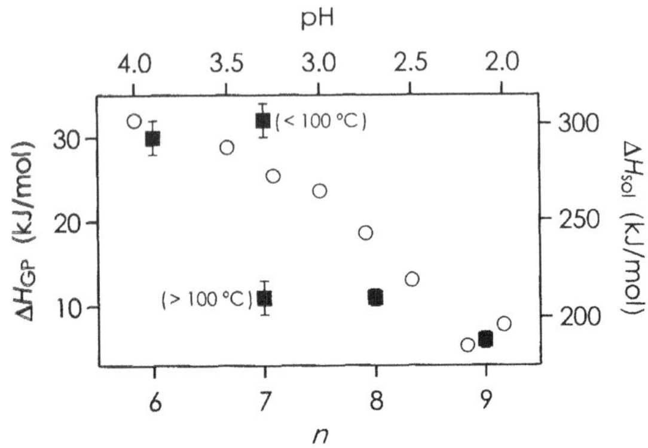

FIGURE 5.5. Unfolding enthalpies for ubiquitin in solution (open circles) versus pH from Wintrode et al. (1994), and in the gas phase (filled squares) versus ion charge state from Breuker et al. (2002).

Site-specific equilibrium melting temperatures for the +6 to +9 ubiquitin ions were obtained from sigmoidal fits of the ECD yield versus temperature for each cleavage site (Breuker et al. 2002). For residues 21–32, the almost uniform melting temperatures for the +6 and +7 ions of ~145°C and ~85°C, respectively, may indicate the partial preservation of the solution helix structure. On the other hand, the low melting temperatures in the β_1/β_2 region (residues 1–17) fully contrast the solution stability data (Cordier and Grzesiek 2002). For the gaseous ubiquitin +7 ions, the last region to unfold is between residues 41 and 50, with melting temperatures of 154°C and 144°C at sites 44 and 47, respectively. The refolding at 25°C after IR laser unfolding is slowest at site 48 ($k = 0.18\ s^{-1}$) (Breuker et al. 2002), illustrating that in the gas phase, high regional thermal stability does not necessarily cause rapid refolding.

More recently, IRPDS was introduced as a method to study the different types of noncovalent bonds that account for the higher order structure of the gaseous protein ions (Oh et al. 2002; Oomens et al. 2005, Lin et al. 2005). The noncovalent bonding in the gaseous ubiquitin $(M + 7H)^{7+}$ ions was probed by photofragmentation of the $(M + 7H)^{6+\bullet}$ ions from ECD that are in fact complemetary $c/z^\bullet$ ion pairs held together by noncovalent bonding (Breuker et al. 2002; Oh et al. 2002). Their IRPDS spectrum showed a broad (<3100–~3500 cm^{-1}) absorption envelope peaking at ~3350 cm^{-1} (Oh et al. 2002). More recent studies that include reference spectra of amino acid dimers indicate that the 3300–3400 cm^{-1} absorptions arise from hydrogen-bonded ammonium functionalities and saltbridge interactions (Oh et al. 2005). With the appearance in the literature of gaseous ion IR spectra of amino acids, amino acid multimers, and peptides (Oh et al. 2002, 2005; Polfer et al. 2005), a more detailed characterization of the stabilizing interactions in gaseous protein structures should soon become possible.

The ECD and H/D exchange data on ubiquitin, in combination with IRPDS experiments, provided the basis for postulated structures of the +5 to +13 ubiquitin ions (Figure 5.6) (Oh et al. 2002). The secondary structure is α-helical for all

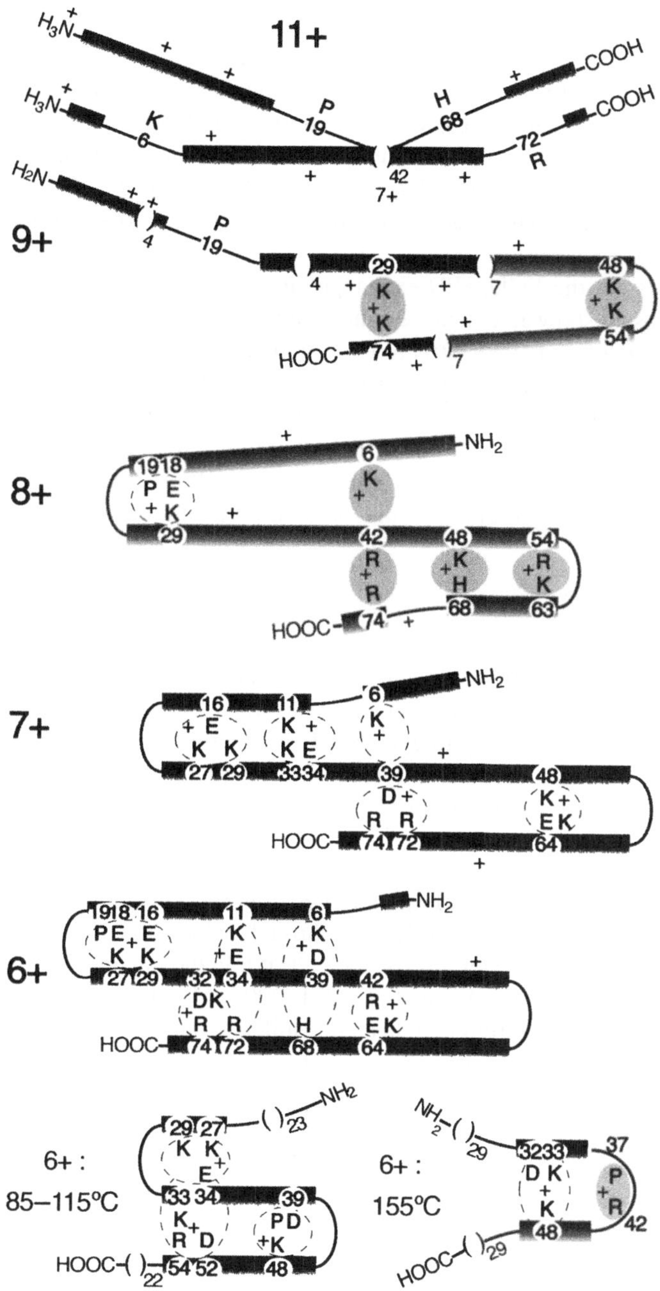

FIGURE 5.6. Proposed structures of the gaseous +6 to +11 ubiquitin ions. [Reprinted with permission from Oh et al. (2002).]

conformers, with the +13 ions representing a single α helix stabilized by charge solvation and charge-dipole interactions. Removing charge introduces bend regions, which allow for the formation of three-helix-bundle structures in the +5 to +7 ions. Hydrogen bonding, saltbridges, and helix dipole–dipole interactions are thought to stabilize the bundle structures (Oh et al. 2002). The proposed structures suggest tertiary structure only for the +5 to +9, but not the +10 to +13, ubiquitin ions. However, molecular dynamics calculations are necessary to support the proposed structures.

In conclusion, the ECD, H/D exchange, and ion mobility data on gaseous +13 ubiquitin ions all show an extensive conformational homogeneity, and are consistent with an extended α-helical structure of high thermal stability. Conformational heterogeneity increases with decreasing charge, and is most pronounced for ions with less than +9 charges. These lower-charge-state ions, which preferentially form by electrospray of aqueous ubiquitin solutions, can exist in a multitude of conformations that include fairly compact as well as extended structures, in surprising contrast to the unique native structure in solution.

5.7.2. Cytochrome *c*

The perhaps most striking feature of cytochrome *c* is its heme group, covalently bound to the protein chain via thioether linkages at C14 and C17. Although the heme can obviously not be part of a protein's secondary structural element, its many hydrophobic and some hydrogen bond interactions with the protein chain considerably stabilize the native cytochrome *c* structure. At present, it is unclear whether or how the heme group also contributes to the stability of the gaseous cytochrome *c* ions.

We have seen in Section 5.6 that the native cytochrome *c* structure becomes unstable after dehydration, and that the dehydration-induced unfolding takes place in the heated capillary for desolvation through which the ions enter the FTMS instrument (Breuker and McLafferty 2003, 2005). In the ion mobility apparatus used by the Clemmer group (Badman et al. 2001, 2005; Myung et al. 2002), ions do not enter the instrument through a heated capillary but are electrosprayed into a differentially pumped desolvation region (1.3–13 mbar, room temperature) instead, from where they are focused into the quadrupole ion trap ($\sim 7 \times 10^{-4}$ mbar helium, 27°C). After variable ion storage times in the trap, the ions are injected into the drift tube for conformational probing. Initial studies by the Clemmer group monitored the structural changes of gaseous cytochrome *c* ions over ~10–200 ms storage times in the trap (Badman et al. 2001). The +7 to +15 ions were from ESI of 49% water/49% methanol/2% acetic acid solutions, in which cytochrome *c* exists in a helical denatured state without any tertiary structure (Kamatari et al. 1996). However, the +7 to +10 ions had rather small collision cross sections at the shortest trapping times (Badman et al. 2001), indicating that these lower-charge-state ions have collapsed into more compact conformers on transfer into the gas phase. As for the +6 to +8 ions of ubiquitin studied by the same group, the compact +7,+8 and partially folded +9,+10 ions of cytochrome *c* underwent unfolding transitions by populating discrete intermediate states. Unfolding started after ~30 ms storage time in the trap for cytochrome *c* (Badman et al. 2001), close to the induction periods of 30 and 25 ms for the +7 and +8 ubiquitin ions, respectively, but substantially shorter than the 100 ms for the +6 ubiquitin ions (Myung et al. 2002).

In another study, the ion storage times were extended to 10 s, during which the +9 cytochrome *c* ions first underwent the rapid unfolding transition observed before (Badman et al. 2001), but then the newly formed extended conformers refolded into structures of intermediate compactness ("*F* state") (Badman et al. 2005). The onsets for unfolding and refolding of the +9 cytochrome *c* ions were ~30 and ~320 ms, respectively, with the fraction of extended conformers continuously decreasing at later times. Collisional heating of the +9 cytochrome *c* ions by application of low-amplitude resonance frequency waveforms to the Paul trap electrodes produced mostly extended conformers that also refolded into the *F* state during the prolonged storage time (Badman et al. 2005). However, the +6, +7, +8, and +10 cytochrome *c* ions from ESI of the same solution showed no evidence for refolding under the exact same experimental conditions (Badman et al. 2005), despite the lower Coulombic energy in the +6, +7, and +8 ions.

As for ubiquitin, the collision cross sections of the gaseous cytochrome *c* ions generally increase with increasing charge state *n* (Figure 5.7) (Badman et al. 2001;

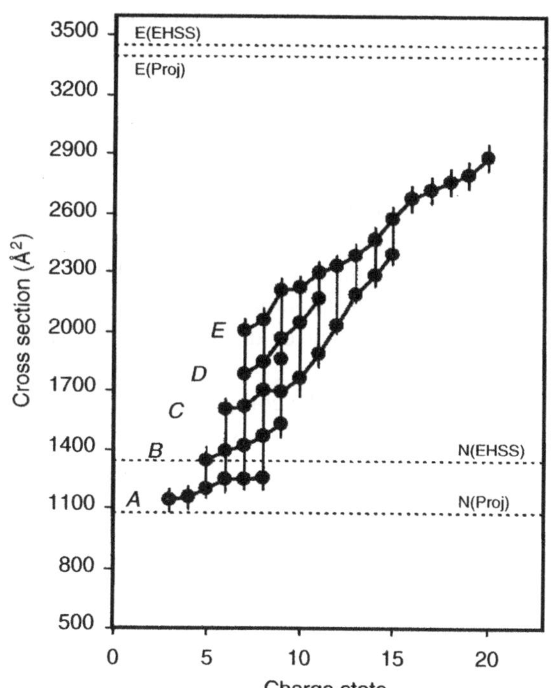

FIGURE 5.7. Collision cross sections of gaseous cytochrome *c* ions versus charge state *n*. [Reprinted with permission from Badman et al. (2001). Copyright (2001) American Chemical Society.]

Clemmer et al. 1995; Hudgins et al. 1997; Jarrold 1999; Shelimov et al. 1997). The conformational heterogeneity is largest for $5 < n < 10$; for $n = 8$, the collision cross sections can be as small as ~1260 Å^2 and as large as ~2070 Å^2 (Badman et al. 2001; Shelimov et al. 1997). Unlike the gaseous ubiquitin ions, however, the extended +10 to +18 cytochrome *c* ions refolded into compact structures in less than 100 ms after extensive charge removal, giving mostly +4 and +5 ions in proton transfer reactions with MTBD (7-methyl-1,5,7-triazabicyclo[4.4.0]dec-5-ene) (Shelimov and Jarrold 1996). Although increasing the kinetic energy with which the ions are injected into the drift tube can result in their unfolding as a result of collisional heating (Jarrold 1999; Mao et al. 1999; Shelimov et al. 1997; Valentine and Clemmer 1997), the +3 to +5 cytochrome *c* ions produced from high-charge-state ions by proton transfer to MTBD actually refolded into even tighter structures after collisional heating and subsequent thermalization (Shelimov et al. 1997; Shelimov and Jarrold 1996), indicating a rugged and shallow folding energy landscape for these gaseous cytochrome *c* ions. However, only the +3 to +5 ions, but neither the +7 ions produced by charge stripping from higher charge states, nor the +7 ions from ESI of unacidified solutions, adopted compact conformations when high injection energies were used (Shelimov and Jarrold 1996). The +5 cytochrome *c* ions from ESI of 75% water/25% acetonitrile solutions also showed little change in collision cross section (~1200 Å^2) when high injection energies were used and the drift tube temperature was increased from ~1°C to ~299°C, whereas the compact +7 ions (~1300 Å^2) injected with 350 eV unfolded into extended structures (~2100 Å^2) over the same temperature range (Mao et al. 1999).

Thus the most stable gas-phase structures of the +3 to +5 cytochrome *c* ions are compact conformations of high thermal stability (Mao et al. 1999; Shelimov et al. 1997; Shelimov and Jarrold 1996; Woenckhaus et al. 1997), and those of the +10 to +20 ions are extended structures (Clemmer et al. 1995; Jarrold 1999; Shelimov et al. 1997). For the +6 to +9 ions, multiple coexisting conformers with a wide range of collision cross sections have been observed, and it is yet unclear which of these have the highest stability (Badman et al. 2001; Shelimov et al. 1997), although for the +9 ions more recent evidence points to a partially folded state (Badman et al. 2005).

The difference in conformation of the compact +5 and extended +7 cytochrome *c* ions is also reflected in their hydration behavior in the drift tube of an ion mobility instrument at 2°C, with the +5 ions having a larger propensity for water adsorption than the +7 ions (Woenckhaus et al. 1997). The free-energy changes for initial hydration of the +5 ions were more negative than those of the +7 ions, in contrast to the expected correlation between water affinity and ion charge state for a hydration process dominated by water interactions with charged groups. A possible explanation for this would be a more effective intramolecular charge solvation in the extended +7 cytochrome *c* ions (Woenckhaus et al. 1997). Note that intramolecular charge solvation was also suggested as an explanation for the decreasing H/D exchange with increasing protein ion charge state in FTMS experiments (McLafferty et al. 1998; Oh et al. 2002). The compact +5 cytochrome *c* ions, along with the extended +9 cytochrome *c* ions, have also been subjected to H/D exchange reactions

(~25 ms) with D_2O (~0.9 mbar) in the ion mobility drift tube (Valentine and Clemmer 2002). The maximum number of exchanged hydrogens increased with increasing temperature from 53 and 63 at 27°C to ~200 (complete exchange) and ~190 (14 hydrogens short of complete exchange) at ~160°C for the +5 and +9 cytochrome *c* ions, respectively. The collision cross sections of the +5 and +9 cytochrome *c* ions did not vary significantly with temperature (1242 Å^2 at 27°C and 1210 Å^2 at 127°C for the +5 ions; 2180 Å^2 at 27°C and 2155 Å^2 at 127°C for the +9 ions). At temperatures below ~100°C, the compact +5 ions exchanged fewer hydrogens than did the extended +9 ions, but at higher temperatures the exchange level of the compact +5 conformers exceeded the level of the extended +9 structures (Valentine and Clemmer 2002). The relatively small differences in exchange level for the compact +5 and extended +9 ions, 10 at both 27°C and ~160°C, shows that gas-phase H/D exchange with D_2O is at the most marginally sensitive to protein tertiary structure.

H/D exchange of the gaseous +6 to +19 cytochrome *c* ions with D_2O (~ 1.3×10^{-7} mbar) in a FTMS instrument for 1800 s indicated as many as seven stable conformers that differ by ~10 in the number of exchanged hydrogens (Figure 5.8) (McLafferty et al. 1998); note that the multiple conformers found by ion mobility also have almost regularly spaced collision cross-sectional values (Figure 5.7). Within a given conformer family, the number of exchanged hydrogens decreased with increasing charge state (McLafferty et al. 1998), similar to the exchange

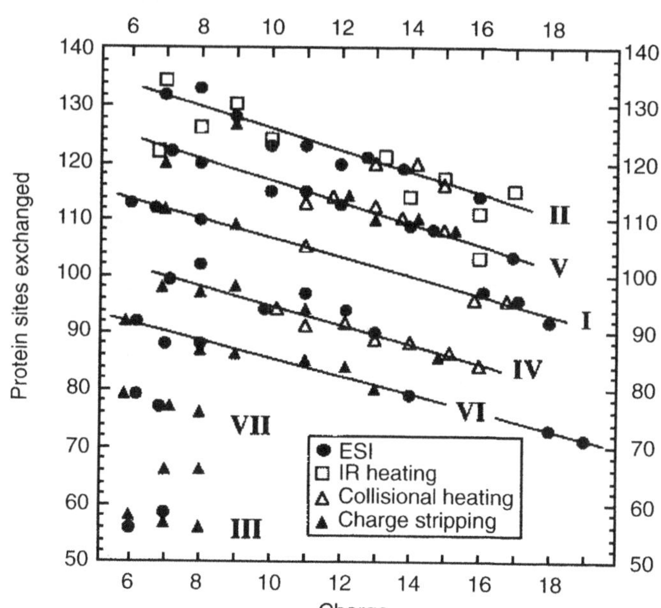

FIGURE 5.8. H/D exchange of gaseous cytochrome *c* ions versus charge state *n*. [Reprinted with permission from McLafferty et al. (1998). Copyright (1998) American Chemical Society.]

behavior observed for ubiquitin (Oh et al. 2002). The decrease in exchanged hydrogens per charge was ~2 for cytochrome *c* and ~4 for ubiquitin, which might be related to the number of basic residues: 24 in cytochrome *c* and 12 in ubiquitin. As for the H/D exchange reactions in the ion mobility apparatus (Valentine and Clemmer 2002), H/D exchange in the FTMS instrument increased with increasing temperature up to a maximum value that differed for the different conformers (but not the ion charge states), between 75 and 110°C. Above these temperatures, H/D exchange in the FTMS cell decreased, in one case even below the exchange value at room temperature (McLafferty et al. 1998). The conformer states populated by ions directly formed by ESI depended on the composition of the cytochrome *c* solution and instrumental conditions, and for a given charge state, some conformer states were observed only after charge stripping reactions, or collisional or IR laser heating (McLafferty et al. 1998; Wood et al. 1995). For example, H/D exchange of the +9 and +15 ions from ESI showed that they exist as states II and V, respectively. Collisional or IR heating in the trapped ion cell did not affect the +9/state II conformer, but converted the +15/state V into a +15/state II conformer (McLafferty et al. 1998).

ECD of the +9 cytochrome *c* ions (the abovementioned state II conformer) stored in the trapped ion cell of an FTMS instrument for ~40 s at 26, 60, 110, 130, and 140°C showed an increase in the number of separated cleavage products from 23 to 44 with increasing temperature (Horn et al. 2001), consistent with a partially folded structure that could be the *F* state observed in ion mobility experiments at extended ion storage times (Badman et al. 2005). Refolding of the +9 cytochrome *c* ions after IR laser unfolding was substantially slower (~120 s) (Horn et al. 2001) than that of the +7 ubiquitin ions (~2 s) (Breuker et al. 2002). The m/z values of the +9 cytochrome *c* and +7 ubiquitin ions are 1374 and 1224, respectively, so that the slower folding of the +9 cytochrome *c* ions is most likely not a result of increased charge but rather due to the larger number of residues (Naganathan and Muñoz 2005). In addition, the bulky heme group could slow down the folding process. ECD of the +15 cytochrome *c* ions (the abovementioned state V conformer) stored in the trapped ion cell for ~40 s showed an increase in the number of separated cleavage products from 41 at 26°C to 60 at 140°C. Unfolding of the +15/state V ions after IR laser exposure took >5 s, and increased the number of separated cleavage products from 41 to 67, with no substantial evidence for refolding during 60 s; ion mobility also found two conformers for the +15 ions (Badman et al. 2001).

The cytochrome *c* +12 to +16 ions from ESI of 29% water/69% methanol/2% acetic acid solutions have also been studied by IRPDS in the range 1420–1790 cm^{-1} (Oomens et al. 2005). Infrared absorption spectra obtained by monitoring the low-energy dissociation reaction $(M + nH + K)^{(n+1)+} \rightarrow (M + nH)^{n+} + K^+$ showed amide I (C=O stretching modes) and II (N—H bending modes) bands, plus another yet unassigned band at ~1480 cm^{-1} whose intensity increased with increasing charge state. The amide I and II band positions are consistent with a mostly α-helical structure of the gaseous +12 to +16 ions, and the bandwidths suggest a conformational flexibility similar to that in solution (Oomens et al. 2005).

5.8. OUTLOOK AND PERSPECTIVES

A considerable number of ion mobility and FTMS experiments on gaseous ubiquitin and cytochrome *c* ions have already provided many detailed structural data. These are the basis for the structural characterization of the gaseous protein ions. However, the combination of data obtained with different experimental strategies is not always straightforward; an example of this is H/D exchange and ion mobility spectrometry. Thus a major future challenge lies in designing experiments that combine existing techniques for the probing of different structural aspects of the same ion population, such as ion mobility spectrometry and ECD. Even though the structural elucidation of gaseous proteins is not yet possible, experiments have provided very detailed thermodynamic and kinetic data on the folding and unfolding of the gaseous protein ions. However, identification of the driving force for folding in the gas phase requires further study. This could involve mutational studies, well established in condensed-phase protein research. Molecular dynamics calculations are another promising approach for the understanding of gaseous protein structures. Despite all the open questions, it is also evident that a clearer picture of protein structure and folding in the gas phase is beginning to emerge.

ACKNOWLEDGMENT

The author acknowledges generous funding from the Austrian FWF, BMBWK (grant T229), and TWF (grant UNI-0404/158), and discussions with Fred W. McLafferty, Ekkehart Breuker, Huili Zhai, Xuemei Han, Mi Jin, Cheng Lin, Harold Hwang, Peppe Infusini, Robert Konrat, Bernhard Kräutler, Marc-Olivier Ebert, Thomas Müller, and Michal Steinberg.

REFERENCES

Babu KR, Moradian A, Douglas DJ (2001): The methanol-induced conformational transitions of β-lactoglobulin, cytochrome *c*, and ubiquitin at low pH: A study by electrospray ionization mass spectrometry. *J. Am. Soc. Mass Spectrom.* **12**:317–328.

Badman ER, Hoaglund-Hyzer CS, Clemmer DE (2001): Monitoring structural changes of proteins in an ion trap over ~10–200 ms: Unfolding transitions in cytochrome *c* ions. *Anal. Chem.* **73**:6000–6007.

Badman ER, Myung S, Clemmer DE (2005): Evidence for unfolding and refolding of gas-phase cytochrome *c* ions in a Paul trap. *J. Am. Soc. Mass Spectrom.* **16**:1493–1497.

Banci L, Bertini I, Gray HB, Luchinat C, Reddig T, Rosato A, Turano P (1997): Solution structure of oxidized horse heart cytochrome *c*. *Biochemistry* **36**:9867–9877.

Banci L, Bertini I, Huber JG, Spyroulias GA, Turano P (1999a): Solution structure of reduced horse heart cytochrome *c*. *J. Biol. Inorg. Chem.* **4**:21–31.

Banci L, Bertini I, Rosato A, Varani G (1999b): Mitochondrial cytochromes *c*: A comparative analysis. *J. Biol. Inorg. Chem.* **4**:824–837.

Barber M, Bordoli RS, Sedgwick RD, Tyler AN (1981): Fast atom bombardment of solids (F.A.B.): A new ion source for mass spectrometry. *J. Chem. Soc., Chem. Commun.* **7**: 325–327.

Beckey HD, Schulten HR (1975): Field desorption mass spectrometry. *Angew. Chem. Int. Ed.* **14**:403–415.

Berkenkamp S, Karas M, Hillenkamp F (1996): Ice as a matrix for IR-matrix-assisted laser desorption/ionization: Mass spectra from a protein single crystal. *Proc. Natl. Acad. Sci. USA* **93**:7003–7007.

Berman HM, Westbrook J, Feng Z, Gilliland G, Bhat TN, Weissig H, Shindyalov IN, Bourne PE (2000): The protein data bank. *Nucleic Acids Res.* **28**:235–242.

Bothner B, Siuzdak G (2004): Electrospray ionization of a whole virus: Analyzing mass, structure, and viability. *ChemBioChem* **5**:258–260.

Breuker K (2004): The study of protein-ligand interactions by mass spectrometry—a personal view. *Int. J. Mass Spectrom.* **239**:33–41.

Breuker K (2006): Segmental charge distributions of cytochrome *c* on transfer into the gas phase. *Int. J. Mass Spectrom.* in press.

Breuker K, McLafferty FW (2003): Native electron capture dissociation for the structural characterization of noncovalent interactions in native cytochrome *c*. *Angew. Chem. Int. Ed.* **42**:4900–4904.

Breuker K, McLafferty FW (2005): The thermal unfolding of native cytochrome *c* in the transition from solution to gas phase probed by native electron capture dissociation. *Angew. Chem. Int. Ed.* **44**:4911–4914.

Breuker K, Oh HB, Horn DM, Cerda BA, McLafferty FW (2002): Detailed unfolding and folding of gaseous ubiquitin ions characterized by electron capture dissociation. *J. Am. Chem. Soc.* **124**:6407–6420.

Breuker K, Oh HB, Lin C, Carpenter BK, McLafferty FW (2004): Nonergodic and conformational control of the electron capture dissociation of protein cations. *Proc. Natl. Acad. Sci. USA* **101**:14011–14016.

Briggs MS, Roder H (1992): Early hydrogen-bonding events in the folding reaction of ubiquitin. *Proc. Natl. Acad. Sci. USA* **89**:2017–2021.

Brutscher B, Brüschweiler R, Ernst RR (1997): Backbone dynamics and structural characterization of the partially folded A state of ubiquitin by 1H, 13C, and 15N nuclear magnetic resonance spectroscopy. *Biochemistry* **36**:13043–13053.

Campbell S, Rodgers MT, Marzluff EM, Beauchamp JL (1995): Deuterium exchange reactions as a probe of biomolecule structure. Fundamental studies of gas phase H/D exchange reactions of protonated glycine oligomers with D_2O, CD_3OD, CD_3CO_2D, and ND_3. *J. Am. Chem. Soc.* **117**:12840–12854.

Cassady CJ, Carr SR (1996): Elucidation of isomeric structures for ubiquitin $[M + 12H]^{12+}$ ions produced by electrospray ionization mass spectrometry. *J. Mass Spectrom.* **31**: 247–254.

Catalina MI, van den Heuvel RHH, van Duijn E, Heck AJR (2005): Decharging of globular proteins and protein complexes in electrospray. *Chem. Eur. J.* **11**:960–968.

Chowdhury SK, Katta V, Chait BT (1990): Probing conformational changes in proteins by mass spectrometry. *J. Am. Chem. Soc.* **112**:9012–9013.

Clemmer DE, Hudgins RR, Jarrold MF (1995): Naked protein conformations: Cytochrome *c* in the gas phase. *J. Am. Chem. Soc.* **117**:10141–10142.

Clemmer DE, Jarrold MF (1997): Ion mobility measurements and their applications to clusters and biomolecules. *J. Mass Spectrom.* **32**:577–592.

Constantopoulos TL, Jackson GS, Enke CG (2000): Challenges in achieving a fundamental model for ESI. *Anal. Chim. Acta* **406**:37–52.

Cooper A (1999). Thermodynamics of protein folding and stability. In Allen G (ed), *Protein: A Comprehensive Treatise*, JAI Press, Vol. 2, pp. 217–270.

Cooper A, Johnson CM, Lakey JH, Nöllmann M (2001): Heat does not come in different colors: Entropy-enthalpy compensation, free energy windows, quantum confinement, pressure perturbation calorimetry, solvation and the multiple causes of heat capacity effects in biomolecular interactions. *Biophys. Chem.* **93**:215–230.

Cordier F, Grzesiek S (2002): Temperature-dependence of protein hydrogen bond properties as studied by high-resolution NMR. *J. Mol. Biol.* **715**:739–752.

Cornilescu G, Marquardt JL, Ottiger M, Bax A (1998): Validation of protein structure from anisotropic carbonyl chemical shifts in a dilute liquid crystalline phase. *J. Am. Chem. Soc.* **120**:6836–6837.

Cox HA, Julian RR, Lee SW, Beauchamp JL (2004): Gas-phase H/D exchange of sodiated glycine oligomers with ND_3: Exchange kinetics do not reflect parent ion structures. *J. Am. Chem. Soc.* **126**:6485–6490.

de la Mora JF (2000): Electrospray ionization of large multiply charged species proceeds via Dole's charged residue mechanism. *Anal. Chim. Acta* **406**:93–104.

de la Mora JF, Van Berkel GJ, Enke CG, Cole RB, Martinez-Sanchez M, Fenn JB (2000): Electrochemical processes in electrospray ionization mass spectrometry. *J. Mass Spectrom.* **35**:939–952.

Dill KA (1990): Dominant forces in protein folding. *Biochemistry* **29**:7133–7155.

Dinner AR, Sali A, Smith LJ, Dobson CM, Karplus M (2000): Understanding protein folding via free-energy surfaces from theory and experiment. *Trends Biochem. Sci.* **25**:331–339.

Dobson CM, Sali A, Karplus M (1998): Protein folding: A perspective from theory and experiment. *Angew. Chem. Int. Ed.* **37**:868–893.

Dole M, Mack LL, Hines RL, Mobley RC, Ferguson LD, Alice MB (1968): Molecular beams of macroions. *J. Chem. Phys.* **49**:2240–2249.

Engen JR, Smith DL (2001): Investigating protein structure and dynamics by hydrogen exchange MS. *Anal. Chem.* **73**:256A-265A.

Englander SW, Krishna MMG (2001): Hydrogen exchange. *Nat. Struct. Biol.* **8**:741–742.

Englander SW, Mayne L, Bai Y, Sosnick TR (1997): Hydrogen exchange: The modern legacy of Linderstrøm-Lang. *Protein Sci.* **6**:1101–1109.

Felitsyn N, Peschke M, Kebarle P (2002): Origin and number of charges observed on multiply-protonated native proteins produced by ESI. *Int. J. Mass Spectrom.* **219**:39–62.

Fenn JB, Mann M, Meng CK, Wong SF, Whitehouse CM (1989): Electrospray ionization for mass spectrometry of large biomolecules. *Science* **246**:64–71.

Filosa A, English AM (2000): Probing local thermal stabilities of bovine, horse, and tuna ferricytochromes *c* at pH 7. *J. Biol. Inorg. Chem.* **5**:448–454.

Franks F (2002): Protein stability: The value of "old literature." *Biophys. Chem.* **96**:117–127.

Freitas MA, Hendrickson CL, Emmett MR, Marshall AG (1999): Gas-phase bovine ubiquitin cation conformations resolved by gas-phase hydrogen/deuterium exchange rate and extent. *Int. J. Mass Spectrom.* **185/186/187**:565–575.

Ge Y, Lawhorn BG, ElNaggar M, Strauss E, Park JH, Begley TP, McLafferty FW (2002): Top down characterization of larger proteins (45 kDa) by electron capture dissociation mass spectrometry. *J. Am. Chem. Soc.* **124**:672–678.

Geller O, Lifshitz C (2005): A fast flow tube study of gas phase H/D exchange of multiply protonated ubiquitin. *J. Phys. Chem. A.* **109**:2217–2222.

Goeringer DE, McLuckey SA (1998): Relaxation of internally excited high-mass ions simulated under typical quadrupole ion trap storage conditions. *Int. J. Mass Spectrom.* **177**:163–174.

Goto Y, Hagihara Y, Hamada D, Hoshino M, Nishii I (1993): Acid-induced unfolding and refolding transitions of cytochrome *c*: A three-state mechanism in H_2O and D_2O. *Biochemistry* **32**:11878–11885.

Guevremont R (2004): High-field asymmetric waveform ion mobility spectrometry: A new tool for mass spectrometry. *J. Chromatogr. A* **1058**:3–19.

Hagihara Y, Tan Y, Goto Y (1994): Comparison of the conformational stability of the molten globule and native states of horse cytochrome *c*. *J. Mol. Biol.* **237**:336–348.

He F, Ramirez J, Lebrilla CB (1999): Evidence for enzymatic activity in the absence of solvent in gas-phase complexes of lysozyme and oligosaccharides. *Int. J. Mass Spectrom.* **193**: 103–114.

Heck AJR, Jørgensen TJD, O'Sullivan M, van Raumer M, Derrick PJ (1998): Gas-phase noncovalent interactions between vancomycin-group antibiotics and bacterial cell-wall precursor peptides probed by hydrogen/deuterium exchange. *J. Am. Soc. Mass Spectrom.* **9**:1255–1266.

Hoaglund CS, Valentine SJ, Sporleder CR, Reilly JP, Clemmer DE (1998): Three-dimensional ion mobility/TOFMS analysis of electrosprayed biomolecules. *Anal. Chem.* **70**:2236–2242.

Hoaglund-Hyzer CS, Counterman AE, Clemmer DE (1999): Anhydrous protein ions. *Chem. Rev.* **99**:3037–3079.

Hoang L, Bedard S, Krishna MMG, Lin Y, Englander SW (2002): Cytochrome *c* folding pathway: Kinetic native-state hydrogen exchange. *Proc. Natl. Acad. Sci. USA* **99**:12173–12178.

Horn DM, Breuker K, Frank AJ, McLafferty FW (2001): Kinetic intermediates in the folding of gaseous protein ions characterized by electron capture dissociation mass spectrometry. *J. Am. Chem. Soc.* **123**:9792–9799.

Horn DM, Ge Y, McLafferty FW (2000): Activated ion electron capture dissociation for mass spectral sequencing of larger (42 kDa) proteins. *Anal. Chem.* **72**:4778–4784.

Hudgins RR, Jarrold MF (1999): Helix formation in unsolvated alanine-based peptides: Helical monomers and helical dimers. *J. Am. Chem. Soc.* **121**:3494–3501.

Hudgins RR, Ratner MA, Jarrold MF (1998): Design of helices that are stable in vacuo. *J. Am. Chem. Soc.* **120**:12974–12975.

Hudgins RR, Woenckhaus J, Jarrold MF (1997): High resolution ion mobility measurements for gas phase proteins: Correlation between solution phase and gas phase conformations. *Int. J. Mass Spectrom.* **165/166**:497–507.

Iavarone AT, Parks JH (2005): Conformational change in unsolvated Trp-cage protein probed by fluorescence. *J. Am. Chem. Soc.* **127**:8606–8607.

Iavarone AT, Williams ER (2003): Mechanism of charging and supercharging molecules in electrospray ionization. *J. Am. Chem. Soc.* **125**:2319–2327.

Ibarra-Molero B, Loladze VV, Makhatadze GI, Sanchez-Ruiz JM (1999a): Thermal versus guanidine-induced unfolding of ubiquitin. An analysis in terms of the contributions from charge-charge interactions to protein stability. *Biochemistry* **38**:8138–8149.

Ibarra-Molero B, Makhatadze GI, Sanchez-Ruiz JM (1999b): Cold denaturation of ubiquitin. *Biochim. Biophys. Acta* **1429**:384–390.

Iribarne JV, Thomson BA (1976): On the evaporation of small ions from charged droplets. *J. Chem. Phys.* **64**:2287–2294.

Jameson CJ (1991): Gas-phase NMR spectroscopy. *Chem. Rev.* **91**:1375–1395.

Jarrold MF (1999): Unfolding, refolding, and hydration of proteins in the gas phase. *Acc. Chem. Res.* **32**:360–367.

Jarrold MF (2000): Peptides and proteins in the vapor phase. *Annu. Rev. Phys. Chem.* **51**: 179–207.

Jockusch RA, Schnier PD, Price WD, Strittmatter EF, Demirev PA, Williams ER (1997): Effects of charge state on fragmentation pathways, dynamics, and activation energies of ubiquitin ions measured by blackbody infrared radiative dissociation. *Anal. Chem.* **69**:1119–1126.

Jourdan M, Searle MS (2001): Insights into the stability of native and partially folded states of ubiquitin: Effects of cosolvents and denaturants on the thermodynamics of protein folding. *Biochemistry* **40**:10317–10325.

Jurchen JC, Williams ER (2003): Origin of asymmetric charge partitioning in the dissociation of gas-phase protein homodimers. *J. Am. Chem. Soc.* **125**:2817–2826.

Kaltashov IA, Eyles SJ (2002): Studies of biomolecular conformations and conformational dynamics by mass spectrometry. *Mass Spectrom. Rev.* **21**:37–71.

Kaltashov IA, Mohimen A (2005): Estimates of protein surface areas in solution by electrospray ionization mass spectrometry. *Anal. Chem.* **77**:5370–5379.

Kamatari YO, Konno T, Kataoka M, Akasaka K (1996): The methanol-induced globular and expanded denatured states of cytochrome *c*: A study by CD fluorescence, NMR and small-angle X-ray scattering. *J. Mol. Biol.* **259**:512–523.

Kaminsky LS, Davison AJ (1969): Effects of organic solvents on cytochrome *c*. *FEBS Lett.* **3**:338–340.

Karas M, Bachmann D, Bahr U, Hillenkamp F (1987): Matrix-assisted ultraviolet laser desorption of non-volatile compounds. *Int. J. Mass Spectrom.* **78**:53–68.

Karas M, Bahr U, Ingendoh A, Hillenkamp F (1989): Laser desorption/ionization mass spectrometry of proteins of mass 100 000 to 250 000 dalton. *Angew. Chem. Int. Ed. Engl.* **6**:760–761.

Karas M, Hillenkamp F (1988): Laser desorption ionization of proteins with molecular masses exceeding 10000 daltons. *Anal. Chem.* **60**:2299–2301.

Kauzmann W (1959): Some factors in the interpretation of protein denaturation. *Adv. Protein Sci.* **14**:1–63.

Kebarle P (2000): A brief overview of the present status of the mechanisms involved in electrospray mass spectrometry. *J. Mass Spectrom.* **35**:804–817.

Kebarle P, Peschke M (2000): On the mechanisms by which the charged droplets produced by electrospray lead to gas phase ions. *Anal. Chim. Acta* **406**:11–35.

Kelleher NL, Lin HY, Valaskovic GA, Aaserud DJ, Fridriksson EK, McLafferty FW (1999): Top down versus bottom up protein characterization by tandem high-resolution mass spectrometry. *J. Am. Chem. Soc.* **121**:806–812.

Kelly MA, Vestling MM, Fenselau CC, Smith PB (1992): Electrospray analysis of proteins: A comparison of positive-ion and negative-ion mass spectra at high and low pH. *Org. Mass Spectrom.* **27**:1143–1147.

Klotz IM (1993): Solvent water and protein behavior: View through a retroscope. *Protein Sci.* **2**:1992–1999.

Kohtani M, Jones TC, Schneider JE, Jarrold MF (2004): Extreme stability of an unsolvated α-helix. *J. Am. Chem. Soc.* **126**:7420–7421.

Konermann L, Douglas DJ (1998a): Equilibrium unfolding of proteins monitored by electrospray ionization mass spectrometry: Distinguishing two-state from multi-state transitions. *Rapid Commun. Mass Spectrom.* **12**:435–442.

Konermann L, Douglas DJ (1998b): Unfolding of proteins monitored by electrospray ionization mass spectrometry: A comparison of positive and negative ion modes. *J. Am. Soc. Mass Spectrom.* **9**:1248–1254.

Konn DO, Murrell J, Despeyroux D, Gaskell SJ (2005): Comparison of the effects of ionization mechanism, analyte concentration, and ion "cool-times" on the internal energies of peptide ions produced by electrospray and atmospheric pressure matrix-assisted laser desorption ionization. *J. Am. Soc. Mass Spectrom.* **16**:743–751.

Koppenol WH, Margoliash E (1982): The asymmetric distribution of charges on the surface of horse cytochrome *c*. *J. Biol. Chem.* **257**:4426–4437.

Krantz BA, Sosnick TR (2000): Distinguishing between two-state and three state models for ubiquitin folding. *Biochemistry* **39**:11696–11701.

Krishna MMG, Englander SW (2005): The N-terminal to C-terminal motif in protein folding and function. *Proc. Natl. Acad. Sci. USA* **102**:1053–1058.

Krishna MMG, Lin Y, Rumbley JN, Englander SW (2003): Cooperative omega loops in cytochrome *c*: Role in folding and function. *J. Mol. Biol.* **331**:29–36.

Laskin J, Futrell JH (2003): Entropy is the major driving force for fragmentation of proteins and protein-ligand complexes in the gas phase. *J. Phys. Chem. A* **107**:5836–5839.

Li J, Taraszka JA, Counterman AE, Clemmer DE (1999): Influence of solvent composition and capillary temperature on the conformations of electrosprayed ions: Unfolding of compact ubiquitin conformers from pseudonative and denaturated solutions. *Int. J. Mass Spectrom.* **185/186/187**:37–47.

Lin C, Infusini G, Oh HB, Hwang H, Breuker K, Carpenter BK, McLafferty FW (2005): Structural characterization of gaseous protein ions by infrared photodissociation spectroscopy, submitted.

Loladze VV, Ibarra-Molero B, Sanchez-Ruiz JM, Makhatadze GI (1999): Engineering a thermostable protein via optimization of charge-charge interactions on the protein surface. *Biochemistry* **38**:16419–16423.

Loo JA, Edmonds CG, Udseth HR, Smith RD (1990): Effect of reducing disulfide-containing proteins on electrospray ionization mass spectra. *Anal. Chem.* **62**:693–698.

Maity H, Maity M, Englander SW (2004): How cytochrome *c* folds, and why: Submolecular foldon units and their stepwise sequential stabilization. *J. Mol. Biol.* **343**:223–233.

Makhatadze GI, Lopez MM, Richardson JM, Thomas ST (1998): Anion binding to the ubiquitin molecule. *Protein Sci.* **7**:689–697.

Mao Y, Woenckhaus J, Kolafa J, Ratner MA, Jarrold MF (1999): Thermal unfolding of unsolvated cytochrome *c*: Experiment and molecular dynamics simulations. *J. Am. Chem. Soc.* **121**:2712–2721.

Marshall AG, Hendrickson CL, Jackson GS (1998): Fourier transform ion cyclotron mass spectrometry: A primer. *Mass Spectrom. Rev.* **17**:1–35.

Marshall AG, Hendrickson CL, Shi SDH (2002): Scaling MS plateaus with high-resolution FT-ICRMS. *Anal. Chem.* 74:252A-259A.

McLafferty FW, Guan ZQ, Haupts U, Wood TD, Kelleher NL (1998): Gaseous conformational structures of cytochrome *c*. *J. Am. Chem. Soc.* **120**:4732–4740.

McLuckey SA, Wells JM, Stephenson JL, Goeringer DE (2000): Novel quadrupole ion trap methods for characterizing the chemistry of gaseous macro-ions. *Int. J. Mass Spectrom.* **200**:137–161.

Mirza UA, Chait BT (1997): Do proteins denature during droplet evolution in electrospray ionization? *Int. J. Mass Spectrom.* **162**:173–181.

Mirza UA, Cohen SL, Chait BT (1993): Heat-induced conformational changes in proteins studied by electrospray ionization mass spectrometry. *Anal. Chem.* **65**:1–6.

Myung S, Badman ER, Lee YJ, Clemmer DE (2002): Structural transitions of electrosprayed ubiquitin ions stored in an ion trap over ~10 ms to 30 ms. *J. Phys. Chem. A* **106**:9976–9982.

Naganathan AN, Muñoz V (2005): Scaling of folding times with protein size. *J. Am. Chem. Soc.* **127**:480–481.

Nesatyy VJ, Suter MJF (2004): On the conformation-dependent neutralization theory and charging of individual proteins and their noncovalent complexes in the gas phase. *J. Mass Spectrom.* **39**:93–97.

Oh HB, Breuker K, Sze SK, Ge Y, Carpenter BK, McLafferty FW (2002): Secondary and tertiary structures of gaseous protein ions characterized by electron capture dissociation mass spectrometry and photofragment spectroscopy. *Proc. Natl. Acad. Sci. USA* **99**:15863–15868.

Oh HB, Lin C, Wang HY, Zhai H, Breuker K, Zabrouskov V, Carpenter BK, McLafferty FW (2005): Infrared photodissociation spectroscopy of electrosprayed ions in a Fourier transform mass spectrometer. *J. Am. Chem. Soc.* **127**:4076–4083.

Oomens J, Polfer NC, Moore DT, van den Meer L, Marshall AG, Eyler JR, Meijer G, van Helden G (2005): Charge-state resolved mid-infrared spectroscopy of a gas-phase protein. *Phys. Chem. Chem. Phys.* **7**:1345–1348.

Pan P, Gunawardena HP, Xia Y, McLuckey SA (2004): Nanoelectrospray ionization of protein mixtures: Solution pH and protein pI. *Anal. Chem.* **76**:1165–1174.

Polfer NC, Paizs B, Snoek LC, Compagnon I, Suhai S, Meijer G, van Helden G, Oomens J (2005): Infrared finderprint spectroscopy and theoretical studies of potassium ion tagged amino acids and peptides in the gas phase. *J. Am. Chem. Soc.* **127**:8571–8579.

Price WD, Schnier PD, Jockusch RA, Strittmatter EF, Williams ER (1996): Unimolecular reaction kinetics in the high-pressure limit without collisions. *J. Am. Chem. Soc.* **118**:10640–10644.

Purves RW, Barnett DA, Ells B, Guevremont R (2000): Investigation of bovine ubiquitin conformers separated by high-field asymmetric waveform ion mobility spectrometry: Cross section measurements using energy-loss experiments with a triple quadrupole mass spectrometer. *J. Am. Soc. Mass Spectrom.* **11**:738–745.

Purves RW, Barnett DA, Ells B, Guevremont R (2001): Elongated conformers of charge states +11 to +15 of bovine ubiquitin studied using ESI-FAIMS-MS. *J. Am. Soc. Mass Spectrom.* **12**:894–901.

Reid GE, McLuckey SA (2002): "Top down" protein characterization via tandem mass spectrometry. *J. Mass Spectrom.* **37**:663–675.

Rinehart KL, Gaudioso LA, Moore ML, Pandey RC, Cook JC, Barber M, Sedgwick RD, Bordoli RS, Tyler AN, Green BN (1981): Structures of eleven zervamicin and two emerimicin peptide antibiotics studied by fast atom bombardment mass spectrometry. *J. Am. Chem. Soc.* **103**:6517–6520.

Robinson EW, Williams ER (2005): Multidimensional separations of ubiquitin conformers in the gas phase: Relating ion cross sections to H/D exchange measurements. *J. Am. Soc. Mass Spectrom.* **16**:1427–1437.

Rohner TC, Lion N, Girault HH (2004): Electrochemical and theoretical aspects of electrospray ionization. *Phys. Chem. Chem. Phys.* **6**:3056–3068.

Rostom AA, Fucini P, Benjamin DR, Juenemann R, Nierhaus KH, Hartl FU, Dobson CM, Robinson CV (2000): Detection and selective dissociation of intact ribosomes in a mass spectrometer. *Proc. Natl. Acad. Sci. USA* **97**:5185–5190.

Samalikova M, Grandori R (2005): Testing the role of solvent surface tension in protein ionization by electrospray. *J. Mass Spectrom.* **40**:503–510.

Sanglier S, Leize E, van Dorsselaer A, Zal F (2003): Comparative ESI-MS study of 2.2 MDa native hemocyanins from deep-sea and shore crabs: From protein oligomeric state to biotope. *J. Am. Soc. Mass Spectrom.* **14**:419–429.

Sanishvili R, Volz KW, Westbrook EM, Margoliash E (1995): The low ionic strength crystal structure of horse cytochrome *c* at 2.1 Å resolution and comparison with its high ionic strength counterpart. *Structure* **3**:707–716.

Schnier PD, Gross DS, Williams ER (1995): On the maximum charge state and proton transfer reactivity of peptide and protein ions formed by electrospray ionization. *J. Am. Soc. Mass Spectrom.* **6**:1086–1097.

Shelimov KB, Clemmer DE, Hudgins RR, Jarrold MF (1997): Protein structure in vacuo: Gas phase conformations of BPTI and cytochrome *c*. *J. Am. Chem. Soc.* **119**:2240–2248.

Shelimov KB, Jarrold MF (1996): "Denaturation" and refolding of cytochrome *c* in vacuo. *J. Am. Chem. Soc.* **118**:10313–10314.

Sheu SY, Yang DY, Selzle HL, Schlag EW (2003): Energetics of hydrogen bonds in peptides. *Proc. Natl. Acad. Sci. USA* **100**:12683–12687.

Smith RD, Loo JA, Edmonds CG, Barinaga CJ, Udseth HR (1990): New developments in biochemical mass spectrometry: Electrospray ionization. *Anal. Chem.* **62**:882–899.

Southall NT, Dill KA, Haymet ADJ (2002): A view of the hydrophobic effect. *J. Phys. Chem. B* **106**:521–533.

Strader MB, VerBerkmoes NC, Tabb DL, Connelly HM, Barton JW, Bruce BD, Pelletier DA, Davison BH, Hettich RL, Larimer FW, Hurst GB (2004): Characterization of the 70S ribosome from Rhodopseudomonas palustris using an integrated "top-down" and "bottom-up" mass spectrometric approach. *J. Proteome Res.* **3**:965–978.

Suckau D, Shi Y, Beu SC, Senko MW, Quinn JP, Wampler FM, McLafferty FW (1993): Coexisting stable conformations of gaseous protein ions. *Proc. Natl. Acad. Sci. USA* **90**:790–793.

Surman DJ, Vickerman JC (1981): Fast atom bombardment quadrupole mass spectrometry. *J. Chem. Soc., Chem. Commun.* **7**:324–325.

Sze SK, Ge Y, Oh HB, McLafferty FW (2002): Top-down mass spectrometry of a 29-kDa protein for characterization of any posttranslational modification to within one residue. *Proc. Natl. Acad. Sci. USA* **99**:1774–1779.

Sze SK, Ge Y, Oh HB, McLafferty FW (2003): Plasma electron capture dissociation for the characterization of large proteins by top down mass spectrometry. *Anal. Chem.* **75**: 1599–1603.

Travaglini-Allocatelli C, Gianni S, Brunori M (2004): A common folding mechanism in the cytochrome *c* family. *TiBS* **29**:535–541.

Valentine SJ, Clemmer DE (1997): H/D exchange levels of shape-resolved cytochrome *c* conformers in the gas phase. *J. Am. Chem. Soc.* **119**:3558–3566.

Valentine SJ, Clemmer DE (2002): Temperature-dependent H/D exchange of compact and elongated cytochrome *c* ions in the gas phase. *J. Am. Soc. Mass Spectrom.* **13**:506–517.

Valentine SJ, Counterman AE, Clemmer DE (1997): Conformer-dependent proton-transfer reactions of ubiquitin ions. *J. Am. Soc. Mass Spectrom.* **8**:954–961.

van Duijn E, Bakkes PJ, Heeren RMA, van den Heuvel RHH, van Heerikhuizen H, van den Vies SM, Heck AJR (2005): Monitoring macromolecular complexes involved in the chaperonin-assisted protein folding cycle by mass spectrometry. *Nat. Meth.* **2**:371–376.

Verkerk UH, Kebarle P (2005): Ion-ion and ion-molecule reactions at the surface of proteins produced by nanospray. Information on the number of acidic residues and control of the number of ionized acidic and basic residues. *J. Am. Soc. Mass Spectrom.* **16**:1325–1341.

Verkerk UH, Peschke M, Kebarle P (2003): Effect of buffer cations and of H_3O^+ on the charge states of native proteins. Significance to determinations of stability constants of protein complexes. *J. Mass Spectrom.* **38**:618–631.

Vijay-Kumar S, Bugg CE, Cook WJ (1987): Structure of ubiquitin refined at 1.8 Å resolution. *J. Mol. Biol.* **194**:531–544.

Wilkinson KD, Urban MK, Haas AL (1980): Ubiquitin is the ATP-dependent proteolysis factor I of rabbit reticulocytes. *J. Biol. Chem.* **255**:7529–7531.

Wilson DJ, Konermann L (2003): A capillary mixer with adjustable reaction chamber volume for millisecond time-resolved studies by electrospray mass spectrometry. *Anal. Chem.* **75**:6408–6414.

Winger BE, Light-Wahl KJ, Rockwood AL, Smith RD (1992): Probing qualitative conformation differences of multiply protonated gas-phase proteins via H/D isotopic exchange with D_2O. *J. Am. Chem. Soc.* **114**:5897–5898.

Wintrode PL, Makhatadze GI, Privalov PL (1994): Thermodynamics of ubiquitin unfolding. *Proteins Struct. Funct. Genet.* **18**:246–253.

Woenckhaus J, Mao Y, Jarrold MF (1997): Hydration of gas phase ions: Folded +5 and unfolded +7 charge states of cytochrome *c*. *J. Phys. Chem. B* **101**:847–851.

Wood TD, Chorush RA, Wampler FM, Little DP, O'Connor PB, McLafferty FW (1995): Gas-phase folding and unfolding of cytochrome *c* cations. *Proc. Natl. Acad. Sci. USA* **92**: 2451–2454.

Wyttenbach T, Bowers MT (1998): Gas phase conformations of biological molecules: The hydrogen/deuterium exchange mechanism. *J. Am. Soc. Mass Spectrom.* **10**:9–14.

Wyttenbach T, Kemper PR, Bowers MT (2001): Design of a new electrospray ion mobility mass spectrometer. *Int. J. Mass Spectrom.* **212**:13–23.

Yang AS, Honig B (1995a): Free energy determinants of secondary structure formation: I. α-helices. *J. Mol. Biol.* **252**:351–365.

Yang AS, Honig B (1995b): Free energy determinants of secondary structure formation: II. Antiparallel β-sheets. *J. Mol. Biol.* **252**:366–376.

Zhai H, Han X, Breuker K, McLafferty FW (2005): Consecutive ion activation for top down mass spectrometry: Improved protein sequencing by nozzle-skimmer dissociation. *Anal. Chem.* **77**:5777–5784.

Zhang X, Cassady CJ (1996): Appararent gas-phase acidities of multiply protonated peptide ions: Ubiquitin, insulin B, and renin substrate. *J. Am. Soc. Mass Spectrom.* **7**:1211–1218.

Zubarev RA, Horn DM, Fridriksson EK, Kelleher NL, Kruger NA, Lewis MA, Carpenter BK, McLafferty FW (2000): Electron capture dissociation for the structural characterization of multiply charged protein cations. *Anal. Chem.* **72**:563–573.

Zubarev RA, Kelleher NL, McLafferty FW (1998): Electron capture dissociation of multiply charged protein cations. A nonergodic process. *J. Am. Chem. Soc.* **120**:3265–3266.

Zubarev RA, Kruger NA, Fridriksson EK, Lewis MA, Horn DM, Carpenter BK, McLafferty FW (1999): Electron capture dissociation of gaseous multiply-charged proteins is favored at disulfide bonds and other sites of high hydrogen atom affinity. *J. Am. Chem. Soc.* **121**:2857–2862.

6

DYNAMICAL SIMULATIONS OF PHOTOIONIZATION OF SMALL BIOLOGICAL MOLECULES

DORIT SHEMESH

Department of Physical Chemistry and the Fritz Haber Research Center
The Hebrew University
Jerusalem, Israel

R. BENNY GERBER*

Department of Chemistry
University of California
Irvine, CA
and
Department of Physical Chemistry and the Fritz Haber Research Center
The Hebrew University
Jerusalem, Israel

*To whom correspondence should be addressed (benny@fh.huji.ac.il).

Principles of Mass Spectrometry Applied to Biomolecules, edited by Julia Laskin and Chava Lifshitz

6.1. INTRODUCTION

The mechanism and dynamics of photoionization of biological molecules are of considerable intrinsic interest and may have potential applications, especially in mass spectrometry. Ionization, carried out by several possible processes (Alexandrov et al. 1984; Yamashita and Fenn 1984a,b; Karas et al. 1987), is obviously a central aspect of mass spectrometry. At the same time, mass spectrometry has already developed into a major tool in the study and characterization of biological molecules, from small to very large (Koster and Grotemeyer 1992; Schlag et al. 1992; Bowers et al. 1996; Burlingame et al. 1998; Lockyer and Vickerman 1998, 2000; Vorsa et al. 1999; Chalmers and Gaskell 2000; Cohen et al. 2000; Aebersold and Goodlett 2001; Griffiths et al. 2001; Jonsson 2001; Mann et al. 2001; Nyman 2001; Aebersold and Mann 2003; Ferguson and Smith 2003; Lin et al. 2003; Mano and Goto 2003; Standing 2003). Little is currently known on the dynamical processes that take place on photoionization of biological molecules. The mass spectra give useful information about the parent molecule (reactant) and the fragments resulting from photoionization (products). In addition, standard calculations can supply energetics of the reactants and products, and the transition states that connect them (Weinkauf et al. 2002). This may serve as an explanation for the abundance of any fragment recorded in the mass spectra. However, mass spectra cannot explain the dynamical evolution of the system after ionization. Molecular dynamics simulation has already been applied for various kinds of processes. Ionization is seldom modeled since appropriate potential energy surfaces are rare. We introduce here a molecular dynamics simulation of ionization processes on a semiempirical potential energy surface. The molecular dynamics simulation can give deep insight into existing conformers and also can show whether there are, for example, unstable intermediate states that cannot be detected by the mass spectra. Molecular dynamics is also able to assign fragment structures to unresolved peaks in the mass spectra. Complementary, statistical theories such as RRK and RRKM can be used for calculating rate constants and for comparing them to the rate constants obtained from the dynamical simulations. Today, RRK and RRKM are a very helpful tool in understanding experimental reaction rates. (Lifshitz 1992, 2001, 2003).

The channels that open up on ionization include internal flow and redistribution of the vibrational energy between the modes, conformational transition of the nascent ion; transfer of hydrogen within the ion (Rodriguez-Santiago et al. 2000; Simon et al. 2002), and fragmentation of the ion (Depke et al. 1984; Rizzo et al. 1985a,b; Elokhin et al. 1991; Ayre et al. 1994; Dey and Grotemeyer 1994; Reilly and Reilly 1994; Belov et al. 1995; Lindinger et al. 1999; Vorsa et al. 1999; Simon et al. 2002; Shemesh et al. 2004; Shemesh and Gerber 2005). The study presented here focuses mostly on the first two types of processes: intramolecular vibrational energy redistribution (IVR) and transitions between different conformers. These are believed to be the fastest and most efficient dynamical processes in such systems. However, chemical processes that involve bond breaking or shifting and require much longer timescales also depend ultimately on the outcome of the IVR and conformational transition events. It is important to know whether a statistical distribution of vibrational energy is indeed obtained and on which timescale it is reached. The issue is how long after the ionization a vibrational distribution compatible with RRKM is obtained. The timescale that will be explored here is short, only 10 ps, but it is useful to know whether the system approaches a statistical distribution. If not, characterization of the patterns of vibrational energy flow is of considerable interest. The issue of conformational transitions is likewise important: Which conformers are populated following ionization, and on which timescale does this take place? This issue is often discussed qualitatively in mass spectrometry, but it seems that it was not studied quantitatively, by dynamics simulations. Another interesting question is whether *some* fragmentation events can take place already on the short timescale studied here. Clearly, the yield for such processes on this short timescale is expected to be very low, but it is interesting to explore whether the event is not too rare to be seen for some of the trajectories in the set (and hopefully be measured experimentally). It will indeed be seen that for the set of trajectories used, some fragmentation events are found. In summary, the present chapter explores the dynamical evolution of the system after ionization, using classical trajectory simulations and a semiempirical potential surface, the choice of which will be discussed later. The chapter is organized as follows. Section 6.2 presents the methodology used. In Section 6.3 the findings are described and analyzed. Section 6.4 introduces future prospects.

6.2. METHODOLOGY

6.2.1. Potential Energy Surface

Biological molecules are most often studied using empirical forcefields such as Amber (Weiner and Kollmann 1981; Weiner et al. 1984), OPLS (Jorgensen and Tiradorives 1988; Jorgensen et al. 1996), MMFF94 (Halgren 1996a,b). The advantages of these forcefields are that they are simple to use, are computationally fast, and give adequate results for neutral molecules. However, these forcefields are most often inapplicable for ionic molecules, since they were not parameterized for such cases. Exceptions are DNA and proteins in their protonated states, which are well described by forcefields.

A more accurate approach is to use ab initio potentials, but these are computationally expensive, even for biological molecules of modest size. In dynamical simulations the potential energy is evaluated thousands of times along the trajectories. This operation is the main computational effort in such simulations, and using ab initio potentials would be very time-consuming. Such ab initio simulations would be limited to only very short timescales and the smallest systems.

Therefore, in this study we use PM3 semiempirical electronic structure theory (Stewart 1989a,b). PM3 is one of several modified semiempirical NDDO approximation (neglect of diatomic differential overlap) methods (Jensen 1999). Rather than performing a calculation of all the integrals required in the formation of the Fock matrix, three- and four-center integrals are neglected in PM3, and one-center, two-electron integrals are parameterized. Thus, in principle, PM3 is closer to ab initio methods than forcefields. Additionally, PM3 has been applied to calculations of small proteins (Lee et al. 1996; Stewart 1997; Daniels and Scuseria 1999). It is not known whether PM3 is capable of correctly describing bond breaking for radical ions. Another problem that may arise is the Hartree–Fock instability and possible degeneracy for open-shell systems. The simulations presented here were carried out with standard PM3. At the same time we also tested some properties of the potential surface (e.g., the barrier for internal rotation of the glycine ion) against ab initio results. This is discussed later.

All calculations were performed using the electronic structure package GAMESS (available from http://www.msg.ameslab.gov/GAMESS/GAMESS.html). The relevant conformers were optimized using PM3 semiempirical electronic structure theory on the neutral surface. For glycine the global minimum and a conformer only slightly higher in energy were chosen (Shemesh et al. 2004; Shemesh and Gerber 2005). For tryptophan the global minimum (Snoek et al. 2001) was computed and optimized by PM3. The second derivative (Hessian) matrix was calculated to ensure that the stationary point is indeed a minimum. Harmonic normal-mode analysis was performed on this geometry. Initial coordinates for glycine were selected according to the following procedure. Each mode was distorted from equilibrium. For each mode 16 different positions were chosen on an equidistant grid, and the Wigner distribution for these geometries was calculated. Using this procedure, 384 initial geometries were found. For tryptophan the same procedure was used for simulating single-photon ionization; 1200 geometries were found in this way. The excess energy of these geometries on the ionic surface (compared to a reference nearby ionic minimum) was computed and the 100 geometries with the highest excess energy were chosen as initial geometries.

For the two-photon ionization process of tryptophan 91 initial geometries were found by the following procedure. Two modes were simultaneously distorted from equilibrium, each on a 16-point grid; 710,400 geometries were found in this way. The excitation energy to the first excited state of tryptophan has been measured by D. H. Levy and coworkers (Rizzo et al. 1985b) and corresponds to 34,873 cm^{-1} (4.32 eV or 416.8 kJ/mol). The two-photon ionization mechanism is presumed to go through this state. The vertical ionization energy therefore is equal to two photons (8.64 eV or 833.6 kJ/mol). Only geometries with ionization energy of 8.64 eV were accepted; 91 geometries were chosen in such a way, that their ionization energy differed by up to 0.001 eV (~0.1 kJ/mol) from the abovementioned value. Note that few of the 710,400 geometries have this property and therefore are not suitable for simulating the two-photon ionization.

For each geometry in the Franck–Condon regime, single-photon ionization was modeled by vertical promotion into the ionic potential energy surface. For two-photon ionization it is assumed that the first photon promotes the system to the first excited state. Immediately, a second photon is absorbed that leads to the desired ionization. It is assumed that two-photon absorption is a very fast process, so that the geometry does not change at all. Evidence for this is provided by studies by Kushwaha and Mishra (2000) and Rizzo et al. (1986a). Kushwaha and Mishra (2000) show in their calculations that the first excited-state geometry of tryptophan is almost equivalent to the neutral geometry. Rizzo et al. (1986a) show in a supersonic beam experiment that the excited-state conformers of tryptophan do not interconvert during the fluorescence lifetime.

Note that the initial geometry sampling for single- and two-photon ionization is different. The simulation assumes a monochromatic source for the single-photon ionization. Therefore each geometry chosen by the procedure described above is suited for the simulation. In contrast, the two-photon ionization process implies that the ionization energy is exactly equal to two photons of certain energy. The assumed procedure of the two-photon ionization process goes through an excited state, which is well known, and measured experimentally. Therefore the initial geometries chosen must fulfill this property. This leads to a small number of accepted geometries.

6.2.2. Classical Trajectory Simulations of the Dynamics

The method used for the simulation is "on the fly" molecular dynamics (Stewart et al. 1987; Maluendes and Dupuis 1990; Taketsugu and Gordon 1995a–c; Gordon et al. 1996; Takata et al. 1998) that is implemented into the electronic structure program package GAMESS. More recently, studies of dynamics on the fly using QM/MM and semiempirical potentials have been pursued by Hase and coworkers (Wang et al. 2003; Wang and Hase 2003). Obviously, some quantum effects are expected to play a role, and these are neglected in the classical approach. Zero-point energy is probably the most important quantum effect neglected in this study. However, the excess internal energy of the ionic states is fairly high [$E > 0.8\,\text{eV}(77.2\,\text{kJ/mol})$], and we assume that the classical description should be reasonable. In on-the-fly molecular dynamics, at each timestep the current potential energy is evaluated and the forces are computed.

The atoms are moved according to these forces to a new position (next timestep) and there again, the forces are calculated from the potential energy and the atoms are moved, and so on. A very demanding self-consistent field (SCF) (ROHF) convergence criterion of 10^{-11} au was employed to ensure in this case accurate force calculations for the timescale of the study. The default value of 10^{-5} au employed in the standard GAMESS code is not sufficiently small for the present case. Calculations with this value have shown that the computed trajectory contains unphysical artifacts. The reason is obvious—a more accurate description of the potential energy surface yields a more precise force calculation. Fewer errors are then accumulated during the simulation, and the calculated trajectory deviates less from the true one. Each trajectory was calculated for 10 ps with a timestep of 0.1 fs (to ensure energy conservation). Ionization was modeled by vertical promotion into the ionic potential energy surface. After ionization the trajectory was propagated in time on the ionic PM3 potential energy surface (using ROHF in the Hartree–Fock part of the code). Trajectories where the ROHF energy calculations failed to converge or for which energy conservation was not satisfied were rejected. Energy was considered to be conserved when the difference between the initial total energy and the total energy (in atomic units) at the current timestep was smaller than 5×10^{-5}. The analysis was carried out completely for the remaining trajectories, which successfully reached the end of the simulation timescale (here 10 ps). Note that the single-photon ionization model used does not correspond to a monochromatic source, but to radiation having frequencies across the range of the photoionization (absorption) spectrum. This means that different trajectories had different energies in the ionic state. In contrast, the two-photon ionization model corresponds to a monochromatic source as can be supplied for example by a laser.

6.2.3. Modeling the Initial State for One- and Two-Photon Ionization

The molecule is assumed to be initially in its vibrational ground state. This is an experimentally realizable (e.g., in low-temperature-beam experiments), well-defined condition. For such an initial state, the classical description is quite unrealistic (classically, the system at $T = 0$ K is initially at rest at the minimum configuration, with no zero-point energy), so there will be only one classical trajectory for these conditions. Classical description becomes closer to reality if the simulations are done for initial temperature $T > 0$, in fact for sufficiently high T. In summary, the ground-state vibrational wavefunction is appropriate for sampling the initial state. Furthermore, for this state the anharmonic corrections are modest, and a harmonic wavefunction seems to be a reasonable wavefunction. To sample trajectories according to the initial state, we use the Wigner distribution function (Wigner 1932).

$$W(r,p) = \left(\frac{1}{2\pi\hbar}\right)^N \int ds \cdot \exp\left[\frac{is \cdot p}{\hbar}\right] \cdot \Psi^*\left(r + \frac{s}{2}\right) \times \Psi\left(r - \frac{s}{2}\right) \qquad (6.1)$$

where Ψ is the wavefunction of the state, r are the coordinates, and p are the momenta.

Each normal mode is treated as a classical harmonic oscillator in its ground vibrational state. Substitution of the harmonic oscillator wavefunction into the Wigner distribution gives the Wigner distribution for an n-dimensional harmonic oscillator:

$$W(q,p) = (\pi\hbar)^{-n} \prod_{i=1}^{n} e^{-p_i^2/\alpha_i} e^{-k_i q_i^2/\alpha_i} \qquad (6.2)$$

where q_i are the normal modes, p_i are the corresponding momenta, k_i is the force constant of the ith normal mode, and α_i is related to the corresponding vibrational frequency ($\alpha_i = \hbar\omega_i$). For the excitation process, we assume, in the spirit of the Franck–Condon principle, vertical promotion to the ionic state. This implies that the initial velocities are zero on the ionic surface; the configurations are those sampled for the neutral species.

6.2.4. Statistical Approximation

RRK is a very well established theory (Baer and Hase 1996) and has been tested in ionization processes for fragmentation and conformational transitions. RRK gives in most of the cases very satisfactory results.

To test the validity of the statistical approximation for the interconversion between conformers, the results of the trajectory calculations will be compared with the classical RRK rates. For a process at energy E, the RRK rate is given by Baer and Hase (1996) and Holbrook et al. (1996)

$$k(E) = A \cdot \left(\frac{E - E_0}{E}\right)^{s-1} \tag{6.3}$$

where A is taken as

$$A = \left(\frac{\prod_{j=1}^{s} \upsilon_j}{\prod_{i=1}^{s-1} \upsilon_i}\right) \tag{6.4}$$

where E is the total energy, E_0 is the energy barrier, s is the number of degrees of freedom, υ_j are the frequencies from the initial minimum geometry, and υ_i are the frequencies from the transition state. The transition state of the interconversion between conformers was found by searching for an extremum along the reaction path, and checking for a configuration having a single negative Hessian frequency eigenvalue. If the initial state corresponds to a distribution of energies of the molecules rather than to be a microcanonical ensemble, the overall RRK rate is

$$\bar{k}_{\text{RRK}} = \int_{E_0}^{\infty} P(E)A\left(\frac{E - E_0}{E}\right)^{s-1} dE \tag{6.5}$$

where $P(E)dE$ is the fraction of initial states having energies between E and $E + dE$. In the statistical approximation, the specific initial conditions for an ensemble are assumed to be "forgotten" on a very short timescale. Thus, $P(E)$ in Eq. (6.5) represents the only effect of a nonmonoenergetic ensemble on the RRK reaction rate. The validity of RRK does not, of course, depend on having a monoenergetic ensemble, and can be tested in principle for any ensemble.

Another way of calculating the rate constant is to use the microcanonical transition state theory (i.e., RRKM theory), wherein the rate constant for a monoenergetic initial state ensemble is given by Baer and Hase (1996)

$$k(E) = \frac{N^{\mathrm{TS}}(E - E_0)}{h\rho(E)} \tag{6.6}$$

where $N^{\mathrm{TS}}(E - E_0)$ is the number of states in the transition state and $\rho(E)$ is the reactant density of states. This includes static quantum effects. We chose to use the RRK rate constant, since it is more compatible with the results of classical dynamics performed in this study.

6.3. APPLICATIONS

6.3.1. Systems

This study focuses on single- and two-photon ionization of biological molecules: glycine and tryptophan. Single-photon ionization of glycine, the simplest amino acid, has been previously extensively studied by the present authors (Shemesh et al. 2004; Shemesh and Gerber 2005). The main aspects of this process have been summarized here. There are a wealth of relevant data on (neutral) glycine, including the structure of its conformers and their spectroscopy (Jensen and Gordon 1991; Godfrey and Brown 1995; Godfrey et al. 1996; Ivanov et al. 1997, 1999; Stepanian et al. 1998; Huisken et al. 1999; McGlone et al. 1999; Chaban et al. 2000; Brauer et al. 2004; Miller and Clary 2004), and many characteristics of the potential energy surface. On the contrary, relatively little is known about the glycine ion produced by photoionization. In particular, little is known about the ionic potential energy surface (Depke et al. 1984; Yu et al. 1995; Rodriguez-Santiago et al. 2000; Simon et al. 2002). Yu et al. (1995) found two ionic conformers at the G2(MP2) level of ab initio electronic structure theory. Both are similar in structure with neutral conformers. The positive charge in both conformers is located on the nitrogen. One ionic conformer has one hydrogen-bonding interaction between one hydrogen of the NH_2 group and the carbonyl oxygen. The hydrogens connected to the nitrogen are in the same plane as the C—C=O backbone. The second conformer has a bifurcated interaction linking the amino hydrogens to the hydroxyl oxygen lone pairs. The energy difference between these two isomers at the G2(MP2) level is predicted to be 5.9 kJ/mol. According to the work of Rodríguez-Santiago et al. (2000) (DFT and MP2 calculations), there is a third ionic conformer that is similar in structure with another neutral conformer. This conformer has an interaction linking the hydroxyl hydrogen to the amino lone pair.

The single- and two-photon ionization of tryptophan has also been studied in this work. Tryptophan is an aromatic amino acid, containing the indole group as the chromophore. Tryptophan has also been well characterized by experiments and calculations (Rizzo et al. 1985a,b, 1986a,b; Elokhin et al. 1991; Ayre et al. 1994;

Dey and Grotemeyer 1994; Reilly and Reilly 1994; Belov et al. 1995; Callis 1997; Lindinger et al. 1999; Vorsa et al. 1999; Kushwaha and Mishra 2000; Piuzzi et al. 2000; Compagnon et al. 2001; Snoek et al. 2001; Bakker et al. 2003; Huang and Lin 2005), in particular, the neutral conformers existing in a supersonic beam and in matrix have been identified using spectroscopy. The excited states of tryptophan are well known (Rizzo et al. 1985b; Callis 1997). In the gas phase, the lowest-lying excited singlet state is denoted by ^1L_b and the second excited singlet state, by ^1L_a (Rizzo et al. 1985b; Callis 1997). The mass spectra of tryptophan have been recorded by different groups (Rizzo et al. 1985a,b, 1986a,b; Elokhin et al. 1991; Ayre et al. 1994; Dey and Grotemeyer 1994; Reilly and Reilly 1994; Belov et al. 1995; Lindinger et al. 1999; Vorsa et al. 1999). Still, there are no available data on the ionic surface such as, for example, stable ionic conformers.

6.3.2. Ultrafast Internal Rotation Effect in Photoionization of Glycine and Tryptophan

6.3.2.1. Single-Photon Ionization of Glycine. Figure 6.1 shows the equilibrium structures of conformers I and II computed from PM3. The energy difference between the conformers is 0.09 eV (8.7 kJ/mol). Conformer I corresponds to the global minimum, and conformer II is another low-energy minimum on the potential energy surface. These conformers are well characterized in the literature by calculations and experiments, including spectroscopy (Stepanian et al. 1998; Huisken et al. 1999; McGlone et al. 1999; Chaban et al. 2000). The main difference between these conformers lies in the arrangement of the NH_2 group. At conformer I the N atom lies in the same plane as the C–COOH atoms. The hydrogens connected to the nitrogen are perpendicular to that plane. This conformer has the C_s symmetry. The two amino hydrogens interact with the carbonyl oxygen. Conformer II has an almost planar arrangement of the C–COOH atoms and a pyramidal arrangement of the C–NH_2 part of the molecule. The nitrogen lies out of the plane, whereas the hydrogens connected to it point into the plane.

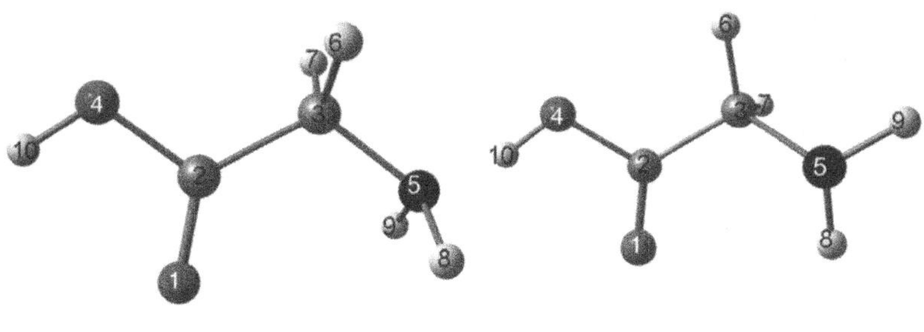

(a) (b)

FIGURE 6.1. Glycine conformers used in this study as optimized by PM3: (a) global minimum, $E = 0\,\text{eV}$; (b) second conformer used in this study, $\Delta E = 0.09\,\text{eV}$. [Reproduced from Shemesh D, Gerber RB (2005): *J. Chem. Phys.* **122**:241104 (Shemesh and Gerber 2005) with permission of the American Institute of Physics.]

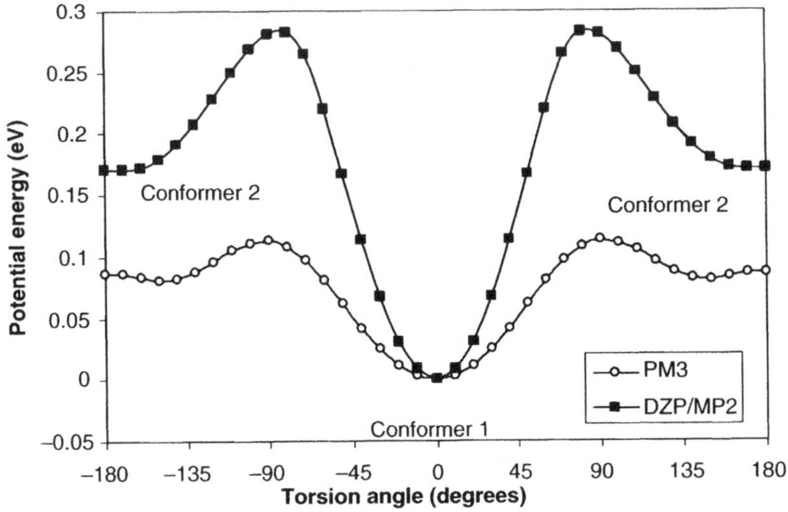

FIGURE 6.2. Potential energy as a function of the torsion angle N5–C3–C2–O1 calculated by PM3 and DZP/MP2. [Reproduced from Shemesh, D et al. (2004): *J. Phys. Chem. A* 108: 11477–11484 (Shemesh et al. 2004) with permission of the American Chemical Society.]

A detailed analysis of conformer II photoionization dynamics was given in the authors' previous studies (Shemesh et al. 2004; Shemesh and Gerber 2005). Important results are summarized here. A total number of 361 trajectories successfully reached 10 ps. Several trajectories show a fast internal rotation about the C–C bond. The potential energy as a function of rotation angle has two unequal wells as can be seen in Figure 6.2. This corresponds to two stable conformers on the ionic surface. The molecule has to overcome a barrier in order to rotate freely. The barrier height is 0.11 eV (10.6 kJ/mol) by PM3 [0.28 eV (27.0 kJ/mol) by MP2/DZP]. For nearly all the trajectories the total energy of the ion is much larger than the barrier, but in some events the energy along the internal rotation is insufficient and the system can become stuck in one of the two wells. As energy flows back into the rotational mode, "hopping" between the conformers can occur. The rotation occurs to a different extent in both conformers. Out of 362 trajectories of conformer I, 13 show the rotation, whereas for conformer II about 3 times more (37 out of 361) show the rotation. The corresponding probabilities from the Wigner distribution are as follows. For conformer I the probability of rotation is 0.79% whereas for conformer II the probability is about twice as high (2.23%).

6.3.2.2. Two-Photon Ionization of Tryptophan. Figure 6.3 shows the global neutral minimum of tryptophan as optimized by PM3. The indole ring is mainly planar. Note that the distance between O14 and H17 in the global minimum is 2.81 Å. Most of the trajectories show that these two atoms become closer during the dynamics and thus form a hydrogen bond (average distance during the dynamics: 1.85 Å). This hydrogen bond prevents the COOH group from rotating, and therefore this rotation is observed in only one trajectory in case of the two-photon ionization.

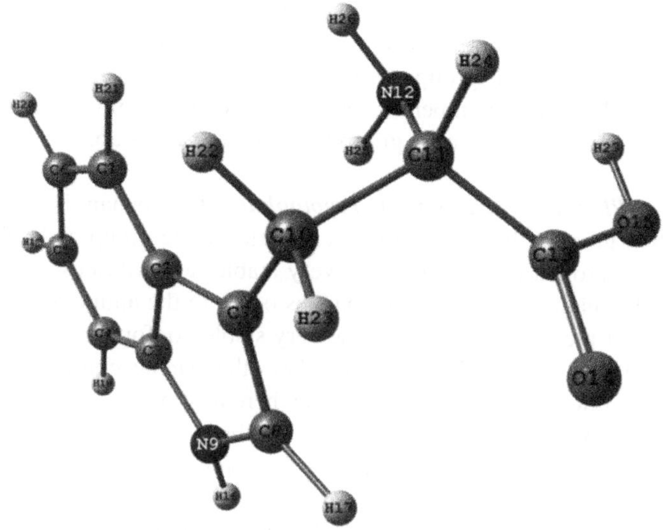

FIGURE 6.3. Global minimum of tryptophan as optimized by PM3.

Even after breaking the hydrogen bond, there is still steric repulsion between the H17 of the aromatic ring and the OH of the carboxylic group. The excess energy after the two-photon ionization process on the ionic surface is up to 4.3 eV (414.9 kJ/mol). In principle, this is sufficient for breaking the hydrogen bond. But one must keep in mind that tryptophan is a relatively large molecule with 75 normal modes. The excess energy can distribute among these modes, and it can take some time before this energy flows into one specific bond.

6.3.2.3. Single-Photon Ionization of Tryptophan. Here the trajectories were chosen in such a way that they have large excess energies [between 1.8 eV (173.7 kJ/mol) and 7.2 eV (694.7 kJ/mol)]. This enables the molecule to undergo more conformational changes. This is indeed the case here. The rotation about the C11–C13 bond (see Figure 6.3 for nomenclature of the atoms) is observed in 3 (out of 94) trajectories. Also, many of the trajectories here show the conformer with the abovementioned hydrogen bond. This seems to be a kinetically stable conformer. Out of 94 trajectories, 3 show a rotation of the amino acid backbone about the C7–C10 bond. This rotation does not show up for the case of two-photon ionization of tryptophan. These three trajectories have excess energy of >3.8 eV (366.6 kJ/mol). Therefore, the C7–C10 rotation requires a large amount of excess internal energy.

6.3.3. Conformational Transitions Induced by Photoionization: Glycine and Tryptophan

6.3.3.1 Single-Photon Ionization of Glycine. The main conformers observed during the dynamics correspond to a geometry that is similar to conformer I (ionic conformer A) and another geometry where the COOH group is rotated by 180° (ionic conformer B). Structural changes during the dynamics are rather small and involve rotation of the NH_2 group and variation of the C–O–H angle. Isomerization is also observed; a small number of trajectories for both initial conformers show hydrogen transfer from the amino group to the carbonyl group.

6.3.3.2. Two-Photon Ionization of Tryptophan. Tryptophan is a much larger molecule than glycine. It consists of an aromatic indole group and an amino acid backbone. The aromatic group itself is very stable, and vibrates only slightly on ionization. The main conformational changes occur in the amino acid backbone. A large number of the trajectories show the very stable conformer with the hydrogen bond between O14 and H17. There are also other stable conformers that occur during the dynamics. One trajectory is chosen here to show the conformers accessed during the simulation. A detailed study of the photoionization dynamics of tryptophan is given in (Shemesh et.al 2006). The timescale of the trajectory is 10 ps. The excess energy for this trajectory is 2.7 eV (260.5 kJ/mol). Geometries of this trajectory were optimized by PM3, and the energy as a function of the time was plotted. Different conformers can be distinguished by their energies. By comparison to the 3D structure (see Figure 6.4), six main conformers could be identified (gray boxes in Figure 6.5). The small differences in the energy of each conformer are due to small changes in the geometry, as, for example, rotation of the NH_2 group. The first conformer in Figure 6.4 corresponds to the optimized starting geometry. Then after a short time, the molecule jumps to the nearby conformer by bringing the C=O group close to one hydrogen of the indole ring. This geometry (second conformer of Figure 6.4) is stabilized by a hydrogen bond between the carbonyl oxygen and a hydrogen of the indole ring. This conformer is especially kinetically stable, as can be seen from Figure 6.5. This is the same conformer as described above. In the first conformer the COOH group is above the plane of the indole ring. During the dynamics it changes so that it is later below the plane of the indole ring. The fifth and sixth conformers are especially stable. Here the COOH group is far from the indole ring, so there is less repulsion between these two groups. This has a stabilizing effect on the conformer. The main changes from the first conformer to the last conformer are due to the rotation about the C11–C10 bond. In summary, the two photon ionization of tryptophan leads to conformational changes, which are exemplified with one trajectory. This provides insight into which different conformers are populated during, for example, a mass spectrometric experiment. The trajectories calculated here show only conformational changes as seen in case the presented above, and no isomerization or fragmentation is found for the two-photon ionization process. This may be due to the fact that the trajectories here do not contain enough energy for these changes.

6.3.3.3. Single-Photon Ionization of Tryptophan. Single-photon ionization of tryptophan leads to higher excess energies than does two-photon ionization. Besides the conformational changes in the sense discussed above, this excess energy is able to isomerize the molecule. Still, fragmentation has not been observed despite

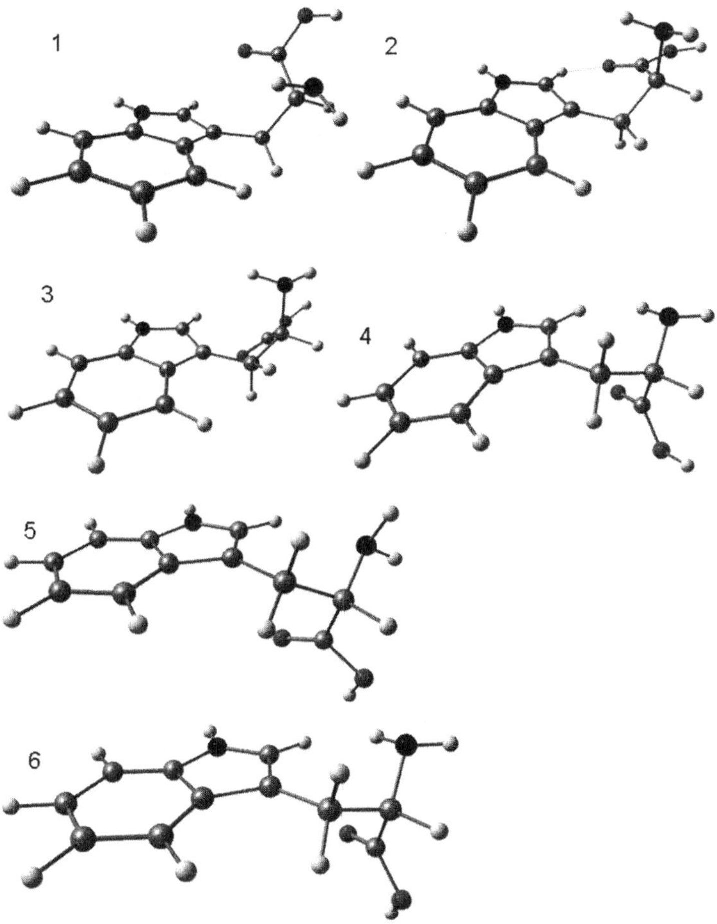

FIGURE 6.4. Stable conformers during one trajectory as optimized by PM3.

the large excess energy in the molecule. Stable intermediates are created during two of the trajectories. Figure 6.6 shows the optimized structures as calculated by PM3. Details on the dynamical evolution that leads to the creation of these conformers are discussed in a separate paper by Shemesh et al (2006). Molecular dynamics simulation therefore gives us important information about the intermediates created during the ionization process. These intermediates can lead to fragments in mass spectrometric experiment. This fragments may be unexpected if one considers the initial geometry only as a precursor.

6.3.4. Internal Energy Flow and Redistribution

6.3.4.1. Single-Photon Ionization of Glycine. Internal energy flow and redistribution have been extensively discussed for the case of glycine (conformer II) in a previous paper by the current authors (Shemesh et al. 2004). It is often assumed that

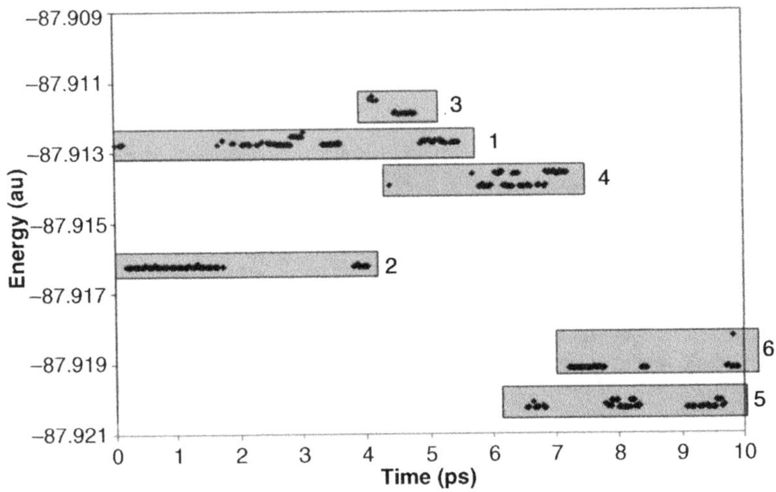

FIGURE 6.5. Energies of optimized structures during one trajectory.

IVR takes place on the timescale of several picoseconds. In terms of temperature, the normal modes should equilibrate on this timescale to a common temperature. The normal modes of the equilibrium structure of the ionic global minimum most similar to the neutral lowest energy conformer were calculated. These modes were used in order to analyze the energy partition into normal modes during the simulation. It should be noted that the normal-mode approximation fails for large displacements from equilibrium. The mean energy partition into the normal modes was calculated

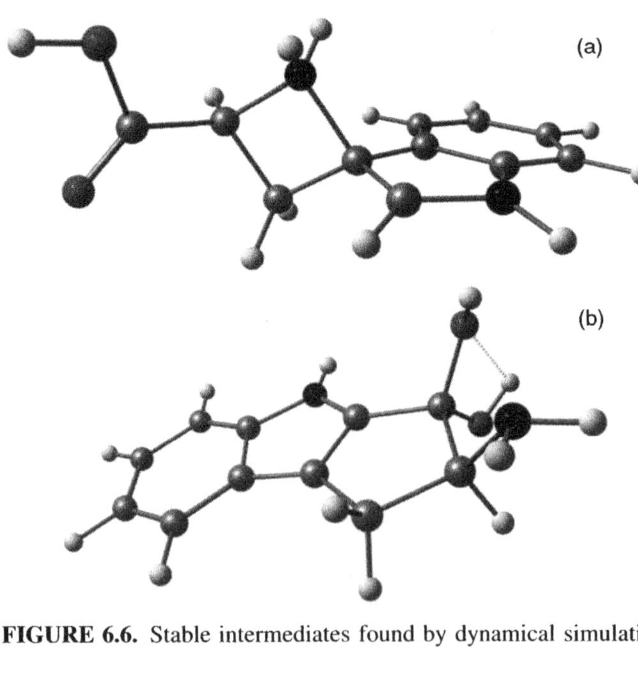

FIGURE 6.6. Stable intermediates found by dynamical simulations.

using all 361 trajectories with their statistical weights obtained from the Wigner distribution.

The kinetic energy of each mode in each trajectory is calculated during the dynamics and averaged using the Wigner distribution. From the average kinetic energy in each mode, the temperature of each mode is obtained. The temperature of each mode at time t is therefore defined by

$$T(t) = \sum_i \frac{2 \cdot w_i \cdot E^i_{\text{kin}}(t)}{k} \tag{6.7}$$

where the sum is over all trajectories sampled, w_i is the statistical weight of trajectory i, k is the Boltzmann factor, and $E^i_{\text{kin}}(t)$ is the kinetic energy of trajectory i at time t.

Fluctuations of the temperature have a high frequency and were averaged over time. The temperatures so obtained are referred to as *effective temperatures.* Figure 6.7 shows the effective temperature of four modes as a function of time. These modes were selected out of the set of 24 modes in order to show how different frequencies and different locations of the normal modes affect the energy flow between the modes.

Two of them correspond to the NH_2 group. These modes are the NH_2 bending (frequency 1609.92 cm^{-1}) and the NH_2 rocking mode (frequency 868.98 cm^{-1}). Another mode depicted here is a skeletal movement mode (432.49 cm^{-1}). The last mode chosen is the OH stretch (3815 cm^{-1}), which has the highest frequency in the molecule. No other modes involve vibration of this group to such an extent. Because this mode's high frequency and the location, it is isolated from other modes and does

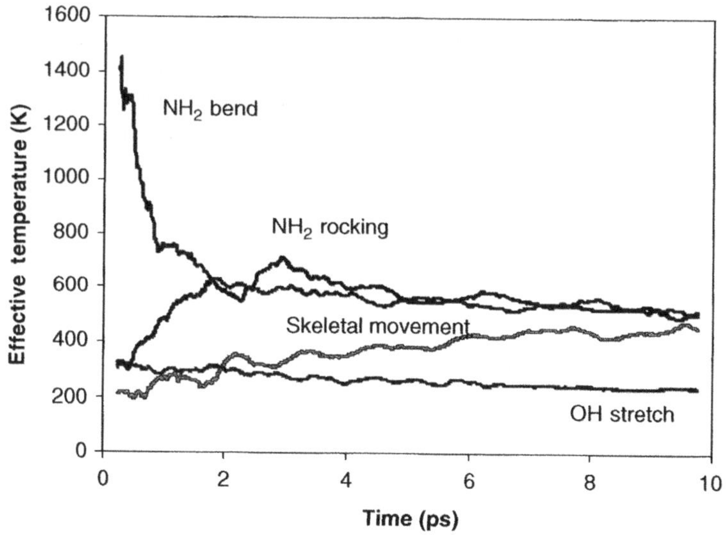

FIGURE 6.7. Effective temperature of four normal modes of glycine as a function of time. [Reproduced from Shemesh, D et al. (2004): *J. Phys. Chem. A* 108: 11477–11484 (Shemesh et al. 2004) with permission of the American Chemical Society.]

not participate effectively in the energy flow. The temperature of this mode remains almost constant during the 10 ps of simulation. The other modes are more excited than the OH stretch. The NH_2 bending mode and the NH_2 rocking mode are in the same part of the molecule, and also their frequencies have a ratio of almost 1 : 2. Therefore the equilibration is faster between these two modes and takes about a few picoseconds. The skeletal movement involves mainly the heavy atoms. It almost reaches equilibration with the NH_2 modes in the timescale of 10 ps. The frequency of this mode is almost half that of the NH_2 rocking mode, which also facilitates the energy flow.

In summary, it has been shown that the timescale of equilibration between all the modes in this system is definitely exceeds 10 ps. Although RRK is a very well established theory (Baer and Hase 1996; Holbrook et al. 1996), in certain cases some of the assumptions break down (Schranz et al. 1991a,b; Sewell et al. 1991; Shalashilin and Thompson 1996; Pena-Gallego, A et al. 1998, 1999, 2000; Martinez-Nuñez and Vazquez 1999a,b, 2000; Martinez-Nuñez et al. 2000; Lee et al. 2000; Rahaman and Raff 2001; Leitner et al. 2003). The system under study clearly does not show fast IVR. The efficiency of the energy flow between two modes depends on two parameters: the geometric proximity of the modes and the ratio between their frequencies. The geometric location of the modes has a stronger influence on the efficiency of the energy flow. Same-location modes couple better, and the energy flow between them is fast. A low-order resonance can strongly couple the modes, but this depends on the mismatch of frequencies. In the present example, the strength of the coupling between modes is most strongly dependent on geometry.

6.3.4.2. Two-Photon Ionization of Tryptophan. The kinetic energy of each normal mode in every trajectory was plotted against time. The system was divided into a subset of normal modes located predominantly on the indole ring (39 normal modes) and a subset of modes located predominantly on the amino acid backbone. The kinetic energies of the indole ring atoms and of the amino acid backbone atoms were separately summed and temperatures for the indole ring and for the amino acid backbone subsets were defined in terms of the corresponding kinetic energies. The temperature of each part at time t is thus

$$T(t) = \frac{2 \cdot E_{\text{kin}}(t)}{k} \tag{6.8}$$

where k is the Boltzmann factor and $E_{\text{kin}}(t)$ is the kinetic energy at time t. The high-frequency fluctuations of the effective temperatures were averaged over time. The effective temperatures versus time are plotted in Figure 6.8, where it can be seen that initially the indole ring is much hotter than the amino acid backbone. The energy exchange between both parts is extremely fast and until 10 ps, an equilibration between both parts has almost been reached. Note that for glycine the energy flow between modes has been discussed, whereas here the energy flow between two parts of the molecule has been emphasized. A closer look at the energy flow between the normal modes could still reveal that some modes do not exchange energy with some

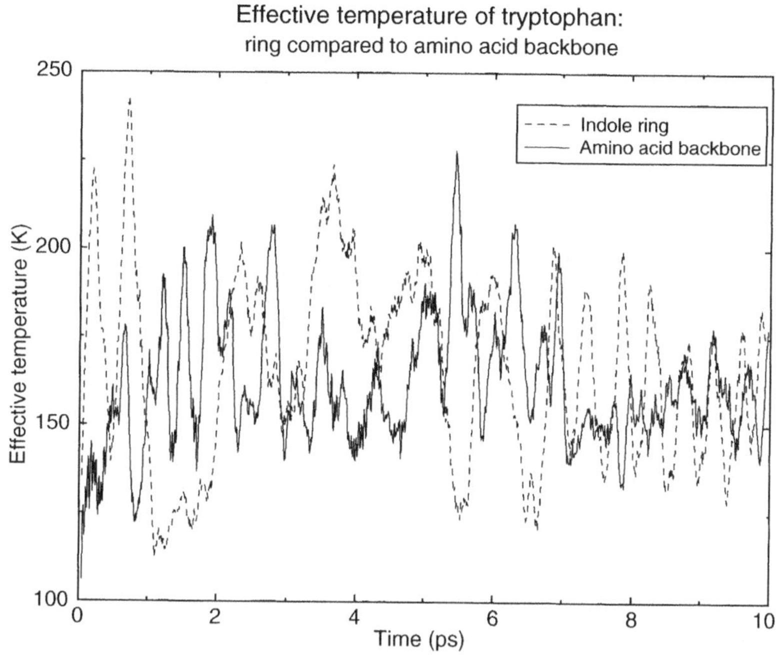

FIGURE 6.8. Effective temperature of ring modes compared to amino acid backbone modes.

other modes. The approach of dividing the molecule into two parts disregards possible bottlenecks.

6.3.5. Breakdown of RRK for Short-Timescale Conformational Transitions

Both glycine conformers show conformational hopping as described above. The first hopping time obtained from molecular dynamics can be compared to the hopping time predicted by RRK for the same excess energy. It has been shown in a previous paper (Shemesh et al. 2004) for conformer II that the hopping time obtained from molecular dynamics deviates from the hopping time predicted by the statistical theory. The RRK results are compared with the hopping rates computed directly from the trajectories. Note that different trajectories correspond in general to different excess energies, depending on the initial geometry for which the trajectory was sampled. Two RRK lifetimes are plotted. One uses $A = 70.08\,\text{cm}^{-1}$. The vibrational motion related to this frequency has been shown to be responsible for the hopping (and rotation). The lower-lying RRK graph uses the ratio between the frequencies as in Eq. (6.4), which is here equal to $85.55\,\text{cm}^{-1}$; s is assumed to be equal to 24. The frequencies have been tabulated in a previous paper by the current authors (Shemesh et al. 2004). It seems from Figure 6.9 that the dynamics results differ qualitative from RRK. Note that we are testing RRK here for a nonmonoenergetic ensemble. At any "slice" of energy E, we obtain $E - \Delta E \leq E \leq E + \Delta E$, where ΔE is small, the results can be compared with the monoenergetic RRK expression. In the ensemble used, we have few trajectories

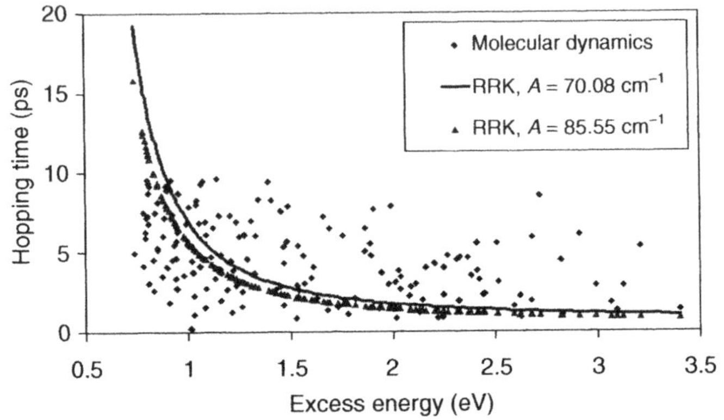

FIGURE 6.9. First hopping time as a function of the excess energy available. The dots distributed on the graph are from the molecular dynamics simulations, whereas the two smoothed lines are calculated by RRK method. [Reproduced from Shemesh D et al. (2004): *J. Phys. Chem. A* **108**:1477–11484 (Shemesh et al. 2004) with permission of the American Chemical Society.]

for each energy E, but there are results for many E values, so the test of RRK here is quite stringent. Some trajectories show faster hopping times; some show slower hopping times as predicted by RRK. RRK assumes statistical distribution of the vibrational energy, which is not the case here, at the short timescales investigated. This is due to different coupling strengths between the modes. Modes that are strongly coupled to the rotational degree of freedom show hopping times faster than predicted by RRK and vice versa. Also the process takes place in highly nonequilibrium conditions and the trajectories do not follow the minimum-energy path between the conformers. Whether hopping in fact occurs does not seem to depend significantly on the excess energy available, which is central to RRK. The minimum excess energy in these trajectories is 0.8 eV (77.2 kJ/mol) for some trajectories. Some of them show hopping, and some do not. This seems to depend on how strongly coupled the initially excited mode is with the rotational mode. It should be noted that a quantitative, rigorous way to show whether the system is statistical is to work with an ensemble of trajectories with constant energy and to express the lifetime probability distribution for this ensemble from the RRK model as

$$P(t) = k(E)e^{-k(E)t} \tag{6.9}$$

where $k(E)$ is the RRK rate and t is the time. This lifetime probability should then be compared to the lifetime probability obtained from molecular dynamics (Steinfeld et al. 1990). It seems, however, that the rates from the trajectories fluctuate so radically from the RRK rates that, qualitatively, the behavior is very non-RRK-like. Therefore, glycine exhibits strong non-RRK behavior. For tryptophan, this has not yet been tested.

The dynamical evolution of two conformers has been compared for glycine. The probability for rotation of conformer II is 2.23%, compared to 0.79% for conformer

I. The probability for conformer hopping for conformer II is 21.55% compared to 2.35% for conformer I [results for conformer II in Shemesh et al. (2004) had a numerical error]. The probability for the rotational events is thus higher by an order of magnitude for conformer II. Rotation is favored if the NH_2 group is already distorted from the plane as it is for conformer II. Statistical theories assume that after a short time, the system forgets the initial conditions. This is not the case here. The processes show conformer-specific dynamics, strongly dependent on the initial state, and therefore pronounced non-RRK behavior is obtained.

6.3.6. Short-Timescale Fragmentation in Single- and Two-Photon Ionization

Fragmentation has been observed only for glycine. Almost all the fragmentations that were computed cleave the C—C bond. It has been shown theoretically that this is the lowest barrier for fragmentation in this system (Depke et al. 1984; Simon et al. 2002; Lu et al. 2004). The main peak in the mass spectra measured shows the immonium ion ($CH_2NH_2^+$), which also corresponds to the C—C cleavage (Vorsa et al. 1999; Polce and Wesdemiotis 2000). The probability of fragmentation in photoionization of conformer II is low, about 2.7×10^{-7}%. This is due to the fact that the corresponding trajectories sample the tails of the initial wavefunction. This gives a low Wigner probability. The probability of fragmentation in photoionization of conformer I is higher by a factor of 1000 (4.8×10^{-4}%).

One fragmentation pathway of conformer I shows dehydration. This fragmentation pathway leads to a water molecule and a glycine fragment of $m/z = 57$. One mass spectrum recorded for glycine (Vorsa et al. 1999) does not show this peak; the other one (Polce and Wesdemiotis 2000) shows a very small peak. Until now, this peak was not assigned to a structure. The absence of the peak in the former mass spectrum could have two explanations: (1) the probability of this pathway is very small, so that the amount of the glycine fragment produced is minimal and cannot be measured; or (2) the glycine fragment can fragment further to yield CO and CH_2NH^+. The excess energy in the simulation is sufficient to yield these fragments, but they are not created in the dynamics. The dynamics stops after the first fragmentation, due to convergence problems. In experiments, the excess energy can be much larger and the fragmentation can proceed further and yield CO, H_2O, and CH_2NH^+. In summary, the new mass peak predicted by our simulation was seen in one experiment but not in another. The excitation energies in these experiments do not correspond to our simulations, and it should be very interesting to test the prediction for excitations that correspond to the model used. In summary, molecular dynamics simulation is able to predict fragmentation pathways, and to assign fragments to unresolved peaks.

6.3.7. Testing RRK for Short-Timescale Fragmentation

RRK has been tested for 11 trajectories of glycine that show fragmentation. It has been observed that RRK predicts a fragmentation timescale shorter than observed in the molecular dynamics simulations. Since only a few trajectories undergo fragmentation during the first 10 ps, the statistics is insufficient for conclusions. However, tentative evidence suggests that RRKM underestimates the fragmentation time for these events.

6.4. CHALLENGES AND POSSIBILITIES FOR DYNAMICAL SIMULATIONS OF MASS SPECTROMETRIC PROCESSES

It has been shown in this chapter that molecular dynamics simulation is a useful tool for predicting conformational changes and fragmentations in ionization processes for large systems. The method enables us to identify reaction mechanisms, intermediates, and products. The products can be compared with existing mass spectra, and unresolved peaks can be identified. Interesting features such as rotation can be predicted by this method. The potential energy surface used is appropriate for biological molecules. This method can be extended to peptides or even small proteins. This may lead to better understanding of mass spectrometric processes and to the identification of unknown compounds.

ACKNOWLEDGMENT

This chapter is dedicated to the memory of Prof. Chava Lifshitz. We have learned much from her unique insights into ionic processes and benefited from her very helpful comments.

REFERENCES

Aebersold R, Goodlett DR (2001): Mass spectrometry in proteomics. *Chem. Rev.* **101**: 269–295.

Aebersold R, Mann M (2003): Mass spectrometry-based proteomics. *Nature* **422**: 198–207.

Alexandrov ML, Gall LN, Krasnov NV, Nikolaev VI, Pavlenko VA, Shkurov VA, Baram GI, Grachev MA, Knorre VD, Kusner YS (1984): Direct coupling of a microcolumn liquid chromatograph and a mass-spectrometer. *Bioorg. Khim.* **10**:710–712.

Ayre CR, Moro L, Becker CH (1994): Effects of desorption method and photoionizing laser characteristics on molecular fragmentation. *Anal. Chem.* **66**:1610–1619.

Baer T, Hase WL (1996): *Unimolecular Reaction Dynamics,* Oxford Univ. Press, Oxford.

Bakker JM, Aleese LM, Meijer G, von Helden G (2003): Fingerprint IR spectroscopy to probe amino acid conformations in the gas phase. *Phys. Rev. Lett.* **91**.

Belov ME, Alimpiev SS, Mlynsky VV, Nikiforov SM, Derrick PJ (1995): Laser-ablation of organic-molecules from frozen matrices. *Rapid Commun. Mass Spectrom.* **9**: 1431–1436.

Bowers MT, Marshall AG, McLafferty FW (1996): Mass spectrometry: Recent advances and future directions. *J. Phys. Chem.* **100**:12897–12910.

Brauer B, Chaban GM, Gerber RB (2004): Spectroscopically-tested, improved, semi-empirical potentials for biological molecules: Calculations for glycine, alanine and proline. *Phys. Chem. Chem. Phys.* **6**:2543–2556.

Burlingame AL, Boyd RK, Gaskell SJ (1998): Mass spectrometry. *Anal. Chem.* **70**:647r–716r.

Callis P (1997): 1La and 1Lb transitions of tryptophan: Applications of theory and experimental observations to fluorescence of proteins. *Meth. Enzymol.* **278**:113–151.

Chaban GM, Jung JO, Gerber RB (2000): Anharmonic vibrational spectroscopy of glycine: Testing of ab initio and empirical potentials. *J. Phys. Chem. A* **104**:10035–10044.

Chalmers MJ, Gaskell SJ (2000): Advances in mass spectrometry for proteome analysis. *Curr. Opin. Biotechnol.* **11**:384–390.

Cohen R, Brauer B, Nir E, Grace L, de Vries MS (2000): Resonance-enhanced multiphoton ionization spectroscopy of dipeptides. *J. Phys. Chem. A* **104**:6351–6355.

Compagnon I, Hagemeister FC, Antoine R, Rayane D, Broyer M, Dugourd P, Hudgins RR, Jarrold MF (2001): Permanent electric dipole and conformation of unsolvated tryptophan. *J. Am. Chem. Soc.* **123**:8440–8441.

Daniels AD, Scuseria GE (1999): What is the best alternative to diagonalization of the Hamiltonian in large scale semiempirical calculations? *J. Chem. Phys.* **110**:1321–1328.

Depke G, Heinrich N, Schwarz H (1984): Mass-spectrometric investigation of nitrogen-compounds. 41. On the gas-phase chemistry of ionized glycine and its enol—a combined experimental and abinitio molecular-orbital study. *Int. J. Mass Spectrom.* **62**:99–117.

Dey M, Grotemeyer J (1994): Multiphoton ionization and photodissociation at the 2nd-order space focus in a time-of-flight mass-spectrometer—the amino-acid tryptophan. *Org. Mass Spectrom.* **29**:659–667.

Elokhin VA, Krutchinsky AN, Ryabov SE (1991): The resonantly enhanced 2-Photon Ionization and Electron Ionization of Tryptophan Molecules in beams formed by laser evaporation of aqueous cryogenic matrix. *Rapid Commun. Mass Spectrom.* **5**:257–259.

Ferguson PL, Smith RD (2003): Proteome analysis by mass spectrometry. *Annu. Rev. Bioph. Biom.* **32**:399–424.

Godfrey PD, Brown RD (1995): Shape of glycine. *J. Am. Chem. Soc.* **117**:2019–2023.

Godfrey PD, Brown RD, Rodgers FM (1996): The missing conformers of glycine and alanine: Relaxation in seeded supersonic jets. *J. Mol. Struct.* **376**:65–81.

Gordon MS, Chaban G, Taketsugu T (1996): Interfacing electronic structure theory with dynamics. *J. Phys. Chem.* **100**:11512–11525.

Griffiths WJ, Jonsson AP, Liu SY, Rai DK, Wang YQ (2001): Electrospray and tandem mass spectrometry in biochemistry. *Biochem. J.* **355**:545–561.

Halgren TA (1996a): Merck molecular force field. 1. Basis, form, scope, parameterization, and performance of MMFF94. *J. Comput. Chem.* **17**:490–519.

Halgren TA (1996b): Merck molecular force field. 3. Molecular geometries and vibrational frequencies for MMFF94. *J. Comput. Chem.* **17**:553–586.

Holbrook KA, Pilling MJ, Robertson SH (1996): *Unimolecular Reactions*, 2nd ed., Wiley, Chichester, UK.

Huang ZJ, Lin ZJ (2005): Detailed ab initio studies of the conformers and conformational distributions of gaseous tryptophan. *J. Phys. Chem. A* **109**:2656–2659.

Huisken F, Werhahn O, Ivanov AY, Krasnokutski SA (1999): The O-H stretching vibrations of glycine trapped in rare gas matrices and helium clusters. *J. Chem. Phys.* **111**:2978–2984.

Ivanov AY, Sheina G, Blagoi YP (1999): FTIR spectroscopic study of the UV-induced rotamerization of glycine in the low temperature matrices (Kr, Ar, Ne). *Spectrochim. Acta A* **55**:219–228.

Ivanov AY, Plokhotnichenko AM, Izvekov V, Sheina GG, Blagoi YP (1997): FTIR investigation of the effect of matrices (Kr, Ar, Ne) on the UV-induced isomerization of the monomeric links of biopolymers. *J. Mol. Struct.* **408**:459–462.

Jensen F (1999) *Introduction to Computational Chemistry,* Wiley, Chichester, UK.

Jensen JH, Gordon MS (1991): The conformational potential-energy surface of glycine—a theoretical-study. *J. Am. Chem. Soc.* **113**:7917–7924.

Jonsson AP (2001): Mass spectrometry for protein and peptide characterisation. *Cell Mol. Life Sci.* **58**:868–884.

Jorgensen WL, Tiradorives J (1988): The OPLS potential functions for proteins—energy minimizations for crystals of cyclic-peptides and crambin. *J. Am. Chem. Soc.* **110**: 1657–1666.

Jorgensen WL, Maxwell DS, TiradoRives J (1996): Development and testing of the OPLS all-atom force field on conformational energetics and properties of organic liquids. *J. Am. Chem. Soc.* **118**:11225–11236.

Karas M, Bachmann D, Bahr U, Hillenkamp F (1987): Matrix-assisted ultraviolet-laser desorption of nonvolatile compounds. *Int. J. Mass Spectrom.* **78**:53–68.

Koster C, Grotemeyer J (1992): Single-photon and multiphoton ionization of infrared laser-desorbed biomolecules. *Org. Mass Spectrom.* **27**:463–471.

Kushwaha PS, Mishra PC (2000): Electronic spectra, excited-state geometries and molecular electrostatic potentials of aromatic amino acids. *J. Photochem. Photobiol. Chem.* **137**:79–86.

Lee IR, Chen WK, Chung YC, Cheng PY (2000): A direct observation of non-RRKM behavior in femtosecond photophysically activated reactions. *J. Phys. Chem. A* **104**:10595–10599.

Lee TS, York DM, Yang WT (1996): Linear-scaling semiempirical quantum calculations for macromolecules. *J. Chem. Phys.* **105**:2744–2750.

Leitner DM, Levine B, Quenneville J, Martinez TJ, Wolynes PG (2003): Quantum energy flow and trans-stilbene photoisomerization: An example of a non-RRKM reaction. *J. Phys. Chem. A* **107**:10706–10716.

Lifshitz C (1992): Recent developments in applications of RRKM-QET. *Int. J. Mass Spectrom.* **118**:315–337.

Lifshitz C (2001): Some recent aspects of unimolecular gas phase ion chemistry. *Chem. Soc. Rev.* **30**:186–192.

Lifshitz C (2003): Editorial: Basic aspects and principles of mass spectrometry applied to biomolecules. *Mass Spectrom. Rev.* **22**:157.

Lin D, Tabb DL, Yates JR (2003): Large-scale protein identification using mass spectrometry. *Bba-Proteins Proteom.* **1646**:1–10.

Lindinger A, Toennies JP, Vilesov AF (1999): High resolution vibronic spectra of the amino acids tryptophan and tyrosine in 0.38 K cold helium droplets. *J. Chem. Phys.* **110**:1429–1436.

Lockyer NP, Vickerman JC (1998): Multiphoton ionization mass spectrometry of small biomolecules with nanosecond and femtosecond laser pulses. *Int. J. Mass Spectrom.* **176**:77–86.

Lockyer NP, Vickerman JC (2000): Single photon and femtosecond multiphoton ionisation of the dipeptide valyl-valine. *Int. J. Mass Spectrom.* **197**:197–209.

Lu HF, Li FY, Lin SH (2004): Theoretical interpretation of the fragments generated from a glycine radical cation. *J. Phys. Chem. A* **108**:9233–9243.

Maluendes SA, Dupuis M (1990): A dynamic reaction coordinate approach to abinitio reaction pathways—application to the 1,5 hexadiene cope rearrangement. *J. Chem. Phys.* **93**:5902–5911.

Mann M, Hendrickson RC, Pandey A (2001): Analysis of proteins and proteomes by mass spectrometry. *Annu. Rev. Biochem.* **70**:437–473.

Mano N, Goto J (2003): Biomedical and biological mass spectrometry. *Anal. Sci.* **19**:3–14.

Martinez-Nuñez E, Vazquez SA (1999a): Classical dynamics study of the unimolecular decomposition of CH3SH+. *J. Phys. Chem. A* **103**:9783–9793.

Martinez-Nuñez E, Vazquez SA (1999b): Further dynamical studies of the dissociation and elimination reactions of methyl nitrite. *J. Chem. Phys.* **111**:10501–10510.

Martinez-Nuñez E, Vazquez SA (2000): Unimolecular decomposition of CH3SH+: An ab initio and RRKM study. *J. Mol. Struct.* (*Theochem*) **505**:109–116.

Martinez-Nuñez E, Pena-Gallego A, Rodriguez-Fernandez R, Vazquez SA (2000): Direct dynamics simulation of the methanethiol cation decomposition. *Chem. Phys. Lett.* **324**: 88–94.

McGlone SJ, Elmes PS, Brown RD, Godfrey PD (1999): Molecular structure of a conformer of glycine by microwave spectroscopy. *J. Mol. Struct.* **486**:225–238.

Miller TF, Clary DC (2004): Quantum free energies of the conformer of glycine on an ab intito potential energy surface. *Phys. Chem. Chem. Phys.* **6**:2563–2571.

Nyman TA (2001): The role of mass spectrometry in proteome studies. *Biomol. Eng.* **18**: 221–227.

Pena-Gallego A, Martinez-Nuñez E, Vazquez SA (1998): Classical trajectory study of the cis-trans isomerization and F-O dissociation of FONO. *J. Phys. Chem. A* **102**:8708–8715.

Pena-Gallego A, Martinez-Nuñez E, Vazquez SA (1999): Nonstatistical effects in the unimolecular dissociation of the acetyl radical. *J. Chem. Phys.* **110**:11323–11334.

Pena-Gallego A, Martinez-Nuñez E, Vazquez SA (2000): Dynamics of the cis-trans isomerization and Cl-O dissociation of chlorine nitrite. Classical trajectory and statistical calculations. *Phys. Chem. Chem. Phys.* **2**:5393–5399.

Piuzzi F, Dimicoli I, Mons M, Tardivel B, Zhao QC (2000): A simple laser vaporization source for thermally fragile molecules coupled to a supersonic expansion: Application to the spectroscopy of tryptophan. *Chem. Phys. Lett.* **320**:282–288.

Polce MJ, Wesdemiotis C (2000): First generation and characterization of the enol of glycine, H2N–CH=C(OH)(2), in the gas phase. *J. Mass Spectrom.* **35**:251–257.

Rahaman A, Raff LM (2001): Theoretical investigations of intramolecular energy transfer rates and pathways for vinyl bromide on an ab initio potential-energy surface. *J. Phys. Chem. A* **105**:2147–2155.

Reilly PTA, Reilly JP (1994): Laser-desorption gas-phase ionization of the amino-acid tryptophan. *Rapid Commun. Mass Spectrom.* **8**:731–734.

Rizzo TR, Park YD, Levy DH (1985a): A molecular-beam of tryptophan. *J. Am. Chem. Soc.* **107**:277–278.

Rizzo TR, Park YD, Levy DH (1986a): Dispersed fluorescence of jet-cooled tryptophan—excited-state conformers and intramolecular exciplex formation. *J. Chem. Phys.* **85**: 6945–6951.

Rizzo TR, Park YD, Peteanu L, Levy DH (1985b): Electronic-spectrum of the amino-acid tryptophan cooled in a supersonic molecular-beam. *J. Chem. Phys.* **83**:4819–4820.

Rizzo TR, Park YD, Peteanu LA, Levy DH (1986b): The electronic-spectrum of the amino-acid tryptophan in the gas-phase. *J. Chem. Phys.* **84**:2534–2541.

Rodriguez-Santiago L, Sodupe M, Oliva A, Bertran J (2000): Intramolecular proton transfer in glycine radical cation. *J. Phys. Chem. A* **104**:1256–1261.

Schlag EW, Grotemeyer J, Levine RD (1992): Do large molecules ionize. *Chem. Phys. Lett.* **190**:521–527.

Schranz HW, Raff LM, Thompson DL (1991a): Statistical and nonstatistical effects in bond fission reactions of Sih2 and Si2h6. *J. Chem. Phys.* **94**:4219–4229.

Schranz HW, Raff LM, Thompson DL (1991b): Nonstatistical effects in bond fission reactions of 1,2-difluoroethane. *Chem. Phys. Lett.* **182**:455–462.

Sewell TD, Schranz HW, Thompson DL, Raff LM (1991): Comparisons of statistical and nonstatistical behavior for bond fission reactions in 1,2-difluoroethane, disilane, and the 2-chloroethyl radical. *J. Chem. Phys.* **95**:8089–8107.

Shalashilin DV, Thompson DL (1996): Intrinsic non-RRK behavior: Classical trajectory, statistical theory, and diffusional theory studies of a unimolecular reaction. *J. Chem. Phys.* **105**:1833–1845.

Shemesh D, Gerber RB (2006): Classical Trajectory Simulations of Photoionization Dynamics of Tryptophan: Intramolecular Energy Flow, Hydrogen Transfer Processes and Conformational Transitions. *J. Phys. Chem. A* in press.

Shemesh D, Gerber RB (2005): Different chemical dynamics for different conformers of biological molecules: Photoionization of glycine. *J. Chem. Phys.* **122**:241104.

Shemesh D, Chaban GM, Gerber RB (2004): Photoionization dynamics of glycine: The first 10 picoseconds. *J. Phys. Chem. A* **108**:11477–11484.

Simon S, Sodupe M, Bertran J (2002): Isomerization versus fragmentation of glycine radical cation in gas phase. *J. Phys. Chem. A* **106**:5697–5702.

Snoek LC, Kroemer RT, Hockridge MR, Simons JP (2001): Conformational landscapes of aromatic amino acids in the gas phase: Infrared and ultraviolet ion dip spectroscopy of tryptophan. *Phys. Chem. Chem. Phys.* **3**:1819–1826.

Standing KG (2003): Peptide and protein de novo sequencing by mass spectrometry. *Curr Opin Struct. Biol.* **13**:595–601.

Steinfeld JI, Francisco JS, Hase WL (1990) *Theory of Unimolecular and Recombination Reactions*, 2nd ed., Prentice-Hall, New York.

Stepanian SG, Reva ID, Radchenko ED, Rosado MTS, Duarte MLTS, Fausto R, Adamowicz L (1998): Matrix-isolation infrared and theoretical studies of the glycine conformers. *J. Phys. Chem. A* **102**:1041–1054.

Stewart JJP (1989a): Optimization of parameters for semiempirical methods. 1. Method. *J. Comput. Chem.* **10**:209–220.

Stewart JJP (1989b): Optimization of parameters for semiempirical methods. 2. Applications. *J. Comput. Chem.* **10**:221–264.

Stewart JJP (1997): Calculation of the geometry of a small protein using semiempirical methods. *J. Mol. Struct. (Theochem)* **401**:195–205.

Stewart JJP, Davis LP, Burggraf LW (1987): Semiempirical calculations of molecular trajectories—method and applications to some simple molecular-systems. *J. Comput. Chem.* **8**:1117–1123.

Takata I, Taketsugu T, Hirao K, Gordon MS (1998): Ab initio potential energy surface by modified Shepard interpolation: Application to the CH3+H-2 → CH4+H reaction. *J. Chem. Phys.* **109**:4281–4289.

Taketsugu T, Gordon MS (1995a): Dynamic reaction-path analysis based on an intrinsic reaction coordinate. *J. Chem. Phys.* **103**:10042–10049.

Taketsugu T, Gordon MS (1995b): Dynamic reaction-path study of Sih4+H–]Sih5- and the Berry pseudorotation mechanism. *J. Phys. Chem.* **99**:14597–14604.

Taketsugu T, Gordon MS (1995c): Dynamic reaction coordinate analysis—an application to Sih4+H–]Sih5-. *J. Phys. Chem.* **99**:8462–8471.

Vorsa V, Kono T, Willey KF, Winograd N (1999): Femtosecond photoionization of ion beam desorbed aliphatic and aromatic amino acids: Fragmentation via alpha-cleavage reactions. *J. Phys. Chem. B* **103**:7889–7895.

Wang JP, Meroueh SO, Wang YF, Hase WL (2003): Efficiency of energy transfer in protonated diglycine and dialanine SID—effects of collision angle, peptide ion size, and intramolecular potential. *Int. J. Mass Spectrom.* **230**:57–64.

Wang YF, Hase WL (2003): Direct dynamics study of N-protonated diglycine surface-induced dissociation. Influence of collision energy. *J. Am. Soc. Mass Spectrom.* **14**:1402–1412.

Weiner PK, Kollmann PA (1981): *J. Comput. Chem.* **2**:287–303.

Weiner SJ, Kollman PA, Case DA, Singh UC, Ghio C, Alagona G, Profeta S, Weiner P (1984): A new force-field for molecular mechanical simulation of nucleic-acids and proteins. *J. Am. Chem. Soc.* **106**:765–784.

Weinkauf R, Schermann JP, de Vries MS, Kleinermanns K (2002): Molecular physics of building blocks of life under isolated or defined conditions. *Eur. Phys. J. D* **20**:309–316.

Wigner E (1932): *Phys. Rev.* **40**:749.

Yamashita M, Fenn JB (1984a): Negative-ion production with the electrospray ion-source. *J. Phys. Chem.* **88**:4671–4675.

Yamashita M, Fenn JB (1984b): Electrospray ion-source—another variation on the free-jet theme. *J. Phys. Chem.* **88**:4451–4459.

Yu D, Rauk A, Armstrong DA (1995): Radicals and ions of glycine—an ab-initio study of the structures and gas-phase thermochemistry. *J. Am. Chem. Soc.* **117**:1789–1796.

7

INTRAMOLECULAR VIBRATIONAL ENERGY REDISTRIBUTION AND ERGODICITY OF BIOMOLECULAR DISSOCIATION

CHAVA LIFSHITZ*

Department of Physical Chemistry and The Farkas Center for Light Induced, Processes
The Hebrew University of Jerusalem
Jerusalem, Israel

* This chapter is a last work of Prof. Chava Lifshitz, who passed away on March 1, 2005. Chava occupied a special place in my life; she was my graduate advisor, my teacher, my mentor, and colleague; she was a person to whom I came to ask for advice. I took the liberty to finalize this work and prepare it for publication, trying to follow Chava's style and way of thinking as much as I could (Julia Laskin).

Principles of Mass Spectrometry Applied to Biomolecules, edited by Julia Laskin and Chava Lifshitz

7.1. WHAT WE HAVE LEARNED FROM ORGANIC MASS SPECTROMETRY

7.1.1. Introduction and Historical Background

Development of mass spectrometry of organic compounds began in the 1950s and 1960s. It was realized rather early on that a mass spectrum of a molecule is the result of a series of parallel and consecutive gas-phase unimolecular reactions. The theoretical approach known as the quasiequilibrium theory (QET) was developed by Rosenstock, Wallenstein, Wahrhaftig, and Eyring in 1952 (Rosenstock et al. 1952) and was also referred to in the early years as *statistical theory of mass spectra* (STMS). Development of the statistical theory of unimolecular reactions in neutral systems known as RRKM (Rice, Ramsperger, Kassel, Marcus) took place independently and was finalized by Marcus and Rice (1951). These theories are based on Eyring's absolute reaction rate theory and define an activated complex—the transition state—located at the barrier along the reaction coordinate, separating the reactants from the products. All detailed statistical theories of unimolecular reactions begin with the calculation of $k(E)$, the rate constant as a function of the internal energy E that is the microcanonical rate constant (Baer and Hase 1996). The RRKM/QET expression is given by

$$k(E) = \sigma W^{\ddagger}(E\text{-}E_0)/h\rho(E) \tag{7.1}$$

where E_0 is the activation energy, $\rho(E)$ is the density of rotational and vibrational states at the energy E, $W^{\ddagger}(E - E_0)$ is the sum of the vibrational states from 0 to E–E_0 in the transition state, h is Planck's constant, and σ is the reaction symmetry factor. The RRKM/QET formulation, expression (1), is the starting point for modern statistical theories (Baer and Hase 1996). There are additional statistical theories, notably phase space theory [PST; see Baer and Hase (1996) and references cited therein], that do not assume the existence of an activated complex. The rate constant $k(E)$ increases with increasing internal energy E and all other things being equal, is smaller for larger molecules, that is, for larger numbers of vibrational degrees of freedom.

At the cornerstone of statistical theories of unimolecular reactions such as RRKM/QET stands the ergodic* hypothesis. The coupling of internal motions is postulated to lead to rapid intramolecular energy randomization on a timescale that is short relative to the mean lifetime of decomposing species (Oref and Rabinovitch 1979). In the case of the ionic species that are of central interest to us, the internal motions involve both electronic and vibrational degrees of freedom. Whereas ionization can lead to any number of electronically excited states of the ion, radiationless transitions to the ground electronic state (or "internal conversion") are fast, so that dissociations take place from a vibrationally excited, ground electronic state of the ion.

*According to *Webster's Dictionary*, the term *ergodic* means (1) of or relating to a process in which every sequence or sizable sample is the same statistically and therefore equally representative of the whole; (2) involving or relating to the probability that any state will recur.

TABLE 7.1. Some Characteristic Times

Process	Time (s)
Radiationless transitions	10^{-14}
IVR	10^{-10}–10^{-12}
Unimolecular reactions	10^{-3}–10^{-12}

Nonergodic (or nonstatistical) behavior pertains to the situation in which only a portion of the available internal energy-coordinate hypervolume is explored during the lifetime of the excited molecules prior to dissociation (Oref and Rabinovitch 1979). Since intramolecular vibrational redistribution (IVR) of the energy normally occurs on the timescale of picoseconds, the experimental examples for RRKM behavior are numerous and "non-RRKM" molecules are scarce. Nearly all organic molecules that have been studied mass-spectrometrically behave statistically (are "ergodic"). Examples given below include some work on neutral gas-phase molecules in addition to ionic systems.

Table 7.1 summarizes some of the important characteristic times for radiationless transitions, IVR, and unimolecular decompositions. The competition between these various processes is what determines whether statistical behavior in unimolecular decompositions is observed. There are some observations in the literature based on spectroscopic measurements indicating times longer than 10^{-10} s for IVR (Lehmann et al. 1994; Keske et al. 2000). Longer unimolecular reaction times than approximately milliseconds are suppressed because radiative decay rates in the infrared are of the order of 100 s^{-1} (Dunbar and Lifshitz 1991). Peptides were observed to have lower radiative decay rates of 10 – 40 s^{-1}, so lower dissociative rates can be achieved (Laskin et al. 2000, 2004). Furthermore, this lifetime limitation does not apply to thermal experiments that determine $k(T)$ rather than $k(E)$.

7.1.2. Early Experiments on Organic Molecules and Agreement with Statistical Theories

Early comparisons between experimental results and theoretical calculations based on statistical theories were limited to mass spectra (Rosenstock et al. 1952). In other words, in these early studies the molecular ions undergoing fragmentation were produced, generally by electron impact ionization, with a broad distribution of internal energies $P(E)$. Expressions had to be assumed for the internal energy distribution functions, for comparison between experiments and theory. Whereas agreement was observed between experiment and theory, it was not until ions with well-defined internal energies were formed experimentally that valuable information could be obtained on the very basic attributes such as $k(E)$ or at least on breakdown diagrams (a *breakdown diagram* is a plot of a set of breakdown curves where the breakdown curve is the relative intensity of each ion formed from energy-selected parent ions as a function of the parent ion internal energy, i.e., sets of mass spectra determined at well-defined internal energies).

Experiments using photoionization in the vacuum ultraviolet (VUV) (Chupka and Berkowitz 1967) were very thorough studies of the alkanes. These were high-energy-resolution experiments with the ability to deduce breakdown diagrams, that is, collections of breakdown curves, from normalized derivative data of the parent and fragment ionization efficiency curves. The experimental breakdown curves were in good qualitative and fair quantitative agreement with the predictions of RRKM/QET. Further QET calculations and comparisons with experiments were carried out (Vestal 1968).

7.1.3. Development of Modern Experiments

The method of chemical activation in neutral systems (Butler and Kistiakowsky 1960; Rabinovitch and Flowers 1964; Rabinovitch and Diesen 1959; Rabinovitch et al. 1963) was the first successful approach to energy selection, to the determination of the microcanonical rate constant $k(E)$ (Baer and Hase 1996), and to the observation of nonrandom energy flow (Baer and Hase 1996; Rynbrandt and Rabinovitch 1970, 1971a,b). In a classical experiment (Rabinovitch et al. 1963), a hydrogen atom undergoes addition to the double bond in butene [Eq. (7.2)] forming a chemically activated *sec*-butyl radical. Since a new C—H bond is being formed, the radical has a quite well-defined internal energy; the minimum energy is 40 kcal/mol in the case of hydrogen atom addition to *cis*-but-2-ene (there is a small superimposed energy spread because of the thermal energy distribution):

$$H + C_4H_8 \rightarrow CH_3CH_2\dot{C}HCH_3^* \xrightarrow{k(E)} \cdot CH_3 + C_3H_6 \qquad (7.2)$$

The radical undergoes dissociation to propylene plus a methyl radical—a clear indication that interchange of energy between vibrational modes occurs. The microcanonical rate constant $k(E)$ for the methyl loss reaction was determined for a series of internal energies by using different butene precursors and the results were in agreement with RRKM calculations. Many other systems have been studied by B. S. Rabinovitch and coworkers by this method, using different bimolecular association reactions to form the chemically activated species. The reaction of hydrogen atoms with butene may be particularly relevant to our central topic of interest, namely, biomolecular ions, because of electron capture dissociation (ECD), discussed below.

Electron impact ionization and photoionization are nonresonant processes because the electrons carry off excess energy. As a result, the parent ions are formed with a distribution of internal energies. The first experimental method developed to circumvent this problem was charge exchange. The recombination energy of an ion is the energy released when it recombines with an electron. This is well defined, particularly for atomic ions that have no contributions from excited electronic states. The method has been developed in tandem (MS/MS) mass spectrometry by Lindholm and coworkers. Mass spectra were determined as a function of internal energy yielding breakdown curves, as, for example, for *n*-butane (Chupka and Lindholm 1963). Experimental breakdown patterns were successfully compared with RRKM/QET calculations (Vestal 1968; Lifshitz and Tiernan 1971). The effect of thermal energy on dissociative charge transfer was found to be in excellent agreement with QET calculations (Lifshitz and Tiernan 1973; Chupka 1971). Charge exchange experiments constitute the first direct measurements of *k*(*E*) for any ionic decomposition (Andlauer and Ottinger 1971, 1972). Two reaction systems were studied: HCN elimination from the benzonitrile ion and $C_6H_5^+$ and $C_4H_4^+$ formation from benzene cations. Although the benzene reactions were concluded not to be in competition, which is a conclusion contrary to common knowledge today, these results, particularly for benzonitrile, constitute a landmark achievement.

The most successful method to date has been photoelectron photoion coincidence (PEPICO). This method has been applied to a large variety of ion fragmentations, and the results were compared with RRKM/QET calculations. In PEPICO, the molecule M is photoionized by a VUV photon of well-defined energy $h\nu$, and the product molecular ions or any of their fragment ions are energy-selected by measuring them in coincidence with energy-selected electrons. Since 1973, dispersed continuum photon sources have been used in which the ions are measured in coincidence with initially zero, or threshold, electrons [see Baer and Hase (1996) and references cited therein]. Since the kinetic energy of the electron is zero, the internal energy of the molecular ion is simply given by the photon energy minus the ionization energy of the molecule: $E = h\nu - IE$ (assuming the thermal energy to be negligibly small). This method is called *threshold* PEPICO or TPEPICO and has been perfected by Tomas Baer and coworkers. More recent studies have involved dispersed synchrotron radiation (Baer et al. 1988). Excellent agreement has been obtained in most cases between experimental dependences of absolute rates *k*(*E*) on the internal energy and RRKM/QET calculations. The agreement between theory and experiment in the case of phenyl ion formation from bromobenzene and deuterobromobenzene ions indicates that the CH or CD vibrational modes participate fully in the energy flow of the isolated molecular ion, even though the CH and CD bonds are not involved in the formation of product $C_6H_5^+$ or $C_6D_5^+$ fragment ions (Baer and Kury 1982). Kinetic energy release distributions (KERDs) in the fragmentation of energy-selected ions were another successful result demonstrating statistical behavior (Mintz and Baer 1976; Baer et al. 1981).

Photoexcitation is a method that leads to dissociation of molecular ions in well-defined internal energy states. Time-resolved photodissociation (TRPD) is a very successful method developed by Dunbar and coworkers (Dunbar 1987, 1989; Faulk ands Dunbar 1991). Ions can be produced by electron impact ionization. They undergo relaxation and thermalization by radiative and collisional decay in an ion trap before being photoexcited by a laser to a well-defined internal energy that leads to unimolecular dissociation. Energy-selected benzene ions have been very thoroughly studied experimentally using resonance enhanced multiphoton ionization, REMPI (Neusser 1989), PEPICO (Baer et al. 1979), TRPD (Klippenstein et al. 1993), and other methods. Experiments were combined with RRKM/QET and variational RRKM calculations. In the variational calculations both the definition of the reaction coordinate and its value were independently optimized (Klippenstein et al. 1993). Figure 7.1 illustrates experimental and theoretically evaluated rate

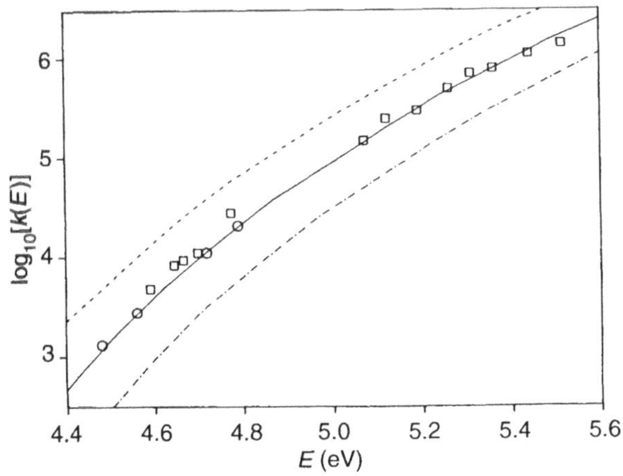

FIGURE 7.1. Plot of experimental and theoretically calculated rate constants for the H-atom atom decay rate constant as a function of the excess internal energy E. The squares denote the experimental REMPI results of Neusser and coworkers (Neusser 1989), and the circles denote TRPD results (Klippenstein et al. 1993). The dashed, solid, and dashed–dotted lines denote the optimized reaction coordinate theoretical results based on assumed dissociation energies of 3.78, 3.88, and 3.98 eV, respectively. [Reproduced from Klippenstein et al. (1993) with permission of the American Physical Society.]

constants for the H-atom decay rate constant as a function of the excess energy. Satisfactory agreement between the theoretical and experimental results was obtained for an assumed dissociation energy of 3.88 eV to the lowest triplet state of $C_6H_5^+$.

It has been concluded (Baer and Hase 1996) that most of the experimental tests of RRKM theory have supported its assumptions. At the same time, only few of the claims of nonstatistical behavior have withstood the test of time. Experimental artifacts have been the major sources of apparent non-RRKM behavior. In fact, it appears that as experiments have become more controlled and refined, the more dramatic has been the validation of the statistical assumptions. The theory has been tested from long times (ms) to short times (ps), from large molecules to the very smallest molecules (Baer and Hase 1996).

7.1.4. IVR Lifetimes and Nonrandom Decompositions

If IVR is very fast compared to the rate of reaction, this guarantees a single exponential decay, because the same distribution of phase space is sampled during the course of the reaction.* Rapid IVR is the basis of the random lifetime assumption, according to which the lifetime is independent of where in phase space the molecule happens to be located (Baer and Hase 1996; Robinson and Holbrook

*A $6N$-dimensional space giving the positions and momenta of each particle in three independent directions is known as the *phase space* of *N particles*.

1972). According to RRKM, the lifetime distribution $P(\tau)$ is given by

$$P(\tau) = k(E)\exp\{-k(E)\tau\} \tag{7.3}$$

Classical trajectory calculations were carried out on various model systems to study the nature of the lifetime distribution (Baer and Hase 1996).

There are cases of apparent non-RRKM behavior and cases of intrinsic non-RRKM behavior that are entirely distinct (Bunker and Hase 1973). Intrinsic non-RRKM behavior originates from "bottlenecks" in the phase space that prevent free energy flow between different vibrational modes, while apparent non-RRKM behavior is observed when dissociation is very fast (on a femtosecond timescale) and the reaction products are formed prior to the complete redistribution of the internal energy in the excited molecule. Apparent non-RRKM behavior can in principle be caused by chemical activation or photoexcitation that are site specific processes. Strong internal coupling would yield a $P(\tau)$ not very different from a random one, which is consistent with most of the experimental observations that have been made (Bunker and Hase 1973). However, the system studied by Rynbrandt and Rabinovitch (1970, 1971a,b) is a classic example of apparent non-RRKM behavior. A chemically activated symmetrical bicyclic molecule, hexafluorobicyclopropyl-d_2 (HBC), was produced by the addition of $^1CD_2$ or $^1CH_2$ to hexafluorovinylcyclopropane (HVC) or HVC-d_2, respectively:

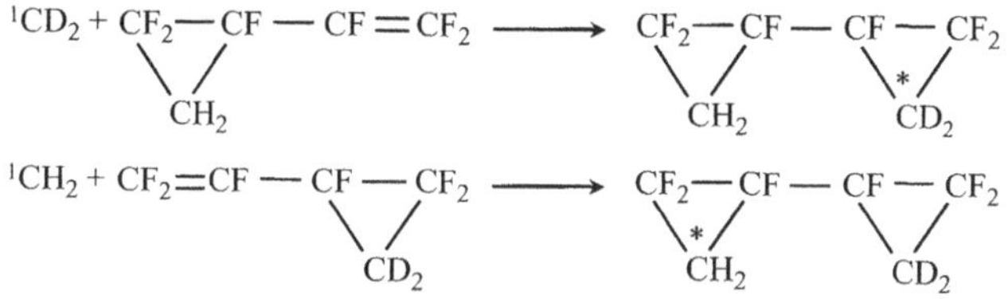

(7.4a)

(7.4b)

The two ring systems of HBC are distinguishable through deuteration. The excess internal energy (marked by an asterisk) resides originally in one of the rings. IVR can randomize the energy over the two-ring system of the whole molecule Unimolecular dissociation can occur in either of the two rings through ring opening and CF_2 elimination. The overall pressure in the system is used as an internal clock. As the pressure is increased, the time between deactivating collisions decreases and only fast dissociations can take place. At low pressures, dissociation of the two rings follows IVR and occurs RRKM-like at a 1 : 1 ratio. However, as the pressure is increased to higher and higher values, dissociation favors that ring that has originally been activated, yielding a nonrandom dissociation that circumvents IVR. The dissociation that follows IVR has a rate constant of $2.3 \times 10^9\ s^{-1}$. With increasing pressure the time between collisions shortens and this RRKM-like reaction is suppressed because there is not enough time available for it to occur. However, the reaction that involves excitation of only one of the rings is much faster, since fewer degrees of freedom are involved. Its rate constant is $3.5 \times 10^{11}\ s^{-1}$, and it can compete with the IVR process (Rynbrandt and Rabinovitch 1970, 1971a,b), since the rate of intramolecular relaxation is $1.1 \times 10^{12}\ s^{-1}$. This result is of historical significance and remains one of the very few examples of incomplete energy randomization in the dissociation of molecules (Baer and Hase 1996).

Decomposition of the enol ion of acetone is an example of chemical activation through isomerization (Lifshitz 1983) and is a well-studied case of apparent non-RRKM behavior in ionic systems. Dissociation occurs to give the acetyl cation plus methyl radical via the chemically activated acetone ion intermediate:

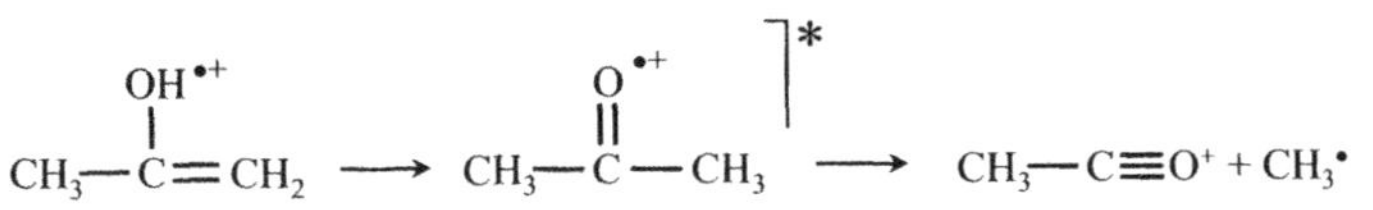

(7.5)

As in the case of HBC, acetone is a chemically symmetrical system having two equivalent groups (the two methyls), one of which can be isotopically labeled either with deuterium (McAdoo et al. 1970; McLafferty et al. 1971) or—in order to avoid possible kinetic isotope effects—with ^{13}C (Depke et al. 1981). Contrary to the HBC case, there are no deactivating collisions in the ionic reaction system taking place in the mass spectrometer. Incomplete energy randomization in the acetone ion intermediate leads to nonrandom decomposition at a ratio different from 1 : 1. The KERDs are bimodal and have been analyzed in terms of their surprisal content,* (Lifshitz 1982, 1983; Levine and Bernstein 1974), which is part of the maximum-entropy method to be discussed below. The high-energy component of the bimodal KERD reflects the contribution of the nonrandomized decay fraction. The low-energy component reflects the decay fraction that samples the bottom of the deep acetone ion well and dissociates ergodically. A quasiclassical trajectory study has been reported recently (Nummela and Carpenter 2002), and the results have been reviewed in what has been termed "nonexponential decay of reactive intermediates" (Carpenter 2003) as well as more recently (Carpenter 2005). The trajectory calculations were in agreement with experiments in terms of giving a nonrandom decomposition ratio in favor of the newly formed methyl. Starting the trajectories in the vicinity of the isomerization transition state yielded a half-life of 238 fs, whereas trajectories started in the vicinity of the acetone radical cation indicated a half-life of 409 fs even though the two sets of trajectories had the same total energy.

Dissociation of acetone radical cation has been also studied using *ab initio* direct classical trajectory calculations at the MP2/6-31G(*d*) level of theory (Anand and Schlegel 2004). It has been demonstrated that once the keto form of acetone radical cation is formed, it dissociates in a few tens of fs. This time corresponds to one or two vibrational cycles of the C–C stretch and CCO bend and is not sufficient for a complete redistribution of energy among all vibrational modes. The branching ratio of the two methyl loss channels obtained in this study is 1.53 ± 0.20, in excellent agreement with the experimental ratio of 1.55 at an excess energy of 8–10 kcal/mol (Osterheld and Brauman 1992). The calculated kinetic energy distribution of the

*The surprisal factor $[I = -\ln(P/P_0)]$ indicates the information content of a distribution function P by determining the extent to which it deviates from a prior expectation of the distribution P_0.

active and the spectator methyl groups is in good agreement with the bimodal KERDs observed experimentally (Lifshitz 1982, 1983).

Intrinsically non-RRKM behavior can arise if transitions between two groups of states are less probable than those leading to products. A group of states not well coupled to the product channel would introduce a component of long lifetimes into $P(\tau)$ (Bunker and Hase 1973). It has been claimed that a heavy atom such as tin at the center of a molecule may prevent energy flow (Rogers et al. 1982); however, this has come under criticism (Wrigley at al. 1984), and similar systems were demonstrated to undergo energy randomization on a subpicosecond timescale. Hindered energy flow was found in loosely bound dimers such as I_2—He, due to a large mismatch in vibrational frequencies that leads to very weak coupling between the I_2 stretch and those of the van der Waals bond between I_2 and helium, or in other words, to a bottleneck in phase space (Davis and Gray 1986; Gray et al. 1986). Neutral dimers as a class of molecules represent the clearest evidence for non-RRKM behavior (Baer and Hase 1996). IVR is slow, and the molecule fragments by transferring the energy from the high-frequency mode originally excited directly to the van der Waals bond in a so-called vibrational predissociation. There are dramatic differences when different parts of the molecule are excited. For example, in the case of the nonsymmetric van der Waals dimer HF—HF (Huang et al. 1986) in which only one of the hydrogen atoms is hydrogen-bonded to a second fluorine atom and the other is not, the outcome depends on which of the two HF monomers that make up the dimer is being excited. In addition, the dissociation rates are orders of magnitude slower than predicted by RRKM.

It has been claimed (Hoxha et al. 1999) that distributions of kinetic energy released during fragmentation processes provide a complementary and more sensitive test for the validity of the statistical theories of mass spectra than the measurement of unimolecular reaction rate constants. The topic of KERDs in mass spectrometry has been reviewed recently (Laskin and Lifshitz 2001). The subtopic that is of interest to us is the efficiency of phase space sampling that can be deduced from analyzing KERDs by the maximum-entropy method (Hoxha et al. 1999; Laskin and Lifshitz 2001). An "ergodicity index" e^{-DS}—which measures the efficiency of phase space sampling, where DS denotes the so-called entropy deficiency associated with incomplete energy randomization—can be extracted from experiments (Lorquet 2000c). Halogen loss reactions from $C_2H_3Br^+$, $C_2H_5I^+$, $C_3H_7I^+$, $C_6H_5Br^+$, and $C_6H_5I^+$ were studied in some detail (Hoxha et al. 1999; Lorquet 2000a,c). Phase space appears to be sampled with an efficiency close to 100% at both very low and very high values of the internal energy. For intermediate values of E, the minimal efficiency is of the order of 75%. The behavior at high internal energies is interpreted as resulting from the conjugated effect of IVR and radiationless transitions among potential energy surfaces. Figures 7.2 and 7.3 represent experimental KERDs for $C_3H_7I^+$ isomers determined via TPEPICO (Brand et al. 1983) and calculations based on predictions of phase space theory (Brand et al. 1983) and on the maximum-entropy method (Lorquet 2000a). The efficiency of phase space sampling was found by the maximum-entropy method to be about 96%.

It was concluded (Lorquet 2000b) that KERDs were more sensitive (i.e., less robust) than were rate constants to incomplete energy randomization. However, for

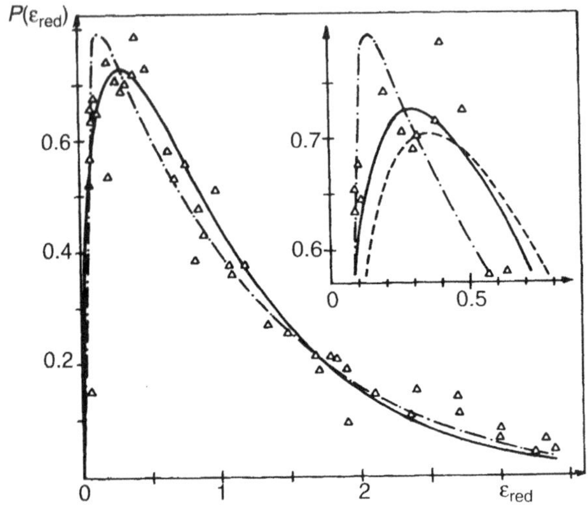

FIGURE 7.2. Normalized KERD in reduced units ($\varepsilon_{red} = \varepsilon/\langle\varepsilon\rangle$, where ε is the kinetic energy release and $\langle\varepsilon\rangle$ is the average release energy) for the reaction $C_3H_7I^+ \rightarrow C_3H_7{}^+ + I$. Triangles are experimental data (Brand et al. 1983) for both 1- and 2-iodopropane at all excess internal energies studied in the TPEPICO experiment. Dashed–dotted line: predictions of phase space theory (Brand et al. 1983). Dashed line in the inset: prior (i.e., most statistical) distribution, corresponding to full phase space sampling. Solid line: maximum entropy calculation leading to an ergodicity index of 0.96. [Reproduced from Lorquet (2000a) with permission of Elsevier Science.]

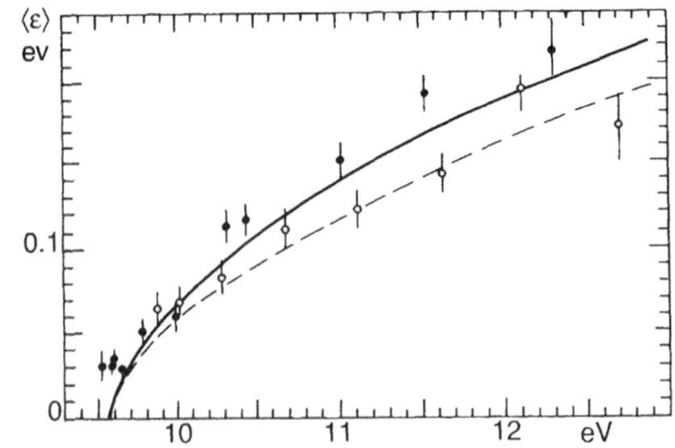

FIGURE 7.3. The average release energy for reaction $C_3H_7I^+ \rightarrow C_3H_7{}^+ + I$, as a function of the absolute 0 K energy (the zero of the absolute energy scale is the energy of the elements in their standard states). Open symbols represent experimental data (Brand et al. 1983) for 1-iodopropane. Closed symbols are experimental data for 2-iodopropane. Dashed line: predictions of phase-space theory (Brand et al. 1983). Solid line: the average release energy of the maximum entropy distribution having a 96% efficiency of phase space sampling. [Reproduced from Lorquet (2000a) with permission of Elsevier Science.]

most molecular ions created in a conventional way (e.g., by electron impact, photoionization or chemical ionization), the assumption of rapid energy randomization leading to nearly complete phase space sampling seems to be a good approximation paving the way for the application of a statistical theory ($e^{-DS} \approx 75$–80%). This success probably results from the nonspecific nature of the conventional ionization, which involves a sequence of initial points (Lorquet 2000b). It is important to point out that the system does not have to explore 100% of the phase space for statistical theories to apply.

7.1.5. Cases of "Isolated" Electronic States

This section deals very briefly with what is known about the assumption of radiationless transitions from excited electronic states and the possibility of direct dissociations from so-called isolated electronic states. Isolated state decay has been first observed by Lifshitz and Long (1965a) for C—F bond cleavages in fluoroethene cations and in $C_2F_6^+$(Lifshitz and Long 1965b). PEPICO experiments (Simm et al. 1973, 1974; Stadelmann and Vogt 1980a, 1980b) have corroborated the early electron impact results. Several additional examples have been found, including CH_3^+ production from CH_3OH^+ (Berkowitz 1978), the fragmentation of the formaldehyde cation (Bombach et al. 1981; Lorquet and Takeuchi 1990), and of acetone and nitromethane cations (Shukla and Futrell 1993).

These topics have been reviewed more recently (Lorquet 2000b). The situation of isolated state decay is a rather rare event. Some of the earliest TPEPICO experiments have already demonstrated the efficiency of radiationless transitions (Stockbauer and Inghram 1975). No correlation between the shape of the breakdown curves and the structure in the photoelectron spectra was observed in the case of ethylene. The breakdown curves were observed to vary smoothly, whereas the threshold photoelectron spectra showed two distinct bands in the same energy range. In other words, there was no correlation between the breakdown curves and the initial electronic states of the molecular ions. The fragmentation of these ions was found to depend only on the total internal energy and not on the initial state of excitation of the ions. This is the general rule, in agreement with assumptions of statistical theories of mass spectra. Isolated state decay is a rare event mainly because of "conical intersections" between potential energy surfaces. These lead to ultrafast relaxation mechanisms that bring about radiationless transitions with lifetimes of the order of 10^{-14} s (Lorquet 2000b). Furthermore, the situation of isolated state decay has been interpreted as indicating a possible failure of the assumption of fast radiationless conversion to the ground electronic state of the ion, related to the bonding character of a particular molecular orbital. However, this has been shown to be the case only for formaldehyde where the rate limiting step is internal conversion and not dissociation (Lorquet and Takeuchi 1990). Other cases have been interpreted (Lorquet 2000b; Galloy et al. 1982) using *ab initio* calculations by showing that internal conversion causes a very specific preparation of the lower electronic state so that the reactive flux branches into two paths. One of them leads directly to the dissociation asymptote and escapes statistical energy redistribution. The other path leads to the bottom of the potential energy surface and gives rise to the RRKM/QET component of the mass spectrum. The conclusion was that "isolated state decay corresponds to dissociation of the nonrandomized fraction of the population and not to isolation." When internal conversion takes place, the vibrational energy remains fixed in one or only a few degrees of freedom (Stadelmann and Vogt 1980a). The branching into two paths can be brought about only by avoided crossings between potential surfaces, which are relatively rare. The much more frequently encountered conical intersections do not lead to branching but to ultrafast relaxation processes and to ergodic nuclear motion (Lorquet 2000b).

7.2. BIOMOLECULES

7.2.1. Introduction

Mass spectrometry has become a success story in the analysis of biomolecules in general and in the sequencing of peptides and proteins in particular. We are at the beginning of a new era in proteomics research, and there is no doubt concerning the central role played by mass spectrometry in this field. It is therefore of importance to understand some of the physics behind the analytical methods.

The term *nonergodic behavior* has been raised quite often in connection with mass spectra of biomolecules, and it is our intention in this second part of the chapter to study this question in depth. We will put the emphasis on peptides and proteins.

We know that biomolecules are special insofar as they constitute the basis for life on earth. Are they also special in their unimolecular reactions that make up their mass spectra? There are several aspects that could lead to a greater extent of nonergodic behavior in biomolecules than in ordinary organic molecules, and we need to check each of them.

7.2.2. Protein Size and Potential for Nonergodic Behavior

Proteins are large molecules with many degrees of freedom; could this lead to nonergodic behavior?

It is possible to derive information on IVR from optical spectra [see Pavlov-Verevkin and Lorquet (2002) and references cited therein]. A complicated spectrum consisting of a series of "clumps" should be a valid assumption for a molecule such as a protein. When such spectra were analyzed it was found that the available phase space is virtually completely explored after a time $10hD$, where h is Planck's constant and D is the average density of optically active vibrational states detected in the spectrum. As the size of the molecule increases, so does the density of states and the time necessary for phase space exploration.

The idea was brought up many years ago in connection with the "degrees of freedom (DOF) effect" (Lin and Rabinovitch 1970; Bente et al. 1975) that for very large polyatomic ions a point should be reached where IVR will not compete with dissociation following a site-specific excitation. For example, in a series of 2-alkanones there are two consecutive reactions, $M^{\bullet +} \rightarrow A^{\bullet +} \rightarrow D^{+}$:

$$C_nH_{2n}O^{\bullet +} \rightarrow C_3H_6O^{\bullet +} \xrightarrow{m^*} C_2H_3O^{+} \tag{7.6}$$

With increasing size of the alkanone chain the internal energy content transferred to the product $A^{\bullet +}$ of the first dissociation step decreases and the relative abundance of the metastable ion m^* decreases accordingly. Theoretically, the logarithm of the normalized m^* intensity log $[m^*]/[A]$ approaches $-\infty$ as 1/DOF approaches zero. Experimentally, DOF plots demonstrate finite intercepts. The model used assumed statistical partitioning (randomization) of the excess energy of $M^{\bullet +}$ between $A^{\bullet +}$ and the neutral fragment, and the disagreement of the calculated results with experiment at high DOF values could be due to failure of this assumption. The conclusion was not clearcut. On one hand, the average internal energy of $C_3H_6O^{\bullet +}$ (i.e., $A^{\bullet +}$) decreases with increasing size of $M^{\bullet +}$ because increasingly large neutral fragments carry away a larger fraction of the $M^{\bullet +}$ internal energy. Offsetting this is an increasing abundance of tight complex fragmentations of $M^{\bullet +}$ that compete with $C_3H_6O^{\bullet +}$ formation at low $M^{\bullet +}$ internal energies, causing the average internal energy of those $M^{\bullet +}$ ions that fragment to give $C_3H_6O^{\bullet +}$ to be higher for larger $M^{\bullet +}$ ions.

Large gas-phase biomolecules fragment on a microsecond timescale. Fragmentation is, of course, a prerequisite for sequencing of proteins. The probability of concentrating the required energy in the reaction coordinate seems intuitively to be very small if the energy is originally randomized over all the degrees of freedom of a protein. Calculated dissociation rates were originally orders of magnitude lower than observed ones, and the question arose as to whether there is a threshold size above which statistical theories fail (Schlag and Levine 1989). The calculations were carried out using the older version of RRKM (or its classical limit) called RRK. In RRK all the vibrational frequencies are considered equal and energy is randomized efficiently. There are additional assumptions made in deriving the RRK reaction rate equation (Baer and Hase 1996; Schlag and Levine 1989). A nonstatistical approach following the original version of Slater (1959) was applied as an alternative, according to which the energy is localized in a moiety rather than randomly distributed. In Slater's theory the vibrational frequencies of the molecule are considered to be harmonic, thus preventing energy flow among them. Starting from either the RRK or the Slater limiting description, the authors (Schlag and Levine 1989) were led to the conclusion that the observed fragmentation of large molecules and its variation with excess energy reflects bottlenecks to intramolecular vibrational energy distribution. However, other calculations for large protein ions demonstrated RRKM unimolecular rate coefficients that agree well with fragmentation lifetimes in a mass spectrometer (Bernshtein and Oref 1994).

7.2.3. Protein Structure and Potential for Nonergodicity

Proteins are biopolymers made up of amino acids bound by amide bonds. Is there something special in the structure that leads to nonergodic behavior?

Classical trajectory simulations of the photoionization dynamics of glycine have demonstrated no significant approach to statistical distribution of the energy throughout the first 10 ps (Shemesh et al. 2004). Furthermore, rates computed from the dynamics for conformational transitions differ considerably from RRK predictions. It was concluded that the system shows strong nonstatistical behavior. Whereas this is of great interest, it has to be pointed out that some organic molecules demonstrate extremely slow IVR rates with relaxation times; τ_{IVR} values for some of the modes are as long as 1–3 ns (Keske et al. 2000; Engelhardt et al. 2001). However, on the other hand, peptides, which are the main topic of interest here, were demonstrated to undergo extremely fast IVR (Hamm et al. 1998). Femtosecond infrared (IR) spectroscopy experiments were used to examine the ultrafast response of the so-called amide I modes, which involve mainly the C=O stretching displacements of the peptide backbone with contributions from the CN and NH motions. Vibrational relaxation of these modes occurs in 1.2 ps. The relaxation is dominated by IVR and reflects an intrinsic property of the peptide group in any environment. An even faster relaxation value of 450 fs was obtained for *N*-methylacetamide, a model for the peptide unit. Picosecond infrared spectroscopy of the amide I band of myoglobin demonstrated that vibrational relaxation from this mode into the hydration shell of the protein occurs in approximately 20 ps (Austin et al. 2005). These experiments support fast IVR and ergodic behavior of large biomolecules.

It has been suggested that peptide cation radicals avoid the "pitfall of large molecule kinetics," namely, vibrational energy redistribution, because the charge scouts for the site of reactivity without energy dissipation (Weinkauf et al. 1995, 1996, 1997). According to this approach, the peptide behaves as a collection of amino acids, like beads on a string rather than a viable supermolecule, namely, an ordinary organic molecule albeit with a large number of degrees of freedom. Electronic energy relaxation does not necessarily precede dissociation, but there is rather site selectivity and charge-directed reactivity (Remacle et al. 1998). These concepts will be discussed further below in the light of experimental findings.

7.2.4. Ionization and Excitation of Biomolecules

Special methods have been developed to introduce biomolecules into the gas phase and induce fragmentation. Could some of these lead to nonergodic behavior?

7.2.4.1. Protonated Peptides. Peptide and protein ions are currently introduced into the gas phase mainly by two soft ionization methods—matrix-assisted laser desorption/ionization [(MALDI); see Karas and Hillenkamp (1988)] and electrospray ionization [(ESI); see Fenn et al. (1989)]. These produce singly or multiply protonated peptides and proteins. Several activation methods are being employed to induce fragmentation. These include collision-induced (or -activated) dissociation [(CID or CAD; see Senko et al. (1994), and Laskin and Futrell (2003a)], surface-induced dissociation [(SID); see Williams et al. (1990a), Dongré et al. (1996a), and Laskin and Futrell (2003b)], UV photodissociation [see Hunt et al. (1987), Williams

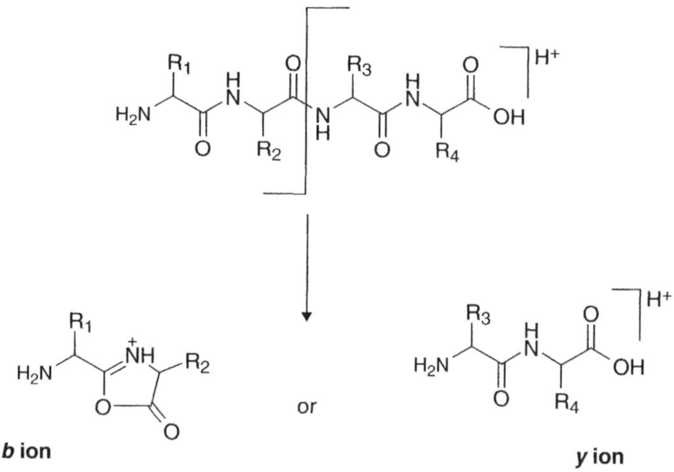

SCHEME 7.1

et al. (1990b), and Thompson et al. (2004)], infrared multiphoton dissociation [(IRMPD); see Woodin et al. (1978), Zimmerman et al. (1991), and Little et al. (1994)], and blackbody infrared radiation [(BIRD); see Price et al. (1996)]. Internally excited multiply protonated peptides undergo fragmentation of amide bonds to produce *N*-terminal *b* and *C*-terminal *y* ions (Scheme 7.1).

The "mobile proton model" has been developed (Burlet et al. 1992; Dongré et al. 1996b) to describe how protonated peptides dissociate by the various activation methods. On excitation the protons added to a peptide will migrate to various protonation sites prior to fragmentation. Less favored protonation sites, like those of the reactive intermediates leading to backbone dissociation and to sequence-informative fragment ions, become more populated, as the internal energy of the ions increases. Most of the common ion types observed by CID, SID, or photodissociation are rationalized as charge-directed fragmentations; nonetheless, most of the dissociation events can fit into a statistical RRKM-like picture. In fact, the mobile proton model has been validated by a combination of quantum-chemical data and a RRKM formalism (Csonka et al. 2000; Paizs et al. 2001; Paizs and Suhai 2001, 2005).

Experimental CID and SID energy-resolved fragmentation efficiency curves (FECs) of protonated dialanine (Laskin et al. 2000), and protonated trialanine and tetraalanine (Laskin et al. 2002a) were successfully modeled by RRKM/QET. In the first step, microcanonical rate constants [$k(E)$] were calculated. Results for the four primary reactions of Ala–AlaH$^+$ are presented in Figure 7.4. The rate–energy dependences $k(E)$ were next employed to calculate fragmentation probabilities (i.e., breakdown curves) as a function of the internal energy of the parent ion and the experimental observation time. The internal energy deposition function was set up as an analytical expression. Experimental and computed FECs were compared and the fitting parameters varied until the best fit was obtained. The fitting parameters

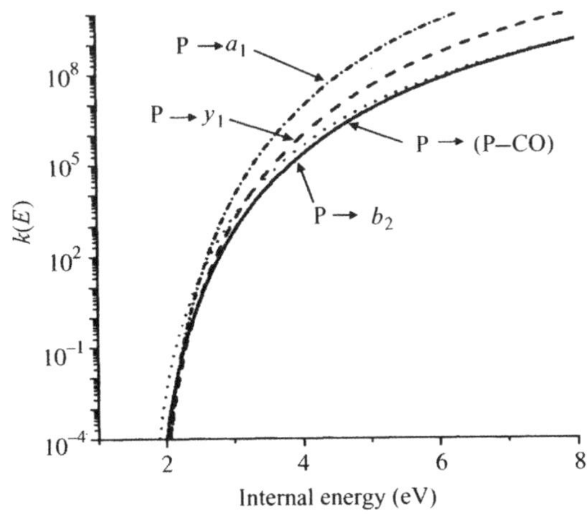

FIGURE 7.4. RRKM rate–energy dependences for the four primary fragmentation pathways of Ala–AlaH^+: (—) b_2 production; (· · · · · · ·) CO elimination; (-----) y_1 production; (-·-·-) a_1 formation. The ion notations conform with the accepted nomenclature (Roepstorff and Fohlmann 1984; Biemann 1988). [Reproduced from Laskin et al. (2000) with permission of the American Chemical Society.]

included the critical energy and activation entropy for dissociations of the precursor ion that affect the $k(E)$ curves and the parameters characterizing the energy deposition function. Experimental and theoretical SID curves are presented in Figure 7.5 demonstrating excellent fits. SID time- and energy-resolved FECs for larger peptides were also modeled by RRKM (Bailey et al. 2003; Laskin et al. 2002b, Laskin 2004). RRKM modeling revealed that addition of a basic residue such as arginine, which is known to sequester the proton, to the C terminus of a peptide has a negligible effect on the dissociation threshold at acidic residues. However, the Arrhenius preexponential factor is reduced by two orders of magnitude by such an addition. SID results for des-Arg^1– and des-Arg^9–bradykinin were rationalized in the light of previous BIRD experiments (Schnier et al. 1996). This required RRKM calculations of $k(E)$ for the dissociations of the two peptides and the use of Tolman's theorem (Tolman 1920) that connects between threshold energies and Arrhenius activation energies.

The evidence presented thus far has been that protonated peptides follow statistical theories when they are dissociated by various activation methods such as CID, SID, and IRMPD. Their reactions can be modeled successfully by RRKM/QET; thus, they behave ergodically. However, there has been in addition accumulating evidence, both from experimental studies (Laskin and Futrell 2003b; Laskin et al. 2003) and from classical trajectory simulations (Meroueh et al. 2002; Wang et al. 2003) that above a certain collision energy peptides undergo "shattering fragmentations" under SID. At low energies the peptide ion activated by collision with the surface, bounces off, and then dissociates in the gas phase after undergoing IVR. The trajectory calculations demonstrate that in the shattering collision the ion fragments instantaneously as it collides with the surface. It has been

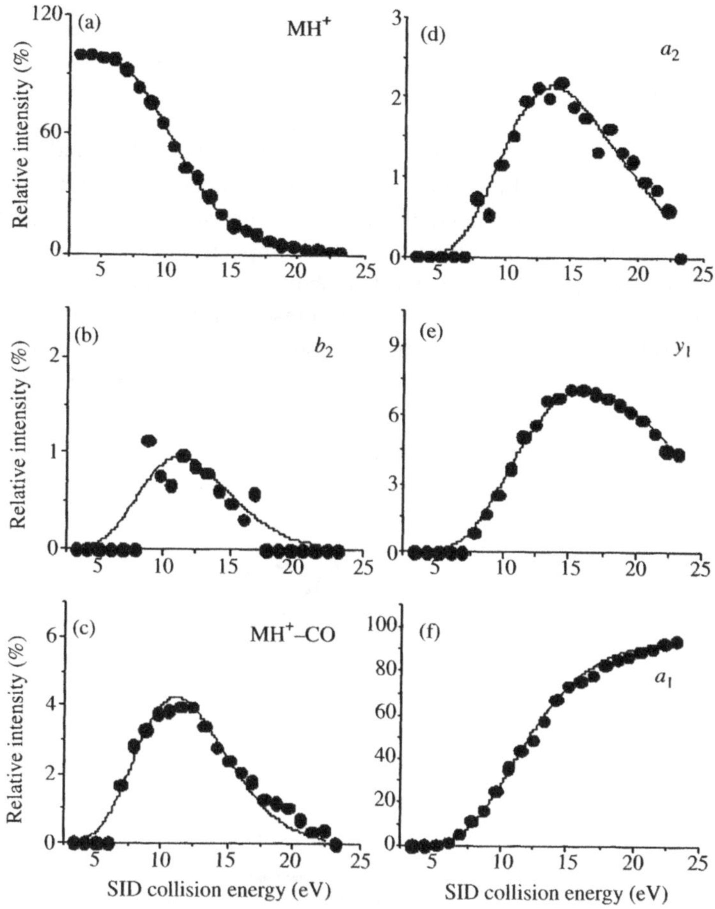

FIGURE 7.5. Collision-energy-resolved SID fragmentation efficiency curves and the results of theoretical modeling for (a) MH^+; (b) b_2; (c) MH^+-CO; (d) a_2; (e) y_1; (f) a_1 ions of Ala–AlaH^+. [Reproduced from Laskin et al. (2000) with permission of the American Chemical Society.]

suggested that as the molecule undergoes deformation during its impact with the surface, it can be driven to a very specific transition state and fragment instantly.

The characteristic results of shattering fragmentations are as follows:

1. There is an abrupt increase in the number of fragments formed. The selectivity observed for slow fragmentations at low energies is lost. This is very useful in protein sequencing because the fragments formed in the shattering regime are due to backbone cleavages of the amide bonds and are sequence informative.
2. The fragments are not due to secondary decompositions of ions formed at lower energies.

3. The last conclusion is based on modeling the experimental results, and it states that the fragments due to shattering are independent of the reaction delay times.

Experimental results for time- and energy-resolved SID of des-Arg[1] and des-Arg[9] on a fluorinated self-assembled monolayer surface were modeled (Laskin et al. 2003). Time-dependent and time-independent fragments were separated and two dissociation rate constants were used for the total ion decomposition to account for the slow, RRKM-like decomposition and for the fast shattering fragmentation, respectively. Experimental FECs and modeling results are presented in Figure 7.6.

For all peptide ions studied experimentally so far, it was found that the shattering transition occurs when the ion internal energy exceeds 10 eV (Laskin and Futrell

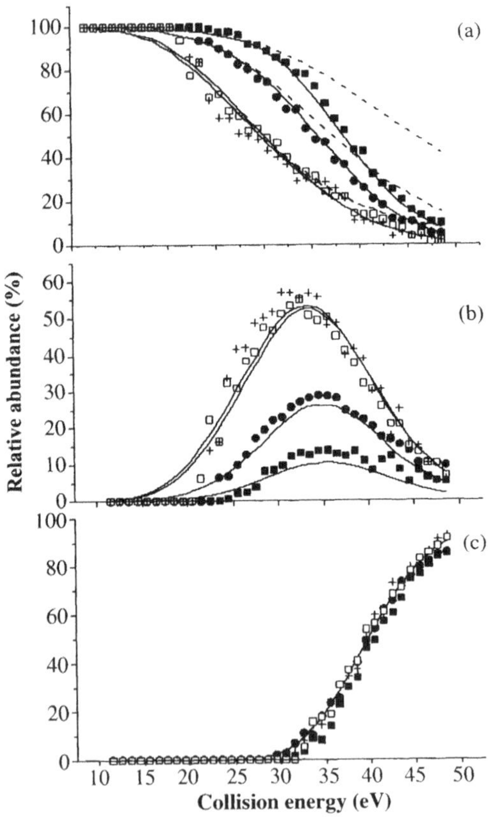

FIGURE 7.6. Experimental FECs for the parent ion (a), time-dependent fragments (b), and time-independent fragments (c) of des-Arg[1]–bradykinin for reaction delays of 1 ms (solid squares), 10 ms (solid circles), 100 ms (open squares), and 1 s (crosses). Solid lines represent the modeling results taking into account slow (RRKM-like) and fast (shattering) dissociation channels. Dashed lines in panel (a) represent the contribution of the slow decay channel to the FEC of the parent ion. [Reproduced from Laskin et al. (2003) with permission of the American Chemical Society.]

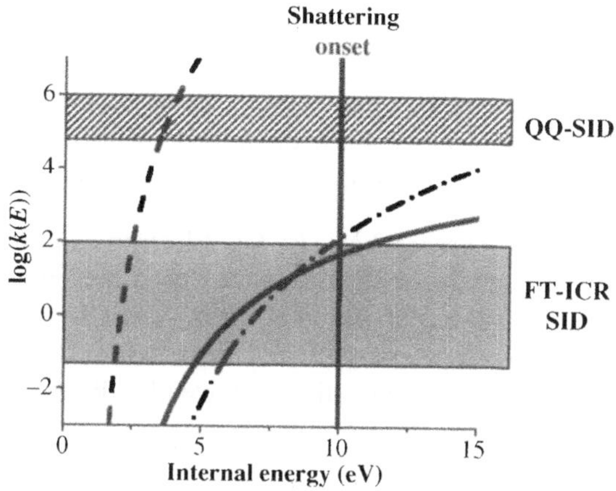

FIGURE 7.7. Semilogarithmic plot of the microcanonical rate-energy dependencies for dissociation of singly protonated pentaalanine (dashed line), des-Arg[1]–bradykinin (solid line) and fibrinopeptide A (dashed–dotted line). Vertical solid line shows the assumed shattering onset. The hatched area corresponds to the experimental time window of a tandem quadrupole QQ-SID setup, while the shadowed area shows the observation window in the FT-ICR SID experiment. [Reproduced from Laskin and Futrell (2003b) with permission of the American Society for Mass Spectrometry.]

2003b). As a result, a small peptide such as pentaalanine does not reach this limit—shattering does not compete with its unimolecular dissociation even on a microsecond timescale, but larger peptides, such as des-Arg[1]–bradykinin and fibrinopeptide A, do undergo shattering dissociations even on the long timescale of an FTICR (see Figure 7.7). The results presented in Figure 7.7 also imply that for many large and medium-sized peptides microsecond dissociation is dominated by shattering.

It is not at all clear that shattering dissociations of the peptides are nonstatistical events (Laskin 2005). It is possible that they reflect a "phase transition" in which, on collision with the surface, a stable conformer isomerizes to a collection of floppy molecules whose dissociation products are a multitude of smaller fragments. A similar phase transition has been described and analyzed quantitatively for C_{60} and C_{60}^{+} (Campbell et al. 1996). If a temperature can be assigned to the fragmenting peptide as has been done to C_{60}, the dissociation may be considered statistical. Furthermore, there is still disagreement concerning the timescale of the fast dissociations. Laskin, Futrell, and coworkers claim on the basis of their FTICR data that dissociation in the shattering regime is instantaneous with assumed typical rate constants in the range 10^{11}–10^{12} s^{-1} (Laskin and Futrell 2003b). On the other hand, Wysocki and coworkers performed experiments using SID of MALDI- generated peptide ions and claim that they observe directly fast dissociations and that these take place ~250 ns after the precursor ion collides with the surface. These results were interpreted as indicating fragment formation in proximity to the surface but not to immediate fragmentation upon surface impact. Rate constants in the range of 10^{6}–10^{7} s^{-1} were suggested for these fast fragmentation processes (Gamage et al. 2004).

Shattering upon SID has been observed previously for other chemical systems particularly clusters (Raz et al. 1995; Raz and Levine 1996; Christen et al. 1998) and the maximum-entropy method has been employed to treat the data theoretically. Shattering of peptides has not yet been quantified using the maximum-entropy method because of the complexity of the systems (Laskin and Futrell 2003b).

7.2.4.2. ***Electron Capture Dissociation (ECD).*** Electron capture dissociation is a very efficient protein-sequencing technique. It has been presented as a nonergodic process already on its initiation (Zubarev et al. 1998). In ECD a multiply protonated peptide or protein captures a low-energy (≤ 0.1 eV) electron leading to reduction of the cation and formation of a radical cation followed by dissociation:

$$[\mathrm{M} + n\mathrm{H}]^{n+} + e^{-} \rightarrow [\mathrm{M} + n\mathrm{H}]^{(n-1)+\bullet} \rightarrow \text{fragments} \tag{7.7}$$

The major fragmentation pathway of peptides on ECD involves cleavage of the backbone *N*-alkyl (N—C_α) bond to form *N*-terminal *c* and *C*-terminal $z^\bullet$ ions (McLafferty et al. 2001; Cooper et al. 2005)(Scheme 7.2).

Whereas other activation methods such as CID or IRMPD introduce the energy as vibrational energy in a random fashion, ECD has been thought to introduce the energy as the "recombination energy" of an electron with a positive ion, in a site-specific manner at the protonation sites. ECD cleaves many more bonds than CID (Zubarev et al. 2002) and has thus been considered to be nonselective whereas CID preferentially cleaves the weak bonds. The energy released on electron capture, the recombination energy, is estimated to be about 6 eV (Turecek and Syrstad 2003; Turecek 2003; Breuker et al. 2004). In order for dissociation to take place, it has therefore been postulated that it circumvents IVR and occurs at or near the site of

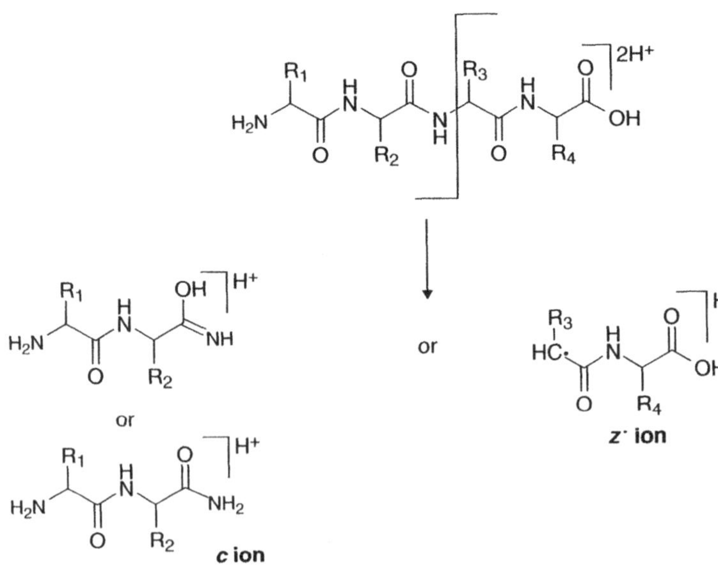

SCHEME 7.2

initial electron capture. Otherwise, the released energy would have to be distributed over a very large number of degrees of freedom and lead to a minor increase in the protein temperature and no dissociation would ensue. In other words, it has been claimed that ECD leads to what has been termed "apparent non-RRKM behavior" (see Section 7.1.4). The available energy is utilized locally in a fast reaction, before it has time to undergo energy randomization. It has been further claimed (Breuker et al. 2004) that the effect should prevail even in smaller peptides. This implies that there is an intrinsic non-RRKM behavior (see Section 7.1.4) because of insufficiently fast dispersal of the excess recombination energy away from the active site, not by how extensively it can then be dispersed. This contradicts previous findings that fast IVR is an intrinsic property of the peptide group (Section 7.2.3).

Several additional unique features of ECD have led to the conclusion that the dissociation is nonergodic. A major factor is the observation that what are considered strong backbone covalent bonds are cleaved but the fragments retain labile groups such as posttranslational modifications. Dissociation of strong bonds in the presence of weak bonding was used to support nonergodic behavior of peptides following electron capture (Zubarev 2003).

The assumption of nonergodic behavior has come under considerable criticism. There has also been some discussion of the mechanism of ECD, and the two aspects of ECD—the mechanism and the question of ergodicity—are tied together. The hydrogen atom capture model (Zubarev et al. 1999, 2002) was able to explain the preferential cleavage of S–S bonds as well as the formation of the c and $z^{\bullet}$ ions (Scheme 7.2). According to this mechanism, electron capture leads to neutralization and formation of a hypervalent species such as $RNH_3^{\bullet}$ that ejects an energetic hydrogen atom.

This hydrogen atom is mobile intramolecularly and is captured by groups having a high $H^{\bullet}$-atom affinity such as the disulfide bond and the carbonyl group of the backbone amide bond. Sufficient excess energy is supplied for near-instantaneous (nonergodic) dissociation.

A model system, 1-hydroxyl-1-(N-methyl)aminoethyl radical (**1**), has been chosen to represent the protein group $-CHR\dot{C}(OH)NH-CHR'-$:

$$CH_3CONHCH_3 + H^{\bullet} \leftarrow \underset{\mathbf{1}}{CH_3\dot{C}(OH)NHCH_3} \rightarrow CH_3C(OH)NH + {}^{\bullet}CH_3 \quad (7.8)$$

RRKM calculations using B3LYP/6-31G(d) geometries and frequencies were carried out. The lifetimes and branching ratios between hydroxyl $H^{\bullet}$ loss and N-substituted ${}^{\bullet}CH_3$ loss were calculated (Zubarev et al. 1999). Very short lifetimes ($\sim 10^{-12}$ s) were calculated favoring the cleavage forming the ${}^{\bullet}CH_3$ radical with increasing internal energy. The calculated energy dependence of the branching ratio between the $H^{\bullet}$ loss and the ${}^{\bullet}CH_3$ loss has been interpreted (Zubarev et al. 2002) as indicating that the observed $N-C_{\alpha}$ cleavage can occur only when the whole recombination energy is released in a small molecular region containing just a few atoms, that is, without energy redistribution over the whole molecule. At lower energies hydrogen loss should prevail. High-level RRKM calculations within the accuracy of the G2 potential energy surface were combined with experimental measurements, and the conclusions were

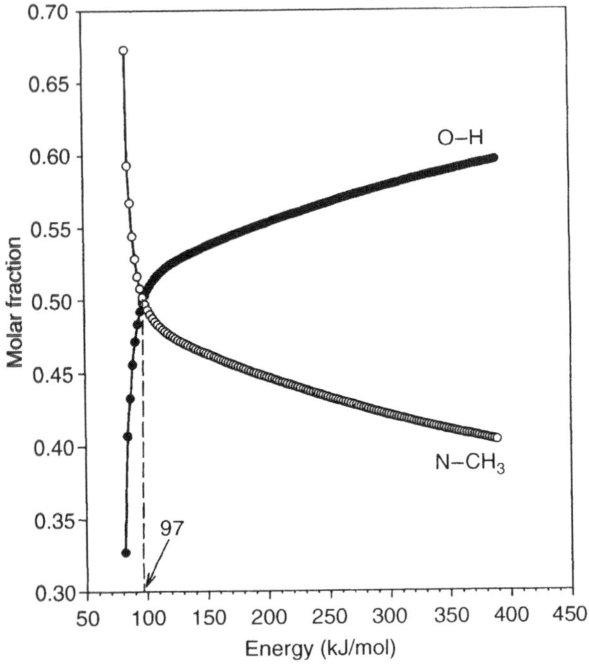

FIGURE 7.8. Energy dependence of the branching ratio for the dissociations of the O–H and N–CH_3 bonds in *N*-methylacetamide. [Reproduced from Syrstad et al. (2003) with permission from the American Chemical Society.]

reversed (Syrstad et al. 2003). The computational results are represented in Figure 7.8. The branching ratio k(O–H)/k(N–CH_3) increases with increasing internal energy. Experimentally, the branching ratio was found to be 1.7 at an internal energy of 190 kJ/mol, in quite good agreement with the RRKM calculation. Hydrogen atom loss is the dominant reaction channel over most of the energy range except at near threshold energies. The conclusion was (Syrstad et al., 2003) that the radical **1** does not represent the best model for ECD, where N–C bond dissociations leading to backbone fragmentation predominate. This conclusion was further supported by a dissociative recombination study (Al-Khalili et al. 2004) showing that hydrogen loss was a dominant dissociation pathway (81.7%) for recombination of protonated *N*-methylacetamide, while N–C_α bond cleavage was a minor pathway accounting for approximately 7% of fragment ions.

The energetics of N–C_α bond cleavage in model aminokethyl radicals representing intermediate species in ECD was studied using density functional theory and ab initio calculations (Turecek 2003). Relatively high dissociation energy of 85–105 kJ/mol was obtained for the smallest model system: β-alanine-*N*-methylamide. However, only 37 kJ/mol was required for N–C_α bond cleavage in a larger system: the N_α-glycylglycine amide radical. RRKM rate constants for dissociations of N–C_α bonds in aminoketyl radicals and cation radicals indicate an extremely facile bond cleavage that occurs with unimolecular rate constants larger than $10^5\ s^{-1}$ at 298 K. Calculations suggested that N—C_α bond cleavage in aminoketyl cation radicals does not necessarily result in formation of the corresponding fragment ions but rather isomerization to ion–molecule complexes that are held together by strong hydrogen bonds was observed. Based on these findings Turecek proposed that it is unnecessary to invoke the hypothesis of nonergodic behavior for ECD intermediates. A subsequent study of the effect of positive charge on electron capture by ammonium and amide groups proposed an efficient pathway for formation of labile aminoketyl radicals that readily dissociate by N—C_α bond cleavage in ECD of protonated peptides (Syrstad and Turecek 2005).

7.2.4.3. Radical Cations Produced by Multiphoton Ionization (MPI). The combination of lasers with mass spectrometry has been utilized to study fragmentation of gas-phase peptide radical cations. Laser desorption (LD) has allowed the transfer of these thermally labile species into the gas phase without breakup. Neutral molecules are transferred into the vacuum region of a mass spectrometer using s supersonic beam, which also provides control over the initial internal energy of the molecule. Strong absorption of an aromatic sidechain of a peptide between 250 and 285 nm is used for localized multiphoton ionization (MPI) at the chromophore. Weinkauf, Schlag, and coworkers conducted the first MPI experiments on small peptides containing a single chromophore (tryptophan or tyrosine). One of the major conclusions from these studies has been that peptides, which are not expected to fragment according to RRKM/QET on the timescale of mass spectrometry because of a very low excess energy $E - E_0$ above the threshold E_0, do fragment, avoiding the pitfall of large molecule kinetics (Weinkauf et al. 1995, 1996, 1997). According to this reasoning, the assumption that the entire available internal energy E is distributed over all degrees of freedom is incorrect. It has been claimed that the charge scouts for the site of reactivity without energy dissipation. These studies were, however, limited to measurements of mass spectra, and no rate–energy $k(E)$ data were reported.

Lifshitz and coworkers used time-resolved photodissociation (TRPD) to explore the extent of statistical versus site-selective fragmentation of small peptide radical cations (Cui et al. 2002; Hu et al. 2003). The objective was to determine how the mode of ion preparation and the initial site of excitation affect the type, degree, and rate of fragmentation. A major question that has been posed is whether a peptide behaves as a collection of amino acids, like beads on a string, or as a viable supermolecule, namely, an ordinary organic molecule, albeit with a large number of degrees of freedom.

TRPD experiments were conducted using a quadrupole ion trap/reflectron time-of-flight (TOF) instrument such as that shown in Figure 7.9. The instrument and the experimental approach have been described in detail elsewhere (Cui et al. 2000, 2002; Hu et al. 2003). Briefly, the TRPD apparatus consists of three vacuum chambers, for (1) laser desorption, (2) the quadrupole ion trap, and (3) a reflectron-TOF analyzer. LD followed by jet cooling is carried out in chamber 1. The desorbed neutrals are entrained and cooled by collisions with CO_2 or Ar gas through the nozzle of a pulsed supersonic valve and transferred into chamber 2 downstream

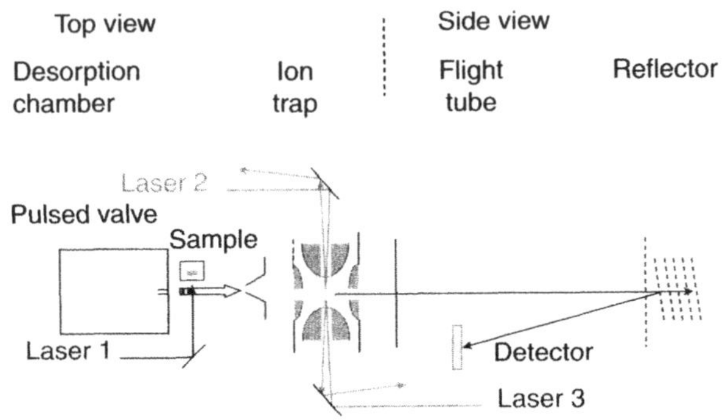

FIGURE 7.9. Quadrupole ion trap/reflectron time-of-flight mass spectrometer. Laser 1 causes desorption of the peptide sample, laser 2 causes ionization, and laser 3 causes dissociation. [Reproduced from Hu et al. (2003) with permission from the American Chemical Society.]

through a skimmer. Ions are formed in chamber 2 and stored in the Paul trap. Pulsed helium buffer gas is used to bring the ions into the center of the trap and ensure collisional thermalization prior to photodissociation. The third laser pulse is used for excitation/photodissociation (PD) of the thermalized ions to well-defined internal energies. This is followed by measurements of the fragment ion buildup as a function of the trapping delay time leading to the TRPD curve. The suitability of this instrumentation for decay time investigations has been demonstrated using perdeuterated benzene and naphthalene (Cui et al. 2000) as model systems, for which TRPD data have been previously reported by others (Grebner and Neusser 1999; Ho et al. 1995).

TRPD of two peptides—LeuTyr and LeuLeuTyr—was studied using this experimental apparatus (Cui et al. 2002; Hu et al. 2003). Both peptides were ionized at the aromatic chromophore of the *C* terminus using a 280.5 or 266 nm laser light and excited for PD in the visible. The internal energy of the ion was varied by varying the wavelength of the PD laser. PD of both peptides resulted in formation of the *N*-terminal immonium ion at m/z 86 (Scheme 7.3). No other fragments were observed in the range of internal energies between 2.3 and 3.7 eV.

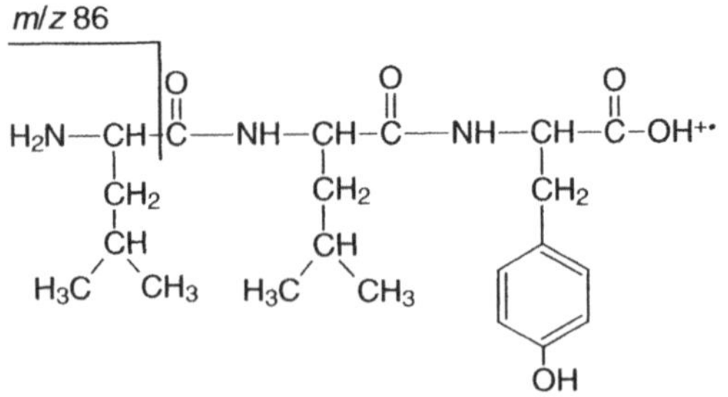

SCHEME 7.3. The α-cleavage reaction resulting in formation of the immonium ion at m/z 86 from the LeuTry and LeuLeuTyr radical cations.

Typical TRPD curves obtained for these peptides are shown in Figure 7.10. The major findings of this study can be summarized as follows: (1) the rate constants deduced for the same species (LeuTyr or LeuLeuTyr) at different PD wavelengths are different, increasing with decreasing wavelength (i.e., with increasing internal energy); and (2) the rate constants at the same PD wavelengths are one to two orders of magnitude lower for LeuLeuTyr than for LeuTyr. At the same total energy of 2.73 eV (photon energy plus average thermal energy), the rate constant for LeuLeuTyr is nearly three orders of magnitude lower than that for LeuTyr. The very different rate constants for LeuTyr and LeuLeuTyr provide a clear indication that the peptide length (i.e., its number of degrees of freedom) strongly correlates with the dissociation rate. The assumption that the entire available internal energy E is distributed over all degrees of freedom is seen to be correct.

Further support for this conclusion was obtained by comparison of RRKM/QET calculated rate constants with the experimental values shown in Figure 7.11. The model that fits the data is $E_0 = 1.4\,\text{eV}$ and $15.2 \leq \Delta S^{\ddagger} \leq 18.4$ cal/mol. The E_0 found as a fit is very nearly equal to the value of Weinkauf et al. (1995). The Arrhenius preexponential factor (A) calculated from $\Delta S^{\ddagger}$ is $1 \times 10^{17} \leq A \leq 6.2 \times 10^{17}\,\text{s}^{-1}$ at 1000 K. The transition state is quite loose with a high positive activation entropy and high A factor. TRPD studies of small peptides demonstrated that these peptides do not circumvent IVR. Electronic excitation (or the charge) may be the scout because charge transfer is a very fast and efficient process (Weinkauf et al. 1996), but the pitfall of large molecule kinetics is not avoided. The charge only directs the dissociation as α-cleavage is the preferred low-energy dissociation route of amine cation radicals leading to the immonium ion (Scheme 7.3).

Although the main conclusion of this study was that peptide radical cations do not circumvent IVR, non-RRKM behavior could not be completely ruled out (Hu et al. 2003). The TRPD curves (Figure 7.10) have finite intercepts at zero delay times. This indicates a contribution from a reaction component that is fast on the timescale of these experiments. Nonzero intercept could be attributed to the dissociation that avoids IVR. Alternatively, fast fragmentation could to be due to multiphoton dissociation processes. In other words, whereas the TRPD curve corresponds to a single-photon dissociation, the intercept is due to a much faster two-photon dissociation but one that is still within the domain of RRKM.

7.2.4.4. Photodissociation of Protonated Peptides. TRPD of small protonated peptides $LysTrpLysH^+$ and $LysTyrLysH^+$ following excitation with a UV laser (266 nm, 4.66 eV) was reported by Andersen et al. (2004). This study utilized an electrostatic storage ring equipped with an electrospray source. TRPD data obtained on a timescale between 0 and 20 ms showed a single-component, nearly exponential decay of photoexcited peptide ions, indicating that the photon energy is converted rapidly to vibrational excitation. Furthermore, the decay rate of $LysTrpLysH^+$ showed a significant increase with decrease in the PD laser wavelength from 266 to 260 and 243 nm. These experiments provide a further support for the statistical nature of dissociation of gas-phase peptide ions.

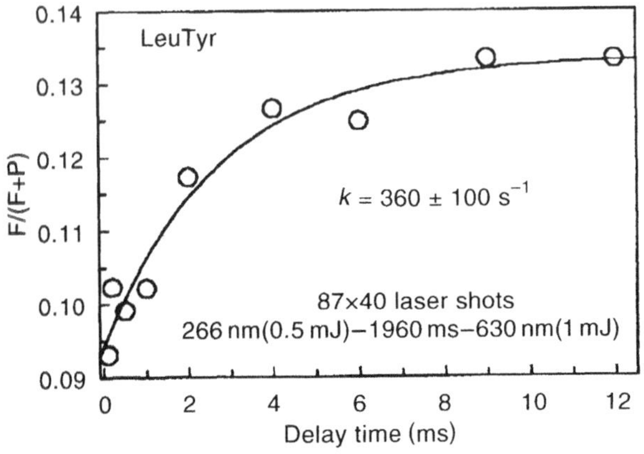
LeuTyr
F/(F+P)
$k = 360 \pm 100\ s^{-1}$
87×40 laser shots
266 nm(0.5 mJ)–1960 ms–630 nm(1 mJ)
Delay time (ms)

(a)

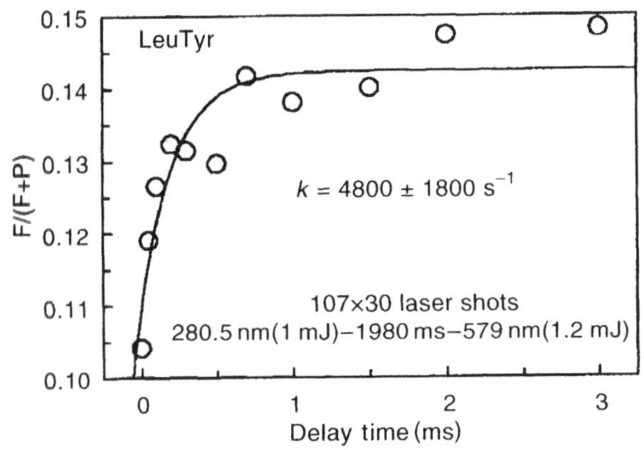
LeuTyr
F/(F+P)
$k = 4800 \pm 1800\ s^{-1}$
107×30 laser shots
280.5 nm(1 mJ)–1980 ms–579 nm(1.2 mJ)
Delay time (ms)

(b)

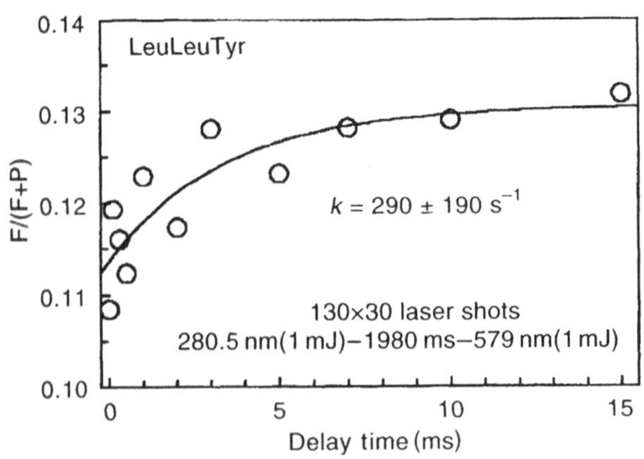
LeuLeuTyr
F/(F+P)
$k = 290 \pm 190\ s^{-1}$
130×30 laser shots
280.5 nm(1 mJ)–1980 ms–579 nm(1 mJ)
Delay time (ms)

(c)

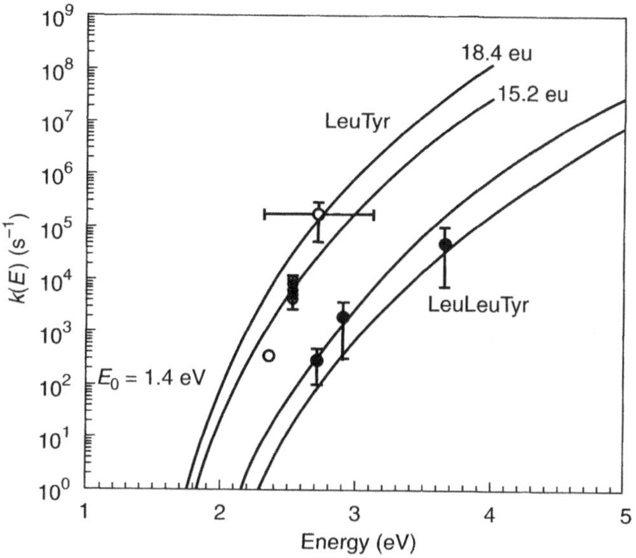

FIGURE 7.11. Microcanonical rate–energy dependences for the reactions producing the immonium ion from $\text{LeuTyr}^{+\bullet}$ and $\text{LeuLeuTyr}^{+\bullet}$. The rate constant $k(E)$ is plotted on a logarithmic scale as a function of internal energy E in the reactant ion. The open circles (for LeuTyr), filled circles (for LeuLeuTyr), and error bars are experimental data, and the lines are calculated using RRKM/QET and a model with $E_0 = 1.4\,\text{eV}$ and $\Delta S^{\ddagger} = 15.2$–$18.4$ cal/mol. The horizontal error bar shown for one LeuTyr point demonstrates the width of the thermal energy distribution. [Reproduced from Hu et al. (2003) with permission from the American Chemical Society.]

Reilly and coworkers proposed that PD of protonated peptides using 157-nm VUV light circumvents IVR (Thompson et al. 2004; Cui et al. 2005). In contrast to slow activation techniques producing *a*-, *b*- and *y*-type ions in MS/MS of protonated peptides, 157 nm PD results in formation of *x*-, *v*-, and *w*-type fragments when a basic arginine residue is located at the *C* terminus of the peptide and *a*- and *d*-type ions for peptides with *N*-terminal arginine. This unusual fragmentation was

FIGURE 7.10. Normalized signals of the immonium ion (F, m/z 86) with increasing trapping time from formed $\text{LeuTyr}^{+\bullet}$ (P, m/z 294) and $\text{LeuLeuTyr}^{+\bullet}$ (P, m/z 407): (a) one-color (266 nm, 0.5 mJ), two-photon ionization, thermalization for 1960 ms, and excitation at 630 nm (1 mJ) of LeuTyr; (b) one-color (280.5 nm), two-photon ionization, thermalization for 1980 ms, and excitation at 579 nm of LeuTyr; (c) one-color (280.5 nm), two-photon ionization, thermalization for 1980 ms, and excitation at 579 nm of LeuLeuTyr. The experimental points are fitted with single exponential curves (solid lines) yielding inverse time constants (i.e., rate constants): (a) $(3.6 \pm 1) \times 10^2\,\text{s}^{-1}$; (b) $(4.8 \pm 1.8) \times 10^3\,\text{s}^{-1}$; (c) $(2.9 \pm 1.9) \times 10^2\,\text{s}^{-1}$. [Reproduced from Hu et al. (2003) with permission from the American Chemical Society.]

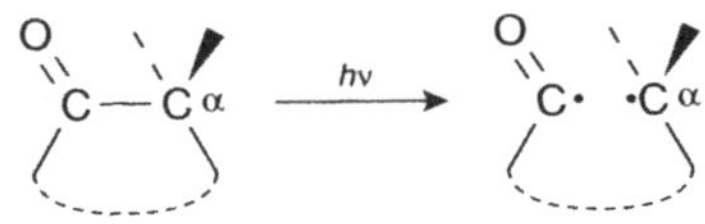

SCHEME 7.4 Norrish type I fragmentation.

rationalized using a mechanism, which involves homolytic radical cleavage via a Norrish type I process (Scheme 7.4) followed by radical elimination processes from the primary radical cations (Cui et al. 2005). According to this mechanism *a*- and *x*-type ions are formed by elimination of a hydrogen atom from the corresponding $a + 1$ and $x + 1$ radical precursors.

The nonergodic nature of peptide PD was proposed on the basis of results reported by Zewail and coworkers for femtosecond PD of acetone (Kim et al. 1995) and cyclic ketones (Diau et al. 1998). In both cases Norrish type I cleavage results in loss of CO and requires cleavage of two C—C bonds. In acetone the first cleavage occurs on a timescale of 50 fs, that is, prior to IVR (Kim et al. 1995). Loss of CO from cyclic ketones [CH_2—$(CH_2)_{n-2}$—C=O, $n = 4,5,6,10$] is characterized by very short decay rates of 100, 125, 180, and 180 fs for $n = 4,5,6,10$, respectively. Reaction rate decreased by a factor of only 2 with increase in the number of vibrational degrees of freedom from 27 to 81. Comparison of experimental decay rates with RRKM calculations suggested that Norrish type I reaction displays nonergodic, nonstatistical behavior (Diau *et. al.* 1998). Zewail and coworkers concluded that because the reaction occurs on the femtosecond timescale, IVR is restricted to the modes near the reaction coordinate and that complete statistical redistribution of internal excitation does not occur prior to dissociation. It is unclear whether the results obtained for small molecules can be used to predict the behavior of large molecules on photoexcitation. TRPD experiments in a storage ring discussed earlier could be used to confirm or challenge the proposed nonergodic PD of peptide ions.

7.2.5. Summary: The Pros and Cons of IVR and Ergodic Behavior in Biomolecules

The development of mass spectrometry of biomolecules has again raised some old questions, such as whether IVR precedes fragmentation and whether electronic and vibrational energy relaxation take place, or whether there is site selectivity and charge-directed reactivity of excited gas-phase biomolecules. Studies of gas-phase ion chemistry of large floppy molecules such as peptides and proteins revealed a variety of very interesting phenomena, some of which (e.g., electron capture dissociation and photodissociation) were described as nonergodic processes that circumvent IVR. However, it seems that in most cases experimental findings can also be explained without assuming non-ergodic behavior. Because of the complexity of biomolecules, theoretical studies describing these phenomena are currently limited to relatively small model systems. In addition, the large numbers of conformers that exist even for relatively small peptides and the uncertainties in the location and degeneracy of the transition state render very accurate RRKM calculations impossible at present.

There is a need for new experimental approaches as well as theoretical treatments for large biomolecules. TRPD has shown an unprecedented potential for rate–energy determination of gas-phase peptide ions. Detailed modeling of time- and energy-resolved SID data and understanding of the shattering transition in collisions of peptide ions with surfaces will address the possibility of nonergodic dissociation during ion–surface collisions. Mechanistic understanding of electron capture dissociation that explains experimental observations is crucial for establishing the interplay between IVR and dissociation in this process.

ACKNOWLEDGMENT

During preparation of this chapter for publication Chava had very helpful discussions with Professors Frank Tureĉek, Jean-Claude Lorquet, Tom Baer, Benny Gerber, Fred McLafferty, Dr. Julia Laskin, and other colleagues. This work was supported by the Israel Science Foundation, founded by the Israel Academy of Sciences and Humanities. The Farkas Research Center is supported by the Minerva Gesellschaft für die Forschung GmbH, München.

REFERENCES

Al-Khalili A, Thomas R, Ehlerding A, Hellberg F, Geppert WD, Zhaunerchyk V, af Ugglas M, Larsson M, Uggerud E, Vedde J, Adlhart C, Semaniak J, Kaminska M, Zubarev RA, Kjeldsen F, Andersson PU, Osterdahl F, Bednarska VA, Paal A (2004): Dissociative recombination cross section and branching ratios of protonated dimethyl disulfide and N-methylacetamide. *J. Chem. Phys.* **121**:5700–5708.

Anand S, Schlegel HB (2004): Dissociation of acetone radical cation ($CH_3COCH_3^{+\bullet} \rightarrow CH_3CO^+ + CH_3^\bullet$): An ab initio direct classical trajectory study. *Phys. Chem. Chem. Phys.* **6**:5166–5171.

Andersen JU, Cederquist H, Forster JS, Huber BA, Hvelplund P, Jensen J, Liu B, Manil B, Maunoury L, Nielsen SB, Pedersen UV, Rangama J, Schmidt HT, Tomita S, Zettergren H (2004): Photodissociation of protonated amino acids and peptides in an ion storage ring: Determination of Arrhenius parameters in the high-temperature limit. *Phys. Chem. Chem. Phys.* **6**:2676–2681.

Andlauer B, Ottinger Ch (1971): Unimolecular ion decompositions: Dependence of rate constants on energy from charge exchange experiments. *J. Chem. Phys.* **55**:1471–1472.

Andlauer B, Ottinger Ch (1972): Dissociation lifetimes of molecular ions produced by charge exchange. *Z. Naturforsch.* **27a**:293–309.

Austin RH, Xie AH, van der Meer L, Redlich B, Lindgard PA, Frauenfelder H, Fu D (2005): Picosecond thermometer in the amide I band of myoglobin. *Phys. Rev. Lett.* **94**:128101.

Baer T, Willett GD, Smith D, Phillips JS (1979): The dissociation dynamics of energy selected $C_6H_6^+$. *J. Chem. Phys.* **70**:4076–4085.

Baer T, Büchler U, Klots CE (1981): Kinetic energy release distributions for the dissociation of internal energy selected $C_2H_5I^+$ Ions. *J. Chim. Phys.* **77**:739–743.

Baer T, Kury R (1982): Random energy flow in the dissociation of energy selected bromobenzene and bromobenzene-d_5 ions. *Chem. Phys. Lett.* **92**:659–662.

Baer T, Dutuit O, Mestdagh H, Rolando C (1988): Dissociation dynamics of *n*-butylbenzene ions: The competitive production of *m/z* 91 and 92 fragment ions. *J. Phys. Chem.* **92**:5674–5679.

Baer T, Hase WL (1996) *Unimolecular Reaction Dynamics: Theory and Experiments,* Oxford Univ. Press, New York.

Bailey TH, Laskin J, Futrell JH (2003): Energetics of selective cleavage at acidic residues studied by time- and energy-resolved surface-induced dissociation in FT-ICR MS. *Int. J. Mass Spectrom.* **222**:313–327.

Bente PF III, McLafferty FW, McAdoo DJ, Lifshitz C (1975): Internal energy of product ions formed in mass spectral reactions: The degrees of freedom effect. *J. Phys. Chem.* **79**: 713–721.

Berkowitz J (1978): Photoionization of CH_3OH, CD_3OH, and CH_3OD: Dissociative ionization mechanisms and ionic structures. *J. Chem. Phys.* **69**:3044–3054.

Bernshtein V, Oref I (1994): Unimolecular dissociation of very large polyatomic molecules. *J. Phys. Chem.* **98**:136–140.

Biemann K (1988): Contributions of mass spectrometry to peptide and protein structure. *Biomed. Environ. Mass Spectrom.* **16**:99–111.

Bombach R, Dannacher J, Stadelmann J-P, Vogt J (1981): The fragmentation of formaldehyde molecular cations: The lifetime of CD_2O^+ (A^2B_1). *Chem. Phys.Lett.* **77**:399–402.

Brand WA, Baer T, Klots CE (1983): Kinetic energy release distributions in the fragmentation of energy-selected iodopropane ions. *Chem. Phys.* **76**:111–120.

Breuker K, Oh HB, Carpenter BK, McLafferty FW (2004): Nonergodic and conformational control of the electron capture dissociation of protein cations. *Proc. Natl. Acad. Sci. USA* **101**:14011–14016.

Bunker DL, Hase WL (1973): On non-RRKM unimolecular kinetics: Molecules in general, and CH_3NC in particular. *J. Chem. Phys.* **59**:4621–4632.

Burlet O, Yang CY, Gaskell SJ (1992): Influence of cysteine to cysteic acid oxidation on the collision-activated decomposition of protonated peptides: Evidence for intraionic interactions. *J. Am. Soc. Mass Spectrom.* **3**:337–344.

Butler JN, Kistiakowsky GB (1960): Reactions of methylene: IV. Propylene and cyclopropane. *J. Am. Chem. Soc.* **82**:759–765.

Campbell EEB, Raz T, Levine RD (1996): Internal energy dependence of the fragmentation patterns of C_{60} and C_{60}^+. *Chem. Phys. Lett.* **253**:261–267.

Carpenter BK (2003): Nonexponential decay of reactive intermediates: New challenges for spectroscopic observation, kinetic modeling and mechanistic interpretation. *J. Phys. Org. Chem.* **16**:858–568.

Carpenter BK (2005): Nonstatistical dynamics in thermal reactions of polyatomic molecules. *Annu. Rev. Phys. Chem.* **56**:57–89.

Christen W, Even U, Raz T, Levine RD (1998): The transition from recoil to shattering in cluster-surface impact: An experimental and computational study. *Int. J. Mass Spectrom. Ion Proc.* **174**:35–52.

Chupka WA, Lindholm E. (1963): Dissociation of butane molecule ions formed in charge exchange collisions with positive ions. *Arkiv Fysik* **25**:349–361.

Chupka WA, Berkowitz J (1967): Photoionization of ethane, propane, and *n*-butane with mass analysis. *J. Chem. Phys.* **47**:2921–2933.

Chupka WA (1971): Effect of thermal energy on ionization efficiency curves of fragment ions. *J. Chem. Phys.* **54**:1936–1947.

Cooper HJ, Haakansson K, Marshall AG (2005): The role of electron capture dissociation in biomolecular analysis. *Mass Spectrom. Rev.* **24**:201–222.

Csonka IP, Paizs B, Lendvay G, Suhai S (2000): Proton mobility in protonated peptides: A joint molecular orbital and RRKM study. *Rapid Commun. Mass Spectrom.* **14**:417–431.

Cui WD, Hadas B, Cao BP, Lifshitz C (2000): Time-resolved photodissociation (TRPD) of the naphthalene and azulene cations in an ion trap/reflectron. *J. Phys. Chem. A* **104**:6339–6344.

Cui W, Hu Y, Lifshitz C (2002): Time resolved photo dissociation of small peptide ions—combining laser desorption with ion trap/reflectron TOF mass spectrometry. *Eur. Phys. J. D* **20**:565–571.

Cui WD, Thompson MS, Reilly JP (2005): Pathways of peptide ion fragmentation induced by vacuum ultraviolet light. *J. Am. Soc. Mass Spectrom.* **16**:1384–1398.

Davis MJ, Gray SK (1986): Unimolecular reactions and phase space bottlenecks. *J. Chem. Phys.* **84**:5389–5411.

Depke G, Lifshitz C, Schwarz H, Tzidony E. (1981): Non-ergodic behaviour of excited radical cations in the gas phase. *Angew. Chem. Int. Ed. Engl.* **20**:792–793.

Diau EWG, Herek JL, Kim ZH, Zewail AH (1998): Femtosecond activation of reactions and the concept of nonergodic molecules. *Science* **279**:847–851.

Dongré AR, Somogyi Á, Wysocki VH (1996a): Surface-induced dissociation: An effective tool to probe structure, energetics and fragmentation mechanisms of protonated peptides. *J. Mass Spectrom.* **31**:339–350.

Dongré AR, Jones JL, Somogyi A, Wysocki VH (1996b): Influence of peptide composition, gas-phase basicity, and chemical modification on fragmentation efficiency: Evidence for the mobile proton model. *J. Am. Chem. Soc.* **118**:8365–8374.

Dunbar RC (1987): Time-resolved photodissociation of chlorobenzene ion in the ICR spectrometer. *J. Phys. Chem.* **91**:2801–2804.

Dunbar RC (1989): Time-resolved photodissociation study of relaxation processes in gas-phase styrene ion. *J. Chem. Phys.* **91**:6080–6085.

Dunbar RC, Lifshitz C (1991): Slow time-resolved photodissociation of p-iodotoluene ion. *J. Chem. Phys.* **94**:3542–3547.

Engelhardt C, Keske JC, Rees FS, Self-Medlin YB, Yoo HS, Pate BH (2001): Very slow intramolecular vibrational energy redistribution (IVR) for molecules in planar conformations. *J. Phys. Chem. A* **105**:6800–6807.

Faulk JD, Dunbar RC (1991): Time-resolved photodissociation of three *tert*-butylbenzene ions. *J. Phys. Chem.* **95**:6932–6936.

Fenn JB, Mann M, Meng CK, Wong SF, Whitehouse CM (1989): Electrospray ionization for mass-spectrometry of large biomolecules. *Science* **246**:64–71.

Galloy C, Lecomte C, Lorquet JC (1982): Unimolecular reaction paths of electronically excited species. III. Production of CH_3^+ ions from CH_3OH^+ as an example of isolated state dissociation. *J. Chem. Phys.* **77**:4522–4528.

Gamage CM, Fern·ndez FM, Kuppannan K, Wysocki VH (2004): Submicrosecond surface-induced dissociation of peptide ions in a MALDI TOF MS.*Anal. Chem.* **76**:5080–5091.

Gray SK, Rice SA, Davis MJ (1986): Bottlenecks to unimolecular reactions and an alternative form for classical RRKM theory. *J. Phys. Chem.* **90**:3470–3482.

Grebner TL, Neusser HJ (1999): Dissociation rates of energy-selected benzene cations: Comparison of experimental results from ion cyclotron resonance and cylindrical ion trap time-of-flight mass spectrometry. *Int. J. Mass Spectrom.* **187**:517–532.

Hamm P, Lim M, Hochstrasser RM (1998): Structure of the amide I band of peptides measured by femtosecond nonlinear-infrared spectroscopy. *J. Phys. Chem. B* **102**:6123–6138.

Ho YP, Dunbar RC, Lifshitz C (1995): C-H bond strength of naphthalene ion—a reevaluation using new time-resolved photodissociation results. *J. Am. Chem. Soc.* **117**:6504–6508.

Hoxha A, Locht R, Lorquet AJ, Lorquet JC, Leyh B (1999): Unimolecular dynamics from kinetic energy release distributions. V. How does the efficiency of phase space sampling vary with internal energy? *J. Chem. Phys.* **111**:9259–9266.

Hu YJ, Hadas B, Davidovitz M, Balta B, Lifshitz C (2003): Does IVR take place prior to peptide ion dissociation? *J. Phys. Chem. A* **107**:6507–6514.

Huang ZS, Jucks KW, Miller RE (1986): The vibrational predissociation lifetime of the HF dimer upon exciting the "free-H" stretching vibration. *J. Chem. Phys.* **85**:3338–3341.

Hunt DF, Shabanowitz J, Yates JR III (1987): Peptide sequence analysis by laser photodissociation Fourier transform mass spectrometry. *J. Chem. Soc. Chem. Commun.* 548–550.

Karas M, Hillenkamp F (1988): Laser desorption ionization of proteins with molecular masses exceeding 10,000 Da. *Anal. Chem.* **60**:2299–2301.

Keske J, McWhorter DA, Pate BH (2000): Molecular rotation in the presence of intramolecular vibrational energy redistribution. *Int. Rev. Phys. Chem.* **19**:363–407.

Kim SK, Pedersen S, Zewail AH (1995): Direct femtosecond observation of the transient intermediate in the alpha-cleavage reaction of $(CH_3)_2CO$ to $2CH_3 + CO$—resolving the issue of concertedness. *J. Chem. Phys.* **103**:477–480.

Klippenstein SJ, Faulk JD, Dunbar RC (1993): A combined theoretical and experimental study of the dissociation of benzene cation. *J. Chem. Phys.* **98**:243–256.

Laskin J (2004): Energetics and dynamics of peptide fragmentation from multiple-collision activation and surface-induced dissociation studies. *Eur. J. Mass Spectrom.* **10**:259–267.

Laskin J, Denisov E, Futrell JH (2000): A comparative study of collision-induced and surface-induced dissociation. 1. Fragmentation of protonated dialanine. *J. Am. Chem. Soc.* **122**: 9703–9714.

Laskin J, Lifshitz C (2001): Kinetic energy distributions in mass spectrometry. *J. Mass Spectrom.* **36**:459–478.

Laskin J, Denisov E, Futrell JH (2002a): Fragmentation energetics of small peptides from multiple-collision activation and surface-induced dissociation in FT-ICR MS. *Int. J. Mass Spectrom.* **219**:189–201.

Laskin J, Bailey TH, Denisov EV, Futrell JH (2002b): On the relative stability of singly protonated des-Arg^1- and des-Arg^9-bradykinins. *J. Phys. Chem. A* **106**:9832–9836.

Laskin J, Futrell JH (2003a): Collisional activation of peptide ions in FT-ICR mass spectrometry. *Mass Spectrom. Rev.* **22**:158–181.

Laskin J, Futrell JH (2003b): Surface-induced dissociation of peptide ions: Kinetics and dynamics. *J. Am. Soc. Mass Spectrom.* **14**:1340–1347.

Laskin J, Bailey TH, Futrell JH (2003): Shattering of peptide ions on self-assembled monolayer surfaces. *J. Am. Chem. Soc.* **125**:1625–1632.

Laskin J, Bailey TH, Futrell JH (2004): Fragmentation energetics for angiotensin II and its analogs from time- and energy-resolved surface-induced dissociation studies. *Int. J. Mass Spectrom.* **234**:89–99.

Laskin J (2005): Personal communication, Jan. 2005.

Lehmann KK, Scoles G, Pate BH (1994): Intramolecular dynamics from eigenstate-resolved infrared spectra. *Annu. Rev. Phys. Chem.* **45**:241–274.

Levine RD, Bernstein RB (1974): Energy disposal and energy consumption in elementary chemical reactions. Information theoretic approach. *Acc. Chem. Res.* **7**:393–428.

Lifshitz C, Long FA (1965a): Appearance potentials and mass spectra of fluorinated ethylenes. III. Calculations based on the statistical theory of mass spectra. *J. Phys. Chem.* **69**: 3737–3740.

Lifshitz C, Long FA (1965b): Some observations concerning the positive ion decomposition of C_2F_6 and C_3F_8 in the mass spectrometer. *J. Phys. Chem.* **69**:3746–3751.

Lifshitz C, Tiernan TO (1971): Dissociation of cyclobutane molecule-ions formed in charge exchange collisions. *J. Chem. Phys.* **55**:3555–3561.

Lifshitz C, Tiernan TO (1973): Effect of thermal energy on dissociative charge transfer. *J. Chem. Phys.* **59**:6143–6146.

Lifshitz C (1982): A surprisal analysis of the dissociation dynamics of $C_3H_6O^+$ cations. *Int. J. Mass Spectrom. Ion Phys.* **43**:179–193.

Lifshitz C (1983): Intramolecular energy redistribution in polyatomic ions. *J. Phys. Chem.* **87**:2304–2313.

Lin YN, Rabinovitch BS (1970): Degrees of freedom effect and internal energy partitioning upon ion decomposition *J. Phys. Chem.* **74**:1769–1775.

Little DP, Speir JP, Senko MW, O'Connor PB, McLafferty FW (1994): Infrared multiphoton dissociation of large multiply charged ions for biomolecule sequencing. *Anal. Chem.* **66**:2809–2815.

Lorquet JC (2000a): Unimolecular reaction dynamics from kinetic energy release distributions. VI. Energy-selected ions. *Int. J. Mass Spectrom.* **201**:59–67.

Lorquet JC (2000b): Landmarks in the theory of mass spectra. *Int. J. Mass Spectrom.* **200**:4356.

Lorquet JC (2000c): Unimolecular reaction dynamics from kinetic energy release distributions. 7. Average translational energy release. *J. Phys. Chem A* **104**:5422–5429.

Lorquet JC, Takeuchi T (1990): Nonadiabatic unimolecular reactions. 4. Isolated state decay in the fragmentation of the formaldehyde cation. *J. Phys. Chem.* **94**:2279–2283.

Marcus RA, Rice OK (1951): The kinetics of the recombination of methyl radicals and iodine atoms. *J. Phys. Colloid Chem.* **55**:894–908.

McAdoo DJ, McLafferty FW, Smith JS (1970): Metastable ion characteristics. XVI. Ketonization of gaseous enol ions. *J. Am. Chem. Soc.* **92**:6343–6345.

McLafferty FW, McAdoo DJ, Smith JS, Kornfeld R (1971): Metastable ions characteristics. XVIII. Enolic C3H6O+ ion formed from aliphatic ketones. *J. Am. Chem. Soc.* **93**:3720–3730.

McLafferty FW, Horn DM, Breuker K, Ge Y, Lewis MA, Cerda B, Zubarev RA, Carpenter BK (2001): Electron capture dissociation of gaseous multiply charged ions by Fourier-transform ion cyclotron resonance. *J. Am. Soc. Mass Spectrom.* **12**:245–249.

Meroueh SO, Wang Y, Hase WL (2002): Direct dynamics simulations of collision- and surface-induced dissociation of *N*-protonated glycine: Shattering fragmentation. *J. Phys. Chem. A* **106**:9983–9992.

Mintz DM, Baer T (1976): Kinetic energy release distributions for the dissociation of internal energy selected CH3I + and CD3I + ions. *J. Chem. Phys.* **65**:2407–2415.

Neusser HJ (1989): Lifetimes of energy and momentum selected ions. *J. Phys. Chem.* **93**: 3897–3907, and references cited therein.

Nummela JA, Carpenter BK (2002): Nonstatistical dynamics in deep potential wells: A quasiclassical trajectory study of methyl loss from the acetone radical cation. *J. Am. Chem. Soc.* **124**:8512–8513.

Oref I, Rabinovitch BS (1979): Do highly excited reactive polyatomic molecules behave ergodically? *Acc. Chem. Res.* **12**:166–175.

Osterheld TH, Brauman JI (1992): Infrared multiple photon dissociation of acetone radical cation—an enormous isotope effect with no apparent tunneling. *J. Am. Chem. Soc.* **114**:7158–7164.

Paizs B, Csonka IP, Lendvay G, Suhai S (2001): Proton mobility in protonated glycylglycine and *N*-formylglycylglycinamide: A combined quantum chemical and RRKM study. *Rapid Commun. Mass Spectrom.* **15**:637–650.

Paizs B, Suhai S (2001): Combined quantum chemical and RRKM modeling of the main fragmentation pathways of protonated GGG. I. *Cis-trans* isomerization around protonated amide bonds. *Rapid Commun. Mass Spectrom.* **15**:2307–2323.

Paizs B, Suhai S (2005): Fragmentation pathways of protonated peptides. *Mass Spectrom. Rev.* **24**:508–548.

Pavlov-Verevkin VB, Lorquet JC (2002): Intramolecular vibrational relaxation seen as expansion in phase space. 4. Generic relaxation laws for a spectroscopic clump profile. *J. Phys. Chem. A* **106**:6694–6701.

Price WD, Schnier PD, Williams ER (1996): Tandem mass spectrometry of large biomolecule ions by blackbody infrared radiative dissociation. *Anal. Chem.* **68**:859–866.

Rabinovitch BS, Diesen RW (1959): Unimolecular decomposition of chemically activated *sec*-butyl radicals from H atoms plus *cis*-butene-2. *J. Chem. Phys.* **30**:735–747.

Rabinovitch BS, Kubin RF, Harrington RE (1963): Unimolecular decomposition in nonequilibrium systems. *sec*-Butyl and *sec*-butyl-d_1 radicals produced by chemical activation at different levels of vibrational excitation. *J. Chem. Phys.* **38**:405–417.

Rabinovitch BS, Flowers MC (1964): Chemical activation. *Quart. Rev.* **18**:122–167.

Raz T, Even U, Levine RD (1995): Fragment size distribution in cluster-impact-shattering versus evaporation by a statistical approach. *J. Chem. Phys.* **103**:5394–5409.

Raz T, Levine RD (1996): On the shattering of clusters by surface impact heating. *J. Chem. Phys.* **105**:8097–8102.

Remacle F, Levine RD, Ratner MA (1998): Charge directed reactivity: A simple electronic model, exhibiting site selectivity, for the dissociation of ions. *Chem. Phys. Lett.* **285**:25–33.

Robinson PJ, Holbrook KA (1972): *Unimolecular Reactions*, Wiley-Interscience, Chichester, UK.

Roepstorff P, Fohlmann J (1984): Proposal for a common nomenclature for sequence ions in mass spectra of peptides. *Biomed. Mass Spectrom.* **11**:601.

Rogers P, Montague DC, Frank JP, Tyler SC, Rowland FS (1982): Non-RRKM decomposition in a chemical activation system: Thermal atomic fluorine plus tetraallyl tin. *Chem. Phys. Lett.* **89**:9–12.

Rosenstock HM, Wallenstein MB, Wahrhaftig AL, Eyring H (1952): Absolute rate theory for isolated systems and the mass spectra of polyatomic molecules. *Proc. Natl. Acad. Sci. USA* **38**:667–678.

Rynbrandt JD, Rabinovitch BS (1970): Intramolecular energy relaxation. A novel and direct test of the RRK-RRKM postulate. *J. Phys. Chem.* **74**:4175–4176.

Rynbrandt JD, Rabinovitch BS (1971a): Intramolecular energy relaxation. Nonrandom decomposition of hexafluorobicyclopropyl. *J. Phys. Chem.* **75**:2164–2171.

Rynbrandt JD, Rabinovitch BS (1971b): Direct demonstration of nonrandomization of internal energy in reacting molecules. Rate of intramolecular energy relaxation. *J. Chem. Phys.* **54**:2275–2276.

Schlag EW, Levine RD (1989): On the unimolecular dissociation of large molecules. *Chem. Phys. Lett.* **163**:523–530.

Schnier PD, Price WD, Jockusch RA, Williams ER (1996): Blackbody infrared radiative dissociation of bradykinin and its analogues: Energetics, dynamics, and evidence for salt-bridge structures in the gas phase. *J. Am. Chem. Soc.* **118**:7178–7189.

Senko MW, Speir JP, McLafferty FW (1994): Collisional activation of large multiply charged ions using Fourier transform mass spectrometry. *Anal. Chem.* **66**:2801–2808.

Shemesh D, Chaban GM, Gerber RB (2004): Photoionization dynamics of glycine: The first 10 picoseconds. *J. Phys. Chem. A* **108**:11477–11484.

Shukla AK, Futrell JH (1993): Collisional activation and dissociation of polyatomic ions. *Mass Spectrom. Rev.* **12**:211–255.

Simm IG, Danby CJ, Eland JHD (1973): Direct observation of an isolated state in the ion C2F6+: A violation of the quasi-equilibrium theory of mass spectra. *Chem. Commun.* 832.

Simm IG, Danby CJ, Eland JHD (1974): The fragmentation of $C_2F_6^+$ ions studied by photoelectron—photoion coincidence spectrometry. *Int. J. Mass Spectrom. Ion Phys.* **14**:285–293.

Slater NB (1959): *Theory of Unimolecular Reactions*, Cornell Univ. Press, Ithaca, NY.

Stadelmann J-P, Vogt J (1980a): Isolated state dissociations from electronically excited radical cations of fluoroethenes studied by photoelectron-photoion-coincidence spectroscopy. In Quayle A (ed), *Advances in Mass Spectrometry*, Vol. 8A, Heyden, Inst. Petroleum, London, pp. 47–55.

Stadelmann J-P, Vogt J (1980b): A photoelectron-photoion coincidence study of *cis*- and *trans*-difluoroethene. *Int. J. Mass Spectrom. Ion Phys.* **35**:83–99.

Stockbauer R, Inghram MG (1975): Threshold photoelectron-photoion coincidence mass spectrometric study of ethylene and ethylene-d_4. *J. Chem. Phys.* **62**:4862–4870.

Syrstad EA, Stephens DD, Turecek F (2003): Hydrogen atom adducts to the amide bond. Generation and energetics of amide radicals in the gas phase. *J. Phys. Chem. A* **107**: 115–126.

Syrstad EA, Turecek F (2005): Toward a general mechanism of electron capture dissociation. *J. Am. Soc. Mass Spectrom.* **16**:208–224.

Thompson MS, Cui W, Reilly JP (2004): Fragmentation of singly charged peptide ions by photodissociation at $\lambda = 157$ nm. *Angew. Chem. Int. Ed.* **43**:4791–4794.

Tolman RC (1920): Statistical mechanics applied to chemical kinetics. *J. Am. Chem. Soc.* **42**:2506–2528.

Turecek F (2003): N-C-alpha bond dissociation energies and kinetics in amide and peptide radicals. Is the dissociation a non-ergodic process? *J. Am. Chem. Soc.* **125**:5954–5963.

Turecek F, Syrstad EA (2003): Mechanism and energetics of intramolecular hydrogen transfer in amide and peptide radicals and cation-radicals. J. Am. Chem. Soc. **125**: 3353–3369.

Vestal ML (1968): Ionic fragmentation processes. In Ausloos P (ed), *Fundamental Processes in Radiation Chemistry*, Wiley, New York, pp. 59–118.

Wang Y. Hase WL, Song K (2003): Direct dynamics study of *N*-protonated diglycine surface-induced dissociation. Influence of collision energy. *J. Am. Soc. Mass Spectrom.* **14**: 1402–1412.

Weinkauf R, Schanen P, Yang D, Soukara S, Schlag EW (1995): Elementary processes in peptides: Electron mobility and dissociation in peptide cations in the gas phase. *J. Phys. Chem.* **99**:11255–11265.

Weinkauf R, Schanen P, Metsala A, Schlag EW, Bürgle M, Kessler H (1996): Highly efficient charge transfer in peptide cations in the gas phase: Threshold effects and mechanism. *J. Phys. Chem.* **100**:18567–18585.

Weinkauf R, Schlag EW, Martinez TJ, Levine RD (1997): Nonstationary electronic states and site-selective reactivity. *J. Phys. Chem. A* **101**:7702–7710.

Williams ER, Henry KD, McLafferty FW, Shabanowitz J, Hunt DF (1990a): Surface-induced dissociation of peptide ions in Fourier-transform mass spectrometry *J. Am. Soc. Mass Spectrom.* **1**:413–416.

Williams ER, Furlong JJP, McLafferty FW (1990b): Efficiency of collisionally-activated dissociation and 193-nm photodissociation of peptide ions in Fourier transform mass spectrometry. *J. Am. Soc. Mass Spectrom.* **1**:288–294.

Woodin RL, Bomse DS, Beauchamp JL (1978): Multiphoton dissociation of molecules with low power continuous wave infrared laser radiation. *J. Am. Chem. Soc.* **100**: 3248–3250.

Wrigley SP, Oswald DA, Rabinovitch BS (1984): On the question of heavy-atom blocking of intramolecular vibrational energy transfer. *Chem. Phys. Lett.* **104**:521–525.

Zimmerman JA, Watson CH, Eyler JR (1991): Multiphoton ionization of laser desorbed neutral molecules in a Fourier transform ion cyclotron resonance mass spectrometer. *Anal. Chem.* **63**:361–365.

Zubarev RA (2003): Reactions of polypeptide ions with electrons in the gas phase. *Mass Spectrom Rev.* **22**:57–77.

Zubarev RA, Kelleher NL, McLafferty FW (1998): Electron capture dissociation of multiply charged protein cations: A nonergodic process. *J. Am. Chem. Soc.* **120**:3265–3266.

Zubarev RA, Kruger NA, Fridriksson EK, Lewis MA, Horn DM, Carpenter BK, McLafferty FW (1999): Electron capture dissociation of gaseous multiply-charged proteins is favored at disulfide bonds and other sites of high hydrogen atom affinity. *J. Am. Chem. Soc.* **121**:2857–2862.

Zubarev RA, Haselmann KF, Budnik B, Kjeldsen F, Jensen F (2002): Towards an understanding of the mechanism of electron-capture dissociation: A historical perspective and modern ideas. *Eur. J. Mass Spectrom.* **8**:337–349.

PART II

ACTIVATION, DISSOCIATION, AND REACTIVITY

8

PEPTIDE FRAGMENTATION OVERVIEW

VICKI H. WYSOCKI, GUILONG CHENG, QINGFEN ZHANG, KRISTIN A. HERRMANN, RICHARD L. BEARDSLEY, AND AMY E. HILDERBRAND

Department of Chemistry
University of Arizona
Tucson, AZ

8.1. INTRODUCTION

With the worldwide daily use of mass-spectrometry based-proteomics, there is a demand to improve protein identification algorithms. At this point, it is not clear what types of improvement will be most effective. One possibility is that a detailed knowledge of peptide fragmentation mechanisms in the gas phase might be used to improve the algorithms, by either improving the confidence of matches after the initial algorithm identification (e.g., applying chemical knowledge in pre- or postfiltering) or incorporating chemical knowledge into algorithms (e.g.,

Principles of Mass Spectrometry Applied to Biomolecules, edited by Julia Laskin and Chava Lifshitz

by making spectral predictions using different fragmentation models (Huang et al. 2005) for different candidate sequences). This chapter gives an overview of selected, currently accepted, peptide fragmentation pathways, experimental factors influencing peptide fragmentation patterns, and methods to study peptide fragmentation. Peptide gas-phase secondary structure, a factor that is gaining attention with respect to peptide fragmentation, will also be discussed briefly. Finally, efforts to incorporate current knowledge of peptide fragmentation into protein identification algorithms will be mentioned.

8.2. OVERVIEW OF ACCEPTED PEPTIDE FRAGMENT ION STRUCTURES

Fragmentation studies of protonated peptides have been ongoing since the early 1980s (Biemann 1986; Hunt et al. 1986, Roepstorff 1984), and a recent review of the fragmentation pathways has been published (Paizs and Suhai 2005). The nomenclature that has been used to describe different major MS/MS ion types of protonated peptides is shown in Scheme 8.1. These major fragment ion types, as well as internal and immonium ions, are summarized in Table 8.1, along with references to representative literature studies that suggested their structures. The relative abundances of different ion types in MS/MS spectra depend on many factors, including instrumentation, peptide structure, and collision energy. Ions of types *b*, *a*, *y*, internal, and immonium are more common in lower-energy multistep activation spectra, while higher-energy activation leads to these ion types plus *c*, *d*, *v*, *w*, *x*, and *z* ions.

For several ion types, different structures might be drawn for the product ion, depending on the amino acid residues present at the cleavage site or the involvement of radical-induced cleavage. The b-type fragment ions, for example, are typically thought to have the structure of protonated oxazolones (Yalcin et al. 1995, 1996), but the proton might be located on the *N* terminus (Polfer et al. 2005). In addition, the *b* ion may have an acylium ion structure (Biemann 1988). Other possible *b* ion structures include a diketopiperazine structure (b_2 ion) (Eckart et al. 1998), an anhydride structure (terminal acidic residue in *b* ion) (Gu et al. 2000), and a bicyclic ring structure (His terminal residue in *b* ion) (Tsaprailis et al. 2004; Wysocki et al.

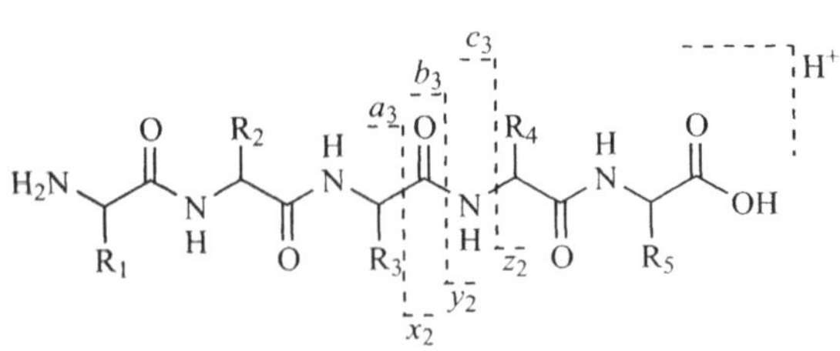

SCHEME 8.1. Nomenclature of common ion types.

TABLE 8.1. Structures and Nomenclature for Fragment Ions of Protonated Peptides, Illustrated for a Pentapeptide

Ion Type	Structure	Ref.	Ion Type	Structure	Ref.
b	b_3 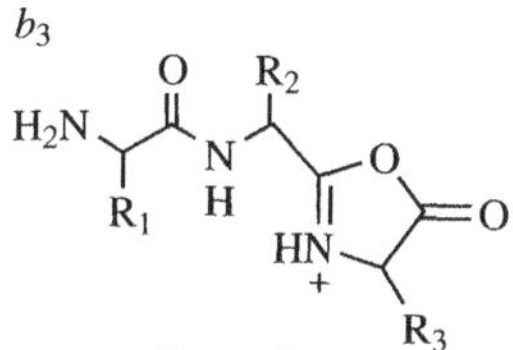	1	*y*	y_2 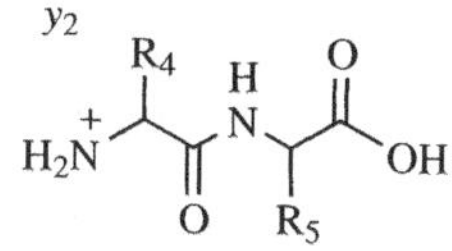	2
a	a_3 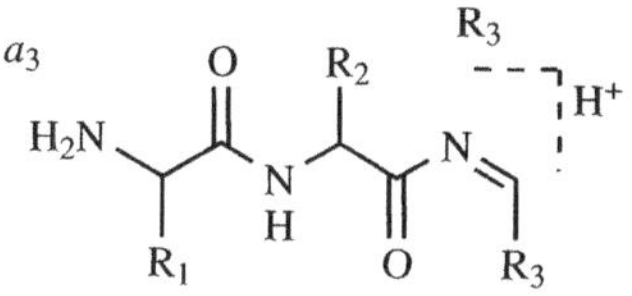	3	*x*	x_2 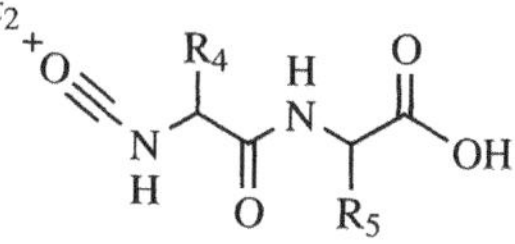	4,5
c	c_3 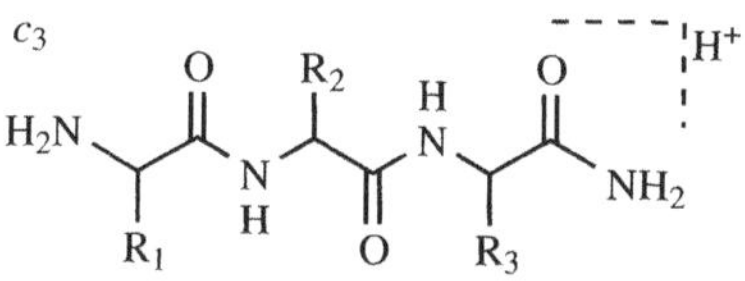	6,7	*z*	z_2 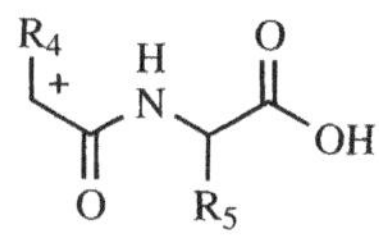	6,7
d	d_3 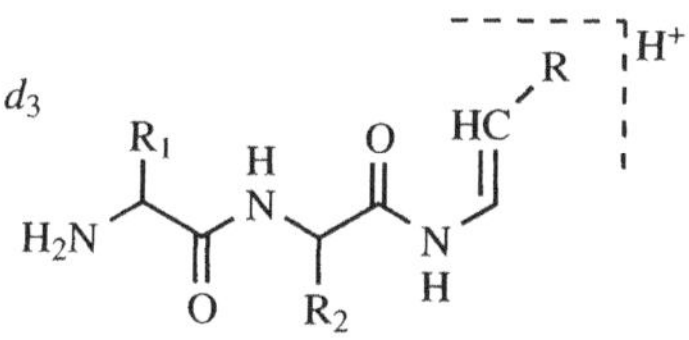	8	*v*	v_2 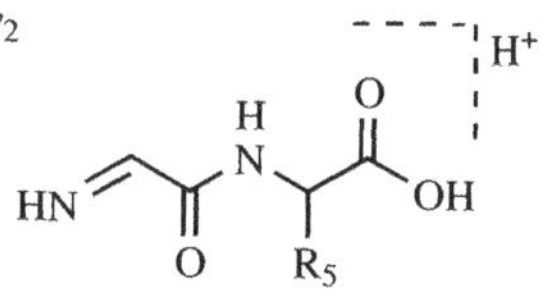	5
Immonium	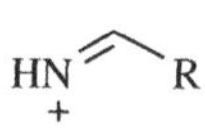	9	*w*	w_2 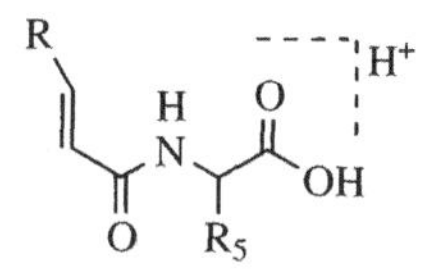	10
Internal fragment	b_3y_4 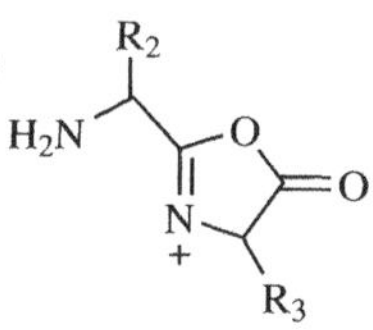	4,5	Internal fragment	a_4y_3 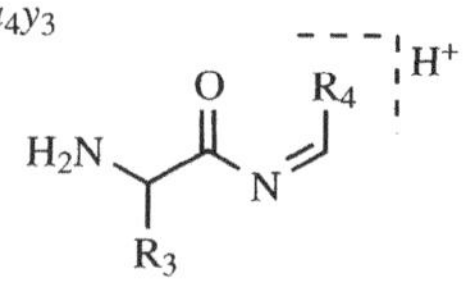	4,5

References:

1. Yalcin et al. (1995, 1996).
2. Kenny et al. (1992).
3. Sadagopan and Watson (2001).
4. Biemann (1988).
5. Johnson et al. (1988).
6. Kruger et al. (1999).
7. Zubarev et al. (1998).
8. Summerfield et al. (1997).
9. Ambihapathy et al. (1997).
10. Wee et al. (2002).

b_3: Anhydride structure; third amino acid is aspartic acid (Gu et al. 2000)

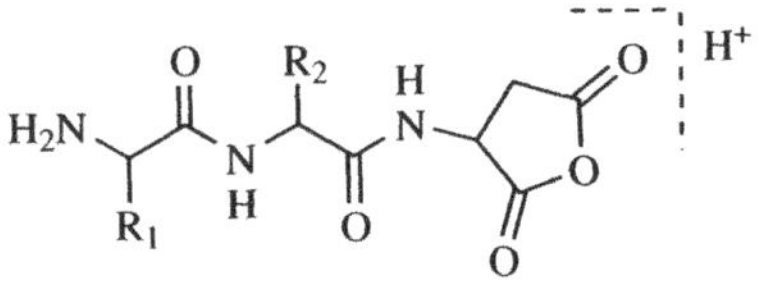

b_3: Bicyclic structure; third amino acid is histidine (Wysocki et al. 2000; Tsaprailis et al. 2004)

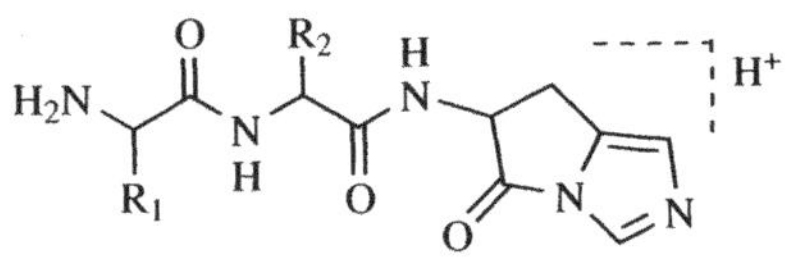

b_3: Acylium (Biemann 1988)

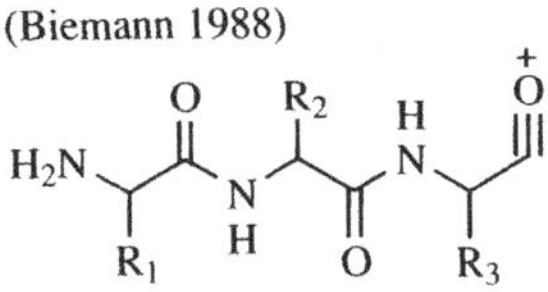

b_2: Diketopiperazine (Eckart et al. 1998)

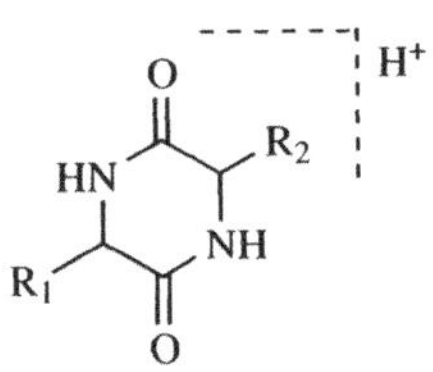

SCHEME 8.2. Representative structures of *b* ions.

2000), as shown in Scheme 8.2. The first direct spectroscopic evidence for the oxazolone *b* ion was obtained in 2005 and involved the use of IR photodissociation (Polfer et al. 2005). The a-type fragment ions are typically thought to be immonium ions, but the a_2 ion of GGG was more recently suggested to be a cyclic structure, protonated 4-imidazolidone (El Aribi et al. 2004). The $z^{\cdot}$ ions produced in ECD are unique and different from the typical *z* ion structure shown in Table 8.1, in that they are odd-electron radical cations. No charge is retained directly at the cleavage site in the ECD $z^{\cdot}$ ion. The charge on the $z^{\cdot}$ ion is acquired from additional charges originally present in the fragment.

8.3. EXPERIMENTAL INFLUENCES ON PEPTIDE FRAGMENTATION

Scheme 8.3 lists some of the parameters that contribute to the fragmentation behavior of peptides in a tandem mass spectrometer. Additional details are provided in Chapter 16. Experimentally, the ion activation method and instrumental configuration (which determines the dissociation reaction observation time window) are two major influencing factors, and have been subjected to extensive studies over

Activation + unimolecular dissociation:
Internal energy distribution
Slow heating vs. energy sudden activation

Molecule:
Different reaction pathways
Different protonation motifs (not single structure)
Energy dependent rate constants

Instrument configuration:
Time window for observation of reaction

SCHEME 8.3. Parameters contributing to MS/MS spectra. [Reprinted from Wysocki et al. (2005) by permission of Elsevier.]

the years. The most widely used ion activation method in MS/MS or MS^n is collision-activated dissociation (CAD). In CAD, an ion collides with a gaseous target, energy is redistributed among different vibrational degrees of freedom within the ion, and fragmentation results. In addition to CAD, other ion activation methods have been developed and studied. These include infrared multiphoton dissociation (IRMPD) (Ballard and Gaskell 1993; Lebrilla 2004, Little et al. 1994; Payne and Glish 2001), blackbody infrared radiative dissociation (BIRD) (Ge et al. 2001; Price et al. 1996; Schnier et al. 1996), surface-induced dissociation (SID) (Cooks et al. 1990; Dongre et al. 1996; Mabud et al. 1985), photodissociation (Martin et al. 1990, Thompson et al. 2004), electron capture dissociation (ECD) (Zubarev et al. 1998, 2002), and electron transfer dissociation (ETD) (Chrisman et al. 2005; Syka et al. 2004). The major motivations behind the development of multiple activation methods are to extract as much structural information from the analytes of interest as possible and to understand the energetics and mechanisms of peptides dissociation. Peptides are larger molecules than the molecules that are typically used to develop the kinetic theory of dissociation in mass spectrometry. Because the various activation methods deposit energy into the ions differently, leading to characteristic fragmentation patterns, the combination of two or more activation methods usually provides more structural and fundamental dissociation information than that from a single technique. For example, when glycopeptides are subjected to CAD in a trapping instrument, carbohydrate moieties of the molecule are typically lost by low-energy charge-directed cleavage in preference to the breakage of peptide bonds, leading to minimal or no information on the locations of carbohydrate (Zhang et al. 2005). ECD, a method involving radical-directed fragmentation, in contrast, leads to backbone cleavage resulting in peptide fragments containing intact carbohydrate moieties, thus indicating the locations of the carbohydrates. For this reason, ECD is gaining popularity in studies of protein posttranslational modification (see Chapter 13). A more recently introduced activation method, ETD, uses a chemical reaction to transfer electron(s) to the analyte, leading to extensive fragmentation of multiply charged peptides (charge $> +2$) and spectra with uniform cleavage across the backbone; thus, it is less sensitive to sequence effects than CAD. While CAD produces y and b backbone ions, ECD and ETD produce c and $z^{\cdot}$ backbone ions.

When investigating peptide dissociation mechanisms, it is important to understand that the fragmentation patterns of the tandem mass spectra of the gaseous protonated peptides are also influenced by the instruments involved in the study. Peptide fragmentation by a particular activation method needs to be examined with knowledge of the observation time windows in the specific instrument in order to understand the fragmentation. Some of the common mass analyzers used in peptide fragmentation studies include quadrupole (Q), time of flight (TOF), quadrupole ion trap (QIT), and Fourier transform ion cyclotron resonance (FTICR). Each instrument, which may contain more than one analyzer type, has its own observation timeframe and its own manner of depositing energy into analyte ions. For example, CAD in trapping instruments leads to slow heating of the ions while keV CAD (seen in TOF-TOF instruments) or eV SID input a larger initial energy step into the ions. In addition, it is not always possible to directly compare results acquired from two different time windows. If fragmentation of a large biomolecule must occur in a short timeframe, a large energy deposition is typically required. Under these higher-energy conditions, however, the fragmenting population may not have the same structures and proton locations as the population into which a much lower average energy was deposited in an instrument with a significantly longer observation timescale, which may affect fragment ions observed.

8.4. APPROACHES USED TO STUDY PEPTIDE FRAGMENTATION

Traditionally, studies on peptide fragmentation have relied heavily on investigation of model peptides. In such studies, the structure of a peptide is systematically altered [e.g., substitution of selected residues (Jonsson et al. 2001; Tsaprailis et al. 2004), blockage of *N* and/or *C* termini (Farrugia et al. 2001, Nair and Wysocki 1998), substitution of D-amino acids for L-amino acids (Wysocki et al. unpublished results; Zhang and Wysocki, unpublished results), and derivatization designed to guide fragmentation or to delineate one ion type from another (Beardsley and Reilly 2004; Beardsley et al. 2005; Keough et al. 1999; Reynolds et al. 2002; Summerfield et al. 1999)]. The resulting changes in peptide fragmentation behavior are used to derive fragmentation mechanism hypotheses.

An example of a study of model peptides involving systematic changes to structure was reported by Tsaprailis and coworkers in 2004 (Tsaprailis et al. 2004). In that study, a series of peptides with sequence RVYI-X-Z-F and VYI-X-Z-F (X = F or H; Z = A, P, or Sar) were employed to study enhanced cleavage *C* terminal to histidine. For doubly charged RVYIHZF, and singly charged VYIHZF, selective cleavage at His was observed. In contrast, the singly charged RVYIHZF did not fragment selectively at His. A mechanism was subsequently proposed in which the backbone carbonyl carbon is attacked by the His sidechain, which leads to formation of an atypical *b* ion structure (see Scheme 8.2).

An example in which a derivatization guides peptide fragmentation was more recently presented by Beardsley and Reilly (Beardsley and Reilly 2004; Beardsley et al. 2005). In this work peptides were modified by generating acetamidine moieties

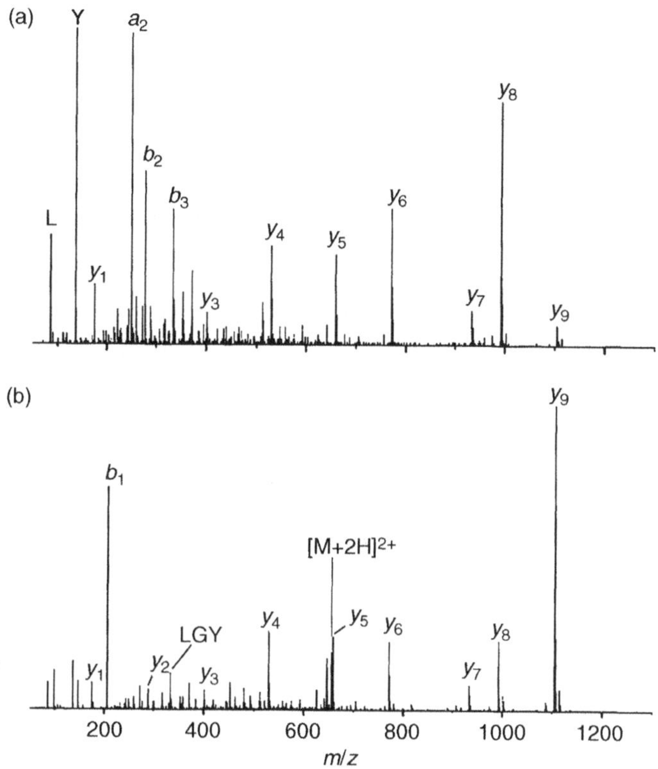

FIGURE 8.1. Q-TOF MS/MS spectra of the $[M+2H]^{2+}$: (a) unmodified and (b) acetamidinated precursor ions of YLGYLEQLLR from α-casein. [Reprinted from Beardsley et al. (2005) by permission of American Chemical Society.]

at the *N*-terminal amines and epsilon amines of Lys sidechains. Compared to their unmodified counterparts, CAD of the derivatized peptides generated significantly enhanced fragmentation of *N*-terminal peptide bonds to produce y_{n-1} fragment ions. This is shown in Figure 8.1, where b_1 and y_9 fragment ions of the acetaminidated peptide YLGYLEQLLR from α-casein are clearly enhanced compared to the fragmentation spectrum of the unmodified peptide. This phenomenon was attributed to the *N*-terminal amidine group facilitating proton transfer to the backbone carbonyl oxygen of the *N*-terminal residue via a hydrogen-bond-stabilized cyclic intermediate. Both this and the previous example illustrate that charge localization in the fragmenting peptide directs selective cleavage.

A structural modification that has been used to identify particular fragment ion types is selective heavy-atom labeling, such as the use of ^{18}O–water to label the *C* terminus of the peptide. In this approach, all *y* ions, which are truncated peptides containing the carboxy terminus, will shift in m/z by the difference between ^{18}O at both positions in the terminal carboxy group (i.e., a singly charged *y* ion will shift by 4 Da compared with the *y* ions from unlabeled peptides) (Reynolds et al. 2002). For labeled and unlabeled compounds, comparison of the spectra immediately reveals the *y* ions and confirms that *y* ions do indeed contain the carboxy termini of the peptides.

Although successful in studies of specific mechanisms, model peptide studies such as those mentioned above are generally slow and time-consuming. It is seldom possible to assess the general applicability of the conclusions from such a study because of the limited number of model peptides used. An alternative to this approach is to statistically investigate fragmentation patterns from a large number of peptide fragmentation spectra. Several studies of this type have appeared in the literature (Breci et al. 2003; Elias et al. 2004; Huang et al. 2002, 2004, 2005; Kapp et al. 2003; Tabb et al. 2003, 2004; van Dongen et al. 1996). These studies have, for example, confirmed that enhanced cleavage at acidic residues does depend on the number and identity of basic residues relative to the number of protons added (Huang et al. 2002; Kapp et al. 2003), that the peak intensity corresponding to cleavage at Pro does depend on the preceding Xxx residue in the Xxx-Pro bond, and that cleavage at Pro is charge-state-dependent. An example of the data from such a study will be provided in the following section. The appeal of these statistical analyses is the large number of "real" peptides (complex sequences experimentally generated, e.g., by tryptic digestion) whose behavior is catalogued.

8.5. INFLUENCE OF CHARGE SITE ON FRAGMENTATION

Two distinct classes of mechanisms have been proposed to account for the relationship between locations of charge and locations of fragmentation in protonated peptides. For the majority of fragment ions, especially those formed by low-energy collision-activated dissociation (CAD) and surface-induced dissociation (SID), it is generally believed that charge-directed fragmentation is the major pathway, with cleavage initiated by the ionizing proton. The other distinct class of fragmentation is "charge-remote," in which cleavage occurs without the direct involvement of the ionizing proton (Gu et al. 2000, Johnson et al. 1988). Both of these types of pathways fall within the scope of a general model for fragmentation, the "mobile proton" model (Scheme 8.4). This model was developed

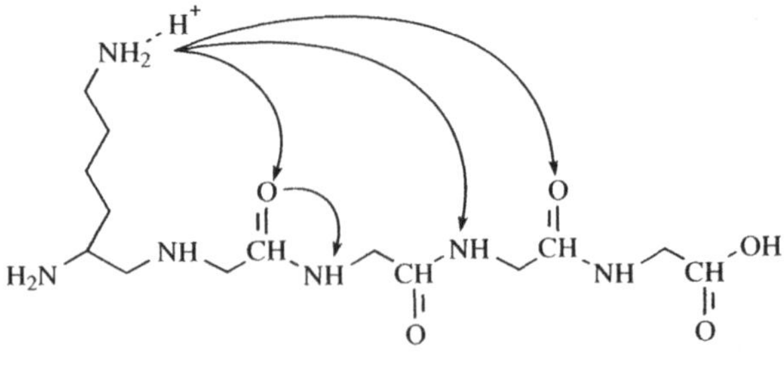

SCHEME 8.4. Mobile proton model. [Reprinted from Dongre et al. (1996) by permission of American Chemical Society.]

over the years by many research groups including Biemann, Gaskell, Wysocki, Harrison, and Paisz (Burlet et al. 1992; Dongre et al. 1996; Gu et al. 2000; Harrison and Yalcin 1997; Johnson et al. 1988; McCormack et al. 1993; Paizs and Suhai 2004).

Simply stated, the mobile proton model predicts that protons are localized at the most basic sites in the peptide prior to activation (Paizs et al. 2004; Wu and Lebrilla 1995). Experimental evidence shows that gas-phase protonated peptides exist with a high degree of internal solvation of the proton by the heteroatoms of the peptide, such as carbonyl oxygens and amino nitrogens (Freitas and Marshall 1999; Wyttenbach and Bowers 1999). Following activation, proton transfers may become possible via the highly internally solvated structure or opened structures leading to a heterogeneous population of structures, with protons at a variety of locations, that fragment at a variety of sites. In situations where proton transfers are less likely, other competitive pathways such as charge-remote fragmentation or cleavage initiated by an acidic hydrogen may open up.

The mobile proton model has been verified by many investigators (Paizs and Suhai 2004), and has also been supported by hydrogen/deuterium scrambling studies (Harrison and Yalcin 1997; Johnson et al. 1995; Mueller et al. 1988; Tsang and Harrison 1976). This model, however, is not intended to be a quantitative model that predicts a full peptide fragmentation spectrum for a given peptide, but rather provides a qualitative framework that allows users to predict the general appearance of a spectrum given the sequence and the known number of protons (or vice versa). Other models that are extensions from the mobile proton model are being developed as detailed predictive models (Paizs and Suhai 2005; Zhang 2004).

The mobile proton model can be illustrated by the published results of a statistical analysis of >28,000 ion trap spectra performed by the Wysocki group in collaboration with biostatisticians, George Tseng and Robert Yuan (Huang et al. 2005). The spectra used for this analysis were acquired by the group of Richard Smith at PNNL, with HPLC-MS/MS ion trap spectra acquired in parallel with HPLC–FTMS accurate mass measurements. Spectra were included in the analysis set only if a MS/MS identification was verified by a matching accurate mass measurement acquired at a corresponding retention time (±5%). Figure 8.2 illustrates that Arg-ending and Lys-ending doubly charged peptides, all containing Pro but no His [···P···noH···], give similar fragmentation patterns, presumably because a mobile proton is available to initiate fragmentation. In contrast, singly charged tryptic Arg-ending and Lys-ending peptides fragment differently from each other (Figure 8.3). Because the proton is strongly sequestered at Arg in singly charged Arg-ending peptides, cleavage tends to occur at acidic residues because the acidic hydrogen of the sidechain can initiate cleavage.

Charge-directed fragmentation is thought to involve a proton-initiated cleavage, which leads to the question of where the proton is located in the fragmenting peptide (Burlet et al. 1992; Dongre et al. 1996; Wysocki et al. 2000). Conflicting reports have appeared in the literature regarding whether the protonation site in fragmenting peptides is the amide nitrogen or carbonyl oxygen (Csonka et al. 2000; McCormack

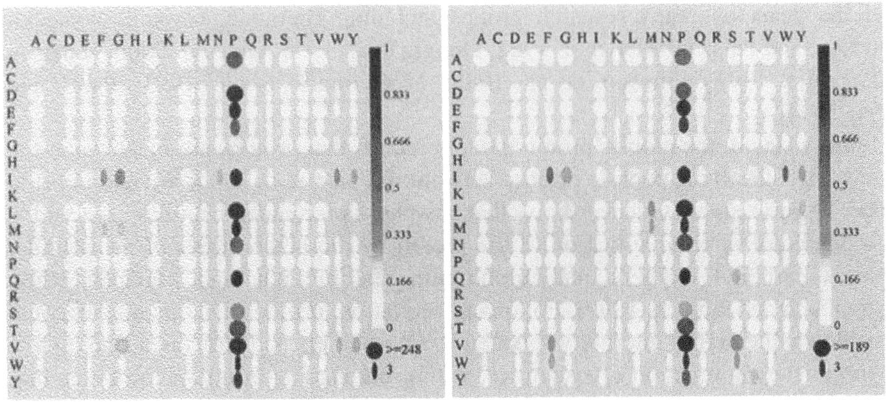

[...P...noH...R]$^{2+}$ [...P...noH...K]$^{2+}$

FIGURE 8.2. Pairwise fragmentation maps showing median bond cleavage intensities for *y* ion formation at specific Xxx–Zzz residue combinations from two sets of doubly charged peptides containing either R or K as the *C*-terminal ending residue and containing proline, but not histidine, $[\cdots P \cdots noH \cdots R]^{2+}$ and $[\cdots P \cdots noH \cdots K]^{2+}$. The ion intensities are normalized to the most abundant *y* ion peak. The single-letter codes of AA residues listed in the leftmost column correspond to the *N*-terminal residue (Xxx) in an Xxx–Zzz pair, while those listed along the topmost row correspond to the *C*-terminal residue (Zzz). The horizontal dimension of each ellipse is proportional to the count of such pairwise cleavages. [Reprinted from Huang et al. (2005) by permission of American Chemical Society.]

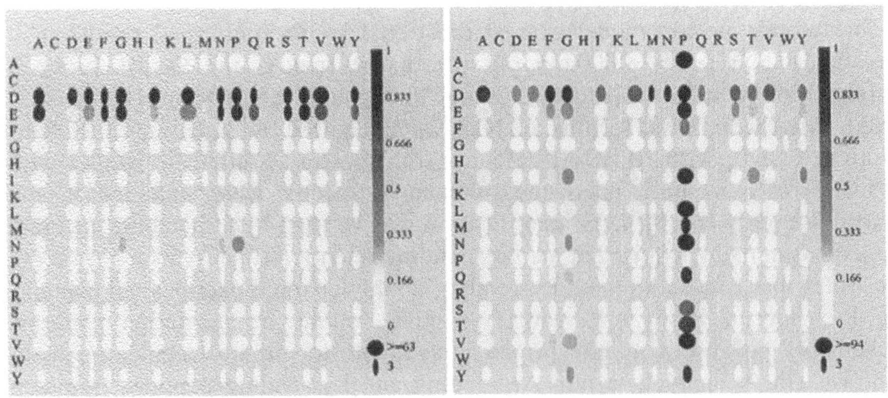

[...P...noH...R]$^{+}$ [...P...noH...K]$^{+}$

FIGURE 8.3. Pairwise fragmentation maps showing median bond cleavage intensities for *y* ion formation at specific Xxx–Zzz residue combinations from two sets of singly charged peptides containing either R or K as the *C*-terminal ending residue and containing proline, but not histidine, $[\cdots P \cdots noH \cdots R]^{+}$ and $[\cdots P \cdots noH \cdots K]^{+}$. The ion intensities are normalized to the most abundant *y* ion peak. The single-letter codes of AA residues listed in the leftmost column correspond to the *N*-terminal residue (Xxx) in an Xxx–Zzz pair, while those listed along the topmost row correspond to the *C*-terminal residue (Zzz). The horizontal dimension of each ellipse is proportional to the count of such pairwise cleavages. [Reprinted from Huang et al. (2005) by permission of American Chemical Society.]

et al. 1993; Paizs and Suhai 2004, 2005; Reid et al. 2000; Wysocki et al. 2000). Because protonation on the carbonyl oxygen is thermodynamically favored, several groups have proposed a mechanism in which carbonyl oxygen is the site of protonation (Hunt et al. 1986; Reid et al. 2000; Wysocki et al. 2000). A more recent computational study has shown the feasibility of intramolecular proton transfers involving two adjacent carbonyls, or via a larger ring, to nonadjacent carbonyls separated by one residue (Kulhanek et al. 2003). Protonation of the carbonyl oxygen increases the electrophilicity of the carbonyl carbon, rendering it more susceptible to nucleophilic attack and formation of, for example, a protonated oxazolone *b* ion product (Scheme 8.5). This mechanism requires that a proton be transferred to the adjacent amide N after nucleophilic attack on the carbonyl carbon and prior to cleavage to form the *N*-terminal departing fragment and the *C*-terminal fragment. A direct O-to-N transfer would require a symmetry-forbidden 1,3-H transfer, known to occur in smaller-model molecules; another possibility is that other heteroatoms downchain assist in this transfer.

Although protonation of the amide N is less likely thermodynamically, partial or complete proton transfer to this site significantly reduces the bond order of the C(O)–N bond and lowers the barrier to dissociation of this protonated form (McCormack et al. 1993; Somogyi et al. 1994). Siu and coworkers have recently calculated the most stable protonated form of GlyGlyGly and found it to be the structure with a proton on the first carbonyl, stabilized by H bonding to the *N*-terminal amino group (Rodriquez et al. 2001). They also showed, via

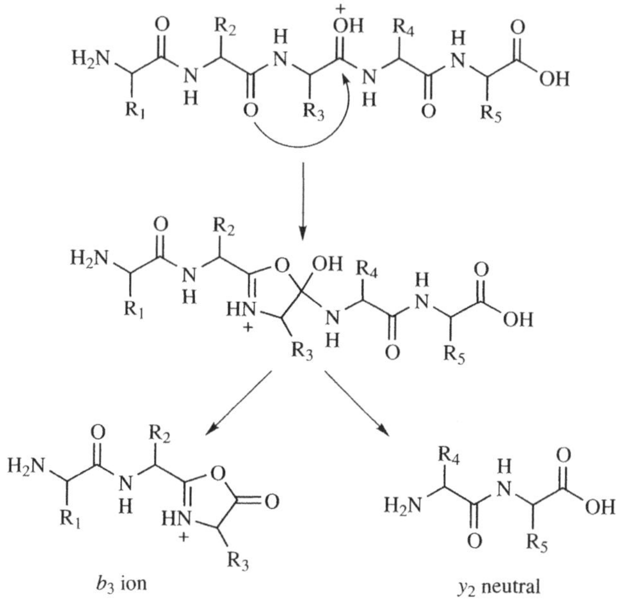

SCHEME 8.5. Protonation of carbonyl oxygen creating protonated oxazolone *b* ion fragment.

experiments and density functional theory formation of the b_2 oxazolone via an amide-protonated structure (Rodriquez et al. 2001; El Aribi et al. 2003). The reaction mechanism elucidated through calculations and experiments is shown in Figure 8.4. More recent calculations for a small model, the *N*-acetylmethyl ester of proline [$CH_3C(O)$—Pro— OMe], show that an adjacent carbonyl can assist in the transfer of a proton from a carbonyl O to an amide N (Komaromi et al. 2005). The only way to definitely distinguish N from O protonation during fragmentation is to observe the fragmentation intermediate spectroscopically during fragmentation. This has not yet been accomplished, although spectroscopic determination that a *b* ion is in part an oxazolone has been achieved (Polfer et al. 2005). It should also be noted that conclusions drawn from experiments and calculations for small-model peptides may not be representative of fragmentation pathways of larger systems.

Although charge-remote fragmentation of peptides has been studied, it has yet to be well characterized. Biemann showed in the late 1980s that *d*, *v*, and *w* ions are produced with greater abundances in Arg-containing peptides and speculated that the ions were produced by charge-remote pathways (Johnson et al. 1988). Burlingame and coworkers more recently showed that MALDI formation of ions from a "cold" matrix, dihydroxybenzoic acid, followed by keV CAD enhanced ions thought to be formed by charge-remote pathways (Stimson et al. 1997). They suggested that the cold matrix helps "fix" the charge at basic residues (i.e., produces a simpler or single population of precursor structures). However, a more recent paper (Luo et al. 2002) characterizes dihydroxybenzoic acid matrix as "hot," not "cold," so these results may need to be reevaluated—a hotter matrix could also explain the larger abundance of higher-energy fragments. Following on the idea that the charge needs to be fixed, several groups have looked at peptides with a fixed charge. Allison and coworkers (Liao et al. 1997) and Gu et al. (2000) both showed that peptides with no acidic residues derivatized with a phosphonium group at the amino terminus produce strong *a* ion fragments, presumably by a charge-remote (thermal) pathway (Figure 8.5). Figure 8.5 shows a comparison of the fragment ion masses for $\varphi_3P^+CH_2C(=O)$—AAAA, a peptide containing exchangeable hydrogens only at amide nitrogens and the *C* terminus, and its H/D exchange product $\varphi_3P^+CH_2$ $C(=O)$—AAAA-d_5. On the basis of the results of Figure 8.5a versus 8.5c the numbers of deuteriums in the *b* (and corresponding *a*) ions of φ_3P^+—CH_2C $(=O)$—AAAA-d_5 [part (c)] are 0, 1, and 2 for b_1, b_2, and b_3 ions, respectively. This indicates that a hydrogen/deuterium at the amide nitrogen migrates away from the *N*-terminal fragment in formation of the charge-remote *b* ions.

Similarly, derivatized peptides that contain acidic residues cleave at the amide bond *C*-terminal to the acidic sidechain rather than producing the *a* ion series (see Figure 8.6, population A, discussed in the next section); that is, the acidic hydrogen of the sidechain directs fragmentation that occurs in preference to formation of the a ion series. Although charge-remote pathways are not as common as charge-directed pathways for protonated peptides, they do occur and are more likely for Arg-containing singly charged peptides.

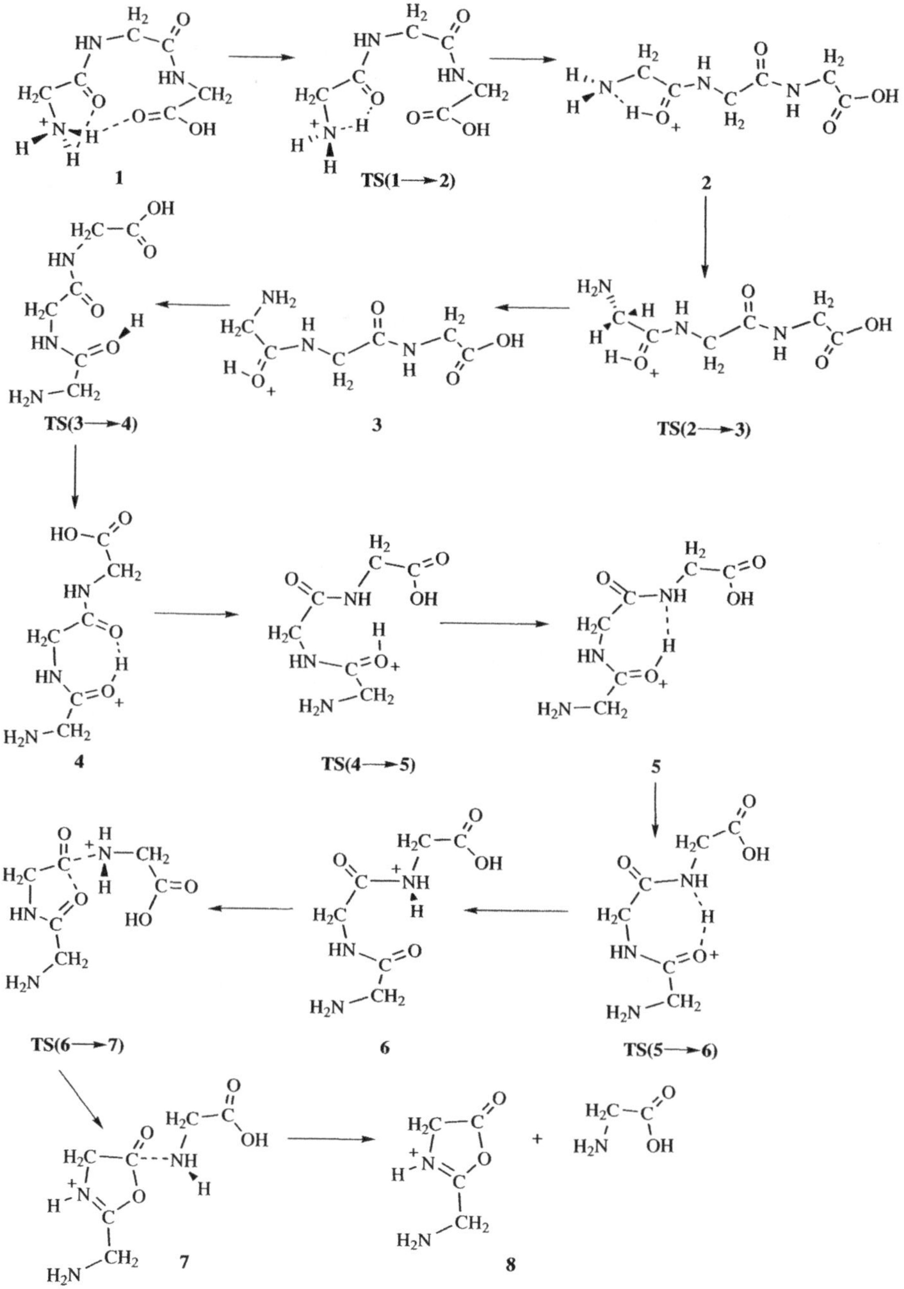

FIGURE 8.4. Fragmentation of GGG to form the b_2 ion. [Reprinted from Rodriquez et al. (2001) by permission of American Chemical Society.]

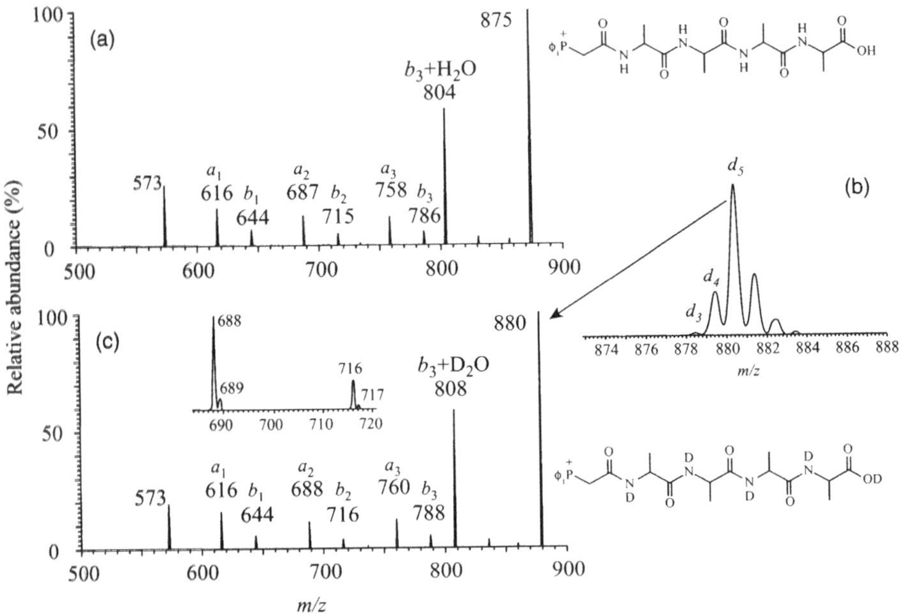

FIGURE 8.5. (a) ESI/CID spectra obtained in an ion trap instrument for singly charged ions of $\varphi_3P^+CH_2C(=O)$-AAAA, (b) molecular ion region of an ESI mass spectrum after H/D exchange for $\varphi_3P^+CH_2C(=O)$-AAAA in deuterated solvent, and (c) ESI/CID spectra of $\varphi_3P^+CH_2C(=O)$-AAAA-d_5 (φ = trimethoxyphenyl). The inset in part (c) shows the expanded region of the b_2/a_2 ions. The minor peaks such as m/z 689 and 717 shown in the inset are likely due to ^{13}C contamination in parent ion selection [^{13}C isotope peak of the peak labeled as d_4 in (b)]. The CID spectra were acquired at ~28% "normalized collision energy." [Reprinted from Gu et al. (2000) by permission of American Chemical Society.]

8.6. INFLUENCE OF SECONDARY STRUCTURE ON PEPTIDE FRAGMENTATION

The influence of secondary structure on peptide fragmentation has been a matter of debate over the years. Some have argued that when a singly or multiply protonated peptide is energized, a number of different conformations are produced, "wiping out" conformational influences on fragmentation. One could argue, however, that any given sequence and charge state has steric restrictions on the conformations that may be formed. Even a different initial placement of a given number of protons might lead to different fragmentation patterns for a particular sequence. An example of this behavior is illustrated in Figure 8.6, which shows different fragmentation patterns for the same m/z, same sequence, and the same charge state—one population of precursors (population A) corresponds to those ions that do not undergo gas-phase H/D exchange with D_2O, and the other population (population B) corresponds to those that undergo fast and extensive gas-phase exchange with D_2O.

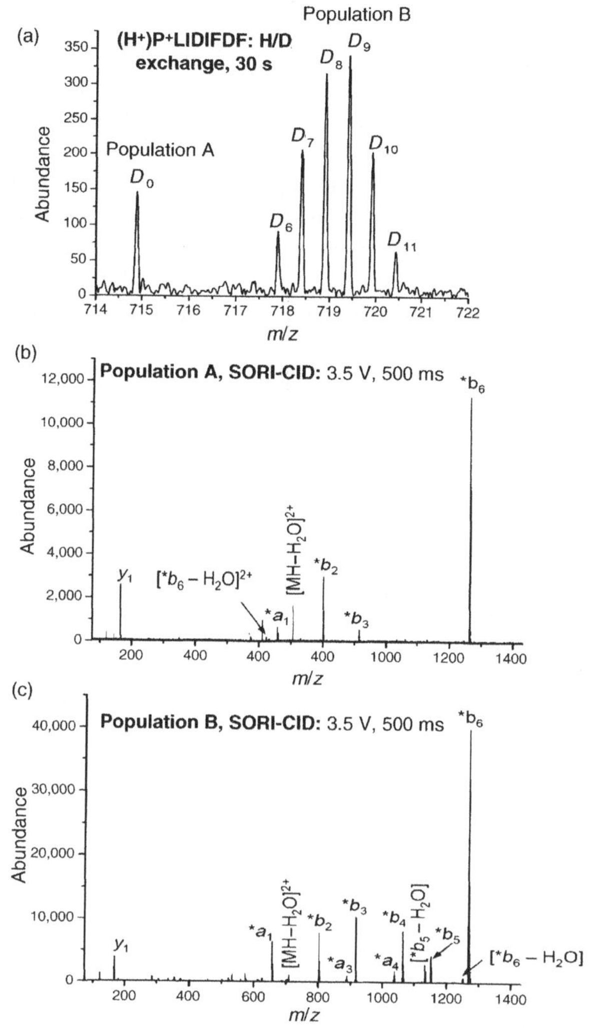

FIGURE 8.6. H/D exchange of $(H^+)P^+$**LDIFSDF** with D_2O (pressure equal to $7 = 10^{-8}$ Torr) and fragmentation of the resulting ion populations. (a) H/D exchange for 30 s after monoisotopic selection of $(H^+)P^+$**LDIFSDF**. Population A represents the precursor ion with no deuterium incorporated. Population B represents the precursor ion with 6–11 deuteriums incorporated. Parts (b) and (c) were obtained without monoisotopic selection prior to H/D exchange for 30 s followed by isolation and SORI of $(H^+)P^+$**LDIFSDF** populations (SORI time equals 500 ms; SORI amplitude equals 3.5 V with argon as the collision gas). (b) Population A (selection and fragmentation of nonexchanging ion along with three non-exchanging carbon-13 isotope peaks, signal not intense enough for monoisotopic selection, but ion populations were well separated). (c) Population B (selection and fragmentation of ions corresponding to 6–11 deuteriums incorporated along with exchanged carbon-13 isotope peaks, signal not intense enough for monoisotopic selection, but ion populations were well separated). The notation "P^+" represents the fixed charge derivative, notated "ϕ_3P^+" in Figure 8.5. [Reprinted from Herrmann et al. (2005) by permission of Elsevier Science.]

Both fragmentation spectra are consistent with the same sequence, but fragmentation of population B is more consistent with charge-directed "mobile proton" fragmentation leading to cleavage along the entire backbone. Fragmentation of population A is consistent with enhanced cleavage at acidic residues that typically occurs when protons are not mobile (Herrmann et al. 2005). One possibility is that, in the nonexchanging population, the added proton is solvated by oxygen-containing functional groups near the *C* terminus, as evidenced by a more abundant y_1 ion. In other work, we have seen different fragmentation patterns for peptides when one of the L-amino amino acids is replaced by a D-amino acid; the atom composition has not changed, only the configuration at one chiral center (Wysocki et al., unpublished results). Clearly, the arrangements of atoms in space affect the fragmentation of the peptide.

8.7. INCORPORATION OF PEPTIDE FRAGMENTATION KNOWLEDGE INTO ALGORITHM DEVELOPMENT

As stated before, the studies of gas-phase peptide fragmentation chemistry have been driven by the development and desire to improve protein identification algorithms. Although currently available protein identification algorithms have made proteomics studies possible, their success rates leave room for improvement (Breci et al. 2003; Resing et al. 2004). Almost all commercially available algorithms ignore the intensity information in the mass spectra. This approach assumes uniform fragmentation behavior for all the amino acids, despite the obvious differences in the 20 sidechains. Although this approach works well for some peptides, it does not perform as well for others. Since 2001, the intensity information missing in existing algorithms has gained some attention. Several groups have started to statistically analyze large sets of peptide fragmentation data (Breci et al. 2003; Elias et al. 2004; Huang et al. 2002, 2004, 2005; Kapp et al. 2003; Tabb et al. 2003, 2004, Tabb et al. 2003; van Dongen et al. 1996), and are exploring the feasibility of quantifying certain fragmentation rules. This work is still in the early stages, and it is not yet clear how intensity information can best be used in protein identification algorithms. The Wysocki group is developing an algorithm that calculates the probability that a candidate sequence matches the experimental tandem mass spectrum. The probability calculation for each candidate is based on pairwise cleavage training data only for peptides whose sequence motif is similar to that of the candidate. Although much work remains, the early success rate of this approach is promising.

8.8. REMAINING CHALLENGES AND FUTURE DIRECTIONS

Several major questions still remain regarding peptide fragmentation. It is not clear how many different protonated forms of a peptide are involved in dissociation. It is still debatable whether the proton is localized at a carbonyl oxygen or amide nitrogen when amide bond cleavage occurs. The kinetics of fast peptide fragmentation, and the nature of the structures that undergo fast fragmentation, are not well established. It remains to be determined whether an improved understanding of peptide dissociation mechanisms of protonated peptides will ultimately lead to improved sequencing algorithms. If this is the case, will whole new algorithmic approaches be developed, or will the mechanistic information simply be used as pre- or postidentification screening tools to increase confidence in identification? Finally, new activation methods continue to be developed and show increasing promise for simplifying data interpretation in the future. Electron transfer dissociation reduces the charge of protonated peptides through a chemical reaction that transfers electrons to the multiply charged peptide, allowing for extensive, more easily interpreted fragmentation for higher charge states of peptides. However, the ETD and ECD methods are inapplicable to singly charged precursor ions, and for reasons that are not yet clear, are less useful for doubly charged precursors than for higher-charged analogs. For some peptides, CAD provides better sequencing information than ETD (Gaskell, unpublished results) because proteomics research is likely to continue for decades to come, studies of peptide fragmentation will continue and will provide fundamental information on gas-phase structure and dissociation, as well as practical information that will increase the utility of mass spectrometry in studies of large molecules.

REFERENCES

Ambihapathy K, Yalcin T, Leung H-W, Harrison AG (1997): Pathways to immonium ions in the fragmentation of protonated peptides. *J. Mass Spectrom.* **32**:209–215.

Ballard KD, Gaskell SJ (1993): Dehydration of peptide [M + H]+ ions in the gas phase. *J. Am. Soc. Mass Spectrom.* **4**:477–481.

Beardsley RL, Reilly JP (2004): Fragmentation of amidinated peptide ions. *J. Am. Soc. Mass Spectrom.* **15**:158–167.

Beardsley RL, Sharon LA, Reilly JP (2005): Peptide de novo sequencing facilitated by a dual-labeling strategy. *Anal. Chem.* **77**:6300–6309.

Biemann K (1988): Contributions of mass spectrometry to peptide and protein structure. *Biomed. Environ. Mass Spectrom.* **16**:99–111.

Biemann K (1986): Mass spectrometric methods for protein sequencing. *Anal. Chem.* **58**: 1288A–1300A.

Breci LA, Tabb DL, Yates JR, III, Wysocki VH (2003): Cleavage *N*-terminal to proline: Analysis of a database of peptide tandem mass spectra. *Anal. Chem.* **75**:1963–1971.

Burlet O, Orkiszewski RS, Ballard KD, Gaskell SJ (1992): Charge promotion of low-energy fragmentations of peptide ions. *Rapid Commun. Mass Spectrom.* **6**:658–662.

Chrisman PA, Pitteri SJ, McLuckey SA (2005): Parallel ion parking: Improving conversion of parents to first-generation products in electron transfer dissociation. *Anal. Chem.* **77**: 3411–3414.

Cooks RG, Ast T, Mabud MA (1990): Collisions of polyatomic ions with surfaces. *Int. J. Mass Spectrom. Ion Proc.* **100**:209–265.

Csonka IP, Paizs B, Lendvay G, Suhai S (2000): Proton mobility in protonated peptides: A joint molecular orbital and RRKM study. *Rapid Commun. Mass Spectrom.* **14**:417–431.

Dongre AR, Jones JL, Somogyi A, Wysocki VH (1996): Influence of peptide composition, gas-phase basicity, and chemical modification on fragmentation efficiency: Evidence for the mobile proton model. *J. Am. Chem. Soc.* **118**:8365–8374.

Eckart K, Holthausen MC, Koch W, Spiess J (1998): Mass spectrometric and quantum mechanical analysis of gas-phase formation, structure, and decomposition of various b2 ions and their specifically deuterated analogs. *J. Am. Soc. Mass Spectrom.* **9**:1002–1011.

El Aribi H, Rodriquez CF, Almeida DRP, Ling Y, Mak WWN, Hopkinson AC, Siu MKW (2003): Elucidation of fragmentation mechanisms of Protonated peptide ions and their products: A case study on glycylglycylglycine using density functional theory and threshold collision-induced dissociation. *J. Am. Chem. Soc.* **125**:9229–9236

El Aribi H, Orlova G, Rodriquez CF, Almeida DRP, Hopkinson AC, Siu MKW (2004): Fragmentation mechanisms of product ions from protonated tripeptides. *J. Phys. Chem. B* **108**:18743–18749

Elias JE, Gibbons FD, King OD, Roth FP, Gygi SP (2004): Intensity-based protein identification by machine learning from a library of tandem mass spectra. *Nat. Biotechnol.* **22**:214–219.

Farrugia JM, O'Hair RAJ, Reid GE (2001): Do all b2 ions have oxazolone structures? Multistage mass spectrometry and ab initio studies on protonated N-acyl amino acid methyl ester model systems. *Int. J. Mass Spectrom.* **210/211**:71–87.

Freitas MA, Marshall AG (1999): Rate and extent of gas-phase hydrogen/deuterium exchange of bradykinins: Evidence for peptide zwitterions in the gas phase. *Int. J. Mass Spectrom.* **182/183**:221–231.

Ge Y, Horn DM, McLafferty FW (2001): Blackbody infrared radiative dissociation of larger(42KD) multiply charged proteins. *Int. J. Mass Spectrom.* **210/211**:203–214.

Gu C, Tsaprailis G, Breci L, Wysocki VH (2000): Selective gas-phase cleavage at the peptide bond C-terminal to aspartic acid in fixed-charge derivatives of Asp-containing peptides. *Anal. Chem.* **72**:5804–5813.

Harrison AG, Yalcin T (1997): Proton mobility in protonated amino acids and peptides. *Int. J. Mass Spectrom.* **165/166**:339–347.

Herrmann KA, Wysocki VH, Vorpagel ER (2005): Computational investigation and hydrogen/deuterium exchange of the fixed charge derivative Tris(2,4,6-trimethoxyphenyl) phosphonium: Implications for the aspartic acid cleavage mechanism. *J. Am. Soc. Mass Spectrom.* **16**:1067–1080.

Huang Y, Triscari JM, Pasa-Tolic L, Anderson GA, Lipton MS, Smith RD, Wysocki VH (2004): Dissociation behavior of doubly-charged tryptic peptides: Correlation of gas-phase cleavage abundance with Ramachandran plots. *J. Am. Chem. Soc.* **126**:3034–3035.

Huang Y, Triscari JM, Tseng GC, Pasa-Tolic L, Lipton MS, Smith RD, Wysocki VH (2005): Statistical characterization of the charge state and residue dependence of low-energy CID peptide dissociation patterns. *Anal. Chem.* **77**:7626–7638.

Huang Y, Wysocki VH, Tabb DL, Yates JR (2002): The influence of histidine on cleavage C-terminal to acidic residues in doubly protonated tryptic peptides. *Int. J. Mass Spectrom.* **219**:233–244.

Hunt DF, Yates JR, Shabanowitz J, Winston S, Hauer CR (1986): Protein sequencing by tandem mass-spectrometry. *Proc. Natl. Acad. Sci. USA* **83**:6233–6237.

Johnson RS, Krylov D, Walsh KA (1995): Proton mobility within electrosprayed peptide ions. *J. Mass Spectrom.* **30**:386–387.

Johnson RS, Martin SA, Biemann K (1988): Collision-induced fragmentation of (M + H)+ ions of peptides: Side chain specific sequence ions. *Int. J. Mass Spectrom. Ion Proc.* **86**:137–154.

Jonsson AP, Bergman T, Jornvall H, Griffiths WJ (2001): Gln-Gly cleavage: A dominant dissociation site in the fragmentation of protonated peptides. *Rapid Commun. Mass Spectrom.* **15**:713–720.

Kapp EA, Schuetz F, Reid GE, Eddes JS, Moritz RL, O'Hair RAJ, Speed TP, Simpson RJ (2003): Mining a tandem mass spectrometry database to determine the trends and global factors influencing peptide fragmentation. *Anal. Chem.* **75**:6251–6264.

Kenny PTM, Nomoto K, Orlando R (1992): Fragmentation studies of peptides: The formation of y ions. *Rapid Commun. Mass Spectrom.* **6**:95–97.

Keough T, Youngquist RS, Lacey MP (1999): A method for high-sensitivity peptide sequencing using postsource decay matrix-assisted laser desorption ionization mass spectrometry. *Proc. Natl. Acad. Sci. USA* **96**:7131–7136.

Komaromi I, Somogyi A, Wysocki VH (2005): Proton migration and its effect on the MS fragmentation of N-acetyl OMe proline: MS/MS experiments and ab initio and density functional calculations. *Int. J. Mass Spectrom.* **241**:315–323.

Kruger NA, Zubarev RA, Horn DM, McLafferty FW (1999): Electron capture dissociation of multiply charged peptide cations. *Int. J. Mass Spectrom.* **187**:787–793.

Kulhanek P, Schlag EW, Koca J (2003): A novel mechanism of proton transfer in protonated peptides. *J. Am. Chem. Soc.* **125**:13678–13679.

Lebrilla CB (2004): Profiling oligosaccharides diversity by infrared multiphoton dissociation. Abstracts of Papers, *227th ACS Nat. Meeting*, Anaheim, CA, March 28–April 1, 2004, PHYS-050.

Liao P-C, Huang Z-H, Allison J (1997): Charge remote fragmentation of peptides following attachment of a fixed positive charge: A matrix-assisted laser desorption/ionization postsource decay study. *J. Am. Soc. Mass Spectrom.* **8**:501–509.

Little DP, Speir JP, Senko MW, Oconnor PB, McLafferty FW (1994): Infrared multiphoton dissociation of large multiply-charged ions for biomolecule sequencing. *Anal. Chem.* **66**:2809–2815.

Luo G, Marginean I, Vertes A (2002): Internal energy of ions generated by matrix-assisted laser desorption/ionization. *Anal. Chem.* **74**:6185–6190.

Mabud MA, Dekrey MJ, Cooks RG (1985): Surface-induced dissociation of molecular ions. *Int. J. Mass Spectrom.* **67**:285–294.

Martin SA, Hill JA, Kittrell C, Biemann K (1990): Photon-induced dissociation with a 4-sector tandem mass-spectrometer. *J. Am. Soc. Mass Spectrom.* **1**:107–109.

McCormack AL, Somogyi A, Dongre AR, Wysocki VH (1993): Fragmentation of protonated peptides: Surface-induced dissociation in conjunction with a quantum mechanical approach. *Anal. Chem.* **65**:2859–2872.

Mueller DR, Eckersley M, Richter WJ (1988): Hydrogen transfer-reactions in the formation of Y + 2 sequence ions from protonated peptides. *Org. Mass Spectrom.* **23**:217–222.

Nair H, Wysocki VH (1998): Are peptides without basic residues protonated primarily at the amino terminus? *Int. J. Mass Spectrom.* **174**:95–100.

Paizs B, Schnoelzer M, Warnken U, Suhai S, Harrison AG (2004): Cleavage of the amide bond of protonated dipeptides. *Phys. Chem. Chem. Phys.* **6**:2691–2699.

Paizs B, Suhai S (2005): Fragmentation pathways of protonated peptides. *Mass Spectrom. Rev.* **24**:508–548.

Paizs B, Suhai S (2004): Towards understanding the tandem mass spectra of protonated oligopeptides. 1: Mechanism of amide bond cleavage. *J. Am. Soc. Mass Spectrom.* **15**:103–113.

Payne AH, Glish GL (2001): Thermally assisted infrared multiphton photodissociation in a quadrupole ion trap. *Anal. Chem.* **73**:3542–3548.

Polfer NC, Oomens J, Suhai S, Paizs B (2005): Spectroscopic and theoretical evidence for oxazolone ring formation in collision-induced dissociation of peptides. *J. Am. Chem. Soc.* **127**:17154–17155

Price WD, Schnier PD, Williams ER (1996.):Tandem mass spectrometry of large biomolecule ions by blackbody infrared radiative dissociation. *Anal. Chem.* **68**:859–866.

Reid GE, Simpson RJ, O'Hair RAJ (2000): Leaving group and gas phase neighboring group effects in the side chain losses from protonated serine and its derivatives. *J. Am. Soc. Mass Spectrom.* **11**:1047–1060.

Resing KA, Meyer-Arendt K, Mendoza AM, Aveline-Wolf LD, Jonscher KR, Pierce KG, Old WM, Cheung HT, Russell S, Wattawa JL, Goehle GR, Knight RD, Ahn NG (2004): Improving reproducibility and sensitivity in identifying human proteins by shotgun proteomics. *Anal. Chem.* **76**:3556–3568.

Reynolds KJ, Yao X, Fenselau C (2002): Proteolytic 18O labeling for comparative proteomics: Evaluation of endoprotease glu-C as the catalytic agent. *J. Proteome Res.* **1**:27–33.

Rodriquez CF, Cunje A, Shoeib T, Chu IK, Hopkinson AC, Siu KW (2001): Proton migration and tautomerism in protonated triglycine. *J. Am. Chem. Soc.* **123**:3006–3012.

Roepstorff P (1984): Proposal for a common nomenclature for sequence ions in mass spectra of peptides. *Biomed. Mass Spectrom.* **11**:601.

Sadagopan N, Watson JT (2001): Mass spectrometric evidence for mechanisms of fragmentation of charge-derivatized peptides. *J. Am. Soc. Mass Spectrom.* **12**:399–409.

Schnier P, Price W, Jockusch R, Williams E (1996): Blackbody infrared radiative dissociation of bradykinin and its analogues: Energetics, dynamics, and evidence for salt-bridge structures in the gas phase. *J. Am. Chem. Soc.* **118**:7178–7189.

Somogyi A, Wysocki VH, Mayer I (1994): The effect of protonation site on bond strengths in simple peptides: Application of ab initio and modified neglect of differential overlap bond orders and modified neglect of differential overlap energy partitioning. *J. Am. Soc. Mass Spectrom.* **5**:704–717.

Stimson E, Truong O, Richter WJ, Waterfield MD, Burlingame AL (1997): Enhancement of charge remote fragmentation in protonated peptides by high-energy CID MALDI-TOF-MS using "cold" matrices. *Int. J. Mass Spectrom.* **169**:231–240.

Summerfield SG, Cox KA, Gaskell SJ (1997): The promotion of d-type ions during the low energy collision-induced dissociation of some cysteic acid-containing peptides. *J. Am. Soc. Mass Spectrom.* **8**:25–31.

Summerfield SG, Steen H, O'Malley M, Gaskell SJ (1999): Phenyl thiocarbamoyl and related derivatives of peptides: Edman chemistry in the gas phase. *Int. J. Mass Spectrom.* **188**:95–103.

Syka JEP, Coon JJ, Schroeder MJ, Shabanowitz J, Hunt DF (2004): Peptide and protein sequence analysis by electron transfer dissociation mass spectrometry. *Proc. Natl. Acad. Sci. USA* **101**:9528–9533.

Tabb DL, Huang Y, Wysocki VH, Yates JR, III (2004): Influence of basic residue content on fragment ion peak intensities in low-energy collision-induced dissociation spectra of peptides. *Anal. Chem.* **76**:1243–1248.

Tabb DL, Smith LL, Breci LA, Wysocki VH, Lin D, Yates JR, III (2003): Statistical characterization of ion trap tandem mass spectra from doubly charged tryptic peptides. *Anal. Chem.* **75**:1155–1163.

Thompson MS, Cui W, Reilly JP (2004): Mass spectrometry: Fragmentation of singly charged peptide ions by photodissociation at l = 157 nm. *Angew. Chem. Int. Ed.* **43**:4791–4794.

Tsang CW, Harrison AG (1976): Chemical ionization of amino-acids. *J. Am. Chem. Soc.* **98**:1301–1308.

Tsaprailis G, Nair H, Zhong W, Kuppannan K, Futrell JH, Wysocki VH (2004): A mechanistic investigation of the enhanced cleavage at histidine in the gas-phase dissociation of protonated peptides. *Anal. Chem.* **76**:2083–2094.

van Dongen WD, Ruijters HF, Luinge HJ, Heerma W, Haverkamp J (1996): Statistical analysis of mass spectral data obtained from singly protonated peptides under high-energy collision-induced dissociation conditions. *J. Mass Spectrom.* **31**:1156–1162.

Wee S, O'Hair RAJ, McFadyen WD (2002): Side-chain radical losses from radical cations allows distinction of leucine and isoleucine residues in the isomeric peptides Gly-XXX-Arg. *Rapid Commun. Mass Spectrom.* **16**:884–890.

Wu J, Lebrilla CB (1995): Intrinsic basicity of oligomeric peptides that contain glycine, alanine, and valine—the effects of the alkyl side chain on proton transfer reactions. *J. Am. Soc. Mass Spectrom.* **6**:91–101.

Wysocki VH, Breci L, Hermann K (no date): Unpublished results.

Wysocki VH, Tsaprailis G, Smith LL, Breci LA (2000): Mobile and localized protons: A framework for understanding peptide dissociation. *J. Mass Spectrom.* **35**:1399–1406.

Wyttenbach T, Bowers MT (1999): Gas phase conformations of biological molecules: The hydrogen/deuterium exchange mechanism. *J. Am. Soc. Mass Spectrom.* **10**:9–14.

Yalcin T, Csizmadia IG, Peterson MR, Harrison AG (1996): The structure and fragmentation of Bn ($n >= 3$) ions in peptide spectra. *J. Am. Soc. Mass Spectrom.* **7**:233–242.

Yalcin T, Khouw C, Csizmadia IG, Peterson MR, Harrison AG (1995): Why are B ions stable species in peptide spectra? *J. Am. Soc. Mass Spectrom.* **6**:1165–1174.

Zhang J, Schubothe K, Li B, Russell S, Lebrilla CB (2005): Infrared multiphoton dissociation of O-linked mucin-type oligosaccharides. *Anal. Chem.* **77**:208–214.

Zhang Q, Wysocki VH (no date): Unpublished results.

Zhang Z (2004): Prediction of low-energy collision-induced dissociation spectra of peptides. *Anal. Chem.* **76**:3908–3922.

Zubarev RA, Haselmann KF, Budnik B, Kjeldsen F, Jensen F (2002): Towards an understanding of the mechanism of electron-capture dissociation: A historical perspective and modern ideas. *Eur. J. Mass Spetrom.* **8**:337–349.

Zubarev RA, Kelleher NL, McLafferty FW (1998): Electron capture dissociation of multiply charged protein cations. A nonergodic process. *J. Am. Chem. Soc.* **120**:3265–3266.

9

PEPTIDE RADICAL CATIONS

Alan C. Hopkinson and K. W. Michael Siu

Centre for Research in Mass Spectrometry and the Department of Chemistry
York University
Toronto, Ontario, Canada

9.1. INTRODUCTION

Electron impact (EI) mass spectrometry (MS) has been used for many years to produce radical cations of organic molecules and to study their fragmentation patterns (McLafferty and Turecek 1993). However, extension of this method to amino acids and peptides was very limited because these biologically important

Principles of Mass Spectrometry Applied to Biomolecules, edited by Julia Laskin and Chava Lifshitz

molecules have very low vapor pressures and decompose on heating, making it difficult if not impossible to create a sufficient gas-phase population for electron impact studies (Jochims et al. 2004). More recently, laser ablation followed by UV photoionization has provided a source of radical cations of peptides (Becker and Wu 1995; Weinkauf et al. 1995, 1996; Grotemeyer and Schlag 1988a, 1988b; Cui et al. 2002, 2005; Sheu et al. 2002; Hu et al. 2003); this technique is restricted to peptides that contain aromatic amino acid residues that act as chromophores and consequently is of limited applicability.

Electrospray ionization (ESI) provides an alternative soft technique for generating ions of amino acids and peptides. Electrospraying solutions of peptides produces high abundances of protonated peptides, and extensive collision-induced dissociation (CID) studies of these ions have led to a general understanding of how protonated peptides fragment. This has proved to be particularly useful in determining the sequences of the amino acid residues in peptides (Aebersold and Goodlett 2001). The dissociation products of protonated peptides are almost entirely even-electron (closed-shell) ions; that is, very few radical cations are produced by this procedure. However, multiprotonated peptides can be converted into radical cations by capturing an electron. The subsequent fragmentation of these hot ions produces mainly c_n and z_m ions, the products of breaking the N—$^{\alpha}$C bond (Zubarev et al. 1998; Zubarev 2003; Iavarone et al. 2004; Turecek 2003; Syrstad and Turecek 2004). This electron capture dissociation (ECD) method will not be discussed further here, as it is the topic of another chapter.

Metal ion complexes containing amino acids or peptides as ligands can also be generated by ESI; CID of these complexes may result in the loss of the metal as a neutral atom, thereby creating a radical cation (Lavanant et al. 1999, El Aribi et al. 2004). For example, one of the minor products in the CID of Ag^+/histidine is the radical cation of histidine (Scheme 9.1, structure **1**) (Shoeib 2002).

Ion **1** readily loses CO_2, a reaction frequently seen in the fragmentation of peptide radical cations, and one that is most easily understood in terms of the histidine having a structure in which the charge is on the aromatic ring and the radical on the carboxy group (Scheme 9.1). Structures that have this type of separation of charge and unpaired spin, as in structures **1** and **2**, are described as *distonic* ions and have been found, in general, to have higher stabilities than their canonical analogs (Holmes et al. 1982; Bouma et al. 1982a, 1982b). Ion **2** can lose a small neutral molecule to give a third distonic ion **3**. The neutral lost is probably aminocarbene,

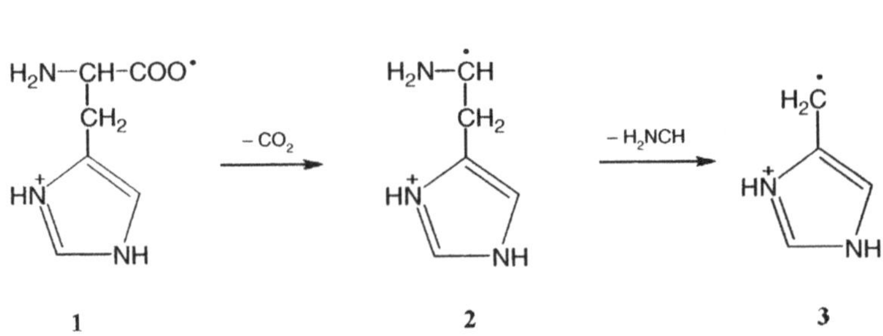

SCHEME 9.1

H_2NCH. Its tautomer, methanimine, $HN{=}CH_2$, is lower in energy by 163 kJ/mol, but the conversion barrier is high at 197 kJ/mol (Pau and Hehre 1986), making the latter species a minor component, if present at all, even under CID conditions.

9.2. GENERATION OF PEPTIDE RADICAL CATIONS FROM METAL ION/PEPTIDE COMPLEXES

The most frequently used method for producing radical cations of peptides is from the CID of doubly charged metal complexes containing both an amine ligand and a peptide. Copper has most commonly been the metal of choice and the process exploits the redox chemistry of this metal. In the initial discovery (Chu et al. 2000), the complex $[Cu^{II}(dien)(TyrGlyGlyPheLeuArg)]^{\bullet 2+}$, where dien is diethylenetriamine ($H_2NCH_2CH_2NHCH_2CH_2NH_2$), gives on CID the radical cation of the peptide and $[Cu(dien)]^+$ (Figure 9.1). Subsequently, complexes with various combinations of peptides, M, and ligands, L (triamines, terpyridines and crown ethers), have been subjected to CID (Chu et al. 2001, 2005; Bagheri-Majdi et al. 2004, Barlow et al. 2004). Several studies on the coordination and chemistry of Cu(II) complexes in the gas phase indicate preference for a square planar arrangement of the ligands

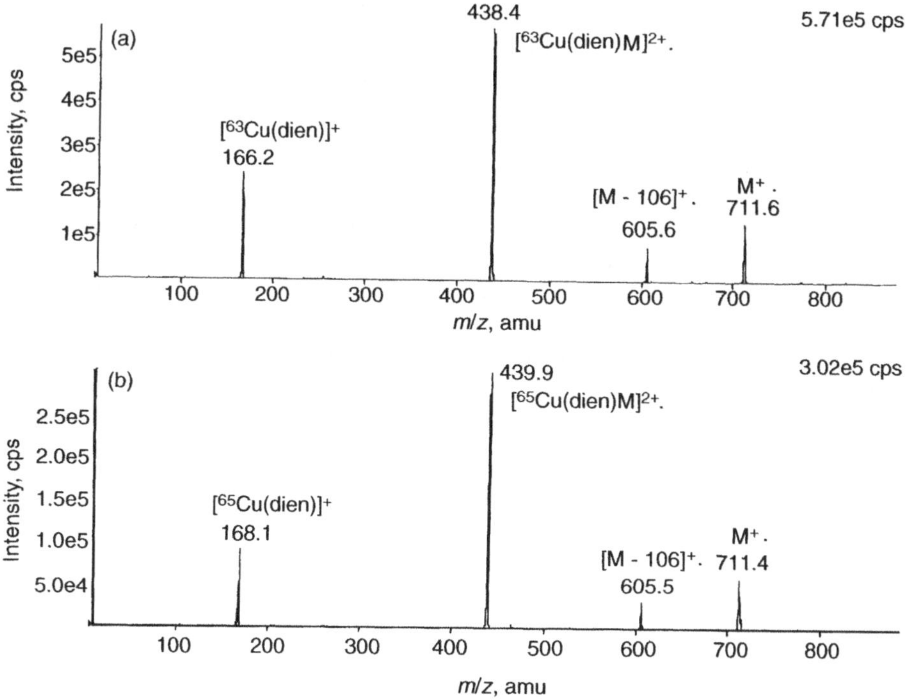

FIGURE 9.1. Product ion spectra of (a) $[^{63}Cu(dien)(M)]^{\bullet 2+}$ and $[^{65}Cu(dien)(M)]^{\bullet 2+}$, where M = TyrGlyGlyPheLeuArg, center of mass (E_{CM}) = 1.3 eV. [Reprinted with permission from Chu et al. (2000): Molecular radical cations of oligopeptides. *J. Phys. Chem. B* **104**:3393–3397. (Chu et al. 2000) Copyright 2000 American Chemical Society.] CPS is counts per second.

surrounding the Cu(II) nucleus (Berces et al. 1999; Wright et al. 2001; Walker et al. 2001; Seymour and Turecek 2002; Seymour et al. 2004). However, alternative binding modes whereby the tridentate amine ligand forces one or two of the coordinating peptide sites into an apical or equatorial position, or into a secondary complexation shell, cannot be excluded (Vachet et al. 1998; Chaparro and Vachet 2003; Combariza and Vachet 2002). Indeed, X-ray crystallographic examination of $Cu^{II}(tpy)Cl_2.nH_2O$ (tpy = 2,2′:6′,2″-terpyridine, $n = 0,1$) reveals a five-coordinate, distorted square pyramidal structure (Henke et al. 1983). Fragmentation pathways, all leading to separation of charge, that have been observed are as follows:

$$[Cu^{II}(L)(M)]^{\bullet 2+} \rightarrow [Cu^{I}(L)]^{+} + M^{\bullet +} \tag{9.1}$$

(peptide radical formation)

$$\rightarrow [Cu^{II}(L\text{–}H)]^{\bullet +} + [M + H]^{+} \tag{9.2}$$

(proton addition to the peptide)

$$\rightarrow [Cu^{II}(M\text{–}H)]^{\bullet +} + [L + H]^{+} \tag{9.3}$$

(proton abstraction from the peptide)

$$\rightarrow [Cu^{II}(L)(M\text{–}b_n)]^{\bullet +} + b_n^{+} \tag{9.4}$$

(peptide fragmentation)

The last reaction can be illustrated by using a tripeptide as M. In this ion, charge can be most easily delocalized by proton migration from the carboxylic group to one of the basic sites along the peptide backbone: a basic side chain, a carbonyl oxygen of one of the amide bonds, or the terminal amino group. However, cleavage of an amide bond is most easily rationalized by protonation at the nitrogen of a peptide bond; this is followed by displacement of the *C*-terminal amino acid by nucleophilic attack by the other carbonyl oxygen and formation of a protonated 2-(aminomethyl)-5-oxazolone, as shown Scheme 9.2.

For complexes $[Cu^{II}(tpy)(M)]^{\bullet 2+}$, where M contains a basic amino acid residue, the peptide is probably zwitterionic with the carboxylate anion attached to the Cu(II)

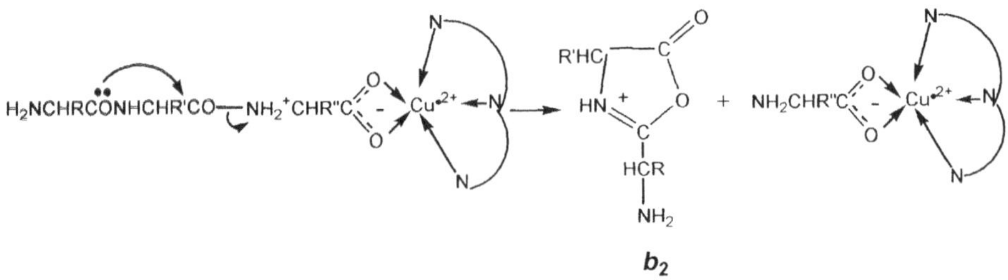

SCHEME 9.2

and the proton on the sidechain. Reaction (9.4) is then facilitated by proton transfer from the sidechain to the amide bond that is to be cleaved. When the bases are the histidine and lysine residues, fragmentation of the peptide bond is more competitive than when it is the arginine residue, and this is attributed to the highest basicity of this last residue, thus rendering proton transfer less facile (Wee 2005).

Reaction (9.1), a dissociative redox reaction, results in production of the radical cation ($M^{\bullet +}$) and in the present context is the desired product. Considerable experimental effort has been expended on maximizing the yield of this step. In the earlier studies, radical cations could be formed from only peptides containing either a tyrosine or tryptophan residue (from the two amino acids with the lowest ionization energies), and ions containing these two residues remain the easiest to produce in high abundances (Bagheri-Majdi 2003). In order to maximize reaction (9.1), ligand L should have a higher binding energy than the peptide, thereby discouraging the former's elimination; in order to prevent reaction (9.2), the ligand should have no acidic hydrogens. Replacing all the amino hydrogens of dien by methyl groups (as in Me_5-dien) was partly successful, but introduced a more basic ligand and, as a consequence, increased the yield of the products in Reaction (9.3) (Bagheri-Majdi 2003). The range of peptide radical cations that can be formed by CID was extended to those containing the basic amino acid residues, derived from lysine, arginine, and histidine, by using Cu(II) complexes in which the ligand is tpy (Wee et al. 2002, 2004; Bagheri-Majdi 2004). 1,4,7,10-Tetraoxacyclododecane (12-crown-4), a cyclic polyether that does not have acidic hydrogens, has also been used as a ligand and this enabled Chu et al. (2004) to generate radical cations GlyGlyXxx, where Xxx is the residue of an aliphatic amino acid (Gly, Ala, Val, Iso, or Leu).

Subsequent systematic studies of $[Cu^{II}(L)(M)]^{\bullet 2+}$ complexes with different ligands—bidentate bipyridines and phenanthrolines as well as tridentate dien derivatives and terpyridines—showed that tridentate ligands are the most effective at preventing reaction (9.3) and that, of these, tpy best prevents reaction (9.2). Chu et al. (2005) have extended the work on tpy by using ligands that are more sterically hindered; complexes $[Cu^{II}(L)(M)]^{\bullet 2+}$, where L is 6-bromo-2,2′:6′,2″-terpyridine, successfully eliminates reactions (9.2) and (9.3), but peptide fragmentation via reaction (9.4) is still significant. Using an even more sterically hindered ligand, 6,6′-dibromo-2,2′:6′,2″-terpyridine, it is possible to generate radical cations of all tripeptides, GlyGlyXxx, where Xxx is (any of the) naturally occurring amino acid residue, except serine, asparagine and aspartic acid.

Copper has a high second ionization energy (20.3 eV), and this is one of the reasons why it is effective in removing an electron from a departing peptide in the CID process. Triply charged metal ions should be more effective in this regard, but formation of complexes having a tripositive charge is difficult. One potential way around this problem is to use triply charged metal ions attached to dinegative ligands as demonstrated by Barlow et al. (2005). They exploited Cr^{3+}, Mn^{3+}, Fe^{3+}, and Co^{3+} complexed with 5,5′-disubstituted salens [*N*,*N*-ethylenebis(salicyldieneaminato), structure **4**] to provide monopositive ions with two open coordination sites that can accommodate peptide ligands.

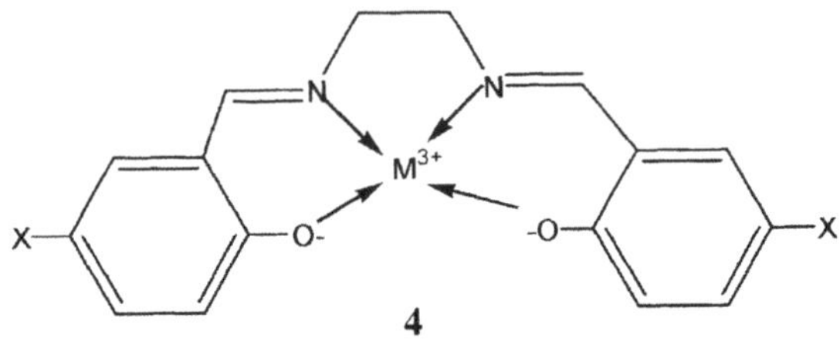

4

Salen dianion ligands, X = H,Cl,NO_2,OMe

Peptides M = TyrGlyGlyPheLeuArg, TrpGlyGlyPheLeuArg, and GlyGlyGlyPheLeuArg were used and CID of all complexes $[metal(salenX)(M)]^{3+}$ produce radical cations $M^{\bullet+}$, with Mn^{3+} and Fe^{3+} giving the highest abundances. The competing fragmentation reactions vary considerably with the metal, and the substituents (X) are sufficiently remote from the metal ion to allow tuning of the dissociation reaction to favor radical cation formation.

Most of the work on the radical cations of peptides has focused on using increasingly complicated ligands in attempts to create high abundances of the peptide ions containing only the amino acid residues that have the higher ionization energies. Rather than using increasingly more complicated ligands, an arguably simpler approach is to use Cu(II) complexes in which the ligand is also a peptide (either the same one, i.e., L = M, or a different one) (Ke et al. 2005). Employing peptides containing amino acid residues that have low ionization energies (tryptophan and tyrosine), we have produced high abundances of radical peptides from complexes $[Cu^{II}(M_a)(M_b)]^{\bullet 2+}$, where M_a and M_b are Xxx, XxxGly, GlyXxx, XxxGlyGly, GlyXxxGly, and GlyGlyXxx (Xxx = Tyr or Trp). When $M_a = M_b$, two reactions are possible: formation of the radical cation [reaction (9.5)] and proton transfer [reaction (9.6)]:

$$[Cu^{II}(M_a)(M_b)]^{\bullet 2+} \rightarrow [Cu^{I}(M_a)]^{+} + [M_b]^{\bullet +} \tag{9.5}$$

$$\rightarrow [Cu^{II}(M_a\text{–}H)]^{\bullet +} + [M_b + H]^{+} \tag{9.6}$$

When both peptides contain a trytophan residue, only radical cation formation was observed, except when $M_a = M_b$ = TrpGlyGly, where some proton transfer was also apparent. Some of the radical cations decompose easily, and it was found that the existence of the complementary ion $[Cu^{I}(M_a)]^{+}$ is a better indicator of reaction (9.5). In mixed complexes where both peptides contain a trytophan residue but one is a dipeptide and the other a tripeptide, the location of the tryptophan residue dictates which peptide will form the radical. Peptides with the tryptophan residue at the *N* terminal are favored and, in the CID of $[Cu^{II}(TrpGlyGly)(TrpGly)]^{\bullet 2+}$, the abundance of TrpGlyGly$^{\bullet +}$ is approximately twice that of TrpGly$^{\bullet +}$.

In the CIDs of $[Cu^{II}(TyrGly)_2]^{\bullet 2+}$ and $[Cu^{II}(GlyTyr)_2]^{\bullet 2+}$, both pathways are observed with similar prevalences. The radical cations fragment easily at the N—$^{\alpha}$C bond of the tryptophan residue, and only decomposition products and the complementary ions $[Cu^{I}(M_a)]^+$ are evident. Complexes with tripeptides that have a tyrosine residue undergo predominantly proton transfer reactions. There is no evidence for the formation of $TyrGlyGly^{\bullet+}$; the other two tripeptide radical cations, $GlyTyrGly^{\bullet+}$ and $GlyGlyTyr^{\bullet+}$, are only formed in low abundances.

9.3. STABILIZING FACTORS IN RADICAL CATIONS

In solution, cations interact strongly with solvent molecules, enabling them to spread the charge over a large network of solvent molecules. In the gas phase, unsolvated cations have to accommodate the whole charge and this is most easily achieved if the molecule has an extended conjugated π system, enabling the charge to be distributed over several nuclei. It is for this reason that amino acids that have the lowest ionization energies, the aromatic ones, most easily form radical cations.

An additional electronic feature that plays an important role in the chemistry of radical cations of amino acids and peptides is that spatial separation of the charge and radical center is stabilizing. As alluded to previously, ions of this type are said to be distonic. The radical cation of methanol provides the simplest example (Holmes et al. 1982; Bouma et al. 1982). Formally, methanol has two lone pairs of electrons on the oxygen atom, and removal of an electron leaves a positive charge on the oxygen atom (structure **5**).

$H_3C{-}O^{\bullet+}H$ **5** $\qquad H_2C^{\bullet}{-}OH_2^+$ **6**

However, this classical structure is ~29 kJ/mol higher in energy than the distonic ion (structure **6**), in which a hydrogen atom has migrated to the oxygen, resulting in the radical center now being located at the carbon, and the oxygen atom again formally carrying the positive charge. Interconversion between isomers **5** and **6** has a significant barrier (108 kJ/mol above **5** at the G2** level of theory).

Three isomers of the glycine radical cation in the gas phase have been studied both experimentally and theoretically (Depke et al. 1984; Yu and Rauk 1995; Polce and Wesdemiotis 1999, 2000). The classical structure (**Gly1**), formed by electron impact on glycine, has the majority of the spin and charge on the amino group (Lu et al. 2004). Ion **Gly2**, formed by the loss of hydrogen atom from protonated glycine, formally has the charge located on the NH_3 group and the spin on the CH, and therefore is distonic (Beranova et al. 1995). This ion is calculated [at CCSD(T)/6-31++G(*d*,*p*)//UB3LYP/6-31++G(*d*,*p*)] to be lower than **Gly1** by 7.0 kcal/mol (Simon et al. 2002). The structure at the global minimum (**Gly3**) is a α radical with a hydrogen having been transferred to the carboxy group, thereby formally creating another distonic ion. Here, there is an additional stabilizing feature; the radical center is attached to a powerful π-electron-donating group (NH_2) and also a powerful π-electron-withdrawing group $C(OH)_2^+$. This arrangement, combining the resonance electron-withdrawing (capto) and donating (dative) groups, creates a planar *captodative* radical in which there is extensive delocalization of both the spin and the charge.

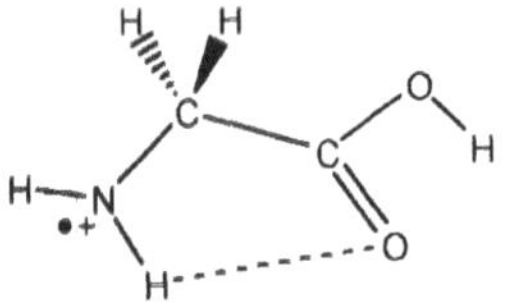

Gly1

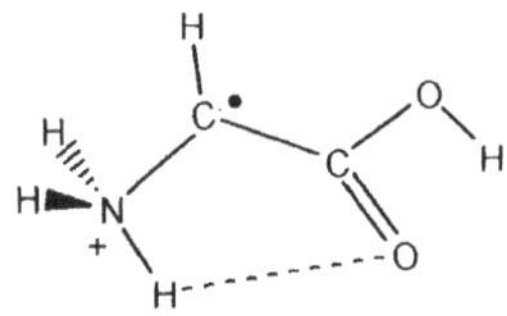

Gly2

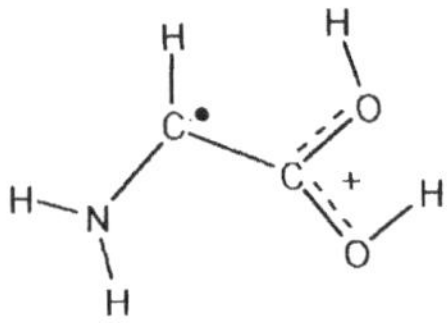

Gly3

The captodative effect has long been recognized as important in stabilizing radicals (Viehe et al. 1985; Sustmann and Korth 1990; Bordwell et al. 1992; Easton 1997). Homolytic cleavage of the $^{\alpha}$C–H bond of an amino acid produces a captodative radical in which the electron-withdrawing COOH group is only moderately effective (Turecek et al. 1999). Furthermore, the stability achieved by the conjugation depends on steric factors that dictate how close to planar the radical can become (Burgess et al. 1989; Rauk and Armstrong 1999). In a theoretical study using an isodesmic reaction [Reaction (9.7)] to compare the resonance-stabilizing energies (RSEs) of groups X on the stability of a methyl radical, it was found that NH_2 is stabilizing by 44.7 kJ/mol and COOH by 21.0 kJ/mol (Croft et al. 2003):

$$XCH_2^{\bullet} + CH_4 \rightarrow CH_3^{\bullet} + XCH_3 \qquad (9.7)$$

When both groups are attached to the *same* center in the neutral glycinyl radical, $H_2NCRCOOH^{\bullet}$ (R = H), the radical stabilization is 95.9 kJ/mol, 30.2 kJ/mol greater than the sum of the individual components. This synergistic stabilization is attributed to the delocalization of the unpaired spin. The alanyl (R = CH_3) and valyl [R = $CH(CH_3)_2$] radicals have larger RSE values, reflecting the abilities of the alkyl sidechains to stabilize a radical center.

Modification of the α-carbon centered radicals by substituting an acetyl group on the amino group, and using the methyl ester instead of the carboxylic acid, has the effect of reducing the RSE, largely due to the reduction in the π-donating ability of the amino group. These changes are often designed to mimic an amino acid residue in a peptide chain. More interestingly, there is a reversal in the relative RSE values; the valine derivative, $H_3CCONHCH(CH(CH_3)_2)COOH^{\bullet}$, now has the lowest RSE, and this is attributed to interaction between the bulky isopropyl sidechain and the carbonyl of the amide, which prevents planarity at the radical center and thereby reducing spin delocalization. Consequently, in a peptide, the glycine residue is one of the most favorable radical sites (Croft et al. 2003).

The $C(OH)_2^+$ group of **Gly3** is much more strongly electron-withdrawing than the neutral COOH; extension of the resonance stabilization energy analysis to the $C(OH)_2^+$ group gives an RSE value of 20.1 kJ/mol, slightly smaller than that of COOH (Zhao et al. 2005). Separately, the sum of the individual stabilizations from NH_2 and $C(OH)_2^+$ is 64.8 kJ/mol; the combined effect of NH_2 and $C(OH)_2^+$ in $H_2NCRCO(OH)_2^{\bullet+}$ (R = H) is 182.0 kJ/mol, 117.2 kJ/mol greater than the individual contributions. Comparing this with the enhancement of only 30.2 kJ/mol attributed to the captodative character in $H_2NCHCOOH^{\bullet}$ shows that the increase in the electron-withdrawing power by introduction of a positive charge has a profound stabilizing effect.

In the case of radical cations of peptides, the location of the spin is less predictable because additional factors involving the charge and the side chain come into play. Delocalization of the positive charge becomes important and for aromatic amino acids either a π-radical structure, where the electron has been removed from the π system and the charge is delocalized over the ring, or a benzyl-type radical structure, where an exocyclic proton has migrated to a basic site and the spin is delocalized over the ring, is possible. Alternatively, the charge and the spin may be located on the α carbon and the two functional groups that are common to all amino acids: the amino and carboxy groups. The three possible types of radical cations derived from an aromatic amino acid are illustrated below using tryptophan.

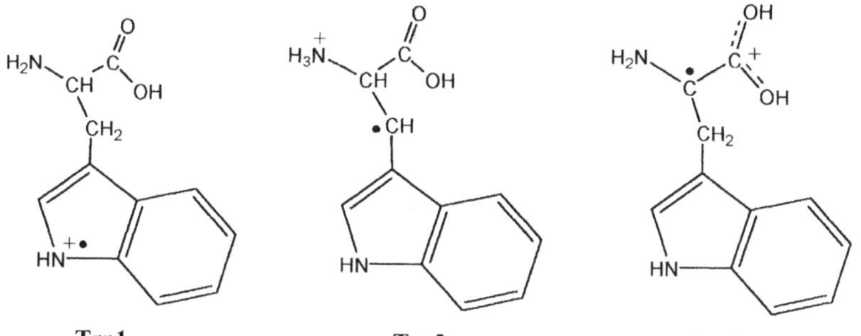

Trp1 **Trp2** **Trp3**

Trp1 is a π-radical with both the charge and spin delocalized over the two rings. **Trp2** is a distonic ion with the unpaired spin delocalized over the ring and the charge localized on the NH_3. **Trp3** is an α radical, the only one of the three that is captodative, with the charge and spin delocalized over the H_2N, the α-C and the $C(OH)_2^{+}$. Density functional calculations at UB3LYP/6-31++G(*d*,*p*) showed all three structures to have very similar energies; the lowest-energy one is **Trp1** and the other two are almost identical in energy, about 25 kJ/mol above **Trp1** (Orlova et al. 2005). Experimentally $Trp^{\bullet+}$ was generated by CID of $[Cu^{II}(dien)(Trp)]^{\bullet 2+}$; mass selection of this ion followed by fragmentation at a higher energy gave protonated indenomethide and the captodative glycyl radical (Bagheri-Majdi et al. 2004). This reaction is most easily rationalized by homolytic fission of the $^{\alpha}C$—$^{\beta}C$ bond in **Trp1** (Scheme 9.3).

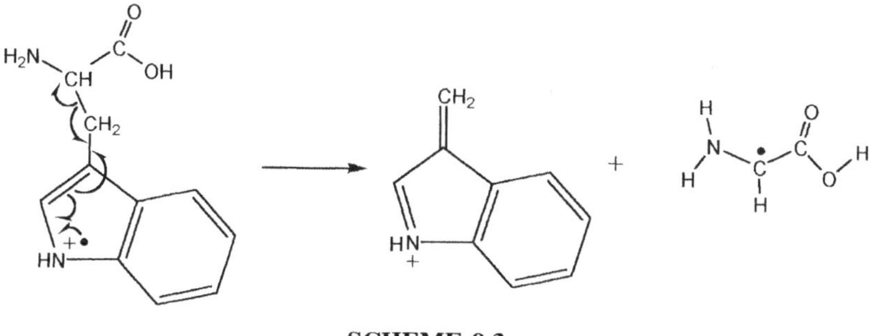

SCHEME 9.3

Cysteine is the most effective amino acid at scavenging radicals in solution, probably because formation of a radical at a sulfur atom is facile (Aliaga and Lissi 2000). However, all attempts at creating $Cys^{\bullet+}$ by CID of $[Cu^{II}(L)(Cys)]^{\bullet 2+}$ failed because of the difficulty in making the complex (Ke et al. 2005). There have been two more recent computational studies of cysteine radical cations. In a comparative study, Simon et al. (2005) examined the structures and fragmentations of the radical cations of glycine, alanine, serine, and cysteine. Unfortunately, only the canonical structure formed by removing an electron from the lowest-energy conformer of the neutral amino acid was considered; potentially lower-energy structures of the type represented by **Gly2** and **Gly3** were not included. In a comprehensive study of the potential energy hypersurface of $Cys^{\bullet+}$, Zhao et al. (2005) found that the distonic ion (**Cys1**), created by transferring the SH hydrogen to the amino group, is 25.9 kJ/mol higher in energy than the structure at the global minimum, the α radical **Cys2**. Additionally, the canonical structure studied by Simon et al. (2005) is 73.2 kJ/mol above **Cys2** at the UB3LYP/6-311++G(*d*,*p*) level of theory.

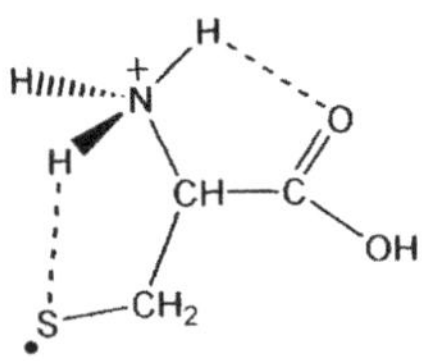

Cys1

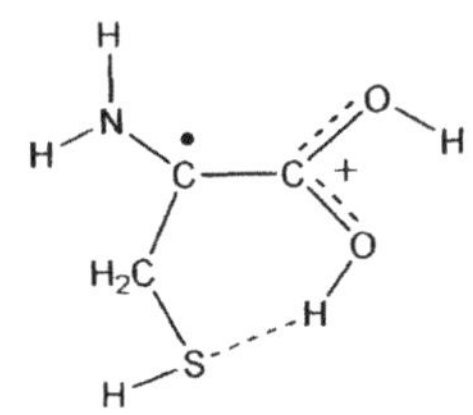

Cys2

9.4. FRAGMENTATION OF RADICAL CATIONS

Proteins have many basic sites. In addition to the nitrogen atoms on the sidechains of histidine, lysine, and arginine residues, there are the terminal amino group and all the amide oxygens along the peptide backbone and on the sidechains of aspargine and glutamine residues. Consequently, electrospraying of proteins typically produces multiply charged protein ions containing a large number of added protons. For small peptides, the number of protons that will add is necessarily fewer, but ions $[M+H]^+$ and $[M+2H]^{2+}$ are usually in high abundance. For sequencing, the doubly protonated ions generally provide more extensive information because they fragment more readily (typically into b_n and y_m ions) as a result of Coulombic repulsion (Aebersold and Goodlett 2001).

Fragmentation of radical peptides $M^{\bullet+}$, while not yet so extensively studied, can provide additional information for sequencing. In general, peptide radical cations show more diverse chemistry than do the closed-shell protonated peptides, as they commonly lose both even-electron and odd-electron neutral molecules. One neutral product frequently lost in the dissociation of peptide radical cations is CO_2, and, as carboxy radicals are known to be relatively unstable in the gas phase (Schroder et al.

2003; Bossio et al. 2003), this has been taken as an indicator that in the complex, the peptide is coordinated to the copper through a carboxylate anion (Gatlin et al. 1995; Seymour and Turecek 2002). Some of the most stable peptide radical cations have a tyrosine or a tryptophan residue; the sidechains of the residues are sometimes lost as *p*-quinomethide or indenomethide. Mechanistically, this is most easily understood if the initially formed radical ion has both the charge and unpaired spin in the aromatic system. This seems likely because on the potential energy hypersurface for $Tyr^{\bullet+}$, although the α radical has the lowest energy, it is only 12.6 kJ/mol lower in energy than the π radical (Zhao et al. 2005). When the aromatic amino acid residue is at a position other than the *N*-terminus of the peptide, the α radical is acylated on the amino nitrogen and is no longer heavily stabilized by a strong π donor. The π radical is unaffected by acylation, and hence is most likely to have the lowest energy.

Starting with a peptide with a tyrosine residue in which the charge and unpaired electron are in the π system, a distonic ion can then be formed by transferring the proton from the OH group of the phenol to either the terminal NH_2 group or an amide oxygen, two sites that have comparable proton affinities (Rodriquez et al. 2001). If the site of protonation is the amide oxygen *C*-terminal to the tyrosine residue, elimination of the tyrosine sidechain produces an α-radical ion that has some captodative character by being attached to a strong electron-withdrawing group on one side (where the proton resides) and to a weakly electron-donating one on the other (Scheme 9.4). Loss of *p*-quinomethide is frequently observed in the

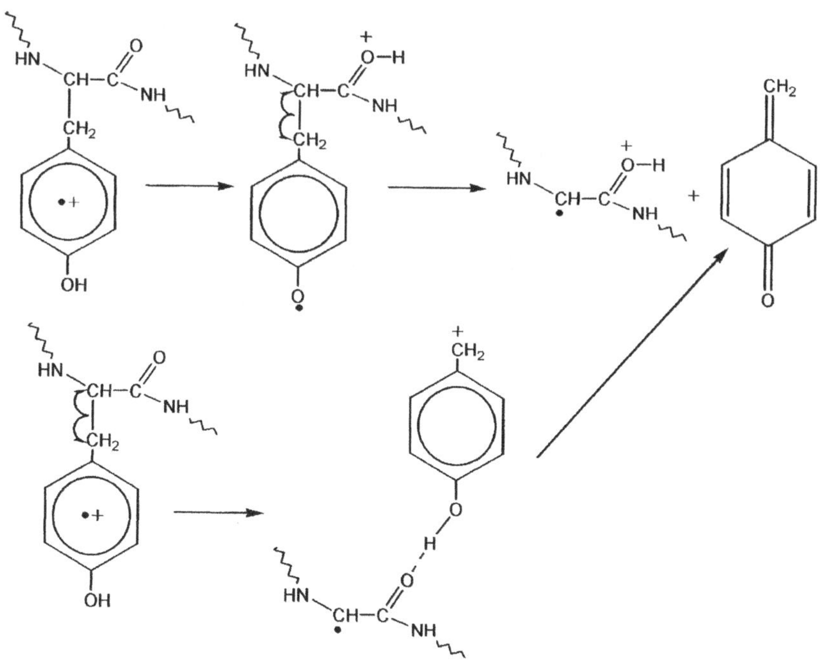

SCHEME 9.4

CIDs of peptides containing a tyrosine residue, but not from radical cations that have the tyrosine residue at the *C* terminus, as in $Tyr^{\bullet+}$ (Bagheri-Majdi et al. 2004) and $ArgTyr^{\bullet+}$ (Wee 2005). This is probably because the rigidity of the sidechain of tyrosine does not permit migration of the hydrogen from the phenolic OH group in these smaller ions. (The alternative mechanism shown at the bottom of Scheme 9.4 involves homolytic cleavage of the $^{\alpha}C$—$^{\beta}C$ bond, formation of an ion–molecule complex, and proton transfer from the phenol to the amide oxygen. This mechanism fits experimental results less well, as $ArgTyr^{\bullet+}$ does not eliminate *p*-quinomethide.) In addition, the carboxylic group is less basic than the amide group. By contrast, in the CID of $TyrArg^{\bullet+}$, *p*-quinomethide is the only product. Here, the initially formed ion probably is a carboxy radical with the sidechain of the arginine residue protonated, and there is sufficient flexibility between the phenolic OH and $COO^{\bullet}$ that hydrogen atom transfer can occur.

The dominant fragmentation pathway for radical peptides of the type $(Ala)_n(Tyr)(Ala)_m^{\bullet+}$ was cleavage at the N—$^{\alpha}C$ of the tyrosine, producing $[c_n + 2H]^+$ ions; little, if any, *p*-quinomethide was produced in the CID spectra of these ions (Ke et al. 2005). Thus the results in totality suggest that the arginine residue plays a role in the generation of *p*-quinomethide. One possibility is that ions that contain an arginine residue tend to be zwitterionic and bind to Cu(II) through the COO^-. CID of these complexes forms a radical cation with $COO^{\bullet}$ that can extract, where sterically possible, the phenolic hydrogen. In the absence of a basic residue, peptides of the type $(Ala)_n(Tyr)(Ala)_m$ may coordinate differently with Cu(II), perhaps through the terminal nitrogen or the oxygens of the peptide backbone. Ions formed in the CID of these complexes will be aromatic π radicals; migration of a proton from the CH_2 of the tyrosine sidechain to the nitrogen initiates the N—$^{\alpha}C$ cleavage.

9.4.1. Radical Cations of Amino Acids

Complexes $[Cu^{II}(M)_2]^{\bullet 2+}$, where M is any of the naturally occurring amino acids except cysteine, can be formed by electrospraying a mixture of a Cu(II) salt and the amino acid in water/methanol (Ke et al. 2005). CIDs of these complexes result in both the dissociative redox reaction to produce the radical cation (or a fragment derived from it) and the dissociative proton transfer reaction (except when M is tryptophan, where no proton transfer occurs). The parent $M^{\bullet+}$ ions are observed directly for three of the aromatic amino acids (tryptophan, tyrosine, and histidine), for two of the basic amino acids (arginine and lysine), methionine, and the two amino acids that have an amide in the sidechain (asparagine and glutamine). The ion that complements the $M^{\bullet+}$ ion—$[Cu^{I}(M)]^+$—is observed for all the complexes.

Most of the $M^{\bullet+}$ ions that have a hydrogen on the carbon in the γ position relative to the carbonyl group undergo McLafferty rearrangement (McLafferty 1955; Djerassi et al. 1965), producing the most stable isomer of the glycine radical cation (**Gly3**) and an unsaturated neutral product (Scheme 9.5). Some of the amino acid radical cations also lose stable neutral molecules from the side chain.

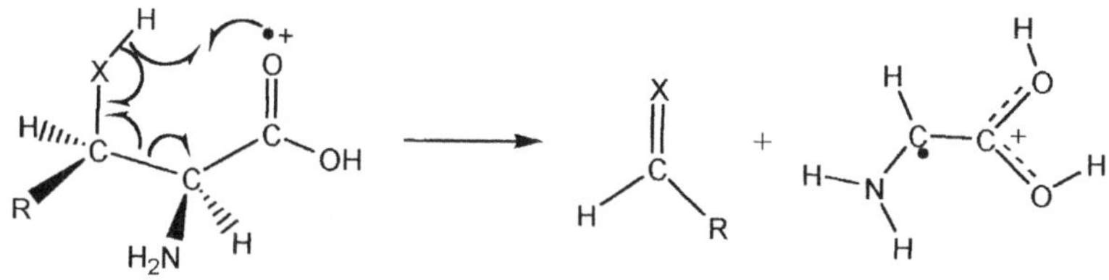

Gly3

Isoleucine:	X = $CHCH_3$, R = CH_3
Leucine:	X = $C(CH_3)_2$, R = H
Serine:	X = O, R = H
Threonine:	X = O, R = CH_3

SCHEME 9.5

9.4.2. Peptide Radical Cations Containing Only Glycine Residues

Loss of the tryptophan sidechain from TrpGlyGly$^{\bullet+}$ produces GlyGlyGly$^{\bullet+}$ in low abundance. Formally, the radical is located on the α carbon of the *N*-terminal residue and the proton is on the oxygen of the *N*-terminal amide bond to give $H_2NC^{\bullet}HC(OH)^{+}NHCH_2CONHCH_2COOH$. This captodative radical fragments into the $[b_2-H]^{\bullet+}$ ion, which can then lose CO to form the $[a_2-H]^{\bullet+}$ ion, as shown in Scheme 9.6 (Bagheri-Majdi et al. 2004).

9.4.3. Peptide Radical Cations Containing Only Glycine and Tryptophan Residues

CIDs of $[Cu^{II}(dien)(TrpGlyGly)]^{\bullet 2+}$ and $[Cu^{II}(TrpGlyGly)_2]^{\bullet 2+}$ give relatively high abundances of TrpGlyGly$^{\bullet+}$ and its decomposition product, $[TrpGlyGly - NH_3]^{\bullet+}$,

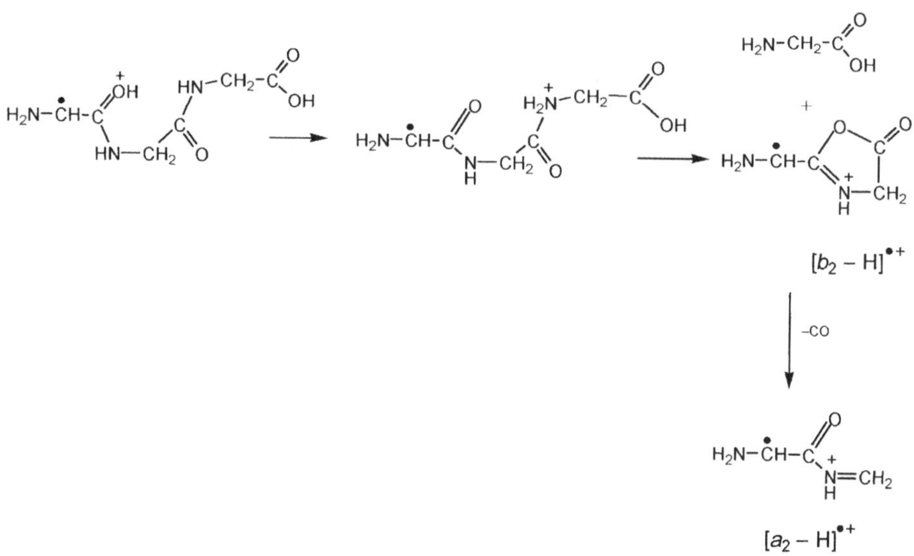

SCHEME 9.6

in moderate abundance; the proton transfer product, [TrpGlyGly + H]$^+$, a less fragile ion, is also produced in significant abundance (Bagheri-Majdi et al. 2004; Ke et al. 2005). In the CID of all other $[Cu^{II}(L)(M)]^{\bullet 2+}$ complexes, where L is either dien or a second M and M is GlyTrp, TrpGly, GlyTrpGly and GlyGlyTrp, *only* the dissociative redox reaction occurs. Why proton transfer is competitive only when the peptide is TrpGlyGly is slightly puzzling. Perhaps it is because protonation is on the terminal NH_2 group and tryptophan is more basic than glycine. However, there is no proton transfer to TrpGly. CIDs of these complexes are unusual in that, with the exception of $[Cu^{II}$(dien)(TrpGlyGly)$]^{\bullet 2+}$, none of them show any evidence of loss of CO_2, suggesting the existence of a canonical (nonzwitterionic) peptide in the complex and the lack of $COO^\bullet$ in the radical cation that is subsequently formed via CID.

The mode by which the peptide attaches to the copper may be important here. Glycine binds to Cu^+ in a bidentate manner through the NH_2 and carbonyl oxygen (structure **7**) (Hoyau and Ohanessian 1997; Marino et al. 2000). By contrast, $Cu^{\bullet 2+}$ prefers to bind with zwitterionic glycine (structure **8**) (Bertran et al. 1999; Pulkkinen et al. 2000; Hoppilliard et al. 2004). When a second glycine molecule binds to structure **8**, the charge on the copper has been sufficiently reduced that the preferred mode of attachment is through the NH_2 and carbonyl oxygen, as in structure **9** (Zhao et al. 2005). If the same type of binding occurs in complexes $[Cu^{II}(M_a)(M_b)]^{\bullet 2+}$, then one peptide will be attached through the carboxy group and the other through a combination of the amino group and the carbonyl oxygens of the amide bonds.

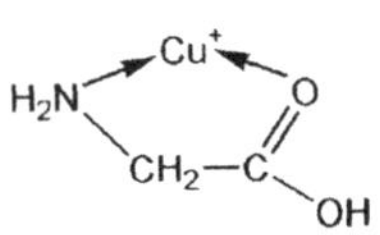

7

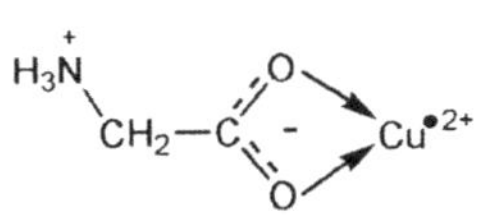

8

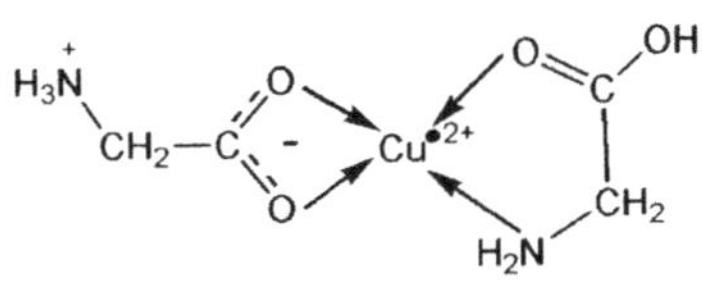

9

Peptides that have the tryptophan residue at the *N* terminal often dissociate as the radical cation. Fragmentation of complexes $[Cu^{II}(M_a)(M_b)]^{\bullet 2+}$, where M_a is TrpGly and M_b is a tripeptide having the various possible combinations of two glycine and one tryptophan residues, yields TrpGly$^{\bullet +}$ in high abundance; only when the tripeptide is TrpGlyGly is there a significant abundance of the other peptide radical cation. If the peptides are attached as in structure **9**, then the M that dissociates most likely will be the canonical one and the dissociation could be accompanied by electron transfer from the ligand to the copper, creating a π radical at the tryptophan residue. Why the possession of a tryptophan residue at the *N* terminal encourages complexation of the peptide in the canonical form is currently unclear. The proton affinity of tryptophan is about 42 kJ/mol higher than that of glycine (Maksic and Kovacevic 1999), and this should also encourage zwitterion formation for peptides with an *N*-terminal tryptophan.

There are some common pathways in the CIDs of radical peptides containing only glycine and tryptophan residues. For four of the five radical cations of the di- and tripeptides examined, the fragment in highest abundance is the $[z_n - H]^{\bullet +}$

ion, resulting from cleavage of the N—$^{\alpha}$C bond of the tryptophan residue. This reaction also occurs for the fifth radical cation, GlyTrpGly$^{\bullet+}$, but the product is in slightly lower abundance than that created by the loss of a glycine molecule from the *C* terminal, thus forming the $[b_2-H]^{\bullet+}$ ion (Bagheri-Majdi et al. 2004; Ke et al. 2005). The corresponding *O*-methyl esters also fragment by this pathway, ruling out the possibility that the reaction is initiated by a proton transfer from the carboxylic acid group. The proposed mechanism for peptides in which the tryptophan residue is *not* at the *C* terminus involves a proton shift from the CH_2 group of the tryptophan sidechain of the π radical to the nitrogen of the amide bond on the *N*-terminal side of the tryptophan residue. The required 1,3-proton shift has a high barrier [>125 kJ/mol at UB3LYP/6-31+G(*d*)], but this is circumnavigated by a 1,4-proton transfer to the oxygen of the amide bond at the *C*-terminal side of the tryptophan residue, followed by a second 1,4-proton shift to the amino nitrogen of the tryptophan residue (Scheme 9.7, illustrated by using an X = Gly, i.e., for ion GlyTrpGly$^{\bullet+}$). A neutral molecule then cleaves from the

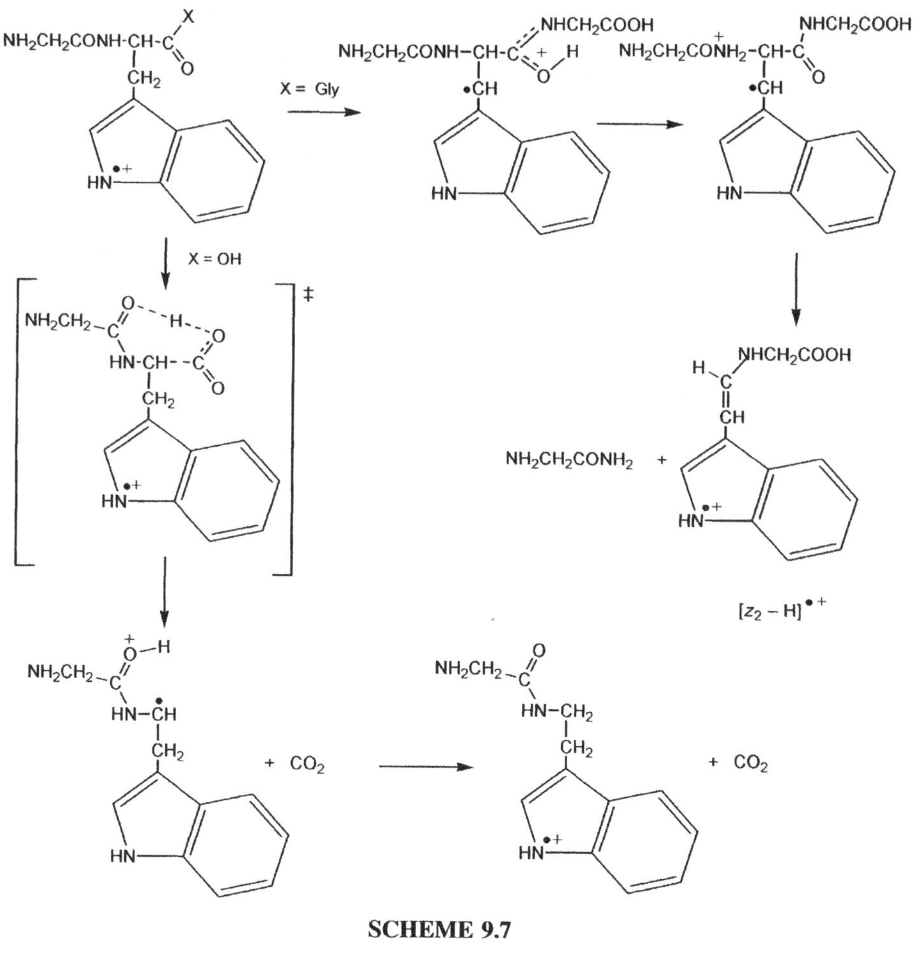

SCHEME 9.7

N terminal as RNH_2 (where R is H, H_2NCH_2CO, or $H_2NCH_2CONHCH_2CO$, depending on the location of the tryptophan residue), resulting in a $[z_n - H]^{\bullet +}$ ion with the charge and unpaired electron delocalized over the tryptophan ring and two exocyclic carbon atoms (Orlova et al. 2005). The $[z_2 - H]^{\bullet +}$ ions formed in the fragmentations of $TrpGly^{\bullet +}$ and $GlyTrpGly^{\bullet +}$ show identical subsequent fragmentation to lose CO_2 and H_2CNH (Bagheri-Majdi et al. 2004).

When the tryptophan residue is at the *C* terminus, the mechanism is slightly different. The carbonyl oxygen of the COOH group has a proton affinity lower than that of an amide, making the 1,4-proton shift a higher-energy process. In $GlyTrp^{\bullet +}$ the proton migrates from the CH_2 of the tryptophan sidechain to the oxygen of the amide bond (a 1,5-proton shift) and, in the rate-determining step [a barrier of 105.9 kJ/mol at UB3LYP/6-31+G(*d*)], the amide bond cleaves, forming glycinamide in its enol form and the $[z_1 - H]^{\bullet +}$ that involves the tryptophan residue (Orlova et al. 2005; Zhao et al. 2005). For radical cations $(Gly)_nTrp^{\bullet +}$, there is only one other dissociation channel: the loss of CO_2. The other three peptide radical cations that do not have the tryptophan residue at the *C* terminal do not lose CO_2. Density functional theory (DFT) calculations at UB3LYP/6-31+G(*d*) on $GlyTrp^{\bullet +}$ showed the mechanism to involve proton migration from the carboxy group to the amide oxygen synchronous with breaking the C—CO_2 bond (see Scheme 9.7). This rate-determining step was calculated to have a barrier of 118.4 kJ/mol at 298 K. The second step in this reaction, with a barrier lower than the first, has a 1,3-hydrogen shift and forms $[a_2 + H]^{\bullet +}$ with the charge and unpaired electron again delocalized over the ring (Bagheri-Majdi et al. 2004).

The CID spectra of all the peptides in which the tryptophan residue is *not* at the *C* terminus show the existence of a_n ions formed by breaking the amide bond at the *C*-terminal side of the tryptophan residue. Protonation of the amide nitrogen usually precedes cleavage of the amide bond. The initial structure in Scheme 9.8 has two electron-withdrawing groups attached to the radical center and, therefore, is likely to be a high-energy species; dissociation of this intermediate forms a captodative radical and an acylium ion that would be unstable to the loss of CO, resulting in an a_n ion (Zhao et al. 2005).

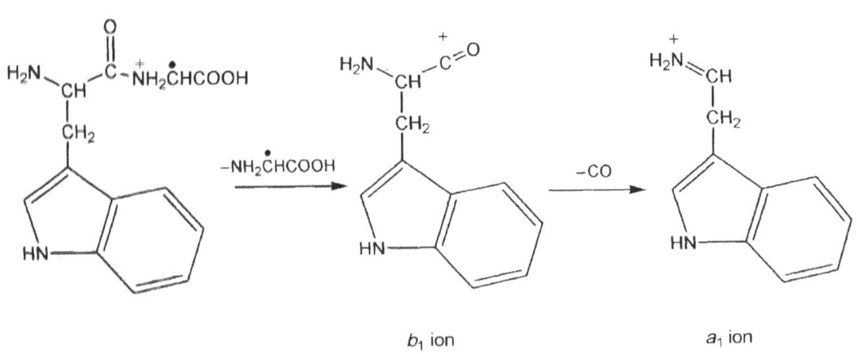

SCHEME 9.8

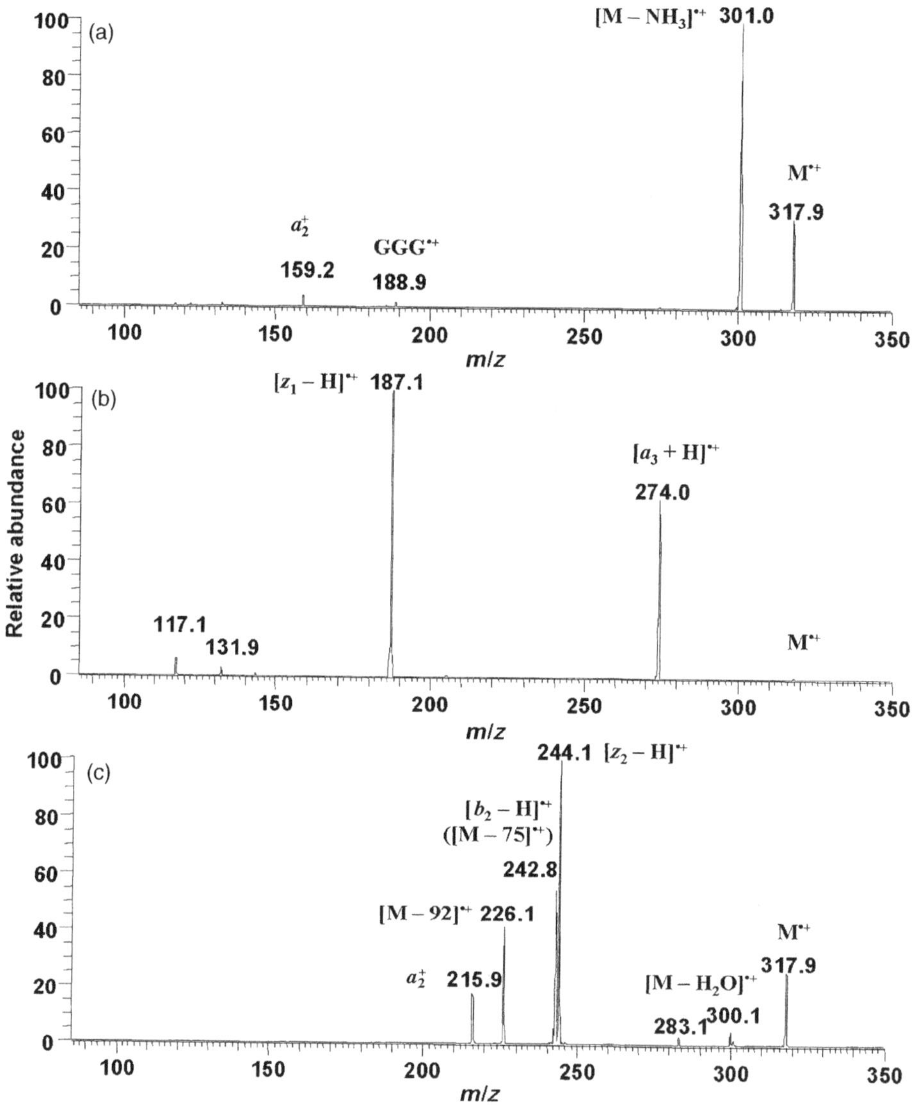

FIGURE 9.2. CID spectra of (a) $TrpGlyGly^{\bullet +}$, (b) $GlyGlyTrp^{+\bullet}$, and (c) $GlyTrpGly^{\bullet +}$ at relative collision energies of 10%. [Ke et al. (2005): Unpublished data.]

Radical cation $GlyTrpGly^{\bullet +}$ has the most complicated CID spectrum. In addition to formation of $[z_2 - H]^{\bullet +}$ and a_2 ions as described above, radical ions $[M - 75]^{\bullet +}$ and $[M-92]^{\bullet +}$ are also in high abundance (Figure 9.2). Both are probably oxazolones (Scheme 9.9).

Ion $[M-75]^{\bullet +}$ can be formed by protonation of the nitrogen of the *C*-terminal amide bond, followed by loss of the *C*-terminal glycine assisted by nucleophilic attack on the carbon of the incipient acylium ion by the oxygen of the other amide

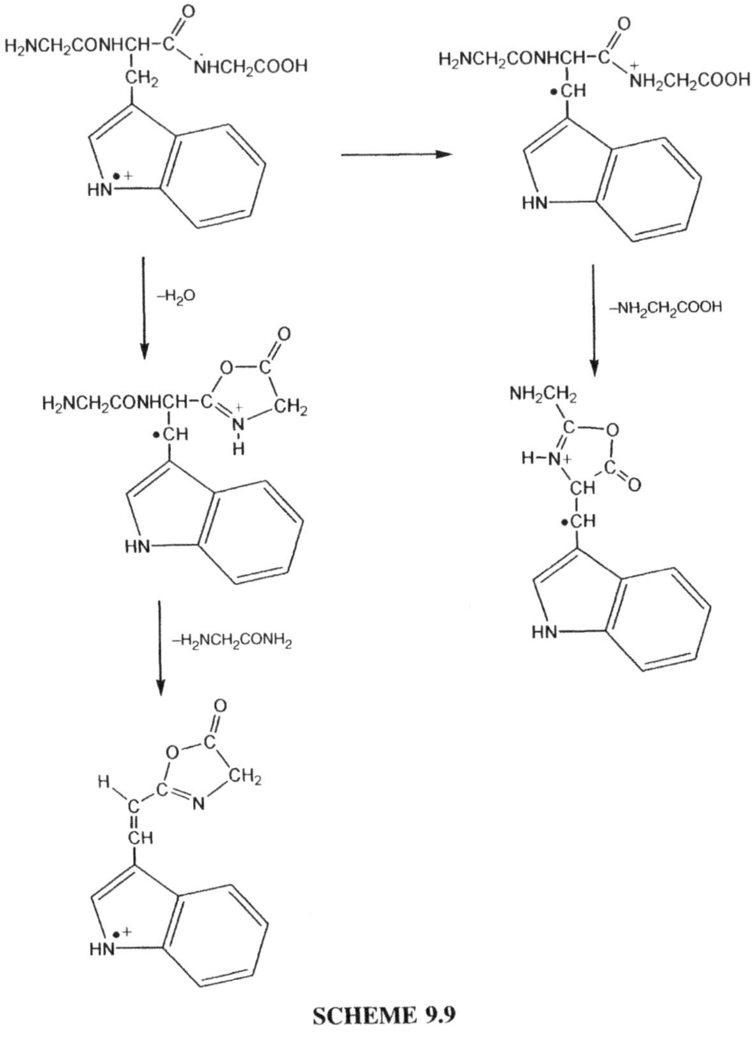

SCHEME 9.9

group. The resulting ion is the $[b_2 - H]^{\bullet+}$ ion (Bagheri-Majdi et al. 2004). The origin of the proton could again be the CH_2 group of the sidechain (as shown in Scheme 9.7), but could also be from the indole nitrogen or from the α carbon of the *N*-terminal glycine residue. The last site would produce the most stable oxazolone, a captodative radical with the formal site of the unpaired electron in conjugation with the terminal amino group and the protonated oxazolone ring. At first glance it would appear that the $[M-92]^{\bullet+}$ ion is formed by the loss of water from $[z_2 - H]^{\bullet+}$, but as alluded to earlier, CID of the latter showed losses of only CO_2 and $H_2C{=}NH$ (Bagheri-Majdi et al. 2004). The CID spectrum of GlyTrpGly$^{\bullet+}$ shows an ion of low abundance at 300 Th, (a Thomson, abbreviated to Th, is the unit of mass/charge) corresponding to the loss of water. Subsequent loss of $H_2NCH_2CONH_2$ (from the *N* terminus) produces the $[M - 92]^{\bullet+}$ ion.

9.4.4. Peptide Radical Cations Containing Only Glycine and Histidine Residues

Histidine deserves special attention because it combines two additional structural features that stabilize radical cations: an aromatic ring that is easily ionized and an imidazole ring that is strongly basic.

9.4.4.1. Histidine Radical Cation. Histidine residues act as ligands in many copper enzymes and proteins, and play a role in copper transportation in living systems. In proteins, copper attaches to the imino nitrogen of the imidazole ring, but in complexes of $Cu^{\bullet 2+}$ with either histidine or small peptides containing a histidine residue, there are other possibilities. As with any amino acid or *N*-terminal residue, coordination with the α-amino group is possible. In this regard, histidine is not unique. An additional interaction, specific to basic amino acids, is the possibility of coordination to an oxygen atom of the carboxylate anion that is part of the zwitterion in which the carboxylic proton has been transferred to a nitrogen atom of the sidechain. In a detailed NMR/DFT computational study of $Cu^{II}(His)_2$ complexes in solution, it was concluded that $Cu^{\bullet 2+}$ is coordinated to one of the histidine anions through both the α-amino and ring imino nitrogens and to the other through one of the carboxylate oxygens, with the fourth coordination site occupied by a water molecule (Baute et al. 2004). The implications of this study for the structure of the $[Cu^{II}(L)(M)]^{\bullet 2+}$ complexes used to generate peptide radical cations is that, assuming that the ligand occupies three coordination sites, the histidine residue may either attach to the Cu^{II} through the ring imino and/or α-amino nitrogen or may prefer to be zwitterionic and attach through the carboxylate anion.

Hu and Loo (1995) examined the CID of several angiotensin derivatives with transition metal dications (Zn^{2+}, Cu^{2+}, Ni^{2+}, and Co^{2+}). All these peptides contain at least one histidine residue, and they observed that the dominant fragmentations were at the amide bond at the *C* terminal of the histidine. From this they concluded that all the complexes had major interactions with the histidine residues. The only complexes that showed significantly different behavior were those of Cu^{2+}, where the most facile reaction was loss of CO_2. Acetylation of the *N*-terminal amino group and removal of aspartic acid residues did not prevent loss of CO_2, but amidation of the *C*-terminal carboxy group did. Furthermore, the $Cu^{\bullet 2+}$ complexes of peptides with a *C*-terminal amide group showed fragmentation patterns almost identical to those of their Zn^{2+}counterparts; specifically, $Cu^{\bullet 2+}$ appeared to interact with the amidated peptides in much the same way as do the other M^{2+} ions. From this, they concluded that in complexes with the underivatized peptides, Cu^{2+} interacts with the carboxylic group.

Complexes $[Cu^{II}(L)(M)]^{\bullet 2+}$, where M = His, HisGly, GlyHis, HisGlyGly, GlyHisGly, and GlyGlyHis and L is one of the tridentate nitrogen-containing ligands, dien, 1,4,7-triazacyclononane (tacn) and tpy, have been subject to CID. The $[Cu^{II}(L)(His)]^{\bullet 2+}$ complexes gave the simplest spectra (Ke et al. 2005) (see Figure 9.3). With dien as ligand, there was no detectable radical cation and only products of proton transfer from the ligand to histidine [reaction (9.2)] were

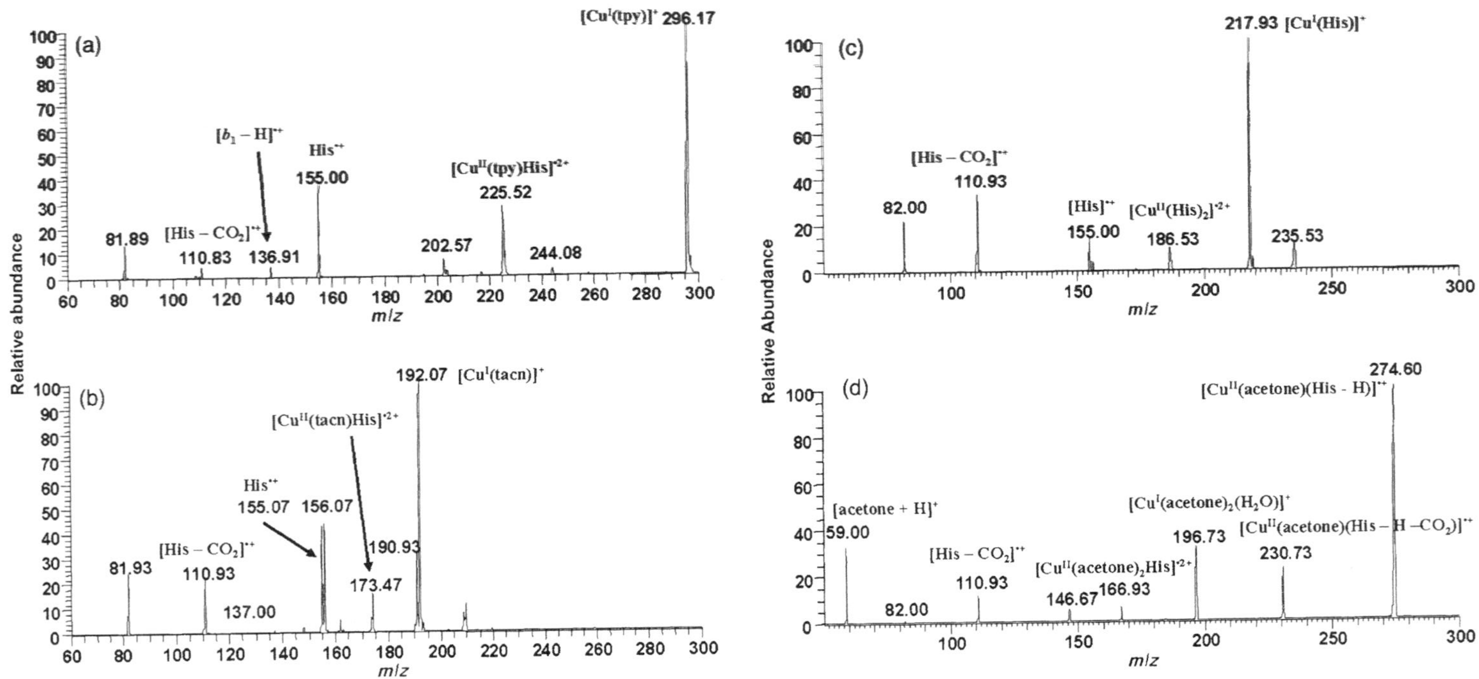

FIGURE 9.3. Product ion spectra of [^{63}Cu^{II}(L)(His)]$^{•2+}$, where L is (a) 2,2′:6′,2″-terpyridine; (b) 1,4,7-triazacyclononane; (c) histidine; and (d) two acetone molecules, at relative collision energies 9%, 8%, 7%, and 6%, respectively. [Ke et al. (2005): Unpublished data.]

observed. With tacn as ligand, both reactions (9.1) and (9.2) occurred and, using the relative abundances of the more stable complementary ions $[Cu^{I}(L)]^{+}$ and $[Cu^{II}(L - H)]^{\bullet +}$ as indicators, formation of $His^{\bullet +}$ is the dominant reaction. The abundances of $His^{\bullet +}$ (155 Th) and $[His + H]^{+}$ were comparable, but two additional ions at 111 and 82 Th, not present in the CID of $[His + H^{+}]^{+}$, were attributed to fragmentation of $His^{\bullet +}$. These correspond to losses of CO_2 followed by aminocarbene H_2NCH. CID of isolated $His^{\bullet +}$ did not lead to the loss of CO_2, but instead that of water, resulting in the $[b_1 - H]^{\bullet +}$ ion at 137 Th. This resulting "acylium" ion is unusually stable, but does lose CO at higher collision energies to form the $[a_1 - H]^{\bullet +}$ ion at 109 Th. With the strongest ligand, tpy, only the dissociative redox reaction (9.1) occurred. The $His^{\bullet +}$ ion at 155 Th was very prominent; the ions at 111 and 82 Th were relatively minor. One other ligand, a weaker one, acetone, was used and CID of $[Cu^{II}(CH_3COCH_3)_2\ (His)]^{\bullet 2+}$ gave the products of the dissociative redox reaction, but *only* ions at 111 and 82 Th.

These results point to the formation of at least two isomeric histidine radical cations, resulting from two different modes of complexation. Assuming that the ligand (tpy) is tridendate and the complex is five-coordinate, the two proposed structures are **10**, in which histidine is canonical and binds to Cu^{II} via the ring imino nitrogen and the α-amino nitrogen atoms, and **11**, in which histidine is zwitterionic and binds to Cu^{II} via a carboxy oxygen and the α-amino nitrogen atoms. Starting with **10**, the dissociative redox reaction would produce **His1**, a radical cation in which the charge and spin are initially both formally localized on the imino nitrogen; starting with complex **11**, the dissociation would give **His2**, a distonic ion.

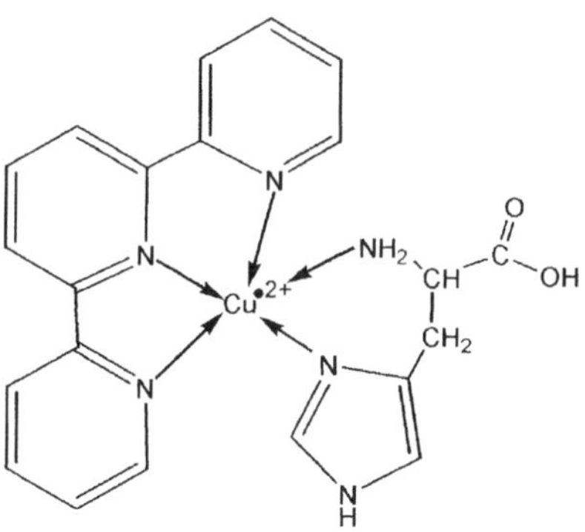

10

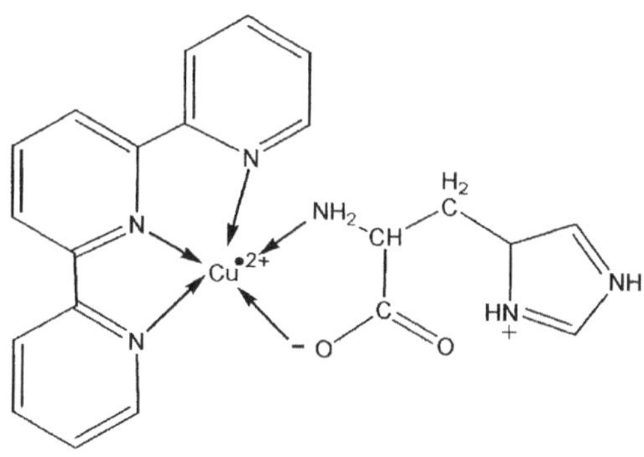

11

Density functional calculations at UB3LYP/6-311++G(*d*,*p*) show **His2** to lie 112 kJ/mol below **His1**. At lower levels of theory, the structure for **His2** has an intact C–CO_2 bond, but at UB3LYP/6-311++G(*d*,*p*) the only structure at a minimum has the CO_2 molecule loosely solvating a distonic ion in which the imidazole is protonated and the radical center is on the α carbon (adjacent to the NH_2 group). Dissociation of this complex is endothermic by only 14 kJ/mol. A further loss of aminocarbene is endothermic by 231 kJ/mol, but if the neutral lost is $H_2C{=}NH$, then it is endothermic by only 88 kJ/mol.

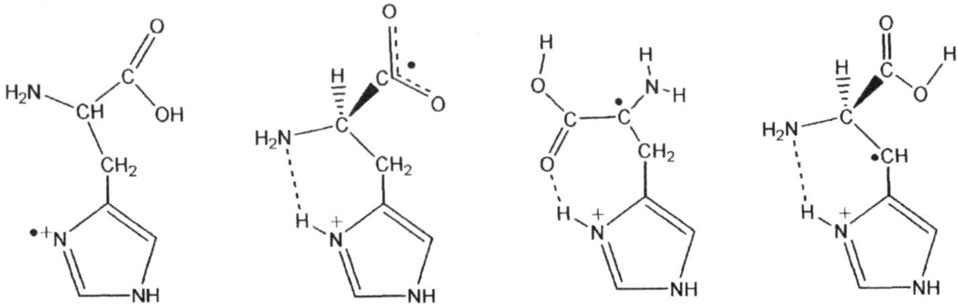

His1 **His2** **His3** **His4**

The dissociation reactions shown in Scheme 9.10 can account for two of the dissociation products at 111 and 82 Th. As the barrier against dissociation to give the 111-Th ion is only 14 kJ/mol, **His2** is probably too fragile to be isolated experimentally. **His1**, formed initially from fragmentation of structure **10**, can convert to **His2**, which requires two internal rotations, one about the CH_2—C_α bond and the other about C–OH. The 1,6 H-shift from this final conformation has a negligible barrier (1.4 kJ/mol), but forming the required conformation to initiate the migration has a barrier of approximately 58 kJ/mol.

His3, the α radical, is at the global minimum, lower in energy than **His2** by 19.2 kJ/mol. This structure benefits from two strongly stabilizing features. The proton is located on the strongly basic imidazole ring and is involved in a strong hydrogen bond to the carbonyl oxygen, effectively transferring a large amount of positive charge onto the carboxy group thereby enriching the captodative character. Conversion of **His1** into **His3** requires a 1,4 H-shift from the α-carbon to the imidazole nitrogen. The barrier against this process is calculated to be only 35.1 kJ/mol, lower than the energy required for conversion into **His2** by 23 kJ/mol.

The $His^{\bullet+}$ ion that is sufficiently stable to isolate and subject to subsequent CID loses water and, at higher collision energies, CO. This combination of losses is common to all protonated amino acids, although the losses are usually concomitant because of the instability of the intermediate acylium ion. The b_1 ion derived from protonated histidine is an exception, and it has been suggested that it is stabilized by cyclization in which the carbonyl group is attached to the imino nitrogen (Farrugia

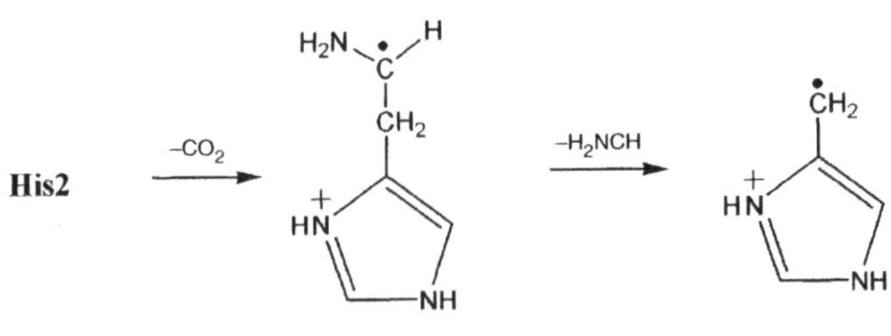

111 Th 82 Th

SCHEME 9.10

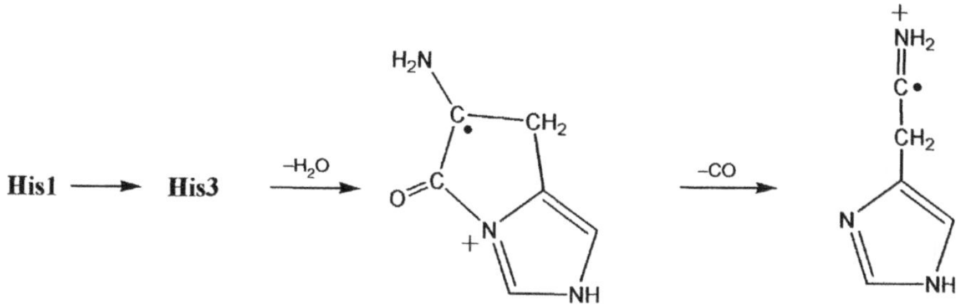

$[b_1 - H]^{\bullet+}$, 137 Th $[a_1 - H]^{\bullet+}$, 109 Th

SCHEME 9.11

et al. 2001). Under similar collision conditions, $His^{\bullet+}$ and $[His + H]^+$ ions both lose water followed by loss of CO (Ke et al. 2005). The second step, however, requires considerably higher energy for $His^{\bullet+}$, implying that the $[b_1 - H]^{\bullet+}$ ion is less fragile than the b_1 ion derived from $[His + H]^+$. Scheme 9.11 depicts a proposed reaction sequence that involves **His1** (from the dissociation of structure **10**) converting into **His3**, which then dissociates to give the $[b_1 - H]^{\bullet+}$ ion and $[a_1 - H]^{\bullet+}$ ion. The extra stability of the $[b_1 - H]^{\bullet+}$ ion is attributed to its captodative character. The carbonyl group attached to the positively charged imidazole is strongly electron-withdrawing, and the amino group is a powerful electron donor. The $[a_1 - H]^{\bullet+}$ ion created by the loss of CO is distonic, but the CH_2 group blocks the conjugation required for it to be captodative. A 1,2 H-shift would create a lower energy benzylic-like radical that is also captodative.

Formation of a fourth possible isomer, namely, **His4**, created by a 1,3 H-shift from the CH_2 group to the ring imino nitrogen of **His1**, is 67.8 kJ/mol exothermic. As from **His3**, the proton on the imidazole can be transferred to the OH of the carboxylic group and lead to the loss of water. The resulting cyclic ion, however, has the radical center between the saturated α carbon and the imidazole (where the charge is formally located) and is, therefore, distonic but not captodative, making it less stable than the $[b_1 - H]^{\bullet+}$ ion shown in Scheme 9.11.

9.4.4.2. Peptides HisGly$_n$. The dominant pathway for fragmentation of ions $[Cu^{II}(tpy)(M)]^{\bullet 2+}$, where M is HisGly or HisGlyGly, is by cleavage of the peptide chain at the histidine residue to create $[Cu^{II}(tpy)(M - b_1)]^{\bullet+}$ and the b_1 ion (Wee 2005; Ke et al. 2005). This requires the addition of a proton to the nitrogen of the *N*-terminal amide bond and is most easily achieved by a 1,6 H^+-shift from the imidazole as shown in Scheme 9.12 (for the complex in which M = HisGly). The precursor ion has the peptide in the zwitterionic form, a structure that delocalizes the positive charge around the complex. Indirect support for this structure is obtained from the CID of the complexes containing the peptide *O*-methyl ester, where binding with the carboxylate anion is not possible and no b_n ions are formed.

In addition, CIDs also produce low abundances of $HisGly^{\bullet+}$ and $HisGlyGly^{\bullet+}$; however, the complementary ion of the dissociative redox reaction, $[Cu^I(tpy)]^+$, is

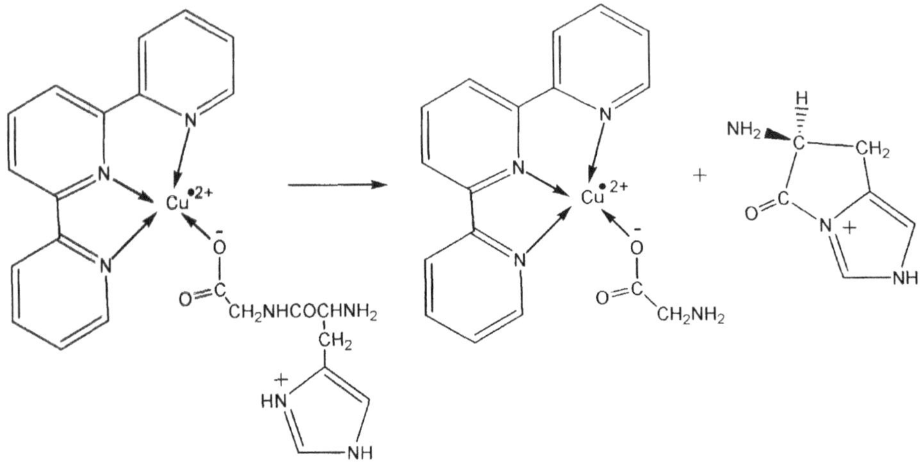

SCHEME 9.12

present in higher abundance, indicating that the peptide radical cations are fragile, as expected for the carboxy radical that would be formed by homolytic fission of the Cu—O bond in the complex shown in Scheme 12. Losses of CO_2 and $H_2C{=}NH$ from the radical cationic peptides produce the same types of ion as in the fragmentation of the radical cation of histidine (Scheme 9.11).

The fragmentation of HisGly$^{\bullet+}$ is uncomplicated, yielding abundant amounts of the $[b_1 - H]^{\bullet+}$ ion by the loss of glycine and lower abundances of $[a_1 - H]^{\bullet+}$. Similarly, fragmentation of HisGlyGly$^{\bullet+}$ produces both $[b_2 - H]^{\bullet+}$ and $[b_1 - H]^{\bullet+}$ in higher abundances, with lower abundances of the corresponding $[a_n - H]^{\bullet+}$ ions. These chemistries are most easily rationalized by formation of α radicals of the **His3** type, followed by the loss of glycine or diglycine. The $[b_2 - H]^{\bullet+}$ (assuming that it is an oxazolone) and $[b_1 - H]^{\bullet+}$ ions both have captodative structures.

9.4.4.3. GlyHisGly. The prevalent reaction in the CID of $[Cu^{II}(tpy)(GlyHisGly)]^{\bullet 2+}$ is cleavage at the amide bond *C*-terminal to the histidine, as in the complexes with peptides HisGly$_n$, yielding in this case the $[b_2 - H]^{\bullet+}$ in high abundance, with a much lower abundance of $[a_2 - H]^{\bullet+}$. This is most easily rationalized in terms of an α radical formed at the *N*-terminal glycine residue, followed by proton transfer from the imidazole imino nitrogen to the *C*-terminal amide nitrogen and cleavage to form $[b_2 - H]^{\bullet+}$, which probably has an oxazolone structure with the radical center on the carbon of the sidechain. Again, the peptide radical cation, GlyHisGly$^{\bullet+}$, is produced in low abundance with the complementary ion, $[Cu^{I}(tpy)]^{+}$, in considerably higher abundance.

Two additional fragment ions are observed in low abundance; both can be formed from the α radical formed at the *N*-terminal glycine residue. Proton transfer from the protonated imidazole ring to the terminal amino group, followed by the loss of NH_3, produces the $[z_1 - H]^{\bullet+}$ ion. Alternatively, proton transfer to the nitrogen of the *N*-terminal amide bond results in cleavage of the C—N bond, yielding HisGly and a

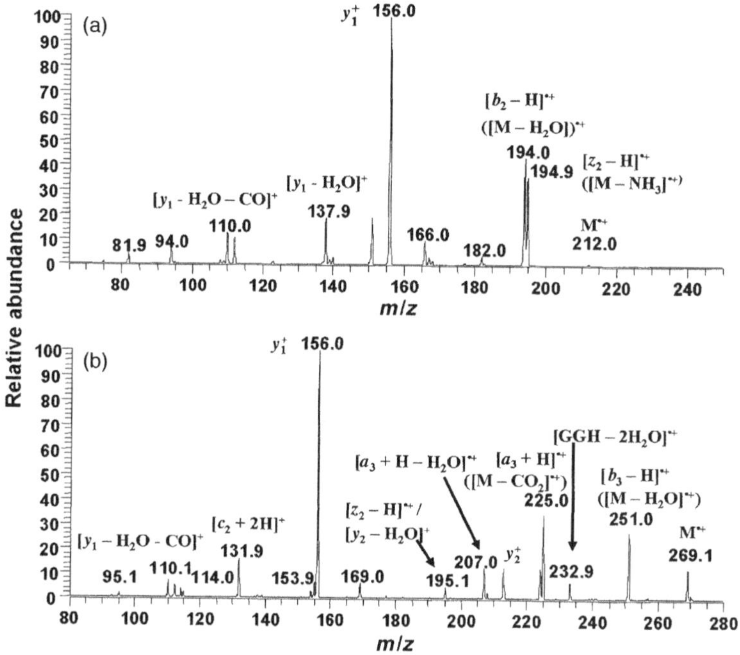

FIGURE 9.4. CID spectra of (a) $GlyHis^{•+}$ and (b) $GlyGlyHis^{•+}$ at relative collision energies of 12%. [Ke et al. (2005): Unpublished data.]

captodative ion, $H_2NCH^•CO^+$. Proton transfer between the incipient species, from $H_2NCH^•CO^+$ to the highly basic HisGly, results in the y_2 ion.

9.4.4.4. Peptides Gly$_n$His. When the histidine residue is at the *C* terminus, the dissociative redox reaction is the dominant channel. The most abundant ion is $[Cu^I(tpy)]^+$, but the abundances of $GlyHis^{•+}$ and $GlyGlyHis^{•+}$ are high (approximately 70%); evidently, these ions are more stable than other isomeric ions in which the histidine residue is not at the *C* terminus.

The fragmentation of $GlyHis^{•+}$ bears some similarities to that of $[GlyHis + H]^+$, forming $[b_2 - H]^{•+}$, $[a_2 - H]^{•+}$, y_1, $[y_1 - H_2O]$, and $[y_1 - H_2O - CO]$ (Figure 9.4). One major difference is that the radical cation loses NH_3 to give the $[z_2 - H]^{•+}$ ion, and also concomitantly NH_3 and CO_2. $[z_2 - H]^{•+}$ ions have been observed in the CIDs of tryptophan-containing peptides (Bagheri-Majdi et al. 2004). Another additional product, in low abundance, is the $[y_1 - CO_2]$ ion. These last products are most easily understood in terms of $GlyHis^{•+}$ being a carboxy radical, while the y_1 ion is most easily formed from the structure in which there is an α radical at the *N*-terminal glycine residue (Scheme 9.13). The mechanism by which $GlyHis^{•+}$ forms the y_1 ion could involve either proton transfer from the imidazole to the amide nitrogen, cleavage of the amide bond, followed by proton transfer back to the incipient histidine from the incipient $H_2NCHCO^{•+}$, or a single step in which $H^•$ transfers from the terminal nitrogen to the amide nitrogen concomitant with cleavage of the amide bond. The

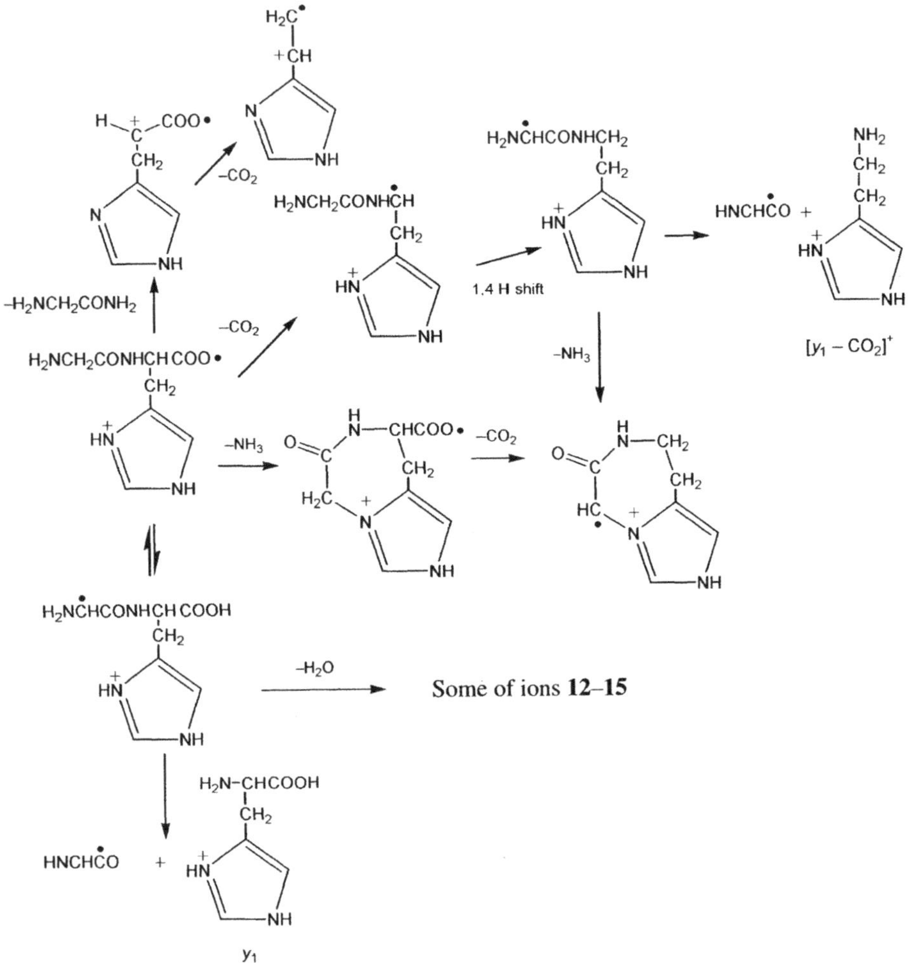

SCHEME 9.13

latter odd-electron process is the more likely in light of the findings of Wee et al. (2004) on the fragmentation of the GlyXxxArg$^{\bullet+}$ ions.

Loss of water from GlyHis$^{\bullet+}$ can occur from two sites, either from the carboxylic group to form a classic $[b_2 - \mathrm{H}]^{\bullet+}$ ion at 194 Th [either a diketopiperazine **12**, as suggested for $[\mathrm{GlyHis} + \mathrm{H}]^+$ (Farrugia et al. 2001), or an oxazolone **13**], or an *N*-acylated imidazole **14** (as suggested for the intermediate formed by the loss of water from protonated histidine); alternatively, the water may be extracted from the peptide bond to form a nitrilium ion (**15**). Isomers **13** and **15** have captodative structures and the other two ions are distonic.

In the CID of GlyGlyHis$^{\bullet+}$, the y_1 ion is again the most abundant, followed by the $[a_3 + \mathrm{H}]^{\bullet+}$ (or $[\mathrm{M} - \mathrm{CO_2}]^{\bullet+}$), $[b_3 - \mathrm{H}]^{\bullet+}$ (or $[\mathrm{M} - \mathrm{H_2O}]^{\bullet+}$), and y_2 ions (Figure 9.4). There is no ion resulting from the loss of only NH_3, but the ion at 195 Th could be

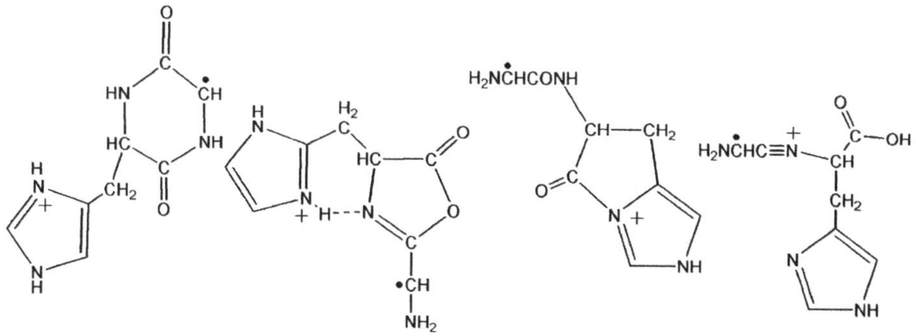

12 13 14 15

assigned as the $[z_2 - H]^{\bullet+}$ ion resulting from the loss of $H_2NCH_2CONH_2$. An alternative assignment is the $[y_2 - H_2O]^+$ ion. The 224 Th ion may arise by the loss of $COOH^\bullet$ from $M^{\bullet+}$, or by the loss of $H^\bullet$ from the $[a_3 + H]^{\bullet+}$ ion. The ion at 132 Th is probably protonated glycylglycinamide or the $[c_2 + 2H]^+$ ion.

9.4.5. Radical Cations of Tripeptides GlyXxxArg

The peptides in the complexes $[Cu^{II}(L)(M)]^{\bullet 2+}$ that led to the initial discovery of this versatile route for formation of peptide radical cations were enkephalin derivatives (Chu et al. 2000, 2001). All had either a tyrosine or tryptophan residue [both known to be radical sites in proteins (Stubbe and van der Donk 1998)] and a basic amino acid residue at the *C* terminal (lysine or arginine). CID of the complex $[Cu^{II}(L)(TyrAlaGlyPheLeuArg)]^{\bullet 2+}$ yielded the radical cation in high abundance, while the similar complex with the pentapeptide TyrAlaGlyPheLeu gave no radical cation. From this, it was initially concluded that two key requirements for formation of the radical cation were that the peptide must contain (1) a residue with low ionization energy and (2) a basic amino acid. The choice of the ligand L seemed less crucial, although it was quickly realized that some ligands reduced or even removed competitive channels.

A consequence of these early discoveries was that much effort was put into examining peptides that contain a basic residue. Wee et al. (2002 and 2004) examined the series GlyXxxArg, where Xxx is a naturally occurring amino acid. One advantage in using this approach is that, because arginine has the highest proton affinity, assessing the location of the positive charge becomes relatively simple thereby making mechanistic considerations easier. Another advantage is that the products of tryptic digests all have either a lysine or arginine residue at the *C* terminus; the $GlyXxxArg^{\bullet+}$ ions in this study provide a good model for the fragmentation of some of these commonly encountered peptides.

In the CID of the $[Cu^{II}(tpy)(GlyXxxArg)]^{\bullet+}$ complexes, the $[Cu^{I}(tpy)]^+$ ion generally had the highest abundance and the complementary ion $GlyXxxArg^{\bullet+}$ frequently had the second highest abundance. There was only one notable exception;

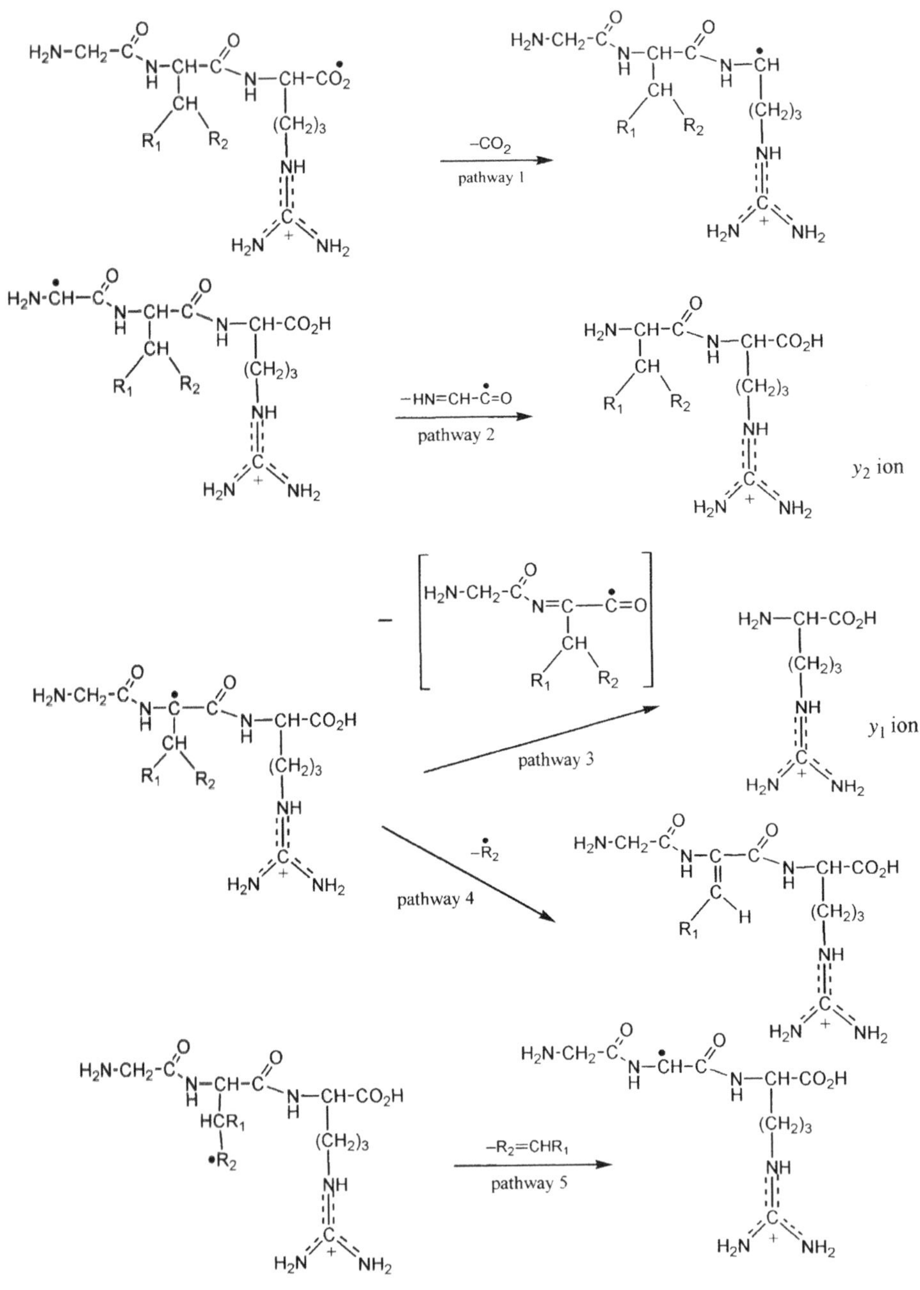

SCHEME 9.14

when Xxx was Arg, no peptide radical cation was formed. Most of the complexes gave some $[GlyXxxArg-CO_2]^{\bullet +}$, perhaps indicating that the peptides in the complexes are zwitterionic and are attached to Cu^{II} through the carboxylate anion with the acidic proton having been transferred to the guanidinyl group at the end of the arginine sidechain. Loss of ammonia also occurred from most complexes; here there were two possible origins: the terminal NH_2 group or the arginine sidechain. As protonated arginine loses NH_3 from its sidechain at relatively low collision energies, this seems to be the more likely origin.

Fragmentation of the GlyXxxArg$^{\bullet+}$ ions follow many different pathways, depending on the composition of the sidechain. Loss of CO_2 is observed from some ions, but only as a minor product (pathway 1 in Scheme 9.14). This indicates that, if the initially formed ion is a carboxy radical with the sidechain of the arginine residue protonated, then intramolecular hydrogen atom migration must occur rapidly, that is, the radical center is "mobile." Most of the products have been rationalized in terms of structures in which the argininine sidechain remains protonated and a hydrogen atom has migrated to the carboxy group from either the sidechain of Xxx or from one of the α carbons of the other two residues. The pathways used to account for the major products are given in Scheme 9.14 (Wee et al. 2004).

Pathway 1. In the CID of GlyXxxArg$^{\bullet+}$, loss of CO_2 occurs only when Xxx is Gly, Ala, Val, Phe, Trp, Asp, His, and Lys, and it is a minor pathway except when Xxx is Trp and Phe (60% and 45%, respectively, of the most abundant ion in the latter two dissociations).

Pathway 2. The most stable captodative radical would be formed, if the α hydrogen of the glycine residue migrates to the carboxy group, and the protonated arginine sidechain forms a hydrogen bond with the oxygen of the *N*-terminal amide bond. This intermediate has a hydrogen missing from the glycine residue and, logically, is the precursor to the y_2 ion, a product observed in the CID of *all* GlyXxxArg$^{\bullet+}$. However, the abundances of the y_2 ions are relatively low (<30% of the most abundant ion), except when Xxx is Gly (40%), Val (38%), Iso (30%), Phe(40%), and Trp(80%). In order to obtain the y_2 ion, it was proposed that a purely radical rearrangement occurs around the *N*-terminal residue of the *N*-terminal α-radical. A 1,4 H-migration from NH_2 to the amide nitrogen accompanied by homolytic cleavage of the amide bond creates a neutral radical, $HN{=}CH{-}C^{\bullet}{=}O$, and the y_2 ion. An alternate mechanism involving proton transfer from the arginine sidechain to the amide nitrogen, followed by amide bond cleavage and proton transfer back to the dipeptide XxxArg, is improbable because the CID of $[GlyXxxArg + H]^+$ does not give any y_2 ions. It is noteworthy that for Xxx = Tyr, this is the only operative pathway, a consequence probably of the high stability of the intermediate phenoxy ion.

Pathway 3. The y_1 ion is a major product in the CID spectra of all GlyXxxArg$^{\bullet+}$ ions, with the exception of [GlyTyrArg]$^{\bullet+}$. In Scheme 9.14, the y_1 ion is shown to be formed from the α radical of the central amino acid residue. Starting from this distonic ion, the same type of radical rearrangement as postulated in the formation of the y_2 in pathway 2 leads to the y_1 ion, again with an acyl radical as the neutral product.

Pathway 4. A second pathway involving the intermediacy of the α radical of the central amino acid residue is used to explain the loss of a radical from the β position of the sidechain, leaving a product ion that has an alkene on the sidechain of the central residue. This dissociation pathway is dominant for ions in which Xxx is Glu,

Gln, Asn, Cys, Met, Leu, and Ile. An apparent anomaly here is that, when the central residue is aspartic acid, there is *no* loss of $CH_2COOH^•$, while for the corresponding amide, asparagine, loss of the isoelectronic radical, $CH_2COONH_2^•$ is the dominant reaction. The difference in reaction paths is explained in terms of the proton of the COOH group in the aspartic acid residue's sidechain being hydrogen-bonded to the nitrogen of the *C*-terminal amide, thereby facilitating the formation of the y_1 ion, which is by far the most abundant product in the dissociation of $GlyAspArg^{•2+}$. In the case of isoleucine, two possible radicals, $CH_3^•$ and $CH_3CH_2^•$, could be lost and both channels are followed. However, loss of $CH_3CH_2^•$ is the dominant pathway, in keeping with the greater stability of this larger radical. Fragmentations of radical cations of this type enable us to distinguish between peptides that contain leucine and the isomeric isoleucine residues. In peptide sequencing using protonated or diprotonated peptides that contain leucine or isoleucine residues, the usual products are b_n and y_m ions in which the sidechains remain intact. From this type of analysis, the products containing a leucine or isoleucine are isobaric and it is impossible to fully characterize a peptide. By contrast, in the CID of $GlyIleArg^{•+}$ the most abundant ion comes from loss of $C_2H_5^•$, while for $GlyLeuArg^{•+}$ the major loss is $(CH_3)_2CH^•$, making distinguishing between the two peptides straightforward (Wee et al. 2004).

Pathway 5. This pathway is observed only when there is a mobile hydrogen at the γ position or further along the sidechain of the central residue. Migration of a hydrogen atom to the initially formed carboxy radical from a heteroatom (when Xxx is Tyr, Cys, Ser, and Thr), from a CH_2 adjacent to a heteroatom (when Xxx is Met and Lys), and from the γ carbons of the leucine and isoleucine residues (creating tertiary and secondary radicals, respectively) forms relatively stable intermediates that dissociate by losses of alkenes, leaving a glycine α radical as the central residue. Here, there are some similarities with the CID of the amino $acid^{•+}$ ions. The radical cations leucine, isoleucine, threonine, and serine all undergo McLafferty rearrangement to form the carbon-centered glycine radical cation, $NH_2CH^•C(OH)_2^+$ and a neutral molecule containing a double bond.

9.5. CONCLUSIONS

Peptide radical cations have rich and fascinating collision-induced dissociation chemistries. These reactions result in product ions that often are very different from their protonated counterparts. It is hoped that the CID of peptide radical cations, once better understood, will offer complementary information to the CID of protonated peptides, which has become the major technique for sequencing peptides in proteomics.

ACKNOWLEDGMENT

In writing this chapter, we drew on considerable unpublished work from the O'Hair group detailed in the PhD thesis of Sheena Wee. Some of Dr. Wee's findings will be published in a special issue of the *International Journal of Mass Spectrometry* in honor of Chava Lifshitz, who was a key proponent of this monograph. We are grateful to Drs. Wee, O'Hair, and McFadyen for making a preprint available to us. We also made extensive use of unpublished materials from our former and current group members: Galina Orlova, Yuyong Ke, and Junfang Zhao. This chapter and our research described herein were made possible by financial support from the Natural Sciences and Engineering Council of Canada (NSERC) of Canada and MDS SCIEX.

REFERENCES

Aebersold R, Goodlett DR (2001): Mass spectrometry in proteomics. *Chem Rev.* **101**: 269–295.

Aliaga C, Lissi EA (2000): Reactions of the radical cation derived from 2,2′-azinobis (3-ethylbenzo-6-sulfonic acid ($ABTS^{\bullet+}$) with amino acids. Kinetics and mechanism. *Can. J. Chem.* **78**:1052–1059.

Bagheri-Majdi E (2003): *Fragmentation of Cu(II)-Peptide Complexes as a Route to Molecular Radical Cations of Peptides*, MSc thesis, York Univ.

Bagheri-Majdi E, Ke Y, Orlova G, Chu IK, Hopkinson AC, Siu KWM (2004): Copper-mediated peptide radical ions in the gas phase. *J. Phys. Chem. B* **108**:11170–11181.

Barlow CK, Wee S, McFadyen WD, O'Hair RAJ (2004): Designing copper (II) ternary complexes to generate radical cations of peptides in the gas phase. Role of the auxiliary ligand. *Dalton Trans.* **2004**:3199–3204.

Barlow CK, McFadyen WD, O'Hair RAJ (2005): Formation of cationic peptide radicals by gas-phase redox reactions with trivalent chromium, manganese, iron and cobalt complexes. *J. Am. Chem. Soc.* **127**:6109–6115.

Baute D, Arielli D, Neese F, Zimmermann H, Weckhuysen BM, Goldfarb D (2004): Carboxylate binding in copper histidine complexes in solution and in zeolite Y: X- and W-band pulsed EPR/ENDOR combined with DFT calculations. *J. Am. Chem. Soc.* **126**:11733–11745.

Becker CH, Wu KJ (1995): On the photoionization of large molecules. *J. Am. Soc. Mass Spectrom.* **6**:883–888.

Beranova S, Cai J, Wesdemiotis C (1995): Unimolecular chemistry of protonated glycine and its neutralized form in the gas phase. *J. Am. Chem. Soc.* **117**:9492–9501.

Berces A, Nukada T, Margle O, Ziegler (1999): Solvation of Cu^{2+} in water and ammonia. insight from static and dynamic density functional theory. *J. Phys. Chem. A* **103**:9693–9701.

Bertran J, Rodriguez-Santiago L, Sodupe M (1999): The different nature of bonding in Cu^{+}-glycine and Cu^{2+}-glycine. *J. Phys. Chem. B* **103**:2310–2317.

Bordwell FG, Zhang X, Alnajjar MS (1992): Effects of adjacent acceptors and donors on the stabilities of carbon-centered radicals. *J. Am. Chem. Soc.* **114**:7623–7629.

Bossio RE, Hudgins RR, Marshall A (2003): Gas-phase chemistry can distinguish different conformations of unhydrated photo-affinity labeled peptide ions. *J. Phys. Chem. B* **107**:3284–3289.

Bouma WJ, Nobes RH, Radom L (1982a): The methylenoxonium radical cation ($CH_2OH_2^{+\bullet}$): A surprisingly stable isomer of the methanol radical cation. *J. Am. Chem. Soc.* **104**:2929–2930.

Bouma WJ, Macleod JK, Radom L (1982b): Experimental evidence for the existence of a stable isomer of $CH_3OH^{+\bullet}$: The methylenoxonium radical cation, $CH_2OH_2^{+\bullet}$. *J. Am. Chem. Soc.* **104**:2930–2931.

Burgess VA, Easton CJ, Hay MP (1989): Selective reaction of glycine residues in hydrogen transfer from amino acid derivatives. *J. Am. Chem. Soc.* **111**:1047–1052.

Chaparro AL, Vachet, RW (2003): Tandem mass spectrometry of Cu(II) complexes: the effects of ligand donor group on dissociation. *J. Mass Spectrom.* **38**:333–342.

Chu IK, Rodriquez CF, Lau TC, Hopkinson AC, Siu KWM (2000): Molecular radical cations of oligopeptides. *J. Phys. Chem. B* **104**:3393–3397.

Chu IK, Rodriquez CF, Hopkinson AC, Siu KWM, Lau TC (2001): Formation of molecular radical cations of enkephalin derivatives via collision-induced dissociation of electrospray generated copper (II) complex ions of amines and peptides. *J. Am. Soc. Mass Spectrom.* **12**:1114–1119.

Chu IK, Siu SO, Lam CNW, Chan JCY, Rodriquez CF (2004): Formation of molecular radical cations of aliphatic tripeptides from their complexes with Cu^{II}(12-crown-4). *Rapid Commun. Mass Spectrom.* **18**:1798–1802.

Chu IK, Lam CNW, Siu SO (2005): Facile generation of tripeptide radical cations in vacuo via intramolecular electron transfer in Cu^{II} tripeptide complexes containing sterically encumbered terpyridine ligands. *J. Am. Soc. Mass Spectrom.* **16**:763–771.

Combariza MY, Vachet RW (2002): Gas-phase ion-molecule reactions of transition metal complexes: The effect of different coordination spheres on complex reactivity. *J. Am. Soc. Mass Spectrom.* **13**:813–825.

Croft AK, Easton CJ, Radom L (2003): Design of radical-resistant amino acid residues: A combined theoretical and experimental investigation. *J. Am. Chem. Soc.* **125**:4119–4124.

Cui W, Hu Y, Lifshitz C (2002): Time resolved photodissociation of small peptide ions. *Eur. Phys. J. D* **20**:565–571.

Cui W, Thompson MS, Reilly JP (2005): Pathways of peptide ion fragmentation induced by vacuum ultraviolet light. *J. Am. Soc. Mass Spectrom.* **16**:1384–1398.

Depke G, Heinrich N, Schwarz H (1984): On the gas phase chemistry of ionized glycine and its enol. A combined experimental and ab initio molecular orbital study. *Int. J. Mass Spectrom. Ion Proc.* **62**:99–117.

Djerassi C, von Mutzenbecher G, Fajkos J, Williams DH, Budzikiewicz H (1965): Mass spectrometry in structural and stereochemical problems. LXV. Synthesis and fragmentation behaviour in 15-keto steroids. The importance of interatomic distances in the McLafferty rearrangement. *J. Am. Chem. Soc.* **87**:817–826.

Easton CJ (1997): Free-radical reactions in the synthesis of α-amino acids and derivatives. *Chem. Rev.* **97**:53–82.

El Aribi H, Orlova G, Hopkinson AC, Siu KWM (2004): Gas-phase fragmentation of protonated aromatic amino acids: Concomitant and consecutive neutral eliminations and radical cation formation. *J. Phys. Chem. A* **108**:3844–3853.

Farrugia JM, Taverner T, O'Hair RAJ (2001): Side-chain involvement in the fragmentation reactions of the protonated methyl esters of histidine and its peptides. *Int. J. Mass Spectrom.* **209**:99–112.

Gatlin CL, Tureček F, Vaisar T (1995): Copper(II) amino acid complexes in the gas phase. *J. Am. Chem. Soc.* **117**:3637–3638.

Grotemeyer J, Schlag EW (1988a): Laser desorbed neutral peptide ionized by multiphoton absorption. *Org. Mass Spectrom.* **23**:388.

Grotemeyer J, Schlag EW (1988b): Biomolecules in the gas phase: Multiphoton ionization mass spectrometry. *Acc. Chem. Res.* **22**:399–406.

Henke W, Kremer S, Reinen D (1983): Cu^{2+} in five-coordination: A case of a second-order Jahn-Teller effect. 1. Structure and spectroscopy of the compounds Cu(terpy)X_2.nH2O. *Inorg. Chem.* **22**:2858–2863.

Holmes JL, Lossing FP, Terlouw JK, Burgers PC (1982): The radical cation $[CH_2OH_2]^{+\bullet}$ and related stable gas-phase ion-dipole complexes. *J. Am. Chem. Soc.* **104**:2931–2932.

Hoppilliard Y, Ohanessian G, Bourcier S (2004): Fragmentation mechanisms of glycine-Cu^+ in the gas phase. An experimental and theoretical study. *J. Phys. Chem. A* **108**:9687–9696.

Hoyau S, Ohanessian G (1997): Absolute affinities of α-amino acids for Cu^+ in the gas phase. A theoretical study. *J. Am. Chem. Soc.* **119**:2016–2024.

Hu Y, Hadas B, Davidovitz M, Balta B, Lifshitz C (2003): Does IVR take place prior to peptide ion dissociation? *J. Phys. Chem. A* **107**:6507–6514.

Hu P, Loo JA (1995): Gas phase coordination properties of Zn^{2+}, Cu^{2+}, Ni^{2+} and Co^{2+} with histidine-containing peptides. *J. Am. Chem. Soc.* **117**:11314–11319.

Iavarone AT, Paech K, Williams ER (2004): Effects of charge state and cationizing agent on the electron capture dissociation of a peptide. *Anal. Chem.* **76**:2231–2238.

Jochims H-W, Schwell M, Chotin J-L, Clemino M, Dulieu F, Baumgartel H, Leach S (2004): Photoion mass spectrometry of five amino acids in the 6-22 eV photon energy range. *Chem. Phys.* **298**:279–297.

Ke,Y, Hopkinson AC, Siu KWM (2005): Unpublished data.

Lavanant H, Hecquet E, Hoppilliard Y (1999): Complexes of L-histidine with Fe^{2+}, Co^{2+}, Ni^{2+}, Cu^{2+}, Zn^{2+} studied by electrospray ionization mass spectrometry. *Int. J. Mass Spectrom.* **185/186/187**:11–23.

Lu H-F, Li F-Y, Lin SH (2004): Theoretical interpretation of the fragments generated from a glycine radical cation. *J. Phys. Chem. A* **108**:9233–9243.

Maksic ZB, Kovacevic B (1999): Towards the absolute proton affinities of 20 amino acids. *Chem. Phys. Lett.* **307**:497–504.

Marino T, Russo N, Toscano M (2000): Gas-phase metal ion (Li^+, Na^+, Cu^+) affinities of glycine and alalnine. *J. Inorg. Biochem.* **79**:179–185.

McLafferty FW (1955): Mass spectrometric analysis. *Anal. Chem.* **31**:82–87.

McLafferty FW, Turecek F (1993): *Interpretation of Mass Spectra*, Univ. Science, Mill Valley, CA.

Orlova G, Hopkinson AC, Siu KWM (2005): Manuscript in preparation.

Pau CF, Hehre WJ (1986): In Hehre WJ, Radom L, Schleyer PvR, Pople, JA (eds), *Ab initio Molecular Orbital Theory*, Wiley.

Polce MJ, Wesdemiotis C (1999): α-Glycyl cation, radical, and anion ($H_2NCH^{+/\bullet/-}COOH$: Generation and characterization in the gas phase. *J. Am. Soc. Mass Spectrom.* **10**:1241–1247.

Polce MJ, Wesdemiotis C (2000): First generation and characterization of the enol of glycine, $H_2N{-}CH{=}C(OH)_2$, in the gas phase. *J. Mass Spectrom.* **35**:251–257.

Pulkkinnen S, Noguera M, Rodriguez-Santiago L, Sodupe M, Bertran J (2000): Gas-phase intramolecular proton transfer in cationizes glycine and chlorine-substituted derivatives

(M-Gly, M = Na^+, Mg^{2+}, Cu^+, Ni^+ and Cu^{2+}): Existence of zwitterionic structures? *Chem. Eur. J.* **6**:4393–4399.

Rauk A, Armstrong DA (1999): Comparison of the $^{\alpha}$C-H bond enthalpies of amino acids in a protein model environment. *Biochemistry* **38**:9089–9096.

Rodriquez CF, Cunje A, Shoeib T, Chu IK, Hopkinson AC, Siu KWM (2001): Proton migration and tautomerism in protonated triglycine. *J. Am. Chem. Soc.* **123**:9229–9236.

Schroder D, Soldi-Lose H, Schwarz H (2003): Unimolecular rearrangements of transient carboxy radicals and cations. *Aust. J. Chem.* **56**:443–451.

Seymour JL, Turecek F (2002): Structure, energetics and reactivity of ternary complexes of amino acids with Cu(II) and 2,2′-bipyridine by density functional theory. A combination of radical-induced and spin-remote fragmentations. *J. Org. Mass Spectrom.* **37**:533–540.

Seymour JL, Turecek F, Malkov AV, Kocovsky P (2004): Chiral recognition in solution and the gas phase. Experimental and theoretical studies of aromatic D- and L-amino acid-Cu(II)-chiragen complexes. *J. Mass Spectrom.* **39**:1044–1052.

Sheu S-Y, Yang D-Y, Selzle HL, Schlag EW (2002): Charge transport in a polypeptide chain. *Eur. Phys. J. D* **20**:557–563.

Shoeib T (2002): *Argentinated Peptides; Theory and Experiment*, PhD thesis, York Univ.

Simon S, Sodupe M, Bertran J (2002): Isomerization versus fragmentation of glycine radical cation in gas phase. *J. Phys. Chem. A* **106**:5697–5702.

Simon S, Sodupe M, Gil A, Bertran J (2005): Structure and fragmentation of glycine, alanine, serine and cysteine. *J. Mol. Struct.* (*Theochem*) **727**:191–197.

Stubbe J, van der Donk WA (1998): Protein radicals in enzyme catalysis. *Chem. Rev.* **98**:705–762.

Sustmann R, Korth HG (1990): The captodative effect. *Adv. Phys. Org. Chem.* **26**:131–178.

Syrstad EA, Turecek F (2004): Toward a general mechanism of electron capture. *J. Am. Soc. Mass Spectrom.* **16**:208–224.

Turecek F, Carpenter FH, Polce MJ, Wesdemiotis C (1999): Glycyl radical is a stable species in the gas phase. *J. Am. Chem. Soc.* **121**:7955–7956.

Turecek F (2003): N-C_{α} bond dissociation energies and kinetics in amide and peptide radicals. Is the dissociation a non-ergodic process? *J. Am. Chem. Soc.* **125**:5954–5963.

Vachet RW, Hartman JAR, Callahan JH (1998): Ion-molecule reactions in a quadrupole ion trap as a probe of the gas-phase structure of metal complexes. *J. Mass Spectrom.* **33**:1209–1225.

Viehe HG, Janousek Z, Merenyi R, Stella L (1985): The captodative effect. *Acc. Chem. Res.* **18**:148–154.

Walker NR, Wright RR, Barran PE, Murell JN, Stace AJ (2001): Comparisons in the behavior of stable copper(II), silver(II) and gold(II) complexes in the gas phase: Are there implications for condensed-phase chemistry? *J. Am. Chem. Soc.* **123**:4223–4227.

Wee S, O'Hair RAJ, McFadyen WD (2002): Side-chain radical losses from radical cations allows distinction of leucine and iosleucine residues in isomeric peptides GlyXxxArg. *Rapid Commun. Mass Spectrom.* **16**:884–890.

Wee S, O'Hair RAJ, McFadyen WD (2004): Comparing the gas-phase fragmentation reactions of protonated and radical cations of the tripeptides GXR. *Int. J. Mass Spectrom.* **234**:101–122.

Wee S (2005): *Studies of the Interaction of Metal Complexes with Ligands and Biomolecules in the Gas Phase Using Mass Spectrometry*, PhD dissertation, Univ. Melbourne.

Weinkauf R, Schanen P, Yang D, Soukara S, Schlag EW (1995): Elementary processes in peptides: Electron mobility and dissociation in peptide cations in the gas phase. *J. Phys. Chem.* **99**:11255–11265.

Weinkauf R, Schanen P, Metsala A, Schlag EW, Burgle M, Kessler H (1996): Highly efficient charge transfer in peptide cations in the gas phase: Threshold effects and mechanism. *J. Phys. Chem.* **100**:18567–18585.

Wright RR, Walker NR, Firth S, Stace AJ (2001): Coordination chemistry of stable Cu(II) complexes in the gas phase. *J. Phys. Chem. A* **105**:54–64.

Yu D, Rauk A, Armstrong DA (1995): Radicals and ions of glycine: An ab initio study of the structures and gas-phase thermochemistry. *J. Am. Chem. Soc.* **117**:1789–1796.

Zhao J, Hopkinson AC, Siu KWM (2005): Manuscript in preparation.

Zubarev RA, Kelleher NL, McLafferty FW (1998): Electron capture dissociation of multiply charged cations. A non-ergodic process. *J. Am. Chem. Soc.* **120**:3265–3266.

Zubarev RA (2003): Reactions of polypeptide ions with electrons in the gas phase. *Mass Spectrom. Rev.* **22**:57–77.

10

PHOTODISSOCIATION OF BIOMOLECULE IONS: PROGRESS, POSSIBILITIES, AND PERSPECTIVES COMING FROM SMALL-ION MODELS

ROBERT C. DUNBAR
Chemistry Department
Case Western Reserve University
Cleveland, OH

Principles of Mass Spectrometry Applied to Biomolecules, edited by Julia Laskin and Chava Lifshitz

10.1. INTRODUCTION

Aside from a few exotic techniques, the two major information sources for characterizing ions coming from a mass spectrometer source are fragmentation, and ion–molecule reactions. Focusing attention in this chapter on fragmentation-based approaches, we will be interested here in photodissociation experiments: those that initiate dissociation by shining light on the ions.

On one hand, photodissociation has long played a subsidiary role as a means of heating ions and inducing fragmentation patterns in ways similar to other fragmentation techniques. This role continues to be important in the mass spectrometry of biomolecules. In particular, IR lasers (mostly CO_2 lasers) are now routinely deployed on ion trapping mass spectrometers looking at biomolecules, in the approach often called IRMPD (infrared multiphoton dissociation) in the literature. This serves as an alternative to collision-induced dissociation (CID), electron capture dissociation (ECD), surface-induced dissociation (SID), and similar methods for inducing fragmentation of trapped ions. Such approaches are well covered in other chapters of this volume. On the other hand, there are a number of ways in which photoexcitation can lead to results that are uniquely useful in some way. Applications of this sort will be the focus of the present chapter.

Discussion of this topic divides itself naturally, as does the present chapter, according to two different points of view: (1) the spectroscopic perspective involves varying the wavelength of the photodissociating light, and viewing the resulting spectrum as a reflection of the optical spectrum of the ion; (2) the activation perspective considers photons as a way of activating molecules in particularly incisive ways and looks at the information gained from the subsequent dissociation process. From this latter perspective, photodissociation takes advantage of precise energy deposition via monochromatic photons, well-defined timing of photoactivation via pulsed lasers, and localization of the activation site in the molecule at the position of the optical chromophore.

The reader might refer to the *Encyclopedia of Mass Spectrometry* for more complete discussion of aspects of this topic, for instance, in Volume 1, article on photodissociation spectroscopy (Dunbar 2003a), article on infrared photodissociation (Riveros 2003), article on vibrational predissociation (Johnson 2003), and article on photodissociation studies of ion thermochemistry (Dunbar 2003b); and Volume 4, article on IRMPD (Fridgen and McMahon 2005), and article on photochemistry and spectroscopy of organic ions (Dunbar 2005). Other reviews of interest include discussions of photodissociation and bond activation in smaller ions (Metz 2004a), multiply charged ions (Metz 2004b), and a slightly older but comprehensive review of the spectroscopic aspects (Duncan 2000).

10.2. THE SPECTROSCOPIC PERSPECTIVE

10.2.1. Distinctive Electronic Chromophores

It is very difficult to do an optical absorption spectroscopy experiment on mass-selected gas-phase ions, because the concentration of molecules in the sample is too low to give measurable light absorption. However, when the wavelength of light driving a photodissociation is varied, a plot of the extent of dissociation versus wavelength creates a spectrum [the photodissociation spectrometry (PDS) spectrum] that mirrors the true optical absorption spectrum of the ion. This is a form of "action spectroscopy" that works because only those light wavelengths that are absorbed by the ions can lead to dissociation. This approach has exquisite sensitivity, because the dissociation of even a single ion is often readily detectable.

Just as with visible/ultraviolet (vis/UV) spectroscopy of neutrals, we can think of PDS in the vis/UV wavelength region in terms of peaks at particular characteristic wavelengths that will correlate with interesting functional groups or bonding patterns in the ion. Much early gas-phase PDS work from this perspective focused on radical cations, first because the alternative condensed-phase spectroscopic study of these highly reactive molecules is difficult, and second because their intense absorptions at visible wavelengths match well with the availability of powerful, tunable light sources in the visible. Radical cations tend to be brightly colored because the HOMO is half-filled, giving rise to "hole promotion" electronic excitations in which a lower-lying valence electron promotes into the hole in the HOMO, which "promotes" the hole to an inner valence orbital. These hole promotion transitions often give intense peaks in the vis/near-UV range.

10.2.1.1. Radical Cation Conjugated π Systems

10.2.1.1.1. $C_9H_{10}^{+\cdot}$. PDS viewed in this way is an excellent means of characterizing the structures of ions with unsaturated electronic systems, since the corresponding π systems are sensitive in predictable ways to the bonding topology. A good early illustration of this is shown in Figure 10.1, where the PDS spectra are compared for two isomers of $C_9H_{10}^{+\cdot}$ (Fu and Dunbar 1978). The two peaks seen in each of the spectra are characteristic of the aromatic π system in radical cations, corresponding

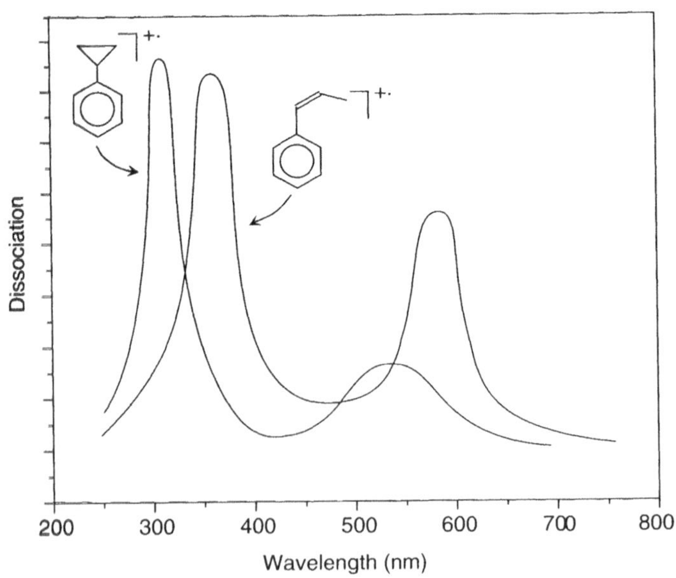

FIGURE 10.1. Photodissociation spectra of two isomeric radical cations, showing the characteristic spectral pattern for the benzene ring system, contrasted with the spectral pattern of the styrene framework with its more extended conjugated system. [Adapted from results displayed in Fu and Dunbar (1978).]

to the hole promotion transition (~500 nm) in the visible, and the electron promotion transition (~300 nm) in the UV. The conjugated π system of the methylstyrene ion is more extensive than that of the cyclopropylbenzene ion, so the peaks in the former ion are significantly farther to the red, making it easy to distinguish these isomers. The 300- and 530-nm peaks seen for cyclopropyl benzene ion are characteristic of the benzene ring radical cation system having no electronically active substituent, while the redshifted peaks of the methylstyrene ion offer a spectroscopic diagnosis of a substituent having a strong conjugative interaction with the ring.

In the same study, the nonconjugated isomer of methylstyrene was also looked at

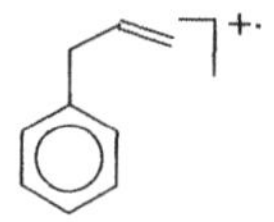

The spectrum observed for this compound was identical with that of methylstyrene ion, showing that ionization of the nonconjugated neutral compound resulted in its rearrangement to the more stable conjugated isomer. This illustrates the capability of PDS for characterizing both ion structures and also ion rearrangements, in this case using the characteristic spectral pattern of the benzene radical cation chromophore.

10.2.2. UV/Visible Study of Coordination and Solvation

When the spectrum of a small molecule is obtained at extremely high resolution, resolving large numbers of individual rotational lines, detailed analysis gives structure parameters with very high accuracy (Duncan 2000). This works as well for ion PDS spectra as for neutral-molecule spectra, and is actively interesting to some groups working mostly on diatomics and triatomics. Since this approach is remote from any obvious application to biomolecules, we will not describe such applications further but will focus on structure-characterizing applications relevant to more complex molecules.

Transition metal complexes have always been spectroscopically interesting, and this is no less true for their study as gas-phase ions. Advanced ion sources have made these molecules more readily available, and much of the more recent activity in vis/UV PDS of larger systems has been directed to them. The mass-selective character of PDS makes it possible to look at the spectroscopy of complexes having selected numbers of ligands in the inner shell, as well as selected numbers of solvent molecules outside the inner shell.

10.2.2.1. Coordination Number of Inner Ligand Shell

10.2.2.1.1. Co^{2+}(methanol)$_n$ and Co^{2+} (water)$_n$. Using PDS to characterize the ligand arrangement around a transition metal core ion has obvious interest for metal-containing peptides, enzymes, and other compounds. Metz' study of some

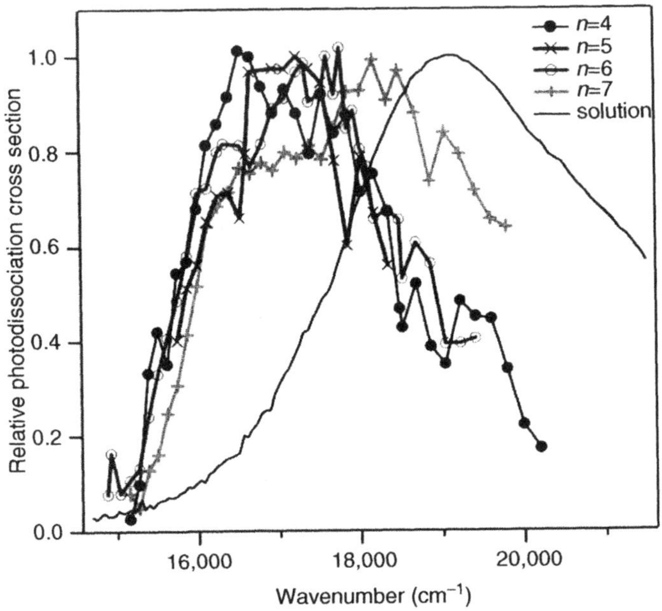

FIGURE 10.2. PDS spectra of Co^{+2}(methanol)$_n$ complexes. [From Thompson et al. (2005) reproduced by permission of the PCCP Owner Societies.]

model systems from this perspective is some of the most suggestive recent work. Figure 10.2 shows spectra for the methanol clusters of Co^{2+} with four to seven methanols (Thompson et al. 2005). The authors consider that the similarity of the $n = 4,5,6$ spectra indicates that all of these have the same four-coordinate first solvation shell, with the fifth and sixth methanols, respectively, added in a second, outer solvation shell in the $n = 5$ and $n = 6$ cases. On the other hand, the $n = 7$ spectrum is significantly different, which they believe reflects a changeover to a six-coordinate first solvation shell, with only the seventh methanol occupying the second solvation shell. Support for this idea comes from the fact that the $n = 7$ spectrum has shifted closer to the spectrum of Co^{2+} in liquid methanol, whiere it is known that the first solvation shell is six-coordinate.

In contrast, the spectrum of $Co^{2+}(H_2O)_6$ shown in Figure 10.3 (Faherty et al. 2001) is very similar to that of $Co^{2+}(H_2O)_7$, suggesting that in the case of water, six waters are accommodated in the first solvation shell even in the $n = 6$ complex. Further supporting this picture is the comparison of the gas-phase molar absorptivities ε derived from the photodissociation cross sections. In the $Co^{2+}(H_2O)_6$ case ε has the very low value of 16 M^{-1} cm^{-1}, typical of *d–d* transitions in a symmetric octahedral ligand field. [Note that the extinction coefficients given by Faherty et al. (2001) and Thompson et al. (2005) need to be adjusted downward by a factor of 2.3.] On the other hand, $Co^{2+}(CH_3OH)_6$ has a larger value of ε, 52 M^{-1} cm^{-1}, as would be expected for the noncentrosymmetric tetrahedral four-coordinate first solvation shell.

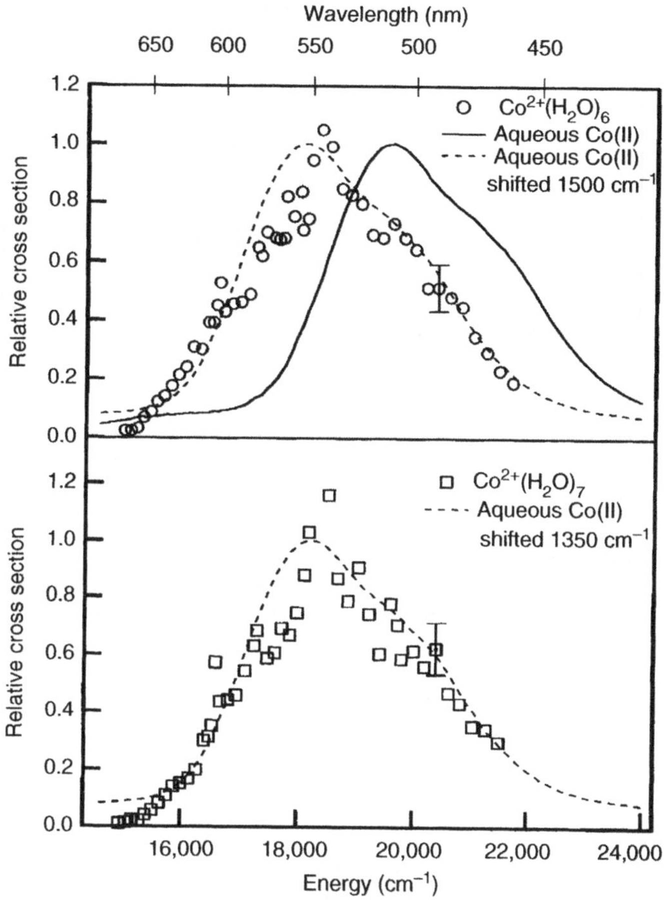

FIGURE 10.3. PDS spectra of $Co^{+2}(H_2O)_n$ complexes, compared with the fully solvated (aqueous) spectrum of six-coordinate Co^{2+}. [Reprinted with permission from Faherty et al. (2001). Copyright 2001 American Chemical Society.]

10.2.2.2. Solvation Effects

10.2.2.2.1. Solvated Fe^{2+}(terpyridyl)$_2$. A theme of more recent interest is the effect of solvation on the spectroscopy, and the comparison of gas-phase and solution-phase spectra. A classic study in this vein was that of Posey's group on Fe^{2+}(terpyridyl)$_2$ complexes solvated with between one and eight DMSO molecules (Spence et al. 1998). The spectrum, shown in Figure 10.4, is similar to the solution phase spectrum even for $n = 1$, but as solvent molecules are added, the strong charge transfer peak near 550 nm shifts progressively to longer wavelength and moves closer to the solution-phase position. Shown in the figure are three points along this route, with the number n of DMSO solvent molecules equal to 1, 8, and ∞. The overall solvent shift is about 20 nm (700 cm^{-1}) going from $n = 1$ to $n = \infty$.

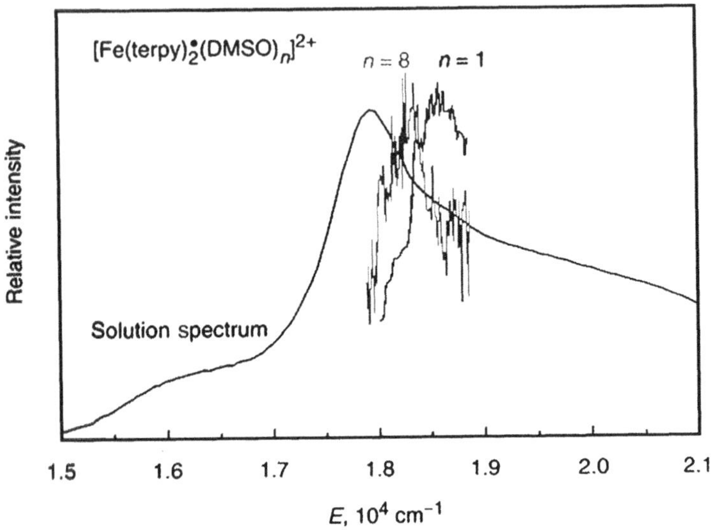

FIGURE 10.4. PDS spectra of Fe(terpyridyl)$_2$•(DMSO)$_n^{2+}$ with $n = 1$ and $n = 8$, along with the absorption spectrum of Fe(terpyridyl)$_2$(PF$_6$)$_2$ in DMSO solution. The observed spectral feature is the metal-to-ligand charge transfer peak, shifted by the perturbation of the solvating dimethylsulfoxide (DMSO) molecules. [Adapted from Spence et al. (1998), reprinted with permission from (Spence et al, 1998). Copyright 1998 American Chemical Society. Reprinted from Dunbar (2003a), with permission from Elsevier.]

10.2.2.2.2. Solvation of $Co^{2+}(L)_n$. In addition to the information about metal ion coordination discussed above, the results from Metz' group on water and methanol complexes of Co^{2+} also gave information about the effects of solvation on the spectroscopy of the complexes (Thompson et al. 2005; Faherty et al. 2001; Metz 2004). We can refer back to Figure 10.3, which compares the PDS spectrum with the solution absorption spectrum for the $Co^{2+}(H_2O)_6$ complex. This is a weak *d–d* transition of the Co^{2+} ion ($\varepsilon = 16\,M^{-1}\,cm^{-1}$) arising from the ligand field splitting in the octahedral ligand shell. The gas-phase and solution spectra are virtually superimposable, except that the peak shifts to shorter wavelength by 1500 cm^{-1} on addition of the solvent (solvation blueshift). Figure 10.2 suggests that there is a slightly smaller solvation blueshift (around 1000 cm^{-1}) for the presumably octahedral $Co^{2+}(CH_3OH)_7$ complex. These systems provide an interesting contrast with the solvation redshift of the charge transfer transition noted in the preceding paragraph for Fe^{2+}(terpyridyl)$_2$. As discussed quantitatively by Spence et al., the highly polar charge transfer excited state is preferentially stabilized electrostatically by solvation in the latter system, leading to the solvation redshift. In contrast, the metal-localized *d–d* transition in the former systems does not have such charge separation effects in the excited state, but instead presumably has increased ligand field splitting when the octahedral core complex is solvated, giving a solvation blueshift.

In addition to displaying the solvation shift directly, this work of Metz' group showcased another advantage of gas-phase spectroscopic study of the complexes.

In solution, uncertainty surrounded the interpretation of the spectrum of aqueous Co^{2+}. Questions arose about respective spectroscopic contributions of the equilibrium mixture of $Co^{2+}(H_2O)_6$, $Co^{2+}(H_2O)_5$, and $Co^{2+}(H_2O)_4$, which was believed to describe aqueous Co(II). The mass-resolved gas-phase PDS spectra obtained by Metz' group for all three of these ions showed their unambiguous spectra, removing all questions about their contributions to the observed aqueous spectrum. Not only did the gas-phase $Co^{2+}(H_2O)_6$ spectrum match nicely with the aqueous spectrum (allowing for the expected solvation shift), confirming the identity of this as the major spectroscopic contributor in solution; further, the PDS spectrum of $Co^{2+}(H_2O)_4$ was found to match well with additional solution spectral features observed at elevated temperature, confirming the hypothesis that these arise from a trace equilibrium concentration of $Co^{2+}(H_2O)_4$ in solution under these conditions.

10.2.2.2.3. Solvation of Cu^{2+}(pyridine)$_n$. Another group studying transition metal cation complexes is that of Stace, who, for instance, have shown PDS spectra for mass-selected complexes Cu^{2+}(pyridine)$_n$, where $n = 4,5,6$ (Figure 10.5) (Puskar et al. 2003). These complexes all have a ligand field split *d–d* transition around 600–800 nm, which is consistent with the extinction coefficient of $100\,M^{-1}\,cm^{-1}$ that they estimated for Cu^{2+}(pyridine)$_4$. The observation that Cu^{2+}(pyridine)$_5$ has the shortest-wavelength band maximum among these three complexes is an interesting observation stimulating ongoing theoretical speculation. On the figure, we have superimposed the solution spectra of the $n = 4$ and $n = 5$ complexes (Leussing and Hansen 1957). It can be seen that the $n = 4$ complex exhibits virtually no solvation shift of its absorption feature, while the $n = 5$ complex shows a very substantial shift ($\sim$5000 cm^{-1}) to the blue on solvation with water. These effects have not yet been explained. Spectra of the corresponding Ag(II) series, Ag^{2+}(pyridine)$_n$, ($n = 4,5,6$) show an absorption maximum near 460 nm, which does not appear to vary much with ligand number. The authors suggest that this might be a metal–ligand charge transfer band. This suggestion is supported by the larger extinction coefficient ($\sim$500 $M^{-1}cm^{-1}$) observed for the silver $n = 4$ species (in this case with methylpyridine ligands).

10.2.2.2.4. Solvation of IrX_6^{2-}. Among the few studies of anion spectroscopy is the work of the Kappes group (Friedrich et al. 2002) on the $IrCl_6^{2-}$ and $IrBr_6^{2-}$ dianions. The $IrBr_6^{2-}$ spectrum, reproduced in Figure 10.6, is particularly rich, having six assignable electronic peaks. The challenging process of assigning the peaks is greatly aided by the good correlation of the peak widths, relative positions and intensities with the aqueous absorption spectrum, which has been extensively analyzed in the literature. This assignment shows a blueshift of all of the peaks on water solvation, by amounts varying from 0.01 to 0.15 eV. This is in accord with the simplest picture of solvent shifts, which attributes the greatest degree of stabilization of the electronic states to those states placing the largest amount of charge on atoms near the solvent shell. In this case, the ground state concentrates charge on the outer bromide atoms, while all of the excited states correspond to a transfer of charge

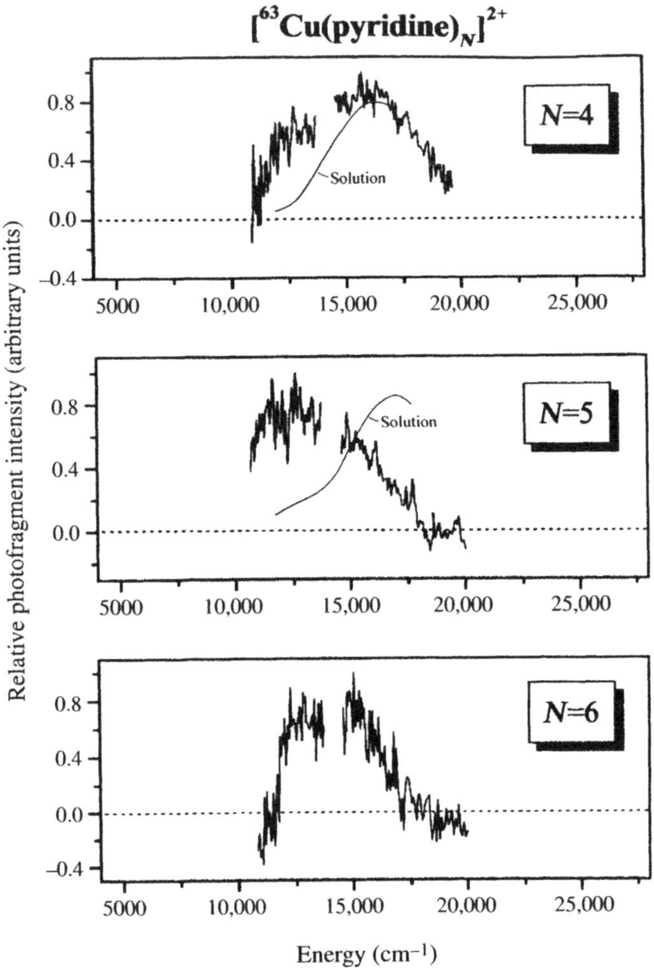

FIGURE 10.5. PDS spectra of three electrosprayed complexes of $Cu(II)(pyridine)_n^{2+}$, obtained by tunable laser irradiation in the flight tube of a fast-ion-beam instrument. For $n = 4$ and $n = 5$ we have superimposed the solution-phase absorption spectra of the same complexes fully water solvated. [Puskar et al. (2003) – Reproduced by permission of The Royal Society of Chemistry.]

toward the Ir atom. This corresponds to a differentially greater solvent stabilization of the ground state, and thus a solvation blueshift.

However, this attractive simple picture is considered to be too simple, both because it fails to account for the wide range of solvent shifts for the different electronic states and also because the $IrCl_6^{2-}$ ion, which should be closely similar to the bromide case, fails to show any discernible solvent blueshift of its two intense peaks. The authors consider that a second solvent effect, namely, the change in bond lengths on solvation, alters the electronic state energy pattern significantly and contributes importantly to the observed solvent shifts.

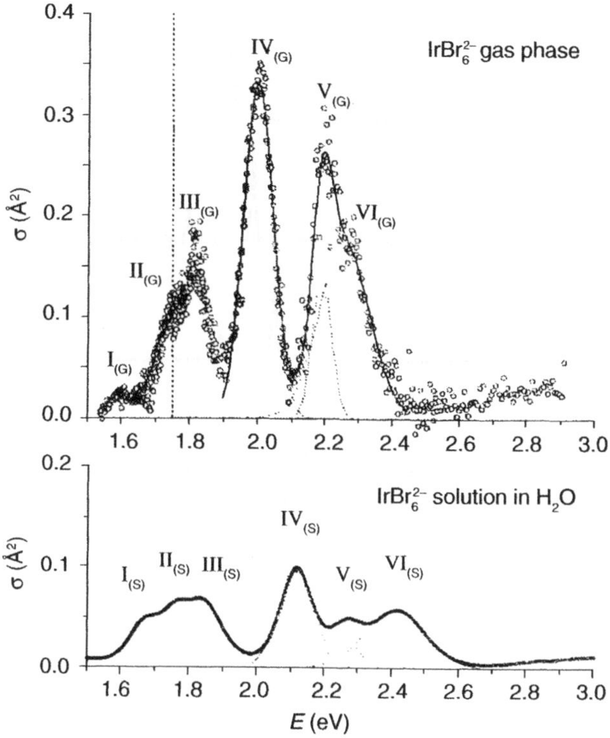

FIGURE 10.6. The gas-phase PDS spectrim of $IrBr_6^{2-}$ compared with the aqueous absorption spectrum. [Reprinted with permission from Friedrich et al. (2002). Copyright 2002, American Institute of Physics.]

10.2.3. Identification of Distinctive Vibrational Modes

Vibrational spectroscopy in the IR region offers another fruitful approach to characterizing ions. Since many photons are usually needed to bring about dissociation, IR PDS spectra are usually obtained via an infrared multiple-photon dissociation (IRMPD) experiment. One useful strategy for probing ion properties is to identify a prominent IR peak corresponding to a distinctive vibrational mode and follow its variations in position and intensity as it appears in different environments. This is well suited to work with lasers having restricted tuning ranges, like CO_2 lasers around 1000 cm^{-1} or OPO lasers around 2800–3600 cm^{-1}. A slightly different perspective is opened up by the recent emergence of laser sources readily tunable accross the IR fingerprint region [notably the free-electron laser (FEL)], which make it convenient to characterize ionic molecules by the global features of the IR spectrum, in which certain modes can emerge as diagnostic for interesting structural, spin, or other properties of the ions. First we will note three prominent vibrational mode types lending themselves to detailed study, and then we will broaden this to viewing a selection of broadband IR spectra from a more global perspective.

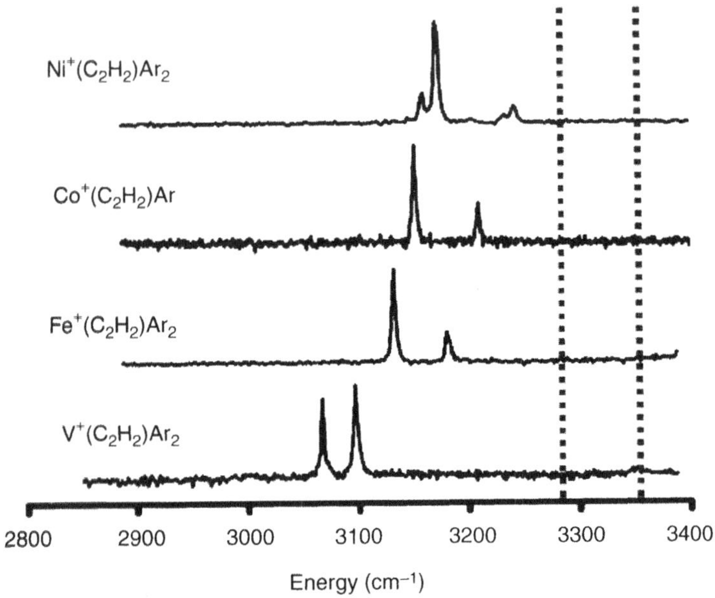

FIGURE 10.7. OPO laser PD spectra of M^+(acetylene) complexes, with varying number of argon atom tags to allow one-photon photodissociation. [Reprinted with permission from Walters et al. (2005). Copyright 2005 American Chemical Society.]

10.2.3.1. Hydrogen Stretches. Examination of the hydrogen stretching vibrational region is well matched to the wavelength range of OPO lasers, and potentially carries much useful information about structures and conformations of biomolecules, H-bonded ion clusters, and so on.

10.2.3.1.1. Metal Ion/Acetylene Complexes. A nice example of the trends of C–H stretching vibrations was shown in the Duncan group's study of metal ion/acetylene complexes, $M^+(C_2H_2)$, for V, Fe, Co and Ni (Walters et al. 2005). Figure 10.7 shows the shift of the pair of peaks (symmetric and antisymmetric C–H stretches of acetylene) as a function of the metal ion.

The dashed lines in Figure 10.7 indicate the peak positions for bare acetylene. The progressively larger shift of the vibrations to lower frequency in going from right to left on the first transition metal row is related to the backdonation of electrons from the metal ion into the antibonding π^* orbitals of acetylene, and is found to correlate nicely with the calculated charge on the metal atom. Ni^+ has the smallest amount of backdonation, the smallest positive charge on the metal, and induces the smallest peak shift.

10.2.3.1.2. Ubiquitin. An early exploitation of the information to be gained from the hydrogen stretches of biomolecules was the OPO laser study of highly protonated charge states of ubiquitin by McLafferty's group (Oh et al. 2002). Spectra of the $(M+7H)^{6+}$ and $(M+8H)^{7+}$ ions were compared with model dipeptide and

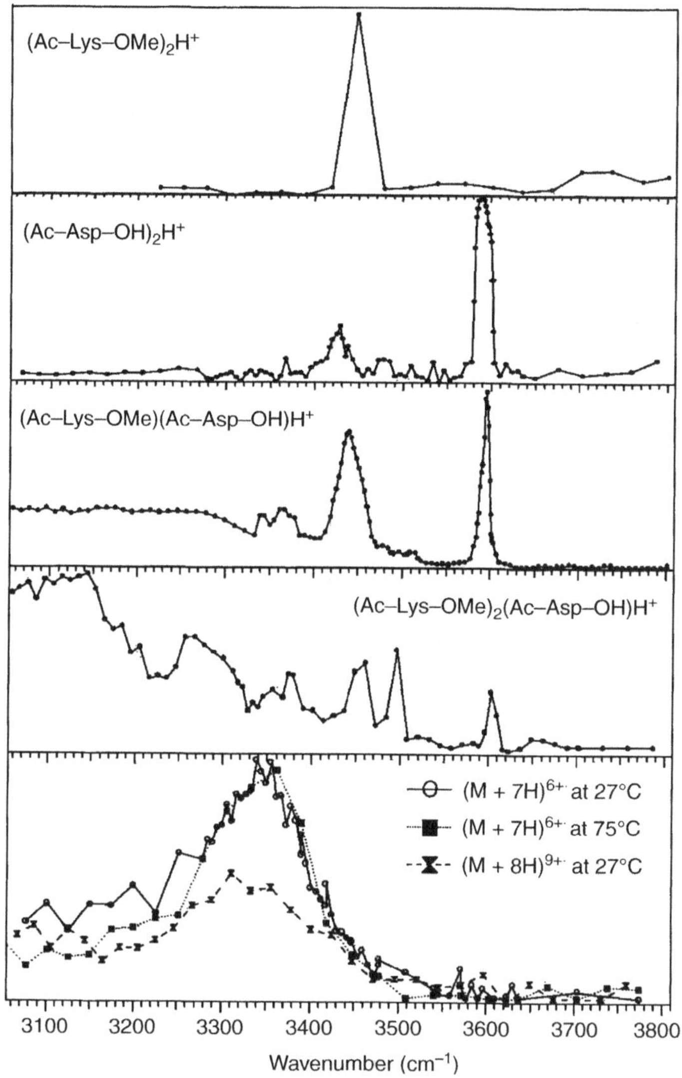

FIGURE 10.8. IR photodissociation spectra of ubiquitin (several protonated charge states, bottom panel), and several small-molecule model systems of proton-bound dimers and trimers of amino acid derivatives (top four panels). The protein case is distinguished from the model systems by the disappearance of the sharp peaks around 3450 and 3600 cm^{-1} that probably reflect non-hydrogen-bonded O–H or N–H vibrations. [Reprinted from Oh et al. (2002). Copyright (2002) National Academy of Sciences, U.S.A.]

tripeptide systems over the range of approximately 3050–3800 cm^{-1} (Figure 10.8). Contrasting with the dipeptide models, which showed mainly sharp features tentatively identified with N–H and COOH stretches (3450 and 3600 cm^{-1}, respectively), the peptide spectrum consisted largely of a broad peak at longer wavelength (3300–3400 cm^{-1}), which was considered as reflecting extensive hydrogen bonding.

10.2.3.1.3. Proton-Bound Amino Acid Dimers. A more extensive study of smaller proton-bound clusters of amino acid systems by the same group revealed much detail (Oh et al. 2005). As an example of the structural conclusions they were able to draw, the $(\text{serine})_2\text{H}^+$ ion showed a clear aliphatic free-OH peak at $3700\,\text{cm}^{-1}$, which ruled out the possibility that this proton-bound dimer ion has a structure in which both of the sidechain OH groups are hydrogen bonded.

10.2.3.2. C=O Stretches. The free electron laser, as an intense IR light source easily tunable across the entire IR fingerprint region (~400–2200 cm^{-1}) has very recently opened up new possibilities for considering various other characteristic vibrational modes. C=O stretching modes in the 1600–1700 cm^{-1} range are a useful example of this.

10.2.3.2.1. Metal Ion/Acetophenone Complexes. Some more recent work (Dunbar et al. 2005) on transition metal complexes graphically illustrates how the intense, characteristic stretching vibration of the ketone C=O group can be used to signal whether a metal ion is bound at that site in the ligand. The ligand in this case was acetophenone (acet), which offers a choice of binding sites. The metal ion may bind to the carbonyl oxygen using the basic oxygen lone-pair electrons, or it may bind in a cation-π configuration over the benzene ring. Several complexes $\text{M}^+(\text{acet})_2$ were studied, including those with M = Cr,Co,Ni. Thermochemical DFT calculations suggested that the two types of binding sites are nearly equal energetically for Co^+, while the oxygen site is substantially preferred for Cr^+ and Ni^+. IRMPD spectra of these three complexes are shown in Figure 10.9. The computed spectrum displayed here for the Co complex assumes one of the two ligands to be ring-bound (R/O), and shows the characteristic peak near 1750 cm^{-1} of the uncomplexed C=O stretch. The computed spectra for the Cr and Ni complexes assume that both ligands are oxygen-bound (O/O), showing the characteristic absorption near 1600 for the metal-complexed C=O stretch. The experimental spectra for the latter two complexes show no spectral indication of uncomplexed C=O, supporting the assignment of predominantly oxygen-bound ligands. The Co complex, on the other hand, shows a prominent peak near 1730 cm^{-1}, assigned as showing at least one R-bound ligand. The latter complex also shows clear spectral features for O-bound ligands, indicating that this population contains comparable amounts of R- and O-bound ligands (perhaps, but not necessarily, a homogeneous population of R/O complexes). Among transition metal ions, the special affinity of Co^+ for π binding to benzene rings, which is well known, is nicely illustrated by these results on competitive binding on a two-site ligand.

10.2.3.3. Amide Bands

10.2.3.3.1. Cytochrome c. New work of great interest for the spectroscopy of gas-phase proteins in general was described in a free-electron laser study of the mid-IR spectrum of cytochrome *c* (Oomens et al. 2005). Electrospray was used to introduce multiply protonated ions into the FTICR ion trap, where the IRMPD spectra of

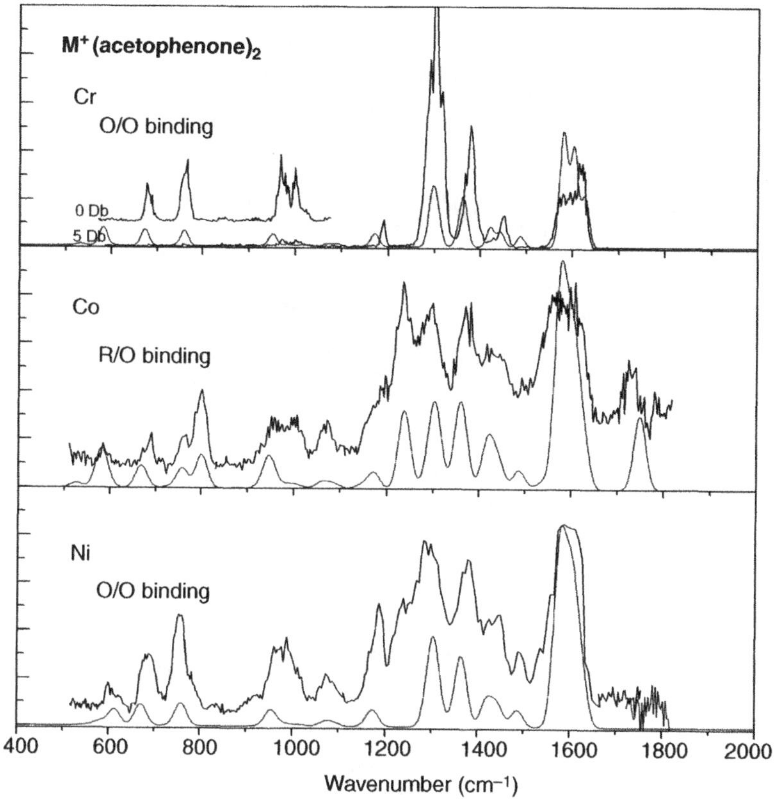

FIGURE 10.9. IRMPD spectra using the FELIX free-electron laser for M^+(acetophenone)$_2$ complexes. The experimental spectrum is the bold line, and the calculated spectrum of the best-fit isomer is shown as a light line. [From Dunbar et al. (2005) by permission.]

charge states from +12 to +16 were observed. As shown in Figure 10.10, three distinct bands were resolved between 1400 and 1800 cm^{-1}, with intensities and positions varying in interesting ways between charge states. The amide I band near 1660 cm^{-1} was discussed as being the most likely to carry interpretable information about the architecture of the protein ion. This band is well known as a valuable signal of the extent of α-helical content of the structure (Barth and Zscherp 2002). The band is blueshifted from its expected solution-phase position for all the charge states, corresponding to the lower hydrogen-bonding conditions of the gas-phase environment. It is interesting that the blueshift of this band is largest for the most highly charged states, which was considered to be consistent with other evidence from ion mobility suggesting more extended conformations for the higher charge states. The amide II band was also identified at 1535 cm^{-1}, which is redshifted relative to solution phase, which is considered to reflect an environment with less hydrogen bonding. The additional band at 1485 cm^{-1} band showed the greatest sensitivity to charge state of the ion, which makes it obviously interesting, but more work is needed to work out the nature and significance of this vibration. The

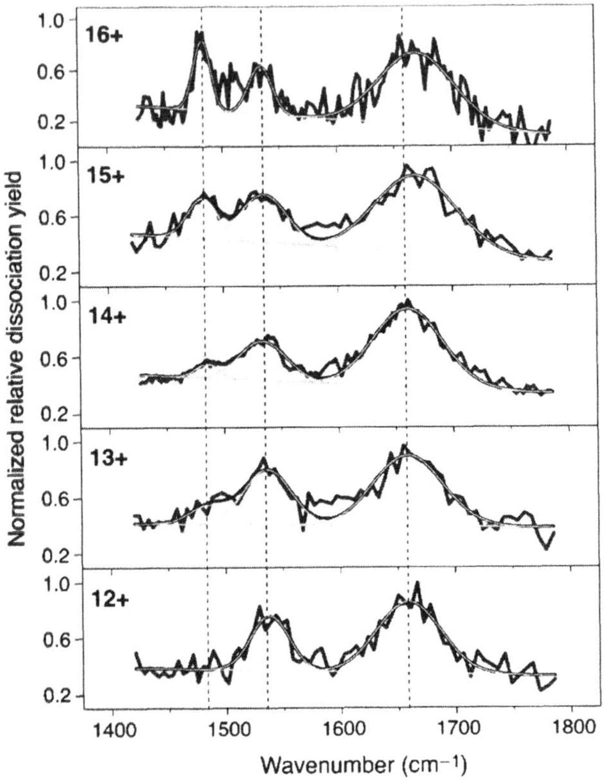

FIGURE 10.10. Free-electron laser spectra of the region containing amide peaks for several charge states of electrosprayed cytochrome *c* ions. [From Oomens et al. (2005), reproduced by permission of the PCCP Owner Societies.]

bandwidths of the amide bands were similar to those in solution, suggesting a similar range of conformational variation and/or flexibility in the condensed-phase and gas-phase environments.

10.2.3.3.2. Protonated Dialanine. The spectrum of protonated dialanine is particularly interesting as a model for peptide ion spectroscopy. Dialanine has a number of possible protonation sites, and numerous conformational variations. The study by Lucas et al. with the CLIO FEL (Lucas et al. 2004) gave a nice example of the sensitivity of the amide bands, particularly the amide I (C=O stretch) band, to the site of protonation. As is seen in Figure 10.11, the spectral predictions for various conformers protonated on the terminal nitrogen all share the expectation of amide I spectral structure between 1700 and 1800 cm^{-1}, in agreement with the IRMPD spectrum. In contrast, protonation of the oxygen of the amide linkage (conformer TransO1) corresponds to the amide I feature shifted to 1670 cm^{-1}, in clear disagreement with the observed spectrum. The amide I peak thus clearly indicates nonprotonated amide oxygen, and in conjunction with spectral modeling, indicates terminal nitrogen protonation.

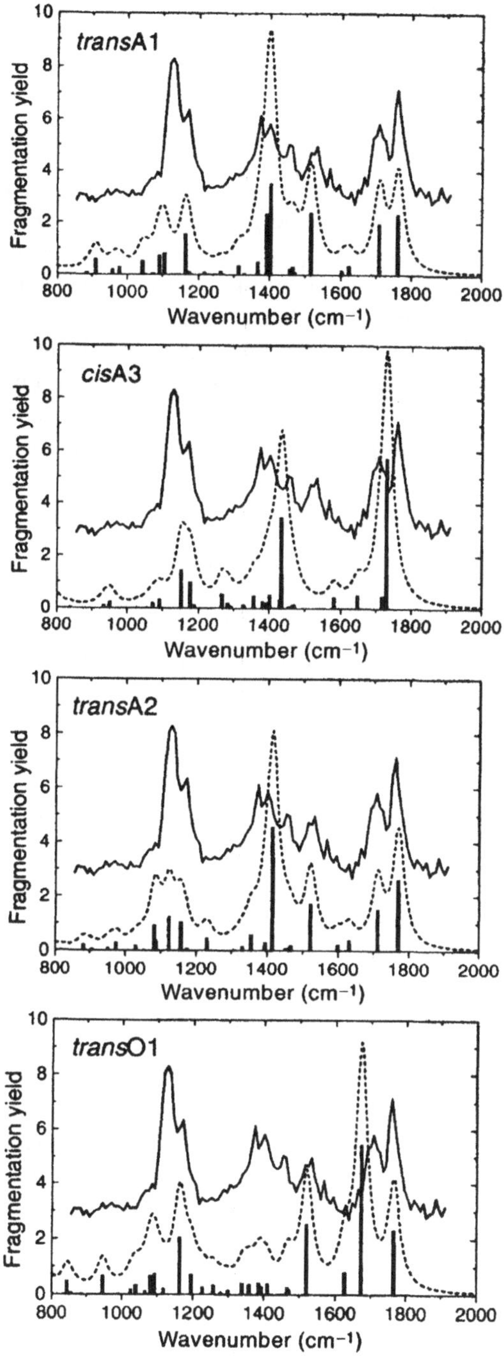

FIGURE 10.11. IRMPD spectroscopy of protonated dialanine. The experimental spectrum is reproduced in each panel, and is compared with the calculated spectra of four likely conformers (the stick spectra are the calculated transitions, which have been convoluted with a linewidth of 40 cm^{-1} to give the dashed spectra). [From Lucas et al. (2005), reproduced by permission of the PCCP Owner Societies.]

10.2.4. Broadband IR Spectra

10.2.4.1. Isomers and Binding Sites. In considering the use of IRMPD spectra for characterizing gas-phase ions, we have so far focused on using a particularly informative diagnostic normal-mode frequency to resolve a particular structural question about the ion. However, the emerging availability of IRMPD spectra across wide stretches of the IR fingerprint region gives a more powerful tool, allowing the integration of the pieces of spectral information coming from an assortment of normal modes of the ion. Some of the first work showing the power of broadband IR spectroscopy using the FEL for distinguishing and identifying structures of isomerically differing ions was the work of the CLIO group at Orsay on Fe^+ complexes with C_4 hydrocarbons (Lemaire et al. 2002; Simon et al. 2004). That group and the group at FELIX in the Netherlands have pursued these possibilities for a number of systems.

10.2.4.1.1. Cr^+/Aniline Complex. An illustration of how the IR spectrum in the fingerprint region can do this has been described for the Cr^+/aniline complex, using the FELIX FEL facility coupled to an FTICR mass spectrometer (Oomens et al. 2004). Aniline offers two sites for the attachment of the metal ion; one is the π site over the benzene ring, and the other is the lone-pair site on the nitrogen. The two sites were found computationally to be almost equally attractive to Cr^+, and in fact DFT calculations using different functionals gave contradictory predictions of the preferred site for the metal ion. Figure 10.12 shows the FELIX spectrum, along with computed spectra of the two possible isomeric complexes. It can be seen at once that the ring-bound prediction gives a much better overall fit to the observed spectrum. In particular, the most diagnostic features from the calculations are the peak near $1070\,cm^{-1}$ for the N-bound structure, and the peak near $1310\,cm^{-1}$ for the ring-bound structure. The experimental spectrum shows a strong feature at $1300\,cm^{-1}$, but nothing near $1070\,cm^{-1}$, clearly indicating a ring-bound geometry. The $1070\,cm^{-1}$ peak in the N-bound geometry is due to the frustrated inversion motion of the NH_2 group (where the attached Cr^+ blocks the usual nitrogen inversion mode). The 1310-cm^{-1} peak in the ring-bound spectrum is a mode with a large component of C–N stretching, mildly blueshifted from a similar mode in the calculated spectrum of bare aniline at $1266\,cm^{-1}$; this mode frequency is drastically altered if the metal ion attaches instead to the nitrogen. Clearly the ring-bound complex is the favored isomer, without any significant component of the N-bound complex in the population. Note that the relative intensities of the peaks do not match the prediction as well as the peak positions. This is typical, and reflects the fact that the calculation of intensities is much more difficult and uncertain than frequencies. Moreover, the experimental peak intensities reflect the nonlinear and complex dynamics of the IRMPD process, giving no more than a semiquantitative mirror of the IR absorption intensity pattern. For instance, the high predicted intensity of the peak at $1620\,cm^{-1}$ in the ring-bound spectrum in Figure 10.12 is probably a reflection of computational peculiarities, and is not a cause of concern about this assignment.

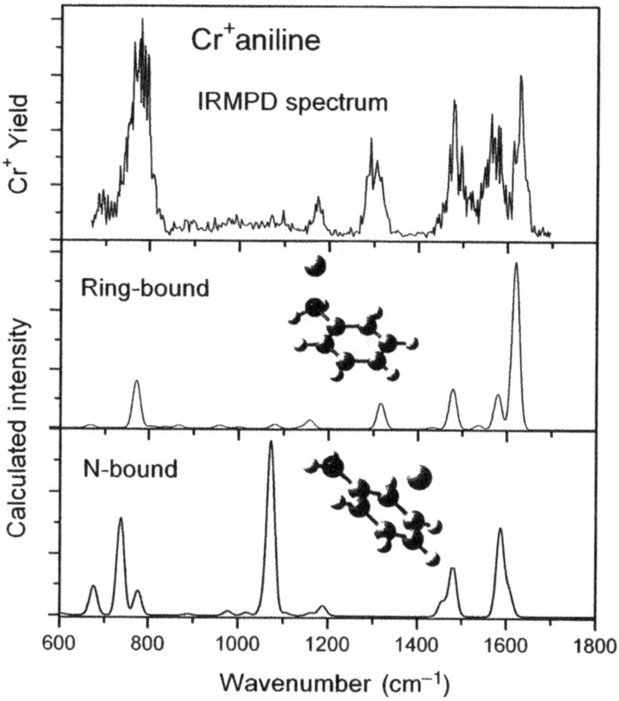

FIGURE 10.12. IRMPD spectrum of the Cr^{+}/aniline complex, compared with the calculated spectra for binding of the metal at the two possible basic sites in the sextet spin state. [Adapted from Moore et al. (2005).]

10.2.4.1.2. Protonated Dialanine. In discussing the protonated dialanine case shown in Figure 10.11, we emphasized the ability of the amide I peak to indicate whether the amide oxygen is protonated. However, looking at the rest of the spectral information in this figure, in conjunction with the calculated spectra for the different conformers, stronger conclusions can be drawn. The conclusion that the amide oxygen is not protonated is strengthened by the generally poor match of the TransO1 prediction with the observed spectrum. Furthermore, the authors concluded that two of the lowest-energy structures, both terminal-nitrogen-protonated, gave such superior fits relative to the spectrum that the actual structure of the ion is likely dominated by these two conformations (Lucas et al. 2004).

10.2.4.2. Computing IR Frequencies. It is interesting and important to know how confidently one can calculate IR frequencies for various trial structures for comparison with observed IRMPD photodissociation spectra. One survey of several neutral ligands of the order of 15 atoms gave a residual average deviation of around 10 cm^{-1} (or ~1%) after applying a linear 4% scaling factor to the raw DFT output values, and suggested that deviations of as much as 25 cm^{-1} (3%) would be unusual

(Moore et al. 2005). However, for "pathological" potential functions for which the restoring potential of a particular motion is highly anharmonic, the harmonic prediction coming from simple harmonic normal-mode analysis can be much more seriously in error. For instance, the potential energy surface for motion involving hydrogen bonding can be pathological, and in the extreme, can split into a double-well potential (which gives the appearance of a reaction coordinate for transfer of the hydrogen between the two heavy atoms). For instance, in an MP2 calculation directly comparing harmonic and anharmonicity-corrected values for the ammonia proton-bound dimer ion, Oh et al. (2005) found that the ordinary H stretching vibrations were generally lowered by 7% (which is not out of line with the empirical scaling factor usually applied to MP2 calculations), but the antisymmetric N–H–N stretch, which is the anharmonic H-bonding mode, went from 1850 (harmonic approximation) to 577 (anharmonicity-corrected). The nitrogen inversion mode of amines is similarly pathological and not adequately treated by harmonic approximations, as was found in the study of the Cr^+/aniline complex noted above (Oomens et al. 2004).

10.2.4.3. Spin State Discrimination. It is a traditional concern in transition metal complexes whether the ligand field of the surrounding ligands pushes the *d* electrons into a low-spin state. This question is often answered spectroscopically in condensed-phase work, and it would be nice to make similar conclusions for the more exotic complexes now accessible in gas-phase ionic systems. For gas-phase complexes several systems have been reported where the IR spectroscopy could be used to infer the spin state.

Among the most clearcut examples of spin state discrimination is the recent work on Cr^+ complexes of the π ligands aniline and anisole (Moore et al. 2005). The free-ion ground state for Cr^+ is the high-spin sextet, where each *d* orbital is singly occupied. However, electron pairing can take place to give a quartet (or a doublet) low-spin excited state in which one (or two) of the *d* orbitals are empty. In the cases of one or two benzene rings π-complexed to this metal ion, the ligand field strongly destabilizes two of the *d* orbitals by Pauli repulsion, and it can become advantageous to empty one or both of these orbitals by going to a low-spin configuration. For the monomer complexes Cr^+(aniline) and Cr^+(anisole) the single ligand cannot push the Cr^+ into a low-spin configuration, and accordingly the spectra are observed to agree with the high-spin expectation. However, two ligands are sufficient to push the system to low spin, and the dimer complexes $Cr^+(\text{aniline})_2$ and $Cr^+(\text{anisole})_2$ are accordingly low-spin complexes, as shown by the IRMPD spectra.

Another nice example of the spectroscopic determination of spin character the study of acetonitrile complexes of Nb^+ by broadband IRMPD using the CLIO FEL. Particularly clear was the $Nb(CH_3CN)_4^+$ complex, which was found to have the high-spin quintet spin state rather than the alternative possible triplet or singlet states, based on the comparison of calculations with the observed position of the CN stretch near 2300 cm^{-1}.

10.2.4.4. Charge State Effects. The charge state of an ionized molecule like a protein, which can accommodate numerous positive or negative charges, has important effects on the gas-phase conformation of the ion. Since electrospray of such systems typically yields a mixture of charge states, each of which can be independently characterized by the powerful arsenal of available tools, there has been growing interest in observing and trying to understand these effects. Spectroscopic tools have only just begun to contribute to this story, but some more recent results point to a promising future.

We have already noted the signficant charge state trends seen for cytochrome *c* ions in the FEL study of Oomens et al. (2005), as displayed in Figure 10.10. The effect seen there was progressive frequency shifts of the bands with increasing charge.

A different manifestation of charge state effects was evident in initial results of McLafferty's group (Oh et al. 2002) on multiply protonated ubiquitin, displayed above in Figure 10.8. In this case, the intensity of the broad feature near 3300–3400 cm^{-1} was compared between the +7 and +8 charge states. Relative to the room-temperature spectrum of the +7 ion, it was seen that the intensity decreased by a factor of 2 on going to either the +8 charge state or high-temperature (75°C) conditions for the +7 state. This was interpreted as showing increased denaturation of the structure resulting either from higher temperature, or from the increased Coulomb effects in the more highly charged species.

10.2.5. IR Studies of Solvation Effects (Hydrogen Stretching Modes)

10.2.5.1. Alkali Ions. The first extensive IRMPD studies using the OPO laser to probe the hydrogen stretching region were by Lisy's group, who explored many details of the buildup of solvation spheres around alkali metal ions. [The article by Vaden et al. (2004) is a recent study in this series.] An interesting example, motivated by the desire to elucidate the interplay of metal ion interactions and hydrogen bonds of water in protein ion channels, was a study of Na^+ and K^+ solvated by combinations of water and benzene ligands (Cabarcos et al. 1999). Figure 10.13 shows the series of spectra for one water molecule with varying numbers of benzenes, and illustrates graphically the use of this probe to show hydrogen-bonding interactions. The authors focus attention on the comparison of $Na^+(C_6H_6)_3(H_2O)_1$ and $K^+(C_6H_6)_3(H_2O)_1$. In the K^+ complex, all four ligands are accommodated in the first solvation sphere, and the two peaks represent the symmetric and antisymmetric O–H stretches of metal-bound water. However, in the Na^+ complex the first solvation shell is unable to accommodate all four ligands, and one of the benzenes moves out to the second solvation shell, forming a hydrogen bond with the water. This greatly enhances the photodissociation intensity, and also gives the new O–H stretching peak near 3550 cm^{-1} that is characteristic of the O–H group involved in a π-hydrogen bond with the benzene ring. This characteristic peak can be followed through the spectra of other species as a specific indicator of the solvation nature of the complexes.

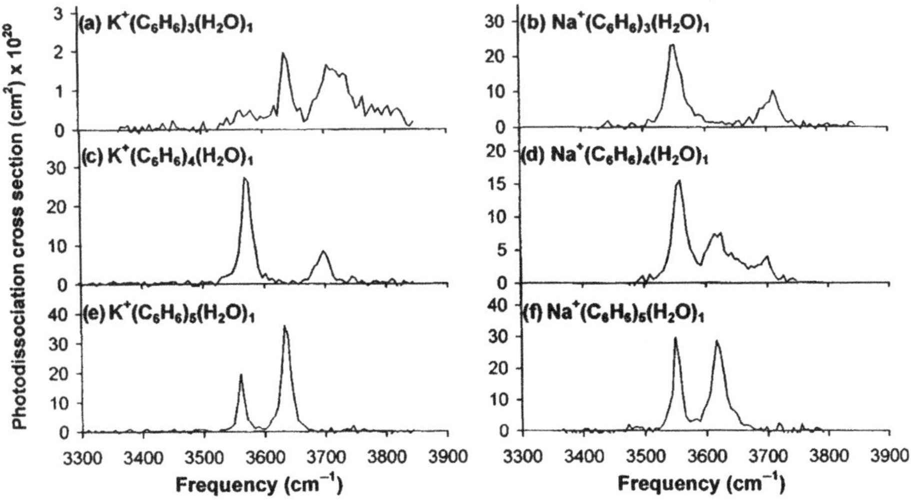

FIGURE 10.13. OPO spectra of the O–H stretching region of alkali ion complexes of benzene, exploring the π-hydrogen-bonding effects. The peak near 3550 cm^{-1} is characteristic of an outer-shell benzene ligand hydrogen bonded to the water OH. [Reprinted with permission from Cabarcos et al. (1999). Copyright 1999, American Institute of Physics.]

10.3. THE ACTIVATION PERSPECTIVE

10.3.1. Threshold Wavelength

Because photoexcitation can be achieved with accurately controlled photon energy and light-pulse timing, it has been attractive to apply photodissociation in highly quantitative ways, often as a route to measuring accurate thermochemical properties of ions. Given the ability to control and tune the excitation energy input precisely, the most obvious approach to measuring dissociation thermochemistry is to find the longest wavelength capable of initiating dissociation. This *threshold approach* has been applied to many systems, and has yielded useful information, but there are numerous pitfalls in actually applying it, and there are many cautionary examples of inaccurate or even erroneous thermochemical conclusions from threshold observations. When it works, it can offer high intrinsic accuracy, but when it goes wrong, the error is often large and hard to recognize.

At the same time as we show some successful examples of threshold thermochemical study, we can also briefly note some of the pitfalls, appearing either as unrecognized artifacts or errors not amenable to accurate corrective measures. One class of effects leads to thresholds lower than the true dissociation energy:

1. Thermal internal energy—small "Boltzmann tail" of ions in the thermal population having internal energies considerably above the average internal energy.

2. The population may not be completely thermalized, thus containing excessively energetic ions even above those constituting the Boltzmann tail.
3. Contribution of weak two-photon dissociation processes.

(Note that high-sensitivity threshold experiments magnify all of problems 1–3.) A second class of effects leads to thresholds higher than the true dissociation energy:

4. Kinetic shifts (dissociation too slow to allow observation on the timescale of the experiment). Much of our understanding of these effects comes from the work of Lifshitz (2002). [An example is the metallocene ions $M^+(C_5H_5)_2$ (Lin and Dunbar 1995; Faulk and Dunbar 1992).]
5. Intrinsic kinetic shift [dissociation slower than the infrared radiative deexcitation of the photoexcited ions (Lifshitz 2002)]. [Examples are $C_{11}H_9^+$ (Huang and Dunbar 1990) and $C_7H_8^+$ (Huang and Dunbar 1991).]
6. Dynamical barriers to near-threshold dissociation [V_3^+ (Fu et al. 2001)].
7. Spectroscopic limitations (no light absorption near the threshold wavelength).
8. Dissociation by the desired channel suppressed by competition from a different dissociation product channel [$RhCH_2^+$, (Hettich and Freiser 1987)].

As an example of the difficulties of the threshold approach, Ag^+(benzene) is an ion where attempted photodissociation threshold measurements have had several of these problems, giving dissociation energies differing by many kJ/mol from the accepted best value (Chen and Armentrout 1993). PDS thresholds have been reported as both too high (Afzaal and Freiser 1994) (probably spectroscopic limitations) and too low (Willey et al. 1991, 1992) (perhaps two-photon dissociation, or a nonthermalized ion population).

On the other hand, we can note some more recent examples of convincing photodissociation threshold energy determinations that are fully consistent with results from other techniques. It has become clear that PDS thresholds are valuable in conjunction with other sources of thermochemical information, but are seldom definitive for assigning dissociation thermochemistry. Most of the more recent literature acknowledges that a threshold can only at best give an upper limit to the thermochemical bond strength, and is cautious about making a direct identification of an observed PDS threshold with the corresponding bond strength.

10.3.1.1. Example Threshold Measurements

10.3.1.1.1. $Zr(C_2H_2)^+$. The Freiser group led the early development of the threshold strategy in the ICR ion trap to give dissociation threshold values for a large variety of mostly metal-containing ions. Some of their early results have been reassessed and revised in the light of later fuller understanding of the pitfalls noted above. A fairly recent study from this group (Ranatunga and Freiser 1995) in which these

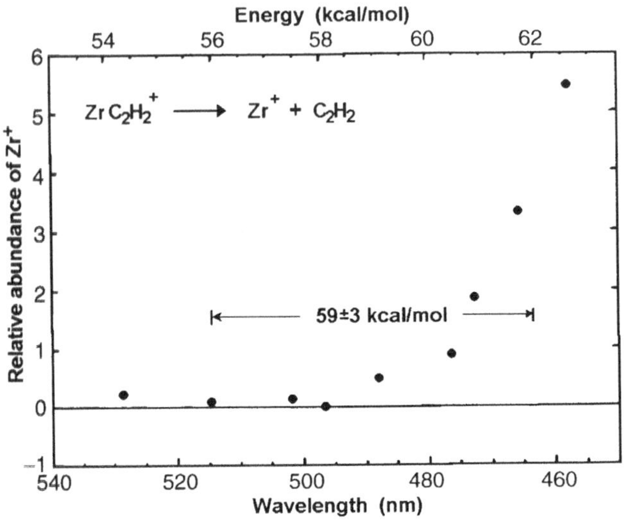

FIGURE 10.14. PDS spectrum observed for $Zr(C_2H_2)^+$. Ions were trapped in the ICR spectrometer, and were irradiated with the argon ion laser for 4 s following the ion formation–cooling sequence. The threshold is assigned with generous error limits as 247 ± 12 kJ/mol^{-1}. [Reprinted from Ranatunga and Freiser (1995), with permission from Elsevier.]

issues were carefully considered was the dissociation of $Zr(C_2H_2)^+$ in the FTICR ion trap:

$$Zr(C_2H_2)^+ \xrightarrow{h\nu} Zr^+ + C_2H_2 \qquad (10.1)$$

Thermalized, trapped ions were irradiated with available spectral lines from an argon ion laser for 4 s following the ion formation and ion cooling sequence. The threshold spectrum shown in Figure 10.14 appears to have a clean threshold, giving a Zr–C bond energy of 247 ± 12 kJ/mol. This is in good agreement with computational expectations.

10.3.1.1.2. $CoCH_2^+$. Another convincing threshold measurement was the $Co-CH_2^+$ bond strength corresponding to the reaction

$$CoCH_2^+ \rightarrow Co^+ + CH_2 \qquad (10.2)$$

This case is an excellent illustration of the value of cooling the ions to a low temperature, in this experiment achieved by supersonic jet expansion. An earlier spectrum of room temperature (and perhaps not fully thermalized) ions (Hettich and Freiser 1986) had a long, gradually disappearing tail to the red side of 340 nm, finally disappearing before 390 nm. A threshold of 340 nm was suggested, considering the dissociation in the long-wavelength tail to be due to hot ions. The new jet-cooled spectrum from Metz' group (Husband et al. 2000) shown in Figure 10.15 clearly shows a step at 361 nm that looks like a good threshold and

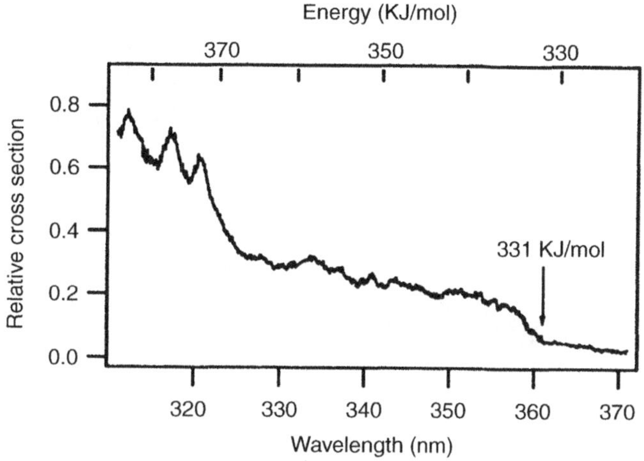

FIGURE 10.15. Photodissociation spectrum and threshold for $CoCH_2^+$. [Reprinted with permission from Husband et al. (2000). Copyright 2000 American Chemical Society.]

definitely shows that the older estimate was at too short a wavelength. From this result the authors assigned a firm new upper limit of 331 kJ/mol for the Co–C bond strength, consistent with other measured and computed values.

10.3.1.1.3. $TiMnO^+$. A final example of a successful PDS threshold measurement again illustrates the high precision possible when the spectroscopic behavior of a small ion is favorable and the ions are cooled by supersonic jet expansion. This is the observation of the triatomic ion $TiMnO^+$ in a collaboration of the Morse and Armentrout groups (Fu et al. 2001; Russon et al. 1994). The PDS spectrum in Figure 10.16 shows a sharp rise in TiO^+ photofragment at 14,219 cm^{-1} (703 nm). The authors concluded that this was probably the true thermochemical onset for dissociation. This gave an assignment of 170.2 ± 0.1 kJ/mol for the TiO^+–Mn bond. Making use of thermochemical cycles with other known thermochemical data, this leads to bond strengths for several other bonds as well, such as TiO–Mn^+ and Ti^+–OMn.

10.3.2. Energy Dependence of IRMPD

Looking to the possibilities for thermochemical study of ions by photodissociation in the infrared region [see, e.g., Riveros (2003)], we find a situation different from what was described above for vis/UV photon excitation. In the IR, photodissociation almost always requires several or many photons to drive the multiphoton dissociation (MPD) process, and it is usually impossible to determine how many photons correspond to a given observed dissociation process. In this situation, a spectroscopic threshold strategy such as that described in the previous section is useless. However, there has been considerable effort to determine the activation energy for dissociation from the rate of dissociation as a function of laser power.

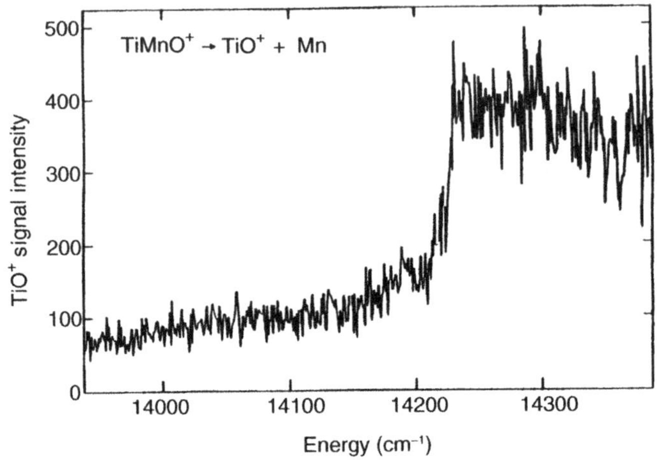

FIGURE 10.16. Threshold PDS of $TiOMn^+$. Parent ions were made in a supersonic-jet-cooled laser desorption source and measured by time-of-flight mass spectrometry. Most or all of the signals at wavelengths longer than the threshold at 14,219 cm^{-1} were attributed to collision-induced dissociation fragments. [Reprinted from Fu et al. (2001) with permission from Elsevier.]

This IRMPD approach uses a continuous-wave laser, usually a CO_2 laser in practice, so that the photoexcitation occurs slowly and sequentially, one photon at a time, over a timescale usually of milliseconds or seconds. In a sense, this strategy is an alternative, to the BIRD (blackbody IR radiative dissociation) approach, in which the "power" of the irradiation is varied by varying the temperature of the surrounding blackbody radiation field. As will be discussed, the MPD process is more difficult to interpret than BIRD, because the "temperature" of the dissociating ions is not easily determined. The BIRD approach, which is fully described elsewhere in this volume, and has also been reviewed in detail by the present author (Dunbar 2003c, 2004), benefits from the well defined temperature of that experiment. However, the MPD approach is an interesting alternative because it can easily access higher-energy dissociation processes, even going up to effective ion temperatures of thousands of kelvins, that are impractical for BIRD experiments.

When the laser is turned on, the MPD dissociation process starts with an induction period during which the ions absorb photons without dissociation and build up enough internal energy for dissociation to begin. Then the system moves into a steady-state phase in which energy deposition by the photons is balanced by removal of energy by three processes: radiative energy relaxation, collisional energy relaxation, and removal of the most energetic members of the ion population by dissociation. A steady-state distribution of ion energies is established in this latter phase, and this distribution persists reasonably unchanged as the parent ion population gradually disappears by dissociation.

In a low-pressure ion trapping mass spectrometer (usually FTICR) it is easy to irradiate the ions and follow the extent of dissociation as a function of time at different laser powers. There has been a lot of study of how to interpret these observations, usually with the goal of obtaining a measure of the dissociation energy or bond strength for the observed fragmentation reaction. This is more difficult than might be expected, because there is no obvious signal in the observations indicating how many photons are being absorbed by the dissociating ions, and it is seldom possible to estimate this crucial number with any degree of confidence. As a consequence of this uncertainty, it has been typical in such studies to find that the observations can be fitted within acceptable tolerances over a distressingly wide range of assumed activation energies. Nonetheless, careful interpretation can extract useful thermochemical information from such data. Most incisive, and most convincing in their outcomes, are detailed master equation modeling procedures, in which all of the microscopic processes of photon absorption, photon emission, collisional energy removal, and dissociation are assigned rate constants; then the resulting kinetics of the dissociative depletion of the ion population are modeled by numerically integrating the evolution of the ion internal energy distribution and the dissociation process through time. By combining the information given by the induction time with that given by the steady-state dissociation rate, and fitting to a range of laser intensity values, the dissociation energy giving the best overall fit can be assigned. This was illustrated in the smaller-ion situation for styrene ion (Dunbar 1992), and it was more recently shown by the Williams group (Jockusch et al. 2000) that such detailed modeling is even feasible for larger bioions; specifically illustrated in the latter reference was a pentapeptide with about 100 atoms.

An appealing, more approximate approach to IRMPD interpretation draws on the close analogy between slow laser IRMPD and blackbody-radiation-driven (BIRD) dissociation (Dunbar 1991). The key assumption and approximation in this approach is that the ion, in the steady-state phase of the IRMPD process, absorbs and emits light only close to the laser wavelength. (This is reasonable for photon absorption, since most of the radiation that the ion sees comes from the laser itself. It is a more serious approximation for the radiative emission. It is argued, but is justifiable to only a limited extent, that much of the fluorescent photon emission from typical ions at steady state is in the vicinity of the CO_2 laser wavelength near 1000 cm^{-1}.) Given this assumption, it is found that the steady-state population of laser-irradiated ions comes to a Boltzmann distribution of internal energies, [more exactly, a truncated Boltzmann distribution (Dunbar 1991)], just as if it were sitting in a blackbody radiation field. Then the temperature of the ions is directly related to the laser intensity, and it becomes possible to write an Arrhenius-type equation giving an activation energy E_a as a function of the laser intensity dependence

$$E_{\mathrm{a}} = -\frac{d \ln k_{\mathrm{diss}}}{d(1/k_{\mathrm{B}} T_{\mathrm{ions}})} = q h\nu \frac{d \ln k_{\mathrm{diss}}}{d \ln I} \tag{10.3}$$

where E_a is the activation energy for dissociation, k_{diss} is the dissociation rate constant in the steady-state kinetic regime, k_B is the Boltzmann constant, h is Planck's constant, ν is the laser frequency, I_{laser} is the light intensity (including the background blackbody radiation at the laser frequency ν, along with the laser light), and q is a factor derived from the theory whose numerical value is near unity. Within the assumptions of this theoretical framework, the last term on the right of this equation gives a very simple route to approximating the dissociation energy from the set of k_{diss} values as a function of laser intensity.

Because of the various approximations used in this treatment, the simple application of Eq. (10.3) gives E_a values with no more than semiquantitative significance. These have been considered most useful for comparing E_a values of similar ions. In the most ambitious approach to improving this situation, the Williams group (Jockusch et al. 2000) showed that the approximations can be removed and detailed master equation modeling can be applied to large biosystems, taking leucine enkephalin as an example. This detailed treatment yielded activation parameters agreeing with BIRD results. However, the complexity of this brute-force modeling approach to data analysis has deterred its widespread application. Two strategies have been followed toward improving the much simpler data analysis offered by Eq. (10.3).

One strategy is to assume that the approximations inherent in the simple analysis are similar and cancel out for similar systems, so that useful comparisons of dissociation energies can be drawn from comparative E_a values derived by direct application of Eq. (10.3). Marshall's group laid out this idea in detail for large biomolecules in the technique termed FRAGMENT (Focused vadiation for gaseous energy transfer) (Freitas et al. 1999). Various studies have applied this approach to comparing dissociation energies. In addition to the protonated bradykinin ions reported by Freitas et al. (1999), other example systems studied in this way have included +5 and +11 charge states of polyprotonated ubiquitin (Freitas et al. 2000), and −3 charge states of unmodified and modified oligonucleotides (Hannis and Muddiman 2002).

The second strategy based on Eq. (10.3), pursued by Williams' group (Paech et al. 2002), is to introduce an empirically adjusted parameter in the derivation of Eq. (10.3). The approximation that is removed in their derivation is the assumption that the ion absorbs and emits IR photons only at the irradiating laser wavelength. These authors carry through a reanalysis taking into account the full vibrational spectrum of the ion, and justify using the empirically adjusted equation

$$E_a = sk_B \frac{d \ln k_{uni}}{d \ln I} \tag{10.4}$$

where the empirical parameter s is fitted to systems known from BIRD measurements, or to results of master equation modeling. They considered that the parameter s would be consistent across a set of molecules having similar chemical composition, and assigned the value $sk_B = 0.208\,\text{eV}$ (or $s = 2420\,\text{K}$) as appropriate for polypeptides [compared with the corresponding value of 0.123 eV used in the unmodified form of Eq. (10.3)]. Among the effects that are still not accounted for in this adjusted treatment are the additional light intensity contributed by the background blackbody radiation field, as well as the dissociation-related perturbations of the Boltzmann distribution of ion energies. These complicating effects are all subsumed into the empirical assignment of the numerical parameter s.

The value of s of 2420 K assigned by Paech et al. (2002) for polypeptides is thus appropriate for a chamber background temperature of 300 K, and for the slow dissociation regime in which the timescale of dissociative ion loss is much longer than the timescale of energy exchange with the surroundings. Comparison of their results with BIRD results for leucine enkephalin (Schnier et al. 1997), bradykinin (Schnier et al. 1996), and ubiquitin (Jockusch et al. 1997) showed excellent agreement. There was less satisfactory agreement of their results with activation energies derived by Marshall's group from IRMPD data for bradykinin (Freitas et al. 1999, 2000), and with high-pressure thermal dissociation measurements for melittin (Busman et al. 1992) and leucine enkephalin (Meot-Ner et al. 1995).

Putting perspective on this promising approach to data analysis, we can note that detailed modeling reported previously by Williams' group (Jockusch et al. 2000) made it clear that both the chamber temperature and also the choice of fast dissociation versus slow dissociation data regimes have significant effects on the results, which shows the need for care in applying a single invariant value of s across a series of experiments. It seems clear that further refinement of this strategy is necessary before it can be adopted for determining general and reliable absolute E_a values for unknown systems.

10.3.3. Time-Resolved Photodissociation

The discussion in Section 10.3.1 described a difficult and uncertain approach to determination of dissociation thermochemistry via UV/vis thresholds. A better approach is to observe and interpret the kinetics of unimolecular dissociation of the ion following the input of a known excitation energy by photoexcitation [see, e.g., Dunbar (1996)]. In this time-resolved photodissociation (TRPD) approach, most of the problems of threshold methods are removed or alleviated, and thermochemical results of good accuracy and reliability are possible. This is one of several promising dissociation-kinetics-based methods for determining dissociation energetics through analysis of the dissociation kinetics. Dissociative photoionization of gas-phase neutral molecules is another, to which Lifshitz made notable contributions (Lifshitz 1997); yet another that is now emerging is surface-induced dissociation (SID), described by Laskin and Futrell (2005) as well as in Chapter 16 of the present volume.

The overall strategy is to excite the ions to a known internal energy E^* above the dissociation threshold, and measure the unimolecular dissociation rate constant k_{diss}. The interpretation always goes through the intermediary of a kinetic model relating k_{diss} and E^* [e.g., see Gilbert and Smith (1990) and Holbrook et al. (1996)]. It is always assumed that the internal energy is statistically spread through the molecule, and some form of transition state theory (TST) is applied, often RRKM theory, or more advanced methods with the same basic theme like variational transition state theory (VTST).

One of the great advantages of dissociation kinetic methods such as TRPD is that the ions do not need to be cold (although working with cold ions always simplifies the analysis when it can be done). A room-temperature population of ions spans a spread of internal energies centered about an average internal energy E_{thermal}. The usual TRPD analysis simply takes the total internal energy of the dissociating ion to be $E^* = E_{\text{photon}} + E_{\text{thermal}}$, and uses the kinetic theory to calculate $k_{\text{diss}}(E^*)$. In favorable circumstances where it seems necessary, the dissociation kinetics can even be calculated more precisely by a convolution of $k_{\text{diss}}(E)$ over the distribution of ion energies in the population. This is all possible because the internal energy characteristics of the thermalized ion population are precisely known (in a statistical way).

For very large ions the situation is a bit different, because the thermal energy E_{thermal} is much larger than the photon energy, so that the photon absorption gives only a small percentage increase in the overall internal energy. For instance, the thermal energy of a model protein molecule of mass 8.4 kDa is ~30 eV at 430 K (Price et al. 1996), compared with a typical UV photon energy of 5 eV. Moreover, the kinetic shifts in these large systems can be much larger than the activation energy E_0 for dissociation, which makes the analysis yet more difficult.

The treatment of TRPD data has not been addressed in detail for such very large ions. However, Laskin and Futrell have successfully worked out a parallel theoretical and operational framework in the context of the SID experiment, where energy input is by surface collisions rather than photons. This situation is closely analogous to the TRPD case, so that their approach to data analysis should carry over to large-molecule TRPD experiments. They have illustrated their approach with analyses of SID fragmentations of peptides (Bailey et al. 2003; Laskin and Futrell 2005). As an example, for several eight-residue peptides (masses of ~1000 Da) having dissociation energies of the order of 1.3 eV, the internal energy required to put k_{diss} in the experimentally accessible range is of the order of 8 eV. Yet the steepness of the rate–energy curves (Figure 10.17) leads to a simplification of the

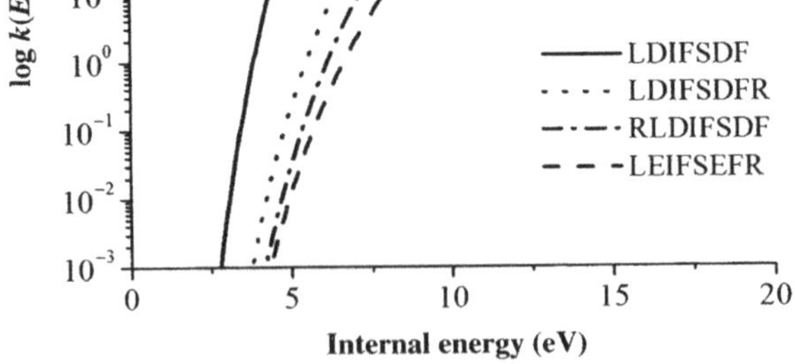

FIGURE 10.17. Rate–energy curves calculated for several polypeptide dissociations. [Reprinted from Bailey et al. (2003) with permission from Elsevier.]

analysis, and makes it possible to extract quite precise E_0 values from observations of dissociation rates versus energy deposition. For instance, the pair of peptides LDIFSDFR and RLDIFSDF in Figure 10.17 have the same number of degrees of freedom, and differ by only 0.07 eV in E_0 (1.24 vs. 1.31 eV), but in the region around 8 eV they differ in dissociation rate by a factor of >5. This large rate difference is a manifestation of the "amplification" effect discussed in more detail by Laskin and Futrell (2005).

Translation of this successful time-resolved SID treatment to the closely parallel situation of TRPD is attractive, in part because photoexcitation intrinsically has higher certainty and precision of energy input than any collisional excitation method. Against this is the negative feature that convenient laboratory light sources are limited in energy to photon energies of the order of 6 eV or less, which complicates work with even moderately large biosystems like those illustrated in Figure 10.17, in which ≥8 eV of energy input is needed to reach the useful dissociation kinetics regime. One way to supersede this limitation and achieve higher energy deposition values in TRPD is through two-photon excitation. Two examples of time- and energy-resolved studies can illustrate this possibility.

10.3.3.1. Ferricinium Ion. One example of two-photon TRPD is the dissociation of ferrocene ion observed by laser TRPD in the FTICR ion trap (Faulk and Dunbar 1992):

$$Fe(C_5H_5)_2^+ \xrightarrow{h\nu} Fe(C_5H_5)^+ + C_5H_5 \qquad (10.5)$$

The experiment ultimately showed the actual critical energy for this dissociation to be 3.7 ± 0.3 eV, but this system, although relatively small, has a large kinetic shift, and even with a photon energy as high as 5.4 eV (240 nm) no one-photon dissociation could be observed ($k_{diss} < 50\,s^{-1}$). However, irradiation at 355 nm (giving an internal energy of 7.24 eV, including thermal energy) gave a fine TRPD curve with $k = 7.5 \times 10^{-4}\,s^{-1}$.

The constant problem with multiphoton excitation such as this is to know exactly how many photons are involved. However, the TRPD experiment in this situation has a built-in selection capability, in that one photon at 355 nm has insufficient energy to give any observable dissociation, while three-photon dissociation is so fast ($k \gg 10^6\,s^{-1}$) that it contributes only a nonzero intercept at zero delay time. Thus, all the dissociation observed with finite dissociation lifetime can be attributed to those ions that absorb exactly two photons.

10.3.3.2. Gold Cluster Ions. The other example of two-photon excitation (Vogel et al. 2001) comes from the powerful experimental approaches being developed by the group in Mainz for photodissociation study of large cluster ions. The goal of the TRPD analysis carried out by Vogel et al. is to obtain the E_0 for the dissociation reaction

$$Au_{17}^+ \rightarrow Au_{16}^+ + Au \qquad (10.6)$$

In these experiments, dissociation was initiated using a two-photon excitation that leaves a large residual internal energy in the Au_{16}^{+} fragment, which proceeds to lose another gold atom in the sequence

$$Au_{17}^{+} \xrightarrow{h\nu} Au_{16}^{+*} \rightarrow Au_{15}^{+} \tag{10.7}$$

Separately, a one-photon photoexcitation was used to measure the dissociation reaction

$$Au_{16}^{+*} \xrightarrow{h\nu} Au_{15}^{+} + Au \tag{10.8}$$

The two TRPD curves obtained are shown in Figure 10.18. It can be seen that removing the energy of reaction (10.8) from that of reaction (10.7) will yield a measurement of the desired reaction (10.6). This is achieved graphically in Figure 10.18, by measuring the displacement (3.47 eV) of the TRPD curves for the two reactions. The two curves are displaced on the energy axis by precisely the energy needed to drive reaction (10.6), so that the desired E_0 for reaction (10.6) is about 3.47 eV. A small correction for differential thermal energies of the cluster ions gives a final value of 3.37 eV for reaction (10.6). The authors point out that this graphical subtraction procedure removes any need for kinetic modeling of the dissociation kinetics, giving a "model-free" determination of the binding energy represented by reaction (10.6). It seems likely that this strategy will be useful for a variety of systems, both cluster ions and molecular dissociations, in which both steps of a sequential two-step dissociation can be observed.

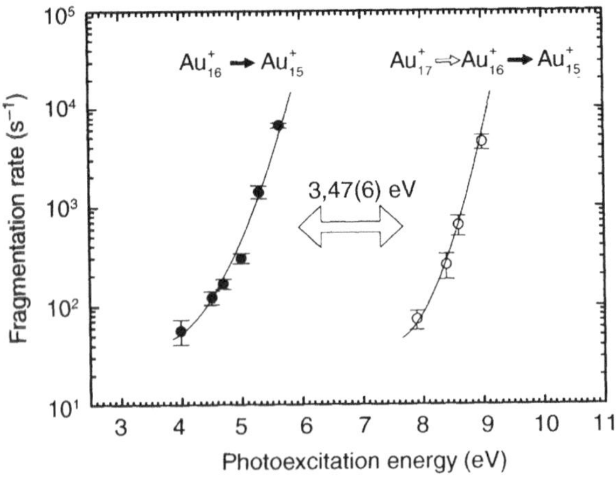

FIGURE 10.18. TRPD plots for two Au_n^+ cluster ions formed by laser vaporization and studied in a Penning trap with time-of-flight mass analysis. A pulsed dye laser tunable in the 2–6 eV photon energy range was the light source. [Reprinted with permission from Vogel et al. (2001). Copyright 2001 by the American Physical Society.]

10.3.4. Distinctive Visible/UV Photofragmentation Patterns

We can regard the pattern of fragmentation products as another way to exploit photodissociation to characterize ions. We want to continue the theme of seeking dissociation chemistry where photoexcitation is distinctively different from other methods. From this perspective, IRMPD fragmentation patterns seem to have no distinctive advantages over other methods, because the continuous-wave (CW) IR laser, like CID or SID excitation, appears to deposit energy essentially in the form of heat in the vibrational degrees of freedom. On the other hand, vis/UV PD is more promising, in that the initial photoexcitation is electronic, and it is common in photochemistry for specific excited electronic states to lead to individual and characteristic reactions. However, unlike the situation with neutral-molecule photochemisty, the literature of ion photofragmentations does not yet provide many verified cases of unique fragmentation patterns associated with particular wavelengths of light. Two promising areas have been described where a vis/UV photon actually does appear to drive unique and distinctive fragmentation chemistry.

10.3.4.1. Internal Charge Transfer Dissociation. Some transition metal complexes have been reported in which photodetachment of the metal leads to the charge ending up on the "wrong" (i.e., less energetically favorable) fragment. These have been called "internal charge transfer processes." Silver ion (which has a relatively high ionization energy) seems prone to such distinctive photochemistry, as for example the photoformation of pyridine$^{+\cdot}$ from the Ag^+-pyridine complex at wavelengths shorter than 341 nm (Yang and Yeh 1999) and benzene$^{+\cdot}$ from the Ag^+–benzene complex at wavelengths shorter than 400 nm (Willey et al. 1991). These cases still seem to be exceptions to the common behavior (at least for organic ions), that photodissociation yields the thermodynamically favored products by a stepwise process: the energy of the photon is initially deposited in an excited electronic state, and then rapidly ($<10^{-9}$ s) undergoes internal conversion into a highly vibrationally excited form of the electronic ground state. This vibrational excitation randomizes through the molecule, leading to dissociation in a quasithermal manner, favoring channels of low activation energy (often referred to as a statistical, or RRKM, pattern of dissociation). There is not enough experience to judge whether other types of systems, such as transition metal complexes or large bioions, will follow this same pattern in general, or whether the "wrong products" will be a common observation in photodissociation chemistry of such systems.

10.3.4.2. Vacuum UV Excitation. One might regard large bioions as a promising place to seek specific dissociation chemistry resulting from photon excitation. The difficulty of putting enough energy into fairly large ions to cause dissociation through an RRKM type of process was discussed above. However, for very large ions, energy randomization through the molecule may not be possible, and fragmentations near the light-absorbing chromophore may be specifically driven. Along these lines, a development of exceptional interest for peptide sequencing was described by Reilly's group (Thompson et al. 2004). This is the induction of a fundamentally new pattern of fragmentation by photoexcitation in the vacuum–ultraviolet wavelength region at 157 nm (8 eV). As background for appreciating the signficance of this observation, two patterns of fragmentation are induced by currently used excitation methods. One is observed for those methods that act essentially by deposition of vibrational energy (CID, metastable decay, SID, IRMPD, low-energy UV/vis PD) and tend to produce *y*- and *a*-type ions. In contrast, electron capture dissociation (ECD) of multiply protonated peptides, inducing dissociation by a charge neutralization mechanism, tends to produce *z*- and *c*-type ions. It was initially hoped that ultraviolet photon PD in the amide band near 190 nm might yield cleavage directly from the excited electronic state, giving a product pattern different from either of these, but in fact no very distinctive pattern has been observed at this wavelength. However, Reilly's group found that excitation at 157 nm, thought perhaps to excite an n–σ^* state, gives a strikingly novel dissociation pattern with a predominance of *x*-, *v*- and *w*-type ions, which is distinctively different from either of the previous patterns.

10.4. FUTURE APPLICATIONS TO BIOMOLECULES

Experience with smaller, simpler systems has lessons for what may or may not be usefully carried over to biomolecules. Many of these have been touched on and explored above, but some summary remarks can be useful.

10.4.1. Spectroscopic Prospects

The emerging body of results on IR spectroscopy of ions using new light sources is highly promising for future gas-phase study of protein ions. The fact that these spectra are readily interpretable in the same terms as for solution phase brings to the gas phase the power of IR spectroscopy for probing features of protein structures and conformations. The significant dependence of the spectra on charge state shows this to be a promising new entry into sorting out charging effects on protein conformation, complementary to the existing approaches such as ion mobility and H/D exchange.

It is both a strength and a weakness of hydrogen stretch spectroscopy (~2800–4000 cm^{-1}) in polymers like proteins that all the monomer subunits contain multiple hydrogens that span many of the possible environments. The spectrum reports hydrogen stretching vibrations corresponding to all of them, and it will be rare to find a situation where an interesting H gives a peak at a unique wavelength not confounded with other uninteresting peaks in the same region. The same problem faces C=O stretching spectroscopy, but is somewhat less severe because there is generally only one C=O group per monomer unit. Most useful for spectroscopic analysis of proteins will be those vibrations, for instance, aromatic ring vibrations, which are localized on, and characteristic of, sidechains of less common amino acid constituents.

IRMPD spectroscopy might be expected to run into difficulty with driving very large systems to dissociation, since such large amounts of energy (as much as hundreds of eV) are required to overcome the kinetic shifts. However, in practice, the spectroscopy of cytochrome *c* described in section 10.2.3.3 shows this not to be a fatal problem. An intense pulsed laser can deliver a very large number of photons into a large molecule, which add to the already large thermal energy, and achieving dissociation is actually quite feasible.

One can expect similarly that the UV/vis spectroscopic approaches illustrated above will continue to be useful for large biomolecules. They should be especially useful for probing characteristic chromophoric groups, analogous to the radical cations discussed in Section 10.2.1, and transition-metal-containing sites that can provide localized chromophores within a large molecular framework.

10.4.2. Activation Perspective

10.4.2.1. Multiphoton Dissociation Experiments. There is an important shift in perspective in considering thermodynamic study of dissociation in small versus large ions. In small molecules, the goal is usually to find the dissociation energy (D_0 or E_0), which is the energy difference of ground-state (zero-*K*) reactants and products. The most admired experiments in this domain are those that cool the reactant ions to essentially zero *K*, and look at the energetics of dissociation to products also at zero *K*. Equivalent to this are similar ideal experiments where the reactant and/or the products are not at zero *K*, but rather in exactly known, monoenergetic excited quantum states. With large molecules, however, the ideal condition of zero-*K* reactants or products is neither achievable nor interesting. The molecules always have a distribution of internal energies, and the ideal situation is where this energy is characterized by a precise Boltzmann thermal distribution. The ideal experiment considers the enthalpy and entropy of dissociation from thermal reactants to thermal products at the same temperature.

The experiments that measure these quantities rigorously are equilibrium experiments, where true thermal equilibrium is measured (as a function of temperature) between reactant molecules and dissociation products. This can sometimes be achieved for ion dissociation processes (Ryzhov and Dunbar 1999). The next best thing is thermal dissociation kinetics (BIRD or thermal CID). In a thermal dissociation kinetics experiment, one is effectively measuring an "equilibrium" between the thermal parent ions and the transition state, giving the activation parameters $\Delta H^{\ddagger}$ and $\Delta S^{\ddagger}$. If the transition state is product-like (as in "barrierless" ion dissociations), these activation parameters can be assumed similar to the true ΔH and ΔS of dissociation. Many of the ionic dissociations that have been worked on have this character. It is from this perspective that the thermodynamic numbers obtained from BIRD or high-pressure thermal dissociation kinetics experiments should be interpreted.

Multiphoton photodissociation experiments on large molecules can be interpreted in a similar spirit, where the laser irradiation is considered to bring about an approximate thermal distribution of the ion population at a high temperature. The fact that this temperature cannot generally be measured, and must be estimated through the intermediary of some theoretical model, is the main obstacle in the way of quantitative thermodynamic studies using this MPD strategy.

The concepts and approaches described above for the intensity dependence analysis of IRMPD results are very relevant to photodissociation studies of truly large biomolecule ions using visible/UV excitation. Because of the massive kinetic shifts and resultant very high effective dissociation thresholds in these systems, many vis/UV photons (typically 4 eV) will be absorbed. Thus the many-photon MPD kinetics analysis needed will be very similar to the IRMPD analysis that was discussed above in the context of medium-sized ions using IR photon energies of the order of 0.1 eV.

10.4.2.2. Threshold Dissociation. Threshold PD is not a very promising approach to thermodynamic measurements as one goes to larger ions, due to the combined problems of large thermal energies and large kinetic shifts. For medium-sized ions (20 atoms), average Boltzmann thermal energies are of the order of 25 kJ/mol at 300 K, so that Boltzmann tail threshold shifts can be of the order of 100 kJ/mol. Kinetic shift errors are even larger. For biomolecules with hundreds or thousands of atoms, threshold measurements of dissociation energies will certainly be meaningless, since effects due to both thermal internal energy and kinetic shifts will be enormous and intractable.

10.4.2.3. Time-Resolved Photodissociation. Dissociation kinetics measurements have more promise of remaining a useful source of thermochemical information, at least for small-biomolecule ions. One particular issue for TRPD of large molecules is the necessity for the population to be thoroughly thermalized at a known temperature before photoexcitation. For large systems collisional cooling and thermalization becomes inefficient. On the other hand, thermalization of the vibrational degrees of freedom by exchange of infrared photons with the cell walls remains efficient with increasing ion size. This is useful for ion trapping techniques where the ions are stored for times of seconds or longer. BIRD studies have shown that large bioions become radiatively equilibrated with the surroundings in a few seconds at temperatures near or above room temperature. On the other hand, temperatures far below room temperature, such as one would like to achieve by supersonic jet cooling, are expected to equilibrate only very slowly by radiative energy exchange, and cooling large molecules to low temperatures would be very difficult.

10.5. CONCLUSION

A number of facets of photodissociation have been explored that have potential relevance and utility for study of large biomolecules. The spectroscopic methods emerging for gas-phase ions in both the infrared and the UV/visible wavelength regions seem eminently transferrable to the domain of very large ions. Thermochemical measurements using the techniques involving discrete photon input are somewhat more problematic, although these methods can and certainly will be extended to small biomolecules. For truly large ions, thermochemistry is most naturally studied by BIRD-related approaches where possible. Beyond this, there is some promise to thermochemical approaches using either IR or UV/visible lasers in an intensity dependence analysis of the multiple-photon dissociation behavior.

ACKNOWLEDGMENTS

The author acknowledges the support of the donors of the Petroleum Research Fund, administered by the American Chemical Society, during the preparation of this chapter.

REFERENCES

Afzaal S, Freiser BS (1994): Gas-phase photodissociation study of Ag(benzene)$^+$ and Ag(toluene)$^+$. *Chem. Phys. Lett.* **218**:254–260.

Bailey TH, Laskin J, Futrell JH (2003): Energetics of selective cleavage at acidic residues studied by time- and energy-resolved surface-induced dissociation in FT-ICR MS. *Int. J. Mass Spectrom.* **222**:313–327.

Barth A, Zscherp C (2002): What vibrations tell us about proteins. *Quart. Rev. Biophys.* **35**:369–430.

Busman M, Rockwood AL, Smith RD (1992): Activation energies for gas-phase dissociations of multiply charged ions from electrospray ionization mass spectrometry. *J. Phys. Chem.* **96**:2397–2400.

Cabarcos OM, Weinheimer CJ, Lisy JM (1999): Size selectivity by cation-pi interactions:-solvation of K^+ and Na^+ by benzene and water. *J. Chem. Phys.* **110**:8429–8435.

Chen Y-M, Armentrout PB (1993): Collision-induced dissociation of $Ag(C_6H_6)^+$. *Chem. Phys. Lett.* **210**:123–128.

Dunbar RC (1991): Kinetics of low-intensity infrared laser photodissociation. The thermal model and applications of the Tolman theorem. *J. Chem. Phys.* **95**:2537–2548.

Dunbar RC (1992): Infrared radiative cooling of gas-phase ions. *Mass. Spectrom. Rev.* **11**: 309–339.

Dunbar RC (1996): New approaches to ion thermochemistry via dissociation and association. In Babcock LM, Adams NG (eds), *Advances in Gas Phase Ion Chemistry*, Vol 2, JAI Press, Greenwich, CT.

Dunbar RC (2003a): Photodissociation spectroscopy. In Armentrout PB (ed), *Encyclopedia of Mass Spectrometry: Chemistry and Physics of Gas-Phase Ions*, Vol. 1, Elsevier, Oxford, pp. 249–261.

Dunbar RC (2003b): Photodissociation studies of ion thermochemistry. In Armentrout PB (ed), *Encyclopedia of Mass Spectrometry: Chemistry and Physics of Gas-Phase Ions*, Vol. 1, Elsevier, Oxford, pp. 403–416.

Dunbar RC (2003c): Black-body infrared radiative dissociation. In Armentrout PB (ed), *Encyclopedia of Mass Spectrometry: Chemistry and Physics of Gas-Phase Ions*, Vol. 1, Elsevier, Oxford, pp. 371–380.

Dunbar RC (2004): BIRD (blackbody infrared radiative dissociation). *Mass. Spectrom. Rev.* **23**:127–158.

Dunbar RC (2005): Photochemistry and spectroscopy of organic ions. In Nibbering NMM (ed), *Encyclopedia of Mass Spectrometry: Fundamentals of and Applications to Organic (and Organometallic) Compounds*, Vol. 4, Elsevier, Oxford, pp. 337–349.

Dunbar RC, Moore DT, Oomens J (2005): IR-spectroscopic characterizatrion of acetophenone complexes with Fe^+, Co^+ and Ni^+ using free-electron-laser IRMPD. *J. Phys. Chem. A* in press.

Duncan MA (2000): Frontiers in the spectroscopy of mass-selected molecular ions. *Int. J. Mass Spectrom.* **200**:545–569.

Faherty KP, Thompson CJ, Aguirre F, Michne J, Metz RB (2001): Electronic spectroscopy and photodissociation dynamics of hydrated Co^{2+} clusters: $Co^{2+}(H_2O)_n$ ($n = 4$–7). *J. Phys. Chem. A* **105**:10054–10059.

Faulk JD, Dunbar RC (1992): Time-resolved photodissociation of gas-phase ferrocene cation: Energetics of fragmentation and radiative rate at near-thermal energies. *J. Am. Chem. Soc.* **114**:8596.

Freitas MA, Hendrickson CL, Marshall AG (1999): Gas phase activation energy for unimolecular dissociation of biomolecular ions determined by focused radiation for gaseous multiphoton energy transfer (FRAGMENT). *Rapid Commun. Mass Spectrom.* **13**:1639–1642.

Freitas MA, Hendrickson CL, Marshall AG (2000): Determination of relative ordering of activation energies for gas-phase ion unimolecular dissociation by infrared radiation for gaseous multiphoton energy transfer. *J. Am. Chem. Soc.* **122**:7768–7775.

Friedrich J, Gilb S, Ehrler OT, Behrendt A, Kappes MM (2002): Electronic photodissociation spectroscopy of isolated IrX_6^{2-} ($X = Cl,Br$). *J. Chem. Phys.* **117**:2536–2544.

Fridgen TD, McMahon TB (2005): IRMPD. In Nibbering NMM (ed), *Encyclopedia of Mass Spectrometry: Fundamentals of and Applications to Organic (and Organometallic) Compounds*, Vol. 4, Elsevier, Oxford, pp. 327–337.

Fu EW, Dunbar RC (1978): Photodissociation spectroscopy and structural rearrangements in ions of cyclooctatetraene, styrene, and related molecules. *J. Am. Chem. Soc.* **100**: 2283.

Fu Z, Russon LM, Morse MD, Armentrout PB (2001): Photodissociation measurements of bond dissociation energies: $D_0(Al_2\text{-}Al)$, $D_0(TiO^+\text{-}Mn)$, and $D_0(V_2{}^+\text{-}V)$. *Int. J. Mass Spectrom.* **204**:143–157.

Gilbert RG, Smith SC (1990): *Theory of Unimolecular and Recombination Reactions*, Blackwell, Oxford.

Hannis JC, Muddiman DC (2002): Tailoring the gas-phase dissociation and determining the relative energy of activation for dissociation of 7-deaza purine modified oligonucleotides containing a repeating motif. *Int. J. Mass Spectrom.* **219**:139–150.

Hettich RL, Freiser BS (1986): Gas-phase photodissociation of $FeCH_2^+$ and $CoCH_2^+$—determination of the carbide, carbyne and carbene bond energies. *J. Am. Chem. Soc.* **108**:2537–2540.

Hettich RL, Freiser BS (1987): Determination of carbide, carbyne, and carbene bond energies by gas-phase photodissociation of $RhCH_2^+$, $NbCH_2^+$ and $LaCH_2^+$. *J. Am. Chem. Soc.* **109**:3543–3548.

Holbrook KA, Pilling MJ, Robertson SH (1996): *Unimolecular Reactions*, Wiley, New York.

Huang F-S, Dunbar RC (1990): Time-resolved photodissociation of methylnaphthalene ion. An illustration of kinetic shifts in large-ion dissociations. *J. Am. Chem. Soc.* **112**:8167–8169.

Huang F-S, Dunbar RC (1991): Time-resolved photodissociation of toluene ion. *Int. J. Mass Spectrom. Ion. Proc.* **109**:151–170.

Husband J, Aguirre F, Thompson CJ, Laperle CM, Metz RB (2000): Photofragment spectroscopy of $FeCH_2^+$, $CoCH_2^+$ and $NiCH_2^+$ near the M^+–CH_2 dissociation threshold. *J. Phys. Chem. A* **104**:2020–2024.

Jockusch RA, Paich K, Williams ER (2000): Energetics from slow infrared multiphoton dissociation of biomolecules. *J. Phys. Chem. A* **104**:3188–3196.

Jockusch RA, Schnier PD, Price WD, Strittmatter EF, Demirev PA, Williams ER (1997): Effects of charge state on fragmentation pathways, dynamics, and activation energies of ubiquitin ions measured by blackbody infrared radiative dissociation. *Anal. Chem.* **69**:1119–1126.

Johnson MA (2003): Vibrational predissociation. In Armentrout PB (ed), *Encyclopedia of Mass Spectrometry: Chemistry and Physics of Gas-phase Ions*, Vol. 1, Elsevier, Oxford; pp. 271–276.

Laskin J, Futrell JH (2005): Activation of large ions in FT-ICR mass spectrometry. *Mass Spectrom. Rev.* **24**:135–167.

Lemaire J, Boissel P, Heninger M, Mauclaire G, Bellec G, Mestdagh H, Simon A, LeCaer S, Ortega JM, Glotin F, Maître P (2002): Gas phase infrared spectroscopy of selectively prepared ions. *Phys. Rev. Lett.* **89**:273002–1.

Leussing DL, Hansen RL (1957): The copper(II)-pyridine complexes and their reactions with hydroxide ions. *J. Am. Chem. Soc.* **79**:4270–4273.

Lifshitz C (1997): Energetics and dynamics through time-resolved measurements in mass spectrometry: Aromatic hydrocarbons, polycyclic aromatic hydrocarbons and fullerenes. *Int. Rev. Phys. Chem.* **16**:113–139.

Lifshitz C (2002): Kinetic shifts. *Eur. J. Mass Spectrom.* **8**:85–98.

Lin C-Y, Dunbar RC (1995): Time-resolved photodissociation of gas phase nickelocene cation: Determination of bond strength and radiative relaxation rate. *J. Phys. Chem.* **99**:1754–1759.

Lucas B, Gregoire G, Lemaire J, Maître P, Ortega J-M, Rupenyan A, Reimann B, Schermann JP, Desfrancois C (2004): Investigation of the protonation site in the dialanine peptide by infrared multiphoton dissociation spectroscopy. *Phys. Chem. Chem. Phys.* **6**:2659–2663.

Meot-Ner M, Dongre AR, Somogyi A, Wysocki VH (1995): Thermal decomposition kinetics of protonated peptides and peptide dimers, and comparison with surface-induced dissociation. *Rapid Commun. Mass Spectrom.* **9**:829–836.

Metz RB (2004a): Photofragment spectroscopy of covalently bound transition metal complexes: A window into C–H and C–C bond activation by transition metal ions. *Int. Rev. Phys. Chem.* **23**:79–108.

Metz RB (2004b): Optical spectroscopy and photodissociation dynamics of multiply charged ions. *Int. J. Mass Spectrom.* **235**:131–143.

Moore DT, Oomens J, Eyler JR, von Helden G, Meijer G, Dunbar RC (2005): Infrared spectroscopy of gas-phase Cr^+ coordination complexes: Determination of binding sites and electronic states. *J. Am. Chem. Soc.* **127**:7243–7254.

Oh H, Breuker K, Sze SK, Carpenter BK, McLafferty FW (2002): Secondary and tertiary structures of gaseous protein ions characterized by electron capture mass spectrometry and photofragment spectroscopy. *Proc. Natl. Acad. Sci. USA* **99**:15863–15868.

Oh HB, Lin C, Hwang HY, Zhai H, Breuker K, Zabrouskov V, Carpenter BK, McLafferty FW (2005): Infrared photodissociation spectroscopy of electrosprayed ions in a Fourier-transform mass spectrometer. *J. Am. Chem. Soc.* **127**:4076–4083.

Oomens J, Moore DT, von Helden G, Meijer G, Dunbar RC (2004): The site of Cr^+ attachment to gas-phase aniline from infrared spectroscopy. *J. Am. Chem. Soc.* **126**:724–725.

Oomens J, Polfer N, Moore DT, van der Meer L, Marshall AG, Eyler JR, Meijer G, von Helden G (2005): Charge-state resolved mid infrared spectroscopy of a gas-phase protein. *Phys. Chem. Chem. Phys.* **7**:1345–1348.

Paech K, Jockusch RA, Williams ER (2002): Slow infrared laser dissociation of molecules in the rapid energy exchange limit. *J. Phys. Chem. A* **106**:9761–9766.

Price WD, Schnier PD, Jockusch RA, Strittmatter EF, Williams ER (1996): Unimolecular reaction kinetics in the high-pressure limit without collisions. *J. Am. Chem. Soc.* **118**: 10640–10644.

Puskar L, Cox H, Goren A, Aitken GDC, Stace AJ (2003): Ligand-field spectroscopy of Cu(II) and Ag(II) complexes in the gas phase: Theory and experiment. *Faraday Discuss.* **124**:259–273.

Ranatunga DRA, Freiser BS (1995): Gas-phase photodissociation of $MC_2H_2^+$ (M = Zr, Nb)—determination of $D_0(M^+\text{-}C_2H_2)$. *Chem. Phys. Lett.* **233**:319–323.

Riveros JM (2003): Infrared photodissociation. In Armentrout PB (ed), *Encyclopedia of Mass Spectrometry: Chemistry and Physics of Gas-phase Ions*, Vol. 1, Elsevier, Oxford, pp. 262–269.

Russon LM, Heidecke SA, Birke MK, Conceicao J, Morse MD, Armentrout PB (1994): Photodissociation measurements of bond dissociation energies—Ti_2^+, V_2^+, CO_2^+, and CO_3^+. *J. Chem. Phys.* **100**:4747–4755.

Ryzhov V, Dunbar RC (1999): Direct associative equilibrium in the FT-ICR ion trap: The hydration equilibrium of protonated 18-crown-6. *J. Am. Soc. Mass Spectrom.* **10**: 862–868.

Schnier PD, Price WD, Jockusch RA, Williams ER (1996): Blackbody infrared radiative dissociation of bradykinin and its analogs: Energetics, dynamics, and evidence for salt-bridge structures in the gas phase. *J. Am. Chem. Soc.* **118**:7178–7189.

Schnier PD, Price WD, Strittmatter EF, Williams ER (1997): Dissociation energetics and mechanisms of leucine enkephalin $(M + H)^+$ and $(2M + X)^+$ ions (X = H, Li, Na, K, and Rb) measured by blackbody infrared radiative dissociation. *J. Am. Soc. Mass Spectrom.* **8**:771–780.

Simon A, Jones W, Ortega J-M, Boissel P, Lemaire J, Maître P (2004): Infrared multiphoton dissociation spectroscopy of gas-phase mass-selected hydrocarbon-Fe^+ complexes. *J. Am. Chem. Soc.* **126**:11666–11674.

Spence TG, Trotter BT, Posey LA (1998): Influence of sequential solvation on metal-to-ligand charge transfer in bis(22′2″-terpyridyl)iron(II) clustered with dimethyl sulfoxide. *J. Phys. Chem. A* **102**:7779–7786.

Thompson CJ, Faherty KP, Stringer KL, Metz RB (2005): Electronic spectroscopy and photodissociation dynamics of Co^{2+}-methanol clusters: $Co^{2+}(\text{methanol})_n$, ($n = 4$–7). *Phys. Chem. Chem. Phys.* **7**:814–818.

Thompson MS, Cui W, Reilly JP (2004): Fragmentation of singly-charged peptide ions by photodissociation at lambda = 157 nm. *Angew. Chem. Int. Ed.* **43**:4791–4794.

Vaden TD, Weinheimer CJ, Lisy JM (2004): Evaporatively cooled $M^+(H_2O)Ar$ cluster ions: Infrared spectroscopy and internal energy simulations. *J. Chem. Phys.* **121**:3102–3107.

Vogel M, Hansen K, Herlert A, Schweikhard L (2001): Model-free determination of dissociation energies of polyatomic systems. *Phys. Rev. Lett.* **87**:013401.

Walters RS, Schleyer PvonR, Corminboeuf C, Duncan MA (2005): Structural trends in transition metal cation-acetylene complexes revealed through the C-H stretching fundamentals. *J. Am. Chem. Soc.* **127**:1100–1101.

Willey KF, Cheng PY, Bishop MB, Duncan MA (1991): Charge-transfer photochemistry in ion-molecule cluster complexes of silver. *J. Am. Chem. Soc.* **113**:4721–4728.

Willey KF, Yeh CS, Robbins DL, Duncan MA (1992): Charge-transfer in the photodissociation of metal ion-benzene complexes. *J. Phys. Chem.* **96**:9106–9111.

Yang Y-S, Yeh C-S (1999): Photodissociative charge transfer in Ag^+-pyridine complex. *Chem. Phys. Lett.* **305**:395–400.

11

CHEMICAL DYNAMICS SIMULATIONS OF ENERGY TRANSFER AND UNIMOLECULAR DECOMPOSITION IN COLLISION-INDUCED DISSOCIATION (CID) AND SURFACE-INDUCED DISSOCIATION (SID)

ASIF RAHAMAN AND WILLIAM L. HASE
Department of Chemistry and Biochemistry
Texas Tech University
Lubbock, TX

KIHYUNG SONG
Department of Chemistry
Korea National University of Education
Chongwon, Chungbuk, Korea

JIANGPING WANG
Department of Chemistry
Wayne State University
Detroit, MI

SAMY O. MEROUEH
Department of Chemistry and Biochemistry
University of Notre Dame
Notre Dame, IN

Principles of Mass Spectrometry Applied to Biomolecules, edited by Julia Laskin and Chava Lifshitz

11.1. INTRODUCTION

Since the early 1960s (Bunker 1962; Blais and Bunker 1962) trajectories, with proper selection of initial conditions, have been used to simulate the dynamics of atomic and molecular collisions, including chemical reactions. The trajectory simulations give an atomic-level description of experiments and provide the type of fundamental information required to develop theoretical models of chemical dynamics. Comparisons between trajectory simulations and both experiments and quantum-dynamics calculations have shown that trajectories often give a correct description of the chemical dynamics (Hase 1998a). For many gas-phase chemical processes classical trajectories give accurate cross sections, rate constants, product energy partitionings, energy transfer probabilities, and other parameters. Quantum effects, contributing to the inaccuracy of classical mechanics, include tunneling through potential energy barriers, transitions between electronic states, and unphysical flow of zero-point energy.

Classical trajectories give accurate results for the energy transfer dynamics associated with collision-induced dissociation (CID) and surface-induced dissociation (SID) (Schultz et al. 1997; Bosio and Hase 1998; Meroueh and Hase 2001; Liu et al. 2003; Martinez-Nuñez et al. 2005). This is expected because of the high collision energy and the large densities of internal states for the projectile ion and surface in CID and SID. In addition, the energy transfer is a direct process, and possible unphysical flow of zero-point energy during the collision is not an important issue. Abrupt, short-time unimolecular dissociation of the projectile, identified as "shattering," is also expected to be accurately represented by the trajectories (Grebenshchikov et al. 2003). For projectile dissociations that occur on a longer timescale, dissociation may be assisted by flow of zero-point energy into the reaction coordinate, resulting in a unimolecular rate constant that is too large and inaccurate (Hase and Buckowski 1982).

In this chapter the use of classical trajectory chemical dynamics simulations to study energy transfer and unimolecular decomposition in CID and SID are reviewed. The methodology of these simulations is described first. Specific simulations of CID and SID are then reviewed, and important findings of this work with respect to experiment and fundamental theories of chemical dynamics are discussed. The chapter ends with some perspectives of future directions of trajectory simulations of CID and SID.

11.2. METHODOLOGY OF CID AND SID SIMULATIONS

In a classical trajectory study the motions of the individual atoms are simulated by solving the classical equations of motion. For chemical dynamics simulations, these equations have traditionally been expressed in the form of Hamilton's equations (Goldstein 1950)

$$\frac{\partial H}{\partial q_i} = \frac{-dp_i}{dt} \quad \text{and} \quad \frac{\partial H}{\partial p_i} = \frac{dq_i}{dt} \tag{11.1}$$

where H, the sum of the kinetic $T(\mathbf{p}, \mathbf{q})$ and potential energies $V(\mathbf{q})$, is the system's Hamiltonian:

$$H = T(\mathbf{p}, \mathbf{q}) + V(\mathbf{q}) \tag{11.2}$$

In the molecular dynamics community (Schlick 2002) the classical equations have usually been expressed in Newton's form

$$m_i \frac{d^2 q_i}{dt^2} = -\frac{\partial V(\mathbf{q})}{\partial q_i} \tag{11.3}$$

which assume Cartesian coordinates. For the most general case, such as for internal coordinates (Wilson et al. 1955), T depends on both the momenta $\mathbf{p}$ and the coordinates $\mathbf{q}$. The index i in the equations above is the number of coordinates or conjugate momenta for the Hamiltonian. If Cartesian coordinates are used, this number is $3N$, where N is the number of atoms.

There are several components to a classical trajectory chemical dynamics simulation. A potential energy function $V(\mathbf{q})$ for the system under study must be chosen. An ensemble of trajectories is calculated, with each trajectory specified by the system's initial set of momenta $\mathbf{p}$ and coordinates $\mathbf{q}$. The initial ensemble of $\mathbf{p}$ and $\mathbf{q}$ is chosen to represent the experiment under investigation or chosen so that a particular dynamical attribute of the system may be studied. Distribution functions are usually sampled randomly in choosing the ensemble of initial conditions, and the methodology of sampling is often called Monte Carlo sampling (Blais and Bunker 1962; Bunker 1962; Peslherbe et al. 1999). An algorithm is required for integrating the classical equations of motion. When the trajectory is completed, the final values of the momenta and coordinates are transformed into properties that may be compared with experiment, such as atomic-level reaction mechanisms, product vibration, rotation, and translation energies, scattering angles, the lifetime of a vibrationally excited molecule, and energies in a molecule's vibrational modes versus time. These components for a classical trajectory simulation are incorporated in the general chemical dynamics program VENUS. The Cartesian coordinate representation is the most general for systems of any size, and VENUS uses Cartesians.

Detailed descriptions have been given previously of the abovementioned components for a classical trajectory simulation (Sun and Hase 2003; Bolton et al. 1998; Hase 1998b,c; Peslherbe, et al. 1999; Bolton and Hase 1998). In the following section a brief description of these components is given, as applied to CID and SID.

11.2.1. Potential Energy Functions

The potential energy function used for the SID simulations is written as

$$V = V_{\text{projectile}} + V_{\text{surface}} + V_{\text{projectile, surface}} \tag{11.4}$$

where $V_{\text{projectile}}$ is the projectile's intramolecular potential, V_{surface} is the surface potential, and $V_{\text{projectile, surface}}$ is the projectile–surface intermolecular potential. For CID the potential used is

$$V = V_{\text{projectile}} + V_{\text{projectile, Rg}} \tag{11.5}$$

where Rg represents a rare-gas atom.

Equation (11.4) is an approximation for the exact SID potential. Although the potential describes projectile dissociation on impact with the surface, possible reaction(s) of the projectile with the surface are not included. Also, a fixed set of parameters is used for the $V_{\text{projectile, surface}}$ intermolecular potential. If the structures of the projectile and/or surface change, the charge distribution on the atoms may change, which may affect the intermolecular parameters. Such changes in the intermolecular potential are not included in this potential model with fixed parameters. There is the same limitation for $V_{\text{projectile, Rg}}$ in Eq. (11.5). The models represented by Eqs. (11.4) and (11.5) are expected to give correct energy transfer results and projectile dynamics for direct impulsive collisions, when the projectile and surface have structures for which the intermolecular parameters were obtained. This is the manner in which the simulations were performed and, as is seen below where the simulation results are discussed, excellent agreement is obtained with many features of the experiments.

If the projectile sticks to the surface or the rare gas, and its conformation changes, the nature of the intermolecular potential may also change. The best way to model these types of collisions is by a direct dynamics quantum-mechanical (QM) calculation for a Rg + projectile system (Liu et al. 2003) and a quantum-mechanical/molecular mechanical (QM/MM) calculation for a projectile + surface system (Li et al. 2000). In these methods the QM theory will directly provide the intermolecular potential and represent any changes in its form.

V_{surface} is an analytic function with parameters so that properties of the surface agree with those from experiment. The $V_{\text{projectile, surface}}$ intermolecular potential is a sum of two-body potentials between the atoms of the projectile and those of the surface, and $V_{\text{projectile, Rg}}$ is a sum of two-body potentials between the Rg atom and atoms of the projectile.

Both analytic and electronic structure theory models were used for the projectile's potential energy function, $V_{\text{projectile}}$. The analytic potential models used for the Al_n and $Cr^+(CO)_6$ clusters properly describe their dissociations. On the other hand, the analytic potentials used for protonated peptides are molecular mechanics (MM) quadratic forcefield and nonbonded potential models, and dissociation pathways are not represented. To study the dissociation dynamics of peptide ions, the AM1 semiempirical QM electronic structure theory model was used for the peptide's potential. The classical trajectory chemical dynamics with AM1 are QM+MM direct dynamics simulations (Sun and Hase 2003; Hase et al. 2003), with MM-type analytic functions for the remaining potential energy terms. Details of the specific models used for $V(\mathbf{q})$ are described below as part of the simulation results. More accurate QM theories such as DFT or MP2 may be used for peptides with one or two amino acids. However, the use of such higher-level theories are not computationally practical for large peptides. For these peptides, a semiempirical method with a floating-occupation molecular orbital configuration interaction wavefunction (Granucci and Toniolo 2000) may be practical and sufficiently accurate QM method.

11.2.2. Trajectory Initial Conditions

11.2.2.1. CID Simulations. To choose initial conditions for a trajectory representing the collision of a rare-gas atom with a polyatomic ion involves first randomly sampling the coordinates and momenta of the ion and then selecting random conditions for the relative properties between the atom and polyatomic ion collision partners. In experiments, the polyatomic ion is prepared at a temperature T. To simulate this condition, the quasiclassical method (Truhlar and Muckerman 1979) is used to randomly sample the ion's rotational and vibrational energy levels and then transform them to the Cartesian coordinates and momenta used for the trajectory. The following are the specific steps performed for this sampling procedure:

1. The normal-mode frequencies $\boldsymbol{\omega} = 2\pi\mathbf{v}$ and normal-mode eigenvector $\mathbf{L}$ for the polyatomic ion are determined by diagonalizing its mass weighted Cartesian force constant matrix (Califano 1976).
2. The harmonic oscillator QM Boltzmann distribution

$$P(n_i) \propto \exp\left[-\left(n_i + \frac{1}{2}\right)h\nu_i/kT\right] \tag{11.6}$$

is randomly sampled for each of the ion's normal modes to select its vibrational quantum number n_i. The energy for the normal mode is

$$E_i = \left(n_i + \frac{1}{2}\right)h\nu_i.$$

3. The polyatomic ion is assumed to be a symmetric top (i.e., this approach is also accurate for spherical tops, but approximate for asymmetric tops) and its angular momentum j and component j_z along the z axis are found by sampling their classical Boltzmann distributions (Bunker and Goring-Simpson 1973). The resulting projection of j into the xy plane [i.e., $(j^2 - j_z^2)^{1/2}$] is randomly assigned onto the x and y axes. The random values of j_x, j_y, j_z, form the rotational angular momentum vector $\mathbf{j}$ for the polyatomic ion.
4. The energy for each normal mode, selected in step 1, is expressed classically as

$$E_i = \frac{P_i^2 + \omega_i^2 Q_i^2}{2} \tag{11.7}$$

where the P_i and Q_i are the normal-mode momenta and coordinates. Random values for P_i and Q_i, with energy E_i, are chosen by giving each normal mode a random phase. This gives random **Q** and **P** for the normal modes. The total energy for the polyatomic ion E° is a sum of its vibration and rotation energies, i.e.

$$E^\circ = \sum_{i=1}^{3N-6} E_i + 3RT/2 \tag{11.8}$$

5. The **Q** and **P** are transformed to Cartesian coordinates **q** and momenta **p** for the polyatomic ion's N atoms using the normal-mode eigenvector **L** (Califano 1976):

$$\mathbf{q} = \mathbf{q}_0 + \mathbf{M}^{(-1/2)}\mathbf{LQ} \tag{11.9a}$$

$$\mathbf{p} = \mathbf{M}^{1/2}\mathbf{LP} \tag{11.9b}$$

where $\mathbf{q}_0$ is a matrix of equilibrium coordinates and **M** is the diagonal matrix whose elements are the atomic masses.

6. Since normal modes are approximate for finite displacements (Califano 1976), a spurious angular momentum $\mathbf{j}_s$ arises following this transformation (Chapman and Bunker 1975; Sloane and Hase 1977). The spurious momentum is found from

$$\mathbf{j}_s = \sum_{i=1}^{N} \mathbf{r}_i \times m_i \dot{\mathbf{r}}_i \tag{11.10}$$

where m_i is the mass of the ith atom and $\mathbf{r}_i$ is its position vector. The desired angular momentum **j** is added to the molecule by forming the vector

$$\mathbf{j}_a = \mathbf{j} - \mathbf{j}_s \tag{11.11}$$

and adding the rotational velocity $\boldsymbol{\omega} \times \mathbf{r}_i$ to each of the atoms, where

$$\boldsymbol{\omega} = \boldsymbol{I}^{-1}\mathbf{j}_a \tag{11.12}$$

and $\boldsymbol{I}^{-1}$ is the inverse of the inertia tensor (Goldstein 1950).

7. The actual internal energy E for the Cartesian coordinates and momenta, chosen from steps 1–6, is calculated using the correct Hamiltonian and compared with the intended energy E°. If they do not agree within some acceptance criterion, the Cartesian coordinates and momenta are scaled by $(E^\circ/E)^{1/2}$. Any spurious center of mass translational energy is subtracted from the molecule, and the procedure loops back to step 6. If E is within the acceptance criterion of E°, the selected coordinates **q** and momenta **p** of the polyatomic ion are saved.

The following steps are carried out to choose relative properties between the polyatomic ion and rare-gas atom:

8. To study energy transfer for ion dissociation, random collision impact parameter b is chosen between 0 and b_{max}, where b_{max} is the largest impact parameter leading to energy transfer in excess of the ion's dissociation threshold. b_{max} is determined by tests. An approximate way to choose b_{max} is to assume a value for the collision cross section, πb_{max}^2.
9. The polyatomic ion's Cartesian coordinates and momenta, selected above, are randomly rotated through Euler's angles (Wilson et al. 1955) to give a random orientation

$$\mathbf{q} = \mathrm{ROT}(\theta, \phi, \chi)\mathbf{q} \qquad \dot{\mathbf{q}} = \mathrm{ROT}(\theta, \phi, \chi)\dot{\mathbf{q}} \tag{11.13}$$

where $\mathrm{ROT}(\theta, \phi, \chi)$ is the Euler rotation matrix (Wilson et al. 1955).
10. Since the polyatomic ion A has a random orientation in a space-fixed coordinate frame, the B atom may be placed in the y,z plane without loss of generality. The x,y,z coordinates of B are then

$$x = 0 \qquad y = 0 \qquad z = (s^2 - b^2)^{1/2} \tag{11.14}$$

where s is the initial separation between B and the A center of mass and b is the impact parameter.
11. The A + B relative velocity v_{rel} is now added along the z axis with the restraint that the A + B center of mass remains at rest. The space-fixed Cartesian momenta are then

$$\mathbf{p} = \mathbf{M}\left(\dot{\mathbf{q}} - \dot{\mathbf{q}}_{\mathbf{rel}}\right) \tag{11.15}$$

The elements of the relative velocity $\dot{\mathbf{q}}_{\mathbf{rel}}$ are zero for the x and y components and equal to $[m_A/(m_A + m_B)]v_{rel}$ for the z component of atom B, and equal to $-[m_A/(m_A + m_B)]v_{rel}$ for the z component of each atom of A.

This concludes the section on random Cartesian coordinates and momenta for a CID trajectory.

11.2.2.2. SID Simulations. For a SID trajectory, initial conditions are chosen for the polyatomic projectile ion, the surface, and relative properties for the projectile–surface collision. Random Cartesian coordinates and momenta are chosen for the projectile ion as described by steps 1–7 and 9 above.

Harmonic quasiclassical (Bosio and Hase 1997), harmonic classical (Song et al. 1995), and anharmonic classical (Meroueh and Hase 2001) approaches have been used to choose random coordinates and momenta for the surface at temperature T_s. Since zero-point energy is added to the surface in the quasiclassical method, the total energy added to the surface is substantially larger for this initial condition sampling approach as compared to either the harmonic or anharmonic classical sampling. Tests of possible effects of this different surface energy for energy transfer in gas–surface collisions has been investigated in collisions of rare-gas atoms with alkyl thiolate self-assembled monolayer (SAM) surfaces at much lower collision energies (Yan and Hase 2002) than those for SID. Within the statistical uncertainties of these simulations, the quasi-classical harmonic and classical anharmonic sampling methods give the same energy transfer efficiencies. The effect of zero-point energy and classical versus quantum T_s is expected to be even less important for the high collision energies of SID. Each of these sampling methods is discussed below, and classical anharmonic sampling is preferred since all potential energy minima and structures are sampled in this approach.

If the model for the surface is sufficiently small, a normal-mode analysis, as described in step 1 above, may be performed to determine the vibrational frequencies and eigenvectors for the surface normal modes of vibration (i.e., phonons). For quasiclassical sampling of the surface's coordinates and momenta at T_s, the above steps 1–7 are performed with the surface's rotational angular momentum $\mathbf{j}$ set to zero. This quasiclassical approach samples a QM temperature for T_s, since Eq. (11.6) is used to sample the quantum numbers for the surface's vibrational modes. To sample a classical T_s temperature for the surface, the energy for each surface mode is randomly selected according to the classical statistical mechanical distribution function

$$P(E_i) = \exp(-E_i/RT_s)/RT_s \tag{11.16}$$

which gives an average mode energy of RT_s. Initial coordinates and momenta are then chosen for the surface, using steps 1–7 above with the preceding classical mechanical algorithm replacing step 2.

The classical mechanical sampling method described above assumes the molecule vibrates about its potential energy minimum with harmonic frequencies. This model is accurate for a surface, such as diamond, with a single potential energy minimum populated at the temperature T_s and whose frequencies are accurately approximated by the harmonic oscillator model. However, for many surfaces, which have soft forces (i.e., low frequencies) and multiple potential energy minima, this model is inappropriate. Liquid and solid hydrocarbon and self-assembled monolayer (SAM) surfaces fall into this class. Classical initial conditions for such surfaces may be selected by performing a standard molecular dynamics (MD) simulation (Allen and Tildesley 1987), which samples the surface's anharmonic potential. Initial random velocities are chosen for the surface atoms by sampling each atom's Maxwell–Boltzmann distribution at T_s. The surface is then equilibrated by a MD trajectory simulation, with velocity rescaling, so there is equilibration between the surface's kinetic and potential energies for T_s resulting in random $\mathbf{p}$ and $\mathbf{q}$ for the surface.

The remaining step in the sampling for SID is to choose the relative properties for the projectile–surface collisions. The projectiles collide with a collision energy E_i and incident angle θ_i. Their impact points with the surface are random. Two approaches have been used to sample random relative properties for the projectile surface collisions. For one (Bosio and Hase 1997), a beam of projectiles, as illustrated in Figure 11.1, is directed toward the surface as in a SID experiment. The

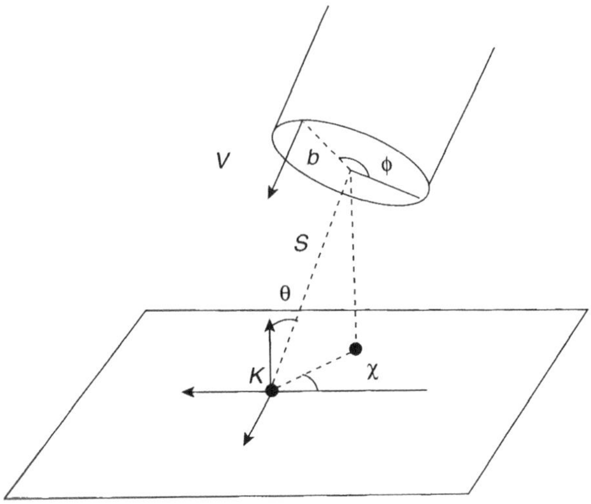

FIGURE 11.1. Definition of coordinates used to sample initial conditions for a SID simulation. [Reprinted with permission from Bosio and Hase (1997).]

projectile ions are randomly placed in a circular area of the beam, with radius r, so that the area spanned by the beam overlaps a unit area on the surface. A projectile is placed randomly in the beam by choosing a random value of b between 0 and b_{max} (see step 8 above for CID simulations) and a random value for the ϕ angle between 0 and 2π. Then incident angle θ_i is fixed and the center of the beam of projectiles is aimed at the middle of the surface's center unit cell (i.e., the aiming point K). The distance S is between the center of the beam and the aiming point K. A shortcoming of this approach is that a single unit cell on the surface is not sampled uniformly.

In experiments, there are often different domains of growth on the surface and each domain may have a different orientation with respect to the beam of projectiles. This orientation is identified by the azimuthal angle χ in Figure 11.1. If it were possible to scatter projectiles off individual domains of growth, for which the orientation is known, it would be possible to perform experiments with χ selected. On the other hand, if the domains on which the projectiles collide is not controlled, there is probably a random distribution of χ for the projectile–surface collisions. To model this latter experimental condition, χ is randomly chosen from a uniform distribution between 0 and 2π.

In a second approach for choosing properties of the projectile–surface collisions, projectiles with E_i and θ_i are randomly aimed at points within the central unit cell of the surface, which provides uniform sampling of impact sites in the unit cell. As described above, the collisions may have either a fixed or a random azimuthal angle χ. Some experiments are performed with large domains of regular, periodic structures on the surface, with a well-defined unit cell (Isa et al. 2004). This is an excellent method for modeling such systems. It should be noted that simulations of peptide–H^+ SID have shown that the energy transfer dynamics is insensitive to the surface site impacted by the center of mass of the peptide ion (Rahaman et al.

2006). Such a result is not unexpected given the size of a peptide ion and the multiple surface sites its atoms strike. Thus, for simulations of peptide–H^+ SID, periodic surface models may be adequate for representing experimental surfaces that may be rough at the molecular level.

11.3. SIMULATIONS OF CID

Classical trajectory simulations of the CID of Al_6 and Al_{13} clusters (de Sainte Claire et al. 1995; de Sainte Claire and Hase 1996) and of protonated polyglycine and polyalanine (Meroueh and Hase 1999, 2000) and $Cr^+(CO)_6$ (Martinez-Nuñez et al. 2005) have provided an atomic-level description of the dynamics of energy transfer in CID and the ensuing unimolecular fragmentation. The simulations have shown how the CID dynamics depend on the structure and vibrational frequencies (i.e., stiffness) of the cluster, the mass of the collision partner, repulsiveness of the intermolecular potential between the cluster and collision partners, and the collision's relative velocity. The simulations are based on accurate intermolecular potentials derived from high-level ab initio calculations. Large collision energies were considered in the simulations, and thus it is the short-range region of the intermolecular potential that is important for the CID dynamics. Obtaining this part of the potential accurately, from ab initio calculations, is much less challenging than determining the attractive potential and its energy minimum, since electron correlation is much less important for the repulsive component as compared to the attractive component. Both analytic functions and electronic structure methods were used to represent the intramolecular potential energy functions of the clusters. In the following these simulations are reviewed and a number of important findings obtained from these simulations are discussed.

11.3.1. Rare-Gas Collisions with Al_6 and Al_{13} Clusters

Chemical dynamics simulations of aluminum cluster CID (de Sainte Claire and Hase 1996; de Sainte Claire et al. 1995) are discussed below.

11.3.1.1. Potential Energy Functions. Two different Al_6 analytic potential energy functions were used in CID. One consists of two-body Lennard-Jones (L-J) terms

$$V_{ij} = \frac{A}{r_{ij}^{12}} + \frac{B}{r_{ij}^{6}} \tag{11.17}$$

and three-body Axilrod–Teller (A-T) terms

$$V_{ijk} = C\frac{1 + 3\cos\alpha_1\cos\alpha_2\cos\alpha_3}{\left(r_{ij}r_{jk}r_{ki}\right)^3} \tag{11.18}$$

with parameters $A = 2{,}975{,}343.77$ kcal · Å^{12}/mol, $B = -17{,}765.823$ kcal · Å^6/mol, and $C = 81{,}286.093$ kcal · Å^9/mol determined by a fit to ab initio calculations for Al_n ($n = 2$–$6, 13$) clusters (Petterson et al. 1987). The r_{ij}, r_{jk}, r_{ki} and $\alpha_1, \alpha_2, \alpha_3$ in Eq. (11.18) represent the sides and angles, respectively, of the triangle formed by the three particles i, j, and k.

The second Al_6 analytic function consists of two-body Morse terms with the exponential β parameter written as a cubic function of r_{ij} (Hase et al. 1987), that is

$$V_{ij} = D_e\{1 - \exp[-\beta(r_{ij} - r_e)]\}^2 - D_e \tag{11.19a}$$

$$\beta = \beta_e + c_2\Delta r^2 + c_3\Delta r^3 \tag{11.19b}$$

where $\Delta r = r_{ij} - r_e$. Parameters in this Morse function were varied to determine how "stiffness" of the Al_6 cluster affects energy transfer. The octahedral (O_h) Al_6 cluster was used for this study. Parameters for the Morse potentials, identified as Morse$^*n$ and Morse/n, were chosen so that they give the same equilibrium geometry and dissociation energy for Al_6 (O_h) as the L-J/A-T potential presented above but give vibrational frequencies for Al_6 (O_h) that are n times larger and smaller, respectively, than those of the L-J/A-T potential. The L-J/A-T potential of Eqs. (11.17) and (11.18), with the parameters given above, was used to represent the potential energy surface for the different Al_{13} clusters.

Equations (11.17) and (11.18) give multiple potential energy minima for Al_6 and Al_{13}, with a wide variety of structures. The structures considered in the simulations are shown in Figure 11.2. They are octahedral O_h and planar C_{2h} for Al_6, and D_{3d} [face-centered cubic (fcc)] and planar D_{2h} and D_{6h} for Al_{13}. The threshold energies for dissociation of these clusters to $Al_5 + Al$ and to $Al_{12} + Al$ are as follows: Al_6 (C_{2h}), 43.8 kcal/mol; Al_6 (O_h), 38.8 kcal/mol; Al_{13} (D_{2h}), 57.5 kcal/mol; Al_{13} (D_{6h}), 42.3 kcal/mol; and D_{3d} (fcc), 16.8 kcal/mol.

Intermolecular potentials for Ne, Ar, and Xe rare-gas (Rg) atom collisions with Al_6 and Al_{13} were written as a sum of two-body potentials between the Rg and Al atoms. Both Rg–Al and Rg–Al^+ potentials were calculated using unrestricted MP2 theory (Head-Gordon et al. 1988; Frisch et al. 1988) with the 6-31G* basis set (Hariharan and Pople 1973, 1974). At long range the Rg–Al^+ potential is more attractive than that for Rg–Al because of the former's ion-induced dipole interaction. However, for the high translational energies of CID, it is the short-range potential that is important, and Rg–Al and Rg–Al^+ have similar short-range potentials. Also, the positive charge is delocalized for the Al_n^+ cluster and, as a result, the Rg–Al two-body potential for this cluster should be intermediate of those for Rg–Al and Rg–Al^+. For the energies of CID, Rg–Al_n and Rg–Al_n^+ are expected to have similar intermolecular potentials.

A very accurate fit to the Ar–Al potential is obtained with

$$V = \frac{a}{r^{12}} + \frac{b}{r^6} + c\exp(-dr) \tag{11.20}$$

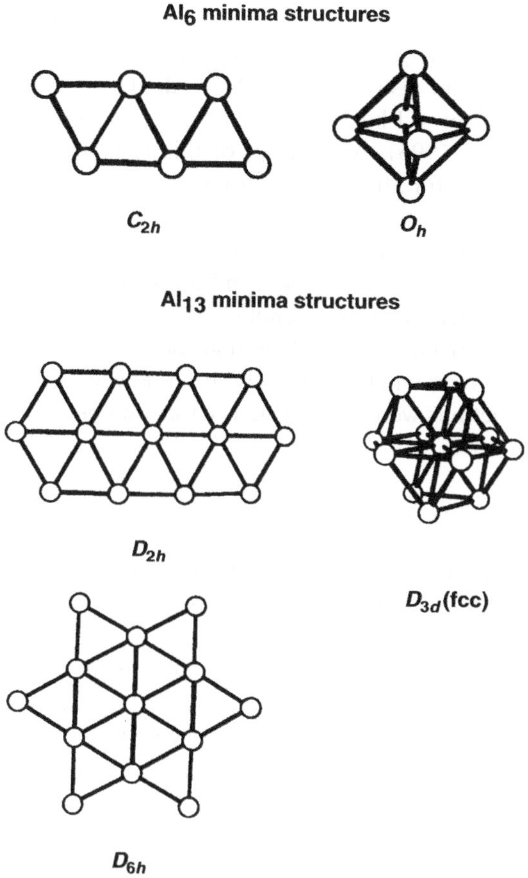

FIGURE 11.2. Al_6 and Al_{13} structures modeled in the CID simulations. [Adapted with permission from de Sainte Claire et al. (1995).]

where r is the distance between Ar and the Al atom. The parameters in this equation are $a = 5212.13281\,\text{kcal}\cdot\text{Å}^{12}/\text{mol}$, $b = 884.959534\,\text{kcal}\cdot\text{Å}^{6}/\text{mol}$, $c = 10,421.443\,\text{kcal/mol}$, and $d = 2.75301790\,\text{Å}^{-1}$. The short-range repulsive regions of the Rg–Al potentials are accurately described by

$$V = V_0 \exp\left(\frac{-r}{L}\right) \tag{11.21}$$

where L is the range parameter for the intermolecular interaction. The Ne–Al, Ar–Al, and Xe–Al potentials were fit with Eq. (11.21), and the V_0 and L parameters are $V_0 = 17642.938$ kcal/mol and $L = 0.2819$ Å for Ne, $V_0 = 17352.530$ kcal/mol and $L = 0.3503$ Å for Ar, and $V_0 = 18479.848$ kcal/mol and $L = 0.4019$Å for Xe.

11.3.1.2. Simulation Results. Initial conditions were chosen for the trajectories to represent CID experiments (Jarrold et al. 1987; Hanley et al. 1987). A 138 K rotation energy of $RT/2$ was added to each principal axis of the cluster, to give a total initial cluster rotational energy E_{rot} of 0.4 kcal/mol. A test showed that the CID cross section and energy transfer distribution for Ar + Al_6 (O_h) are insensitive to whether the zero-point energy of 4.1 kcal/mol is added to Al_6 in the initial conditions (de Sainte Claire et al. 1995). As a result of this test, initially vibrationless Al_n clusters were considered for all the simulations.

11.3.1.2.1. CID Cross Sections and T → V and T → R Energy Transfer. The final cluster internal energy was evaluated for each trajectory, and the difference between the final and initial cluster energies is denoted ΔE_{int}. The cluster is assumed to have sufficient energy to dissociate if its final internal energy is larger than its dissociation threshold, which is the assumption of RRKM unimolecular rate theory (Baer and Hase 1996). Cross sections were calculated as a function of collision energy E_{rel} and fit by

$$\sigma_{CID} = A(E_{rel} + E_{rot} - E_0^{CID})^n / E_{rel} \tag{11.22}$$

Similar CID cross sections were found for Al_6 (O_h) and Al_6 (C_{2h}). For example, at E_{rel} of 120.8 kcal/mol σ_{CID} is 21.7 and 22.8 Å^2 for Al_6 (O_h) and Al_6 (C_{2h}), respectively. Their σ_{CID} fitted parameters for Eq. (11.22) are $A = 40.0$ Å^2, $E_0^{CID} = 52.7$ kcal/mol, and $n = 0.99$ for Al_6 (O_h), and for Al_6 (C_{2h}), $A = 44.6$ Å^2, $E_0^{CID} = 54.9$ kcal/mol, and $n = 0.98$. For $n = 1$, Eq. (11.22) reduces to the well-known hard-sphere model, in which the effective energy is directed along the line of centers and the parameter A equals πb_{max}^2. That the fitted value of n is near unity indicates a hard-sphere and line-of-centers energy transfer model adequately describes both Al_6 (O_h) and Al_6 (C_{2h}). By assuming this model, a b_{max} value of 3.6 Å is extracted from the A parameter for Al_6 (O_h), which is in agreement with the b_{max} values determined directly from the trajectories.

The similar cross sections for Al_6 (O_h) and Al_6 (C_{2h}) arise from different attributes of the collision. The planar cluster has b_{max} of 5.5 Å, that is, 2.0 Å larger than that for the octahedral cluster. Also for $b \leq b_{max}$ a larger fraction of the trajectories for the planar cluster do not transfer sufficient energy for CID to occur. These two effects combine to give similar CID cross sections for the two clusters. The b_{max} value of 3.8 Å extracted from the A parameter for Al_6 (C_{2h}) is considerably smaller than $b_{max} = 5.5$ Å for the trajectories. Trajectories with relatively small impact parameters do not collide with the planar C_{2h} cluster, if the collision's relative velocity is parallel to the cluster's surface plane.

Energy transfer distribution functions for the planar and octahedral Al_6 clusters are compared in Figure 11.3. For both clusters there are no trajectories which transfer all of E_{rel} (i.e., 120.8 kcal/mol) to cluster internal energy. The two distribution functions are quite different, with the increase in the energy transfer probability from zero at large ΔE_{int} much more pronounced for the octahedral cluster. Most of the energy transfer to the Al_6 (O_h) cluster is in the form of vibration. For $E_{rel} = 120.8$ kcal/mol, the average energy transfer to cluster vibration is 35.5 kcal/mol. The average energy transfer to rotation is much lower: only

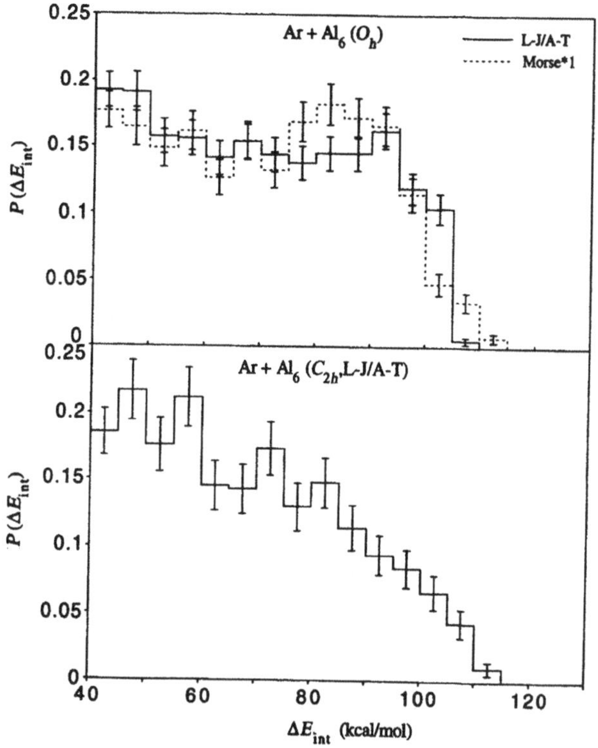

FIGURE 11.3. Ar + Al_6 energy transfer distributions for $E_{rel} = 120.8$ kcal/mol. $P(\Delta E_{int})$ is in units of $(\text{kcal/mol})^{-1}$, and the distributions are normalized so that the total probability is unity for transferring energy in excess of the true dissociation threshold E_0. The probabilities on the y axis have been multiplied by a factor of 10. [Reprinted with permission from de Sainte Claire et al. (1995).]

4.8 kcal/mol. Unfortunately, the distinction between transfer to vibration and rotation was not determined for Al_6 (C_{2h}), but because of its planar structure transfer to rotation is expected to be significant (see discussion below).

CID cross sections were calculated for the D_{3d} (fcc), D_{6h}, and D_{2h} Al_{13} clusters at $E_{rel} = 120.8$ kcal/mol and a cluster rotational temperature T_{rot} of 138 K. The dissociation threshold of the spherical-type D_{3d} (fcc) cluster is substantially smaller than those of the planar D_{2h} and D_{6h} clusters (see numbers above). The former cluster is compact and has a b_{max} for CID equal to 5.0 Å, while b_{max} is 6.5 Å for the planar less compact clusters. The CID cross sections calculated from the trajectories are 54.7, 47.7, and 35.4 Å^2, and the fraction of the trajectories that transfer sufficient energy for CID to occur is 0.70, 0.36, and 0.27 for the D_{3d} (fcc), D_{6h}, and D_{2h} clusters, respectively. These latter fractions are consistent with the different dissociation thresholds for the clusters.

For the planar D_{2h} and D_{6h} clusters a large fraction, 39% and 35%, respectively, of the energy transfer is to rotation. For the spherical D_{3d}, the fraction is much smaller and only 7%. Larger rotational energy transfer is expected for D_{2h} and D_{6h}

clusters in comparison to the D_{3d} cluster, because of the former's planar, anisotropic structures and larger collision impact parameters.

11.3.1.2.2. Comparison of E_0^{CID} and E_0, the True Dissociation Threshold. An important finding from the fits of σ_{CID} versus E_{rel} is that the derived values of E_0^{CID} are larger than the actual E_0 thresholds. For Al_6 (O_h) E_0^{CID} is 52.7 kcal/mol, in contrast to $E_0 = 38.8$ kcal/mol. For Al_6 (C_{2h}) $E_0^{CID} = 54.9$ kcal/mol, while E_0 is 43.8 kcal/mol. As a check of the trajectory fit for Al_6 (O_h), an additional calculation was performed at $E_{rel} = 50$ kcal/mol and, of the 600 trajectories calculated, not one transferred sufficient energy for CID to occur. The maximum ΔE_{int} was 37.3 kcal/mol, which gives a maximum $\Delta E_{int}/E_{rel}$ energy transfer value of 37.3/50.0 = 0.75. The accuracy of the fitted threshold for Al_6 (C_{2h}) was tested by calculating an additional 600 trajectories at each E_{rel} of 53.0, 50.0, and 47.0 kcal/mol. The number of CID events found for these energies are 10, 3, and 0, respectively, which gives cross sections of 1.3, 0.4, and 0.0 Å^2. Including these low-energy cross sections in the fit by Eq. (11.22), gives $E_0^{CID} = 52.7$ kcal/mol. This E_0^{CID} value is 2.2 kcal/mol smaller than the one obtained without the three low-energy points, but still 9 kcal/mol larger than E_0, the true threshold!

The implication from the trajectories is that for Ar + Al_6 collisions, with E_{rel} near the actual dissociation threshold, there are no collisions that transfer all of E_{rel} to the cluster, and the fitted CID threshold is not the true dissociation threshold. This is an important point since CID experiments are used to determine thresholds (Su and Armentrout 1993). The actual and CID dissociation thresholds for Ar + Al_6 may differ because the system is far from the sudden limit (see following Section 5) for small E_{rel} values. As discussed below, changing the mass of the incident atom, the intermolecular potential, and/or the cluster stiffness can increase the efficiency of energy transfer, so that the CID threshold approaches the true threshold. Given the significance of the Ar + Al_6 simulation results, additional work on this system, both experimental and computational, seems warranted. The simulations agree with experimental findings for Ar + Al_6^+ CID. Product branching ratios and cross sections have been measured (Jarrold et al. 1987) for CID of Al_n^+ clusters ($n = 3$–26) by Ar atoms at a center-of-mass collision energy of 121 kcal/mol. Al^+ is observed to be the main product for clusters with fewer than 15 atoms. The trajectory results agree with this finding. For Al_6^+, the experimental cross section is 21 Å^2, which is similar to the trajectory value of 20–25 Å^2.

11.3.1.2.3. Factors Influencing Energy Transfer in Al_6 CID. The effect of varying the incident atom's mass on the energy transfer to Al_6 is illustrated in Table 11.1, where average energies transferred and CID cross sections are listed for $E_{rel} = 120.8$ kcal/mol. As the mass of the incident atom is increased, there is more energy transfer to the cluster and the CID cross section becomes larger. At an incident atom mass of 2–3 times the mass of argon, a limit of 0.45 seems to be reached for the average energy transfer efficiency.

Simulations with the Morse*n and Morse/n potentials show how cluster stiffness affects Ar + Al_6 (O_h) energy transfer. The Morse*1 potential gives the same

TABLE 11.1. Dependence of A + Al_6 (O_h) Energy Transfer on Mass of A

m_A	$\langle \Delta E_{int} \rangle / E_{rel}$[a]	σ_{CID}(Å^2)	Fraction CID
$3m_{Ar}$	0.45	24.8 ± 0.3	0.64
$2m_{Ar}$	0.43	22.7 ± 0.3	0.59
m_{Ar}	0.40	21.7 ± 0.4	0.56
$m_{Ar}/2$	0.33	18.2 ± 0.4	0.47
$m_{Ar}/4$	0.25	12.6 ± 0.3	0.33

[a]The standard deviation of the mean for each value is less than 0.005. Three thousand trajectories were evaluated for each mass of A.

structure and dissociation energy for Al_6 (O_h) as does the L-J/A-T potential, but has slightly different vibrational frequencies. The Morse*n and Morse/n potentials are the same as Morse*1, except their vibrational frequencies are n times larger and smaller, respectively.

Calculations for the Al_6 (O_h) Morse potentials were performed with E_{rel} = 120.8 kcal/mol, and the results are given in Table 11.2. The energy transfer efficiency is strongly dependent on cluster stiffness (i.e., efficient for soft clusters and inefficient for stiff clusters), and there appear to be soft and stiff asymptotic limits. For the Morse*1 and softer clusters, energy transfer is efficient and the CID cross section is large (i.e., >20 Å^2). The average energy transfer to vibration and rotation (i.e., $\langle \Delta E_{vib} \rangle$ and $\langle \Delta E_{rot} \rangle$) are also included in Table 11.2. For the soft clusters the transfer is primarily to vibration, but to rotation for the stiff clusters. Transfers to vibration and rotation were not determined for the Morse/4, Morse/2, and L-J/A-T potentials. However, the result for the L-J/A-T potential should be similar to that for the Morse*1 potential. Rotational energy transfer for the Morse/4 and Morse/2 potentials is expected to be lower than for the Morse*1 potential.

TABLE 11.2. Dependence of Ar + Al_6 (O_h) Energy Transfer on the Cluster Stiffness[a]

Potential Surface	$\langle \Delta E_{int} \rangle / E_{rel}$[c]	$\langle \Delta E_{vib} \rangle$	$\langle \Delta E_{rot} \rangle$	σ_{CID}(Å^2)	Fraction CID
Morse/4[b]	0.48	—[d]	—	26.3 ± 0.3	0.68
Morse/2	0.47	—	—	25.5 ± 0.3	0.66
L-J/A-T	0.40	—	—	21.7 ± 0.4	0.56
Morse*1	0.42	35.5	4.8	23.1 ± 0.3	0.60
Morse*2	0.25	19.4	10.9	13.5 ± 0.3	0.35
Morse*4	0.17	6.3	13.6	6.9 ± 0.3	0.18
Morse*6	0.13	1.5	14.7	4.2 ± 0.2	0.11
Morse*8	0.13	0.6	15.1	3.0 ± 0.2	0.08

[a]Energies are in kcal/mol. The collision energy E_{rel} = 120.8 kcal/mol.
[b]The Morse/n potentials have vibrational frequencies n times smaller than the actual values, while for the Morse*n potentials the frequencies are n times larger.
[c]The standard deviation of the mean for each value is less than 0.005. Three thousand trajectories were evaluated for each potential surface.
[d]$\langle \Delta E_{vib} \rangle$ and $\langle \Delta E_{rot} \rangle$ values were not evaluated.

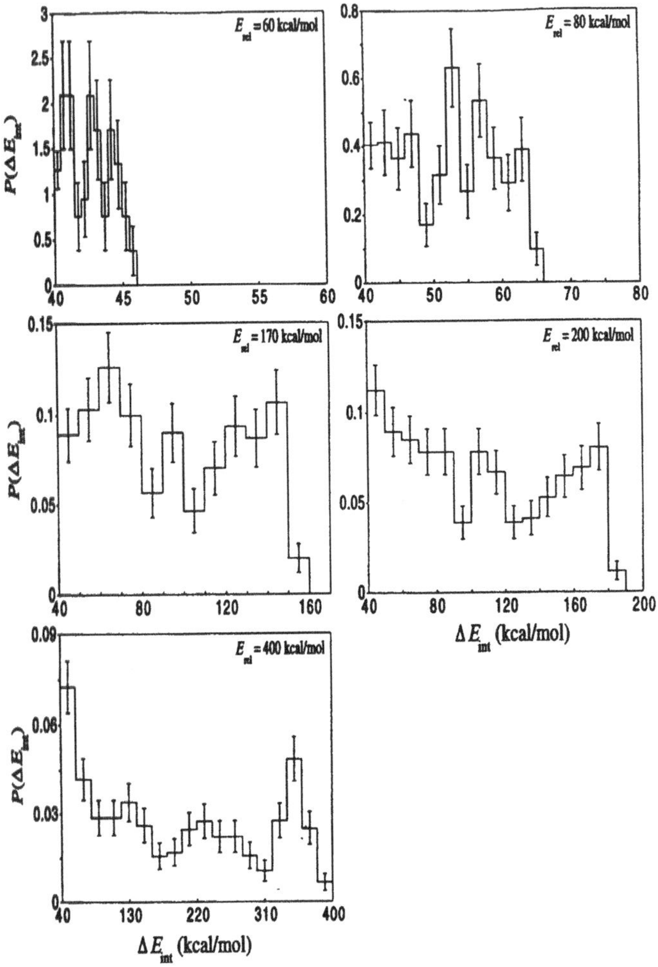

FIGURE 11.4. Energy transfer distributions for Ar + $Al_6(O_h)$ on the L-J/A-T surface for different collision energies. [Reprinted with permission from de Sainte Claire et al. (1995).]

The energy transfer efficiency also depends on the collision energy. Figure 11.3 shows that, for the Al_6 (O_h) cluster and $E_{rel} = 120.8$ kcal/mol, the maximum value of ΔE_{int} transferred divided by E_{rel} is ≈ 0.9. Figure 11.4 shows that, for Ar + Al_6 (O_h) collisions, there is a higher probability of transferring large fractions of E_{rel} as E_{rel} is increased. For $E_{rel} = 400$ kcal/mol, there are collisions in which all of E_{rel} is transferred to ΔE_{int} and the maximum $\Delta E_{int}/E_{rel}$ equals unity. For E_{rel} of 60 and 80 kcal/mol, the maximum $\Delta E_{int}/E_{rel}$ are much lower: approximately 0.77 and 0.82, respectively. As discussed above, for $E_{rel} = 50$ kcal/mol the maximum $\Delta E_{int}/E_{rel}$ is 0.75. The inability to transfer all of E_{rel} to ΔE_{int} as E_{rel} is decreased, is the origin of E_0^{CID} values from the simulations larger than the true thresholds.

The intermolecular potential between the rare-gas atom and the aluminum cluster also affects the energy transfer. This is illustrated in Figure 11.5, where the CID

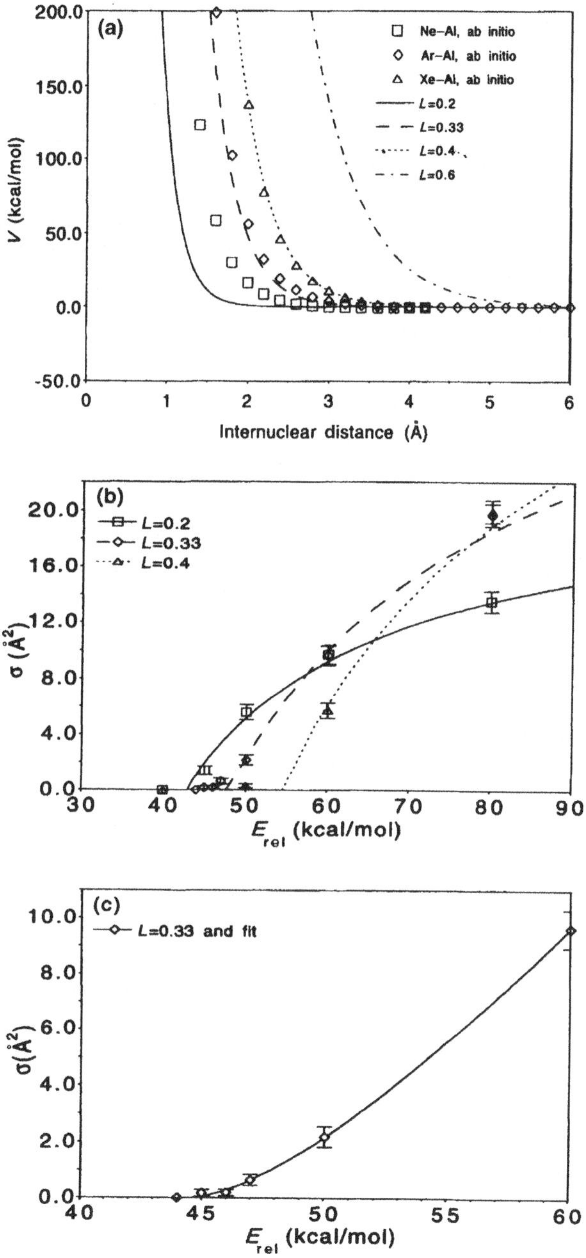

FIGURE 11.5. (a) Different Rg–Al intermolecular potentials. The curves for different L values are plots of Eq. (11.21), with $V_0 = 20{,}257.938$ kcal/mol. (b) Calculated and fitted σ_{CID} versus E_{rel} for Xe + Al_6, using three of the intermolecular potentials in (a). (c) Fit to the σ_{CID} at low E_{rel}, for the curve in (b) with $L = 0.33$. [Adapted with permission from de Sainte Claire and Hase (1996).]

TABLE 11.3. Fits of Eq. (11.22) to Al_6 (O_h) CID Cross Sections versus E_{rel}[a]

Rare Gas	Cluster	Potential: Cluster	Potential: Intermolecular	A	E_0^{CID}	n
Ar	Al_6	L-J/A-T	Eq. (11.20)	40.0	52.7	0.99
Xe	Al_6	L-J/A-T	Eq. (11.21), $L = 0.2$[b]	48.3	43.4	0.86
Xe	Al_6	L-J/A-T	Eq. (11.21), $L = 0.33$[b]	63.3(6.91)[c]	48.5(44.9)	0.91(1.61)
Xe	Al_6	L-J/A-T	Eq. (11.21), $L = 0.4$[b]	81.0	55.0	0.90

[a] A is in units of Å^2 and E_0^{CID} in kcal/mol. The fits are shown in Figure 11.5.
[b] $V_0 = 20{,}257.938$ kcal/mol and L is in units of Å.
[c] The numbers in the parentheses are fit to the cross sections at low E_{rel}.

cross sections versus E_{rel} are plotted for Xe + Al_6 (O_h), using the L-J/A-T potential for the cluster. The Xe–Al_6 intermolecular potential is written as a sum of Xe–Al two body interactions given by Eq. (11.21), with $V_0 = 20257.938$ and a value of either 0.2, 0.33, or 0.4 for the range parameter L. Both the high-energy, asymptotic value of σ_{CID} and CID threshold increase as L is increased. The former is expected, since the Xe + Al_6 collision radius increases as L is increased. Apparently, increasing L decreases the probability of energy transfer, and as a result the CID threshold increases. Values of the parameters in Eq. (11.22), which give fits to the σ_{CID}–E_{rel} curves in Figure 11.5, are listed in Table 11.3. The A parameter increases from 48.3 to 81.0 Å and E_0^{CID} increases from 43.4 to 55.0 kcal/mol, as L is increased from 0.2 to 0.4 Å. The fitted values of n are near unity, which indicates that a hard-sphere and line-of-centers energy transfer model is appropriate.

For the Xe + Al_6 calculation with $L = 0.33$, σ_{CID} was calculated for a range of low E_{rel} values. Fitting the low E_{rel} points by Eq. (11.22), as shown in the bottom graph in Figure 11.5, gives an E_0^{CID} value only 3–4 kcal/mol lower than that found from the complete σ_{CID}–E_{rel} curve, which suggests that a meaningful E_0^{CID} value may be obtained by fitting the complete σ_{CID} curve. On the other hand, a value of n significantly larger than unity is obtained for the fits to σ_{CID} at low E_{rel}. This is because near threshold the collisions are not impulsive (Mahan 1970; Yardley 1980; Shin 1976) and the line-of centers models [i.e., Eq. (11.22) with $n = 1$] no longer applies. Values of n larger than unity have been observed in fits of experimental σ_{CID} curves by Eq. (11.22) (Lian et al. 1992; Su and Armentrout 1993).

11.3.1.3. Qualitative Model for Energy Transfer in CID. Insight into the dynamics of energy transfer in CID can be acquired by considering the refined impulsive model developed by (Mahan 1970; Yardley 1980) for translation to vibration ($T \rightarrow V$) energy transfer in collinear A + BC collisions. The model accounts for the mass terms and interactions between the BC vibration and Fourier components of the intermolecular interaction between A and B. For this model, the fraction of initial relative translational energy transferred to BC vibration is

$$\frac{\Delta E}{E_{rel}} = 4\cos^2\beta \sin^2\beta \left(\frac{\xi}{2}\operatorname{cosech}\frac{\xi}{2}\right)^2 \tag{11.23}$$

where

$$\cos^2\beta = \frac{m_A m_C}{(m_A + m_B)(m_B + m_C)} \tag{11.24}$$

The ξ term in Eq. (11.23) is often called the *adiabaticity parameter* and is given by

$$\xi = 4\pi^2 \nu L / v_{rel} \tag{11.25}$$

where ν is the BC vibrational frequency, v_{rel} is the A + BC initial relative velocity, and L is the range parameter for the A–B interaction, given by Eq. (11.21). In this equation r is the A–B internuclear separation. For small ξ the collisions are in the sudden limit, and $\Delta E/E_{rel}$ reaches its maximum value

$$\frac{\Delta E_{sudden}}{E_{rel}} = 4\cos^2\beta\sin^2\beta \tag{11.26}$$

Thus, according to the definition of ξ in Eq. (11.25), energy transfer reaches its maximum efficiency for a small BC vibrational frequency, short-range parameter, and/or large relative velocity between the collision partners. As discussed above, and reviewed below, these are the dynamical properties that give efficient energy transfer in the trajectory simulations of Al_n CID:

Effect of Al_n Vibrational Frequency. Table 11.2 shows that the CID cross section increases as the cluster frequencies are lowered, which is in accord with the impulsive model.

Effect of Rg–Al_n Intermolecular Potential. Table 11.3 and Figure 11.5 show that the efficiency of energy transfer to the Al_n cluster is increased at low E_{rel} as the range parameter L becomes smaller and the Rg–Al interaction becomes more short-range. This is the effect predicted by the adiabaticity parameter.

Effect of Collision Energy. The adiabaticity parameter becomes smaller, and the collision approaches the sudden limit, as the collision velocity v_{rel} is increased. Thus, more efficient energy transfer is predicted at large collision energies. Figure 11.4 illustrates this effect, where at $E_{rel} = 400$ kcal/mol there are collisions that transfer all of E_{rel} to Al_6 internal energy, while at $E_{rel} = 60$ kcal/mol the maximum fraction of E_{rel} transferred is 0.77.

Overall, Mahan's impulsive model provides a good qualitative understanding of the energy transfer dynamics for Rg + Al_n CID.

11.3.2. Ar Atom Collisions with N-Protonated Glycine and Alanine Polypeptides

Chemical dynamics simulations of N-protonated polyglycine and polyalanine CID (Meroueh and Hase 1999, 2000) are discussed below.

11.3.2.1. Potential Energy Functions. The protonated peptide potential was represented by both the Amber valence forcefield (Cornell et al. 1995) and the AM1 semiempirical electronic structure theory (Dewar et al. 1985; Stewart 1989). For the Amber model the potential is

$$V_{\text{peptide}} = \sum_{\text{bonds}} K_r(r - r_{\text{eq}})^2 + \sum_{\text{angles}} K_\theta(\theta - \theta_{\text{eq}})^2 + \sum_{\text{dihedrals}} \frac{V_n}{2}[1 - \cos(n\phi - \gamma)] + \sum_{i>j} [A_{ij}/r_{ij}^{12} - B_{ij}/r_{ij}^{6} + q_i q_j/(\varepsilon r_{ij})] \tag{11.27}$$

Values for the potential parameters are derived to fit properties of biological molecules (Cornell et al. 1995). For the AM1 model, there is no analytic potential energy function, and instead V_{peptide} and the derivatives of this potential with respect to the Cartesian coordinates are determined at each integration step of the classical trajectory by solving the time-independent Schrödinger equation for the AM1 semiempirical model. Such a calculation is called a direct dynamics simulation (Sun and Hase 2003; Hase et al. 2003). The simulations reported below indicate that either the Amber or AM1 potential energy model may be used to study the dynamics of collisional energy transfer. However, only AM1 may be used to study peptide ion dissociation.

The Ar–peptide intermolecular potential is written as a sum of two-body interactions of the form

$$V(r) = ae^{-br} + \frac{c}{r^9} \tag{11.28}$$

between the Ar atom and the atoms of the peptide. Since high-energy collisions are considered in the CID simulation and the short repulsive region of the intermolecular potential is critical to the energy transfer, no attempt was made to represent the shallow attractive potential energy minima between Ar and the peptide's atoms. The two-body parameters in Eq. (11.28) for the Ar–peptide potential were determined by using the small molecules CH_4, NH_3, NH_4^+, and H_2CO to represent the functional groups for the N-protonated polyglycine and polyalanine peptides. Ab initio calculations, at the QCISD(T)/6-311++G** level of theory, were used to calculate an intermolecular potential between Ar and each of these model molecules. These ab initio potentials were then fit by a sum of the two-body function in Eq. (11.28) to derive parameters for Ar interacting with the atoms of the model molecules; for example, a, b, and c parameters for Ar–H and Ar–C two-body interactions were determined from the Ar–CH_4 ab initio intermolecular potential.

11.3.2.2. Simulation Results. Quasiclassical normal mode sampling was used to select initial coordinates and momenta for the peptide ions. Energies for the peptides normal modes of vibration were selected from their 300 K Boltzmann distribution. A 300 K rotational energy of *RT*/2 was added to each of the peptide's principal axis of

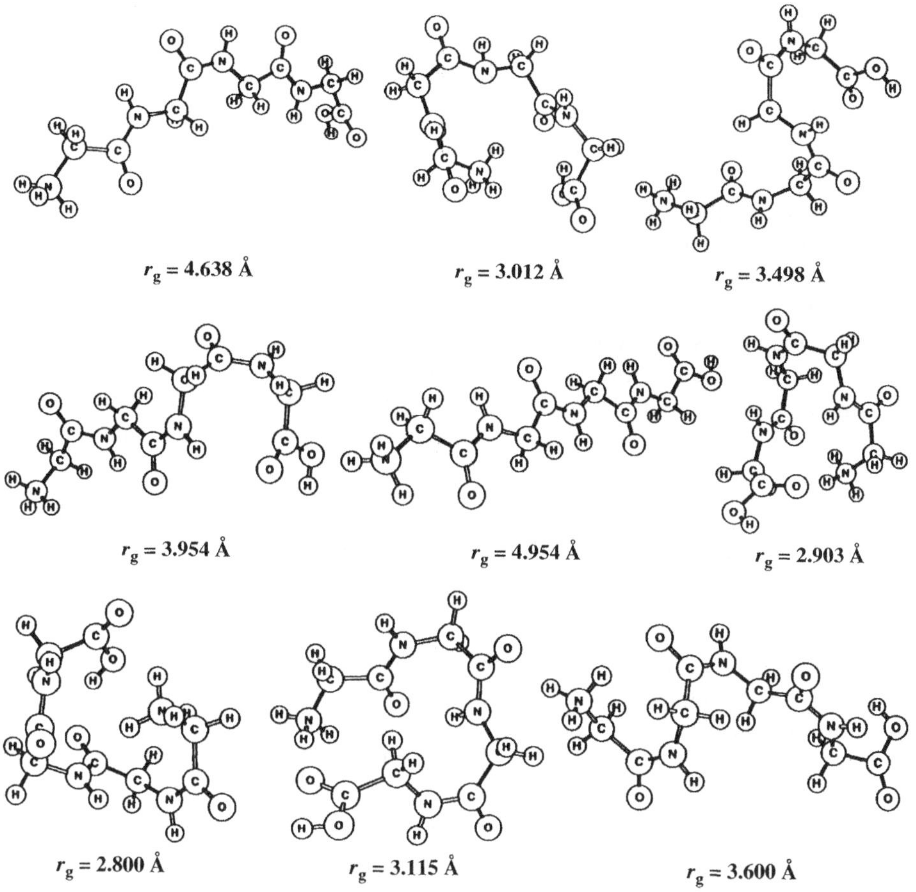

FIGURE 11.6. Structures of gly$_4$-H$^+$ investigated for energy transfer efficiency. The radius of gyration is given for each structure. [Reprinted with permission from Meroueh and Hase (1999).]

rotation. Argon–peptide collision energies of 100, 500, and 1000 kcal/mol were considered with the energy transfer dynamics studied versus impact parameter *b*. Different polypeptide structures were investigated and those for protonated tetraglycine (gly$_4$-H$^+$) are shown in Figure 11.6. The results of the simulations are described in the following text.

11.3.2.2.1. Comparison of Amber and AM1 Peptide Intramolecular Potentials. Amber represents the vibrational motion of the peptide ions and, thus, may be used to model collisional energy transfer to the peptide. However, since Amber is a molecular mechanics (MM) model, it does not describe the peptide ion's dissociation pathways. An important question is whether this incompleteness in the Amber potential affects the collisional energy transfer in the simulations. This question was addressed by also using the AM1 semiempirical quantum chemistry model for the peptide ion's potential in direct dynamics simulations of Ar + peptide–H^+ and peptide–H^+ + surface collisional energy transfer (Meroueh et al. 2002; Wang et al. 2003b). The peptide ions considered are gly-H^+ and gly_2-H^+. What was discovered is that the Amber and AM1 models for the peptide–H^+ intramolecular potential give the same efficiencies for collisional energy transfer. The comparison for Ar + gly-H^+ at $E_i = 70$ eV, $\theta_i = 45°$, and $b = 0$ is shown in Figure 11.7. For the Amber potential the percent energy transfers to ΔE_{int}, ΔE_{surf}, and E_f are 11, 37, and 52, respectively. Using AM1 potential, these percentages are nearly identical: 12, 38, and 50. A similar result is found for collisional energy transfer to the gly_2-H^+ peptide. Thus, these calculations show that the Amber potential may be used to study collisional energy transfer in peptide–H^+ CID, and the results described below were determined using AMBER for the peptide-H^+ intramolecular potential. The agreement between the AMBER and AM1 energy transfer efficiencies for both gly-H^+ and gly_2-H^+, indicates the agreement is not coincidental and both models adequately represent the peptide–H^+ intramolecular potential. Thus Amber, which is more computationally efficient, may be used to study the dynamics of energy transfer.

11.3.2.2.2. Role of the Collision Impact Parameter. The percent energy transfer to the internal degrees of freedom of β-sheet polyglycine peptides [i.e., β-(gly_n-H^+)] are shown in Figure 11.8 as a function of the collision impact parameter b for $n =$ 2–7 and a collision energy of 100 kcal/mol. As one would expect, the average fraction of energy transfer decreases with increase in impact parameter, with the most rapid decline for the smallest peptide. At $b = 0$ the peptides have average energy transfer efficiencies in the range ~50–60 %. The largest b at which measurable energy transfer occurs varies from 7 to 16 Å in going from β-(gly_2-H^+) to β-(gly_7-H^+).

Because of a finite energy transfer for all impact parameters, unambiguously defining b_{max} remains a difficult problem if one wishes to determine the complete energy transfer distribution curve $P(\Delta E_{int})$, starting with $\Delta E_{int} = 0$. In contrast, if the $P(\Delta E_{int})$ that is of interest is the one with $\Delta E_{int} > E_0$, where E_0 is the ion's dissociation threshold, an unambiguous b_{max} may be determined. A quantity that may be converged from the trajectory simulations is the average energy transfer versus impact parameter $\langle \Delta E(b) \rangle$ integrated over the differential cross section $2\pi b\, db$ (Peslherbe et al. 1999):

$$\langle \Delta E \rangle_\sigma = \int_0^\infty \langle \Delta E(b) \rangle 2\pi b\, db \tag{11.29}$$

Values for $\langle \Delta E \rangle_\sigma$ versus the size of the β-(gly_n-H^+) peptide increase nearly linearly from 1000 kcal Å^2/mol for β-(gly_2-H^+) to 7169 kcal Å^2/mol for β-(gly_7-H^+), for $E_{rel} = 100$ kcal/mol. A value for $\langle \Delta E \rangle$, the average energy transfer averaged over b, may be determined by dividing $\langle \Delta E \rangle_\sigma$ by an assumed collision cross section πb_{max}^2. The collision cross section is often equated to the Lennard-Jones cross section (Whyte et al. 1988).

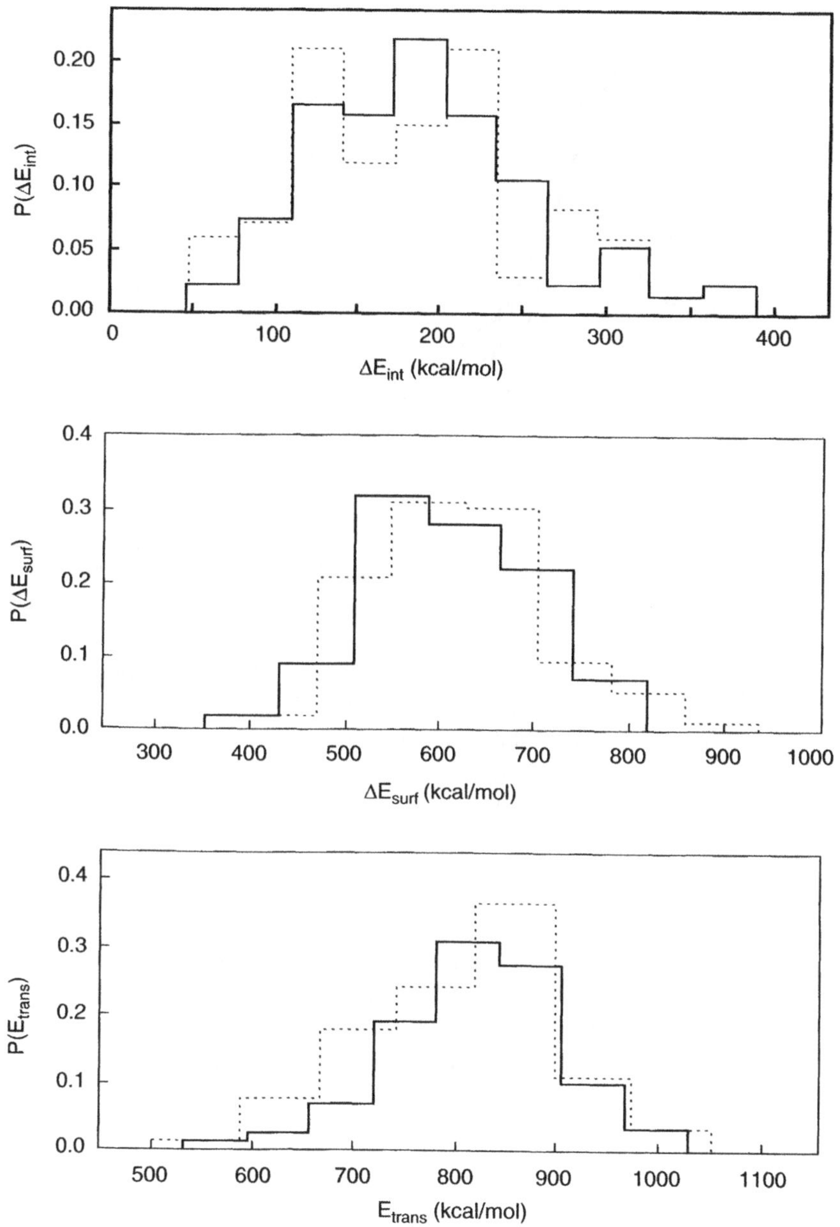

FIGURE 11.7. Distributions of the energy transfer to (gly-H^+) vibration/rotation ($E_i \rightarrow E_{\text{int}}$), the surface ($E_i \rightarrow E_{\text{surf}}$), and translation ($E_i \rightarrow E_{\text{trans}}$) for the (gly-$H^+$) intramolecular potential represented by the AMBER force field (—) and AM1 (···). Results for (gly-H^+) colliding with diamond {111} at an initial energy and angle of 70 eV and 45°. [Reprinted with permission from Meroueh et al. (2002).]

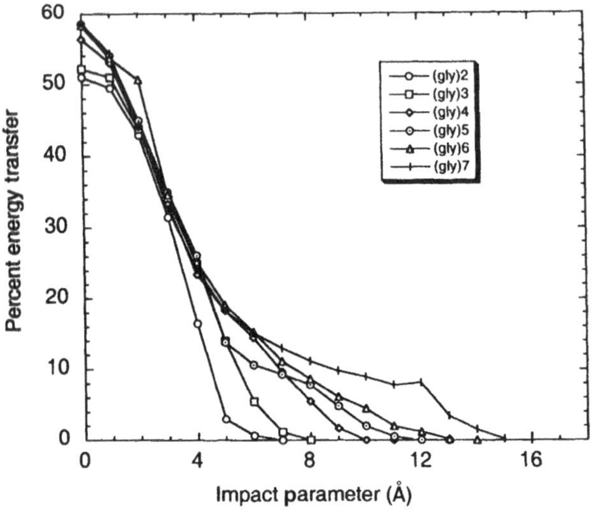

FIGURE 11.8. Percent energy transfer to β-(gly_4-H^+) structures versus impact parameter at E_{rel} = 100 kcal/mol. [Reprinted with permission from Meroueh and Hase (1999).]

11.3.2.2.3. Peptide Size and Collision Energy. The effect of peptide size and collision energy on collision activation was studied by determining the average energy transfer for the collision of Ar with both polyglycines and polyalanines at E_{rel} of 100, 500, and 1000 kcal/mol. Extended β-sheet structures were considered for the polyglycines, while these structures as well as folded, α-helix structures were considered for the polyalanines. The structures are given in Meroueh and Hase (1999). To circumvent the need to choose a value for b_{max}, the simulations were performed for $b = 0$.

The percent energy transfers are shown in Figure 11.9. As is discussed in Section 11.3.1.3. in the impulsive sudden limit the percent energy transfer would be independent of the peptide and E_{rel}. Overall the percent energy transfer values are inconsistent with this model. At E_{rel} = 100 kcal/mol there is a linear-like increase in the percent energy transfer versus the number of atoms for each of the polypeptides. The compounds β-(gly_n-H^+) and α-(ala_n-H^+), with the same number of atoms, have similar energy transfer efficiencies. The energy transfers are somewhat smaller for β-(ala_n-H^+). Different energy transfer patterns are observed at E_{rel} of 500 and 1000 kcal/mol as compared to E_{rel} = 100 kcal/mol. At these higher E_{rel}, the percent energy transfer to the different size β-sheet peptides shows no apparent trend, with a variation of only 4%. On the other hand, for α-(ala_n-H^+) the percent energy transfer retains a near-linear increase. Another identifier in the difference between the energy transfer dynamics, at E_{rel} of 100 kcal/mol as compared to 500 and 1000 kcal/mol, is the similar range of percent energy transfer values at 500 and 1000 kcal/mol, which are significantly larger than those at 100 kcal/mol. The picture from these simulations is that energy transfer to the β-sheet peptides attains the sudden limit between E_{rel} of 100 and 500 kcal/mol, while an E_{rel} larger than 1000 kcal/mol

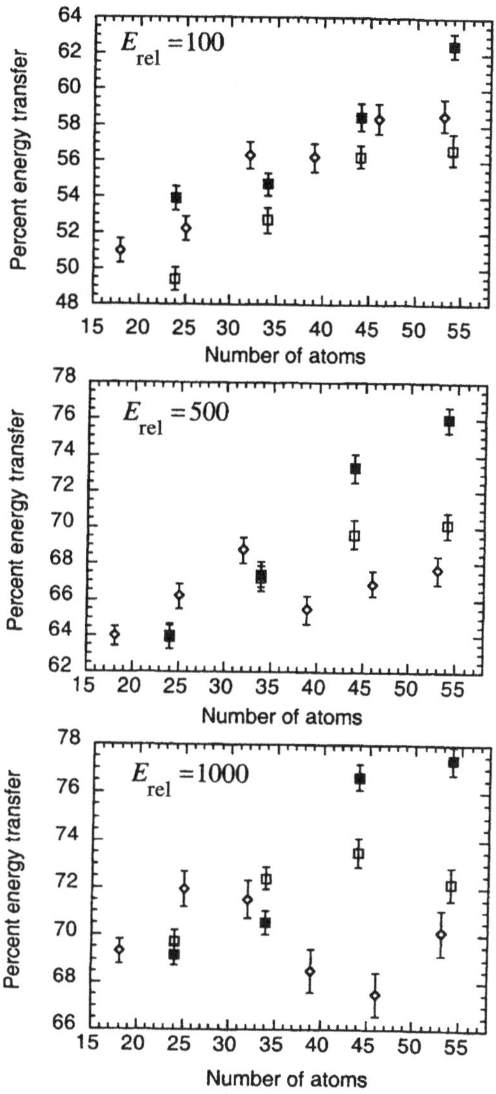

FIGURE 11.9. Percent energy transfer at $b = 0$ versus size of gly$_n$-H$^+$ and ala$_n$-H$^+$ polypeptides for E_{rel} of 100, 500, and 1000 kcal/mol: (■) β-(gly$_n$-H$^+$) and (□) α-(ala$_n$-H$^+$); (◇) β-(gly$_n$-H$^+$). [Reprinted with permission from Meroueh and Hase (1999).]

is required to reach the sudden limit for α-(ala$_n$-H$^+$). At $E_{rel} = 3000$ kcal/mol α-(ala$_2$-H$^+$) and α-(ala$_5$-H$^+$) have statistically the same energy transfer efficiencies of 67% and 69%, respectively.

11.3.2.2.4. Peptide Structure. The role of the peptide structure in the collision activation was investigated in detail by considering a range of structures for gly$_4$-H$^+$, shown in Figure 11.6. Each structure is characterized by its radius of gyration r_g, defined as

$$r_g = \left(\frac{\sum r_i^2}{n+1} \right)^{1/2} \tag{11.30}$$

where the r_i are the distances of the n atoms from the peptide's center of mass. Collisions of Ar with these gly$_4$-H$^+$ structures were studied for $b = 0$ and $E_{rel} = 100$ kcal/mol and 1000 kcal/mol.

The simulation results are given in Figure 11.10 and it is seen that, even though the collisions are head-on with $b = 0$, the structure of the peptide affects the efficiency of energy transfer. For the $E_{rel} = 100$ kcal/mol collision, there is a near-linear 15% decrease in the percent energy transfer as r_g increases from 3 to 5 Å. At $E_{rel} = 1000$ kcal/mol this linear trend is not observed. However, there is still a predominant decrease in the energy transfer efficiency as r_g is increased. The general result from these calculations is that the more compact, folded structures give more efficient energy transfer.

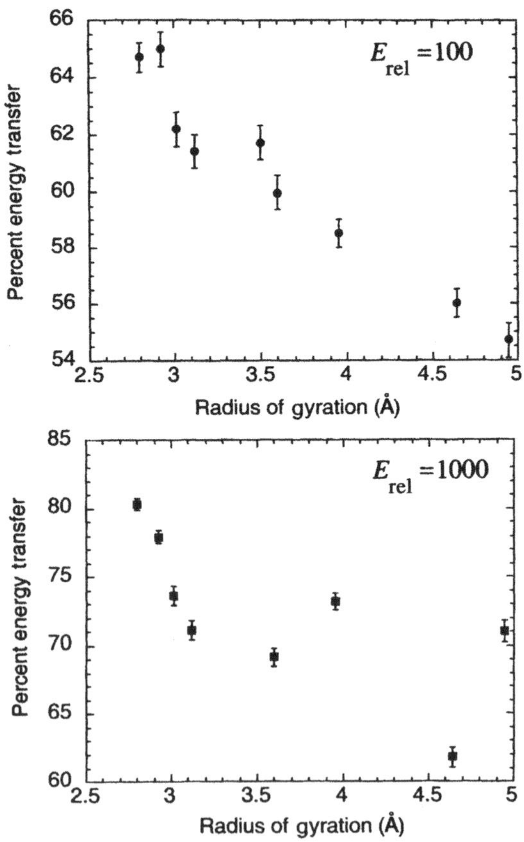

FIGURE 11.10. Percent energy transfer at $b = 0$ and E_{rel} of 100 and 1000 kcal/mol for gly$_4$-H$^+$ structures in Figure 11.6. [Reprinted with permission from Meroueh and Hase (1999).]

The trends in the energy transfer efficiencies versus r_g for E_{rel} of 100 and 1000 kcal/mol are consistent with the discussion in the previous section concerning the sudden limit for energy transfer. At $E_{rel} = 100$ kcal/mol, the energy transfer is not in the sudden limit and it depends on the peptide structure. At the larger E_{rel} of 1000 kcal/mol some of the peptides are in the sudden limit and have similar energy transfers.

Although the $b = 0$ energy transfer efficiencies depend on peptide structure, the quantity $\langle \Delta E \rangle_\sigma$ in Eq. (11.29) appears to be relatively independent of peptide structure (Meroueh and Hase 2000). $\langle \Delta E \rangle_\sigma$ increases near linearly with increase in E_{rel} for β-(gly_4-H^+) and α-(gly_4-H^+), suggesting b_{max} and percent energy transfer values that are only weakly dependent on E_{rel} (see Figure 11.9). At $E_{rel} = 100$ kcal/mol, these two peptides structures have nearly identical $\langle \Delta E \rangle_\sigma$ values, while for $E_{rel} = 1000$ kcal/mol $\langle \Delta E \rangle_\sigma$ is 20% larger for β-(gly_4-H^+).

11.3.2.2.5. $T \rightarrow V$ versus $T \rightarrow R$ Energy Transfer. In Section 11.3.1.2.1, it is shown that the energy transfer is primarily translation to vibration ($T \rightarrow V$) for the spherically shaped Al_n clusters, with ~40% of the transfer is translation to rotation ($T \rightarrow R$) for planar Al_n clusters. A similar effect is found for the polypeptides. A comparison of $T \rightarrow V$ and $T \rightarrow R$ energy transfers, for β-(gly_4-H^+) and α-(gly_4-H^+), is shown in Figure 11.11 as a function of b for $E_{rel} = 100$ kcal/mol. At small impact parameters the energy transfer in primarily to vibration, with transfer to rotation becoming more important as the impact parameter is increased. For the extended β-(gly_4-H^+) structure, energy transfer to rotation and vibration become similar at the larger impact parameters. For the extended peptide energy transfer becomes negligible at b of ~10 Å, whereas for the more compact α-(gly_4-H^+) peptide it becomes negligible at ~7 Å. These are approximate values for b_{max}.

A more quantitative value for b_{max} may be determined from $\langle \Delta E \rangle_\sigma$ in Eq. (11.29), which equals 3638 and 3207 kcal Å^2/mol for extended and folded gly_4-H^+, respectively. Decreasing the upper limit of the integral in Eq. (11.29) will give $\langle \Delta E \rangle_\sigma$ values that are smaller than the limiting value. Tests show that there is a value for the upper limit that, when varied by a very small amount, gives values of $\langle \Delta E \rangle_\sigma$, which are in the range of 1–10^{-4} % of the limiting value. This upper limit identifies a value of b_{max}. Setting it so that the value of $\langle \Delta E \rangle_\sigma$ is within 10^{-4} % of the limiting value gives b_{max} of 10.2 and 7.6 for β-(gly_4-H^+) and α-(gly_4-H^+), respectively. The average values of the energy transfer are then $\langle \Delta E \rangle_\sigma / \pi b_{max}^2$ and equal 11.1 and 17.5 kcal/mol for the extended and folded gly_4-H^+ peptides, respectively. Writing $\langle \Delta E(b) \rangle$ in Eq. (11.29) as the sum $\langle \Delta E_{vib}(b) \rangle + \langle \Delta E_{rot}(b) \rangle$ and taking $\langle \Delta E_{vib}(b) \rangle$ and $\langle \Delta E_{rot}(b) \rangle$ from Figure 11.11, the average transfers to vibration and rotation, averaged over b, are found to be $\langle \Delta E_{vib}(b) \rangle = 7.7$ kcal/mol and $\langle \Delta E_{rot}(b) \rangle =$ 3.4 kcal/mol for the extended β-(gly_4-H^+) peptide and $\langle \Delta E_{vib}(b) \rangle = 14.5$ kcal/mol and $\langle \Delta E_{rot}(b) \rangle = 3.0$ kcal/mol for the folded α-(gly_4-H^+) peptide. Thus 31% and 14% of the energy transfer is to rotation for the extended and folded gly_4-H^+ structures, respectively. These numbers are quite precise given the large number of trajectories calculated. Their accuracy depends on the accuracy of the model for the potential energy function.

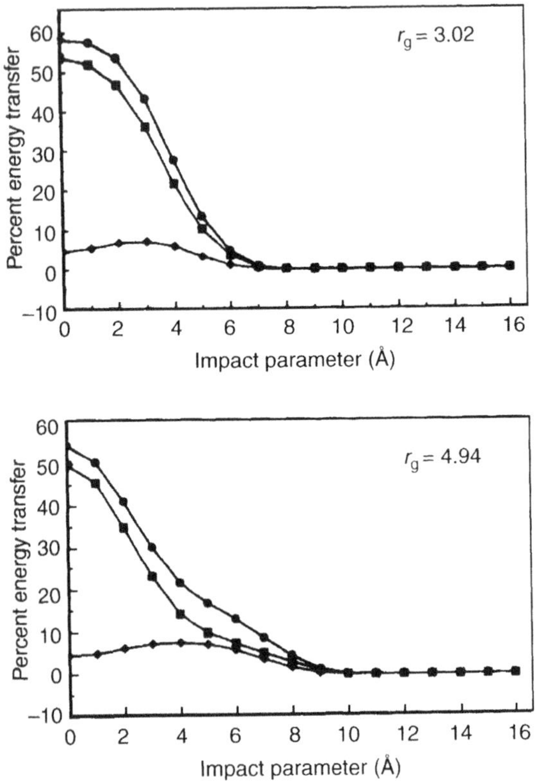

FIGURE 11.11. Total percent energy transfer (circles) and percent energy transfer to vibration (squares) and rotation (diamonds) versus impact parameter for collisions with $E_{rel} = 100$ kcal/mol and $T_{peptide} = 300$K. Results for the folded ($r_g = 3.02$Å) and extended ($r_g = 4.94$ Å) gly$_4$-H$^+$ peptides. [Reprinted with permission from Meroueh and Hase (2000).]

11.3.2.2.6. Pathways for Energy Transfer. Two important components of the mechanism of energy transfer for the peptide–H$^+$ are (1) the peptide modes that receive the energy and (2) the nature of the collision between the rare-gas atom and the peptide. The efficiency of energy transfer to specific modes of β-(gly$_4$-H$^+$) was studied by constraining different sets of internal coordinates. This was accomplished by increasing the force constants for a set of internal coordinates so that their vibrations are in the "stiff limit" and do not accept energy. Using H to identify a "heavy" atom such as C, N, or O and L to identify the "light" hydrogen atom, the internal coordinates were grouped into the following sets: HL stretches, HH stretches, LHL bends, HHH bends, HHHL torsions, and HHHH torsions. Calculations were performed for $E_{rel} = 100$ kcal/mol and $b = 0$, using the β-(gly$_4$-H$^+$) peptide initially in its classical potential energy minimum with no rotational energy. The percent energy transfers with different sets of internal coordinates constrained are listed in Table 11.4.

Constraining the stretches and bends has a small effect on the energy transfer. With no constraints, 58% of the collision energy is transferred to β-(gly$_4$-H$^+$)

TABLE 11.4. Percent Energy Transfer to Extended Gly_4-H^+ with Specific Internal Coordinates Constrained[a]

Coordinates Constrained[b]	$E_{rel} = 100$[c]
None	58 (4)[d]
HL stretches	56 (4)
HH stretches	56 (4)
All stretches	57 (4)
LHL bends	57 (5)
HHL bends	56 (5)
HHH bends	55 (5)
All bends	53 (5)
All bends and stretches	49 (6)
HHHL torsions	37 (9)
HHHH torsions	31 (9)
All bends and dihedrals	11 (11)
All	13 (13)

[a]The calculations are for $E_{rel} = 100$ kcal/mol and $b = 0$. The peptide is initially in its classical potential energy minimum.
[b]H corresponds to a heavy atom (i.e., C, N, or O); L corresponds to the light hydrogen atom.
[c]Energy is in kcal/mol.
[d]The standard deviation in the mean percent energy transfer is approximately 1%. The percent energy transfer to rotation is given in parentheses.

internal energy. With all the stretches constrained, this percent is lowered to 57%. With all the bends constrained, the energy transfer is 53%. Constraining both stretches and bends, with only the torsions available for receiving the collision energy, lowers the energy transfer to 49%. This shows that $0.49/0.58 \times 100 = 84$ of the initial energy transfer is to the torsional modes. More energy is transferred to peptide rotation as modes are constrained, and thus there is less energy transfer to vibration. With no constraints the overall energy transfer is 58% with only 4% to rotation. However, with all the modes constrained there is no energy transfer to vibration, but 13% to rotation.

In the simplest model for collision energy transfer (discussed below), there is an impulsive collision between Ar and the peptide with only one inner turning point in their relative motion. Some of the collisions are of this type, but the dominant collisions are indirect events with multiple Ar–peptide encounters. The nature of the collisions may be monitored by calculating the internal energy of the peptide during the collision. Two typical collisions are depicted in Figure 11.12 for Ar + α-(ala_5-H^+), with an impact parameter of zero and a collision energy of 100 kcal/mol. One of the collisions is direct with only one Ar–peptide encounter. The other is indirect with multiple encounters, identified by the sharp changes in the slope of the peptide's internal energy versus time. The direct collision has an encounter time of ~100 fs, while the indirect collision with multiple encounters has a total encounter time of ~180 fs. The indirect collision has as many as five encounters.

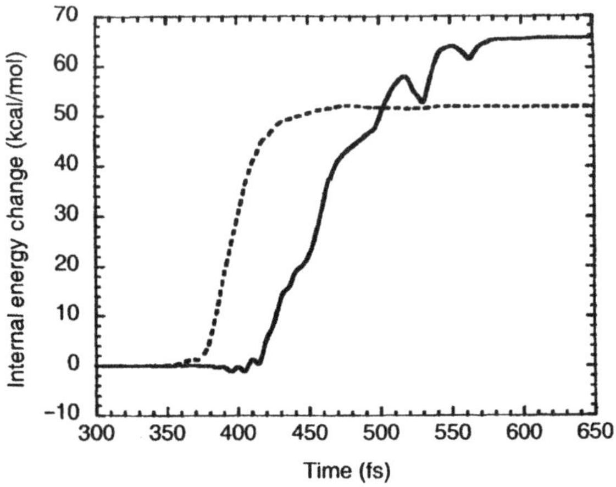

FIGURE 11.12. Examples of a direct trajectory (- - -) and an indirect trajectory (——) with multiple encounters for Ar + α-(ala$_5$-H$^+$) at E_{rel} = 100 kcal/mol and an impact parameter of zero. The internal energy change is that for the peptide. [Reprinted with permission from Meroueh and Hase (1999).]

The relative number of trajectories with a direct encounter, two encounters, and multiple encounters were determined for ten trajectories calculated at different collision energies for three protonated peptides. The results are listed in Table 11.5. Although there are no extraordinary differences in the nature of the encounters for the three peptides, β-(gly)$_7$ appears to have more direct encounters. Such a possible effect needs to be investigated by studying more trajectories. The encounter times range from 40 to 300 fs, with average total encounter times ranging from 80 to 180 fs. More than 50% of the trajectories for each peptide and collision energy have two or more encounters. This effect must be incorporated into any theoretical model describing the efficiency of energy transfer in peptide

TABLE 11.5. Relative Number of Encounters and Average Total Encounter Time for Ar–Protonated Peptide Collisions[a]

E_{rel}	Relative Number of Encounters[b]	Range of Encounter Times[c]	Average Total encounter Time[b]
		Ar + β-(gly)$_4$	
100	1 : 3 : 6	70–135	100
		Ar + β-(gly)$_7$	
100	4 : 1 : 5	55–300	145
		Ar + α-(ala)$_5$	
100	1 : 2 : 7	120–250	180
1000	1 : 6 : 3	40–125	80

[a]The collision impact parameter is zero. Ten trajectories were calculated for the Ar–peptide collisions at each collision energy E_{rel}.

[b]Relative number of trajectories with (one encounter): (two encounters): (multiple encounters).

[c]The encounter time is given in fs.

collisional activation. In comparing this rather incomplete analysis for the three peptides and the two different collision energies, the total encounter time tends to increase as the peptide size is increased or E_{rel} is decreased. The results for α-(ala$_5$-H$^+$) and β-(gly$_7$-H$^+$) suggest that the folded peptide may have more multiple encounters and larger total encounter times. Clearly, more work needs to be done to investigate the mechanism for collisional activation of protonated peptide ions. Multiple encounters are expected to enhance the efficiencies of collisional energy transfer.

11.3.2.3. Comparisons with Mahan's Impulsive Energy Transfer Model. Equations (11.23)–(11.26) represent the impulsive model developed by Mahan for $T \rightarrow V$ energy transfer in A + BC collisions. In this model, energy transfer is controlled by the adiabaticity parameter ξ in Eq. (11.25), with the efficiency of energy transfer increasing as ξ becomes smaller. Decreasing the BC vibrational frequency, increasing the collision relative velocity, or decreasing the intramolecular range parameter L, and making the collision less repulsive, decreases ξ and increases energy transfer. As discussed above, this model provides a meaningful interpretation of energy transfer in collisions of Al_n^+ clusters with rare gas atoms and it is of interest to investigate its applicability to Ar+ peptide–H$^+$ collisions.

As the size of a gly$_n$-H$^+$ or ala$_n$-H$^+$ peptide increases, the distribution of vibrational frequencies for the peptide extends to lower values and, in an average sense, ξ decreases if E_{rel} held constant. Thus, if the collisions are not in the sudden limit, more efficient energy transfer is expected as n is increased. This is the behavior observed for both the β-sheet and α-helix peptides at E_{rel} of 100 kcal/mol (see Figure 11.9), where there is a near-linear increase in the percent energy transfer as n is increased. In the high-collision-energy sudden limit, when $\xi \ll 1$, the energy transfer should be nearly independent of peptide size, and this is the behavior seen for the β-(ala$_n$-H$^+$) and β-(gly$_n$-H$^+$) peptides at $E_{rel} = 1000$ kcal/mol. However, as discussed above, a higher E_{rel} is required to attain the sudden limit for the α-(ala$_n$-H$^+$) peptides.

We have seen above that Mahan's model provides at least a qualitative interpretation of the efficiency of energy transfer with respect to peptide size and the collision energy. However, since peptides with the same vibrational frequencies, mass, and intermolecular potential are predicted to have the same energy efficiency according to Mahan's model, this model does not provide a qualitative interpretation of the dependence of the energy transfer efficiency on peptide structure (see Figure 11.10). This is exemplified by the results for the different gly$_4$-H$^+$ structures. The energy transfer efficiency to the peptide varies by up to 15% as its structure is changed. At E_{rel} of 1000 kcal/mol, Figure 11.10 shows that peptides with very different structures may have similar energy transfer efficiencies, which suggests that some of the peptide structures may have reached the sudden limit.

The manner in which a change in the Ar/peptide intermolecular potential affects energy transfer is qualitatively explained by Mahan's model. The simulations presented above were performed with an accurate ab initio Ar+ peptide–H$^+$

intermolecular potential, as described in Section 11.3.2.1. If a molecular-mechanics (MM)-based empirical potential is used, substantially different energy transfer efficiencies are found (Meroueh and Hase 1999). This empirical potential has range parameters L that are ~2.5 times smaller than the ab initio values and gives energy transfer efficiencies for Ar + β-(gly$_n$-H$^+$), where $n = 2$–7, as much as 30% larger than those discussed above for the ab initio potential. As shown by Eqs. (11.23) – (11.26), this increase in energy transfer efficiency with decrease in L agrees with Mahan's model.

Finally, it must be emphasized that Mahan's impulsive model does not provide a means to quantitatively interpret the energy transfer dynamics for collisions with multiple encounters as illustrated in Figure 11.12. Considerable work needs to be done to develop a theoretical model that provides a quantitative understanding of how the peptide size, structure, and amino acid constituents, the collision's intermolecular potential, and the multiple encounters during a collision affect the energy transfer dynamics for peptide–H$^+$ CID. Simulations such as those discussed here, which address all of these issues, should provide the information needed to develop an accurate theoretical model for energy transfer in peptide–H$^+$ CID.

11.4. SIMULATIONS OF SID

Trajectory simulations of $Cr^+(CO)_6$ (Bosio and Hase 1998; Meroueh and Hase 2001; Song et al. 2003) and protonated peptide ion (Meroueh and Hase 2002; Meroueh et al. 2002; Wang et al. 2003a,b) collisions with surfaces have provided fundamental information concerning the energy transfer and fragmentation dynamics of surface-induced dissociation (SID). The results of these simulations are reviewed here. We begin by describing the construction of potential energy surfaces needed for the simulations.

11.4.1. Potential Energy Surfaces for SID Simulations

11.4.1.1. $Cr^+(CO)_6$. Collisions of $Cr^+(CO)_6$ with both diamond {111} and alkyl thiolate self-assembled monolayer (H-SAM) surfaces have been simulated. The potential energy function for each of these simulations is represented by Eq. (11.4). The V_{surface} potential for diamond {111} consists of harmonic stretches and bends, with parameters chosen to fit the diamond phonon spectrum (Hass et al. 1992). The surface potential for the SAM is written as

$$V_{\text{SAM}} = V_{\text{inter}} + V_{\text{intra}} \tag{11.31}$$

and is an extension of the potential by Mar and Klein (Mar and Klein, 1994) to represent the 300 K structure of alkyl thiolate SAMs on Au{111}. The SAM's intramolecular potential V_{intra} is a molecular mechanics potential, [Eq. (11.27)] and includes all the stretch, bend, dihedral, and torsional motions of the alkyl thiolate chains. The intermolecular part of the SAM potential is written as the following sum of two-body terms

$$V_{\text{inter}} = \sum_{\text{C,H}} V_{i,j}(r) + \sum_{\text{C,S,H}} V_{i,\text{S}}(r) + \sum_{\text{CH}_2,\text{CH}_3} V_{i,\text{Au}}(z) \qquad (11.32)$$

where r is the distance between atom pairs and z is a generalized coordinate representing the distance of a CH_2 or CH_3 group from the Au surface. For the Mar–Klein SAM potential, the Buckingham function

$$V(r) = A\exp(-Br) - C/r^6 \qquad (11.33)$$

is used for the first two terms in Eq. (11.32). Buckingham-type potentials have the disadvantage of becoming attractive at sufficiently small internuclear separations, which approaches negative infinity as the separation decreases. For the C–H and H–H Buckingham potentials, the barriers for the transition from the repulsive short-range interaction to the unphysical short-range attractive interaction were sufficiently low that they were surmounted by the high-energy SID simulations. To correct this shortcoming, switching functions were used to connect the long-range (LR) C–H and H–H Buckingham potentials to the short-range (SR) Buckingham potentials determined from ab initio calculations (Meroueh and Hase 2001).

The potential energy function for $Cr^+(CO)_6$ is given by (Meroueh and Hase 2001)

$$\begin{aligned} V = &\sum_{\text{CrC}} D(\Omega)\{1 - \exp[-\beta(r - r_e)]\}^2 \\ &+ \sum_{\text{CrC}} \Delta D(\Omega) + \sum_{\text{CO}} k_R(R - R_e)^2 \\ &+ \sum_{\text{CrCO}} \frac{k_\phi}{2}(\phi - \phi_e)^2 + \sum_{\text{CCrC}} \frac{k_\theta}{2}(\theta - \theta_e)^2 \end{aligned} \qquad (11.34)$$

where a Morse function is used for the Cr–C stretches and harmonic terms are used for the other degrees of freedom. The force constant for each ϕ and θ bending potential is attenuated, so that the force constant goes to zero as a Cr–C bond defining the bend angle ruptures.

The $D(\Omega)$ and $\Delta D(\Omega)$ terms in Eq. (11.34) are determined as follows. Experiments have shown that the Cr–CO bond dissociation energy depends on the extent of dissociation (Khan et al. 1993) and varies according to

$$\begin{aligned} &Cr^+(CO)_6 \rightarrow Cr^+(CO)_5 + CO && D_e = 1.40(0.08) \\ &Cr^+(CO)_5 \rightarrow Cr^+(CO)_4 + CO && D_e = 0.66(0.03) \\ &Cr^+(CO)_4 \rightarrow Cr^+(CO)_3 + CO && D_e = 0.59(0.08) \\ &Cr^+(CO)_3 \rightarrow Cr^+(CO)_2 + CO && D_e = 0.59(0.06) \\ &Cr^+(CO)_2 \rightarrow Cr^+(CO) + CO && D_e = 0.98(0.04) \\ &Cr^+(CO) \rightarrow Cr^+ + CO && D_e = 0.95(0.04) \end{aligned}$$

where D_e is in units of eV and the values in the parentheses are the uncertainties. To represent these changes in the Cr–C dissociation energy, D_e is written as a function of the extent of dissociation defined by

$$\Omega = \sum_{i=1}^{6} \frac{1}{r_i} \tag{11.35}$$

The value of Ω for a $Cr^+(CO)_n$ complex is evaluated by setting r_i for the ruptured bonds to ∞ and r_i for the intact bonds to their equilibrium value. The Cr–C bond dissociation is written as a function of Ω to fit the values listed above and give $D(\Omega)$. As additional Cr–C bonds rupture, the dissociation energy of the ruptured bonds must be shifted to maintain the proper total potential energy for the ruptured bonds. This is accomplished by $\Delta D(\Omega)$.

Two-body intermolecular potentials, for the Cr^+, C, and O atoms of $Cr^+(CO)_6$ interacting with the H and C atoms of the diamond {111} and H-SAM surfaces, were determined from ab initio calculations (Meroueh and Hase 2001) as described in Section 11.3.2.1 for the Ar + peptide–H^+ potentials. The ab initio calculations were performed for Cr^+ and CO interacting with CH_4 as a representative of the H and C atoms on the surface. The Cr^+–H and Cr^+–C potentials are given by

$$V(r) = A\exp(-Br) + C/r^6 + D/r^4 \tag{11.36}$$

The D/r^4 term is included to model the charge-induced dipole interaction. The same potential function is used for the interactions of the C and O atoms of $Cr^+(CO)_6$ with the C and H atoms of the surfaces, except the D/r^4 term is not included. Also, C/r^6 is replaced by the more general term C/r^d. The ab initio calculations for the Cr^+/CH_4 system were carried out at the UMP2/6-311+G(3*df*) level of theory. While those for CO/CH_4 used the MP2/aug-cc-pVTZ level of theory.

11.4.1.2. Peptide–H$^+$. For the peptide–H^+ SID simulations, the potentials used for diamond {111} and the H-SAM are the same as those described above for $Cr^+(CO)_6$. Both the empirical Amber and the AM1 semiempirical electronic structure theory models were used for the peptide ion's intramolecular potential (see Section 11.3.2.1): $V_{\text{projectile}}$ in Eq. (11.4). An intermolecular potential, for the peptide ion interacting with the surface, was developed in the same manner as described in Section 11.3.2.1 for the Ar/peptide–H^+ intermolecular potential. The peptide/surface intermolecular potential is modeled by a sum of two-body potentials between the atoms of the peptide and surface. The two-body potentials are given by the Buckingham function in Eq. (11.33).

To determine the parameters for the two-body potentials, ab initio potential energy curves were calculated between CH_4, as a model for the C and H atoms of diamond {111} and *n*-hexyl thiolate SAM surfaces, and CH_4, NH_3, NH_4^+, H_2CO, and H_2O, as models for the different types of atoms and functional groups representing the protonated polyglycine and polyalanine peptides. The ab initio calculations are carried out at the MP2 level theory with the 6-311+G(2*df*,2*pd*) basis set. The molecules were held fixed in their optimized geometries and intermolecular potential energy curves for different orientations of CH_4/CH_4, CH_4/NH_3, CH_4/NH_4^+, CH_4/H_2CO, and CH_4/H_2O systems were calculated. These curves were then fit to determine the two-body potential parameters.

11.4.2. SID Simulation Results

The results obtained from simulations of $Cr^+(CO)_6$ and peptide–H^+ SID on H-SAM and diamond {111} surfaces are summarized in Table 11.6. The simulations provide

TABLE 11.6. Dynamics of SID from Trajectory Simulations

Projectile(s)	Chemical Dynamics Result	Ref.
$Cr^+(CO)_6$	Shattering dissociation with H-SAM and diamond surfaces; shattering on the H-SAM requires a higher E_i	Meroueh and Hase (2001), Song et al. (2003)
$Cr^+(CO)_6$	Negligible energy transfer to CO vibration during collision with or after rebounding off surface	Song et al. (2003)
Gly_3-H^+, $Cr^+(CO)_6$	Very different energy transfer distributions for collisions with H-SAM and diamond surfaces; for gly_3-H^+ collisions at $E_i = 30\,eV$, the average percent transfers to ΔE_{int}, ΔE_{surf}, and E_f are 20, 8, and 72 for diamond and 8, 54, and 38 for H-SAM.[a]	Meroueh and Hase (2001, 2002); Song et al. (2003)
Gly_n-H^+	Energy transfer is very similar for gly_3-H^+ and gly_5-H^+	Meroueh and Hase (2002)
Gly_3-H^+	Folded and extended structures for gly_3-H^+ give similar energy transfer efficiencies	Meroueh and Hase (2002)
Gly_n-H^+, $Cr^+(CO)_6$	Percent energy transfer to ΔE_{int} is only weakly dependent on E_i, while percent transfer to ΔE_{surf} and E_f increases and decreases, respectively, with increase in E_i	Meroueh and Hase (2002); Song et al. (2003); Wang et al. (2003a).
Gly_3-H^+	In collisions with diamond at 30 eV, 80% of ΔE_{int} goes to the dihedrals	Meroueh and Hase (2002)
Gly-H^+, gly_2-H^+	Shattering fragmentation in collisions with a diamond surface. The shattering contribution to the dissociation increases with E_i	Meroueh et al. (2002) Wang et al. (2003b)
Gly-H^+, gly_2-H^+	Amber and AM1 potentials for the peptide give the same energy transfer efficiencies	Meroueh et al. (2002) Wang et al. (2003b)
Gly_2-H^+, ala_2-H^+	These two peptides have similar energy transfer efficiencies	Wang et al. (2003a)

[a]The results are for extended gly_3-H^+.

an interpretation of SID experiments and fundamental insight into SID energy transfer and fragmentation dynamics. The dynamics found from the simulations are reviewed in the following.

11.4.2.1. Comparison with Experiments. Excellent agreement is found between experiments and the simulations. Plots of the distributions of energy transferred to the projectile's internal degrees of freedom $P(\Delta E_{\text{int}})$ and the surface vibrations $P(\Delta E_{\text{surf}})$ and the energy remaining in the peptide translation $P(E_{\text{f}})$ are given in Figure 11.13 for simulations of $Cr^{+}(CO)_6$ collisions with the diamond {111} and

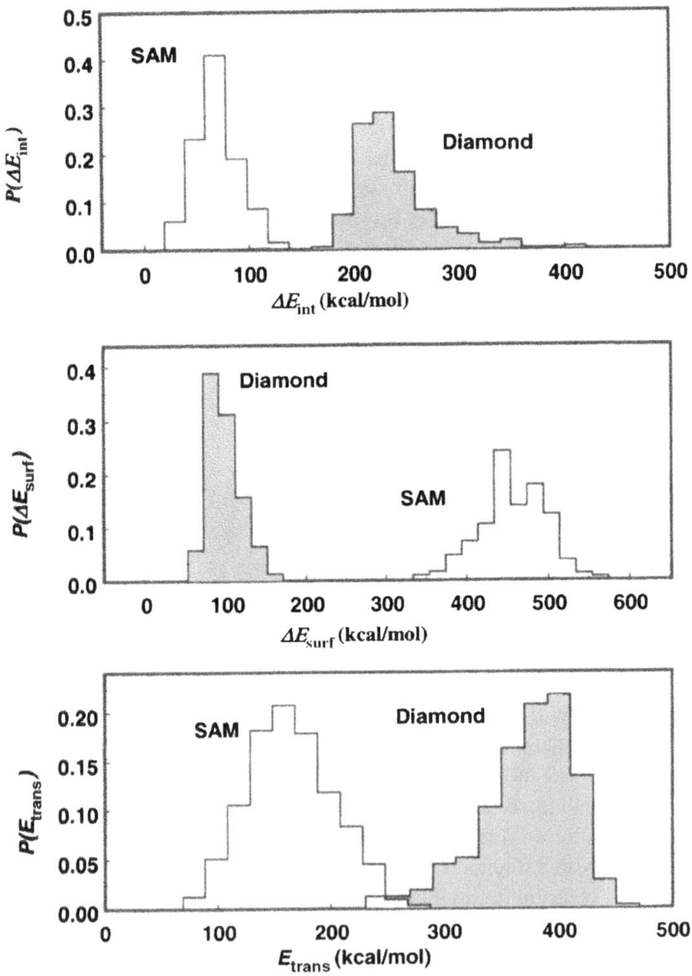

FIGURE 11.13. Distribution of energy transfer to the ion (ΔE_{int}), energy transfer to the surface (ΔE_{surf}), and the translational energy of the recoiling ion (E_{f}) as a result of $Cr^{+}(CO)_6$ collisions with the diamond {111} and H-SAM surfaces at an initial translational energy of 30 eV (692 kcal/mol) and θ_i of 45°. [Reprinted with permission from Meroueh and Hase (2001).]

n-hexylthiolate H-SAM surfaces at an incident collision energy and angle of $E_i = 30$ eV and $\theta_i = 45°$ (Meroueh and Hase 2001). The average energy transfer to ΔE_{int} for collision with the H-SAM is 10%, which agrees well with the value of 11–12% reported by Cooks and coworkers (Morris et al. 1992).

An indirect comparison may be made between the simulations and experiments for peptide-H^+ SID (Laskin et al. 2000; Laskin and Futrell 2003a). Laskin and coworkers find that, for collisions of ala$_2$-H^+ and the protonated octapeptide des-Arg^1–bradykinin with a fluorinated F-SAM surface at $\theta_i = 0°$, the percent energy transfer to ΔE_{int} is independent of E_i for the respective ranges of 3–23 eV and 10–55 eV. For collisions of the octapeptide with diamond {111}, the energy transfer to ΔE_{int} is 19.2% and a similar value is expected for ala$_2$-H^+, since the octapeptide and ala$_2$-H^+ have similar energy transfer efficiencies when colliding with the F-SAM. Simulations have shown that energy transfer for ala$_2$-H^+ and gly$_2$-H^+ are statistically the same (Wang et al. 2003a). From this work, the suggested value for energy transfer to ΔE_{int} for ala$_2$-H^+ + diamond {111} collisions is 24% at $E_i = 30$ eV and $\theta_i = 0°$. This value is close to that suggested by the above experiments.

11.4.2.2. Amber and AM1 Models for the Peptide–H^+ Intramolecular Potential. Energy transfer efficiencies given by the Amber and AM1 models for the peptide–H^+ intramolecular potential were compared in simulations of gly-H^+ and gly$_2$-H^+ collisions with the diamond {111} surface. The results are given in Table 11.7, and it is seen that there is excellent agreement between the Amber and AM1 results. In Section 11.3, the Amber potential is used to investigate the energy transfer dynamics, since the chemical dynamics simulations are much faster using Amber instead of AM1 for the peptide–H^+ intramolecular potential. The AM1 potential is used to study the fragmentation dynamics in Section 11.4.

TABLE 11.7. Comparison of Amber and AM1 Peptide-H^+ Intramolecular Potentials for Simulations of Energy Transfer in SID

	Average Percent Energy Transfer[a]		
Potential	ΔE_{int}	ΔE_{surf}	E_f
		Gly-H^+, $E_i = 70$ eV and $\theta_i = 45°$	
AMBER	11	37	52
AM1	12	38	50
		Gly$_2$-H^+, $E_i = 70$ eV and $\theta_i = 45°$	
AMBER	15	25	60
AM1	13	26	61
		Gly$_2$-H^+, $E_i = 70$ eV and $\theta_i = 0°$	
AMBER	20	40	40
AM1	20	40	40

[a]The standard deviation of the mean for the percent energy transfer is of the order of 1%.

11.4.2.3. Energy Transfer Dynamics

11.4.2.3.1. Pathways for Energy Transfer. It is important to know which modes of the projectile are initially excited when it collides with the surface. Such a property is discussed above, in Section 11.3.2.2.6 for Ar^+ peptide–H^+ collisions. The simulations of $Cr^+(CO)_6$ collisions with the diamond {111} and *n*-hexyl thiolate H-SAM surfaces show there is very little energy transfer to the CO vibrations during the collision of $Cr^+(CO)_6$ with the surface or later, as a result of intramolecular vibrational energy redistribution (IVR), after the ion scatters off the surface (Song et al. 2003). Thus, of the ion's 33 vibration degrees of freedom, 6 are apparently inactive. The inactivity of the CO vibrations for receiving and transferring energy is not unexpected, given their high frequencies and the much lower frequencies for the remaining $Cr^+(CO)_6$ vibrations. No attempt was made to determine which modes of the $Cr^+(CO)_6$ ion are initially excited on its impact with the surface.

A simulation was performed for folded gly_3-H^+ collisions with diamond {111}, similar to the Ar + gly_4–H^+ simulation discussed above, where modes of gly_4-H^+ are constrained to determine the pathways for collisional excitation of the peptide. With none of the modes of gly_3-H^+ constrained, 18, 9, and 73% of $E_i = 30$ eV, for $\theta_i = 45°$ collisions, is transferred to E_{int}, E_{surf}, and E_f. With all the modes of the gly_3-H^+ peptide constrained, except the dihedrals, the respective energy transfers are 14%, 16%, and 70%. This result suggests that $\frac{14}{18} \times 100 = 78\%$ of the internal energy transfer is to the peptide dihedrals, a result similar to that found for Ar + gly_4-H^+ collisions.

11.4.2.3.2. Surface Properties. The simulations show that the properties of the surface have a profound effect on the energy transfer efficiencies. This is illustrated in Table 11.8 for collisions of $Cr^+(CO)_6$ and folded gly_3-H^+ with the diamond {111}

TABLE 11.8. Comparisons of Trajectory Simulations of Energy Transfer (%) for Folded Gly_3-H^+ and $Cr^+(CO)_6$ Collisions with Diamond {111} and H-SAM Surfaces[a]

	ΔE_{int}		ΔE_{surf}		E_f	
E_i^b	$(Gly)_3$	$Cr(CO)_6^+$	$(Gly)_3$	$Cr(CO)_6^+$	$(Gly)_3$	$Cr(CO)_6^+$
			Diamond {111}			
30	18	30	9	14	73	56
70	17	17	21	29	62	54
110	14	16	29	38	57	46
			H-SAM[c]			
30	7	10	63	69	30	21
70	—	9	—	81	—	10
110	—	8	—	81	—	11

[a]The collision angle is 45°.
[b]The collision energy is in eV.
[c]The H-SAM is *n*-hexyl thiolate.

and n-hexyl thiolate H-SAM surfaces. Collision with diamond {111} transfers a factor of 2–3 more energy to ΔE_{int} as compared to colliding with the H-SAM. Collisions with the diamond {111} surface retain the majority of the collision energy in projectile translation, while collisions with the H-SAM deposit the majority of the energy in ΔE_{surf}. These results concur with the experiments (Laskin and Futrell 2003a), which show that for collisions of the singly protonated octapeptide des-Arg[1]–bradykinin with different surfaces the percent $E_i \rightarrow \Delta E_{int}$ transfer increases in the order H-SAM (10%), LiF (12.0%), diamond (19.2%), and F-SAM (20.5%).

Another important difference between collisions with the diamond {111} and H-SAM surfaces is the greater breadth of the $P(\Delta E_{int})$ distribution for collision with diamond {111}. This is illustrated in Figure 11.13 for $Cr^+(CO)_6$ collision, at 30 eV. At the higher E_i of 70 and 110 eV, $P(\Delta E_{int})$ appears to become bimodal (Song et al. 2003). A broader $P(\Delta E_{int})$ for collision with the diamond {111} surface, is also seen for peptide–H^+ projectiles, as shown in Figure 11.14 for folded gly$_3$-H^+.

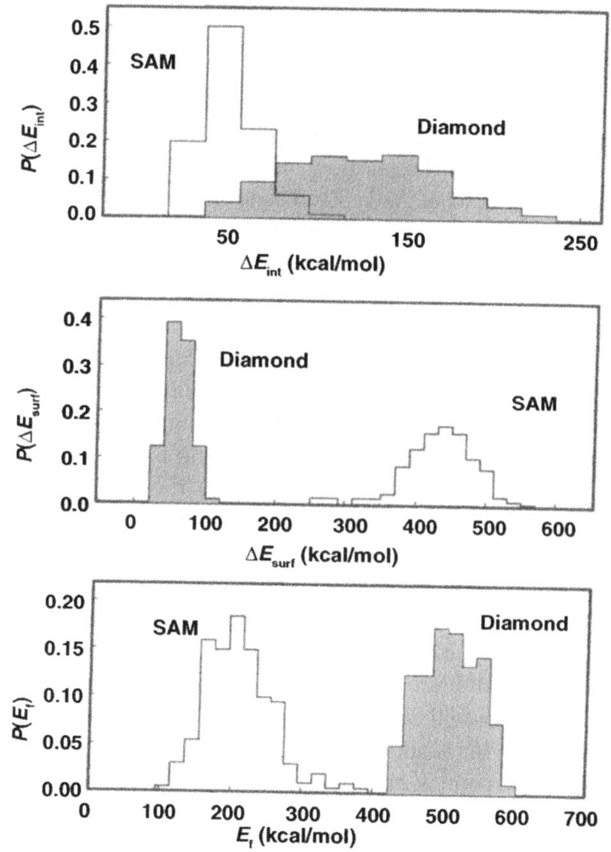

FIGURE 11.14. Distribution of the energy transfer to the ion (ΔE_{int}) and to the surface (ΔE_{surf}) and the translational energy of the recoiling ion (E_f) following folded gly$_3$-H^+ collisions with diamond and H-SAM surfaces at an initial translational energy of 30 eV (692 kcal/mol) and θ_i of 45°. [Reprinted with permission from Meroueh and Hase (2002).]

The origin of the difference in the energy transfer dynamics for collisions with the diamond and H-SAM surfaces is unclear. It has been suggested from simulations that the stiffness of the diamond {111} surface enhances energy transfer to the projectile ion (Meroueh and Hase 2001). From experiments (Laskin and Futrell 2003a) it has been inferred that the surface stiffness has a major effect on the width of $P(\Delta E_{\text{int}})$, while the average energy deposited into the ion is affected mainly by the mass of the chemical moiety representing an immediate collision partner for the ion impacting the surface. The interpretation of the experiments is similar to that of the simulations, since the effective mass of the surface impact site depends on the surface stiffness (Grimmelmann et al. 1980; de Sainte Claire et al. 1995).

11.4.2.3.3. Peptide Size, Structure, and Amino Acid Constituents. An important finding is that the size, structure, and amino acid constituents of the peptide–H^+ ions do not have substantial effects on the energy transfer efficiencies. For collision with diamond {111} at $E_i = 30\,\text{eV}$ and $\theta_i = 45°$, the respective energy transfer to ΔE_{int}, ΔE_{surf}, and E_f are 18%, 9%, and 73% for gly$_3$-H^+ and 23%, 5%, and 72% for gly$_5$-H^+, (Meroueh and Hase 2002). For $E_i = 70\,\text{eV}$ and $\theta_i = 45°$, and collision with the diamond surface, the transfer to ΔE_{int} is 11%, 15%, and 17% for gly-H^+, gly$_2$-H^+, and gly$_3$-H^+, respectively (Wang et al. 2003a). The energy transfer to ΔE_{int} is similar for these different size gly$_n$-H^+ peptides.

The structure of the peptide has at most only a small effect on the energy transfer. For collisions with diamond at $E_i = 30\,\text{eV}$ and $\theta_i = 45°$, the transfer to ΔE_{int} is 18%, and 20%, respectively, for folded and extended gly$_3$-H^+. For collision with the n-hexyl thiolate H-SAM, these respective energy transfers are 7%, and 8%. Thus variations in the structure of the peptide have an insignificant effect on the energy transfer. Energy transfer has been studied for both gly$_n$-H^+ and ala$_n$-H^+ peptides, and, for the same n, both peptides have similar energy transfer efficiencies (Wang et al. 2003a). Thus, alanine and glycine constituents are equally efficient in receiving energy.

The independence of the energy transfer on peptide size, observed in the simulations and amino acid constituents, agrees with experimental results (Laskin and Futrell 2003a). For ala$_2$-H^+ collisions with a F-SAM surface 21% of E_i is transferred to ΔE_{int}, while this transfer is 20.5% for des-Arg[1]–bradykinin collisions with the same surface.

11.4.2.3.4. Projectile Incident Energy and Angle. Experimental studies of ala$_2$-H^+ SID on a F-FAM surface for E_i in the range of 2–23 eV (Laskin et al. 2000) and protonated des-Arg[1]–bradykinin SID on a variety of surfaces for E_i in the range of 10–100 eV (Laskin and Futrell 2003a) have shown that the percent energy transfer to ΔE_{int} is independent of E_i. A similar result is found from the simulation. For gly$_2$-H^+ SID at $\theta_i = 45°$, the percent energy transfer to ΔE_{int} changes from 16 to 13% as E_i is increased from 5 to 110 eV (Wang et al. 2003a). For ala$_2$-H^+ the energy transfer changes from 19 to 15%. These small changes in the transfer to ΔE_{int}, over a large range of E_i, may be difficult to observe experimentally.

There are changes in the energy transfer efficiencies with change in collision angle. This is illustrated in Table 11.7 for gly$_2$-H^+ collisions with the diamond surface. There is a small increase in the percent transfer to ΔE_{int} as θ_i is changed from 45° to 0°. In contrast, the change in the projectile translation energy is much larger, decreasing from 60 to 40%.

11.4.2.3.5. Projectile Orientation and Surface Impact Site. Simulations of $gly_2\text{-}H^+$ colliding with the diamond {111} surface, at E_i of 70 eV and θ_i of 0° and 45°, were performed to determine how the orientation angle of the peptide ion and the surface impact site affect the collisional energy transfer (Rahaman et al. 2006). The peptide ion was randomly rotated in the trajectory initial conditions, which establishes the initial orientation of the ion with respect to the surface. To ensure the peptide ion approaches the surface in this orientation, the rotational temperature of the ion was set to 0 K. The initial orientation of the peptide ion is determined by defining a vector from the nitrogen atom of the protonated amino group to the hydroxyl oxygen atom of the carboxylic group. The angle between this vector and the vector normal to the diamond surface determines the orientation of the peptide ion. There are two unique impact sites on the diamond {111} surface at which the center of mass of $gly_2\text{-}H^+$ may collide. One is a hydrogen atom on the outer layer of the surface. The other is a carbon atom in the second surface layer that is bonded to a carbon atom just below the surface hydrogen atom. The first impact site is denoted a H site and second, a C site. In order to determine possible effects of the surface impact point on the efficiency of energy transfer to $gly_2\text{-}H^+$, one ensemble of classical trajectories was calculated with the center of mass of the diglycine ion directed toward the H site and another with the ion directed towards the C site. An ensemble of trajectories was also calculated with random impact sites for the colliding ion. The results of the simulations are shown in Figure 11.15.

The top graphs in the figure show that the energy transfer to ΔE_{int} is independent of the impact site. The same is found for the $P(\Delta E_{surf})$ and $P(E_f)$ distributions, which are not shown. However, the middle graphs show that energy transfer to ΔE_{int} is enhanced if the backbone of the peptide is perpendicular to the surface during the collision. Concurrently, the transfer to ΔE_{surf} and the energy remaining in projectile translation are decreased for the perpendicular orientation. For these perpendicular collisions, more energy is transferred to the peptide if the *C* terminus ($\Psi = 180°$) instead of the *N* terminus first strikes the surface. For an incident angle θ_i of 45°, instead of 0°, the role of peptide orientation on the energy transfer efficiency is less pronounced. This is shown in the bottom graphs in Figure 11.15.

11.4.2.4. Fragmentation Mechanisms. Simulations for $Cr^+(CO)_6$, $gly\text{-}H^+$, and $gly_2\text{-}H^+$ show that the projectile ion may fragment by two different mechanisms (Meroueh and Hase 2001; Song et al. 2003, Meroueh et al. 2002; Wang et al. 2003b). One is the traditional RRKM model in which the projectile ion is activated by collision with the surface, "bounces off", and then dissociates after undergoing intramolecular vibrational energy redistribution (IVR). The other mechanism is shattering, which dominates at high collision energy. For shattering the ion fragments as it collides with the surface (Raz and Levine 1996).

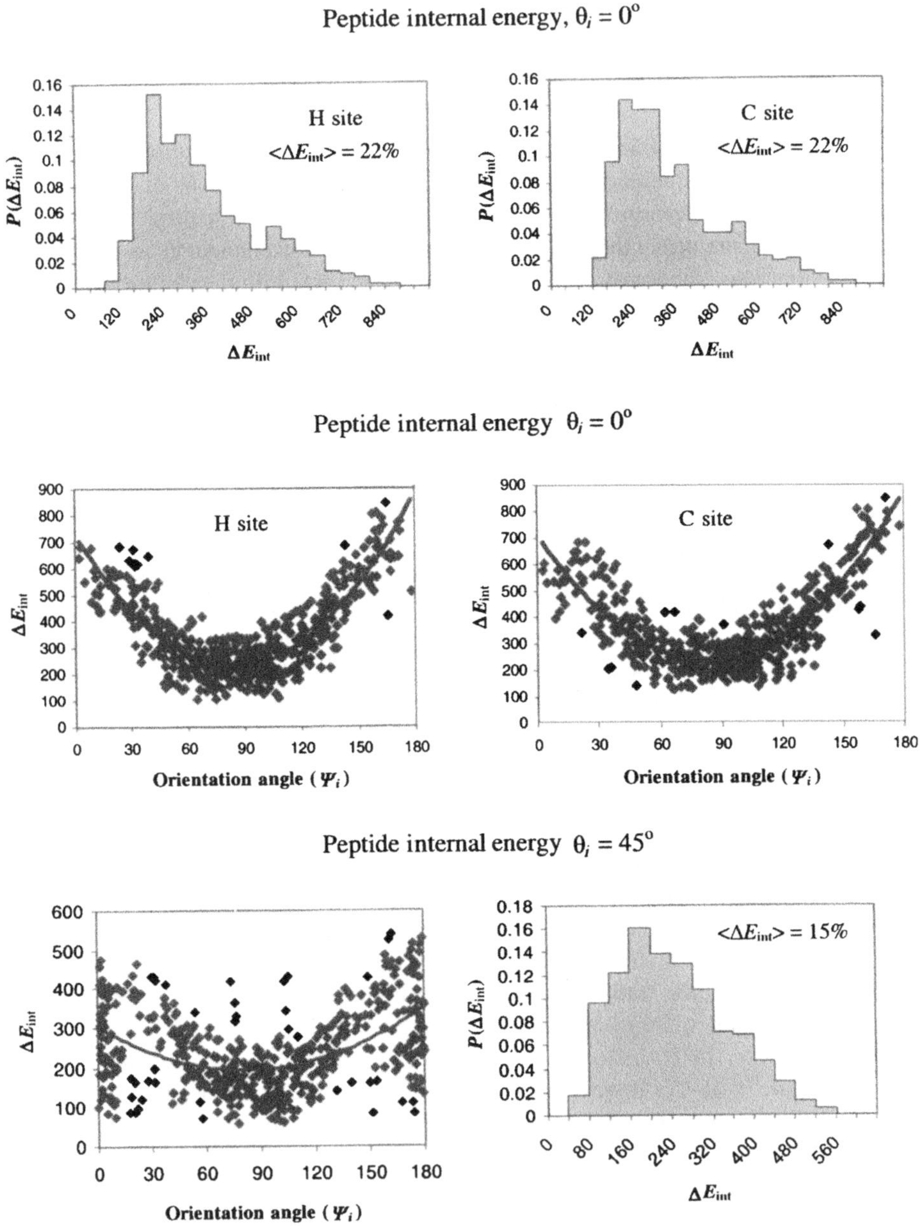

FIGURE 11.15. Upper graphs: distributions of ΔE_{int} for gly$_2$-H$^+$ collisions with the H-atom and C-atom sites of diamond {111}. Energy in the plots is in kcal/mol. $E_i = 70\,\text{eV}$ and $\theta_i = 0°$. Middle graphs: scatterplots of the energy transfers in the upper plots, versus the gly$_2$-H$^+$ orientation angle. The curves are parabolic fits. Energy in the plots is in kcal/mol. Lower graphs: simulation results for $E_i = 70\,\text{eV}$ and $\theta_i = 45°$. Both energy transfer distributions and scatter-plots of the energy transfer versus gly$_2$-H$^+$ orientation angle are given. Energy in the plots is in kcal/mol. [Reprinted with permission from Rahaman et al. (2006).]

11.4.2.4.1. $Cr^+(CO)_6$ Dissociation. Shattering is observed in collisions of $Cr^+(CO)_6$ with the diamond {111} surface at E_i of 30, 70, and 110 eV (Song et al. 2003). For $E_i = 30$ eV, the $n = 4$–6 $Cr^+(CO)_n$ ions dissociate by shattering, while the dissociation for $n = 1$–3 is RRKM-like. For $E_i = 70$ eV, the $n = 4$–6 dissociate by shattering, the dissociation of the $n = 1,2$ ions is RRKM-like, and the dissociation of $n = 3$ ions exhibits a transition from shattering to RRKM behavior. At the highest E_i studied of 110 eV, each of the $n = 4$–6 ions and the majority of the $n = 1$–3 ions dissociate by shattering. The importance of shattering is illustrated in Figure 11.16, where scatterplots are given of the height (h) of the ion above the surface at the time dissociation occurred versus the ion's dissociation lifetime (τ).

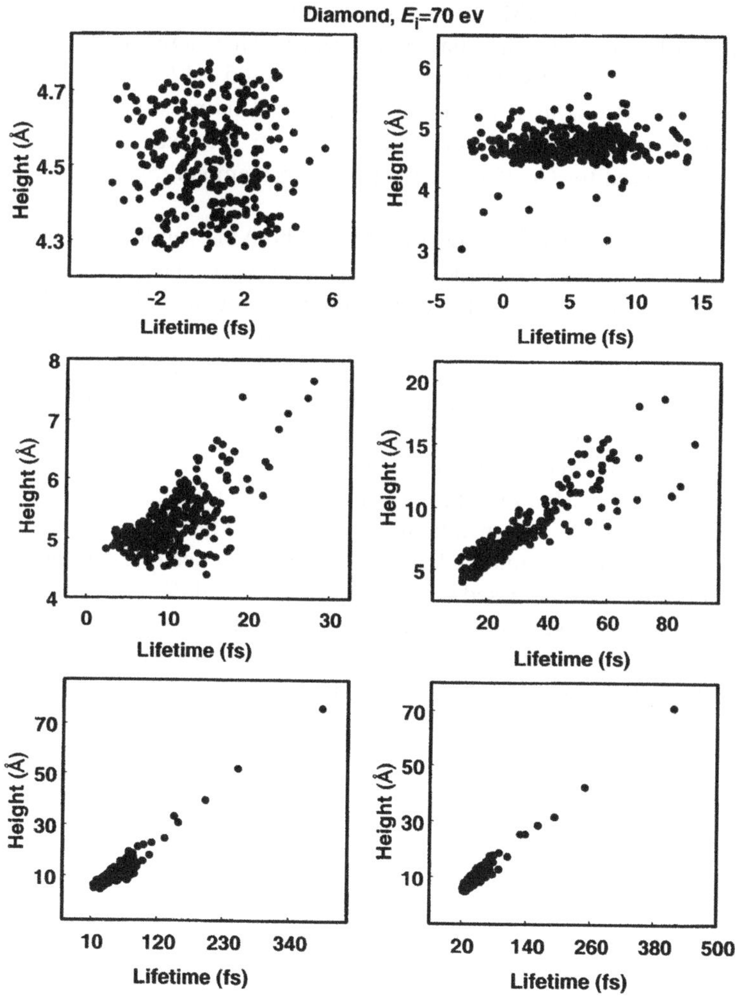

FIGURE 11.16. Scatterplots of the height of the $Cr^+(CO)_n$ ions above the diamond {111} surface when $Cr^+(CO)_n \rightarrow Cr^+(CO)_{n-1} + CO$ dissociation occurs versus the dissociation lifetime. Results are for $E_i = 70$ eV and $\theta_i = 45°$. The H atoms define the top of the diamond surface. [Reprinted with permission from Song et al. (2003).]

For $Cr^{+}(CO)_6$ collision with a H-SAM surface, at $\theta_i = 45°$ and E_i of 30 and 70 eV, there is no shattering and dissociation of each of the $Cr^{+}(CO)_n$ ions is RRKM-like. However, at the higher collision energy 110 eV, the $n = 4$–6 ions dissociate by shattering, with some shattering also for the $n = 3$ ions. Dissociation of the $n = 1,2$ ions remains RRKM-like. For some of the collisions with the H-SAM, at E_i of both 70 and 110 eV, the dissociations occur at heights of zero or less and thus occur while the ion has penetrated the surface.

Shattering requires direct $T \rightarrow V$ energy transfer into the rupturing Cr^{+}–CO bond(s) as $Cr^{+}(CO)_6$ impacts the surface. The probability of this energy transfer is expected to increase as the collision energy E_i is increased and this is the observed result. The greater likelihood of shattering for collisions with the hard diamond, as compared to the softer H-SAM surface, is greater with more efficient $T \rightarrow V$ energy transfer for collision with the diamond surface. That some collisions lead to shattering and others to RRKM-like dissociation, for a fixed E_i, may result from the orientation of $Cr^{+}(CO)_6$ as it impacts the surface. This possible effect should be investigated in future work. It may be related to the near bimodal $P(\Delta E_{int})$ distribution observed for $Cr^{+}(CO)_6$ collisions with diamond (Song et al. 2003).

11.4.2.4.2. Gly-H^{+} and Gly$_2$-H^{+} Dissociation. The AM1 semiempirical model for the peptide–H^{+} intramolecular potential was used to study the fragmentation mechanism for gly-H^{+} and gly$_2$-H^{+} dissociation (Meroueh et al. 2002; Wang et al. 2003b). The diamond {111} surface was used for these simulations. For gly-H^{+} + diamond {111}, at $E_i = 70$ eV and $\theta_i = 45°$, 23% of the collisions resulted in fragmentation by shattering. Within the 1.5 ps of the trajectories, another 19% dissociated by IVR and a RRKM-like mechanism; 58% of the trajectories did not dissociate during the timescale of the simulation. The SID fragmentation dynamics of gly$_2$-H^{+} + diamond {111} collisions was simulated at E_i of 30, 50, 70, and 100 eV for $\theta_i = 0°$, perpendicular collisions. The trajectories were only integrated for a maximum time of 1 ps and not all the trajectories fragmented. The percent that fragmented increased with E_i and were 21%, 44%, 66%, and 83% for E_i of 30, 50, 70, and 100 eV, respectively. The fragmenting trajectories were analyzed to determine the site where the initial bond rupture occurred. These sites are identified by the lettering scheme in Figure 11.17.

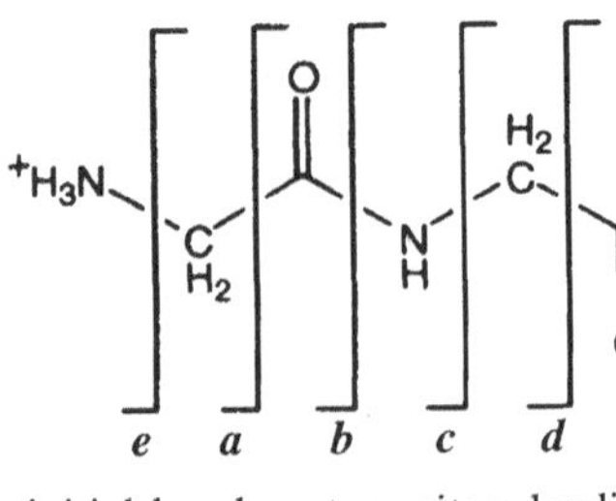

FIGURE 11.17. Possible initial bond rupture sites, leading to fragmentation, for gly$_2$-H^{+}. [Reprinted with permission from Wang et al. (2003b).]

TABLE 11.9. Fraction of Gly_2-H^+ + Diamond {111} Fragmentation Occurring by Different Initial Bond Ruptures[a]

E_i^b	Number of Trajectories	Fragmenting Fraction[c]	Fragmentation Fraction Initiated at Different Bond Rupture Sites				
			a	*b*	*c*	*d*	*e*
30	100	0.21	0.18	0.02	0	0	0.01
50	100	0.44	0.31	0.01	0	0.06	0.06
70	122	0.66	0.43	0.12	0	0.09	0.02
100	106	0.83	0.56	0.03	0	0.15	0.09

[a]The initial bond rupture sites are shown in Figure 11.17.
[b]E_i is in units of eV.
[c]Number of trajectories that fragment divided by the total number of trajectories.

The number of trajectories calculated at each E_i and the fraction of the fragmentation occurring by the different initial bond rupture sites are given in Table 11.9. The fraction of the gly$_2$-H^+ trajectories that dissociate increases with increase in E_i. Site *a* is the most probable position for initial bond rupture to occur, increasing to 56% for $E_i = 100$ eV. In addition, there is no initial bond rupture at site *c*. The absence of dissociation at site *c* is consistent with previous experimental (Klassen and Kebarle 1997) and computational (Paizs and Suhai 2001; Yalcin et al. 1995) studies for diglycine fragmentation. Preference for initial bond rupture at site *a*, namely, the CH_2–CONH bond, concurs with the fragmentation mechanism proposed (Paizs and Suhai 2001) for which the initial step is elimination of CO bonded to the amide nitrogen. As found from CID experiments (Klassen and Kebarle 1997), the ions $NH_2CH_2^+$ and its isomer $NH_3CH_2^+$ are principal fragmentation products.

The number of product channels observed in the gly$_2$-H^+ + diamond {111} simulations increases dramatically as E_i is increased; 6, 23, 44, and 59 different product channels are observed for E_i of 30, 50, 70, and 100 eV, respectively. This increase in the number of product channels is a result of shattering, in which dissociation occurs as gly$_2$-H^+ collides with the surface. The shattering results are summarized in Table 11.10, where the fraction of the trajectories that shatter is listed

TABLE 11.10. Fraction of Gly_2-H^+ + Diamond {111} Trajectories that Fragment by Shattering

E_i^a	Shattering Fraction[b]	Shattering Fraction Initiated at Different Bond Rupture Sites				
		a	*b*	*c*	*d*	*e*
30	0.08	0.06	0.02	0	0	0
50	0.13	0.10	0	0	0	0.03
70	0.44	0.29	0.08	0	0.06	0.01
100	0.71	0.48	0.02	0	0.13	0.08

[a]Collision energy in eV.
[b]Number of trajectories that shatter divided by the total number of trajectories.

as well as the fraction that shatter for the different initial bond rupture sites: sites *a*, *b*, *c*, *d*, and *e* in Figure 11.17. The fraction of the trajectories that shatter increases from 0.08 to 0.71 as E_i is increased from 30 to 100 eV. For all E_i the most important initial bond rupture site for shattering is site *a*. At low E_i of 30 and 50 eV shattering is also observed with initial dissociation at sites *b* and *e*. Shattering occurs with initial dissociation at site *d* for E_i of 70 eV and this becomes the second most important initial dissociation site for shattering at E_i of 100 eV.

At each E_i the principal product channels are those with initial rupture of site *a* in Figure 11.17 and formation of the immonium ion $NH_2CH_2^+$. Initial rupture at site *a* and formation of $NH_2CH_2^+$ constitute 81%, 36%, 35%, and 11% of the fragmentation channels at E_i of 30, 50, 70, and 100 eV, respectively. The decrease in the yield of this ion as E_i is increased, results from the onset of the multitude of shattering pathways. From experiments (Klassen and Kebarle 1997), $NH_2CH_2^+$ is expected to be the principal product ion in CID.

The energy transfer efficiencies of E_f, ΔE_{int}, and ΔE_{surf} are very similar for shattering and nonshattering trajectories, and shattering is not concomitant with a significantly larger transfer of ΔE_{int} than for nonshattering events. Initial studies suggest that shattering is influenced by the orientation of gly$_2$-H$^+$ as it collides with the surface. This is consistent with a suggestion (Meroueh et al. 2002) that shattering is promoted by collisions that direct gly$_2$-H$^+$ to a dissociation transition state on impact with the surface. An incomplete analysis (Wang et al. 2003b) suggests shattering is enhanced when the peptide chain is oriented perpendicular to the surface during collision, so that one "end" of the chain first strikes the surface. A number of important questions may be addressed by an analysis of peptide–H$^+$ orientation and shattering in the simulations. What fraction of ions has the proper orientation to decompose by a shattering mechanism? Does orientation select particular transition states? How do the shattering transition states differ from transition states involved in RRKM type dissociation?

11.4.2.4.3. Peptide–H$^+$ Shattering in Simulation and Experiments. The simulations presented above illustrate the importance of shattering for gly-H$^+$ and gly$_2$-H$^+$ SID during collisions with the diamond {111} surface. Shattering is exemplified by many product channels and the relative product yields for these channels are unrelated to their dissociation thresholds; thus, an RRKM analysis does not predict the presence of such a large number of dissociation channels, nor the observed relative product yields. Shattering becomes increasingly important as E_i is increased.

From experimental studies (Laskin et al. 2003; Laskin and Futrell 2003b,c) of des-Arg^1– and des-Arg^9–bradykinin and fibrinopeptide SID, shattering has been suggested for high-energy collisions. As in the simulations, shattering becomes more important as the collision energy is increased and the appearance of many more product channels, than predicted by unimolecular thresholds and RRKM theory is attributed to shattering. The dissociation of peptide–H$^+$ ions in the experiments may be modeled by two decay rates (slow and fast), with the later modeled by the "sudden death" approximation, in which the molecule fragments instantaneously if its internal energy reaches a certain threshold. It is stated (Laskin and Futrell 2003b) that "instantaneous fragmentation of large molecules at high collision energies occurs on or very near the surface rather than in the gas phase and indicates a transition in the dynamics of ion–surface interaction—namely, the shattering transition." This is the picture of the shattering dynamics observed in the simulations.

An unresolved issue in comparing the simulation and experimental studies of shattering is that in the experiments the ions are observed on a 1 ms or larger timescale. Additional information needed, to compare with experiment, are the translational energies of ions formed by shattering. This will be determined in future simulations.

11.5. FUTURE DIRECTIONS

The presentations and discussions in this chapter show that classical trajectory chemical dynamics simulations give accurate results for the atomic-level details of CID and SID. They may be used to study the dynamics of both the energy transfer and unimolecular dissociation processes, and may be compared directly with experiment. They also provide the fundamental information needed to develop accurate theoretical models for CID and SID.

Although much has been achieved from the simulations, a considerable amount of additional work needs to be done. To date only the SID of protonated peptides with alanine and glycine constituents have been simulated, and it is important to consider additional polypeptides with a variety of amino acids. Also, different surfaces need to be investigated. The reported simulation results are for diamond and H-SAM surfaces, while experiments have also considered surfaces with varying degrees of fluorination and hydroxylation (Smith et al. 2002, 2003). It would also be of interest to consider organic surfaces with additional types of functionalization (Ferguson et al. 2004).

For the potential energy surfaces used in the simulations, the intermolecular potentials between the projectile ions and surfaces have been represented by analytic functions fit to ab initio calculations. Furthermore, these fits have focused on the short-range repulsive regions of the intermolecular potential, so that the energy transfer during the projectile's initial impact with the surface is correctly described. These models do not accurately represent the projectile–surface attractive interaction, nor represent possible projectile + surface chemical reactions. The inaccurate attractive potentials preclude the possibility of simulating soft-landing experiments (Miller et al. 1997), in which projectile ions are captured intact on either solid or liquid surfaces. Soft landing may be simulated by fitting analytic potential energy functions to high-level ab initio calculations that give accurate representations of both the short-range and long-range regions of the intermolecular potentials (Yan and Hase 2002). Chemical reaction between the projectile and the surface may be simulated by QM/MM direct dynamics trajectory calculations in which the potential energy for the projectile and the projectile's interaction with the surface are represented by a quantum mechanics (QM) electronic structure theory (Li et al. 2000). In the simulation only the part of the surface directly interacting with the projectile is represented by the QM model. The remaining surface atoms are represented by analytic molecular mechanics (MM) potentials, with a QM/MM interface connecting the QM and MM surface regions.

The previous direct dynamics simulations of peptide–H^+ surface collisions have used QM+MM models, in which the peptide–H^+ potential is the semiempirical AM1 QM model and the peptide–H^+/surface and surface potentials are analytic MM-type potentials. These QM+MM models do not allow reaction with the surface. Only the gly-H^+ and gly$_2$-H^+ peptides have been considered in these simulations and it is important to extend these simulations to larger peptides with a range of amino acid constituents. In addition, higher level QM methods such as MP2 and DFT need to be investigated. It is important to determine if these more accurate QM models also show that shattering is an important peptide–H^+ fragmentation mechanisms. The MP2 and DFT models may also be used in the QM/MM studies of peptide–H^+ + surface collisions, from which both shattering and reaction with the surface may be investigated.

ACKNOWLEDGMENTS

The authors' CID and SID simulations were supported by the National Science Foundation. Special thanks are expressed to Jean Futrell, Julia Laskin, and Vicki Wysocki for valuable discussions concerning the dynamics of peptide–H^+ SID.

REFERENCES

Allen MP, Tildesley DJ (1987): *Computer Simulation of Liquids*, Oxford, New York.

Baer T, Hase WL (1996): *Unimolecular Reaction Dynamics—Theory and Experiments*, Oxford Univ. Press, New York.

Blais NC, Bunker DL (1962): Monte Carlo calculations. II. The reactions of alkali atoms with methyl iodide. *J. Chem. Phys.* **37**:2713–2720.

Bolton K, Hase WL, Peslherbe GH (1998): Direct dynamics simulations of reactive systems. In Thompson DL (ed), *Modern Methods for Multidimensional Dynamics Computations in Chemistry*, World Scientific, London, pp. 143–189.

Bolton K, Hase WL (1998): Integrating the classical equations of motion. In Allinger NL (ed), *Encyclopedia of Computational Chemistry*, Vol. 2, Wiley, New York, pp. 1347–1360.

Bosio SBM, Hase WL (1997): Energy transfer in rare gas atom collisions with self-assembled monolayers. *J. Chem. Phys.* **107**:9677–9686.

Bosio SBM, Hase WL (1998): Simulations of energy transfer in $Cr^+(CO)_6$ surface induced dissociation. *Int. J. Mass Spectrom. Ion Proc.* **174**:1–9.

Bunker DL (1962): Monte Carlo calculations of triatomic dissociation rates. I. N_2O and O_3. *J. Chem. Phys.* **37**:393–403.

Bunker DL, Goring-Simpson EA (1973): Alkali-methyl iodide reactions. *Faraday Discuss. Chem. Soc.* **55**:93–99.

Bunker DL, Chapman S (1975): An exploratory study of reactant vibrational effects in $CH_3 + H_2$ and its isotopic variants. *J. Chem. Phys.* **66**:1523–1533.

Califano S (1976): *Vibrational States*, Wiley, New York.

Cornell WD, Cieplak P, Bayley CI, Gould IR, Merz KM, Ferguson DM, Spellmeyer DC, Fox T, Cladwell JW, Kollman PA (1995): A second generation force field for the simulation of proteins, nucleic acids, and organic molecules. *J. Am. Chem. Soc.* **117**:5179–5197.

de Sainte Claire P, Hase WL (1996): Thresholds for the collision-induced dissociation of clusters by rare gas impact. *J. Phys. Chem.* **100**:8190–8196.

de Sainte Claire P, Peslherbe GH, Hase WL (1995): Energy transfer dynamics in the collision-induced dissociation of Al_6 and Al_{13} clusters. *J. Phys. Chem.* **99**:8147–8161.

Dewar MJS, Zoebisch EG, Healy EF, Stewart JJP (1985): The development and use of quantum mechanical molecular models. 76. AM1: A new general purpose quantum mechanical molecular model. *J. Am. Chem. Soc.* **107**:3902–3909.

Ferguson MK, Lohr JR, Day BS, Morris JR (2004): Influence of buried hydrogen-bonding groups within monolayer films on gas-surface energy exchange and accommodation. *Phys. Rev. Lett.* **92**:073201.

Frisch MJ, Head-Gordon M, Pople JA (1988): A direct MP2 gradient method. *Chem. Phys. Lett.* **166**:275–280.

Goldstein H (1950): *Classical Mechanics*, Addison-Wesley, London.

Granucci G, Toniolo A (2000): Molecular gradients for semiempirical CU wavefunctions with floating occupation molecular orbitals. *Chem. Phys. Lett.* **325**:79–85.

Grebenshchikov SY, Schinke R, Hase WL (2003). State-specific dynamics of unimolecular dissociation. In Green NJB (ed), *Comprehensive Chemical Kinetics*, Vol. 39, *Unimolecular Kinetics*, Part 1. *The Reaction Step*, Elsevier, New York, pp. 105–242.

Grimmelmann EK, Tully JC, Cardillo MJ (1980): Hard-cube model analysis of gas-surface energy accommodation. *J. Chem. Phys.* **72**:1039–1043.

Hanley L, Ruatta SA, Anderson SL (1987): Collision-induced dissociation of aluminum cluster ions: Fragmentation patterns, bond energies, and structures for Al_2^+-Al_7^+. *J. Chem. Phys.* **87**:260–268.

Hariharan PC, Pople JA (1973): Influence of polarization on MO hydrogenation energies. *Theor. Chim. Acta* **28**:213–222.

Hariharan PC, Pople JA (1974): Accuracy of AH_n equilibrium geometries by single determinant molecular orbital theory. *Mol. Phys.* **1**:209–214.

Hase WL (ed), (1998a): *Advances in Classical Trajectory Methods*, Vol. 3, *Comparisons of Classical and Quantum Dynamics*. JAI Press, London.

Hase WL (1998b): Classical trajectory simulations: Final conditions. In Allinger NL (ed), *Encyclopedia of Computational Chemistry*, Vol. 1, Wiley, New York, pp. 399–402.

Hase WL (1998c): Classical trajectory simulations: Initial conditions. In Allinger NL (ed), *Encyclopedia of Computational Chemistry* Vol. 1, Wiley, New York, pp. 402–407.

Hase WL, Buckowski DG (1982): Dynamics of ethyl radical decomposition. II. Applicability of classical mechanics to large molecule unimolecular reaction dynamics. *J. Comput. Chem.* **3**:335–343.

Hase WL, Mondro SL, Duchovic RJ, Hirst, DM (1987): Thermal rate constant for $H + CH_3 \rightarrow CH_4$ recombination. III. Comparison of experiment and canonical variational transition state theory. *J. Am. Chem. Soc.* **109**:2916–2922.

Hase WL, Song K, Gordon MS (2003): Direct dynamics simulations. *Comput. Sci. Eng.* **5**:36–44.

Hass KC, Tamor MA, Anthony TR, Banholzer WF (1992): Lattice dynamics and Raman spectra of isotopically mixed diamond. *Phys. Rev. B* **45**:7171–7182.

Head-Gordon M, Pople JA, Frisch MJ (1988): MP2 energy evaluation by direct methods. *Chem. Phys. Lett.* **153**:503–506.

Isa N, Gibson KD, Yan T-Y, Hase WL, Sibener SJ (2004): Experimental and simulation study of neon atom collision dynamics with a 1-decanthiol monolayer. *J. Chem. Phys.* **120**:2417–2433.

Jarrold MF, Bower EJ, Krauss JS (1987): Collision-induced dissociation of metal cluster ions: Bare aluminum clusters, Al_n^+ ($n = 3$–26). *J. Chem. Phys.* **86**:3876–3885.

Khan FA, Clemmer DE, Schultz RH, Armentrout PB (1993): Sequential bond energies of $Cr(CO)_x$, $x = 1$–6. *J. Chem. Phys.* **97**:7978–7987.

Klassen JS, Kebarle P (1997): Collision-induced dissociation threshold energies of protonated glycine, glycinamide, and some related small peptides and peptide amino amides. *J. Am. Chem. Soc.* **119**:6552–6563.

Laskin J, Denisov E, Futrell JH (2000): A comparative study of collision-induced and surface-induced dissociation. 1. Fragmentation of protonated dialanine. *J. Am. Chem. Soc.* **122**: 9703–9714.

Laskin J, Bailey TH, Futrell JH (2003): Shattering of peptide ions on self-assembled monolayer surface. *J. Am. Chem. Soc.* **125**:1625–1632.

Laskin J, Futrell JH (2003a): Energy transfer in collisions of peptide ions with surfaces. *J. Chem. Phys.* **119**:3413–3420.

Laskin J, Futrell JH (2003b): Surface-induced dissociation of peptide ions: Kinetics and dynamics. *J. Am. Soc. Mass Spectrom.* **14**:1340–1347.

Laskin J, Futrell JH (2003c): Collisional activation of peptide ions in FT-ICR mass spectrometry. *Mass Spectrom. Rev.* **22**:158–181.

Li G, Bosio SBM, Hase WL (2000): A QM/MM model for $O(^3P)$ reaction with an alkyl thiolate self-assembled monolayer. *J. Mol. Struct.* **556**:43–57.

Lian L, Su C-X, Armentrout PB (1992): Collision-induced dissociation of Ni_n^+ ($n = 2$–18) with Xe: Bond energies, geometrical structures, and dissociation pathways. *J. Chem. Phys.* **96**:7542–7554.

Liu J, Song K, Hase WL, Anderson SL (2003): Direct dynamics study of energy transfer and collision-induced dissociation: Effect of impact energy, geometry, and reactant vibrational modes in H_2CO^+-Ne collisions. *J. Chem. Phys.* **119**:3040–3050.

Mahan BH (1970): Refined impulse approximation for the collisional excitation of the classical harmonic oscillator. *J. Chem. Phys.* **52**:5221–5225.

Mar W, Klein M (1994): Molecular dynamics study of the self-assembled monolayer composed of $S(CH_2)_{14}CH_3$ molecules using an all-atoms model. *Langmuir* **10**: 188–196.

Martinez-Nuñez E, Fernandez-Ramos A, Vazquez SA, Marquez JMC, Xue M, Hase WL (2005): Quasiclassical dynamics simulation of the collision-induced dissociation of $Cr(CO)_6^+$ with Xe. *J. Chem. Phys.* **123**:154311/1–9.

Meroueh O, Hase WL (1999): Collisional activation of small peptides. *J. Phys. Chem. A* **103**:3981–3990.

Meroueh O, Hase WL (2000): Energy transfer pathways in the collisional activation of peptides. *Int. J. Mass Spectrom.* **201**:233–244.

Meroueh O, Hase WL (2001): Effect of surface stiffness on the efficiency of surface-induced dissociation. *Phys. Chem. Chem. Phys.* **3**:2306–2314.

Meroueh O, Hase WL (2002): Dynamics of energy transfer in peptide-surface collisions. *J. Am. Chem. Soc.* **124**:1524–1531.

Meroueh SO, Wang Y, Hase WL (2002): Direct dynamics simulations of collision- and surface-induced dissociation of N-protonated glycine. Shattering fragmentation. *J. Phys. Chem. A* **106**:9983–9992.

Miller SA, Luo H, Pachuta S, Cooks RG (1997): Soft-landing of polyatomic ions at fluorinated self-assembled monolayer surfaces. *Science* **275**:1447–1450.

Morris MR, Riederer Jr DE, Winger BE, Cooks RG, Ast T, Chidsey CED (1992): Ion/surface collisions at functionalized self-assembled monolayer surfaces. *Int. J. Mass Spectrom. Ion Proc.* **122**:181–217.

Paizs B, Suhai S (2001): Theoretical study of the main fragmentation pathways for protonated glycylglycine. *Rapid Commun. Mass Spectrom.* **15**:651–663.

Peslherbe GH, Wang H, Hase WL, (1999): Monte Carlo sampling for classical trajectory simulations. *Adv. Chem. Phys.* **105**:171–202.

Pettersson LGM, Bauschlicher, Jr. CW, Halicioglu T (1987): Small Al clusters. II. Structures and binding in Al_n ($n = 2$–6, 13). *J. Chem. Phys.* **87**:2205–2213.

Rahaman A, Zhou JB, Hase WL (2006): Effect of projectile orientation and surface impact site on the efficiency of projectile excitation in surface-induced dissociation. Protonated diglycine collisions with diamond {111}. *Int. J. Mass Spectrom.* **249–250**:321–329.

Raz T, Levine RD (1996): On the shattering of clusters by surface impact heating. *J. Chem. Phys.* **105**:8097–8102.

Schlick T (2002): *Molecular Modeling and Simulation*, Springer, New York.

Schultz DG, Weinhaus SB, Hanley L, de Sainte Claire P, Hase WL (1997): Classical dynamics simulations of $SiMe_3^+$ ion surface scattering. *J. Chem. Phys.* **107**:9677–9686.

Shin HK (1976): Vibrational energy transfer. In Miller WH (ed), *Dynamics of Molecular Collisions*, Part A, Plenum Press, New York, pp. 131–210.

Sloane CS, Hase WL (1977): On the dynamics of state selected unimolecular reactions. Chloroacetylene dissociation and predissociation. *J. Chem. Phys.* **66**:1523–1533.

Smith DL, Wysocki VH, Colorado Jr. R, Shmakova OE, Graupe M, Lee TR (2002): Low-energy ion-surface collisions characterize alky- and fluoroalkyl-terminated self-assembled monolayers on gold. *Langmuir* **18**:3895–3902.

Smith DL, Selvan R, Wysocki VH (2003): Reactive ion scattering spectrometry of mixed methyl- and hydroxy-terminated alkanethiolate self- assembled monolayers. *Langmuir* **19**:7302–7306.

Song K, de Sainte Claire P, Hase WL, Hass KC (1995): Comparison of molecular dynamics and variational transition state theory calculations of the rate constant for H-atom association with the diamond {111} surface. *Phys. Rev. B* **52**:2949–2958.

Song K, Meroueh O, Hase WL (2003): Dynamics of $Cr(CO)_6^+$ collisions with hydrogenated surfaces. *J. Chem. Phys.* **118**:2893–2902.

Stewart JJP (1989): Optimization of parameters for semiempirical methods. *J. Comput. Chem.* **10**:209–220.

Su C-X, Armentrout PB (1993): Collision-induced dissociation of Cr_n^+ ($n = 2$–21) with Xe: Bond energies, dissociation pathways and structures. *J. Chem. Phys.* **99**:6506–6516.

Sun L, Hase WL (2003): Born-Oppenheimer direct dynamics classical trajectory simulations. In Lipkowitz KB, Larter R, Cundari TR (eds), *Reviews in Computational Chemistry*, Vol. 19, Wiley-VCH, Hoboken, NJ, pp. 79–146.

Truhlar DG, Muckerman JT (1979): Reactive scattering cross sections III. Quasiclassical and semiclassical methods. In Bernstein RB (ed), *Atom-Molecule Collision Theory*, Plenum, New York, pp. 505–566.

Wang J, Meroueh SO, Wang Y, Hase WL (2003a): Efficiency of energy transfer in protonated diglycine and dialanine SID. Effects of collision angle, peptide ion size, and intramolecular potential. *Int. J. Mass Spectrom.* **230**:57–64.

Wang Y, Hase WL, Song K (2003b): Direct dynamics study of N-protonated diglycine surface-induced dissociation. Influence of collision energy. *J. Am. Soc. Mass Spectrom.* **14**: 1402–1412.

Whyte AR, Lim KF, Gilbert RG, Hase WL (1988): The calculation and interpretation of average collisional energy transfer parameters. *Chem. Phys. Lett.* **152**:377–381.

Wilson Jr. EB, Decius JC, Cross PC (1955): *Molecular Vibrations*, McGraw-Hill, New York.

Yalcin T, Khouw C, Csizmadia IG, Peterson MR, Harrison AG (1995): Why are B ions stable species in peptide spectra? *J. Am. Chem. Soc. Mass Spectrom.* **6**:1165–1174.

Yan TY, Hase WL (2002): Comparison of models for simulating energy transfer in Ne-atom collisions with an alkyl thiolate self-assembled monolayer. *J. Phys. Chem. B* **106**: 8029–8037.

Yardley JT (1980): *Introduction to Molecular Energy Transfer*, Academic Press, London, pp. 95–129.

12

ION SOFT LANDING: INSTRUMENTATION, PHENOMENA, AND APPLICATIONS

Bogdan Gologan, Justin M. Wiseman, and R. Graham Cooks

Department of Chemistry
Purdue University
West Lafayette, IN

Principles of Mass Spectrometry Applied to Biomolecules, edited by Julia Laskin and Chava Lifshitz

12.1. INTRODUCTION

In the early twentieth century, Rutherford (1911), using MeV ion beams impinging on thin metal films, provided the first experimental basis for the theory of atomic structure. Much later, this phenomenon was developed into an analytical method of surface and near-surface analysis known as *Rutherford backscattering spectrometry* (RBS). A related binary elastic scattering experiment in the keV rather than the MeV energy range was developed in the 1960s into *ion scattering spectrometry* (ISS), a method for the analysis of the elemental composition of the outermost layers of solid surfaces (Smith 1967). This method has subsequently grown to include recoil and shadowing processes, the latter depending on angular relationships involving the positions of surface adsorbates. Another keV collision phenomenon is sputtering, the process on which secondary-ion mass spectrometry (SIMS) (Honig 1957; Benninghoven et al. 1982; Pachuta and Cooks 1987; Winograd 2005) is based. A series of binary elastic collisions leads to desorption of ions, neutral molecules, and clusters from a surface struck by a primary ion of keV energy; the secondary ions leave with relatively low translational energies and a range of internal energies and are mass-analyzed to provide elemental and molecular structural information on surface constituents. In the following sections ion–surface collision phenomena are introduced and discussed individually in some detail. Ion–surface collision studies can be categorized into four energy regimes based on the collision energy: (1) the thermal range with ions having kinetic energies below 1 eV, (2) the hyperthermal range involving ions having energies 1–100 eV, (3) the low-energy range from 0.1 to 10 keV, and (4) the high-energy range covering ion energies into the MeV regime. This chapter exclusively deals with the hyperthermal energy regime where a number of important chemical events occur.

12.1.1. Ion–Surface Collision Phenomena

The phenomena observed following the collision of polyatomic, organic projectile ions with a surface include (1) simple elastic scattering leading to reflection; (2) surface-induced dissociation resulting from ion excitation in the course of an inelastic collision; (3) chemical sputtering, a process in which surface molecules are ejected into the gas phase as a result of low-energy chemical reactions; (4) reactive collisions leading to chemical transformation of the ion, the surface, or both; and (5) ion soft landing. These processes, which occur competitively in the hyperthermal energy regime, are summarized in Figure 12.1. The likelihood of chemical reactions involving the atoms of the ion and the surface is maximized at collision energies that are neither too high (where the loss of chemical information results from complete dissociation of the projectile at the surface) nor too low (where there is insufficient center-of-mass energy to drive chemical reactions). In most cases, hyperthermal energy collisions provide highly useful chemical information on the projectile ions and the outermost monolayer of the surfaces or surface adsorbates. The resulting information leads to insights into structures and reactivities of both the projectiles and the surfaces. Fundamental aspects of ion–surface collision processes include

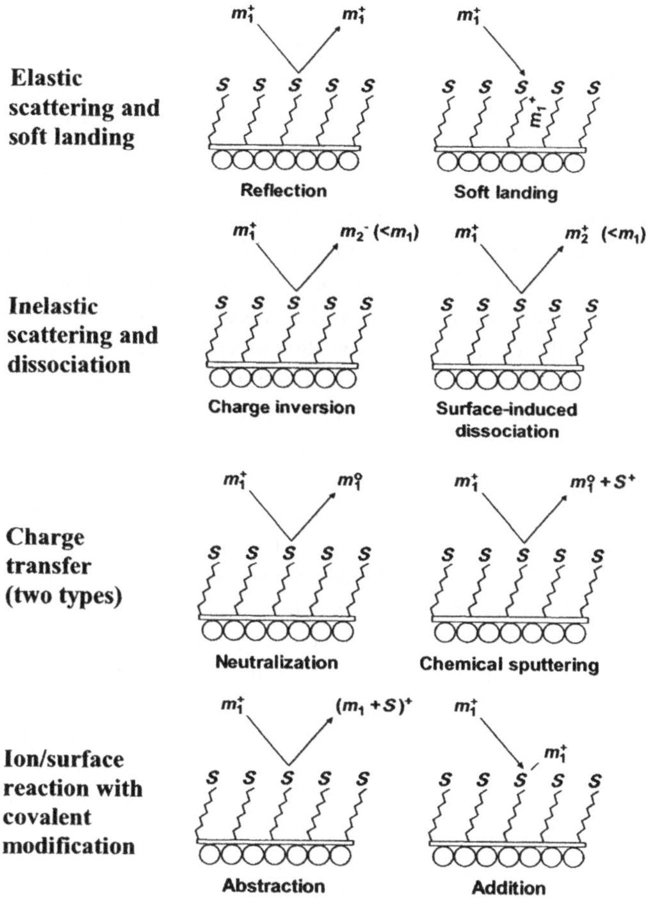

FIGURE 12.1. Processes associated with collisions of polyatomic ions at surfaces. There are other types of processes and categorizations. [Reprinted from Gologan et al. (2005) with permission from the Royal Society of Chemistry. Copyright 2005.]

studies of energy partitioning during inelastic collisions, in particular the transfer of translational energy to vibrational energy ($T \rightarrow V$)(Kubista et al. 1998; Jo and Cooks 2003), the recently recognized nonstatistical dissociation mechanism known as *ion shattering* (Hendell et al. 1995; Schultz and Hanley 1998; Laskin et al. 2003), and comparisons of reactivity at interfaces with gas-phase and solution-phase analogs. The various ion/surface scattering phenomena are discussed in the following sections.

12.1.1.1. Inelastic Scattering. During inelastic scattering a fraction of the initial kinetic energy of the projectile ion is transferred to the outermost surface monolayer and/or converted into internal energy of the scattered ion. If sufficient internal energy is acquired, the ion may dissociate either immediately at the surface [a nonstatistical process known as *shattering* (Hendell et al. 1995; Kang et al. 1998; Schultz and Hanley 1998; Laskin et al. 2003)], or it may do so following recoil, at some distance from the surface (Cooks, Ast et al. 1990). If fragmentation is delayed it occurs by standard gas phase unimolecular dissociation mechanisms. (Robinson and Holbrook 1972; Forst 1973; Baer and Hase 1996) The activation/dissociation process is collectively known as surface-induced dissociation (SID), by analogy with collision-induced dissociation (CID), except that a solid surface is used as a collision target instead of an inert gas molecule. Kinetic energy of the projectile ion is transferred into potential energy upon collision with a solid surface, with resulting activation and subsequent dissociation of the precursor ion.

Surface-induced dissociation was first developed in the 1970s and most of the pioneering studies were performed in this laboratory at Purdue University (Cooks et al. 1975, 1990b). In advancing the field notable results were obtained by several other groups (Williams et al. 1990; Chorush et al. 1995; Dongre et al. 1996; Nair et al. 1996; Schaaff et al. 1998; Laskin et al. 2002, 2003, 2004). The main idea behind this activation method stems from the fact that energy transfer in CID is limited by the energy available in the center-of-mass reference frame (E_{com}), which is dependent on the mass of the target gas. By increasing the mass of the target, E_{com} becomes larger and thus energy transfer can be improved, provided that the mass of the surface target is larger than that of the target gas. Energy conversion should therefore be much more efficient in SID if a substantial part of the surface acts as target. In certain instances, terminal groups on the surface appear to be primarily responsible for energy conversion. In this case, the center-of-mass collision energy is dependent on the mass of these groups. In general, similar products are observed in SID and CID, and the mechanism of activation for SID was therefore rationalized as a two-step process. Initially, the incident ion collides inelastically with the solid surface, forming an internally excited ion that then undergoes unimolecular dissociation. An additional step defined by the period of time spent by the ion on the surface can be distinguished (Wainhaus et al. 1997). The interaction time with the surface is of the order of 10^{-12} s, a short period compared with the dissociation time required for polyatomic ions (Cooks et al. 1990b).

Although not nearly as widely practiced, SID has some advantages over CID, including the simplicity conferred on the experiment by obviating the need to introduce gas into the vacuum system as well as a somewhat narrower internal energy distribution deposited into the activated ions. In addition, SID is a fast activation method for large ions and, as such, it opens up all the dissociation channels available to the molecule at each internal energy (while CID is discriminatory) (Laskin and Futrell 2003). This feature presents an important advantage for some MS/MS (tandem MS) applications. Another feature is that the internal energy deposited in SID is very readily varied as one varies the collision energy. Energy transfer, the key step in the SID process, has been studied in some detail by a number of investigators (DeKrey et al. 1986; Pradeep et al. 1993; Burroughs et al. 1994; Kubista et al. 1998). As an example, Figure 12.2 shows the internal energy of fragmenting ions on modified gold surfaces with fluorinated self-assembled monolayers (F-SAM) and hydrogen-terminated self-assembled monolayers (H-SAM)

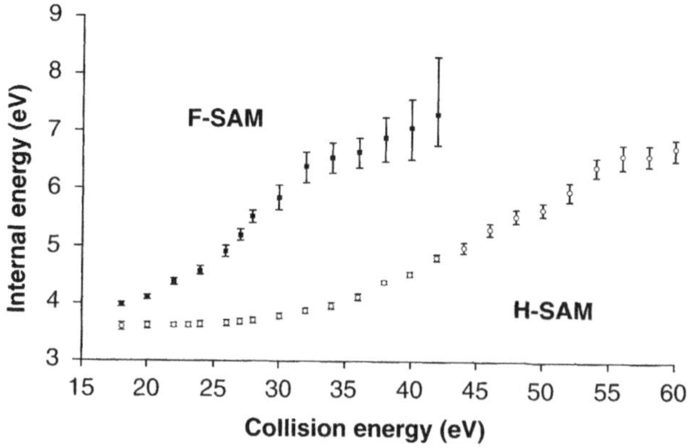

FIGURE 12.2. Plot of internal energy versus laboratory collision energy for *n*-butylbenzene ion. [Reprinted from Jo (2004).]

as estimated by their dissociation behavior, versus their collision energy in the laboratory frame (Jo 2004). This is essentially a plot of the average vibrational energy of the activated ion versus the collision energy at a particular angle of incidence (55° to the normal). The slope of this plot is therefore a measure of translational to vibrational ($T \rightarrow V$) energy conversion. In systems where small organic ions collide with organic surfaces, the internal energy transferred increases with collision energy. This increase is roughly in proportion to the translational energy, and the energy transfer efficiency (% T–V) is greater for the F-SAM surfaces (20%) than H-SAM (12%) surfaces. (Jo and Cooks 2003).

In parallel with experimental studies of ion–surface collision phenomena, there has been an increasingly strong effort in simulations. Noteworthy early work was Sigmund's elucidation (Sigmund 1973) of the higher-energy ion–surface collisions that lead to sputtering in SIMS and the use of a binary collision model (Heiland and Taglauer 1976) that correctly predicted energy loss in binary and ternary scattering in ISS. More relevant to the hyperthermal energy range is the work of Hase, in which the dynamics of collisions of polyatomic ions was simulated (Meroueh et al. 2002; Song et al. 2003). In these studies, computer simulations were used to investigate the SID dynamics of collisions with self-assembled monolayers (SAMs) and diamond surfaces; the findings of these studies illustrate that collisions with SAMs are softer than those with diamond, which translates into more surface penetration and longer residence times for the same ion–surface collision energy. Another noteworthy finding of the simulations is that ions activated by collisions with SAM surfaces have average lifetimes in overall good agreement with the prediction of RRKM theory and, for the most part, that these ions dissociate via intramolecular vibrational energy redistribution after collisional activation. A full description of ion–surface collision interactions at the molecular level remains unavailable largely because of the difficulties of adequately calculating and simulating the dynamics of interactions at a surface. Nevertheless, excellent progress has been made for very simple systems and more recently for larger ions and more complex surfaces with organic adsorbates (Krantzman et al. 2003; Postawa et al. 2003; Zhigilei et al. 2003).

12.1.1.2. Charge Inversion. Among the many phenomena associated with gas-phase collisions at keV energies is charge inversion. Charge inversion occurs from electron transfer and charge stripping processes in which incident positively charged ions are converted into negatively charged ions and vice versa. This and related gas-phase charge changing phenomena, including charge stripping of singly charged ions to give doubly charged ions, has proved to be of considerable value in the determination of thermochemical properties of gaseous ions (Danell and Glish 2001; Hayakawa 2004). Charge inversion reactions of positive ions of hydrogen and some simple radicals, such as oxygen, carbon, and chlorine (Thomson 1913) along with dissociative charge inversion reactions of C_2H^+(Aston 1919; Cooks 1995) were observed to be caused by collisions with residual gases due to the modest vacuum technology available at the time. Further improvements in vacuum systems significantly minimized collisions with residual gases and made it possible to collide ions with selected targets. In the case of charge inversion mass spectrometry, mass-selected precursor ions have been made to collide with a gaseous target, metal vapors (Bursey 1990; Bowie 2001; Danell and Glish 2001; Hayakawa 2001) or with a target surface (Douglas and Shushan 1982; Bier et al. 1987; Vincenti and Cooks 1988; Cooks et al. 1990b, 1994). The resulting product ions of opposite polarity to the precursor ion, formed on two-electron transfer, are mass-analyzed and detected. This type of MS/MS experiment is also known as *charge permutation* (CP) or *charge reversal* (CR) mass spectrometry. Most charge inversion reactions have been performed in the keV energy range, mainly because at these energies electron stripping and electron transfer reactions have sufficiently high cross sections. Charge inversion mass spectrometry using alkali metal gaseous targets has provided information about the dissociation pathways of energy selected neutral species (Hayakawa et al. 1995; Hayakawa 2001). Furthermore, differences in the chemical nature of positive and negative ions have allowed charge inversion mass spectrometry to be successfully applied to the problem of isomeric ion differentiation in the gas phase.

Neutralization-reionization (NR) and charge reversal (CR) mass spectrometric experiments can be combined to investigate the fragmentation of neutrals generated in high energy collisions. Provided that the species under study can exist as a stable anion, neutral, and cation, the reactions of neutral molecules can be distinguished from those of projectile and recovery ion signals recorded by taking neutral–ion decomposition difference (NIDD) mass spectra (Hornung et al. 1997; Schalley et al. 1998a,b; Schroder et al. 1999). Although most charge inversion studies involve collisions with gaseous targets, charge inversion experiments using solid targets have also been reported to reverse the polarity of precursor ions (Douglas and Shushan 1982; Bier et al. 1987; Vincenti and Cooks 1988; Cooks et al. 1990, 1994). As expected, the results show that the charge inversion process is strongly dependent on the nature of the surface and the collision energy of the projectile ions. Ionic collisions at surfaces can lead to charge changing, although their usefulness in estimating thermochemical values remains to be explored. In cases in which the scattering of projectile ions from self-assembled monolayers results in charge permutation, the scattering process can sometimes be accompanied by fragmentation of recoiled ions. This can occur at eV collision energies, in contrast to the keV energies typically used to record the corresponding data for gas-phase charge-changing processes (Hayakawa 2004). Charge-changing collisions at a surface are strongly affected by the thermochemistry associated with the particular reaction and since ions do not penetrate the surface in this energy range, these reactions take place at the outermost atomic layers of the surface.

12.1.1.3. Chemical Sputtering. The powerful surface analytical technique of secondary-ion mass spectrometry (SIMS) (Honig 1957; Benninghoven et al. 1982; Pachuta and Cooks 1987) uses keV energy projectiles to effect desorption/ionization of molecules present on the surface. This technique provides information on the elemental, isotopic and, less often, molecular composition of the surface. Energy transfer from the projectile to the surface results in a dynamic excited region that is in temporal and spatial nonequilibrium, and from which neutral and ionic components of the surface are released into the gas phase. Specifically, the projectile's kinetic energy is converted into a translationally and later a vibrationally excited interfacial region through collision cascades. This description makes it clear that in SIMS the projectile plays a physical role in the sputtering process, although it also has a chemical role in promoting the ionization of the sputtered atoms and molecules (Garrison et al. 2003; Cooks et al. 2004).

Ionic collisions in the hyperthermal energy range also result in sputtering of surface material but in this case, instead of the material being released by momentum transfer, ions are liberated in this energy regime as a result of a chemical reaction. Consequently, it has become common to refer to these low-energy processes as *chemical sputtering* (Busharov et al. 1976; Tu et al. 1981; Ast et al. 1993) since in this lower-energy regime the ion beam acts as a chemical sputtering reagent (Vincenti and Cooks 1988) and the surface molecules are both ionized and removed from the surface by chemical reactions with the hyperthermal projectile ion. Simple charge transfer occurs most efficiently when the recombination energy of the ion matches or exceeds the work function in the case of a metal surface, or the ionization energy of the adsorbate in the case of collisions at a molecular surface. If this process occurs with deposition of sufficient energy into the surface, ionic fragments derived from the surface or an adsorbate may be released into the gas phase, and their mass spectrum is then a useful characteristic of the nature of the outermost monolayers of the surface. Overall, the greater the reaction exothermicity, the greater is the internal energy deposition while, by contrast, kinetic energy is less effective in activation. In chemical sputtering, high secondary ion yields are recorded at quite low impact energies. In regard to the fundamental characteristics of the process, work from this laboratory (Grill et al. 2001) has shown that the energetics of charge transfer between the incoming projectile ion and the surface dramatically influences the degree of surface fragmentation observed. Dissociation accompanying endothermic or thermoneutral charge transfer (e.g., Br^+ reagent ions colliding with a fluorocarbon SAM) is not extensive and is strongly influenced by the collision energy. Exothermic charge transfer reagents such as $Ar^{+\cdot}$ give chemical sputtering mass spectra of the fluorocarbon surface that show extensive fragmentation and these spectra that are only weakly dependent on the projectile ion collision energy. A specific example is found in the case of $Xe^{+\cdot}$ collisions with desorption of a surface molecule like benzene and many others; neutralization of the projectile ion leaves a charged surface moiety that, if given sufficient energy, can be desorbed into the vacuum as the ion. This occurs with or without further fragmentation, depending on the difference in ionization energies of Xe and the target (Winters 1982; Vincenti and Cooks 1988). At low energies, some odd-electron fluorinated ions undergo collision with hydrocarbon covered surfaces but no secondary ions are emitted from the surface. In these cases, the parent ions apparently are neutralized, but without sufficient energy transfer to cause hydrocarbon ion desorption. Nonfluorinated organic ions yield fragment ions and ion–surface reaction products without causing significant desorption of hydrocarbon ions from the target surfaces. Energy made available in the form of translational energy of the projectile is much less effective than that provided by the reaction exothermicity in causing dissociation.

12.1.1.4. Reactive Scattering. Dynamic studies of elementary gas-phase bimolecular reactions have progressed significantly in the past following advances in molecular-beam and laser techniques as well as in theoretical methodologies. The study of elementary chemical reactions at the level of individual atoms and molecules constitutes the essence of chemical reaction dynamics and is now one of the major areas of inquiry in chemical kinetics (Smith 1980; Herschbach 1987; Lee 1987; Levine and Bernstein 1987; Polanyi 1987). Molecular beam and laser spectroscopic techniques permit experiments that are able to give detailed information about the dynamics of elementary gas-phase chemical reactions. Fundamental questions concerned with product energy partitioning, product angular distribution, the effect of reagent quantum state distribution, and the evolution from reactants to products can be addressed, and in most cases detailed answers can be provided. The history of the field of reaction dynamics goes back to 1928, when detailed descriptions of the microscopic rearrangement of atoms leading to chemical reactions were made possible (London 1929). Since then, the fundamental idea in reaction dynamics has been that reactive collisions can be described and understood by considering the motion of the system over a potential energy surface. The first classical trajectory study of the dynamics of a chemical reaction was carried out as long ago as 1936 by Hirschfelder and coworkers (Hirschfelder et al. 1936) on the hydrogen exchange reaction $H + H_2 \rightarrow H_2 + H$ (London 1929), and interpretation of most reaction dynamic experiments has been based on the concept of adiabatic motion on a single electronically adiabatic potential energy surface. Experimentally, because of technological difficulties, it was not until the 1960s and 1970s that the field received was propelled forward with the work of Polanyi and coworkers on IR chemiluminescence (Polanyi and Tardy 1969; Polanyi 1987), the work of Parker and Pimentel (1969) on chemical lasers, and that of Herschbach (1966, 1987) Lee and coworkers (Schafer et al. 1970; Lee 1987), and others (Gillen et al. 1969) on molecular beams. This early experimental work was devoted mostly to atom–diatom reactions.

The experimental techniques usually employed for gas-phase reaction dynamics studies fall essentially in two categories: crossed molecular-beam (CMB) methods and laser-based spectroscopic methods. The CMB method, introduced in the 1950s (Taylor and Datz 1955) and widely applied in the 1960s to reactions of alkali metals using the hot-wire surface-ionization detector (London 1929; Herschbach 1966, 1987; Parker and Pimentel 1969; Polanyi and Tardy 1969; Farrar and Lee 1974; Polanyi 1987; Herman 2001), reached its maturity in the 1970s with the development of universal instruments (Lee et al. 1969), which, using a mass spectrometer as a detector, rendered the method generally applicable. The CMB method allows for the direct observation of the consequences of single reactive collisions of well-controlled reagents. In these experiments, the internal states and the velocities of the reactants can be selected, usually by supersonic expansion and seeding techniques (Herman 2001). Since the initial relative velocity vector is well defined in magnitude and direction, the measured quantities of scattered products can be related to the mechanics of single collisions. The experimental observables that provide information on the reaction dynamics are the angular and velocity distribution (i.e., the doubly differential cross section) of reaction products.

As part of the reactive scattering phenomena, ion–molecule reactions have long been a significant field of science, with applications to flame, plasma, atmospheric, and interstellar chemistry. In contrast to the situation in the keV energy range, reactive collisions are common in the hyperthermal energy regime, whether they involve gaseous or solid targets. These processes can involve transfer of not only electrons but also protons or other chemical species and are strongly dependent on the chemical nature of the projectile. Reactive scattering from surfaces is a younger field than gas-phase collisions, but one with potentially wide significance. Hyperthermal ion–surface collisions can lead to formation of new covalent bonds, often as a result of transfer or abstraction of an atom or group of atoms to or from the outermost layers of the surface by the projectile ion (Ada et al. 1998; Shen et al. 1999a). A systematic study of different projectiles under varying experimental conditions revealed that often odd-electron ions abstract hydrogen from the surface while even-electron species do not undergo reactive scattering as readily (Ast et al. 1988). Justification for this general observation is the difference in stability of the two groups; odd-electron ions typically will abstract a hydrogen from a surface in order to form the thermodynamically more stable and kinetically more accessible even-electron species, $(M + H)^+$. For example, nitrogen containing ions (nitrogen centered radical cations) were found to be the most likely to abstract hydrogen atoms from a surface on collision while ions that were solely hydrocarbon in nature showed the weakest tendency; the behavior of oxygen and sulfur containing ions was found to be intermediate. Aromatic and heteroaromatic species were also observed to much more readily form $(M + H)^+$ ions than the corresponding saturated ions. Exceptions to these trends do arise and can be explained as the result of particular stability of a given product ion. There are also qualitative thermochemical correlations that can be made with the heats of formation for the hydrogen abstraction reaction; ions that give mainly $(M + H)^+$ as the base peak in the mass spectra recorded after collision with the surface typically have negative heats of reaction while ions that give relatively weak or no signals for the $(M + H)^+$ ion following surface collisions will have positive heats of reaction (Rosenstock and Draxl 1977). The nature of chemical adsorbates on the collision surface controls the degree to which reactive ion–surface collisions occur. Experiments performed using the BQ instrument (Sec.12.2.2) fitted with a succession of metal surfaces showed that the SID spectra observed were independent of the substrate composition (Mabud 1987). Further evidence for the importance of adsorbates on the types of ions produced during ion–surface reactions was given by experiments in which an aluminum target was bombarded with 1.5-keV projectiles over a period of 1 min during which the intensity of the $(M + H)^+$ ion diminishes and the $M^{+\cdot}$ ion becomes the dominant peak in the spectrum. This observation is the result of sputter cleaning of the metal substrate over the course of the experiment (Ast et al. 1990).

Ion–surface reactions are not limited to single hydrogen atom abstraction; heteroaromatics, quinones, and other classes of compounds have been observed to undergo multiple hydrogenation on collision with a surface (Detter et al. 1988; Hand et al. 1989). Intermolecular transalkylation is also a common reaction path in desorption–ionization events. In addition to altering the chemical nature of the projectile, ion–surface reactions can be used to modify the top layers of the surface with a desired chemical reagent provided one selects reagent ions of appropriate mass and velocity (Wijesundara et al. 2000). An example is the silylation of hydroxyl-terminated SAMs (HO-SAMs) (Wade et al. 2000; Evans et al. 2002), which demonstrates the possibility of performing multistep synthesis/modification using low-energy ion–surface collisions. Generally, the advantage of the surface modification approach is the versatility and chemical control acheived, a feature not present in more traditional plasma and other approaches (Grill et al. 2001).

12.1.1.5. Ion Soft Landing. Ion soft landing is an ion–surface collision phenomenon occurring at hyperthermal energies, namely, in the range of laboratory energies of 1–100 eV. It particularly emphasizes collisions of large organic ions with organic, metal, or liquid surfaces without adverse effects on their structure and/or bioactivity. It is important to distinguish two processes that are both described as soft landing: one where the landed ion is trapped as the ion itself and the other, where the landed ion remains structurally intact but the trapped product is neutralized, often by proton transfer to surface adsorbates.

Molecular ion deposition was described in 1977 (Franchetti et al. 1977) and two decades later (Miller et al. 1997) was demonstrated unambiguously in the case of organic ions deposited onto self-assembled monolayers surfaces using the BEEQ mass spectrometer. This method of preparing modified surfaces by gently landing intact polyatomic ions from the gas phase into a monolayer surface at room temperature is referred to as *ion soft landing* (or just *soft landing*) and is the basis for more recent novel experiments in preparative mass spectrometry (Ouyang et al. 2003). During these experiments (Miller et al. 1997; Luo et al. 1998; Shen et al. 1999b) ions are trapped in a fluorocarbon, hydrocarbon and other functionalized matrices for many hours and then released, intact, on sputtering at low or high energy or by

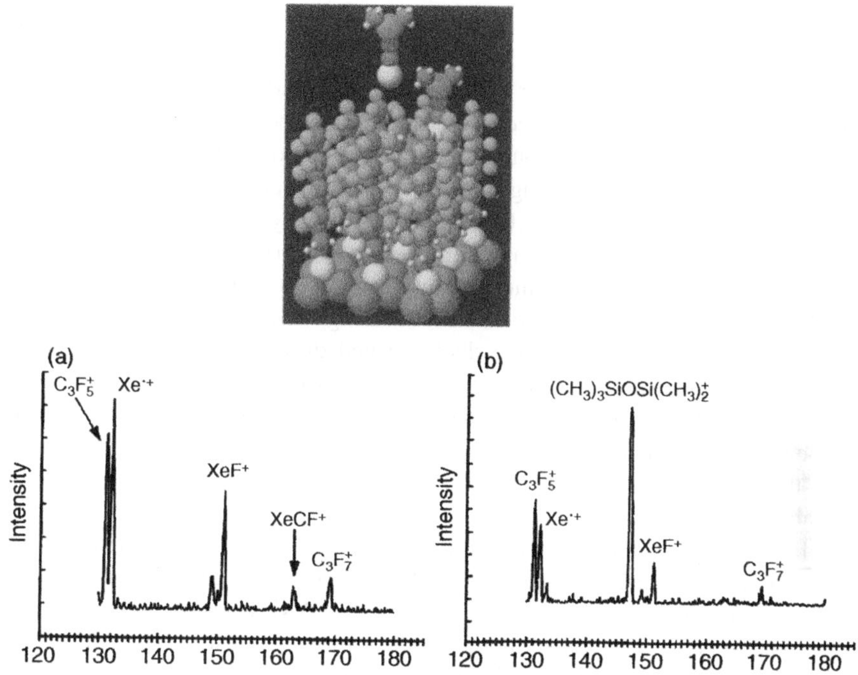

FIGURE 12.3. Three-dimensional molecular modeling representation of the soft-landing process for $(CH_3)_2SiNCS^+$ projectile ions impinging on an F-SAM monolayer surface. Mass spectrum recorded by 60-eV $^{132}Xe^{\cdot+}$ chemical sputtering of (a) an F-SAM surface and (b) the same surface after treatment for 1 h at 5 eV collision energy with $(CH_3)_3SiOSi(CH_3)_2^+$ ions (m/z 147), at a total dose corresponding to ~7% of a monolayer. [Reprinted from Miller et al. (1997) with permission from AAAS. Copyright 1997.]

thermal desorption. Confirmation of molecular composition is usually achieved by isotopic labeling and high-resolution mass measurements.

Figure 12.3 depicts an example of the soft-landing process (Miller et al. 1997) and illustrates the deposition of $(CH_3)_2SiNCS^+$ projectile ions into an F-SAM surface [the monolayer is actually constructed using $CF_3(CF_2)_7(CH_2)_2SH$–Au)]. Two $(CH_3)_2SiNCS^+$ ions are shown penetrating the surface to different depths, and a third is approaching the F-SAM surface. The sterically bulky and covalently bound silyl ether ion, $(CH_3)_3SiOSi(CH_3)_2^+$ (m/z 147), was generated by electron impact on vapor-phase hexamethyldisiloxane, mass-selected, decelerated to 10 eV, and allowed to collide with a monolayer surface of the thiol, self-assembled onto a polycrystalline gold substrate. The F-SAM surface was examined before and after 1 h of deposition, in both cases with the use of 60-eV $^{132}Xe^{\cdot+}$ ions to allow chemical sputtering surface analysis. When compared to the background spectrum of an unmodified F-SAM surface (Figure 12.3a), the chemically sputtered mass spectrum of the modified surface (Figure 12.3b) showed a prominent new peak due to the ion at m/z 147. After storage of the treated surface in laboratory air for 1 day, the signal at m/z 147 decreased by only 30% and was still observable after 4 days. This is in spite of the fact that it is present in the SAM matrix as the free ion.

Through this and several other experiments, it was demonstrated that F-SAM surfaces can be selectively modified via soft landing of polyatomic ions at collision energies of about 10 eV. In addition to the intact deposition of the projectile ion into the surface, two other soft landing phenomena were recognized: dissociative soft landing and reactive soft landing. Projectiles that readily dissociate on collision at the surface yield fragments that may be trapped in the surface. Similarly, projectiles that react readily with atoms or functional groups present at the surface can undergo ion/molecule reactions in the course of soft landing. In considering these ancillary phenomena, Luo et al. (1998) have distinguished dissociation that occurs in the analysis stage of the experiment from processes that occur during soft landing per se. The degree of surface modification was determined to be controlled by the total ion dose, the projectile ion and the energy it carries. The extent of modification achieved was on the order of 0.1% up to a few percent of the available surface sites, and remarkably, the modified surfaces prepared by soft landing retained the deposited material for relatively long periods, inside the vacuum or in the ambient environment. As described in the previous paragraph, the most striking result was the ability to directly deposit intact polyatomic ions into F-SAM surfaces at low collision energies. The results, especially those for the variously substituted pyridine ions, showed that the combination of steric hindrance in the polyatomic ions and the inert and ordered matrix formed by the fluorinated alkylthiolate long chains of the F-SAM surface, play an important role in successful soft landing. Electrostatic interaction between the soft-landed ions and induced electric dipoles in the surface substrate may contribute significantly to their strong binding to the F-SAM surface in addition to the role of steric hindrance. The systems examined led to the conclusion that intact soft landing is more successful with closed-shell rather than open-shell ions, presumably because of the ease of neutralization of the latter (Luo et al. 1998; Shen et al. 1999b). The loss of the soft-landed ions may also involve chemical reactions with adventitious chemical reagents, such as water. These types of chemical reactions appear to be responsible for the observation of water adducts. Several lines of evidence showed that the deposited polyatomic ions were present at the surface as charged species. This evidence includes the following: (1) the projectile ions were often liberated intact by low-energy sputtering or by thermal desorption, (2) the secondary-ion yields in the sputtering process showed only a small dependence on the recombination energy of the sputtering ion, and (3) there was no evidence for products other than those readily explained as the result of ionic dissociation or reaction. The remarkable fact that intact ions can be held at surfaces for long periods was explained by the fact that these ions are trapped in potential wells near metal surfaces and, in a system like F-SAM, where the ions may be $\sim$7 Å from the surface, the electrostatic binding is 2 eV or more. Additionally, ions were sterically trapped in the F-SAM matrix. The failure of the deposited ions to be removed rapidly by reaction was ascribed to the fact that strongly hydrophobic F-SAM matrix helped exclude reagents such as water at least in the two-dimensional domain parallel to the surface and to the steric bulk of the projectile ion itself that helped screen the reactive charged site from attack. The authors expected that the "charge down" configuration of the deposited projectile ion would be especially stable, both because of the stronger electronic interaction because of the closer distance to the gold substrate and because of more effective steric protection from reaction with reagents approaching from the top of the monolayer. It was emphasized that the matrix may be highly disturbed and that the bulky steric substituents play several distinct roles such as to increase the cross section of collisions during landing that help remove translational energy from the ion and allow its trapping, to help lock the ion into the matrix through steric interactions and to protect the reactive site from reagents. The ability to directly deposit intact polyatomic ions into F-SAM surfaces at low collision energies is an extremely striking result, although the deposition occurs into only a small fraction of the available surface sites (Luo et al. 1998).

In a separate and subsequent study at Pacific Northwest National Laboratories (PNNL) (Alvarez et al., 2005) the soft landing of hyperthermal peptide ions on collision with functionalized SAMs was examined. The surfaces used for soft landing and analysis were similar or identical to those used previously for studying this phenomenon using smaller polyatomic ions as projectiles (Miller et al. 1997; Luo et al. 1998; Shen et al. 1999b). Peptides of interest were ionized by electrospray ionization, mass-selected using a quadrupole mass analyzer, and deposited onto F-SAM surfaces by soft landing in a Fourier transform ion cyclotron resonance (FTICR) instrument specially configured for studying ion–surface interactions. Both in situ and ex situ analyses of modified surfaces using FTICR SIMS and time-of-flight SIMS confirmed that a significant number of soft-landed peptide ions remained charged on the surface, even when exposed to air for several hours after deposition. SIMS analysis of a surface on which there were deposited doubly charged ions of the peptide substance P, showed a signal characteristic of the singly charged ion on the surface. This signal was orders of magnitude stronger than that seen when the same amount of the corresponding neutral peptide was applied to the surface, consistent with the expectation that precharged ions at surfaces provide much higher ion yields in SIMS (Busch and Cooks 1982). Peptide ion soft landing on F-SAM surfaces gave much greater sputtered ion signals than on hydrocarbon self-assembled monolayer (H-SAM) surfaces, which can be attributed again to the fact that the ion-induced dipole interaction potential is stronger for the F-SAM surface because of its greater polarizability, which results in weaker physisorption of ions from the surface and therefore better retention of the charged species.

During SIMS analysis of soft-landed peptides (Alvarez et al., 2005) fragment ions were observed together with ions due to the intact molecule. Fragmentation observed in the SIMS analysis was deemed to be the result of either "crash landing" [i.e., projectile ion dissociation on impact with some of the fragments being retained by the organic monolayer, a known process for small organic ions (Shen et al. 1999b)] or the result of internal energy deposition onto the intact peptide (whether ion or neutral) by keV ion desorption during the SIMS analysis. When the kinetic energy of the peptide ions was varied over a range of collision energies (0–150 eV), these alternatives could be distinguished. If dissociation occurred during ion–surface impact an increase in fragment ion abundance with soft-landing energy would be expected. However, soft landing of doubly charged bradykinin at 30 and 150 eV collision energy resulted in a similar amount of fragmentation and, in fact, for two of the peptides investigated, bradykinin and substance P, the amount of fragmentation was independent of the energy of the soft-landing event, providing no evidence of "crash landing" in these experiments (Alvarez et al., 2006).

In the experiments done by Laskin and coworkers (Alvarez et al., 2006) no significant differences in the fragmentation pattern of deposited peptides at various soft-landing energies were observed on static SIMS analysis. The amount of singly protonated peptide sputtered from the surface on 2 kV Cs^+ SIMS analysis varied significantly with the soft-landing energy, indicating a significant decrease in the soft-landing efficiency with increase in the kinetic energy of the projectile ion. Similarly, soft-landing studies of small organic molecules have demonstrated that at low energies the soft-landing process is favored since other ion–surface collision channels are suppressed. In addition, as the energy increases the probability of soft landing decreases (Luo et al. 1998) because of competition with other processes. Interestingly, the decrease in the soft-landing efficiency with increasing collision energy can be rationalized (in spite of the questionable transfer of a gas-phase model to a surface environment) by the corresponding decrease in the Langevin capture cross section, which determines the probability of ion capture by the polarizable medium. The general phenomenon is also supported by an investigation into the interaction of positively charged antimony clusters Sb_n^+ with highly oriented pyrolythic graphite (HOPG) as a function of cluster size ($2 < n < 13$) and cluster kinetic energy (<600 eV) by tandem time-of-flight mass spectrometry and scanning tunneling microscopy (Kaiser et al. 1999). On the basis of the reported observations, the processes taking place when the cluster hits the surface can be classified into three main categories according to kinetic energy. As illustrated in Figure 12.4, at very low kinetic energies ($\leq$13 eV), soft landing should be feasible, since no fragmentation of the cluster takes place and the result of the impact may lead to deposited clusters without any or with only a weak deformation of their original structures. With increasing kinetic energy—up to ~150–180 eV—two competing processes are found: (1) scattering and fragmentation (SID) of the cluster into the most stable products and (2) neutralization. The efficiency of the latter process increases with increasing particle energy from about 85% to almost 100% at 150–180 eV kinetic energy. At 150 eV and above, the interaction between clusters and surfaces leads to ultrafast heating and shattering of the projectile cluster into very small fragments. In the same energy regime (~110 eV), implantation of the clusters into the surface is also observed. During this process the cluster is presumably strongly deformed and the surface atoms are displaced from their original positions, leaving the impact area in a highly amorphous state. The authors suggest that controlled nanostructuring of surfaces using soft-landing and implantation is promising (Kaiser et al. 1999).

In one study the relative amounts of sputtered peptides were investigated after soft landing on different surfaces, including H-SAM, F-SAM, carboxyl-terminated SAM (HOOC-SAM), and gold surfaces (Alvarez et al., 2006). The results showed

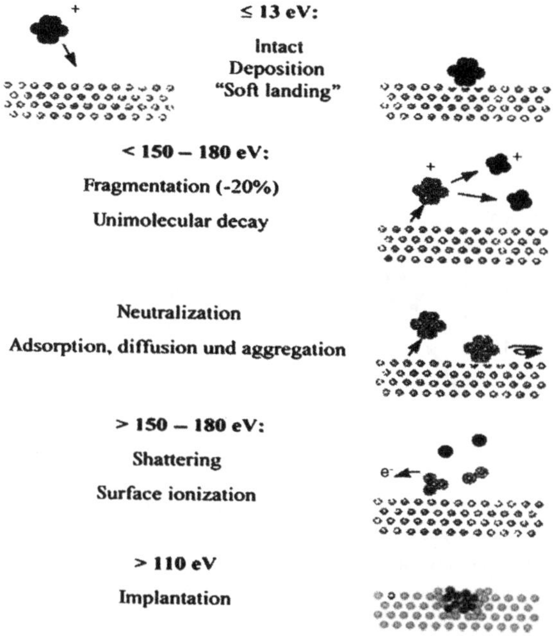

FIGURE 12.4. Overview of the different interaction mechanisms found for the collision of antimony clusters with HOPG. [Reprinted from Kaiser et al. (1999) with permission from Elsevier. Copyright 1999.]

that SIMS sputtering yields of soft-landed peptides are strongly dependent on the surface, a result that parallels other ion–surface collision experiments since neutralization is such a dominant effect (Ast et al. 1990). For example, angiotensin III showed a relatively high sputtering yield from F-SAM surfaces; however, the sputtering yield was reduced to about 25% when the experiment was performed using an H-SAM surface. Further reduction to about 4% was observed on bare gold surfaces and only about 1% relative yield was obtained on analysis from a HOOC-terminated SAM surface. The results are interpreted in terms of neutralization, not differences deposition efficiency.

One question that arises (Alvarez et al., 2005) is the number of ions that can be deposited on a surface before it is saturated and all the subsequent ions are repelled by the surface charge. Given the evidence that soft-landed small organic and peptide ions retain their charge to at least some extent, this phenomenon should be observable. Surface saturation has been observed previously on deposition of small organic ions onto F-SAM surfaces (Luo et al. 1998) where the saturation plateau was reached when about 1×10^{13} ions (7% of a monolayer) had been deposited. More recently, in the case of peptide soft landing (Alvarez et al., 2005), it was shown that saturation due to Coulombic repulsion also occurs before deposition of one monolayer. Accumulation experiments were performed on an F-SAM surface for various periods, and the signal observed in subsequent SIMS analysis was compared to the fraction of a monolayer deposited through ion soft landing (Figure 12.5). For

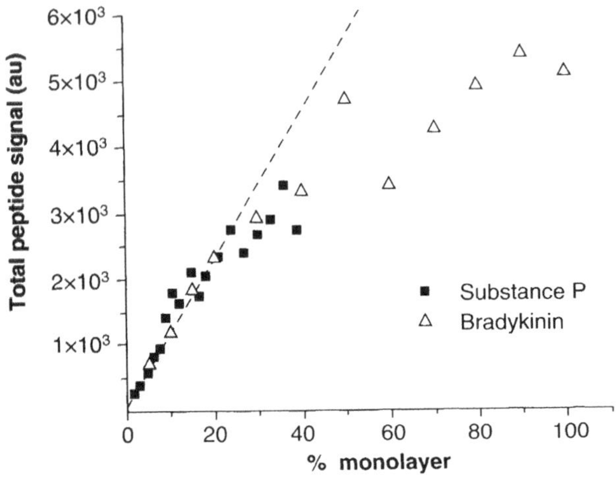

FIGURE 12.5. Saturation plots showing the total peptide signal sputtered from the surface as a function percent of ion exposure for doubly protonated bradykinin and substance P. The exposure is expressed in percent of an equivalent of a monolayer. [Reprinted with permission from Alvarez et al. (2005). Copyright 2005 American Chemical Society.]

doubly protonated bradykinin and substance P saturation occurred on deposition of about 30% of a monolayer.

It is worth mentioning that an unusual application of ion soft landing is its use in the study of homochiral serine octamers, where, starting from nonracemic serine solutions and taking advantage of the special properties of serine, chiral enrichment was achieved (Nanita et al. 2004). In a study done by Nanita et al. serine octamers were generated by means of electrospray and sonic spray ionization of aqueous solutions of d_3-L-serine (108 Da) and D-serine (105 Da) having different molar ratios of enantiomers. A sequence of processes allowed the formation of chirally enriched octameric cluster ions, their mass selection in the gas phase and their dissociation back to the monomer: $Ser_1 \rightarrow Ser_8 \rightarrow Ser_1$. The regenerated serine monomers were shown by isotopic labeling to have an increased enantiomeric excess, that is, to be chirally purer than the starting material. This experiment was performed in two ways:

1. Chiral enrichment in serine was observed in MS/MS/MS experiments in a quadrupole ion trap in which the entire distribution of serine octamers formed from nonracemic solutions was isolated, collisionally activated, and fragmented. Monomeric serine was regenerated with increased enantiomeric excess on dissociation of the octamers when compared with the enantiomeric composition of the original solution.
2. Chiral enrichment was observed in the products of soft landing of mass-selected protonated serine octamers. These ions were generated by means of electrospray ionization, and a single quadrupole mass filter, modified for ion soft landing, was used to isolate the entire isotopic distribution of

protonated serine octamers (m/z 841–865) and to soft-land these ions on a gold surface.

Chiral enrichment of the soft-landed serine was verified by redissolving the recovered material and comparing the intensities of protonated molecules of d_3-L-serine and D-serine after atmospheric-pressure CI (chemical ionization) analysis. Both experiments showed comparable results (20% enantiomeric excess originally, 50% enantiomeric excess after the processing), suggesting that formation of serine octamers depends on the enantiomeric composition of the serine solution and that the magnitude of the chiral preference is intrinsic to octamers formed from solutions of given chiral composition.

12.2. PAST AND CURRENT INSTRUMENTATION USED FOR ION SOFT LANDING

12.2.1. Manhattan Project and Isotope Separation by Mass Spectrometry

The separation and collection of ions by means of mass spectrometry was first demonstrated using the *calutron*, (Smith et al. 1947), an instrument developed in the laboratory of E. O. Lawrence at the University of California at Berkeley. At the time of its construction, the calutron represented the only means of preparing enriched uranium for the construction of a fission bomb that was accomplished under the Manhattan Project. The effort was successful enough that every atom of the 42 kg of ^{235}U used in the first atomic bomb had passed through at least one stage of calutron separation (Rhodes 1986). The classic calutron is an 180° sector homogeneous magnetic field instrument with the source and the collector lying within the magnetic field. It employs a side extraction pulse, produces a line image at the collector, and has multibeam capabilities. Figure 12.6 shows a schematic drawing of a second-stage instrument, also known as the β unit separator. The schematic shows trajectories for two positively charged isotopic ions and one of the several pairs of collectors, also known as *receivers*. In practice, each ion source assembly employed either two or four pairs of receivers. Each vacuum system assembly, also known as a "tank," held the ion sources and their associated collectors, and the main portion of each ion source contained the number of slit assemblies required to produce separate ion beams for each receiver. Some concept of the size of these devices can be obtained from the diffusion pump dimensions. Each tank was evacuated using a 20-in. diffusion pump backed by an 8-in. diffusion pump backed in turn by a rough pump. The α units were first-stage separators that enriched natural uranium to >10 at% ^{235}U. Materials obtained from this approximately 20-fold enrichment from the natural abundance level were used as feed for the second-stage separators. The β units yielded ^{235}U of weapons-grade enrichment, >88 at%. Operating conditions for the calutrons for uranium separation are summarized in Table 12.1. Under typical conditions, each α tank operated at a uranium beam intensity at the collectors of approximately 20 mA and each β tank at a beam intensity of approximately 215 mA.

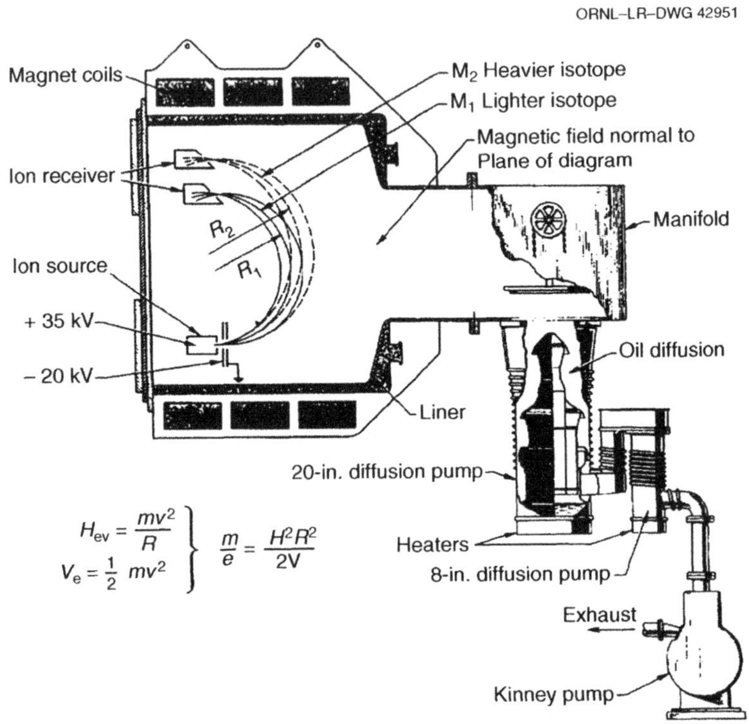

FIGURE 12.6. Schematic of the second-stage, β-unit, separator. ORNL drawing 42951. [Reprinted from Yergey and Yergey (1997) with permission from Elsevier. Copyright 1997.]

Bulk separation of isotopes for bomb production ceased in 1945, and since then calutrons have been used to separate stable isotopes, including therapeutically important materials, but on a much more limited scale. The use of the calutrons for medical purposes is a fitting conclusion to the history of these devices.

12.2.2. Separation of Small Polyatomic Molecules Using Sector Instruments

Instrumentation for the study of hyperthermal ion–surface collisions and especially for ion soft landing has been developed in a targeted fashion, following advances in tandem mass spectrometry for analytical chemistry. The simplest tandem mass spectrometry experiment involves mass selection of the ion of interest in the first MS stage, excitation of the ion by collision leading to its dissociation, and then finally mass analysis of the resulting fragment ions in the second MS stage. Because a number of different types of mass analyzers are widely used, many combinations of two analyzers ("hybrid" instruments) are considered when designing a tandem mass spectrometer. Numerous mass analyzer combinations have been successful in mass spectrometers utilizing collision-induced dissociation (CID) for structural analysis of molecules, and some of these combinations have been also used for ion–surface collisions in the hyperthermal energy range.

TABLE 12.1. Calutron Operating Conditions and Estimated Calutron Efficiency

	r (cm)	B (G)	$2\Delta r$ (mm)	n
Calutron Operating Conditions				
α	122	3200	15.6	864/96 (5α − 1, 4α − 2)
β	61	6400	7.8	216/34
Accelerating voltage = 35 kV				
Magnet power/tank ~ 4500 kW, 7500 A at 600 V				
Pumping capacity ~7.5 m^3/s				
n = number of units = tanks/tracks				
Estimated Calutron Efficiency				
Change ~ 100-g UCl_4				
Sample rate 1–4 g/h				
Peak production rate ~200 g/day of ~88 at% ^{235}U (42 kg, 88 at% ^{235}U in 6 months)				
~20 mA beam current in α unit				
~215 mA beam current in β unit				
Only 10% of charge reaches collectors				

Source: Reprinted from Geiger et al. (1999) with permission from Elsevier. Copyright 1999.

The first instrument used to study low-energy ion–surface reactions and ion soft landing was a Colutron ion-beam kit model BK-1-D (Figure 12.7) (Franchetti et al. 1977). The ion beam was produced with a plasmatron-type ion source which was operated with filament currents of 15–17 A and discharge currents of 200 mA. The source was able to generate both positive and negative ions. The ions were extracted and focused with a three-element Einzel lens. A set of vertical deflection plates

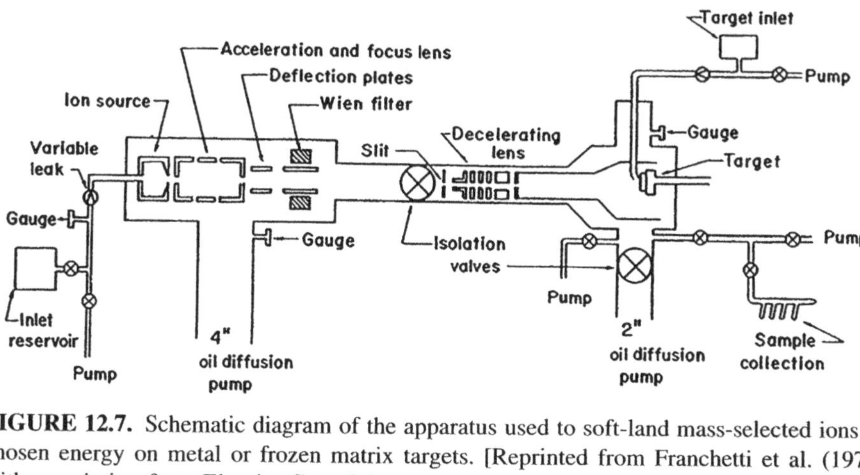

FIGURE 12.7. Schematic diagram of the apparatus used to soft-land mass-selected ions of chosen energy on metal or frozen matrix targets. [Reprinted from Franchetti et al. (1977) with permission from Elsevier Copyright 1977.]

aligned the beam prior to mass selection by a Wien velocity filter. The mass-selected ions were decelerated with a six-element decelerator, as shown in the Figure 12.7. The soft-landing experiments done with this instrument involved sulfur-containing ions such as CS_2^+, CS_2^{2+}, CS^+, S^+, and S_2^+ deposited on and reacted with lead-based surfaces (Franchetti et al. 1977).

The subsequent instruments built for the study of hyperthermal ion–surface collisions included hybrid devices of BQ (magnet, quadruple filter) design. [It should be noted that in these and other instruments, mass spectrometric design features are denoted in the appropriate abbreviation (e.g., QqQ)—capitalized components representing mass analyzers.] These instruments allowed the study of polyatomic ion–surface collisions at variable collision energy and fixed incident and scattering angles. Typical of later instruments, which allowed more detailed examination of ion–surface collision phenomena in the hyperthermal energy regime, is the BEEQ mass spectrometer (Winger et al. 1992) (Figure 12.8), where B = magnetic

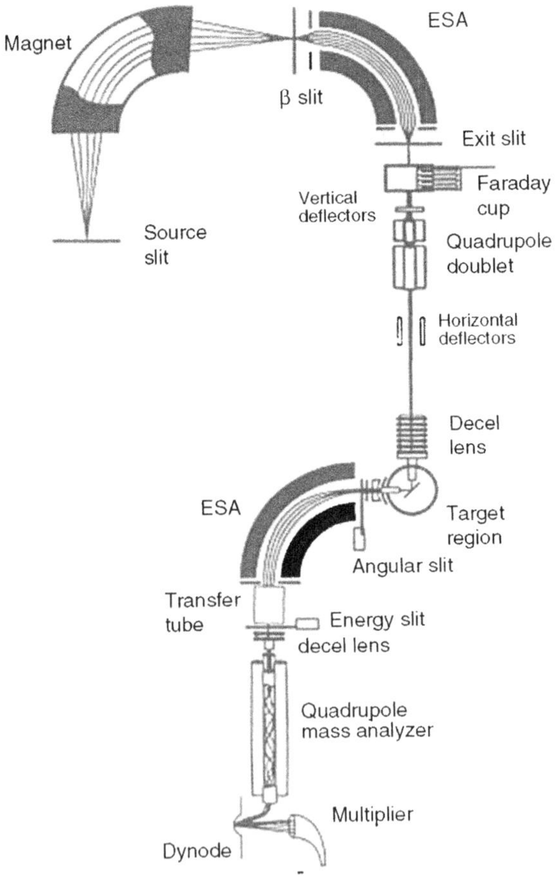

FIGURE 12.8. Hybrid BEEQ tandem mass spectrometer used for the study of ion–surface collision phenomena. [Reprinted with permission from Winger et al. (1992). Copyright 1992 American Institute of Physics.]

sector, E = electric sector, and Q = quadrupole. In many of the initial ion soft-landing experiments (Miller et al. 1997; Luo et al. 1998; Shen et al. 1999b) surface modification experiments were carried out using this custom-built, hybrid mass spectrometer. This instrument allowed momentum, velocity, and angular selection of the impinging ions with velocity, mass/charge ratio, and angular analysis of scattered and desorbed ions. A DC quadrupole doublet focused the m/z and energy-selected ion beam onto the surface, and an appropriate deceleration lens system allowed selection of particular collision energies in the eV or keV range. An electrically floated, EQ postcollision analyzer and detection system was mounted on a rotating rail capable of acquisition of angle- and energy-resolved SID spectra (Winger et al. 1992). In an example of a typical experimental setup (Luo et al. 1998), projectile ions were formed by 70 eV electron impact, accelerated to 2 keV, mass-selected using the first two sectors (BE) of the BEEQ, and directed onto a surface in a high-vacuum scattering chamber. The entire system was differentially pumped with the scattering chamber maintained at a pressure of the order of 2–4 $\times$ 10^{-9} Torr. Ion-beam transport from the ion source to the target surface involved deflection through 180°, preventing fast neutrals formed at the ion source and the magnet from entering the scattering chamber. The mass-selected ion beam was decelerated using a deceleration lens located in the high-vacuum scattering chamber and then focused onto the target surface. Secondary ions were accelerated from the surface and directed into the EQ section of the instrument. The second electric sector was set to pass scattered ions with kinetic energies corresponding to the maximum abundance in their distribution. The transmitted ions were finally mass-analyzed using the quadrupole mass filter. The EQ portion of the instrument was mounted on a rotating rail, allowing study of the angular distribution of the scattered ions. The incident angle could also be varied independently by rotating the surface with respect to the primary ion beam while maintaining the scattering angle of 90°. The incident angle with respect to the surface normal was varied to maximize the efficiency of surface modification (0–55°) and set to 55° for the sputtering experiments. Overall, in these early experiments, one can notice not only the elegance of the apparatus required to nondestructively direct mass-selected ions onto a surface but also the intrinsic complexity associated with designing and completing a meaningful experiment.

12.2.3. Clusters, Polymers, and Experiments Using Non-Mass-Selected Ions in Quadrupole Mass Spectrometers

In 1996 Bromann et al. (1996) used a variable-temperature scanning tunneling microscope to study the effect of cluster kinetic energy and rare-gas buffer layers on the deposition process of size-selected silver nanoclusters on a platinum (111) surface. The Ag clusters were produced by sputtering of a silver target in a differentially pumped secondary-ion source, energy-filtered and subsequently mass-selected by a quadrupole. The resulting clusters with impact energies of $<$1 eV per atom were landed nondestructively, with preservation of the cluster's spatial resolution, on the bare substrate, whereas at higher kinetic energies fragmentation and substrate damage were observed. Clusters with higher impact energy were soft-landed on the platinum substrate through an argon buffer layer, which efficiently dissipated the kinetic energy. Nondestructive cluster deposition represents a promising method to produce monodispersed nanostructures at surfaces. In a *Science* report (Bromann et al. 1996) the authors detailed the investigation of the deposition of size-selected Ag_n clusters ($n = 1,7,19$) of varying kinetic energy (1–14 eV per cluster atom) onto a Pt(111) substrate in ultrahigh vacuum (UHV). Deposition took place either onto the bare surface at 80 or 90 K or into a preadsorbed Ar buffer layer at 26 K, which was subsequently evaporated at 90 K. The surface and cluster morphologies were characterized in situ in the same UHV chamber by variable-temperature scanning tunneling microscopy (STM), as schematically shown in Figure 12.9. The aim of the study was to obtain controlled soft landing through energy dissipation in a rare-gas buffer layer, a possibility suggested by both the matrix deposition experiments, which demonstrated that size-selected Ag clusters codeposited with rare gases do not fragment, and by the molecular dynamics simulations focused on the deposition dynamics of Cu nanoclusters on bare and rare-gas-covered Cu(111). The Bromann experiments illustrated the possibility of soft landing of nanoclusters through energy dissipation into a rare-gas buffer layer. Clusters, which under otherwise identical conditions decayed during the deposition process and created substrate defects, were landed nondestructively by the use of such layers. The opposite

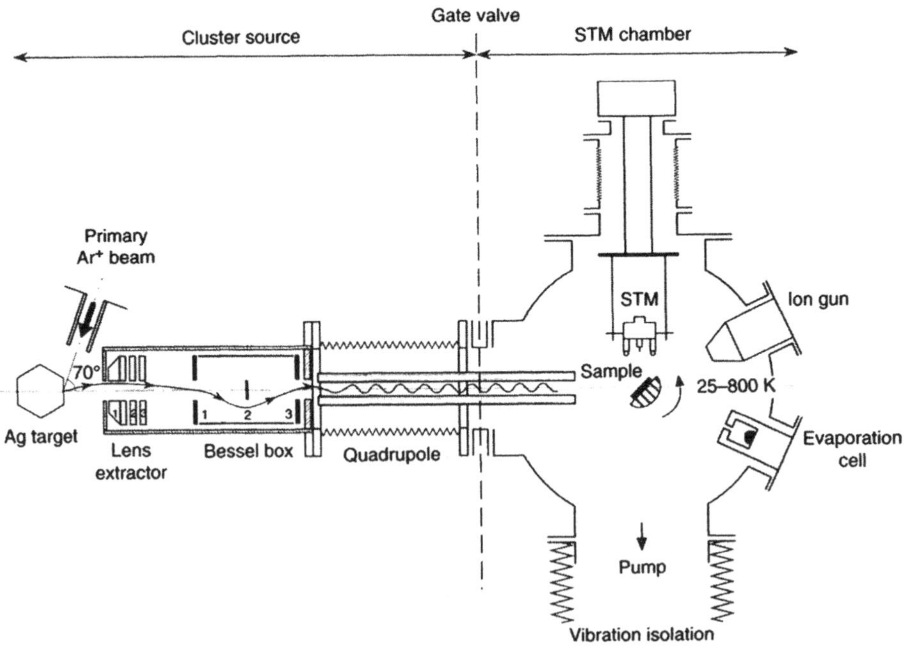

FIGURE 12.9. Apparatus used for the cluster deposition experiment consisting of two UHV chambers separated by a gate valve. [Reprinted with permission from Bromann et al. (1996). Copyright 1996 AAAS.]

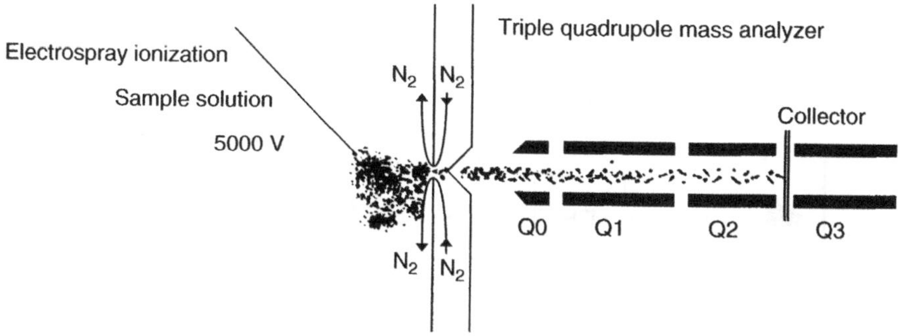

FIGURE 12.10. A modified electrospray ionization quadrupole mass spectrometer. Virus is ionized and electrostatically directed through four sets of quadrupoles, Q0, Q1, Q2, and Q3, which allow for ion focusing. [Reprinted from Siuzdak et al. (1996) with permission from Elsevier. Copyright 1996.]

case of a hard landing is also of potential interest for the nanostructuring of surfaces. In these experiments, otherwise thermally unstable structures might be stabilized by the use of defects created in the violent deposition process.

The gap between biology and mass spectrometry was further bridged by the soft-landing experiments of Siuzdak et al. (1996), who showed that the rice yellow mottle and tobacco mosaic viruses retained their structure and biological activity even after ionization and vacuum exposure. In these experiments a commercial Perkin Elmer SCIEX API-3 triple-quadrupole mass spectrometer that was modified by placing a brass plate as an ion collector in the flight path of the ions between the second (Q2) and third quadrupoles (Q3) (Figure 12.10). Additional experiments were also performed with the collector placed directly behind the orifice within the vacuum of the mass spectrometer. In both cases the collector was coated with a thin layer of methanol : glycerol (50 : 50) and dried for 3 min, leaving a thin, gelatinous surface. It should be noted that, due to the mass limitations of their instrument, no mass separation was employed. To compensate for the inability to scan within the m/z range of the viral ions (>1 000 000 Da), the authors used the radiofrequency (RF)-only mode for quadrupoles Q0, Q1, and Q2, which allowed only ions of high m/z to pass through the quadrupoles.

In 1999, Busch and coworkers employed a modification to a hybrid geometry (EBqQ) mass spectrometer for soft landing of ions onto stainless-steel surfaces (Figure 12.11).(Geiger et al. 1999). This hybrid geometry tandem mass spectrometry instrument was developed specifically for tandem mass spectrometry experiments to combine the performance of a double-focusing, sector-based instrument with the capabilities of a multiple-quadrupole instrument for tandem mass spectrometry. In the case of the instrument used by Busch, a VG 70-SEQ model, the electric and magnetic sectors were followed by an RF-only quadrupole collision cell and a final mass-analyzing quadrupole. As shown in Figure 12.11, this hybrid instrument was fitted with two detectors; the final detector was used for tandem mass spectrometry experiments and located at the end of the beam line after the mass-analyzing

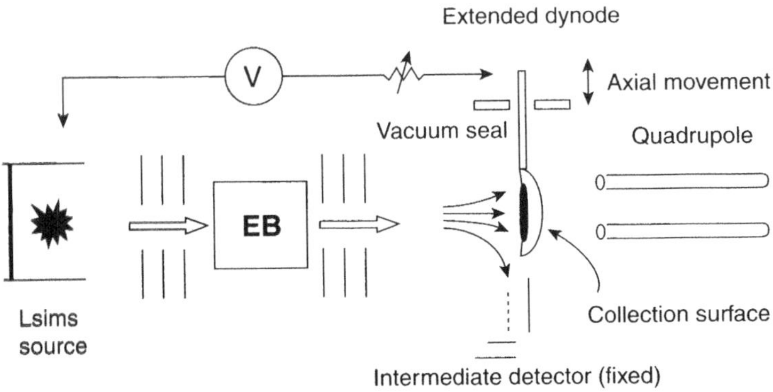

FIGURE 12.11. A modified VG70-SEQ instrument for the soft landing of metal containing and organic ions. [Reprinted from Geiger et al. (1999) with permission from Elsevier. Copyright 1999.]

quadrupole. The intermediate detector was positioned after the magnetic sector, but before ion deceleration into the RF-only quadrupole collision cell and was employed when the instrument was used as a double-focusing mass spectrometer. When imposing a voltage on a deflection dynode, the intermediate detector diverts the ion beam from its straight-line path and directed it toward a scintillator, resulting in the detection of emitted light. The axis of the deflection dynode and scintillator detector is orthogonal to the path of the ion beam. The deflection dynode was mounted to a support attached to a standard flange that was also used for placement of a collection surface to intercept the mass-analyzed beam that exits from the sector portion of the mass spectrometer (Figure 12.11). The pressure in the region of the intermediate detector was approximately 5×10^{-7} Torr. From an ion flux perspective the authors rationalized that a larger number of ions can be obtained with a continuous-flow sample introduction if a positive liquid secondary-ion (LSIMS) source is combined with repetitive introduction of aliquots of the sample solutions. This provided a convenient means to maintain a high flux of sample ions from the source. Using this approach, Busch and coworkers were able to soft-land metal ion clusters. Organic cations were also soft-landed at the collector surface and ESI tandem mass spectrometry analysis of the recovered material was used to confirm preservation of the original organic structure.

As a follow-up to the initial 1996 study, Siuzdak and coworkers further tested whether electrosprayed ions could be collected following mass separation (Siuzdak et al. 1999). A quadrupole mass spectrometer (Perkin-Elmer SCIEX API100), as schematically shown in Figure 12.12, was used to separate gas-phase ions generated by ionizing polypropylene glycol solutions using ESI. Once the ions had been generated and separated, they were collected on a vacuum grease-coated brass plate designed as a collector with an orifice on the surface to allow some ions to pass through and be detected by an electron multiplier detector. The design was innovative since it allowed simultaneous collection and detection of the ions. Preparative mass spectrometric experiments were also performed on a polydispersed polymer,

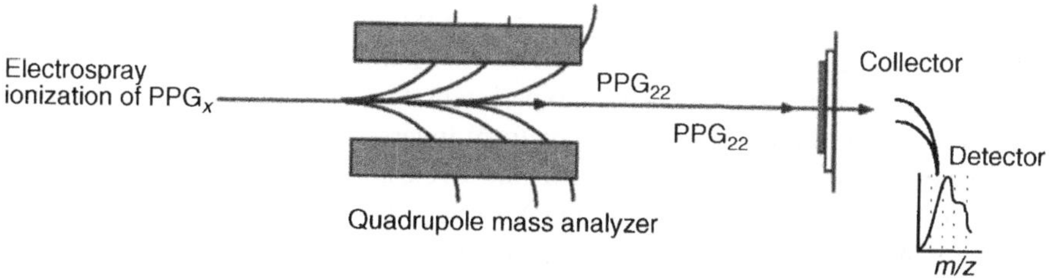

FIGURE 12.12. A Perkin-Elmer SCIEX API100 electrospray ionization single-quadrupole mass analyzer used for preparative experiments. The 22mer of PPG was separated from the polydisperse polymer and collected on a brass plate coated with vacuum grease. The brass plate contained an orifice that allowed some ions to pass to the electron multiplier detector, thus allowing for simultaneous detection and collection. [Reprinted with permission from Siuzdak et al. (1999). Copyright 1999, John Wiley & Sons.]

polypropylene glycol (PPG), to create a monodispersed subpopulation. The collected polymer was recovered from the target by solvent extraction and analyzed by matrix-assisted laser desorption/ionization (MALDI). The results indicated that only monodispersed PPG was present on the collection plate, which successfully demonstrated separation and purification of large molecules using a mass spectrometer.

12.2.4. DNA, Peptides Studied Using FTICR Instruments

A short communication by Feng and colleagues (Feng et al. 1999) reported the use of a modified Fourier transform ion cyclotron resonance mass spectrometer coupled with electrospray ionization (ESI-FTICR) to achieve ion soft landing for the high-resolution analysis, separation, and selective collection of oligonucleotides. The apparatus used in this experiment resembled that depicted in Figure 12.13. The soft-landing apparatus, which included a 40-in.-long probe with a surface attached at

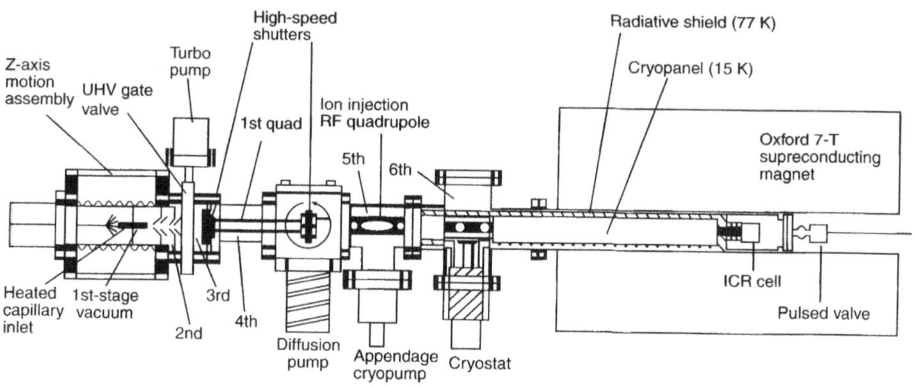

FIGURE 12.13. Schematic representation of the Fourier transform ion cyclotron resonance mass spectrometer used for ion soft landing of DNA. [Reprinted from Winger et al. (1992) with permission from Elsevier. Copyright 1993.]

one end, was coupled to a 7-T ESI-FTICR. A mixture of two 50mer single-strand DNA segments, differing in mass by 40 Da, was ionized and injected into the ion cyclotron resonance cell. An FTICR experimental sequence, which included ion injection, cooling, SWIFT (selected waveform inverse fourier transform) isolation, and soft-landing steps, was repeated for 300 cycles, enabling all charge states of the selected lower-molecular-weight DNA to be soft-landed at low energy on the membrane. An electrically biased metal grid in front of the membrane prevented unselected species from reaching the surface during ion injection and trapping. The membrane was then retrieved and subjected to PCR (polymerase chain reaction) to amplify the soft-landed DNA. The PCR amplification products were examined by agarose gel electrophoresis, and the experimental results showed that approximately 1 attomol of DNA was successfully soft-landed. With this study, the authors demonstrated for the first time that oligonucleotides can be retrieved from the mass spectrometer for further enzymatic manipulation after mass analysis and selection.

Peptides have been studied extensively as models of the soft-landing process. This work has used various instruments, including the modified FTICR instrument described in Figure 12.14. Details of the studies are given in Sec. 12.1.5.

12.2.5. Multiplex Sector Instruments

Turecek and coworkers introduced a sector-based mass spectrometer (Figure 12.15) designed for simultaneous separation and collection of low-molecular-weight

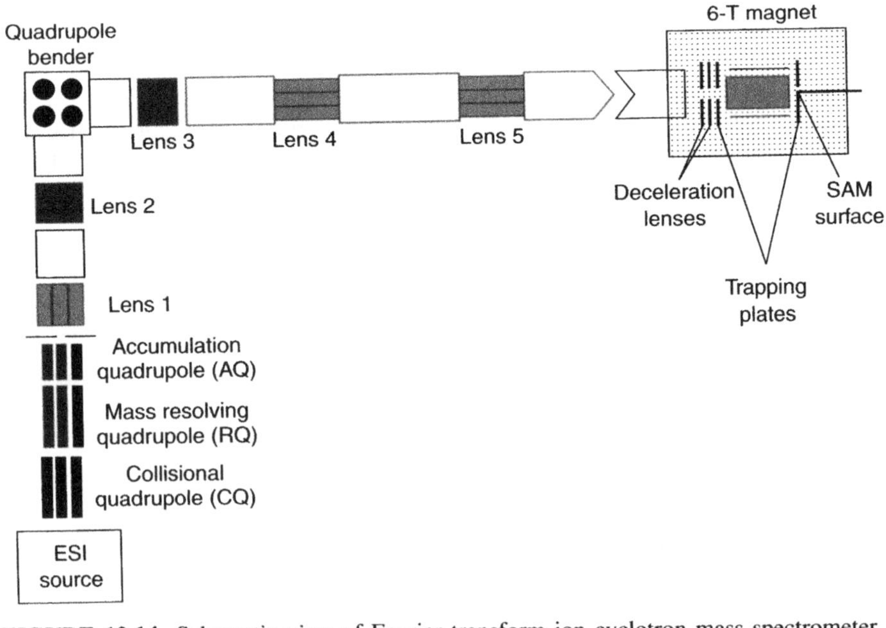

FIGURE 12.14. Schematic view of Fourier transform ion cyclotron mass spectrometer at the Pacific Northwest National Laboratory. [Reprinted with permission from Alvarez et al. (2005). Copyright 2005 American Chemical Society.]

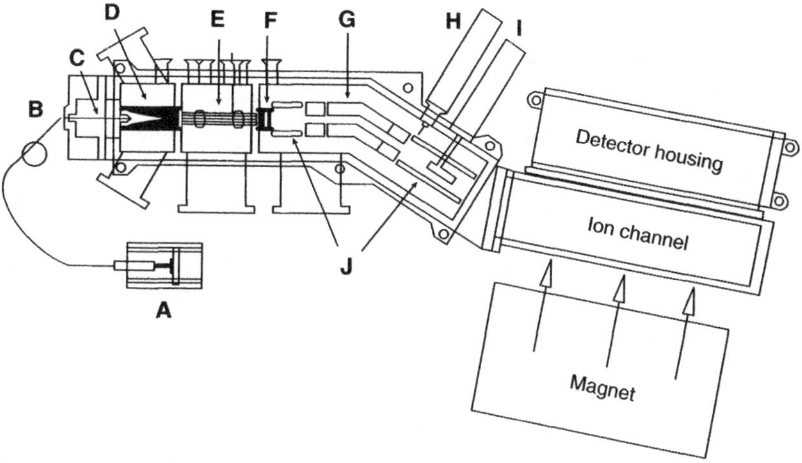

FIGURE 12.15. Magnetic-sector-based mass spectrometer used for multichannel ion separation: A—syringe pump; B—electrospray needle; C—Glass-lined transfer capillary; D—funnel lens; E–octopole; F—acceleration lens; G—electrostatic sector with shunts; H—movable slit mounted on a linear feedthrough; I—Faraday cup ion collector mounted on a linear feedthrough; J—HV-floated Faraday cages. [Reprinted with permission from Mayer et al. (2005). Copyright 2005 American Chemical Society.]

compounds (Mayer et al. 2005). The advantages of this mass spectrometer were its ability to reduce space charge effects at high ion currents and to allow simultaneous separation of all ions. This was achieved by dispersing the ions in a magnetic field at high velocity. The mass spectrometer consisted of an electrospray ion source, ion transfer optics, and a confocal magnetic sector. With appropriate differential pumping, an operating pressure of 1.5×10^{-6} Torr was achieved. Ions generated in the source were directed toward a funnel lens and then an octopole ion guide. An acceleration lens provided the ions with well-defined, tunable kinetic energy (1–3 keV). Faraday cages provided a drift space of well-defined potential that allowed the accelerated ions to reach the electrostatic analyzer and linear dispersion magnet without gain or loss of kinetic energy. The electrostatic analyzer dispersed the ion beam by kinetic energy and a specially designed magnet with an inhomogeneous field provided linear dispersion along the 38-cm focal plane. The ions were spaced equidistantly as a linear function of m/z values and a deceleration lens lowered their kinetic energy to allow for nondestructive soft landing on the ion collector array. The ion collector was composed of a retractable array of bins, each containing a collector electrode and a counterelectrode. The collected components were recovered solvent extraction and subsequently analyzed by mass spectrometry. Although the design of a sector instrument for manufacturing arrays of various compounds through ion soft landing is highly innovative, the limitations imposed by the focal plane dimensions of the sectors have not yet allowed the authors to separate components larger than a mixture of rhodamine B (m/z 443) and gramicidin S (m/z 571).

12.2.6. Quadrupole Mass Filters and Linear Ion Traps for Protein Separations

In 2003, we introduced two soft-landing mass spectrometers; one was a commercial instrument modified to accommodate protein soft-landing experiments and the other, a prototype of a future commercial instrument dedicated to separation of proteins by ion soft landing. These two instruments were used to demonstrate the capability of mass spectrometry to separate proteins from simple and complex mixtures. The soft landing of proteins was achieved using a single stage quadrupole mass spectrometer (SSQ-710c, ThermoElectron, Corp., San Jose, CA) and a prototype linear ion trap (LIT) mass spectrometer (ThermoElectron, Corp., San Jose, CA). The SSQ instrument shown schematically in Figure 12.16 was modified to allow a surface to be positioned in the vacuum chamber. Ions were generated in an electrospray ion source and transmitted for mass selection into a 160-mm hyperbolic quadrupole mass analyzer by an octopole ion guide. The voltage on the surface was set to 0 V, resulting in approximately 5 eV collision energy, which was mostly residual kinetic energy and attributed to the free-jet expansion in the atmospheric interface. The surface was exposed to the ion of interest for 1 h and resulted in the deposition of 10^{10} ions.

The prototype preparative mass spectrometer, described in detail elsewhere (Blake et al. 2004), consisted of an atmospheric pressure ionization source and vacuum interface, ion optics, a mass analyzer and associated electronics made from the adapted components of a prototype commercial linear ion trap mass spectrometer (LIT), as well as a surface positioning system housed in a single-vacuum manifold (Figure 12.17). Using a separate vacuum system for the loading chamber reduces array preparation time and also helps maintain constant experimental conditions in the mass analyzer.

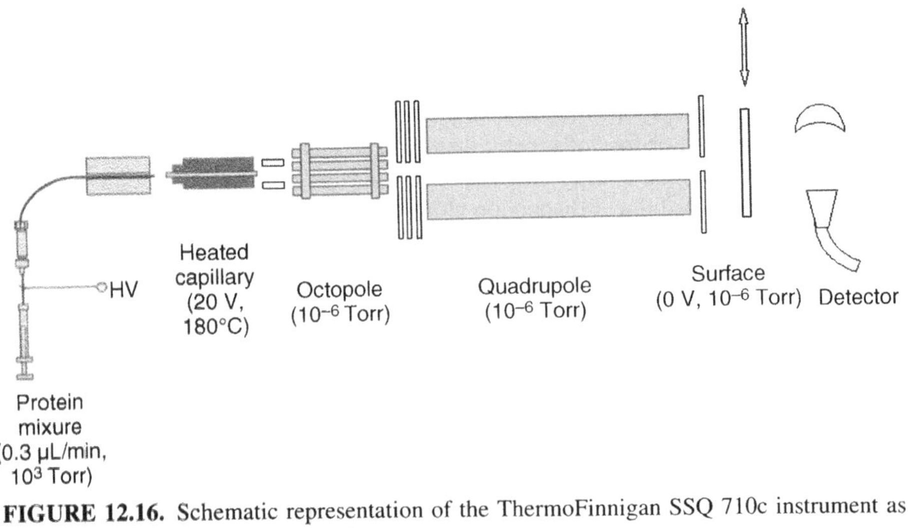

FIGURE 12.16. Schematic representation of the ThermoFinnigan SSQ 710c instrument as modified and used for ion soft landing.

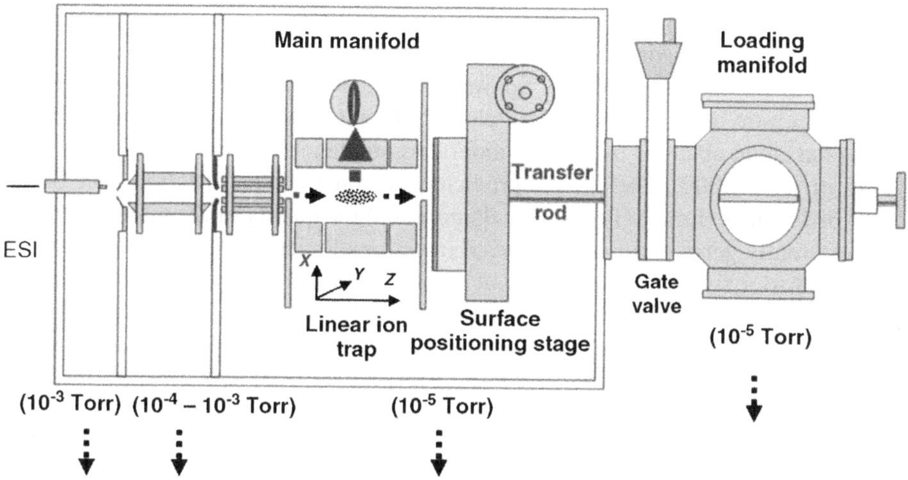

FIGURE 12.17. Top view schematic of the soft-landing instrument including the linear ion trap and the surface loading mechanism. Note that the image depicts ions being ejected radially from the trap along the *x*-axis for detection purposes and that the surface is aligned along the *z*-axis (the *y*-axis comes out of the plane of the figure). The ESI source, heated capillary atmospheric interface, tube lens, skimmer, square quadrupole ion guide, intermultipole gate lens, and octopole ion guide as well as the other labeled components are shown. [Reprinted with permission from Blake et al. (2004). Copyright 2004 American Chemical Society.]

The most significant differences between the single-stage-quadrupole-based instrument and the linear ion trap are outlined in Table 12.2. The improvements provided by an ion-trap-based instrument, compared to the mass-filter-based instrument, include higher resolving power, allowing for a narrow ion isolation window in the linear trap and the ability to perform tandem mass spectrometry experiments on selected ions. The LIT is also equipped with an advanced automated moving stage that allows precise positioning of the ions on various spots. More importantly, in comparison with a classic 3D Paul ion trap, the 2D linear ion trap provides increased ion storage capacity and enhanced ion injection efficiency for externally generated ions, and has the ability to eject ions both axially and radially (Schwartz et al. 2002). The linear ion trap also retains the ability for a user to conduct tandem MS^n experiments and to perform product ion analysis of ion–molecule reactions. These characteristics are of particular interest to preparative mass spectrometry

TABLE 12.2. Comparison of Mass Spectrometers Used in Ion Soft Landing

Mass Analyzer	Maximum m/z	Operating Pressure	MS/MS	m/z Resolution
Quadrupole mass filter	4000 Th	10^{-4}–10^{-6} Torr	No	10^3
Linear ion trap	4000 Th	10^{-3} Torr	Yes	10^4

experiments in which ion injection efficiency and trapping capacity are requirements. Increased ion storage capacity in the linear ion trap is accomplished by an increase in ion trapping volume. Schwartz and coworkers developed a model that involves a ratio of the respective ion cloud radii within the trap using an estimation of the ion cloud obtained from ion tomography studies (Hemberger and Nogar 1992; Schwartz et al. 2002). Their model predicted an increase in the spectral space charge limit of 15 times (limit before severe degration of the quality of the mass spectrum). For externally generated ions, enhanced injection efficiency of the 2D ion trap was attributed to a significant reduction in the RF field in the axial dimension on ion injection. In order for ions to be trapped within a 3D ion trap, the ions must penetrate a significant RF field in the axial dimension that is phase-dependent. As a result, trapping ions from externally generated sources has been shown to be relatively inefficient, (McLuckey et al. 1989). Another advantage of the linear ion trap is the ability to eject ions using the mass-selective instability mode (Stafford and Kelly 1984) and resonance ejection in the radial dimension, as well as the ability to eject ions in the axial dimension by a DC gradient from the center section to the detector or landing surface. Hager (2002) demonstrated the ability to utilize normally detrimental fringe fields between the quadrupole electrodes and the exiting lens to eject ions axially to an external detector. This concept is applied here and is very attractive for ion soft-landing experiments in which ions are trapped in the radial dimension, scanned out radially by resonance ejection or pulsed out the z dimension by a DC pulse.

12.3. APPLICATIONS

12.3.1. Applications and Considerations for Ion Soft Landing as a Preparative Technique

Growing demands from the proteomics and biotechnology fields for increased throughput in biological analyses have produced increased interest in the production of protein microarrays from biological materials. Typically, fabrication of these arrays involves sample purification using various chromatographic techniques and microdrop deposition methods (Martin et al. 1998; Laurell et al. 1999; Morozov and Morozova 1999; MacBeath and Schreiber 2000; Roda et al. 2000). The complementary separatory power of mass spectrometry to that of chromatography might allow preparative-scale experiments using mass spectrometry, or experiments that combine separation by mass spectrometry and by chromatography, to address a broad range of problems in proteomics and related areas. Preparative-scale mass spectrometry using ion soft landing therefore appears to be a possible technology for the separation of biological compounds from mixtures and their storage in the array format for convenience of later analysis. In a *Science* paper, Ouyang et al. (2003) successfully demonstrated this approach.

The use of mass spectrometry as a separation method is expected to provide selectivity since, in principle, any components having different molecular formulas, including isotopomers, can be separated and individually soft-landed. This would include separate glycoforms of proteins, individual polysulfonated forms, and other posttranslationally modified variants. In principle, the technique should also exhibit good spatial resolution since ion-beam spots of μ dimensions are attainable with appropriate ion optical methods. While the traditional methods may fail to separate species having highly similar chemical structures, mass spectrometry should be successful as long as the molecular weights are different. Although the ionization of complex protein mixtures can suffer from suppression effects (Pan and McLuckey 2003) in either ESI or MALDI, these will be minimized when the analytes are highly similar, as in the case of mixtures of synthetic (Siuzdak et al. 1999) and biological (Fuerstenau et al. 2001) polymer congeners such as polyclonal antibodies, glycoproteins, and polysulfonated or phosphorylated products of posttranslationally modified species. Collection of "fractions" in a soft-landing experiment can be performed by moving either the target [as demonstrated by Siuzdak et al. (1999)] or the ion beam on the surface to yield an array of separated components of the initial mixture. These arrays provide tools for combinatorial-chemistry-based methods like identification of drug candidate molecules.

Although a potentially powerful preparative technique in the biological sciences, soft landing has disadvantages. Some of these include the necessity for the sample to be introduced as an ion so limiting the amount of material which can be injected in a mass spectrometer at one time. Many potential applications of preparative soft landing require retention of the activity of the biomolecules, especially those that are difficult to separate by traditional means based on chemical properties rather than mass. While this may be difficult to achieve in all cases, successes have been achieved by soft landing into biocompatible liquid films (see text below). In addition, applications in some areas, as in glycopeptide separation and analysis, are simpler since they only require that the structure remain intact during the separation and landing process. Other potential applications include drug interaction assays, protein purification, and identification of the forms of microheterogeneous proteins responsible for bioactivity.

Preparation of microarrays by ion soft landing faces a number of requirements:

1. The ionization method should be efficient with minimal suppression effects.
2. Biological activity should not be lost in the ionization process, in the mass selection step or in the soft-landing step.
3. Mass spectrometric performance must be adequate to separate the components of interest from other components and to achieve the resolution and mass range necessary for the compounds of interest.

These requirements mean that the mass spectrometer and operating conditions should be such as to minimize the time ions spend in the gas phase, minimize the possibility of collisional activation that might lead to unfavorable conformational changes, and yet still maximize the efficiency of ion transport and collection.

Proteins that have been shown to retain at least some biological activity include lysozyme, trypsin, hexokinase, and protein kinase A catalytic subunit. In the case of the kinases, the soft-landed enzymes were shown to phosphorylate D-fructose and LRRASLF oligopeptide, respectively. Overall, in these prior studies (Ouyang et al. 2003; Blake et al. 2004; Gologan et al. 2004), retention of biological activity for proteins purified through soft landing was shown to be possible with a much higher success rate when thin liquid films were used to capture the ions. The nature of the surface controls the amount and the chemical form of the soft-landed material, and will either facilitate or inhibit retention of the biological activity of the landed materials, therefore limiting the types of post-soft-landing characterization experiments that can be performed. Some surfaces promote the neutralization of charged species, while others allow charge to build up, and this will have a direct impact on the type and thickness of the layers of proteins that can be collected. An extended discussion of surface effects on soft landing can be found elsewhere (Gologan et al. 2004).

The overall efficiency of soft landing is dependent on the ability to minimize losses that occur when creating ions and transferring them from atmosphere to vacuum. First and foremost, the compounds of interest, during ionization, may acquire a different number of protons resulting in a charge state distribution. For example, a protein with 10 basic amino acids, depending on the pH of the solution and on its p*I*, will generally exhibit multiple characteristics peaks in an electrospray ionization mass spectrum. This phenomenon is useful in determining the molecular weight of an unknown protein. However, when one desires to separate such a protein from a mixture, the experiment is currently performed by selecting only one of the protein's charge states. In addition, electrospray ionization is well known to be prone to suppression effects when salts, glycerol, or other additives are present in the sample and the nature of the solvent used (typically an aqueous–organic mixture) has a denaturizing effect on the structure of proteins resulting in loss of structural information and biological activity. Important features of an ideal ionization method for soft-landing of proteins include (i) high efficiency (ii) the method is "soft" (iii) tolerance of high salt concentrations (iv) the ion current is retained in one ionic species and not distributed over multiple charge states. Some of the requirements can be satisfied by using low flow rates, such as those used in nanospray ionization, but this is not an option in an experiment where larger quantities of purified materials are required.

12.3.2. Soft Ionization for Preparative Mass Spectrometry of Biomolecules

Electrosonic spray ionization (ESSI) (Takats et al. 2004a), a variant of electrospray ionization (ESI), employs a traditional micro-ESI source with the addition of a high-velocity nebulizing sheath gas. The high linear velocity of the nebulizing gas provides efficient pneumatic spraying of the charged liquid sample. The variable electrostatic potential can be tuned to allow efficient and gentle ionization. This ionization method has been successfully applied to aqueous solutions of various proteins at neutral pH, and its performance has been compared to that of the nanospray and micro-ESI techniques. Evidence for efficient desolvation during ESSI is provided by the fact that the peak widths for various multiply charged protein ions are an order of magnitude narrower than those for ESI and similar to those observed with nanospray. Narrow charge state distributions compared to other ESI techniques are observed as well; for most of the proteins studied, more than 90% of the protein ions can be accumulated in one charge state using ESSI under optimized conditions. The fact that the abundant charge state is normally as low as or lower than that recorded by ESI or nanospray indicates that folded protein ions are generated since fewer charge sites are available in a compact conformation than in an open conformation. The sensitivity of the ionization technique to high salt concentrations is comparable to that of nanospray, but ESSI is considerably less sensitive to high concentrations of organic additives such as glycerol or 2-amino-2-(hydroxymethyl)-1,3-propanediol (Tris base). The ESSI mass spectrum shown in Figure 12.18 of bovine serum albumin (66 kDa) at neutral pH is a typical example of the kind of performance obtained from ESSI. The mass spectrum reveals predominantly a single charge state and contributions from solvent adducts are minimal for such a large protein. The ion source also produces high ion currents that correspond to 90–100% ionization efficiency measured at atmospheric pressure provided one works under ESSI conditions. The overall experimental evidence indicates that an electrospray/electrosonic spray ionization method from an aqueous or buffered solution will maintain the bioactivity of a compound of interests without increasing the loss in the number of ions directed toward the mass spectrometer.

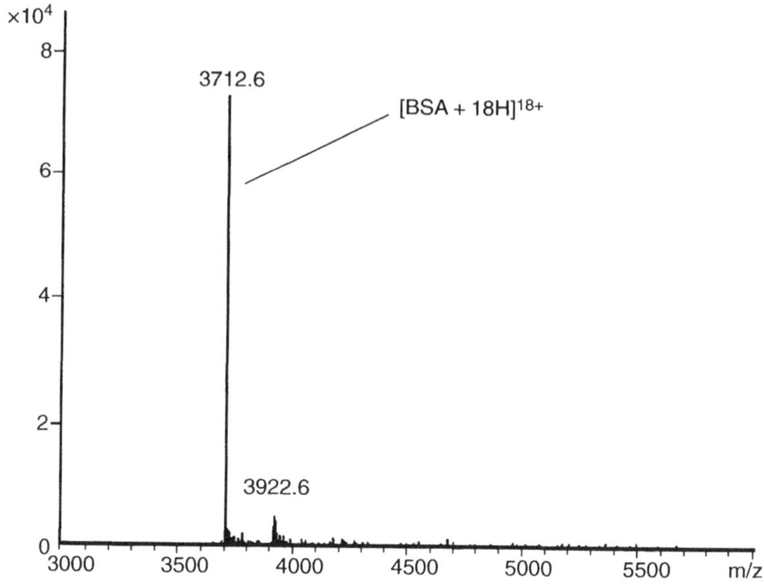

FIGURE 12.18. Electrosonic spray ionization (ESSI) mass spectrum of bovine serum albumin.

12.3.3. Desorption Electrospray Ionization and Applications to Soft-Landing Analysis

Equally important to how well the ions of interest are being isolated and collected in a future preparative soft-landing apparatus is what kind of sample can be separated. If, as would be ideal, no sample preparation steps are to be used before introduction into the mass spectrometer, current ionization methods limit the experiment to a small set of sample types. Instead, one would like to envision the possibility of separating the compounds of interest directly from a natural surface or a tissue, without significant sample processing or separation. The development of desorption ionization techniques provided perhaps the first breakthrough in the mass spectrometric analysis of fragile, non-volatile compounds. MALDI, along with ESI, has revolutionized bioanalytical mass spectrometry by making the analysis of practically any kind of biochemical species feasible. More recently, a novel approach has been introduced to extend these advantages of the desorption ionization methods to direct surface analysis at atmospheric pressure under ambient conditions; the experiment is termed *desorption electrospray ionization* (DESI) (see Figure 12.19) (Takats et al. 2004b). The experiment involves the interaction of low-kinetic-energy cluster ions generated from solvent colliding with ordinary surfaces bearing compounds of interest. DESI is carried out by directing charged droplets and ions of solvent toward the surface to be analyzed. The impact of the charged particles on the surface produces gaseous ions of material originally present on the surface. As an example, the resulting mass spectra of biomolecules are similar to normal ESI mass spectra in that they show mainly singly or multiply charged molecular ions of these analytes (Figure 12.19). The DESI phenomenon was observed both in the case of conductive and insulator surfaces and for compounds ranging from nonpolar small molecules such as lycopene, the alkaloid coniceine, and small drugs, through polar compounds to biopolymers such as peptides and proteins. The mechanisms of DESI are not well

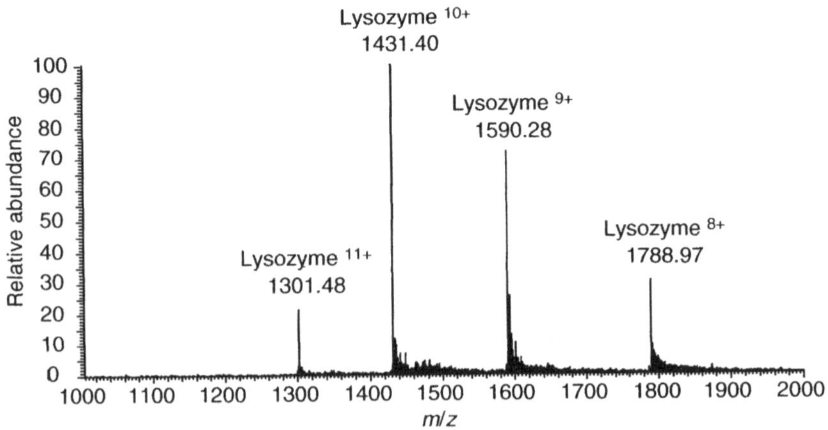

FIGURE 12.19. Desorption electrospray ionization (DESI) mass spectrum of hen eggwhite lysozyme (50 ng/cm^2) present on a PTFE surface.

known, but in the case of small molecules there is evidence that chemical sputtering due to free gas-phase ions is involved while in the case of large biopolymers, a droplet pick-up mechanism has been adduced. Changes in the solution that is sprayed can be used to selectively ionize particular compounds, including those in biological matrices. In vivo analysis was demonstrated using a nitrogen-assisted ethanol–water spray to detect the presence of carnitine and acetylcarnitine in blood.

Many applications (Takats et al. 2004b) are made possible from collisions of ions with surfaces at atmospheric pressure. For example, it is possible to examine protein digests from a Teflon surface, enzyme–substrate molecular recognition is achieved for systems when the kinetics of the enzymatic reaction allow the complex to be observed in the mass spectrum, and chiral analysis directly from a limestone surface containing an amino acid and drug metabolites from a person's skin; these are all achieved. Remarkable new types of chemical analysis are possible using this experiment including direct analysis and chemical imaging of biological tissue (Wiseman et al. 2005, Takats et al. 2004b) which demonstrates just how powerful a method mass spectrometry might become once its vacuum constraints are lifted. Furthermore, when coupled to a soft-landing instrument, DESI may provide the answer to protein and peptide separations directly from an abundant in vivo environment, without any additional sample preparation.

12.4. CONCLUSIONS

Soft landing offers new ways of interrogating and recognizing biomolecules in pure form with the possibility of long-term storage and future analysis of samples. These experiments will lead to highly sensitive detection/identification, such as activity assays, using surface-based spectroscopic methods, including Raman spectroscopy. Note that separation by mass spectrometry from complex mixtures (e.g., serum, plasma) is particularly advantageous for closely related groups of compounds (e.g., glycosylated forms of the same protein). The advantages of ion soft landing extend to minor protein/peptide constituents of mixtures, especially when used in conjunction with other orthogonal separation methods. It is possible to foresee related substance analysis on recombinant and posttranslationally modified proteins as well as high-throughput experiments, including drug receptor screening.

Other potential applications include, but are not limited to, reactions of extremely pure proteins with affinity and other reagents (including enzyme–substrate and receptor–ligand reactions), binding experiments (ligand/receptor identification, small-molecule drug/target pair identification), resolution of multiple modified forms of a protein, effective analysis of biopsy materials, and determination of the effects of posttranslational modifications on protein function.

Exciting possibilities exist for preparative mass spectrometry, performed by soft landing, for purification and storage of biological compounds. While mass spectrometry was a technique used almost exclusively by organic chemists as recently as the mid-1980s to analyze "small molecules," today, with the advances provided by new ionization methods and revolutionary instrument developments, mass spectrometry may significantly impact the quality of life through discovery of new targets for drug development. By combining the powerful capabilities of analytical mass spectrometry within a rugged, high-throughput instrument with a simplified interface and dedicated use, mass spectrometry may yet become a simple yet powerful preparative technology with wide applications.

REFERENCES

Ada ET, Kornienko O, Hanley L (1998): Chemical modification of polystyrene surfaces by low-energy polyatomic ion beams. *J. Phys. Chem. B* **102**:3959–3966.

Alvarez J (2005): *Distinct Collision Processes in Mass Spectrometry: Charge Inversion Reactions and Peptide Ion Soft-Landing*, Purdue University.

Alvarez J, Cooks RG, Barlow SE, Gaspar DJ, Futrell JH, Laskin J (2005): Preparation and in situ characterization of surfaces by soft landing in a Fourier transform ion cyclotron resonance mass spectrometer, *Anal. Chem.* **77**:3452–3460.

Alvarez J, Futrell JH, Laskin J (2006): Soft-landing of peptides on to self-assembled monolayer surfaces, *J. Phys. Chem.* **110**:1678–1687.

Ast T, Mabud MA, Cooks RG (1988): Reactive collisions of polyatomic ions at solid surfaces, *Int. J. Mass Spectrom. Ion Proc.* **82**:131–150.

Ast T, Riederer DE Jr, Miller SA, Morris M, Cooks RG (1993): Collisions of fluorocarbon ions at solid surfaces: Electronic excitation, surface-induced dissociation, and chemical sputtering. *Org. Mass Spectrom.* **28**:1021–1033.

Ast T, Schey KL, Cooks RG (1990): Target effects in ion/surface collisions. *J. Serb. Chem. Soc.* **55**:247–256.

Aston FW (1919): Neon lamps for stroboscopic work. *Proc. Cambridge Phil. Soc.* **19**:300–306.

Baer T, Hase WL (1996): *Unimolecular Reaction Dynamics: Theory and Experiments*.

Benninghoven A, Rudenaurer FG, Werner HW (1982): *Secondary Ion Mass Spectrometry: Basic Concepts, Instrumental Aspects, Applications, and Trends*. Wiley, New York.

Bier ME, Vincenti M, Cooks RG (1987): Ion/surface collisions which lead to charge permutation. *Rapid Commun. Mass Spectrom.* **1**:92–94.

Blake TA, Ouyang Z, Wiseman JM, Takats Z, Guymon AJ, Kothari S, Cooks RG (2004): Preparative linear ion trap mass spectrometer for separation and collection of purified proteins and peptides in arrays using ion soft landing. *Anal. Chem.* **76**:6293–6305.

Bowie JH (2001): Charge-stripping of anions in the gas phase: The formation of interesting neutrals and cations. *Int. J. Mass Spectrom.* **212**:249–265.

Bromann K, Felix C, Brune H, Harbich W, Monot R, Buttet J, Kern K (1996): Controlled deposition of size-selected silver nanoclusters. *Science* **274**:956–958.

Burroughs JA, Wainhaus SB, Hanley L (1994): Impulsive excitation of $Cr(CO)_6^+$ during surface-induced dissociation at organic monolayers. *J. Phys. Chem.* **98**:10913–10919.

Bursey MM (1990): Charge inversion of negative ions in tandem instruments. *Mass Spectrom. Rev.* **9**:555–574.

Busch KL, Cooks RG (1982): Mass spectrometry of large, fragile, and involatile molecules. *Science* **218**:247–254.

Busharov NP, Gorbatov EA, Gusev VM, Guseva MI, Martynenko YV (1976): Chemical sputtering of graphite by H+ ions. *J. Nucl. Mater.* **63**:230–234.

Chorush RA, Little DP, Beu CS, Wood TD, McLafferty FW (1995): Surface-induced dissociation of multiply-protonated proteins. *Anal. Chem.* **67**:1042–1046.

Cooks RG (1995): Collision-induced dissociation: Readings and commentary. *J. Mass Spectrom.* **30**:1215–1221.

Cooks RG, Ast T, Beynon JH (1975): Anomalous metastable peaks. *Int. J. Mass Spectrom. Ion Physics* **16**:348–352.

Cooks RG, Ast T, Kralj B, Kramer V, Zigon D (1990a): Internal energy distributions deposited in doubly and singly charged tungsten hexacarbonyl ions generated by charge stripping, electron impact, and charge exchange. *J. Am. Soc. Mass Spectrom.* **1**:16–27.

Cooks RG, Ast T, Mabud A (1990b): Collisions of polyatomic ions with surfaces. *Int. J. Mass Spectrom. Ion Proc.* **100**:209–265.

Cooks RG, Ast T, Pradeep T, Wysocki V (1994): Reactions of ions with organic surfaces. *Acc. Chem. Res.* **27**:316–323.

Cooks RG, Evans C, Wade N, Jo SC (2004): Collisions of organic ions at surfaces. *Appl. Surf. Sci.* **231–232**:13–21.

Danell AS, Glish GL (2001): Charge permutation reactions in beam type mass spectrometers. *Int. J. Mass Spectrom.* **212**:219–227.

DeKrey M, Kenttamaa HI, Wysocki VH, Cooks RG (1986): Energy deposition in $[Fe(CO)_5]^+$ upon collision with a metal surface. *Org. Mass Spectrom.* **21**:193–195.

Detter LD, Hand OW, Cooks RG, Walton RA (1988): *Mass Spectrom. Rev.* **7**:465–502.

Dongre AR, Somogyi A, Wysocki VH (1996): Surface-induced dissociation: An effective tool to probe structure, energetics and fragmentation mechanisms of protonated peptides. *J. Mass Spectrom.* **31**:339–350.

Douglas DJ, Shushan B (1982): Observation of charge inversion spectra with a triple quadrupole mass spectrometer system. *Org. Mass Spectrom.* **17**:198–199.

Evans C, Wade N, Pepi F, Strossman G, Schuerlein T, Cooks RG (2002): Surface modification and patterning using low-energy ion beams: Si—O bond formation at the vacuum/adsorbate interface. *Anal. Chem.* **74**:317–323.

Farrar JM, Lee YT (1974): Chemical dynamics. *Ann. Rev. Phys. Chem.* **25**:357–385.

Feng BB, Wunschel DS, Masselson CD, Pasa-Tolic L, Smith RD (1999): Retrieval of DNA using soft-landing after mass analysis by ESI-FTICR for enzymatic manipulation. *J. Am. Chem. Soc.* **121**:8961–8962.

Forst W (1973): *Physical Chemistry: a Series of Monographs: Theory of Unimolecular Reactions.*

Franchetti V, Solka BH, Baitinger WE, Amy JW, Cooks RG (1977): Soft landing of ions as a means of surface modification. *Int. J. Mass Spectrom. Ion Phys.* **23**:29–35.

Fuerstenau SD, Benner WH, Thomas JJ, Brugidou C, Bothner B, Siuzdak G (2001): Mass spectrometry of an intact virus. *Angew. Chem. Int. Ed.* **40**:542–544.

Garrison BJ, Delcorte A, Zhigilei L, Itina TE, Krantzman KD, Yingling YG, McQuaw CM, Smiley EJ, Winograd N (2003): Big molecule ejection—SIMS vs. MALDI. *Appl. Surf. Sci.* **203**:69–71.

Geiger RJ, Melnyk MC, Busch KL, Bartlett MG (1999): Modifications to an analytical mass spectrometer for the soft-landing experiment. *Int. J. Mass Spectrom.* **183**:415.

Gerrity DP, Valentini JJ (1985): *J. Chem. Phys.* **82**:1323–1331.

Gillen KT, Riley C, Bernstei R (1969): Reactive scattering of K by Hbr, Dbr in crossed molecular beams—angular and velocity distributions of Kbr in laboratory and Cm systems. *J. Chem. Phys.* **50**:4019–4223.

Gologan B, Green JR, Alvarez J, Laskin J, Cooks RG (2005): Ion/surface reactions and ion soft-landing. *Phys. Chem. Chem. Phys.* **7**:1490–1500.

Gologan B, Takats Z, Alvarez J, Wiseman JM, Talaty N, Ouyang Z, Cooks RG (2004): Ion soft-landing into liquids: Protein identification, separation, and purification with retention of biological activity. *J. Am. Soc. Mass Spectrom.* **15**:1874–1884.

Grill V, Shen J, Evans C, Cooks RG (2001): Collisions of ions with surfaces at chemically relevant energies: Instrumentation and phenomena. *Rev. Sci. Instrum.* **72**:3149–3179.

Hager JW (2002): A new linear ion trap mass spectrometer. *Rapid Commun. Mass Spectrom.* **16**:512–526.

Hand OW, Detter LD, Lammert SA, Cooks RG, Walton RA (1989): *J. Am. Chem. Soc.* (111):5577–5583.

Hayakawa S (2001): Internal energy distribution in charge inversion mass spectrometry using alkali metal targets. *Int. J. Mass Spectrom.* **212**:229–247.

Hayakawa S (2004): Charge inversion mass spectrometry: Dissociation of resonantly neutralized molecules. *J. Mass Spectrom.* **39**:111–135.

Hayakawa S, Endoh H, Arakawa K, Morishita N, Sugiura T (1995): Discrimination of C3H4+ isomeric ions by charge inversion mass spectrometry using an alkali metal target. *Int. J. Mass Spectrom. Ion Proc.* **151**:89–95.

Heiland W, Taglauer E (1976): Low-energy ion scattering—elastic and inelastic effects. *Nucl. Instrum. Meth.* **132**:535.

Hemberger PH, Nogar NS (1992): Laser photodissociation probe for ion tomography studies in a quadrupole ion-trap mass spectrometer. *Chem. Phys. Lett.* (191): 405–410.

Hendell E, Even U, Raz T, Levine RD (1995): Shattering of clusters upon surface impact—an experimental and theoretical study. *Phys. Rev. Lett.* **75**:2670–2673.

Herman Z (2001): The crossed-beam scattering method in studies of ion-molecule reaction dynamics. *Int. J. Mass Spectrom.* **212**:413–443.

Herman Z (2003): *J. Am. Soc. Mass Spectrom.* **14**:1360–1372.

Herschbach DR (1966): *Adv. Chem. Phys.* **10**:309.

Herschbach DR (1987): Molecular-dynamics of elementary chemical-reactions—Nobel lecture, 8 December 1986. *Chem. Scripta* **27**(3):327–347.

Hirschfelder J, Eyring H, Topley B (1936): *J. Chem. Phys.* **4**:170–177.

Honig RE (1957): Sputtering of surfaces by positive ion beams of low energy. *Bull. Am. Phys. Soc.* **2**:34–35.

Hornung G, Schalley CA, Dieterle M, Schroder D, Schwartz H (1997): A study of the gas-phase reactivity of neutral alkoxy radicals by mass spectrometry: a-Cleavages and Barton-type hydrogen migrations. *Chem. Eur. J.* **3**:1866–1883.

Jo SC (2004): *Surface-Induced Dissociation of Hyperthermal Ions at Self-Assembled Monolayers. Chemistry*, PhD thesis, Purdue Univ. West Lafayette, IN.

Jo SC, Cooks RG (2003): Translational to vibrational energy conversion during surface-induced dissociation of n-butylbenzene molecular ions colliding at self-assembled monolayer surfaces. *Eur. J. Mass Spectrom.* **9**:237–244.

Kaiser B, Bernhardt TM, Rademann K (1999): Interaction of mass selected antimony clusters with HOPG. *Nucl. Instrum. Meth B.* **157**(1–4):155–161.

Kang H, Lee HW, Cho WR, Lee SM (1998): Molecular shape effect in polyatomic ion-surface scattering. Surface-induced dissociation of Fe(C5H5)(2)(+) and FeC5H5+ at Si(100). *Chem. Phys. Lett.* **292**:213–217.

Krantzman KD, Fenno R, Decorte A, Garrison BJ (2003): Theoretical simulations of atomic and polyatomic bombardment of an organic overlayer on a metallic substrate. *Nucl. Instrum. Meth. B.* **202**:201–205.

Kubista J, Dolejsek Z, Herman Z (1998): Energy partitioning in collisions of slow polyatomic ions with surfaces: Ethanol molecular ions on stainless steel surfaces. *Eur. Mass Spectrom.* **4**:311–319.

Laskin J, Bailey TH, Futrell JH (2003): Shattering of peptide ions on self-assembled monolayer surfaces. *J. Am. Chem. Soc.* **125**:1625–1632.

Laskin J, Beck KM, Hache JJ, Futrell JH (2004): Surface-induced dissociation of ions produced by matrix-assisted laser desorption/ionization in a Fourier transform ion cyclotron resonance mass spectrometer. *Anal. Chem.* **76**:351–356.

Laskin J, Denisov EV, Shukla AK, Balow SE, Futrell JH (2002): Surface-induced dissociation in a Fourier transform ion cyclotron resonance mass spectrometer: Instrument design and evaluation. *Anal. Chem.* **74**:3255–3261.

Laskin J, Futrell JH (2003): Collisional activation of peptide ions in FT-ICR mass spectrometry. *Mass Spectrom. Rev.* **22**:158–181.

Laurell T, Wallman L, Nilsson JJ (1999): Design and development of a silicon microfabricated flow-through dispenser for on-line picolitre sample handling. *Micromech. Microeng.* **9**:369–376.

Lee YT (1987): Molecular-beam studies of elementary chemical processes. *Science* **236**: 793–798.

Lee YT, Mcdonald JD, Lebreton PR, Herschba D (1969): Molecular beam reactive scattering apparatus with electron bombardment detector. *Rev. Sci. Instrum.* **40**:1402–1408.

Levine RD, Bernstein RB (1987): *Molecular Reaction Dynamics and Chemical Reactivity*, Oxford Univ. Press, New York.

London F (1929): 2. *Elektrochem. Angew. Phys. Chem.* **35**:552–555.

Luo H, Miller SA, Cooks RG, Pachuta SJ (1998): Soft landing of polyatomic ions for selective modification of fluorinated self-assembled monolayer surfaces. *Int. J. Mass Spectrom. Ion Proc.* **174**:193–217.

Mabud A (1987). PhD thesis, Purdue Univ., West Lafayette, IN.

MacBeath G, Schreiber SL (2000): Printing proteins as microarrays for high-throughput function determination. *Science* **289**:1760–1763.

Macdonald RG, Liu KP (1990): A crossed-beam study of the state-resolved dynamics of Ch(X 2-Pi) + D2.2. The isotopic exchange channel. *J. Chem. Phys.* **93**:2443–2459.

Martin BD, Gaber BP, Patterson CH, Turner DC (1998): Direct protein microarray fabrication using a hydrogel "stamper." *Langmuir* **14**:3971–3975.

Mayer PS, Turecek F, Lee HN, Scheidmann AA, Olney TN, Schumacher F, Strop P, Smrcina M, Patek M, Schilin D (2005): Preparative separation of mixtures by mass spectrometry. *Anal. Chem.* **77**:4378–4384.

McLuckey SA, Glish GL, Asano KG (1989): *Anal. Chim. Acta.* **225**:25–35.

Meroueh SO, Wang YF, Hase WL (2002): Direct dynamics simulations of collision- and surface-induced dissociation of N-protonated glycine. Shattering fragmentation. *J. Phys. Chem. A* **106**:9983–9992.

Miller SA, Luo H, Pachuta SJ, Cooks RG (1997): Soft-landing of polyatomic ions at fluorinated self-assembled monolayer surfaces. *Science* **275**:1447–1450.

Morozov VN, Morozova TY (1999): Electrospray deposition as a method for mass fabrication of mono- and multicomponent microarrays of biological and biologically active substances. *Anal. Chem.* **71**:3110–3117.

Nair H, Somogyi A, Wysocki VH (1996): Effect of alkyl substitution at the amide nitrogen on amide bond cleavage: Electrospray ionization/surface-induced dissociation fragmentation of substance P and two alkylated analogs. *J. Mass Spectrom.* **31**: 1141–1148.

Nanita SC, Takats Z, Cooks RG, Myung SM, Clemmer DE (2004): Chiral enrichment of serine via formation, dissociation, and soft-landing of octameric cluster ions. *J. Am. Soc. Mass Spectrom.* **15**:1360–1365.

Neuhauser D, Judson RS, Kouri DJ, Aldeman DE, Shafer NE, Kliner DAV, Zare RN (1992): State-to-state rates for the D+H2(V = 1,J = 1)-]Hd(V′,J′)+H reaction—predictions and measurements. *Science* **257**:519–522.

Ouyang Z, Takats Z, Blake TA, Gologan B, Guymon AJ, Wiseman JM, Oliver JC, Davisson VJ, Cooks RG (2003): Preparing protein microarrays by soft-landing of mass-selected ions. *Science* **301**:1351–1354.

Pachuta SJ, Cooks RG (1987): Mechanisms in molecular SIMS. *Chem. Rev.* **87**:647–669.

Pan P, McLuckey SA (2003): Electrospray ionization of protein mixtures at low pH. *Anal. Chem.* **75**:1491–1499.

Parker JH, Pimentel GC (1969): Vibrational energy distribution through chemical laser studies. I. Fluorine atoms plus hydrogen or methane. *J. Chem. Phys.* **51**:91–96.

Polanyi JC (1987): Some concepts in reaction dynamics. *Science* **236**:680–690.

Polanyi JC, Tardy DC (1969): Energy distribution in exothermic reaction F+H2 and endothermic reaction Hf+H. *J. Chem. Phys.* **51**:5717–5719.

Postawa Z, Ludwig K, Piaskowy J, Krantzman K, Winograd N, Garrison BJ (2003): Molecular dynamic simulations of the sputtering of multilayer organic systems. *Nucl. Instrum. Meth. B* **202**:168–174.

Pradeep T, Miller SA, Cooks RG (1993): Surface-induced dissociation from a liquid surface. *J. Am. Soc. Mass Spectrom.* **4**:769–773.

Rhodes R (1986): *The Making of the Atomic Bomb*, New York.

Robinson PJ, Holbrook KA (1972): *Unimolecular Reactions*.

Roda A, Guardigli M, Russo C, Pasani P, Baraldini M (2000): Protein microdeposition using a conventional ink-jet printer. *BioTechniques* **28**:492–496.

Rosenstock HM, Draxl K (1977): *J. Phys. Chem. Ref. Data* (6).

Rutherford E (1911): *Proc. Manchester Lit. Phil. Soc. IV* **55**:18–20.

Schaaff TG, Qu Y, Farrell N, Wysocki VH (1998): Investigation of the trans effect in the fragmentation of dinuclear platinum complexes by electrospray ionization surface-induced dissociation tandem mass spectrometry. *J. Mass Spectrom.* **33**:436–443.

Schafer TP, Siska PE, Parson JM, Tully FP, Wong YC, LeeYT (1970): Crossed molecular beam study of F + D2. *J. Chem. Phys.* **53**:3385–3387.

Schalley CA, Hornung G, Schroder D, Schwartz H (1998a): Mass spectrometric approaches to the reactivity of transient neutrals. Chem. Soc. Rev. **27**:91–104.

Schalley CA, Hornung G, Schroder D, Schwartz H (1998b): Mass spectrometry as a tool to probe the gas-phase reactivity of neutral molecules. *Int. J. Mass Spectrom. Ion Proc.* **172**:181–208.

Schroder D, Schwarz H, Dua S, Blanksby SJ, Bowie JH (1999): Mass spectrometric studies of the oxocarbons CnOn ($n = 3$–6). *Int. J. Mass Spectrom.* **188**:17–25.

Schultz DG, Hanley L (1998): Shattering of SiMe3+ during surface-induced dissociation. *J. Chem. Phys.* **109**:10976–10983.

Schwartz JC, Senko MW, Syka JEP (2002): A two-dimensional quadrupole ion trap mass spectrometer. *J. Am. Soc. Mass Spectrom.* **13**:659–669.

Shen J, Evans C, Wade N, Cooks RG (1999a): Ion-ion collisions leading to formation of C—C bonds at surfaces: An interfacial Kolbe reaction. *J. Am. Chem. Soc.* **121**:9762–9763.

Shen J, Yim YH, Feng B, Grill V, Evans C, Cooks RG (1999b): Soft landing of ions onto self-assembled hydrocarbon and fluorocarbon monolayer surfaces. *Int. J. Mass Spectrom.* **182/183**:423–435.

Sigmund P (1973): Mechanism of surface micro-roughening by ion-bombardment. *J. Mater. Sci.* **8**:1545–1553.

Siuzdak G, Bothner B, Yeager M, Brugidou C, Fauquet CM, Hoey K, Chang C (1996): Mass spectrometry and viral analysis. *Chem. Biol.* **3**:45–48.

Siuzdak G, Hollenbeck T, Bothner B (1999): Preparative mass spectrometry with electrospray ionization. *J. Mass Spectrom.* **34**:1087–1088.

Smith DP (1967): Scattering of low-energy noble gas ions from metal surfaces. *J. Appl. Phys.* **38**:340–347.

Smith IWM (1980): *Kinetics and Dynamics of Elementary Gas Reactions*, Butterworths, London.

Smith LP, Parkins WE, Forrestor AT (1947): On the separation of isotopes in quantity by electromagnetic means. *Phys. Rev.* **72**:989–1002.

Song KY, Meroueh O, Hase WL (2003): Dynamics of Cr(CO)(6)(+) collisions with hydrogenated surfaces. *J. Chem. Phys.* **118**:2893–2902.

Stafford GC, Kelly PE (1984): Recent improvements in and analytical applications of advanced ion trap technology. *Int. J. Mass Spectrom. Ion Phys.* **60**:85–98.

Takats Z, Wiseman JM, Gologan B, Cooks RG (2004a): Electrosonic spray ionization. A gentle technique for generating folded proteins and protein complexes in the gas phase and for studying ion–molecule reactions at atmospheric pressure. *Anal. Chem.* **76**:4050–4058.

Takats Z, Wiseman JM, Gologan B, Cooks RG (2004b): Mass spectrometry sampling under ambient conditions with desorption electrospray ionization. *Science* **306**:471–473.

Taylor EH, Datz S (1955): Study of chemical reaction mechanisms with molecular beams—the reaction of K with Hbr. *J. Chem. Phys* **23**:1711–1718.

Thomson JJ (1913): *Rays of Positive Electricity and Their Application to Chemical Analysis*. Longmans Green, London.

Tu YY, Chuang TJ, Winters HF (1981): Chemical sputtering of fluorinated silicon. *J. Vac. Sci. Technol.* **18**:357–358.

Vincenti M, Cooks RG (1988): Desorption due to charge exchange in low-energy collisions of organofluorine ions at solid surfaces. *Org. Mass Spectrom.* **23**:317–326.

Wade N, Evans C, Jo SC, Cooks RG (2002): Silylation of an OH-terminated self-assembled monolayer surface through low-energy collisions of ions: A novel route to synthesis and patterning of surfaces. *J. Mass Spectrom.* **37**:591–602.

Wade N, Evans C, Pepi F, Cooks RG (2000): Collisions of silylium cations with hydroxyl-terminated and other self-assembled monolayer surfaces: Reactions, dissociation, and surface characterization. *J. Phys. Chem. B* **104**(47):11230–11237.

Wainhaus SB, Gislason EA, Hanley L (1997): *J. Am. Chem. Soc.* **119**:4001–4007.

Winograd N (2005): The magic of cluster SIMS, *Anal. Chem.* **77**:143A–149A.

Wijesundara MBJ, Hanley L, Ni BR, Sinnott SB (2000): Effects of unique ion chemistry on thin-film growth by plasma-surface interactions. *Proc. Natl. Acad. Sci. USA* **97**:23–27.

Williams ER, Henry KD, McLafferty FW, Shabanowitz J, Hunt DF (1990): Surface-induced dissociation of peptide ions in Fourier-transform mass spectrometry. *J. Am. Soc. Mass Spectrom.* **1**:413–416.

Winger BE, Laue HJ, Horning SR, Julian RK, Lammert SA, Riederer DE, Cooks RG (1992): Hybrid Beeq tandem mass-spectrometer for the study of ion surface collision processes. *Rev. Sci. Instrum.* **63**:5613–5625.

Winters HF (1982): Chemical sputtering, a discussion of mechanisms. *Rad. Effects* **64**:79–80.

Wiseman JM, Puolitaival SM, Takats Z, Cooks RG, Caprioli RM (2005): Mass spectrometric profiling of biological tissue by using desorption electrospray ionization. *Angew. Chem. Int. Ed.* **44**:7094–7097.

Yergey AL, Yergey AK (1997): Preparative scale mass spectrometry: A brief history of the calutron. *J. Am. Soc. Mass Spectrom.* **8**:943–953.

Zare RN, Dagdigia PJ (1974): Tunable laser fluorescence method for product state analysis. *Science* **185**:739–747.

Zhigilei LV, Leveugle E, Garrison BJ, Yingling YG, Zeifman MI (2003): Computer simulations of laser ablation of molecular substrates. *Chem. Rev.* **103**:321–347.

13

ELECTRON CAPTURE DISSOCIATION AND OTHER ION–ELECTRON FRAGMENTATION REACTIONS

ROMAN ZUBAREV

Laboratory for Biological and Medical Mass Spectrometry
Uppsala University
Uppsala, Sweden

Principles of Mass Spectrometry Applied to Biomolecules, edited by Julia Laskin and Chava Lifshitz

13.1. INTRODUCTION: HISTORY OF DISCOVERY OF ION–ELECTRON REACTIONS

Fragmentation of positive ions following electron capture is historically known as *dissociative recombination* (DR). In 1931, Kaplan attributed the oxygen green line in the night sky and in auroras to the formation of the excited atom $O(^1S)$ through the dissociative recombination of electrons with $O_2^{+\bullet}$ ions. DR is believed to be responsible for a range of diverse phenomena in ion–electron plasmas, including formation of water and organic molecules in outer space. One of the DR pioneers, Bates proposed what has became known as *the direct DR mechanism,* in which electron capture by the molecular ion AB^+ occurs into an excited neutral state $(AB\bullet)''$ that lies above the ion state, in the vicinity of its equilibrium position (Figure 13.1a). This state subsequently undergoes dissociation to $A\bullet + B$, where the recombination energy is transferred to the kinetic energy of the products. This picture was subsequently modified by Bardsley, who has suggested *the indirect DR*

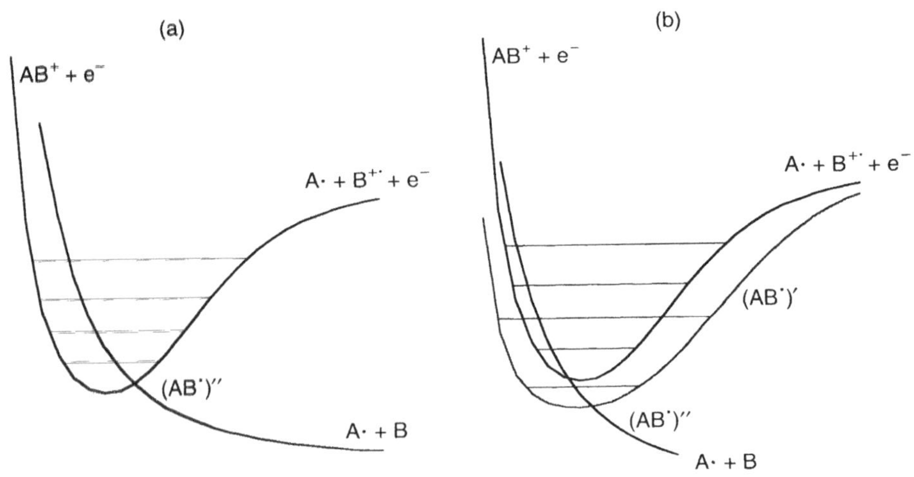

FIGURE 13.1. Dissociative recombination mechanisms: (a) direct; (b) indirect.

mechanism in which electron capture results in formation of a vibrationally excited Rydberg state (AB•)′ (Bardsley and Biondi, 1970). This state would subsequently decay by predissociation via a suitable intersecting neutral state (Figure 13.1b). The indirect process has a considerable influence on the recombination with electrons of polyatomic molecular ions.

DR is an efficient process, especially for complex molecular ions. Recombination rates for some cluster ions reach the value of $10^{-5}\,cm^3\,s^{-1}$ (Bardsley and Biondi 1970). Since bond breakage occurs prior to conversion and redistribution of the recombination energy (i.e. non-ergodically), quasiequilibrium-type theories such as RRKM cannot account for the DR branching ratios; the latter have to be determined experimentally.

In mass spectrometry, ion–electron recombination processes were extensively studied in the 1970s–early 1990s using the neutralization–reionization technique (Wesdemiotis and McLafferty 1984). Porter and coworkers have demonstrated that protonated ammonium NH_4^+ fragments on neutralization on a very short ($\ll 10^{-7}$s) timescale (Gellene and Porter 1984). In 1986, McLafferty made a remarkable prediction that one-electron neutralization of multiply charged proteins should produce nonergodic cleavage of a protonated peptide bond (McLafferty 1986). A decade later, N–C_α backbone bond cleavage in multiply protonated polypeptides was observed, first with α 193-nm UV laser as a source of secondary electrons (photoelectrons emitted presumably from metal surfaces) (Guan et al. 1996), and then with a filament-based electron source (Zubarev et al. 1998). These studies opened the era of electron capture dissociation (ECD), a phenomenon based on capture of low-energy electrons by multiply charged polypeptides (usually protonated):

$$[M + nH]^{n+} + e^- \rightarrow ([M + nH]^{(n-1)+\bullet})_{\text{transient}} \rightarrow \text{fragments} \qquad (13.1)$$

The nature of the unstable intermediate species $([M + nH]^{(n-1)+\bullet})_{\text{transient}}$ varies in different mechanisms suggested to explain ECD.

13.2. ELECTRON CAPTURE DISSOCIATION

13.2.1 Cross Section of Electron Capture

Despite the superficial similarity to DR, the ECD mechanism may or may not be similar to that acting in DR of small cations. One way to learn more about the mechanism is to study the electron capture cross sections and their dependence on the electron energy. Unfortunately, the instrumentation used in DR for very accurate cross-sectional measurements has low mass resolution and is not suitable for large molecules, while mass spectrometers equipped for ECD studies typically have low electron-energy resolution. This may explain why, although DR cross sections are often rich with fine features, all polypeptide ECD cross sections measured so far are smooth (no sharply resolved features). However, both DR and ECD cross sections demonstrate a strong increase with decrease in electron energy. Thus it is usually recommended that electron energy in ECD be maintained as low as possible, preferably below 0.2 eV (Zubarev et al. 2000a). Electron capture cross sections also increase with the ionic charge (Zubarev et al. 1998), reaching considerable values for highly protonated species. For instance, electron capture cross section of cytochrome $c+15$ ions measured at typical ECD conditions exceeded the ion-neutral collisional cross section by two orders of magnitude (Zubarev et al. 2000a). That means that the initial electron capture is primarily Coulombic, and the point-charge model could be useful for describing the process. In that model, the capture distance (Thomson radius) is proportional to the ionic charge n, with the cross section proportional to n^2. The experimental data directly testing this model are scarce, but seem to agree with this prediction (Zubarev et al. 2000a). Dependence of the capture cross section on other parameters has not been reported, although subtle dependences on the charge distribution, ionic composition, structure, and other properties can be expected. Deviations from the prediction of point-charge model are expected to be especially pronounced for larger proteins possessing secondary and higher-order structures in the gas phase.

The introduction of large electrostatic traps has brought the DR and ECD worlds closer together. DR of a small protonated peptide models dimethyl disulfide and 2,2,2-deuterosubstituted *N*-methylacetamide has shown a smooth, monotonically decreasing dependence of the cross section on the electron energy in the region below 0.3 eV (Figure 13.2), and the presence of several competing fragmentation channels following electron capture (Al-Khalili et al. 2004). In another DR study (Tanabe et al. 2003), the relative neutralization cross sections of full-size singly protonated peptides was studied for the energy range above 0.5 eV (Figure 13.3). In both these studies, the recombination cross sections of singly charged ions behaved similarly to the ECD cross section of multiply charged polypeptides (Kjeldsen et al. 2002). Despite these results, the question of the underlying relationship between DR and ECD remains open: on electron capture, multiply charged polypeptides may or

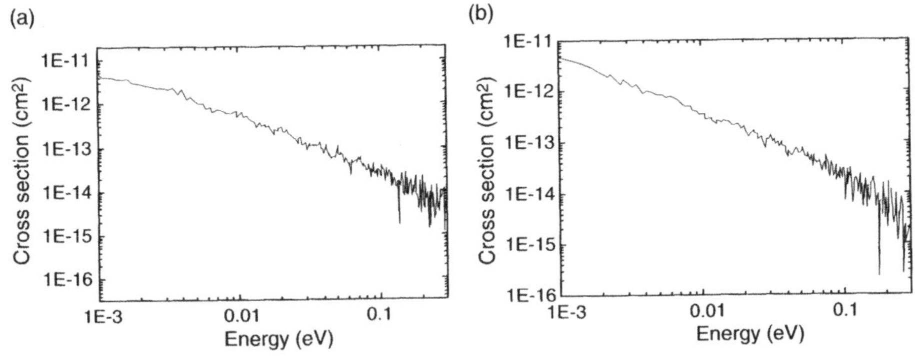

FIGURE 13.2. The measured DR cross section of protonated dimethyldisulfide (a) and protonated 2,2,2-D3 *N*-methylacetamide (b) for center-of-mass collision energies between 1 meV and 0.3 eV. [Reproduced from Al-Khalili et al. (2004) with permission.]

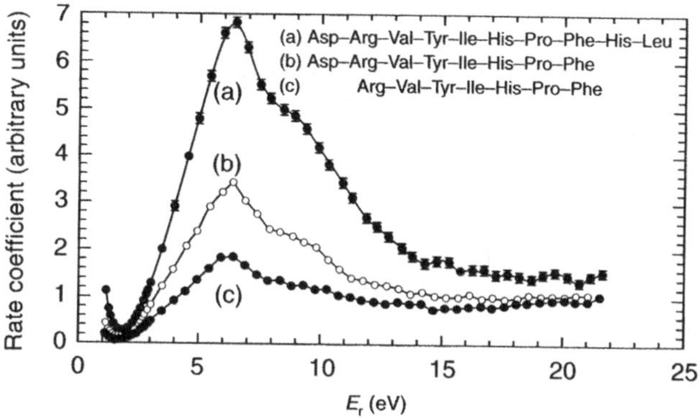

FIGURE 13.3. Neutral-particle production rate as a function of the relative energy in collisions of electrons with singly protonated angiotensin: (a) I; (b) II; and (c) III. [Reproduced from Tanabe et al. (2003) with permission.]

may not follow the same fragmentation pathway as small molecules. The difference is that, in most DR studies, the recombination energy is much greater than the strength of any covalent bond, and the number of degrees of freedom is small; thus rapid fragmentation on electron capture is guaranteed. In ECD of large polypeptides, recombination energy release is often much smaller than the average internal energy content; thus fragmentation is not assured. For instance, in a typical peptide with $N = 200$ atoms, ≈ 6 eV of the recombination energy is distributed over $3N - 6 \approx 600$ vibrational degrees of freedom, corresponding to an increment of just 10 meV per degree, compared to ≈ 27 meV that every degree contains at room temperature. For comparison, at $N = 3$ (typical for DR), each of $3N - 6 = 3$ degrees of freedom will on average receive 2 eV of vibrational energy.

The following section discusses the main fragmentation processes occurring in ECD. The nomenclature used to denote the type of fragment is conventional in mass spectrometry (Kjeldsen et al. 2002): the peptide backbone —$C_\alpha H(R_n)$—(CO)—NH—$C_\alpha H(R_{n+1})$—can be fragmented by cleaving three bonds, C—CO, C—N, and N—C_α. Each cleavage gives a *N*-terminal and a *C*-terminal fragments, which for the C—CO bonds are denoted *a* and *x*, for C—N *b* and *y* and for N—C_α bond *c* and *z* fragments. Moreover, the hydrogen atom transfer compared to homolytic cleavage is monitored. For instance, if the C—N bond cleavage were homolytic, it would result into *b*• and *y*• radical fragments. However, in practice the fragments are formed as if there were a hydrogen atom transfer from the *N*-terminal fragment to the *C*-terminal fragment (the actual cleavage mechanism is different), giving *b* and y' fragments. Here, the prime sign means H• transfer to the radical species; the absence of both radical and prime signs means loss of hydrogen atom from the radical species.

13.2.2. N—C_α-Bond Cleavage

Figure 13.4 presents the ECD mass spectrum of substance P dications, the ions frequently used as a test species for ion–electron reactions. The striking feature of

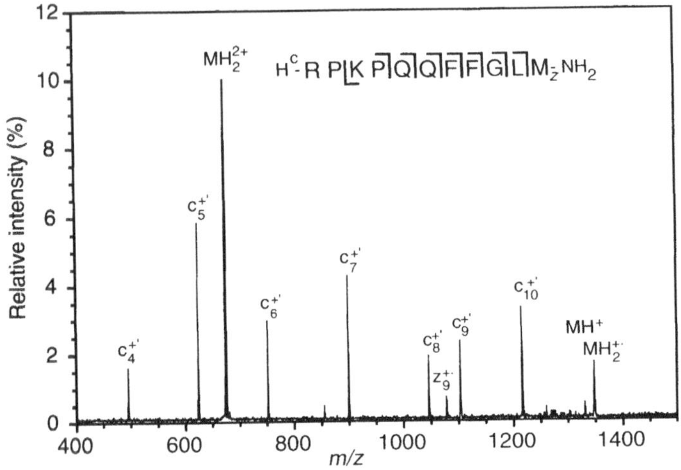

FIGURE 13.4. ECD mass spectrum of substance P +2 ions.

this spectrum, as well as any ECD mass spectrum of polypeptide polycations, is the absence of *b* and *y′* ions typical for collisional and infrared activation mass spectra, and the dominance of *c′* and *z*• fragments. Another major feature of ECD (not observed in Figure 13.4) is the prevalence of the N—C_α backbone bond cleavage even in the presence of much weaker bonds in the molecule, such as bonds to labile chemical groups or noncovalent bonding (Kelleher et al. 1999). The third ECD major feature is the disulfide bond rupture at faster rates than any given N—C_α bond, while disulfide bonds are usually considered strong (Zubarev et al. 1999).

The general explanation for these phenomena may be found in the fact that the fragmenting species in ECD are not even-electron ions as in other fragmentation techniques, but hydrogen-abundant radical cations ($([M + nH]^{(n-1)+\bullet})_{\text{transient}}$. Thus a radical-initiated reaction is a natural explanation for the N—C_α fragmentation. The earliest detailed explanation involved radical-site-induced cleavage following the capture of an electron on protonated carbonyl oxygen (Zubarev et al. 1998):

$$\mathrm{R_1{-}C(OH^+){-}NH{-}CHR_2 + e^- \rightarrow R_1{-}{}^{\bullet}C(OH){-}NH{-}CHR_2}$$
$$\rightarrow \mathrm{R_1{-}C(OH){=}NH}(c) + {}^{\bullet}\mathrm{CHR_2}(z\bullet) \qquad (13.2)$$

Mechanism (13.2) was postulated to be nonergodic, that is, occurring more rapidly than intramolecular vibrational energy conversion and redistribution on the recombination energy release (Zubarev et al. 1998). The nonergodicity of ECD is one of the most attractive features of the phenomenon. It is also the most frequently disputed one. A brief overview of the nonergodicity controversy is discuss below.

13.2.2.1. ECD Nonergodicity. A strong argument for the nonergodic mechanism is the temperature dependence in the 25–125°C range observed for ions fully

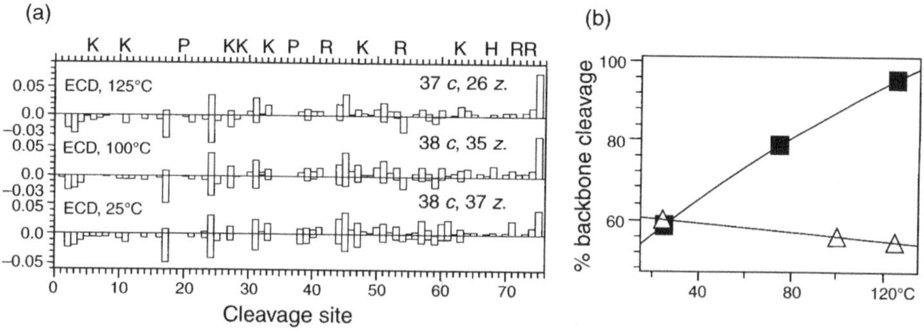

FIGURE 13.5. (a) Cleavage sites of ECD spectra of MH_{13}^{13+} ubiquitin ions as a function of ion cell temperature (*c* ions above line, *z*• below); (b) effect of temperature on the backbone fragmentation yield from subjecting ubiquitin +13 ions to IRMPD (squares); ECD (triangles). [Reproduced from Breuker et al. (2004) with permission.]

"unfolded" (with no or little tertiary structure while secondary structure is often present) in the gas phase (Breuker et al. 2004). The dependence of the *c*,*z* fragment yield on temperature is presented in Figure 13.5, showing the trend almost absent or even opposite that normally seen in vibrational excitation, such as infrared multiphoton dissociation (IRMPD). At freezing temperatures, the overall N—C_α cleavage efficiency diminishes (Mihalca et al. 2004), but the few remaining channels appear to have the same abundance as at room temperature as judged from the signal-to-noise ratios (SNRs) (Figure 13.6).

Generally speaking, a fragmentation pattern produced by any ergodic process can be replicated by any other activation method that increments the internal energy of the same precursors by the same amount, while a nonergodic fragmentation is difficult to reproduce by other means. As another argument for nonergodicity of

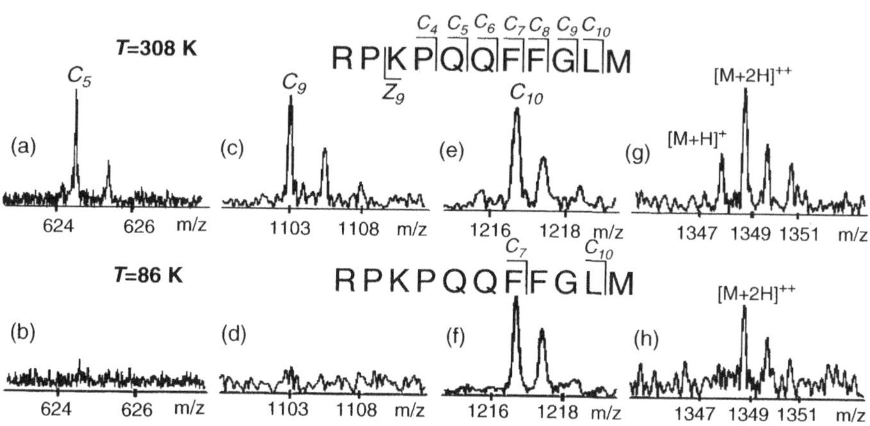

FIGURE 13.6. Partial ECD mass spectra of substance P +2 ions at different temperatures. [Reproduced from Mihalca et al. (2004) with permission.]

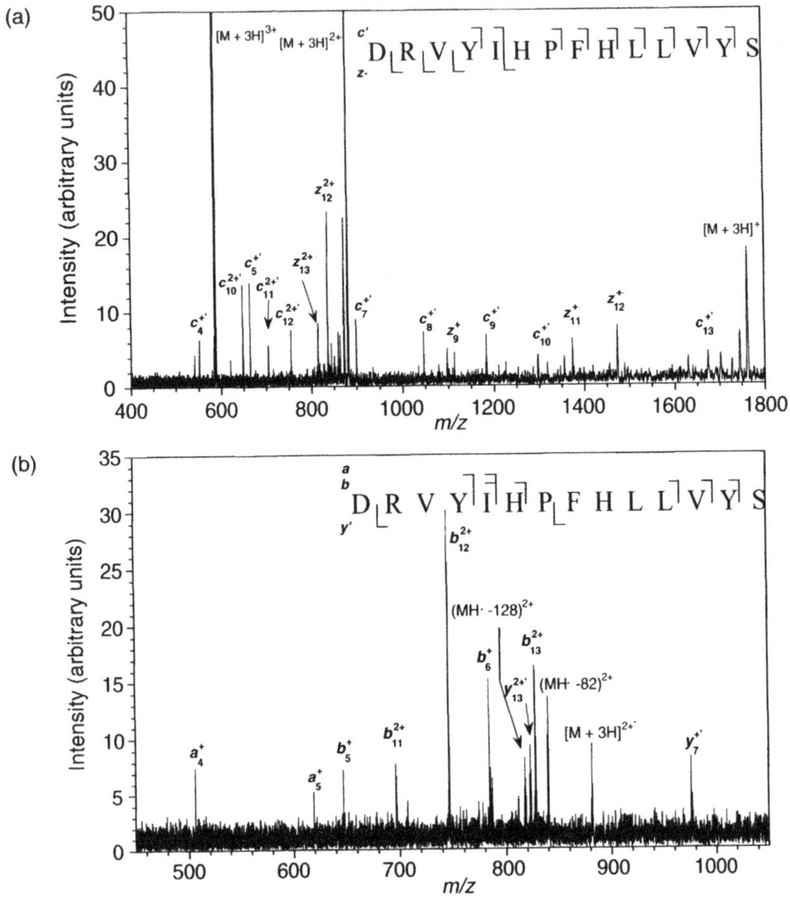

FIGURE 13.7. (a) ECD of +3 of renin substrate; (b) CAD of the reduced species $[M + 3H]^{2+\bullet}$.

ECD, various methods of activation have so far failed to yield the dominant N$-$C$_\alpha$ bond cleavage, although occasional *c,z* ions can be present in tandem mass spectra among a variety of other fragment ions. One could argue that this failure is due to the difficulty to prepare the same intermediates as in ECD, i.e. hydrogen-abundant radical cations, $R_1-\bullet C(OH)-NH-CHR_2$ in (13.2). These species (e.g., $MH_2^{+\bullet}$ ions in Figure 13.7) can be obtained in what appears to be nondissociative electron capture by even-electron precursors. Thus vibrational excitation of charge-reduced species can test the nonergodicity of ECD; if ECD is ergodic, such an excitation should yield predominantly *c* and *z* ions.

It turns out that the outcome of such an experiment depends on the charge state of the precursor radical cations and the size of the peptide. For relatively small polypeptide renin substrate, vibrational activation of $+\bullet 2$ ($[M + 3H]^{2+\bullet}$) reduced species obtained from +3 ions by electron capture produce, preferentially, loss of H•

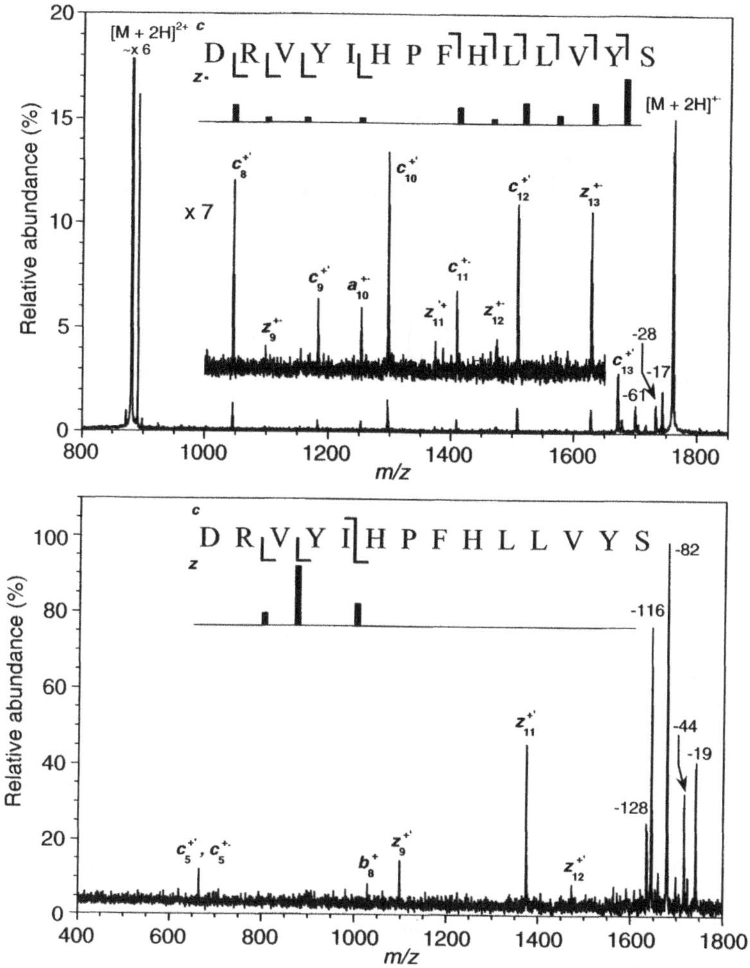

FIGURE 13.8. (a) ECD of +2 of renin substrate; (b) CAD of the reduced species $[M + 2H]^{+\bullet}$.

with subsequent C—N bond cleavage (*b* and *y′* ions in Figure 13.7), while +•1 reduced species yield abundant *c* and *z* ions (Figure 13.8). Both reduced species also yield losses of small radicals and amino acid side chains. Larger molecules, such as ubiquitin, behave in a similar manner. While hydrogen atom loss from the reduced +•11 ions of ubiquitin obtained through electron capture by +12 ions appears to be the lowest-energy channel with no *c*,*z* fragments observed (Breuker et al. 2004), for lower charge states of the reduced species of the same molecule (e.g., +•7 as in Figure 13.9), N—C_α bond cleavage is the dominant fragmentation channel (Zubarev et al. 2000). This latter N—C_α bond cleavage requires usually very lower excitation levels and often leads to unusual *c*•, *z′* ions. The relative abundances of *c*,*z* ions obtained through vibrational excitation of the reduced species are, however, very different from, even complementary to, the "direct" ECD cleavage abundances

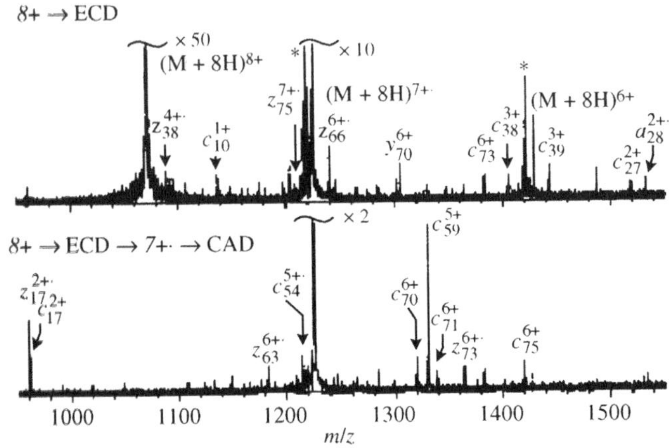

FIGURE 13.9. Top—ECD spectrum of bovine ubiquitin +8 ions; *bottom*—CAD spectrum of SWIFT-isolated, ECD-reduced $[M + 8H]^{7+\bullet}$ radical ions. [Reproduced from Zubarev et al. (2000a) with permission.]

(compare *c,z* ion abundances in the upper and lower panels of Figures 13.8 and 13.9). Thus the general conclusion is that ECD fragmentation patterns are not reproduced by vibrational excitation of charge-reduced radical species, which strengthens the non-ergodic hypothesis.

If N–C_α bond cleavage is nonergodic, then why do some *c,z* fragments still form in vibrational excitation of the reduced species? The nonergodic hypothesis explains this as follows. On electron capture, N–C_α bonds are cleaved, but the complementary *c* and *z* fragments do not always separate and may stay together in the form of the "intact" reduced species as a result of noncovalent bonding. Vibrational excitation applied to these "intact" reduced species destroys this bonding, and the fragments appear in the mass spectrum as individual species. Consistent with delayed fragment separation, many of the N–C_α bond cleavage products observed in vibrational excitation of "intact" reduced species are of the *c*•, *z*′ type, as opposed to the conventional *c*′, *z*• ECD fragments (delayed separation can lead to H• transfer from *c*′ to *z*• fragments; see discussion below).

To be fair, there is still room for alternative explanations to the observations mentioned above. The "reversed" ECD temperature dependence in Figure 13.5 could also be due to the role of neutral hydrogen bonding in N–C_α cleavages and not to ECD nonergodicity. Assuming that the N–C_α bond cleavage frequency relates to the probability of a given carbonyl to be involved in neutral hydrogen bonding (see discussion of the neutral H-bonding ECD mechanism below), one can expect that at higher temperatures, when neutral hydrogen network "melts," ECD cleavages are reduced. The absence of N–C_α bond cleavage in vibrational excitation of unfolded charge-reduced species, as in Figure 13.7, also has an alternative explanation. Perhaps the species resistant to dissociation on electron capture differ in structure from those that immediately fragment; for instance, they can be more stable products of isomerization. Vibration-induced *c,z* fragmentation of reduced species of folded, lower-charge-state ions could be due to ergodic fragmentation induced by an H• atom or another radical migrating under vibrational excitation to a "weak spot" to produce N—C_α bond cleavage. These and other considerations formed the prevailing opinion that, while some of the effects may be best explained by nonergodic cleavage, there is a contribution from slower processes.

13.2.3. ECD Mechanisms

All detailed mechanisms of ECD suggested so far can be divided into two groups, distinguished by the assumed main agent inducing the bond cleavages. The first group favors the electron itself, while the second one suggests an intermediate agent, such as the hydrogen atom or a radical site created by H• attachment. This classification is conditional; for instance, *intra*molecular transfer of a hydrogen atom can under certain conditions be viewed as a coordinated motion of two coupled but physically separated particles: a proton and an electron (Turecek 2003). The main difference between the two groups of mechanisms is that the electron action models assumes that charge neutralization occurs simultaneously with or even after the bond cleavage, while the second group of mechanisms implies prompt charge neutralization followed by a slower bond rupture.

It is currently difficult to name a single model that would account for all experimental observations. There is a growing understanding that several mechanisms are effective in ECD, and the observed mass spectra are the integral result of their competition. Most mechanistic studies focus, therefore, on specific ECD features, first of all on the N—C_α bond cleavage.

To begin understanding this cleavage, let us first consider its main features. The first feature is the role of the chemical context, which has been realized only relatively recently. Consider ECD of doubly protonated α-peptide $(Ahx)_6K$–OH, where Ahx is aminohexanoic acid (Figure 13.10). Note that only one N—C_α bond

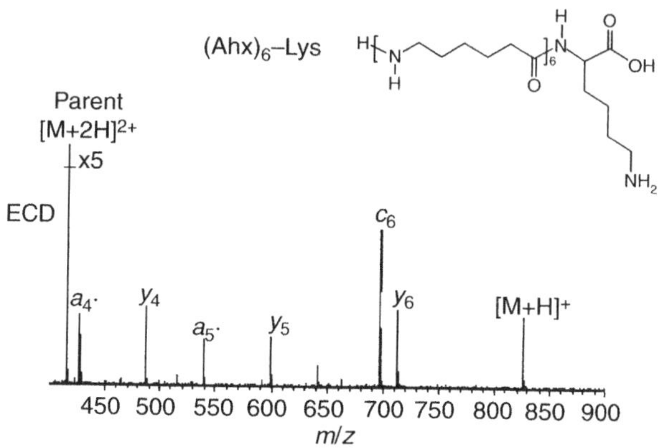

FIGURE 13.10. ECD mass spectrum of the ε-peptide $(Ahx)_6K$ (Ahx = aminohexanoic acid). *Inset*: Structure of $(Ahx)_6K$. [Reproduced from Cooper et al. (2004) with permission.]

cleavage is observed here despite the presence of six peptide bonds (Cooper et al. 2004). For the bonds between the ε-amino acid residues Ahx, the dominating fragmentation channel is C_α–(CO) cleavage ($a\bullet$, y' fragments). The conclusion is that the presence of an adjacent amino acid residue may be required for N–C_α bond cleavage. Note also an unusual for ECD feature, the absence of small losses such as the ubiquitous NH_3 loss (except the hydrogen atom loss) from the reduced species.

13.2.4. Fragment Abundances

Another important aspect of N–C_α bond cleavage is that the range of abundances for different bond cleavages is compressed compared to the abundances of C–N bond cleavages in vibrational excitation (Kruger et al. 1999). At the same time, these abundances are well reproduced and vary modestly with the ion temperature and electron energy (Breuker et al. 2004; Budnik et al. 2002). One of the most intriguing issues is what determines the N–C_α cleavage probabilities. A priori, the following factors may play a role: (1) proximity to the protonated site; (2) chemical nature of the neighboring amino acids, that is, local sequence; (3) local neutral hydrogen bonding, such as found in α helices and β sheets; (4) total charge state and global charge distribution; and (5) global amino acid sequence. The last parameter does not seem to be very important for peptides; as long as c,z ions from a particular N–C_α bond cleavage are observed, their relative abundances are not significantly affected by distant chemical groups (Budnik et al. 2002). Coinciding sequence stretches in otherwise different polypeptides show similar ECD fragmentation patterns. This is in stark contrast to vibrational excitation, where a small chemical modification can induce dramatic changes in the b,y fragmentation pattern (Budnik et al. 2002). Thus we can conclude that N–C_α cleavage is mostly a local phenomenon.

13.2.5. Charge Neutralization

The preceding conclusion seem to contradict another important observation, namely, the preference in ECD for neutralization of one particular charge in multiply charged polypeptides (Adams et al. 2004). Usually, the $(n-1)$ charges located at more basic sites remain intact. The preference for neutralization of the least basic site can be explained by adiabatic exothermicity of electron attachment to n-protonated ions. From the thermodynamic cycle in Scheme 13.1 it follows that

$$\mathrm{RE}(n+) \approx \mathrm{IE}(\mathrm{H}\bullet) - \mathrm{PA}([n-1]+) + \mathrm{HA}([n-1]+) \tag{13.3}$$

where PA is the proton affinity of the neutralized site, HA is its hydrogen atom affinity, and $\mathrm{IE}(\mathrm{H}\bullet) = 13.6\,\mathrm{eV}$ is the ionization energy of the hydrogen atom. Since $\mathrm{HA} \approx 0.6\,\mathrm{eV}$ for amide oxygen (Zubarev et al. 2002; Turecek 2003), $\mathrm{RE} \approx 14.2\,\mathrm{eV} - \mathrm{PA}$; thus, recombination of the least basic sites releases maximum energy, and thus electrons follow the most exothermic path.

Since arginine is the most intrinsically basic amino acid in the gas phase, neutralization of other protonated sites is often preferred when arginine is present.

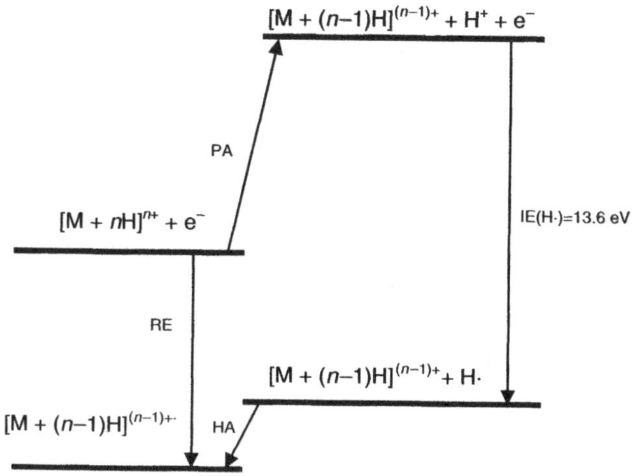

SCHEME 13.1. Thermodynamic cycle of ECD: PA—proton affinity of $[M + (n-1)H]^{(n-1)+}$; HA—hydrogen atom affinity of $[M + (n-1)H]^{(n-1)+}$; RE—recombination energy release; IE—ionization energy.

Thus, doubly charged tryptic peptides with a basic amino acid (either Lys or Arg) at the *C* terminus should, on electron capture, retain the charge on that terminus, preferentially producing *z*• ions. Indeed, in ECD of +2 ions of tryptic peptides, there is a 3 : 1 excess of *z*• ions over *c*′ ions, while in nontryptic polypeptides this ratio is close to 2 : 3 (Kjeldsen et al. 2006). In these peptides, neutralization of the least basic site accounts for >90% of the observed *c,z* ion abundances (Kjeldsen et al. 2006). While the charge at the least basic site is neutralized, the positions of the remaining charges can be determined. Figure 13.11 shows the charge distribution of ECD fragments of +3 ions of the smallest known protein Trp cage. The *c*′ ion series starts

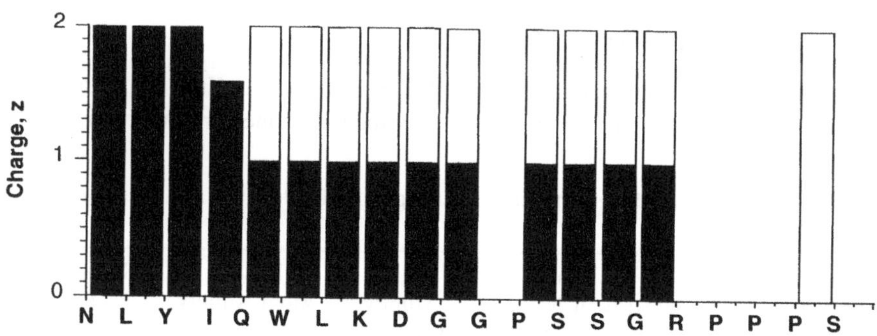

FIGURE 13.11. Distribution of average charge states of ECD fragments of +3 ions of native Trp cage: open columns—*c* ions; filled columns—*z* ions. The shifts in charge states of both types of fragments after Q (Gln_5) and R (Arg_{16}) singles them out as the protonation sites. The position of the third charge remains obscure. [Reproduced from Adams et al. (2004) with permission.]

after Gln_5, while the $z\bullet$ ion series changes the charge state from +1 to +2 past this residue. Similarly, the charge state shift is observed after Arg_{16}. Thus the two nonneutralized charges in the +3 Trp cage are located on Gln_5 and Arg_{16}. These are also charge locations in +2 of the Trp cage (Adams et al. 2004).

How does the electron know which charge's neutralization gives the highest recombination energy? Surely the electron has a means to "test" or "feel" all available charges. That means that charge neutralization in ECD is a global phenomenon; that is, it involves the whole molecule and not just immediate vicinity of the neutralized site. The contrast between the global nature of proton–electron recombination and the local nature of N—C_α bond cleavage is an argument in favor of mechanisms that decouple these two processes.

13.2.6. Cationized Molecules

Another issue important for understanding the N—C_α bond cleavage is the role of ionizing protons. Because of the tendency for gaining the highest possible recombination energy (thermochemical tendency), in mixed species $[M + n\text{Ct} + m\text{H}]^{(n+m)+}$, where Ct is a singly charged cation, protons are preferentially neutralized followed by N—C_α bond cleavage (Iavarone et al. 2004). This is because the recombination energy for proton neutralization is usually higher than that for cations (see Table 13.1). Williams et al. showed that ECD of fully cationized mixed $(M + Li + Cs)^{2+}$ ions there is a 10-fold preference for Li^+ neutralization over that of Cs^+ (Iavarone et al. 2004).

These and other similar data may be interpreted in such a way that N—C_α bond cleavage does not require a presence of a proton. But in the absence of apparent protonation, one of the N—C_α cleavage products (usually the *c* ion) still contains an extra hydrogen atom, suggesting that hydrogen atom transfer takes place, and that hydrogen atoms play some role in ECD of cationized molecules.

Now that the essential features of N—C_α bond cleavages have been reviewed, one can consider various mechanisms proposed to explain this cleavage in ECD.

13.2.7. Hot-Hydrogen-Atom Model

Mechanism (13.2) proposed in the first ECD publication (Zubarev et al. 1998) assumes that the ionizing proton is solvated by the backbone carbonyl oxygens, thus

TABLE 13.1. Calculated Recombination Energies of Protonated and Cationized Glycine

Cationizing Agent	Ionization Energy of Neutral Atom (kcal/mol)	Cation Affinity (kcal/mol)	Neutral Atom Affinity (kcal/mol)	Recombination Energy (kcal/mol)
H^+	314	212	14	116
Li^+	124	51	12	85
H^+	90	21	≈3	72

Source: Iavarone et al. (2004).

linking the charge solvation site and the place of N—C_α cleavage. This model has been amended to accommodate S—S bond cleavage results (see text below) in favor of distant charge solvation and, following neutralization by electrons, migration of a hot H• in search of the site with the highest H• affinity (Zubarev et al. 1999). This hot-hydrogen-atom model experienced a number of difficulties. One was the low recoil energies of H• observed in DR of protonated clusters (Gellene and Porter 1984), far below the level required by the model to account for a preferential rupture of a distant N—C_α bond. Another difficulty was the failure to obtain N—C_α bond cleavage, and even as much as H• capture, in irradiation of gas-phase peptide cations by $\approx$1 eV H atoms (Demirev 2000). Finally, high-level molecular dynamics simulations (MDSs) failed to produce N—C_α bond cleavage by bombardment in silico of amide-containing model molecules with H• atoms moving at different energies (Bakken et al. 2004). Extensive high-level single-point energy calculations revealed that, while adiabatic capture of H• atom by carbonyl is 60–80 kJ/mol exothermic, the H atom must overcome an $\approx$50 kJ/mol energy barrier (Turecek 2003). The H• atom has also to find a way to dissipate the excess energy quickly enough to avoid bouncing off the oxygen atom—the most frequent outcome observed in MDS.

Instead of N—C_α bond cleavage, the high-level MDS produced H• capture by the C_α atom followed by formation and loss of an H_2 molecule (Bakken et al. 2004). This process has not been observed experimentally, probably because of the low cross section. Other calculations showed a relatively high probability of another process: addition of H• to carbonyl carbon (Syrstad and Turecek 2005). This process has activation enthalpy and transition state energy similar to those of the H addition to the carbonyl oxygen, but addition to carbon should lead to a different fragmentation (*a*•, *y*′), which in experiments exhibit a much lower rate of formation compared to N—C_α bond cleavage. Thus, randomly traveling H• atoms in the reduced molecular species can be ruled out as the dominant agents inducing N—C_α bond cleavage. On the other hand, if H• specifically targets carbonyl oxygens, then aminoketyl radicals will be formed, as in (13.2). These radicals are labile with respect to N—C_α bond cleavage, which is either thermoneutral or slightly endothermic (or exothermic, depending on the level of theory and the model molecule) and requires overcoming an energy barrier (Bakken et al. 2004; Syrstad and Turecek 2005). Different levels of theory suggest different barrier heights, ranging between 30 and 120 kJ/mol (Bakken et al. 2004). The upper estimates are too high for fast cleavage, but they are less probable because are usually obtained at lower theory levels. In high-level MDS, addition of a hydrogen atom to either carbonyl oxygen of the molecule $HCONHCH_2CONH_2$ resulted in a very fast (0.2–0.4 ps) N—C_α bond cleavage (Bakken et al. 2004). The quest is therefore to explain the specific attachment of H• to the carbonyl oxygen.

13.2.8. Charge–Solvation Model

Proton sharing (ionic hydrogen bond) with the backbone carbonyls provides a natural way for the proton transfer to carbonyl oxygens. Indeed, high-level MDS showed N—C_α bond cleavage on a picosecond timescale in a small-peptide model

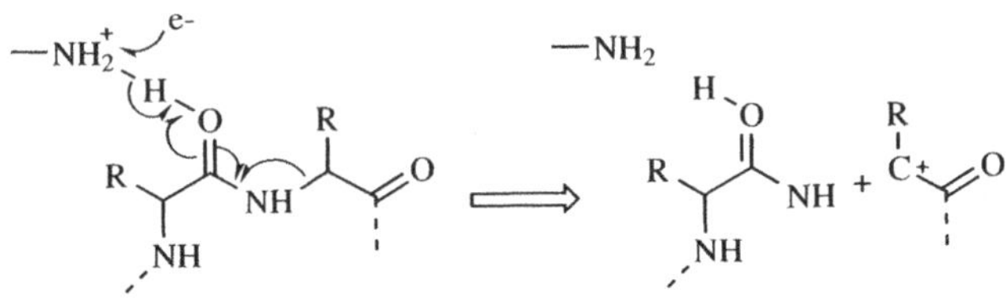

SCHEME 13.2. Charge–solvation model of ECD. [From Leymarie et al. (2003).]

following neutralization of the carbonyl-solvated charge (Bakken et al. 2004). These results supported the following view that has become quite popular (Scheme 13.2). The ionizing protons are located on the most basic sites, which are the sidechains of the arginine, lysine, and histidine residues. In the absence of these, the *N* terminus is protonated as well as less basic amino acids, such as glutamine and tryptophan. Despite Coulombic repulsion with other charges, the protonated sidechains are not pointing outward but instead solvated on the backbone carbonyls (Zubarev et al. 1998). The incoming electron, after being captured at a Rydberg orbit, loses its energy into vibrational and electronic excitation of the ion (the latter rapidly intercoverts to the former), after which it attaches to one of the charges. Charge neutralization leads to a hydrogen-abundant hypervalent group that returns to normal valence by transferring an H· atom to one of the carbonyls participating in charge solvation. The formed aminoketyl radical rapidly dissociates through N–C_α bond cleavage.

The appealing features of the abovementioned model are that it operates with energetically most favorable processes, provides a clear picture in line with mainstream views on gas-phase polypeptides, and possesses formidable predictive power. The charge solvation mechanism predicts that the ECD backbone cleavage abundances reflect the backbone charge solvation pattern. As soon as forcefield MDS became efficient enough, gas-phase structures of multiply charged polypeptides were simulated, and the probabilities of charge solvation on backbone carbonyls were determined (Polfer et al. 2005). The latter, however, agreed poorly with the observed abundances of *c*,*z* ions. The difficulty could be with the choice of model polypeptides. Indeed, even small peptides can acquire so many conformations in the gas phase that statistically valid sampling is very difficult given the existing constraints on simulation time. Moreover, it turned out that in many cases the positions of protons cannot be predicted from intrinsic basicities of individual residues (Schnier et al. 1995). The basicity of each residue must be calculated taking into account both secondary and tertiary structures, as well as the positions of other protons. However, even with well-characterized model polypeptides, little correlation was found between the charge solvation pattern and the N–C_α bond cleavage frequencies. Significant inconsistencies exist between the behavior of the charge solvation pattern and that of the cleavage frequencies of N–C_α bonds (Patriksson et al. 2006). For instance, the pattern of charge solvation is very temperature-sensitive, while experimental abundances of *c* and *z* fragments of "unfolded" ions do not change much with temperature.

13.2.9. Electron Action Mechanisms

A prominent alternative to the charge solvation model is the DR-type fragmentation. This mechanism assumes electron capture either directly into a dissociative electronic state ($AB^+ + e^- \rightarrow A + B\bullet$, direct DR mechanism; Figure 13.1a), or, more likely, on a high-lying, vibrationally excited Rydberg orbital of the peptide cation, followed by crossing to a dissociative state ($AB^+ + e^- \rightarrow AB' \rightarrow AB'' \rightarrow A + B\bullet$, indirect DR mechanism; Figure 13.1b). Alternatively, crossing can occur on a valence state, which can also lead to fragmentation provided sufficient energy is released (Breuker et al. 2004). Understanding of the N—C_α bond cleavage by this mechanism requires high-level quantum-mechanical (QM) calculations of potential energy curves, which are difficult to perform. Experimentally measured DR rates for polyatomic molecules are by orders of magnitude larger than those for small diatomic ions.

One parameter where direct and indirect DR mechanisms provide different predictions is the effect of electron temperature T_e. While direct mechanism predicts ${T_e}^{-1/2}$ dependence of the DR cross section, the indirect model gives ${T_e}^{-3/2}$ scaling. The latter seems to better fit the experimental data for large polypeptides, but not for small-model molecules in Figure 13.2, where the scaling is closer to T^{-1}.

The possible actions of both direct and indirect DR mechanisms are illustrated in Figure 13.12 (Sobczyk et al. 2005). If the incoming electron directly attaches to the σ^* orbital of the carbonyl, the N—C_α bond can be immediately cleaved (Figure 13.12a). This process however requires the electron to have >6 eV kinetic energy, at which the capture probability would be small. It is therefore more likely that the electron first enters the π^* orbital, forming an anion whose potential energy curve has a minimum rather than being purely repulsive (Figure 13.12b). The π^* anion undergoes electronic coupling with the σ^* state, creating an adiabatic surface with a barrier for dissociation (near the crossing point), which must be overcome or tunneled through for N—C_α cleavage to proceed.

The indirect process is also endothermic, albeit by a lesser amount, ~2.5 eV. Therefore the latest trend in the electron action models is to invoke proximity of the electron attachment site to the positive charge for lowering the barrier for dissociation and stabilizing the products (in the hot-hydrogen model, the effect of proximal charge is small). The indirect π^* mechanism becomes exothermic in the presence of strong electric field from which the electron can gain the required 2.5 eV. Since the Coulomb potential energy scales as ≈14.4 eV/r with distance r (Å), and the average distance between C_α atoms of adjacent amino acid residues is 3.63 Å, the positive charge must be present no more than one or two residues away from the cleavage site. Since the indirect π^* mechanism predicts N—C_α cleavages very close to the charge site, its predictions of the *c,z* fragment abundances are very similar to those of the charge solvation model discussed above.

The strong side of the electron action mechanism is the clear and calculable picture of a fast, nonergodic ECD mechanism. However, calculations are still very complex and imprecise.

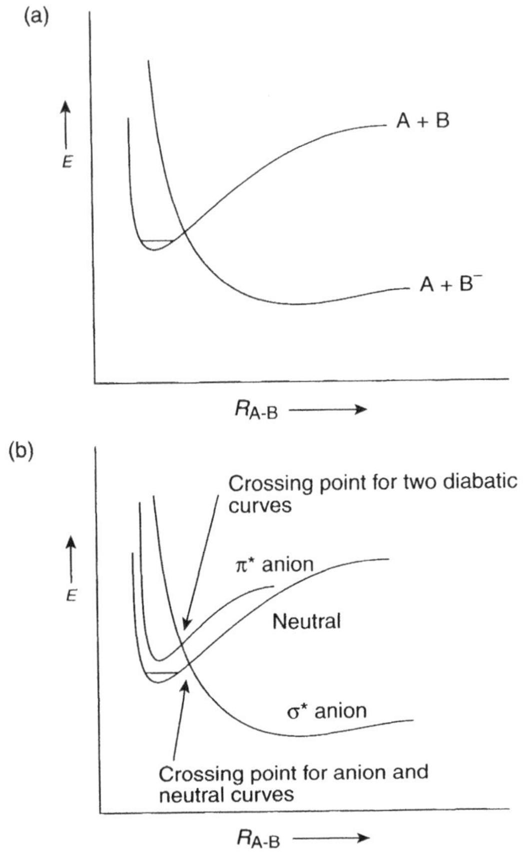

FIGURE 13.12. Neutral and σ* anion curves associated with direct DR (a) and neutral σ* and π* anion curves associated with indirect DR (b). [Reproduced from Sobczyk et al. (2005) with permission.]

Syrstad and Turecek suggested (Syrstad and Turecek 2005) a combination of mechanisms that includes features of both H-mediated and electron-induced models (Figure 13.13). In both cases, electron capture occurs by a cascade process where a number of electronic states are sampled consecutively. In the first (main) mechanism, the ground electronic state of the system is reached, the site with the highest recombination energy is neutralized and hypervalent species (an ammonium or guanidinium radical) are formed. From these hypervalent sites, especially from the ammonium group, a hydrogen atom can be transferred onto a proximate amide carbonyl in competition with hydrogen loss. The transfer is favored from hypervalent sites solvated before charge neutralization at amide carbonyl. In some cases, hydrogen addition can also occur at the carbonyl carbon atom leading to C_α—(CO) cleavage (*a*•, y′ ions). After H• attachment to the carbonyl, the formed aminoketyl radical undergoes N—C_α bond dissociation. The lifetime of the radical

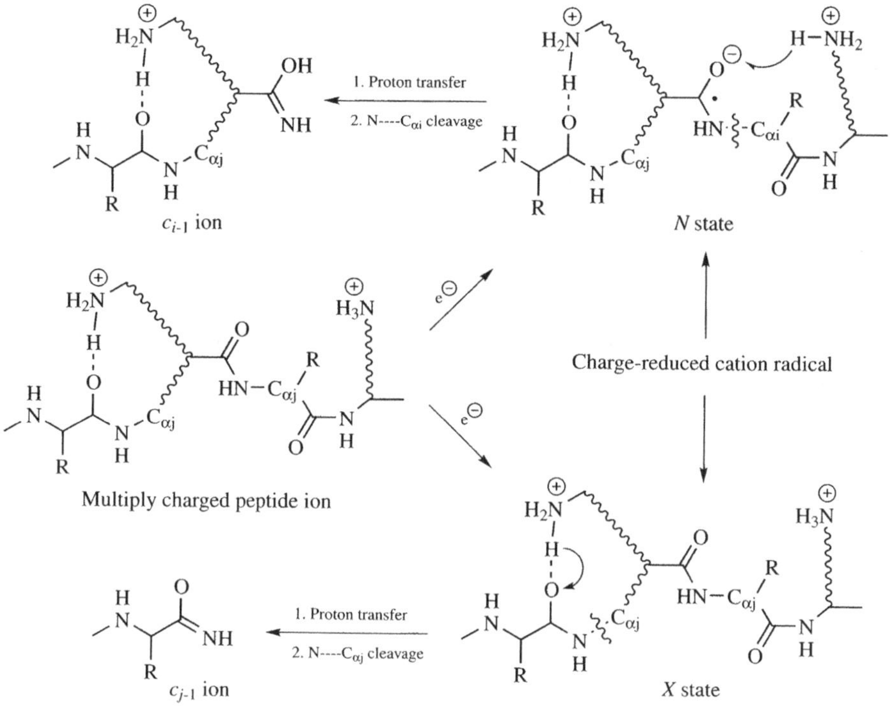

FIGURE 13.13. General mechanism for ECD in ground and excited electronic states of peptide ions. [Reproduced from Syrstad and Turecek (2005) with permission.]

before dissociation is statistically distributed and varies from picoseconds (non-ergodic cleavage) to microseconds and longer (ergodic dissociation).

In parallel with this hydrogen-atom mechanism another mechanism is effective, in which the electron is captured directly on the amide group forming a longlived anion in an excited electronic state (Syrstad and Turecek 2005), as in the model by Simons' group (Sobczyk et al. 2005). Direct N—C_α bond cleavage from such a state is, however, deemed unlikely because of its endothermicity, unless there is a proximal positive charge. But even in the latter case, the N—C_α bond dissociation and transition state energies in the charge-stabilized anions are 20–50 kJ/mol larger than for the corresponding hydrogen atom adduct. It is therefore argued that it is more likely that the amide anion, which is superbasic, abstracts exothermically a proton from an accessible site. The abstracted proton does not necessarily originate from a protonation site (ionizing proton) but from any suitable neutral site, for example, it can be a glycine α-proton. The proton transfer may require a conformational change in the charge-reduced peptide ion. The change can be driven by attractive Coulombic proton–anion or dipole–anion interactions and take nanoseconds to microseconds for completion. Proton transfer onto the amide carbonyl in an excited electronic state produces the same fragile aminoketyl radical as in the hydrogen-atom mechanism, and, driven by the huge exothermicity of anion–proton recombination, facilitates cleavage of the adjacent N—C$_\alpha$ bond (Syrstad and Turecek 2005).

One of the attractive features of this amide-based mechanism is that it does not require the ionizing proton be directly involved, and thus can explain N—C$_\alpha$ bond cleavage in peptide ions with ionizing particles other than protons. The model also addresses the issue of relatively low specificity of N—C$_\alpha$ bond cleavage. The difficulties are those faced by every electron-based mechanism, for example, the need to explain the preference of endothermic electron attachment to a neutral site over exothermic direct charge neutralization. Quantitative predictions of N—C$_\alpha$ bond cleavage frequencies are also difficult, and the reasons for the assumed higher proton mobility compared to the mobility of the anionic site (electron is bound to its site much more loosely than the proton) remain unclear.

13.2.10. H-Bond Mechanism

In this mechanism (Scheme 13.3), the electron is captured on the group involved in *neutral* hydrogen bonding between a backbone nitrogen atom and another backbone carbonyl, as found in α helices and β sheets (Patriksson et al. 2006). Following electron capture by the N—H···O=C group, an anionic site is formed at the nitrogen atom, and a hydrogen atom simultaneously is transferred to the carbonyl forming labile aminoketyl radical:

$$\text{N—H}\cdots\text{O=C} + e^- \rightarrow \text{N}^- - +{\bullet}\text{C(OH)} \qquad (13.4)$$

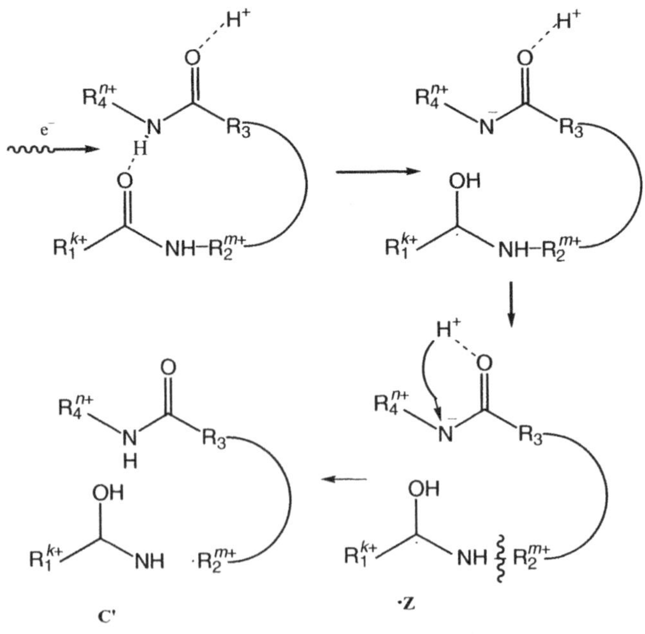

SCHEME 13.3. Neutral hydrogen-bonding mechanism of ECD.

Following this formula, the N—C_α bond in the aminoketyl radical fragments as in the "classical" hydrogen-atom model, and the most loosely bound ionizing proton neutralizes the amide anion, either by a series of proton jumps along the backbone, or by a "throughspace" jump.

This model postulates the necessity of a neutral hydrogen bonding for N—C_α bond cleavage. Otherwise it is compatible with the Simons–Turecek mechanism; reaction (13.4) can be viewed as an initial electron capture on an excited electronic state of carbonyl oxygen followed by proton transfer from the —NH group.

Any neutral hydrogen bond to a carbonyl that can give a stable anion is suitable, and not only the —N–H···O=C-type bonding. Capture on the neutral N–H···O=C hydrogen bond is somewhat more energetically favorable than electron capture to the electronically excited state of carbonyl. Indeed, reaction (13.4) can be presented as successive steps of (1) breakage of the H bonding [≈ -0.2 eV (Hagler et al. 1979)], (2) abstraction of H^+ from —NH— [-15.5 eV (Kjeldsen et al. 2005)], (3) recombination of H^+ and e^- to H• ($+13.6$ eV), and (4) H• attachment to the carbonyl oxygen ($+0.6$ eV). The balance is -1.5 eV, which renders reaction (13.4) 1 eV less endothermic than capture on the π^* C=O orbital. Coulombic attraction to a nearby positive charge assists (13.4) to the same degree as in other electron-action mechanisms.

The —N^--anion is a longlived even-electron entity in the ground electronic state. Note that in Scheme 13.3, the energy release is caused by anion–proton recombination and results in *c,z* fragment separation. This process can be delayed in time compared to the endothermic N—C_α bond cleavage. In the metastable zwitterionic charge-reduced species $[M + nH]^{(n-1)+\bullet}$, intramolecular charge recombination and subsequent *c,z* fragment separation can be triggered by slight vibrational excitation, as observed for ubiquitin +•7 and +•6 ions (Zubarev et al. 2000).

A question faced by any mechanism avoiding direct charge neutralization is "Where does the electron get the required energy (in this case, 1.5 eV)?" The answer is, "from the potential energy of the electrostatic field." In multiply protonated peptides, every new charge increases this energy by on average $\approx$1.1 eV (Budnik et al. 2002a). Thus the kinetic energy gained by an electron resting at infinity and attracted by a doubly protonated peptide is $\approx$2.2 eV. Given the average nature of this figure, more than half of the peptide's N—C_α bonds will be accessible for cleavage with zero-energy electrons. This result is in qualitative agreement with the observed cleavage efficiency for doubly charged peptides.

The role of the neutral hydrogen bonding needs further discussion. On one hand, the presence of multiple H bonding in the molecule prevents fragment separation, thus inhibiting observation of N—C_α bond cleavage. This feature has been used to derive "melting temperature" data on different charge states of gas-phase ubiquitin (Breuker et al. 2002). Furthermore, McLafferty et al. found that the maximum frequency of the N—C_α bond cleavage in α-helical ubiquitin +13 ions is shifted by -(3–4) residues from the assumed protonation sites (Breuker et al. 2004). This distance roughly corresponds to one turn of the α helix, and is consistent with the mechanism in Scheme 13.3—electron capture on the amide nitrogen close to a charged site should result in H• transfer to a carbonyl 3–4 residues toward the

N terminus. Finally, recent force-field MDS on gas-phase dications of the Trp cage protein yielded good correlation of N–C_α bond cleavage abundances with frequencies sufficient to establish single neutral hydrogen bonds for the corresponding carbonyls (Patriksson et al. 2006).

Is the mechanism in Scheme 13.3 nonergodic? The first steps—electron capture and hydrogen atom transfer to the carbonyl—can be very fast, and the fragmentation of the formed aminoketyl radical (without fragment separation) may be a part of a concerted process as a way to dissipate the exothermicity of the hydrogen atom transfer. Because it is nearly thermoneutral, the capture-transfer-cleavage process may also be resonant. Because of the slow anion-proton recombination, fragment separation may be delayed, and other processes may take place before it is complete, such as hydrogen atom transfer and proton transfer between the fragments (see text below).

The abovementioned sensitivity of relative abundances of *c,z* ions to chiral substitution of individual amino acids in polypeptidesh can be used for stereoisomer quantification in a mixture (Figure 13.14) (Adams and Zubarev 2005). This sensitivity is much greater than in vibrational excitation. Given that the chemical properties of D residues are largely the same as L residues, the most logical assumption is that the changes in the H-bonding network are responsible for this effect. Consistent with this suggestion, most changes in N–C_α bond cleavages

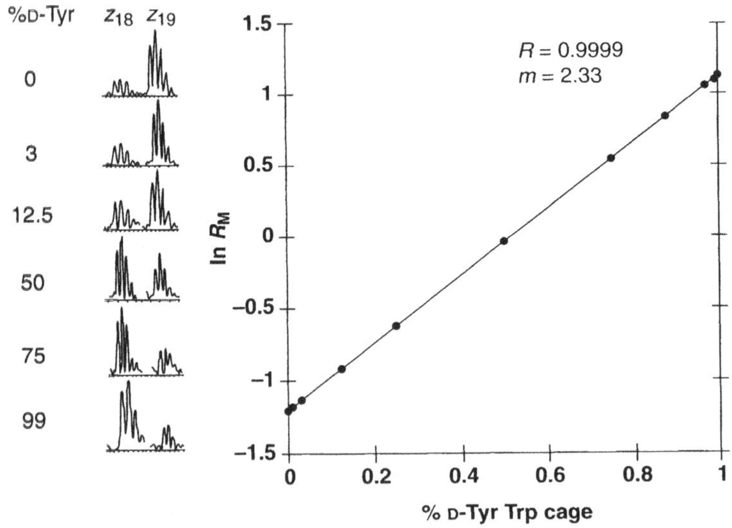

FIGURE 13.14. Shift in z_{18} and z_{19} fragment abundances as a function of D-Tyr_3 content in a stereoisomer mixture (left), a property that is quantifiable as seen in the plot (right). An equation $\ln RM = A + m \cdot r$, where r is the molar content of the D form, $m = \ln(R_{\text{chiral}})$, RM = ratio of the abundances of z_{18} and z_{19} fragments, was fitted to the experimental data points. The detection limit for the admixture of D-Tyr_3 stereoisomers to all-L form is ~1%. [Reproduced from Adams and Zubarev (2005) with permission.]

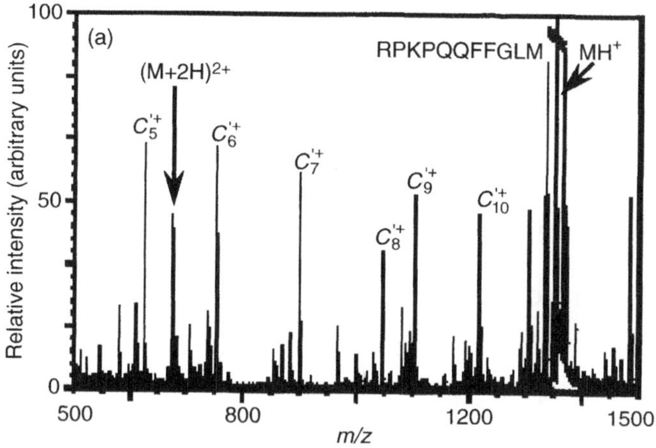

FIGURE 13.15. MALDI ISD mass spectrum of substance P. [Reproduced from Kocher et al. (2005) with permission.]

abundances are observed near the substituted sites (Adams and Zubarev 2005). MDS of neutral hydrogen networks reproduce not only the relative abundances of N—C_α bond cleavage, but also the main features induced by L→D stereoisomeric substitution in Trp cage, such as the larger change in the z_{18}/z_{19} ratio upon D-Tyr_3 substitution (Patriksson et al. 2006).

Additional arguments in favor of the H-bonding ECD mechanism can be found in other processes observed in mass spectrometry. In-source decay (ISD) fragmentation in MALDI (Brown and Lennon 1995) mimics ECD in terms of producing extensive c' ion series. For instance, Figure 13.15 shows an ISD mass spectrum of substance P, which demonstrates a series of c' ions resembling that in Figure 13.4. Much more than in ECD, traditional b,y' ion series are present in ISD and can even predominate in the mass spectra. On the other hand, similarity between ECD and ISD mass spectra may be rather strong: for example, the orders of ion abundances (largest for c_5 and smallest for c_8) are similar in Figures 13.4 and 13.15. The N—C_α bond cleavage in ISD turned out to be due to hydrogen atoms attacking backbone carbonyls (Kocher et al. 2005). Since hydrogen-atom irradiation of gas-phase even-electron peptide cations failed to produce N—C_α bond cleavage (Demirev 2000), the proximity of the hydrogen-atom donor to the backbone carbonyl may be a necessary condition for N—C_α cleavage. In ISD, such a donor may be an electronically excited matrix molecule (Scheme 13.4). Showing the possibility of N—C_α bond cleavage without electrons, ISD lends indirect support to hydrogen-atom models of ECD.

13.2.11. Other Backbone Cleavages

As already mentioned, $a\bullet$ and y' fragments are sometimes observed in ECD mass spectra (Zubarev et al. 1998). The C_α—C cleavage may be explained by several alternative mechanisms, each assuming an intermediate step involving y' fragments

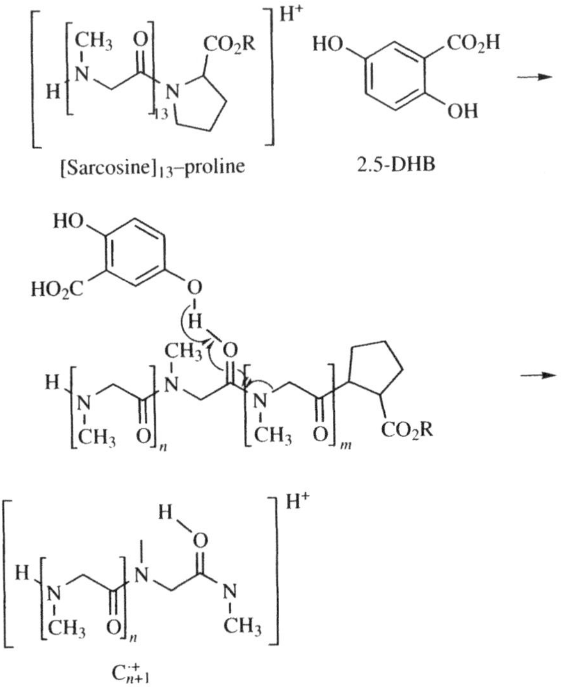

SCHEME 13.4. Mechanism of ISD of the peptide [sarcosine]13–proline. [From Kocher et al. (2005).]

and complementary, unstable *b*• species that further dissociate to *a*• ions via CO loss (this could also be a concerted reaction):

$$\mathrm{RCH{-\!\!-}{\bullet}CO}(b\bullet)+\mathrm{NH_2{-\!\!-}CHR}(y')\rightarrow \mathrm{R(C\bullet)H}(a\bullet)+\mathrm{CO}+\mathrm{NH_2{-\!\!-}CHR}(y') \quad (13.5)$$

The first suggested mechanism was H• attack on the backbone nitrogen followed by amide bond cleavage induced by the $-{\bullet}NH_2-$ hypervalent site (13.6):

$$\mathrm{RCH{-}CO{-}NH{-}CHR + H\bullet \rightarrow RCH{-}CO{-}({\bullet}NH_2){-}CHR \rightarrow RCH{-}{\bullet}CO + NH_2{-}CHR} \quad (13.6)$$

Reactive intermediates may also form via H· capture by the carbonyl carbon (Turecek and Syrstad 2003; Syrstad and Turecek 2005):

$$\mathrm{RCH{-}CO{-}NH{-}CHR + H\bullet \rightarrow RCH{-}({\bullet}CHO){-}NH{-}CHR \rightarrow RCH{-}{\bullet}CO + NH_2{-}CHR} \quad (13.7)$$

The *a*•,*y′* fragmentation may also be due to electron capture by a *b*,*y′* ion complex held together by weak bonding. In such a complex, a fraction of the *b* ions may retain acylium structure (Haselmann et al. 2000), the high recombination energy of which is the prime target for electron attachment:

$$\mathrm{RCH{-}CO^+ \cdots NH_2{-}CHR + e^- \rightarrow RCH{-}{\bullet}CO + NH_2{-}CHR} \qquad (13.8)$$

The CO loss, rarely observed in ECD mass spectra of intact polypeptides, is found to be abundant in ECD of multiply charged *b* ions, which is a way of distinguishing them from their *C*-terminal y' counterparts (Haselmann et al. 2000).

13.2.11.1. S—S-Bond Cleavage. The issue currently debated is whether S—S bonds and N—C_α bonds are cleaved by different mechanisms. This issue could be addressed experimentally by studying the behavior of both types of cleavages as a function of ionic charge state, temperature, and electron energy. If the mechanisms are different, a regime could be found where only one type of cleavage is taking place. Conversely, the absence of such regime would provide evidence in favor of the same mechanism. In the absence of decisive experimental data, several mechanisms are considered plausible.

The hydrogen-atom mechanism suggests that H• will attack any groups with a higher affinity for hydrogen atom than carbonyl groups, such as S—S bonds (Zubarev et al. 1999). An irreversible H• attachment leads to rapid bond dissociation

$$\mathrm{R_1{-}S{-}S{-}R_2H^+ + e^- \rightarrow R_1{-}S{\bullet} + HS{-}R_2,} \qquad (13.9)$$

with unstable R_1—S• fragments undergoing further reactions, such as small losses and backbone cleavage. The RS• radicals can also attach hydrogen or other small radicals, such as •CH_3 and •SH from the complementary fragment (Zubarev et al. 1999). The sulfide radical RS• can, for instance, abstract an H atom from an α carbon in a reaction that is exoergic by ~0.5 eV (Leymarie et al. 2003).

The RS• radical can also attack other disulfides in the molecule, leading to scrambling of the disulfide bonds if more than one is present in the molecule (Zubarev et al. 1999). These secondary processes are facilitated by vibrational excitation of the molecule before or after electron capture. Thus, capture of a single electron can lead to cleavage of more than one disulfide bridge, and for example lead to separation of A and B chains in the insulin molecule linked together by two bonds (Zubarev et al. 1999).

Similarly to N—C_α bonds, the proximity of a hydrogen atom is required for effective H• capture by the disulfide bond. Direct protonation of the disulfide bond is unlikely because of its low proton affinity. Charge solvation of, for instance, protonated lysine sidechain as in Figure 13.16 is more probable (Uggerud 2004). The potential energy diagrams in Figure 13.16 predict rapid dissociation of the disulfide bond. Another prediction that can be made is that of the simultaneous loss from the neutralized amine group of the Lys sidechain *together* with S—S bond cleavage from the reduced species $[M + nH]^{(n-1)+\bullet}$. Indeed, abundant •NH_2 loss from RS• product often appears in ECD mass spectra of disulfide-bound peptide dimers.

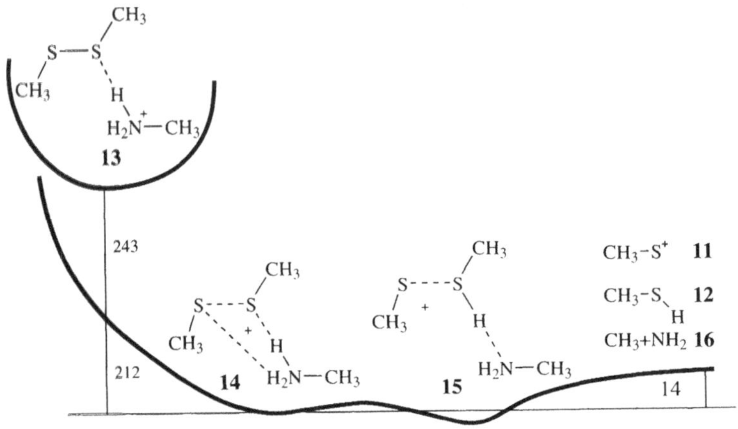

FIGURE 13.16. Potential energy diagrams from B3LYP/6-31G(*d*) calculations of a model constructed to represent a peptide with proton-solvating disulfide bridge. The upper curve shows the initial situation, while the lower curve shows the anticipated behavior on ECD. As indicated, the model predicts straightforward dissociation of the disulfide bridge. [Reproduced from Uggerud (2004) with permission.]

Capture of an electron to a high-*n* Rydberg state with subsequent electron-induced S—S bond cleavage (indirect DR) is an alternative to the hydrogen atom mechanism (Zubarev et al. 1999). Another suggestion is the charge-assisted mechanisms (Sawicka et al. 2003). In the absence of proximal charge, direct vertical dissociative attachment of an electron to an S—S σ^* orbital requires an 0.5 eV electron. At the same time, the presence of two positive charges closer than 40 Å, or a single charge closer to 20 Å to the S—S bond renders direct electron attachment exothermic. Figure 13.17 shows the potential energy diagrams of the charge-assisted dissociation of $MeSSCH_2CH_2{-}NH_3^+$ species as functions of the S—S bond length. While the Rydberg-attached species (curve II, open circles) has a potential minimum at ~2 Å and thus are metastable, the S—S σ^*-attached species (curve III, filled circles) show no minimum and dissociate within one bond vibration (Sawicka et al. 2003).

13.2.12 Small Losses

ECD with low-energy electrons favors cleavage of just one bond per captured electron. This is why *c,z* fragments are seldom observed from the cleavage *N*-terminal to the proline residue, in which nitrogen is tertiary (Zubarev et al. 1998). In rare cases, ($z\bullet$ − 2 Da) fragments are produced from such a cleavage, providing evidence of secondary radical-induced reactions (Cooper et al. 2003a). Such secondary, ergodic reactions are inevitable in ECD, which produces radicals in a vibrationally excited state. Most of the observed secondary reactions are dissociative. If the primary dissociation is a loss of a radical (e.g., H•), the remaining species is a stable even-electron ion, and further dissociation is likely to

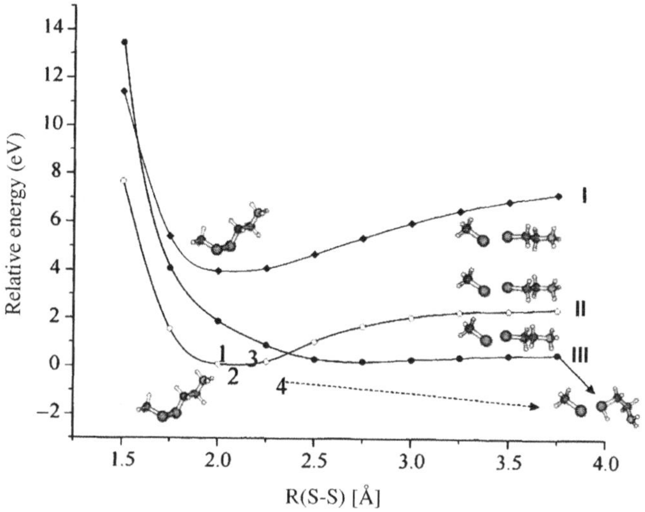

FIGURE 13.17. Energies (curve I, filled diamonds) of the precursor $MeSSCH_2CH_2$-NH_3+ species, (curve II, open circles) the Rydberg attached species, and (curve III, filled circles) the S—S σ^*-attached species, each as a function of the S—S bond length. [Reproduced from Anusiewicz et al. (2005) with permission.]

produce C—N backbone cleavage or losses of stable molecules (H_2O, NH_3, etc.) as in Figure 13.7. If, however, the group lost from the radical ion is a molecule, the radical site remains on the ion, leading to further radical-initiated cleavage [e.g., radical-initiated cascade (Leymarie et al. 2003)].

In many ECD mass spectra, the most abundant product is the reduced $(n-1)^+$ species, typically a mixture of $[M + nH]^{(n-1)+\bullet}$ and $[M + (n-1)H]^{(n-1)+}$ ions. The latter are formed by hydrogen atom loss from the charge-reduced molecular ion, the most trivial fragmentation channel in ECD. Below that ion on the *m/z* scale there is usually a cluster of ions corresponding to other small-molecule losses (<150 Da). Frequent small losses from the reduced species are listed in Table 13.2. Generally, these losses occur either from protonated site (Scheme 13.5), or from groups participating in charge solvation (Cooper et al. 2002; Haselmann et al. 2002), indicating that direct charge neutralization does occur in ECD with a high probability. Some of the losses have a rather strong diagnostic value, but caution is needed—the absence of a loss may mean that the amino acid is not participating in charge solvation rather than it is absent in the sequence.

Small losses can also follow an N—C_α bond cleavage. While even-electron c' ions fragment similarly to the amide forms of truncated *N*-terminal peptides to which they are isomeric, $z\bullet$ ions can lose H• and other small radicals [see Section 13.2.13 (below) and Kjeldsen et al. (2002)]. Hydrogen abstraction (both *intra*- and *inter*-molecular) by a radical fragment can also be a prominent process, especially for vibrationally excited ions with suppressed alternative fragmentation channels, as in cyclic peptides (Leymarie et al. 2003). In *intra*molecular hydrogen abstraction

TABLE 13.2. Losses of Small Groups from Reduced Species of Peptides in ECD

Lost Group	Mass (Da)	Source of Loss	Diagnostic Value
H·	1.0078	Protonated site	N/A
NH_3	17.027	*N* terminus; Lys, Arg	Low
H_2O	18.011	Ser, Thr	Moderate
CO	27.995	Acylium	Distinguishing *b* and *y′* ions
CH_4N_2	44.037	Arg	High
CH_3NO	45.022	Asn/Gln	Moderate–high
C_2H_6O	46.042	Thr	Moderate
C_2H_5NO	59.037	Asn/Gln	Moderate
CH_5N_3	59.048	Arg	Moderate–high
$C_2H_4O_2$	60.021	Asp	Moderate–high
$C_4H_{11}N$	73.089	Lys	Moderate
C_3H_6S	74.019	Met	Moderate
$C_4H_6N_2$	82.053	His	High
$C_4H_{11}N_3$	101.095	Arg	High
C_7H_8O	108.058	Tyr	High
C_9H_9N	131.074	Trp	High

Sources: Cooper et al. (2002), Haselmann et al. (2002a), Cooper et al. (2003a).

(Scheme 13.6), a radical site can propagate inside the molecule and cause a cleavage remote from the original radical site. Such a process may progress via α carbons and lead to scrambling of hydrogen atoms attached to these atoms. It is, however, unlikely that such a radical cascade could be responsible for the dominant N–C_α

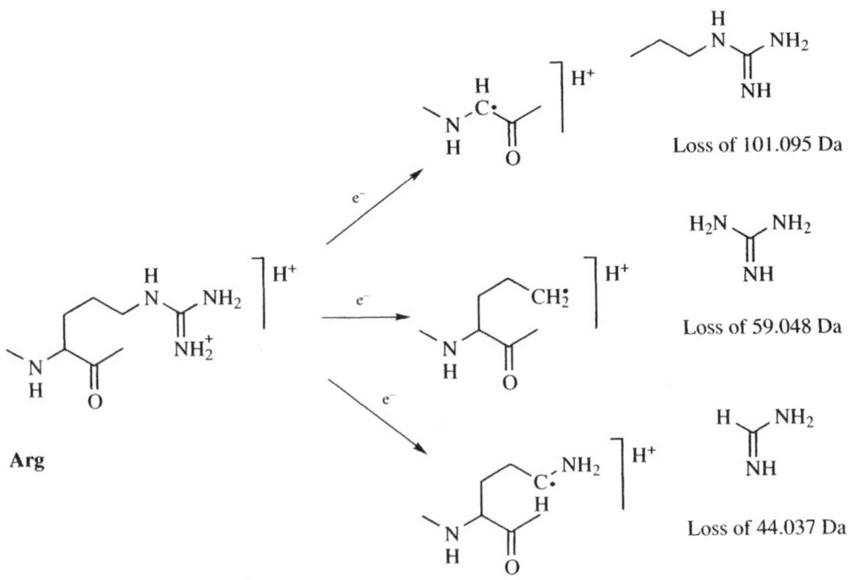

SCHEME 13.5. Mechanism of small losses from protonated Arg residue on electron capture. [From Cooper et al. (2002).]

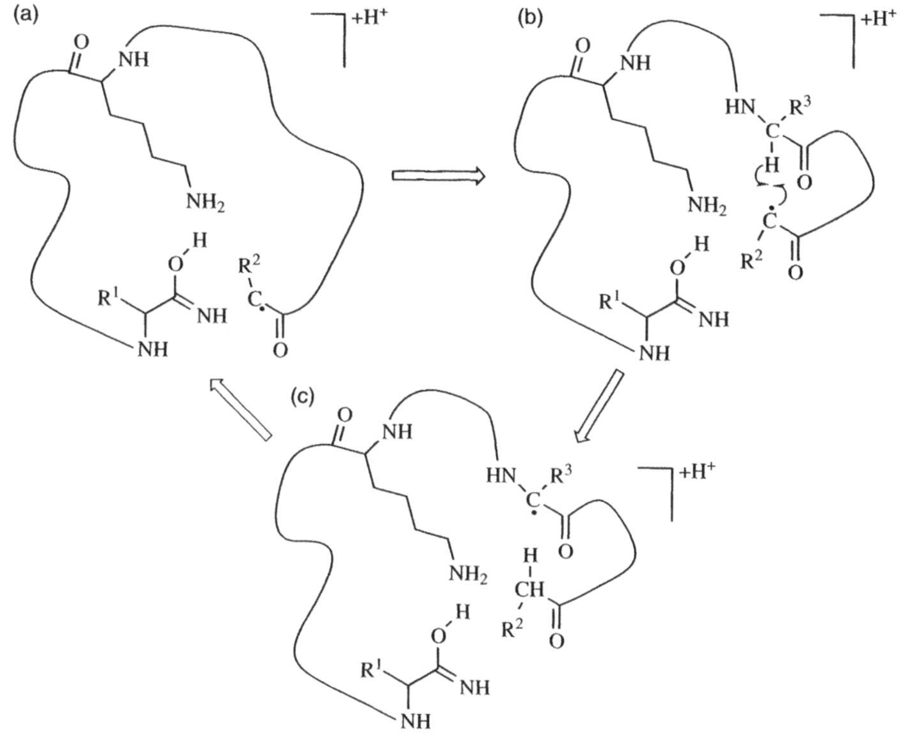

SCHEME 13.6. Propagation of α-carbon radical along the backbone by hydrogen abstraction (B→C). [From Leymarie et al. (2003).]

bond cleavage in ECD. If this were the case, then hydrogen-deficient radical cations $[M + nH]^{(n+1)+\bullet}$ that can be obtained by, for example, >10 eV electron irradiation of $[M + nH]^{n+}$ species, would also fragment preferentially via *c* and *z* ions. Experiments show that these species are rather stable and instead lose small groups and amino acid sidechains (Zubarev et al. 2000a). The same tendency is observed for fragmentation of hydrogen-deficient species $[M + (n-1)H]^{n+\bullet}$ obtained in UV photodissociation of $[M + nH]^{n+}$ species through abundant H• loss.

In the case of *inter*molecular abstraction, the radical fragment can abstract H• from the complementary even-electron fragment after the primary N—C_α bond cleavage and before the fragment separation (which may take a second or longer), which leads to $c\bullet,z'$ ions as opposed to more common in ECD $c',z\bullet$ species (Leymarie et al. 2003). The degree of hydrogen atom rearrangement prior to fragment separation, that is, the ratio between the "classical" $c',z\bullet$ and "rearranged" $c\bullet,z'$ channels, depends on the charge state of the precursor (higher for higher charge states), the cleavage site (higher for unfolded sequence regions), the temperature of the ions, and electron energy. Because these parameters are generally difficult to control, ±1 Da deviation in the mass from their expected values is sometimes observed for ECD fragments.

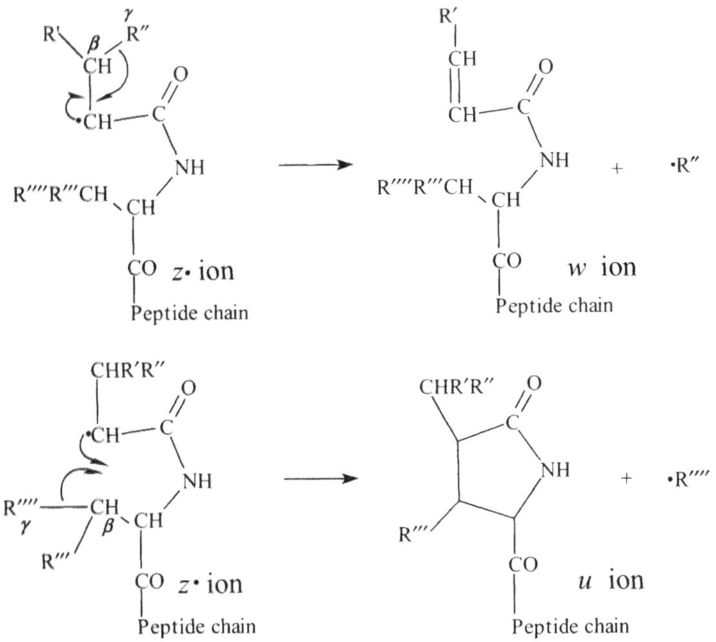

SCHEME 13.7. The mechanisms for *w* and *u* ion formation from the precursor $z^{\bullet}$ radical ions. [From Kjeldsen and Zubarev (2003).]

13.2.13. Hot ECD (HECD)

As already mentioned, HECD differs from ECD by the energy of the electrons. Typically, HECD regime requires >10 eV (Kjeldsen et al. 2002). At such energy, electronic excitation of the precursor ions simultaneously with electron capture becomes an important process. Relaxation of the electronic excitation occurs mostly through conversion into vibrational energy, which leads to secondary fragmentation of radical fragments. Secondary losses from the radical *z*• ions are most frequent in the form of either a hydrogen-atom loss, or partial loss of the sidechain to form even-electron *w* ions (Scheme 13.7). Similar loss from *a*• fragments gives *N*-terminal *d* species. Both *w* and *d* ions are very useful, as they distinguish between the isomeric Leu and Ile residues. With the terminal (closest to the radical site) Leu residue, the mass difference between *z*• and *w* fragments is 43.055 Da ($\bullet C_3H_7$), while with Ile it is 29.039 Da ($\bullet C_2H_5$). The loss can occur not from the terminal but from the neighboring (penultimate) residue instead, forming *u* ions (Kjeldsen and Zubarev 2003). If both terminal and penultimate residues are Leu or Ile, *w* and *u* ions can be differentiated on the basis of their abundances: *w* ions are at least 3 times more abundant.

Formation of *w* ions and other secondary losses may also occur at low electron energies (<1 eV), but becomes much more apparent above the electron energy of 7–10 eV (Figure 13.18). While the electron capture becomes less effective at these energies, the electron current extracted from the source usually greatly increases.

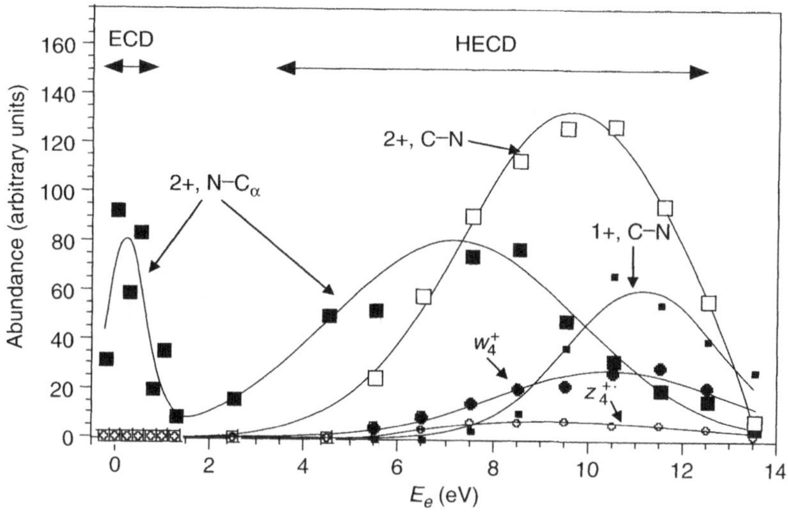

FIGURE 13.18. Fragment ion abundances versus electron energy E_e for 250 ms irradiation of peptide SRP molecular ions, +2: (filled squares) N—C$_\alpha$ bond cleavages, (open squares) C—N bond cleavages, (small open circles) $z_4^{+}\bullet$ fragments, (small filled circles) w_4^+ fragments; +1: (small filled squares) C–N bond cleavages. [Reproduced from Kjeldsen et al. (2002) with permission.]

The net effect can be an overall increase in the efficiency, so that the irradiation time has to be reduced to avoid excessive neutralization of the fragments.

13.2.14. Dissociation of Strong Bonds in Presence of Weak Bonding

The preservation of weak bonding while strong links, including N—C$_\alpha$ and S—S bonds, are preferentially cleaved is one of the most fascinating features of ECD. Even most labile modifications, such as γ-carboxylation and sulfation, and even noncovalent bonding, can be preserved in ECD (Haselmann et al. 2002a). ECD. For example, ECD of multiply charged dimers and higher- order noncovalent oligomers (specific as well as nonspecific) of polypeptides produces small losses as well as N—C$_\alpha$ fragmentation in one of the monomers, while the other one can remain intact and attached to the fragmenting counterpart (Figure 13.19).

Preferential cleavage of a strong bond while preserving much weaker bonding is a physicochemical challenge that can be achieved by two alternative approaches. One is to use ultrafast energy deposition close to the intended cleavage site in attempt to achieve nonergodic fragmentation. The other is to weaken the bond of interest and then apply a slight excess of energy anywhere in the molecule; in the ergodic dissociation process that will follow, the weakest bond will be preferentially cleaved. ECD seems to realize both these approaches—the H• or electron attachment to the backbone carbonyl significantly reduces the strength of the N—C$_\alpha$ bond, while the overall exothermicity of solvated proton recombination (4–7 eV) or, in any case, the deposition inside the ion of the potential energy of coulombic attraction between

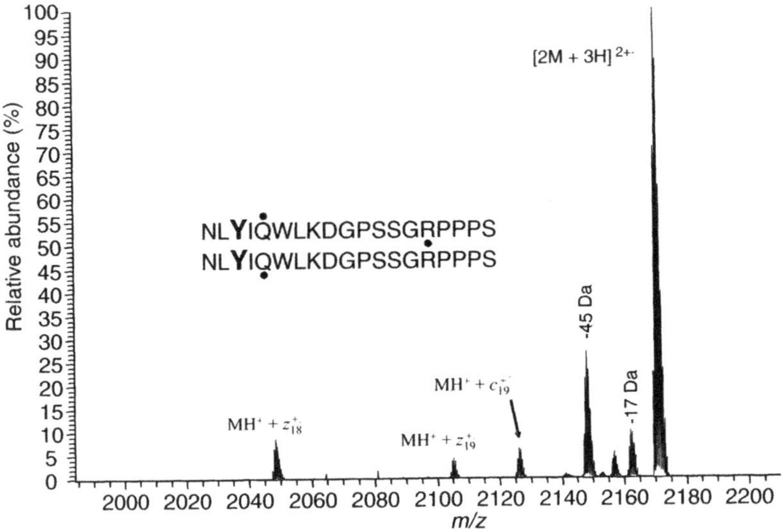

FIGURE 13.19. Part of ECD mass spectrum of +3 ions of noncovalent dimers of Trp cage. Note that the abundance ratio of ($MH^+ + z_{18}^{+\bullet}$) and ($MH^+ + z_{19}^{+\bullet}$) ions reflects that in ECD of +2 monomers (Figure 13.14).

the cation and an electron (~1.1 eV per ionic charge), provides a rapid increment in the internal energy. Both these features seem to be essential for the ECD effect.

13.2.15. Electron Transfer Dissociation

The ability of ECD to preserve weak bonding while cleaving certain strong bonds depends on the internal energy of the system and the electron energy. To obtain even "softer" ECD performance, the energy of both values need to be minimized as to reduce the energy excess that is released in form of vibrational excitation. The internal ion temperature can be lowered well below the freezing point with N–C_α bond cleavage still taking place (e.g., see Figure 13.6), and the electron kinetic energy can be reduced by employing anions instead of free electrons as in electron transfer dissociation (ETD) (Syka et al. 2004). The cation–anion recombination energy is given by

$$RE \approx 14.2\,\text{eV} - PA - EA \tag{13.10}$$

where EA is the electron binding energy in the anion. By a proper choice of the anion the exothermicity of ECD can in principle be made arbitrarily small. In practice, however, the electron affinity should not exceed 1.5–2.0 eV; otherwise the parasitic (nondissociative) proton transfer from the cation to the anion becomes dominant. This parasitic transfer is analogous in its outcome to charge neutralization in ECD with the subsequent H• loss from the reduced species.

Because it is "softer" than ECD, ETD is employed mainly for the analysis of labile posttranslational modifications. A range of anions have been tested, with fluoranthene anions most often used in ETD studies. The electron affinity is the major parameter determining the ETD properties of anions. Anions with high electron affinity cannot induce fragmentation of peptide dications. However, the problem seems to be mostly in fragment separation than in N—C_α bond cleavage; gentle vibrational activation of the reduced species improves the efficiency.

13.3. ELECTRON DETACHMENT DISSOCIATION (EDD)

Charge reduction in deprotonated polypeptides by collisions with >3 eV electrons gives rise to hydrogen-deficient radical anions $[M - nH]^{(n-1)-\bullet}$. These species possess different properties and fragment differently than hydrogen-abundant radical cations $[M + nH]^{(n-1)+\bullet}$ produced in ECD. The oxidized $(n-1)^{-\bullet}$ species tend to rapidly lose CO_2, but remain stable against backbone cleavage even under moderate vibrational excitation. However, if the electron energy exceeds ~10 eV, the probability of backbone cleavage becomes nonnegligible (Budnik et al. 2001). Early reports, based on a limited number of observations, found that *a* and *a*• ions were preferentially formed in this electron detachment dissociation (EDD). Now it is firmly established that EDD yields preferentially *x* and *a*• fragment ions (C_α—C backbone cleavage) with even-electron *x* fragments predominating over radical *a*• ions (Figure 13.20). The backbone cleavage is due to dissociation of —N•— backbone

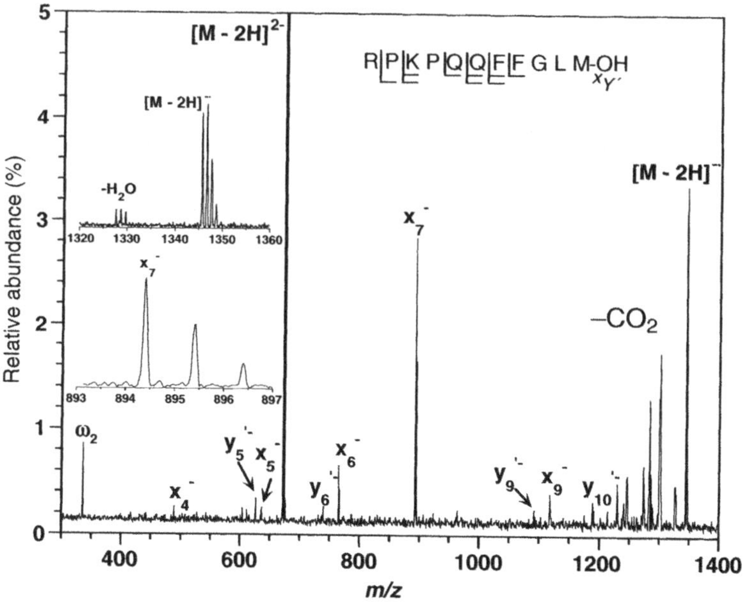

FIGURE 13.20. EDD of −2 ions of the acidic form of substance P. [Reproduced from Kjeldsen et al. (2005) with permission.]

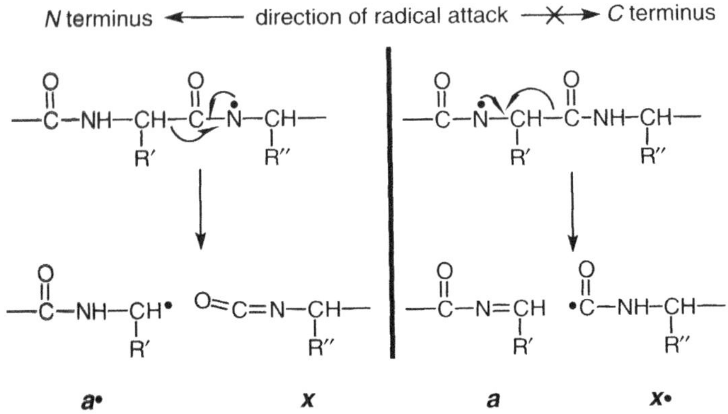

SCHEME 13.8. The mechanism of C_α–C bond cleavage in EDD. [From Kjeldsen et al. (2005).]

radicals (Scheme 13.8) that are formed as a result of electron ejection from $-N^{-}-$ anions. Because of the relatively high acidity of the amide proton (~71 kJ/mol more acidic that the α-carbon proton (Kjeldsen et al. 2005)], amide anions are frequently formed in deprotonated peptides, especially in peptides lacking carboxylic groups, such as the amide form of substance P (Kjeldsen et al. 2005). Ab initio calculations suggest a unidirectional mechanism for EDD (C_α–C cleavage always *N*-terminal to the radical site), favoring *a•*, *x* over *a*,*x•* fragmentation by 74 kJ/mol (Kjeldsen et al. 2005). As in ECD, backbone bonds *N*-terminal to proline residues remain in EDD immune to C_α–C cleavege.

13.4. INSTRUMENTAL REALIZATION

Currently, all manufacturers of Fourier transform ion cyclotron resonance (FTICR) mass spectrometers offer an ECD option with an indirectly heated dispenser cathode as a source of low-energy electrons (Tsybin et al. 2001). The higher electron current (up to 100 μA) of such a cathode traps assists trapping of the positive ions inside the beam that can be up to several millimeters broad, which increases the electron capture rate and the trapping efficiency for ECD fragments. Experience shows that the efficiency of ECD and the quality of the data strongly depend on the alignment of the electron beam in respect to the main axis of the instrument, as well as on the degree of overlap with the ion beam. Good-quality FTICR transients (time-domain signals) last longer, resulting in higher mass resolution and accuracy. When the electron beam is properly aligned, HECD is obtained with a 30–50% drop in efficiency compared to ECD, while in a misaligned one the HECD efficiency can be close to zero. With the dispenser cathode, ECD can be routinely used on LC scale [millisecond irradiation time (Tsybin et al. 2001)], with single-scan spectra of even doubly charged species yielding abundant sequence information.

Several groups have implemented ECD in radiofrequency devices, such as two- and tree-dimensional Paul traps, using one of the three approaches. The first approach is to align a source of free electrons with the main axis of the radiofrequency trap, and to inject the low-energy beam into the trap. Application of magnetic field parallel to the main instrumental axis as well electron injection during the zero-amplitude phase of the radiofrequency potential facilitate ECD (Baba et al. 2004; Silivra et al. 2005). An alternative to that is ETD (Syka et al. 2004), which is easier to implement on commercial instruments than ECD. ETD, however, has two analytical disadvantages compared to ion–electron reactions: (1), ETD efficiency for doubly charged precursors (the most important ionic species in analysis of tryptic peptide mixtures by electrospray ionization) is lower than with ECD; and (2) ETD, which is less energetic, does not induce secondary fragmentation, thus rending impossible distinguishing of the isomeric Leu and Ile residues.

The third approach is to use metastable, electronically excited atoms of, for example, noble gases (Misharin et al. 2005). Colliding with cations, these atoms give away (donate) an electron, producing an effect similar to ECD/ETD. When the same atomic species collide with anions, the resultant fragmentation resembles EDD.

Another method of inducing electron capture is to use high-energy collisions between polypeptide cations accelerated to kinetic energies up to 50 keV per charge, and stationary gaseous targets of alkali metals vapors (Hvelplund et al. 2003). While collisions with Na (ionization energy 5.14 eV) lead to electron capture with subsequent *c* ion formation, collisions with Ne (ionization energy 21.56 eV) produce mostly C—N bond cleavage with subsequent CO losses from *b* ions to form *a* species.

EDD with free electrons has been successfully implemented in Paul ion traps, where it is more easily achieved than ECD due to higher electron energy. For the same reason, EDD efficiency in FTICR MS is usually well below that of ECD. The Coulombic interaction of precursor ions with the space charge produced by electrons is detrimental for anions and their fragments that are repelled from the electron beam. The best EDD results in FTICR MS are obtained with low electron current ($<10\ \mu A$).

13.5. APPLICATION TO STRUCTURAL STUDIES

13.5.1. Primary Sequence Determination

One of the basic requirements for de novo sequencing is the backbone cleavage between each pair of the adjacent amino acids. But even that condition is not sufficient, because the mass difference between two fragments, one *N*-terminal and another *C*-terminal, can accidentally be close to zero or to the mass of an amino acid residue. Application of two complementary fragmentation techniques, such as ECD and CAD or IRMPD, often resolves the confusion, since the mass difference between c' and b fragments is 17.03 Da (mass of NH_3), while that between y' and $z^{\bullet}$ fragments is 16.02 Da (mass of NH_2) (McLafferty et al. 1999). Using two fragmentation techniques and sets of complementary fragments derived from comparison between the respective tandem mass spectra, full sequences (excluding Ile/Leu differentiation) of polypeptides as large as ubiquitin (8.6 kDa) could be determined from MS/MS of the whole molecule (Horn et al. 2000a). McLafferty et al. have shown the utility of ECD for partial sequence determination for molecules up 50 kDa by this "top-down approach" (Ge et al. 2002). In order to break the weak bonding in such large molecules and thus to increase the sequence coverage, mild vibrational excitation prior to ECD ("activated-ion ECD") can be applied (Horn et al. 2000a; Sze et el. 2003). Kelleher et al. showed the potential of ECD for high throughput top-down characterization of small and medium-size proteins (Johnson et al. 2002).

13.5.1.1. Distinguishing Constitutional Isomers. Formation of *w* ions in HECD can be utilized for differentiation of Ile/Leu residues in approximately 80% of the cases (Kjeldsen et al. 2003). Additionally, *u* ions can provide useful information. Application of vibrational activation, ECD and HECD to peptides of a 15 kDa protein PP3 revealed full sequence, including the identities of 23 out of 25 Ile/Leu residues (Kjeldsen and Zubarev 2003).

Isomerization of aspartic and glutamic acids is a frequent phenomenon occurring both in vivo and in vitro. While vibrational activation can distinguish isomers of these residues only by relative abundances of b,y' fragments using internal standards, ECD produces specific $c\bullet$+58 Da and $z - 57$ Da ions for peptides with iso-Asp residues (Cournoyer et al. 2005). Mechanism of the C_α—C_β cleavage (instead of N—C_α bond cleavage) responsible for these fragments proposed by O'Connor et al. is shown in Scheme 13.9. Note that here the cleavage occurs "backward," that is, *N*-terminal to the H· attachment site. A similar cleavage between two carbon atoms occurs in ε-peptides (Figure 13.10).

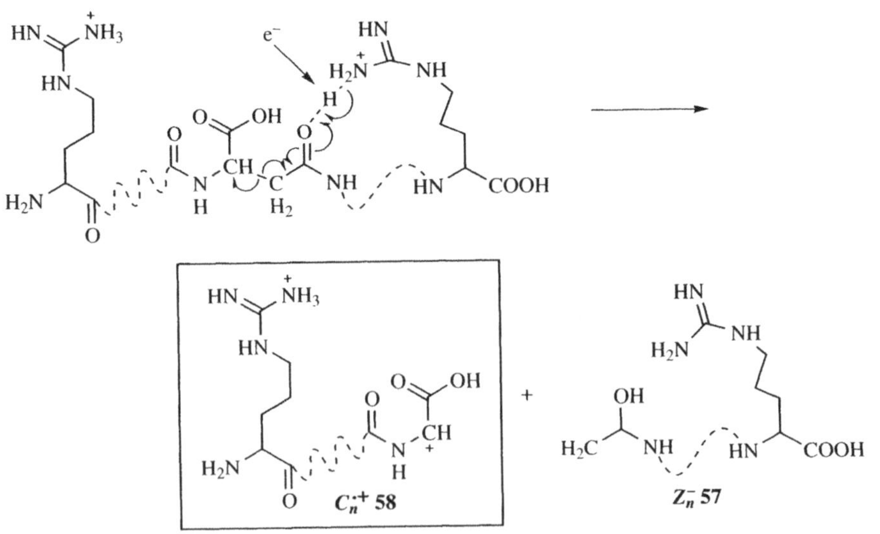

SCHEME 13.9. Proposed fragmentation scheme for the formation of $z_n - 57$ and $c_n^{\bullet} + 58$ ions from ECD of peptides with isoAsp residues. [From Cournoyer et al. (2005).]

13.5.1.2. Distinguishing Stereoisomers. L→D conversion of an amino acid in a polypeptide chain is the most subtle posttranslational modification. Since there can be no characteristic loss distinguishing D-amino acids from L-amino acids, differentiation is based solely on relative abundances of fragment ions (Adams et al. 2004). The abundances of *c,z* ions change quite markedly, especially near the D-substituted residue, with chiral recognition factor as high as 10 (Figure 13.14). Replacement of more than one L-amino acid with a D- counterpart also produces changes in the fragmentation pattern. However, ECD mass spectra of all-L molecules are indistinguishable from those of their all-D counterparts.

13.5.2. Charge Localization

Since one charge is preferentially neutralized in ECD, the remaining $(n-1)$ charges are present as spectators, and their positions can be established by observation of the charge state of the fragments that will increment (or decrement) once the charged residue is included to (or excluded from) the sequence (Kjeldsen et al. 2006). The position of the neutralized *n*th proton remains invisible.

In some cases addition protonation of $+n$ ions does not shift the positions of the n "old" protons, and this newly added proton occupies the least basic site and thus is preferentially neutralized. In these cases, this $(n+1)$st charge becomes invisible in ECD of $+(n+1)$ ions, but the position of the *n*th charge is revealed. Thus in some cases, ECD of $+(n+1)$ ions can reveal all protonation sites in $+n$ species. For instance, ECD of +2 and +3 of Trp cage ions revealed that in +2 of these species the ionizing protons are located at Q_5 and R_{16} (Figure 13.11), and not on K_8 and R_{16} as in solution. One should however remember that charge localization in $+n$ ions by ECD of $+(n+1)$ species is problematic if the protons, after additional protonation, either change their positions or change the basicity order. This usually happens for higher charge states, where Coulombic repulsion between the charges significantly affects the basicity of the sites (Kjeldsen et al. 2006).

13.5.3. Secondary and Tertiary Structure

As already mentioned, ECD is sensitive to neutral *intra*molecular hydrogen bonding, as well as to ionic hydrogen bonding mediated by protonated site solvation. The number of N—C_α bond cleavages in direct ECD shows a temperature dependence, which has been used for deriving melting temperatures and melting enthalpy on gas-phase ubiquitin cations in intermediate charge states (Breuker et al. 2002). For cytochrome *c* gas-phase cations, ECD detected the presence of folding intermediates (Horn et al. 2001). The fragmentation pattern of +15 ions does not change until the temperature reaches 130°C, consistent with the unfolded state of these ions.

13.5.4. Quaternary Structure

ECD shows potential for determination of gas-phase structures of polypeptide–drug and polypeptide–polypeptide complexes. Better understanding of the ECD mechanism is required for reaching this ambitious objective. A more recent approach pioneered by McLafferty et al. is to combine ECD with photofragment spectroscopy that can probe various hydrogen bonds (Oh et al. 2002). The same group suggested that N—C_α bond cleavage can follow *intra*molecular electron capture in electron transfer proteins, such as cytochrome *c* (Breuker and McLafferty 2003). See more on this "native ECD" elsewhere in this book.

13.5.5. ECD of Nonpeptide Molecules

ECD has been applied to a wide variety of (bio)polymers, including DNAs and RNAs (Schultz and Håkansson 2004; Cooper et al. 2005), oligisaccharides (Budnik et al. 2003), peptide nucleic acids (PNAs) (Olsen et al. 2001), synthetic polymers (Cerda et al. 2001, 2002), and dendromers (Koster et al. 2003). Whenever amide bond is present, N—C_α bond cleavage can be expected. In general, the utility of ECD for sequence analysis remains by far most important for polypeptides.

13.6. REMAINING CHALLENGES

Admittedly, possessing detailed knowledge of the action mechanism is not necessary for successful application of a technique in practice and even in some fundamental studies. After all, the mechanisms of both MALDI and ESI are still widely debated. However, solving protein gas-phase structures is a very important goal that can be reached only by establishing a direct, quantitative link between experimental data and molecular dynamics simulations, for which knowledge of fundamental ECD mechanism is essential. Despite recent advances, there is still much to be done here. Particularly, revealing the role of such factors as the proximity to the charged site and the influence of the local neutral hydrogen bonding on ECD/EDD fragment abundances appears to be of utmost value.

REFERENCES

Adams C, Budnik BA, Haselmann KF, Kjeldsen F, Zubarev RA (2004): Electron capture dissociation distinguishes a single D-amino acid in a protein and probes the tertiary structure. *J. Am. Soc. Mass Spectrom.* **15**:1087–1098.

Adams C, Zubarev RA (2005). Distinguishing and quantification of peptides and proteins containing D-amino acids by tandem mass spectrometry. *Anal. Chem.* **77**:4571–4580.

Al-Khalili A, Thomas R, Ehlerding A, Hellberg F, Geppert WD, Zhaunerchyk V, af Ugglas M, Larsson M, Uggerud E, Vedde J, Adlhart C, Semaniak J, Kamińska M, Zubarev RA, Kjeldsen F, Andersson PU, Osterdahl F, Bednarska VA, Paál A (2004): Dissociative recombination cross section and branching ratios of protonated dimethyl disulfide and N-methylacetamide. *J. Chem. Phys.* **121**:5700–5708.

Anusiewicz I, Berdys-Kochanska J, Simons J (2005): Electron attachment step in electron capture dissociation (ECD) and electron transfer dissociation (ETD). *J. Phys. Chem. A* **109**:5801–5813.

Baba T, Hashimoto Y, Hasegawa H, Hirabayashi A, Waki I (2004): Electron capture dissociation in a radio frequency ion trap. *Anal. Chem.* **76**:4263–4266.

Bakken V, Helgaker T, Uggerud E (2004): Models of fragmentations induced by electron attachment to protonated peptides. *Eur. J. Mass Spectrom.* **10**:625–638.

Bardsley J N, Biondi MA (1970). In Bates DR (ed), *Advances in Atomic and Molecular Physics,* Academic Press, New York, pp. 1–57.

Breuker K, Oh HB, Horn DM, Cerda BA, McLafferty FW (2002): Detailed unfolding and folding of gaseous ubiquitin ions characterized by electron capture dissociation. *J. Am. Chem. Soc.* **124**:6407–6420.

Breuker K, McLafferty FW (2003): Native electron capture dissociation for the structural characterization of noncovalent interactions in native cytochrome c. *Angew. Chem. Int. Ed.* **42**:4900–4904.

Breuker K, Oh HB, Lin C. Carpenter BK, McLafferty FW (2004): Nonergodic and conformational control of the electron capture dissociation of protein cations. *Proc. Natl. Acad. Sci. USA* **101**:14011–14016.

Brown RS, Lennon JJ (1995): Sequence-specific fragmentation of matrix-assisted laser-desorbed protein and peptide ions. *Anal. Chem.* **67**:3990–3999.

Budnik BA, Haselmann KF, Zubarev RA (2001): Electron detachment dissociation of peptide di-anions: An electron-hole recombination phenomenon. *Chem. Phys. Lett.* **342**:299–302.

Budnik BA, Nielsen ML, Olsen JV, Haselmann KF, Hörth P, Haehnel W, Zubarev RA (2002a): Can relative cleavage frequencies in peptides provide additional sequence information? *Int. J. Mass Spectrom.* **219**:283–294.

Budnik BA, Tsybin YO, Håkansson P, Zubarev RA (2002b): Ionization energies of multiply protonated polypeptides obtained by tandem ionization in Fourier transform mass spectrometers. *J. Mass Spectrom.* **37**:1141–1144.

Budnik BA, Haselmann KF, Elkin YN, Gorbach VI, Zubarev RA (2003): Applications of electron-ion dissociation reactions for analysis of polycationic chitooligosaccharides in Fourier transform mass spectrometry. *Anal. Chem.* **75**:5994–6001.

Cerda BA, Breuker K, Horn DM, McLafferty FW (2001): Charge/radical site initiation versus coulombic repulsion for cleavage of multiply charged ions. Charge solvation in poly(alkene glycol) ions. *J. Am. Soc. Mass Spectrom.* **12**:565–570.

Cerda BA, Horn DM, Breuker K, McLafferty FW (2002): Sequencing of specific copolymer oligomers by electron-capture-dissociation mass spectrometry. *J. Am. Chem. Soc.* **124**:9287–9291.

Cooper HJ, Hudgins RR, Håkansson K, Marshall AG (2002): Characterization of amino acid side chain losses in electron capture dissociation. *J. Am. Soc. Mass Spectrom.* **13**:241–249.

Cooper HJ, Hudgins RR, Håkansson K, Marshall AG (2003a): Secondary fragmentation of linear peptides in electron capture dissociation. *Int. J. Mass Spectrom.* **228**:723–728.

Cooper HJ, Håkansson K, Marshall AG, Hudgins RR, Haselmann KF, Kjeldsen F, Budnik BA, Polfer NC, Zubarev RA (2003b): The diagnostic value of amino acid side-chain losses in electron capture dissociation of polyepeptides. Comment on: "Can the (M · – X) region in electron capture dissociation provide reliable information on amino acid composition of polypeptides? *Eur. J. Mass Spectrom.* **9**:221–222.

Cooper HJ, Hudgins RR, Marshall AG (2004): Electron capture dissociation Fourier transform ion cyclotron resonance mass spectrometry of cyclodepsipeptides, branched peptides, and ε-peptides. *Int. J. Mass Spectrom.* **234**:23–35.

Cooper HJ, Håkansson K, Marshall AG (2005): The role of electron capture dissociation in biomolecular analysis. *Mass Spectrom. Rev.* **24**:201–222.

Cournoyer JJ, Pittman JJ, Ivleva VB, Fallows E, Waskell L, Costello CE, O'Connor PB (2005): Deamidation: Differentiation of aspartyl from isoaspartyl products in peptides. *Protein Sci.* **14**:452–563.

Demirev PA (2000): Generation of hydrogen radicals for reactivity studies in Fourier transform ion cyclotron resonance mass spectrometry. *Rapid Commun. Mass Spectrom.* **14**: 777–781.

Ge Y, Lawhorn BG, ElNaggar M, Strauss E, Park JH, Begley TP, McLafferty FW (2002): Top down characterization of larger proteins (45 kDa) by electron capture dissociation mass spectrometry. *J. Am. Chem. Soc.* **124**:672–678.

Gellene GI, Porter RF (1984): *J. Phys. Chem.* **88**:6680–6684.

Guan Z, Kelleher NL, O'Connor PB, Aaserud DJ, Little DP, McLafferty FW (1996): 193 nm photodissociation of larger multiply-charged biomolecules. *Int. J. Mass Spectrom. Ion Proc.* **157/158**:357–364.

Hagler AT, Dauber P, Lifson S (1979): Consistent force field studies of intermolecular forces in hydrogen-bonded crystals. 1. Carboxylic acids, amides, and the C=O···H hydrogen bonds. *J. Am. Chem. Soc.* **101**:5111–5121.

Haselmann KF, Budnik BA, Zubarev RA (2000): Electron capture dissociation of b^{2+} ions reveals the presence of the acylium ion structure. *Rapid Commun. Mass Spectrom.* **14**:2242–2246.

Haselmann KF, Polfer NC, Budnik BA, Kjeldsen F, Zubarev RA (2002a): Can the (M • – X) region in electron capture dissociation provide reliable information on amino acid composition of polypeptides? *Eur. J. Mass Spectrom.* **8**:461–469.

Haselmann KF, Jørgensen TJD, Budnik BA, Jensen F, Zubarev RA (2002b): Electron capture dissociation of weakly-bound polypeptide polycationic complexes. *Rapid Commun. Mass Spectrom.* **16**:2260–2265.

Horn DM, Zubarev RA, McLafferty FW (2000a): Automated de novo sequencing of proteins by tandem high-resolution mass spectrometry. *Proc. Natl. Acad. Sci. USA* **97**:10313–10317.

Horn DM, Ge Y, McLafferty FW (2000b): Activated ion electron capture dissociation for mass spectral sequencing of larger (42 kDa) proteins. *Anal. Chem.* **72**:4778–4784.

Horn DM, Breuker K, Frank AJ, McLafferty FW (2001): Kinetic intermediates in the folding of gaseous protein ions characterized by electron capture dissociation mass spectrometry. *J. Am. Chem. Soc.* **123**:9792–9799.

Hvelplund P, Liu B, Nielsen SB, Tomita S (2003): Electron capture induced dissociation of peptide dications. *Int. J. Mass Spectrom.* **225**:83–87.

Iavarone AT, Paech K, Williams ER (2004): Effects of charge state and cationizing agent on the electron capture dissociation of a peptide. *Anal. Chem.* **76**:2231–2238.

Johnson JR, Meng FY, Forbes AJ, Cargile BJ, Kelleher NL (2002): Fourier-transform mass spectrometry for automated fragmentation and identification of 5-20 kDa proteins in mixtures. *Electrophoresis* **23**:3217–3223.

Kelleher NL, Zubarev RA, Bush K, Furie B, Furie BC, McLafferty FW, Walsh CT (1999): Localization of labile posttranslational modifications by electron capture dissociation: The case of γ-carboxyglutamic acid. *Anal. Chem.* **71**:4250–4253.

Kjeldsen F, Budnik BA, Haselmann KF, Jensen F, Zubarev RA, (2002): Dissociative capture of hot (3–13 eV) electrons by polypeptide polycations: An efficient process accompanied by secondary fragmentation. *Chem. Phys. Lett.* **356**:201–206.

Kjeldsen F, Zubarev RA (2003): Secondary losses via γ-lactam formation in hot electron capture dissociation: A missing link to complete de novo sequencing of proteins? *J. Am. Chem. Soc.* **125**:6628–6629.

Kjeldsen F, Sørensen E, Zubarev RA (2003): Distinguishing of Ile/Leu amino acid residues in the PP3 protein by (hot) electron capture dissociation in Fourier transform ion cyclotron resonance mass spectrometry. *Anal. Chem.* **75**:1267–1274.

Kjeldsen F, Silivra OA, Ivonin IA, Haselmann KF, Gorshkov M, Zubarev RA (2005): C_α-C backbone fragmentation dominates in electron detachment dissociation of gas-phase polypeptide anions. *Chem. Eur. J.* **11**:1803–1812

Kjeldsen F, Adams CM, Savitski MM, Zubarev RA (2006): Localization of positive charges in gas-phase polypeptide polycations by tandem mass spectrometry. *Int. J. Mass Spectrom.* X:XXX.

Kocher T, Engström A, Zubarev RA (2005): Fragmentation of peptides in MALDI in-source decay mediated by hydrogen radicals. *Anal. Chem.* **77**:172–177.

Koster S, Duursma MC, Boon JJ, Heeren RMA, Ingemann S, van Benthem RATM, de Koster CG (2003): Electron capture and collisionally activated dissociation mass spectrometry of doubly charged hyperbranched polyesteramides. *J. Soc. Mass Spectrom.* **14**:332–341.

Kruger NA, Zubarev RA, Carpenter BK, Kelleher NL, Horn DM, McLafferty FW (1999): Electron capture versus energetic dissociation of protein ions. *Int. J. Mass Spectrom. Ion Proc.* **182–183**:1–5.

Leymarie N, Costello CE, O'Connor PB (2003): Electron capture dissociation initiates a free radical reaction cascade. *J. Am. Chem. Soc.* **125**:8949–8958.

McLafferty FW (1986). Tandem mass spectrometry of large molecules. In McNeal CJ (ed), *Mass Spectrometry in the Analysis of Large Molecules*, Wiley, New York, pp. 107–120.

McLafferty FW, Fridriksson EK, Horn DM, Lewis MA, Zubarev RA (1999): Biomolecule mass spectrometry. *Science* **284**:1289–1290.

Mihalca R, Kleinnijenhuis AJ, McDonnell LA, Heck AJR, Heeren RMA (2004): Electron capture dissociation at low temperatures reveals selective dissociations. *J. Am. Soc. Mass Spectrom.* **15**:1869–1873.

Misharin AS, Silivra OA, Kjekdsen F, Zubarev RA (2005): Dissociation of peptide ions by fast atom bombardment in a quadrupole ion trap. *Rapid Commun. Mass Spectrom.* **19**:2163–2171.

Oh H, Breuker K, Sze SK, Ge Y, Carpenter BK, McLafferty FW (2002): Secondary and tertiary structures of gaseous protein ions characterized by electron capture dissociation mass spectrometry and photofragment spectroscopy. *Proc. Natl. Acad. Sci. USA* **99**:15863–15868.

Olsen JV, Haselmann KF, Nielsen ML, Budnik BA, Nielsen PE, Zubarev RA (2001): Comparison of electron capture dissociation and collisionally activated dissociation of polycations of peptide nucleic acids. *Rapid Commun. Mass Spectrom.* **15**:969–974.

Patriksson A, Adams CM, Kjeldsen F, Raber J, van der Spoel D, Zubarev RA (2006): Prediction of N–Cα bond cleavage frequencies in electron capture dissociation o Trp-cage dications by force-field molecular dynamics simulations. *Int. J. Mass Spectrom.* X:XXX.

Polfer NC, Haselmann KF, Langridge-Smith PRR, Barran P (2005): Structural investigation of naturally occurring peptides by electron capture dissociation and AMBER force field modeling. *Mol. Phys.* **103**:1481–1489.

Sawicka A, Skurski P, Hudgins RR, Simons J (2003): Model calculations relevant to disulfide bond cleavage via electron capture influenced by positively charged groups. *J. Phys. Chem. B* **107**:13505–13511.

Schnier PD, Gross DS, Williams ER (1995): On the maximum charge-state and proton-transfer reactivity of peptide and protein ions formed by electrospray-ionization. *J. Am. Soc. Mass Spectrom.* **6**:1086–1097.

Schultz KN, Håkansson K (2004): Rapid electron capture dissociation of mass-selectively accumulated oligodeoxynucleotide dications. *Int. J. Mass Spectrom.* **234**:123–130.

Silivra OA, Kjeldsen F, Ivonin IA, Zubarev RA (2005): Electron capture dissociation in a quadrupole ion trap: First results. *J. Am. Soc. Mass Spectrom.* **16**:22–27.

Sobczyk M, Anusiewicz I, Berdys-Kochanska J, Sawicka A, Skurski P, Simons J (2005): Coulomb-assisted dissociative electron attachment: Application to a model peptide. *J. Phys. Chem. A* **109**:250–258.

Syka JEP, Coon JJ, Schroeder MJ, Shabanowitz J, Hunt DF (2004): Peptide and protein sequence analysis by electron transfer dissociation mass spectrometry. *Proc. Natl. Acad. Sci. USA* **101**:9528–9533.

Syrstad EA, Turecek F (2005): Toward a general mechanism of electron capture dissociation. *J. Am. Soc. Mass Spectrom.* **16**:208–224.

Sze SK, Ge Y, Oh HB, McLafferty FW (2003): Plasma electron capture dissociation for the characterization of large proteins by top down mass spectrometry. *Anal. Chem.* **75**:1599–1603.

Takayama M. (2003): In-source decay characteristics of peptides in matrix-assisted laser desorption/ionization time-of-flight mass spectrometry. *J. Am. Soc. Mass Spectrom.* **12**:420–427.

Tanabe T, Noda K, Saito M, Lee S, Ito Y, Takagi H (2003): Resonant neutral-particle emission in collisions of electrons with peptide ions in a storage ring. *Phys. Rev. Lett.* **90**:193–201.

Tsybin YO, Håkansson P, Budnik BA, Haselmann KF, Kjeldsen F, Gorshkov M, Zubarev RA (2001): Improved low-energy injection systems for high rate electron capture dissociation in Fourier transform ion cyclotron resonance mass spectrometry. *Rapid Commun. Mass Spectrom.* **15**:1849–1854.

Turecek F (2003): N–C_α bond dissociation energies and kinetics in amide and peptide radicals. Is the dissociation a non-ergodic process? *J. Am. Chem. Soc.* **125**:5954–5963.

Turecek F, Syrstad EA (2003): Mechanism and energetics of intramolecular hydrogen transfer in amide and peptide radicals and cation-radicals. *J. Am. Chem. Soc.* **125**:3353–3369

Uggerud E (2004): Electron capture dissociation of the disulfide bond—a quantum chemical model study. *Int. J. Mass Spectrom.* **234**:45–50.

Wesdemiotis C, McLafferty FW (1984): Neutralization-reionization mass spectrometry (NR MS). *Chem. Rev.* **87**:485–500.

Zubarev RA, Kelleher NL, McLafferty FW (1998): Electron capture dissociation of multiply charged protein cations. A non-ergodic process. *J. Am. Chem. Soc.* **120**:3265–3266.

Zubarev RA, Kruger NA, Fridriksson EK, Lewis MA, Horn DM, Carpenter BK, McLafferty FW (1999): Electron capture dissociation of gaseous multiply-charged proteins is favored at disulfide bonds and other sites of high hydrogen atom affinity. *J. Am. Chem. Soc.* **121**:2857–2862.

Zubarev RA, Fridriksson EK, Horn DM, Kelleher NL, Kruger NA, Carpenter BK, McLafferty FW (2000a): Electron capture dissociation for structural characterization of multiply charged protein cations. *Anal. Chem.* **72**:563–573.

Zubarev RA, Budnik BA, Nielsen ML (2000b): Tandem ionization mass spectrometry of biomolecules. *Eur. J. Mass Spectrom.* **6**:235–240.

Zubarev RA, Haselmann KF, Budnik BA, Kjeldsen K, Jensen F (2002): Towards an understanding of the mechanism of electron capture dissociation: A historical perspective. *Eur. J. Mass Spectrom.* **8**:337–349.

14

BIOMOLECULE ION–ION REACTIONS

SCOTT A. MCLUCKEY
Department of Chemistry
Purdue University
West Lafayette, IN

Principles of Mass Spectrometry Applied to Biomolecules, edited by Julia Laskin and Chava Lifshitz

14.1. INTRODUCTION

The reaction of ions of opposite polarity in the gas phase has been a subject of interest dating back to the studies of Thomson and Rutherford (Thomson and Rutherford 1896; Rutherford 1897). They are of particular interest in their relevance to interactions such as those in interstellar space, planetary atmospheres, combustion, and plasmas (Bates 1985; Flannery 1982; Mahan 1973, Adams 2003). Most work involving the study of the reactions involving oppositely charged ions has been directed toward relatively small singly charged cations and singly charged anions. The fact that the products of such reactions are neutral has made the study of this chemistry particularly challenging from the experimental standpoint. However, in the case of biological molecules, such as peptides, proteins, and oligonucleotides, the advent of ionization methods that can produce multiply charged ions from such species [the spray ionization methods (Fenn et al. 1990) in particular] has greatly facilitated the study of biomolecule ion–ion reactions. The multiple charging phenomenon enables the facile analysis of reaction products, provided they are not completely neutralized. This condition is readily achieved by adjusting reaction conditions (i.e., reactant number densities and/or reaction times).

By far, the most heavily studied class of bimolecular gas-phase reactions in mass spectrometry has been ion–molecule reactions. While some important analogies can be drawn between ion–molecule and ion–ion reactions, it is important to appreciate that ion–ion reactions represent a distinct category of bimolecular interactions. The dynamics and thermodynamics of ion–ion reactions are quite different from those associated with ion–molecule reactions. Furthermore, the dimensionality associated with ion–ion reactions is potentially much greater than that for ion–molecule reactions. This follows from the fact that, with the current array of ionization methods, a much wider range of ion–ion combinations can be studied than can ion–molecule combinations. The neutral reactant for most ion–molecule reactions must have sufficient volatility to allow for a sufficiently high number density to study the reaction. The volatility requirement is greatly relaxed for ionic reactants because of the ready availability of desorption and spray ionization methods. Furthermore, for many molecules, it is possible to form an array of pseudomolecular ions, such as radical cations, protonated molecules, deprotonated molecules, and multiply charged ions. Hence, it is likely that the range of ion–ion reaction phenomenology examined to date will ultimately prove to be a small (but very important) subset of the entire range of ion–ion chemistry.

Two reviews covering biomolecule ion–ion reactions have appeared that have emphasized instrumentation, applications, and reaction phenomenology (McLuckey and Stephenson 1998; Pitteri 2005a). This chapter does not emphasize applications of gas-phase biomolecule ion–ion reactions, which are still (as of 2006) under active development, and provides only a brief description of the instrumentation used to study ion–ion reactions. Rather, it is devoted to a discussion of the major reaction phenomenologies that have been observed, with emphasis on reaction thermodynamics, along with general reaction dynamics that account for experimental observations made to date. As most detailed studies have been conducted for ion–ion reactions in electrodynamic ion traps, the discussion is most heavily weighted toward conditions that apply in ion trap experiments.

14.2. BIOPOLYMER ION–ION REACTION PHENOMENOLOGY

The most commonly observed general class of ion–ion reactions observed in electrodynamic ion traps involving ions derived from biopolymers has been proton transfer. This is due to the fact that electrospray ionization, the most common means for forming multiply charged bioions, tends to form either protonated (positive-ion-mode) or deprotonated (negative-ion-mode) species. For this reason, protonated and multiply protonated species tend to react as strong Brønsted acids whereas deprotonated and multiply deprotonated species tend to react as strong Brønsted bases. However, as discussed further below, the nature of the reagent ion also plays a major role in the type of reaction that is likely to occur such that reactions other than proton transfer can be observed. The discussion below summarizes the major reaction types noted in ion trap studies along with the relevant thermochemistry and are classified in terms of ion transfer reactions, electron transfer reactions, and condensation reactions. While a variety of ion transfer reactions are possible, the discussion here is restricted to single-proton transfer, multiple-proton transfer, and net metal ion transfer via reactions with anions of metal complexes.

14.2.1. Ion Transfer Reactions

14.2.1.1. Single-Proton Transfer. Most biomolecule ion–ion reaction studies have been conducted with multiply charged bioions in reaction with singly charged ions of opposite polarity. By far, the most commonly observed ion transfer reaction involves a single transfer of a proton. In the case of a multiply protonated biomolecule, $(M + nH)^{n+}$, in reaction with a singly charged anion, A^-, the transfer of a proton leads to neutralization of the anion and the reduction of the charge of the cation by 1, as indicated in the following reaction:

$$(M + nH)^{n+} + A^- \rightarrow (M + (n-1)H)^{(n-1)+} + HA \qquad (14.1)$$

A simple means for representing the relevant portion of the energy surface of a bimolecular reaction is to plot the potential energy of the system as the reactants proceed to products from left to right along a reaction coordinate, as illustrated in Figure 14.1.

The entrance channel is dominated by the long-range Coulomb attraction associated with the oppositely charged reactants (i.e., by a $-Z_1Z_2e^2/r$ potential, where e is the electron charge; Z_1 and Z_2 are the unit charges of the cations and anions, respectively; and r is the interparticle separation) whereas the exit channel follows the potential of an ion–molecule interaction.

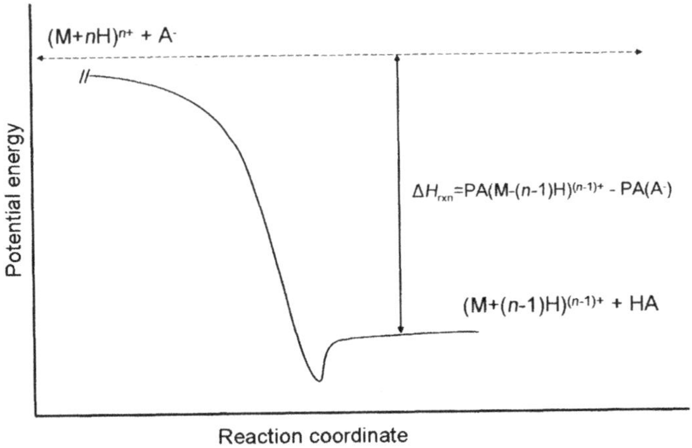

FIGURE 14.1. Simplified energy diagram of an ion–ion single-proton transfer reaction in which one of the products is neutral.

The enthalpy of reaction (14.1) is determined by the difference in the relevant proton affinities (PAs).

$$\Delta H_{\text{rxn}} = \text{PA}[(\text{M} + (n-1)\text{H})^{(n-1)+}] - \text{PA}[\text{A}^-] \tag{14.2}$$

and the free energy of the reaction is determined by the difference in the gas-phase basicity (GPB) of $(\text{M} + (n-1)\text{H})^{(n-1)+}$ and the GPB of A^- (or its equivalent, the gas-phase acidity of HA):

$$\Delta G_{\text{rxn}} = \text{GPB}[(\text{M} + (n-1)\text{H})^{(n-1)+}] - \text{GPB}[\text{A}^-] \tag{14.3}$$

A few quantitative measurements of apparent proton affinities and/or gas-phase basicities of cations have been made using ion–molecule reaction bracketing approaches. The values are indicated as "apparent" because the reactions proceed over an energy surface with a barrier in the exit channel that arises from the fact that the products are of like charge (Williams 1996). Hence, the values obtained via the bracketing method, which is sensitive to the relative difference between the barriers in the entrance and exit channels, do not strictly reflect the thermodynamic values. Nevertheless, deprotonation of even the strongest of gaseous bases by singly charged anions is expected to be highly exothermic because there is a significant gap in energy between the highest proton affinities of neutral bases and the lowest proton affinities of anions. The effect of multiple protonation is to reduce cation proton affinities as charge state increases. Therefore, deprotonation via reaction (14.1) is expected to be exothermic for every value of n, usually by 100 kcal/mol or greater. Furthermore, all anions are expected to have relatively high proton affinities. Hence, reaction (14.1) is expected to be possible, on thermodynamic grounds, for any singly charged anion in reaction with a multiply protonated molecule.

The opposite-polarity case involves a multiply deprotonated species, $(M - mH)^{m-}$, in reaction with a singly protonated species, BH^+, as indicated in (14.4):

$$(M - mH)^{m-} + BH^+ \rightarrow (M - (m-1)H)^{(m-1)-} + B \qquad (14.4)$$

The thermodynamic values associated with this reaction are analogous to those associated with reaction (14.1). In the case of enthalpies, for example, the heat of reaction is determined by the difference in the proton affinities of $(M - mH)^{m-}$ and B:

$$\Delta H_{rxn} = PA[B] - PA[(M - mH)^{m-}] \qquad (14.5)$$

While the ΔH_{rxn} value for any given process represented by either reaction (14.1) or reaction (14.4) is determined by the identities of the reactants, the magnitudes of the relevant proton affinities are such that the range of exothermicities for reactions (14.1) and (14.4) are very similar. In the case of reaction (14.4), the singly charged ion must be capable of donating a proton, and this criterion is not universally met. This stands in contrast with the case of proton transfer from a multiply protonated species [i.e., reaction (14.1)]. Whereas essentially all singly charged anions [see reaction (14.1)] can be expected to have a positive proton affinity, all singly charged cations [see reaction (14.2)] are not necessarily capable of donating a proton. Rare-gas cations, which have been used to extract electrons from $(M - mH)^{m-}$ species (Herron 1995a) (see also text below), are cases in point.

When both reactants of an ion–ion reaction are multiply charged, the complete neutralization of the ion of lesser charge is the favored reaction channel on thermodynamic grounds. Indeed, this is the most commonly observed result. However, partial neutralization reactions are also sometimes noted (Wells et al. 2001), as discussed further in the section dealing with the dynamics of ion–ion reactions (Section 14.3). A generic reaction that represents a single-proton transfer between two multiply charged reactants can be written as follows:

$$(M + nH)^{n+} + (N - mH)^{m-} \rightarrow (M + (n-1)H)^{(n-1)+} + (N - (m-1)H)^{(m-1)-} \qquad (14.6)$$

The energy surface for this reaction differs from that associated with reactions (14.1) and (14.4) in that both the entrance and exit channels are dominated by long-range Coulomb potential attraction $(-Z_1 Z_2 e^2/r)$ (see Figure 14.2). The Coulomb attraction in the exit channel is, of course, of lower magnitude than that of the entrance channel because the charges of each product have been reduced by 1.

The reaction enthalpy is given as follows:

$$\Delta H_{rxn} = PA[(M + (n-1)H)^{(n-1)+}] - PA[(N - mH)^{m-}] \qquad (14.7)$$

Given that multiple charging decreases the proton affinities of cations and increases the proton affinities of anions, the reaction enthalpies associated with a single-proton

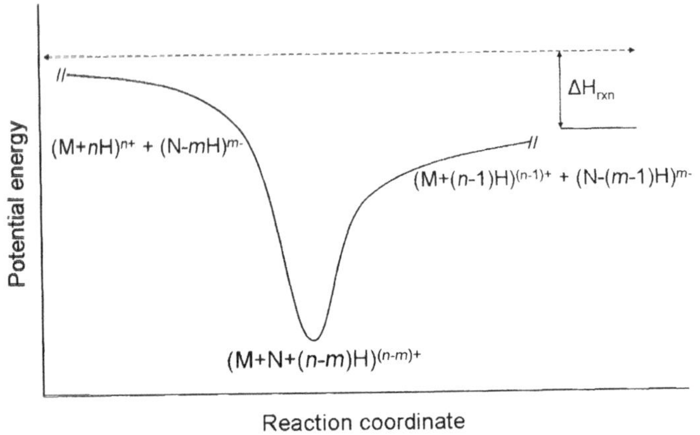

FIGURE 14.2. Simplified energy diagram of an ion–ion single-proton transfer reaction in which both the reactants and products are oppositely charged.

transfer between multiply charged ions can be expected to be higher than a reaction involving ions derived from the same reagent species, M and N, but with one of them singly charged.

14.2.1.2. Multiple-Proton Transfer. When both ionic reactants are multiply charged, multiple-proton transfers in a single ion–ion interaction can occur. Three generic types of reaction involving multiple proton transfer can be classified according to the polarities of the products: (1) *partial neutralization*, in which each product retains the same polarity as its corresponding reactant, although at lower absolute value; (2) *complete neutralization* of one of the reactants; and (3) *charge inversion*, in which the polarity of one of the products differs from that of the corresponding reactant. Partial neutralization involving a single-proton transfer is represented by reaction (14.6). The reaction for the case when $n > m$ can be represented generically as

$$(\mathrm{M} + n\mathrm{H})^{n+} + (\mathrm{N} - m\mathrm{H})^{m-} \rightarrow (\mathrm{M} + (n - x)\mathrm{H})^{(n-x)+} + (\mathrm{N} - (m - x)\mathrm{H})^{(m-x)-} \quad (14.8)$$

where x represents the number of protons transferred and, in the case of multiple-proton transfers, satisfies the criteria that $m > x > 1$. The energy diagram for this generic type of reaction is analogous to that of Figure 14.2 in that both the entrance and exit channels traverse an attractive Coulomb potential. The reaction exothermicity is determined by the sum of the reaction exothermicities of each successive proton transfer:

$$\begin{aligned}\Delta H_{\mathrm{rxn}} = {} & \mathrm{PA}[(\mathrm{M} + (n - 1)\mathrm{H})^{(n-1)+}] + \mathrm{PA}[(\mathrm{M} + (n - 2)\mathrm{H})^{(n-2)+}] \\ & + \cdots \mathrm{PA}[(\mathrm{M} + (n - x)\mathrm{H})^{(n-x)+}] - \mathrm{PA}[(\mathrm{N} - m\mathrm{H})^{m-}] \\ & - \mathrm{PA}[(\mathrm{N} - (m - 1)\mathrm{H})^{(m-1)-}] - \cdots \mathrm{PA}[(\mathrm{N} - (m - x + 1)\mathrm{H})^{(m-x+1)-} \end{aligned} \quad (14.9)$$

Incomplete neutralization of neither of the reactants seldom is the most thermodynamically favored outcome. Nevertheless, as discussed further below, incomplete neutralization has been observed for some combinations of multiply charged reactants. Long-range proton transfer is more plausible than a longlived intermediate mechanism for these reactions (see text below).

The complete neutralization of the reactant of lower initial absolute charge is analogous to reaction (1) in that the exit channel proceeds over an ion–molecule reaction surface (see Figure 14.1). The reaction enthalpy is also represented by (14.9), in which $m = x$:

$$\begin{aligned}\Delta H_{\rm rxn} = {} & \mathrm{PA}[(\mathrm{M}+(n-1)\mathrm{H})^{(n-1)+}] + \mathrm{PA}[(\mathrm{M}+(n-2)\mathrm{H})^{(n-2)+}] \\ & + \cdots \mathrm{PA}[(\mathrm{M}+(n-m)\mathrm{H})^{(n-m)+}] - \mathrm{PA}[(\mathrm{N}-m\mathrm{H})^{m-}] \\ & - \mathrm{PA}[(\mathrm{N}-(m-1)\mathrm{H})^{(m-1)-}] - \cdots \mathrm{PA}[(\mathrm{N}-\mathrm{H})^{-}] \end{aligned} \tag{14.10}$$

This reaction can proceed via a longlived intermediate and can also occur via long-range proton transfer.

When the ionic reactant of lesser charge contains chemical functionalities that can accommodate charges of the polarity of the ionic reactant with greater charge, a net inversion of charge of one of the reactants can take place. An example is illustrated by the data of Figure 14.3.

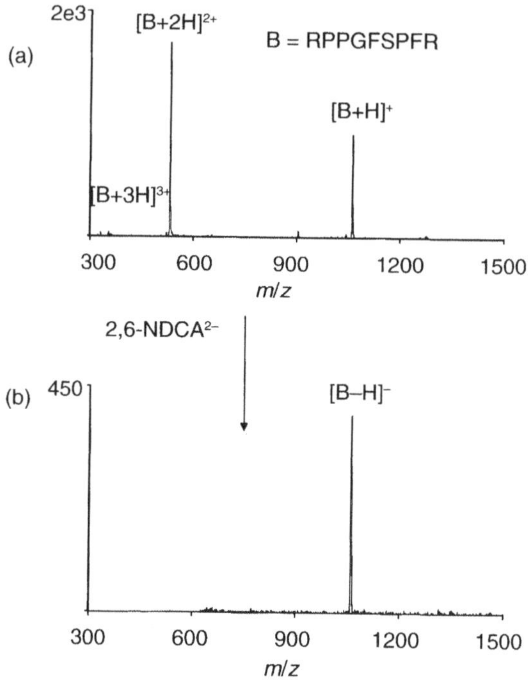

FIGURE 14.3. (a) Positive-ion electrospray mass spectrum of bradykinin, B; (b) negative-ion spectrum obtained after the ions of (a) were subjected to reaction with dianions derived from the negative-ion electrospray of 2,6,-naphthalene di(carboxylic acid), 2,6-NDCA.

When the cations derived from positive-ion electrospray of bradykinin are subjected to ion–ion reaction with dianions derived from the negative-ion electrospray of naphthalene 2,6-dicarboxylic acid (2,6-NDCA), the inversion of the charge of bradykinin is evident from the appearance of the $(B-H)^-$ ion in the negative-ion mode (He et al. 2005). The anions are presumably formed from the $(B+H)^+$ ions and, to a much lesser extent, from sequential reactions involving the triply and doubly charged ions. In the latter case, the first reaction must be a partial neutralization to convert $(B+2H)^{2+}$ to $(B+H)^+$. This particular example is one in which the anion charge is greater than that of the cation but charge inversion has been observed with both combinations (He and McLuckey 2003, 2004). In keeping with the current discussion, which describes the scenario in which positive-ion charge is greater than the charge of the negative reactant (i.e., where $n > m$), the general reaction is given as

$$(M+nH)^{n+} + (N-mH)^{m-} \rightarrow (M+(n-m-x)H)^{(n-m-x)+} + (N+xH)^{x+} \tag{14.11}$$

where x ranges from 1 to $(n-m)$. The overall enthalpy of reaction can be considered to be consist of both ion–ion and ion–molecule reaction components. Thus the ion–ion reaction part is given by relation (14.10), which represents the enthalpy associated with neutralization of the reagent of lesser charge. The ion–molecule part of the overall enthalpy derives from the reaction

$$(M+(n-m)H)^{(n-m)+} + N \rightarrow (M+(n-m-x)H)^{(n-m-x)+} + (N+xH)^{x+} \tag{14.12}$$

The enthalpy associated with reaction (14.12) is given by

$$\begin{aligned}\Delta H_{\mathrm{rxn}} = {} & \mathrm{PA}[(M+(n-m-1)H)^{(n-m-1)+}] + \mathrm{PA}[(M+(n-m-2)H)^{(n-m-2)+}] \\ & + \cdots \mathrm{PA}[(M+(n-m-x)H)^{(n-m-x)+}] - \mathrm{PA}[N] - \mathrm{PA}[(N+H)^+] \\ & - \cdots \mathrm{PA}[(N+(x-1)H)^{(x-1)+}\end{aligned} \tag{14.13}$$

Relation (14.13) represents the enthalpy associated with the transfer of excess protons from $(M+(n-m)H)^{(n-m)+}$ to N, which is expected to take place largely via a longlived complex represented as $[(M+N)+(n-m)H)^{(n-m)+}]$. The overall reaction enthalpy for charge inversion is given by the sum of relations (14.10) and (14.13).

The energy surface associated with a charge inversion reaction that forms products of like charge is distinct from those represented by Figures 14.1 and 14.2. When charge inversion takes place that gives rise to products of like charge, the exit channel is characterized by a Coulomb barrier, in analogy with the exit channel associated with proton transfer ion–molecule reactions that involve multiply protonated species (He et al. 2005). Figure 14.4 shows a representative energy

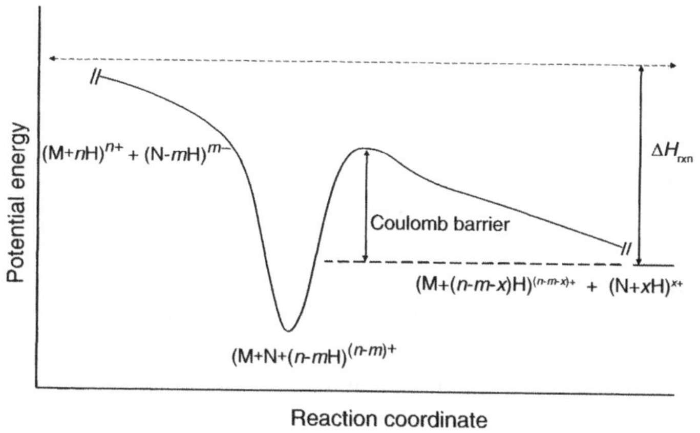

FIGURE 14.4. Simplified energy diagram of a charge inversion reaction that produces products of like charge.

diagram plotted with the entrance channel, characterized by the long-range Coulomb attraction in the entrance channel associated with the reactants and the Coulomb barrier in the exit channel that arises from the formation of products of like charge.

The numbers, strengths, and spacings of the various potential charge sites of the reactants and products determine the relative exothermicities of the various possible charge inversion channels. However, the competition between different product channels may be strongly affected by the Coulomb barrier in the exit channel.

14.2.1.3. Metal Ion Transfer. Strategies employing gas-phase ion–ion reactions as means for incorporating a metal cation into a multiply protonated polypeptide ion have been described (Payne and Glish 2001; Newton and McLuckey 2003; Newton et al. 2004, 2005). The motivation is to develop a capability for manipulating the makeup of the cationizing species associated with a polypeptide ion that is independent of the initial means for forming the ions. A general approach is to incorporate the metal ion in a negative singly charged metal–ligand complex that can be reacted with the polypeptide cation. A general net reaction can be written as

$$[M+nH]^{n+}+[Me_mL_{m+1}]^- \nearrow [M+(n-1)H]^{(n-1)+}+HL+mMeL \quad (14.14)$$

$$[M+nH]^{n+}+[Me_mL_{m+1}]^- \searrow [M+(n-x)H+(x-1)Me]^{(n-1)+}+xHL+(m-x+1)MeL \quad (14.15)$$

where Me represents a metal and L represents a ligand and where x can range from 1 to $(m+1)$. Reaction (14.14) represents a competitive process that does not result in the transfer of a metal ion, whereas reaction (14.15) represents the transfer of one or more metal ions. For simplicity, only the case for the reaction of a doubly protonated species with a metal containing complex of the form MeL_2^-, as indicated below, is discussed further here:

$$[M+2H]^{2+} + MeL_2^- \nearrow [M+H]^+ + HL + MeL \quad (14.16)$$

$$[M+2H]^{2+} + MeL_2^- \searrow [M+Me]^+ + 2HL \quad (14.17)$$

Given that the entrance channel for the reaction is dominated by the attractive Coulomb potential and that the exit channels represent an ionic and two neutral products, the energy diagram of Figure 14.1 most closely represents the energy surface of this type of reaction (any chemical barriers in the exit channels are not depicted in Figure 14.1). The net reactions (14.16) and (14.17) represent cases for proton transfer and metal ion transfer, respectively. However, several intermediate products are sometimes also observed and, in some cases, can be the dominant product ion species. Figure 14.5 summarizes the major possible products from the reactants of reactions (14.16) and (14.17).

An overall kinetic scheme that applies to this reaction is given in Figure 14.6.

Examination of the scheme of Figure 14.6 indicates that there are three distinct pathways for reaction (14.16) and one for reaction (14.17). Reaction (14.16), which represents a proton transfer reaction and does not involve metal ion transfer, can occur via a long-range proton transfer mechanism (indicated as proton hopping in the scheme), as discussed further below. There are two pathways to net proton transfer that can proceed via the breakup of a longlived chemical complex involving loss of HL followed by loss of MeL or loss of MeL followed by loss of HL. Reaction (14.17) is expected to proceed exclusively via a relatively longlived chemical intermediate, represented as $\{[M + 2H]^{2+} + MeL_2^-\}^{*+}$ in the scheme of Figure 14.6, through the losses of two HL molecules. Depending on the stabilities of this intermediate, as well as those formed via loss of either MeL or HL, the product ion spectrum may show significant abundances of the $([M + 2H]–MeL_2)^+$, $(M + 2H + L)^+$, and $(M + H + MeL)^+$ species (see Section 14.2.3). For a given peptide ion, the identity of the ligand plays a major role in the extent to which adducts species are observed, as illustrated in Figure 14.7.

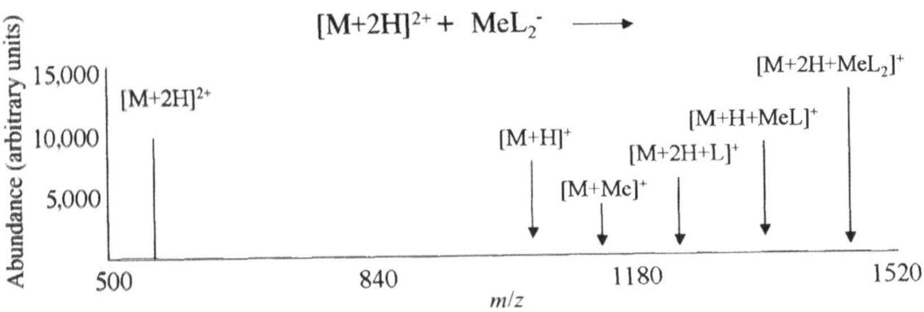

FIGURE 14.5. An illustration of where the major products of an ion–ion reaction involving a doubly protonated peptide (M) and a metal containing anionic comples (MeL_2^-) might fall. [Reprinted from Newton et al. (2005) with permission from the American Chemical Society.]

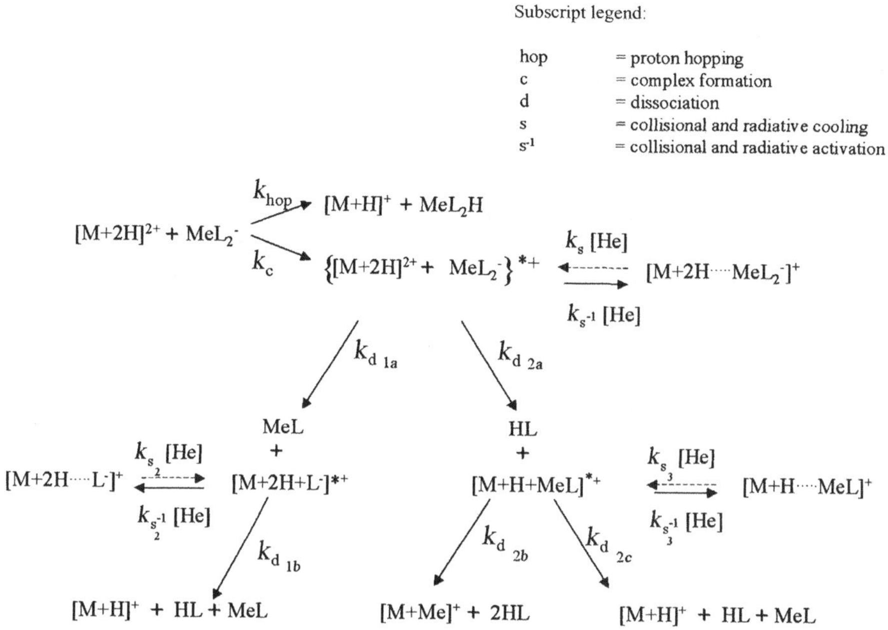

FIGURE 14.6. A kinetic scheme that applies to the reaction $[M+2H]^{2+} + MeL_2^- \rightarrow$ products. [Reprinted from Newton et al. (2005) with permission from the American Chemical Society.]

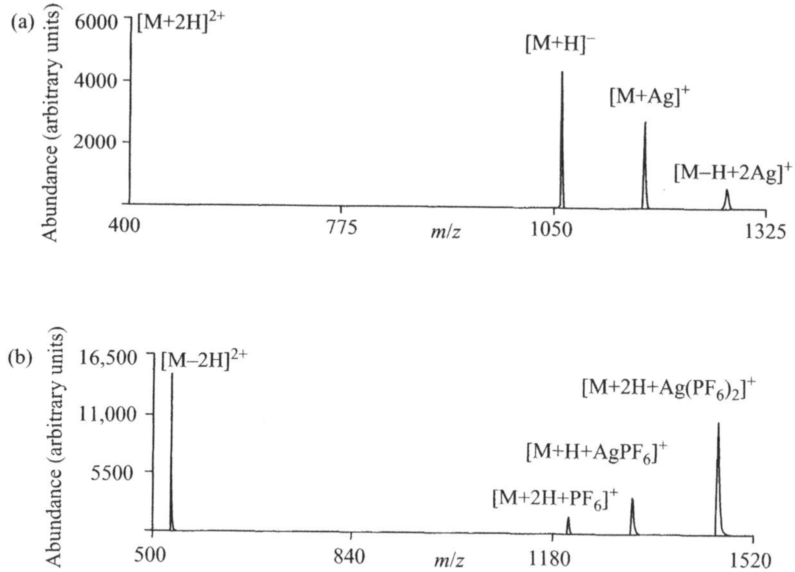

FIGURE 14.7. (a) Product ion spectrum from the reaction of bradykinin $[M+2H]^{2+}$ with silver acetate anions, Ag_nL_{n+1}, where $n = 1$–3; (b) product ion spectrum from the reaction of bradykinin $[M+2H]^{2+}$ with $Ag(PF_6)_2^-$. [Reprinted from Newton et al. (2005) with permission from the American Chemical Society.]

The reactions of multiply protonated peptides with metal acetate anions tends to lead to net proton transfer or metal ion transfer [i.e., reactions (14.16) and (14.17)], whereas reactions with silver hexafluorophosphate anions tend to lead to formation of adduct ions. The tendency for adduct ion formation increases with the strength of the interaction of the ligand with protonation sites in the peptides (Newton et al. 2005). Note that the absence of an abundant $[M+H]^+$ product in Figure 14.7b suggests that the proton hopping channel in the scheme of Figure 14.6 is relatively unimportant, at least for $Ag(PF_6)_2^-$. In any case, it is clear that the identity of L^- plays an important role in the extent to which the net reactions (14.16) and (14.17) are observed.

The identity of L^- also plays an important role in the competition between the net reactions (14.16) and (14.17). While the products are under kinetic control, it is instructive to examine the thermodynamics for these reactions as they can affect relative reaction rates, particularly when reactions proceed over similar energy surfaces (i.e., similar mechanisms apply with similar entropic constraints). For the net competitive reactions (14.16) and (14.17), the difference in the reaction enthalpies is given by

$$\Delta H_{\text{rxn}(14.16)} - \Delta H_{\text{rxn}(14.17)} = \Delta\Delta H_{\text{rxn}} = [\text{PA(M)} + \text{MIA}(\text{L}^-)] - [(\text{MIA(M)} + \text{PA}(\text{L}^-)] \qquad (14.18)$$

where PA(M) and MIA(M) are the proton and metal ion affinities of M, respectively, and $PA(L^-)$ and $MIA(L^-)$ are the respective proton and metal ion affinities of the anionic ligand. As expressed in (14.18), a negative $\Delta\Delta H_{rxn}$ favors formation of $[M+H]^+$ whereas a positive $\Delta\Delta H_{rxn}$ favors formation of $[M+Me]^+$. The relationship in (14.18) can be rewritten as follows:

$$\Delta H_{\text{rxn}(14.16)} - \Delta H_{\text{rxn}(14.17)} = \Delta\Delta H_{\text{rxn}} = [\text{PA(M)} - \text{MIA(M)}] + [(\text{MIA}(\text{L}^-) - \text{PA}(\text{L}^-)] \qquad (14.19)$$

Relation (14.19) makes clear that, because [PA(M) – MIA(M)] is fixed for a given peptide and metal, it is the selection of L^- that determines the relative exothermicities of net proton transfer versus metal ion transfer. Net metal ion transfer is thermodynamically more favored as the $[(MIA(L^-) - PA(L^-)]$ term becomes more positive (or less negative). For the limited set of ligands studied to date, the data are consistent with expectation based on the relative exothermicities of reactions (14.16) and (14.17). For example, the difference between the silver ion affinities and proton affinities of nitrate and acetate anions are −189 and −196 kcal/mol, respectively (Newton et al. 2005). Silver ion transfer to a given peptide ion is therefore expected to be greater when the anionic reagent is $Ag(NO_3)_2^-$ than when it is $Ag(CH_3CO_2)_2^-$, and this is consistent with the data, as can be seen by comparing Figures 14.7a and 14.8. Figure 14.7a shows results for the reaction of doubly protonated bradykinin with silver acetate anions, whereas Figure 14.8 shows the results derived from the reaction of doubly protonated bradykinin with silver nitrate

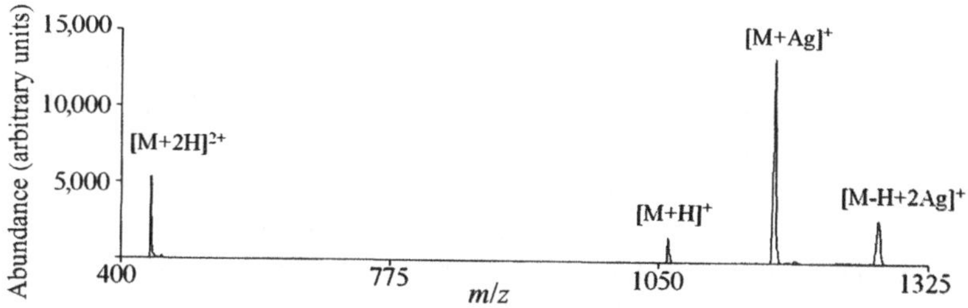

FIGURE 14.8. Product ion spectrum derived from the reaction of doubly protonated bradykinin with silver nitrate anions. [Reprinted from Newton et al. (2005) with permission from the American Chemical Society.]

anions (note that both anion populations also contained significant abundances of $Ag_2L_3^-$ ions, which account for the appearance of $[M - H + 2Ag]^+$ ions).

14.2.2. Electron Transfer Reactions

To date, only single-electron-transfer reactions have been observed in ion–ion reactions of ions derived from biomolecules. The two types of such reactions involve electron transfer *to* a biomolecule cation and electron transfer *from* a biomolecule anion. In both cases, the entrance channel is dominated by long-range Coulomb attraction and the exit channel follows that of an ion–molecule reaction, as depicted qualitatively in Figure 14.1. The first electron transfer reactions to be studied were those of multiply deprotonated oligonucleotides with rare-gas cations (Herron et al. 1995a, McLuckey et al. 1997). More recent work has extended the reactions to include those of multiply deprotonated peptides (Coon et al. 2005) with rare-gas cations. The use of rare-gas cations avoids the complicating possibility of competition between proton transfer and electron transfer such that essentially 100% of the reactions proceed via electron transfer. A reaction involving electron transfer from a biomolecule-ion, represented as a multiply deprotonated species, $(M - nH)^{n-}$, with a radical cation ($A^{+\bullet}$) is given by

$$(M - nH)^{n-} + A^{+\bullet} \rightarrow (M - nH)^{(n-1)-\bullet} + A \qquad (14.20)$$

(Note that the reagent cation need not be odd-electron.) The enthalpy of the reaction is given by

$$\Delta H_{rxn} = EA[(M - nH)^{(n-1)-\bullet}] - RE[A^{+\bullet}] \qquad (14.21)$$

where EA represents the electron affinity and RE represents the recombination energy. Under certain conditions, such reactions have resulted in extensive dissociation of the product anions, as discussed further below.

In the case of electron transfer to a multiply protonated biomolecule, proton transfer to the reagent anion is a competing mechanism because all anions have a positive proton affinity and can serve as a proton acceptor. Most ion–ion reactions involving multiply protonated species have been conducted with reagent anions that tend to behave as Brønsted bases. Electron transfer has not been noted to be a competitive process in these cases. However, electron transfer has been noted to be an important ion–ion reaction channel for an increasing number of reagent anions (Coon et al. 2004). A major motivation for seeking reagents that react more or less exclusively via electron transfer is the observation (Syka et al. 2004) that ion–ion electron transfer to polypeptide ions gives rise to fragmentation that is highly analogous to that observed in electron capture dissociation (ECD) (Zubarev et al. 1998; Zubarev 2003). Such dissociation, denoted *electron transfer dissociation* (ETD), has important applications in biomolecule structure determination via tandem mass spectrometry (MS/MS). A general reaction for electron transfer from a radical anion, $A^{-\bullet}$, to a multiply protonated biomolecule is given by

$$(M + nH)^{n+} + A^{-\bullet} \rightarrow (M + nH)^{(n-1)+\bullet} + A \tag{14.22}$$

Note that the reagent anion need not be odd-electron and, at least for many polypeptide ions, the cationic product fragments as a result of the ion–ion reaction. The enthalpy for the electron transfer reaction is given by

$$\Delta H_{\mathrm{ET}} = \mathrm{EA}[A] - \mathrm{RE}[(M + nH)^{n+}] \tag{14.23}$$

The enthalpy associated with fragmentation that results from electron transfer (i.e., ETD) is given by

$$\Delta H_{\mathrm{ETD}} = \mathrm{EA}[A] + \varepsilon_{\mathrm{crit}}[(M + nH)^{(n-1)+\bullet}] - \mathrm{RE}[(M + nH)^{n+}] \tag{14.24}$$

where $\varepsilon_{\mathrm{crit}}[(M + NH)^{(n-1)+}]$ represents the critical energy associated with a dissociation of the polypeptide product ion.

14.2.3. Condensation Reactions

The reaction implied by the term "condensation" is one in which the two reactants combine to form a single product. Such reactions have proved to be commonplace with bioion/ion reactions in electrodynamic ion traps. The attachment of small anions, such as iodide (Stephenson and McLuckey 1997a) (I^-) to multiply protonated polypeptides, denoted generically as

$$(M + nH)^{n+} + A^{-} \rightarrow (M + A + nH)^{(n-1)+} \tag{14.25}$$

has been noted for reagents that give rise to relatively strong dipole–dipole or ion pairing interactions with protonated sites on the polypeptide. Although less extensively studied, attachment of small cations to multiply deprotonated species has also been observed (Wu and McLuckey 2003). Furthermore, the formation of complexes derived from two or more multiply charged biomolecule–ion reactants, represented as

$$(M + nH)^{n+} + (N - mH)^{m-} \rightarrow (M + N + (n - M)H)^{(n-m)+} \quad (14.26)$$

for the case in which $n > m$, has been noted to be a major ion–ion reaction channel (Wells et al. 2001, 2003; Gunawardena et al. 2004). Figure 14.9 illustrates phenomenology common to the reactions of multiply charged proteins of opposite polarity [e.g., when M and N of reaction (14.26) are both proteins]. Figure 14.9a shows the positive-ion spectrum obtained after the +8 ions of ubiquitin (U^{8+}) were allowed to react with multiply deprotonated cytochrome *c*, C^{5-}. The most abundant product, $(UC)^{3+}$, is formed via reaction (14.26). However, there is also evidence for partial proton transfer [see reaction (14.8)] from the appearance of ions such as U^{5+} and U^{4+}. The appearance of partial proton transfer products has been noted to be dependent on the reactant charge states and identities (Wells et al. 2003). For example, Figure 14.9b shows the products formed with multiply protonated cytochrome *c* ions, C^{8+}, reacting with ubiquitin anions, U^{5-}. In this case, only the condensation product, $(CU)^{3+}$, is observed despite the fact that the charge states of the reactants are the same for both experiments reflected in the Figure 14.9 data. It is also noteworthy that no fragmentation of covalent bonds is noted in either

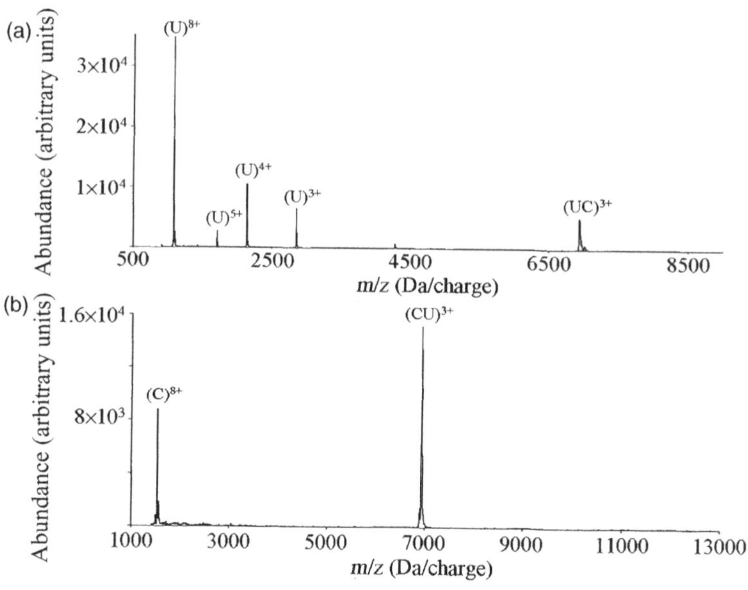

FIGURE 14.9. (a) Positive-ion products noted for the reaction of ubiquitin $(M + 8H)^{8+}$ (denoted as U^{8+}) with cytochrome *c* $(M–5H)^{5-}$ (denoted as C^{5-}); (b) positive-ion products noted for the reaction of cytochrome *c* $(M + 8H)^{8+}$ (denoted as C^{8+}) with ubiquitin $(M–5H)^{5-}$ (denoted as U^{5-}). [Reprinted from Wells et al. (2003) with permission from the American Chemical Society.]

experiment, despite the relatively high exothermicities associated with the condensation of two multiply charged ions.

14.2.4. Fragmentation Arising from Ion–Ion Reactions

Despite the relatively high exothermicities of ion–ion reactions, fragmentation of at least one of the products is observed in only a minority of instances when the reactions take place in an electrodynamic ion trap operated with a bath gas present at roughly 1 mTorr. While this may, on the surface, appear to be a surprising result, it is important to consider the lifetimes of relatively large polyatomic ions that may be formed with excess internal energy; the rates for removal of excess internal energies, both radiatively and via collisions; and the partitioning of reaction exothermicity between internal modes of the products and translational degrees of freedom. Many considerations relevant to ion–ion reactions involving systems with large numbers of states and high state densities have been discussed within the context of ionization of large molecules (Schlag et al. 1992; Schlag and Levine 1992). In general, the factors noted to date that influence the extent to which fragmentation of product ions initially formed as a result of an ion–ion reaction in the quadrupole ion trap environment are as follows: (1) the rate at which excess product ion internal energy is removed by ion–bath gas collisions and emission, (2) reaction exothermicity, (3) the partitioning of reaction exothermicity among all available degrees of freedom, (4) the kinetic stability of the initially formed product ion, (5) the number of degrees of freedom of the initially formed product ion, and (6) the bath gas temperature. All of these factors contribute to the likelihood that fragmentation of a product ion will occur as a result of an ion–ion reaction. Each of these factors is discussed briefly here, in some cases with illustrative data. In the case of ETD, insufficient data have been collected to draw firm conclusions regarding the roles of reaction exothermicitiy, energy partitioning, and cooling rates. Nevertheless, some preliminary observations can be made.

14.2.4.1. Product Ion Cooling Rates. The products of all ion–ion reactions conducted in the presence of a background bath gas are subject to relatively high ion–bath gas collision rates. Therefore, any of the ion–ion reaction exothermicity partitioned into product ion internal energy will tend to be removed from the ion via collisions and infrared photon emission until the ion is essentially thermalized. An inefficient colliders model was used to simulate collisional cooling in the ion trap to determine the order of magnitude of collisional cooling rates that can be expected for bioions in an electrodynamic ion trap environment (i.e., in the presence of a relatively light bath gas at roughly 1 mTorr) (Goeringer and McLuckey 1998). Modeling was done with poly$(AG)_n$ units both for ion cooling and for dissociation using the RRKM parameters published by Griffin and McAdoo (1993). Figure 14.10 compares collisional cooling rates with unimolecular dissociation rates for $(AG)_n$ ions as a function of n for the case of 300-K ions with an input of 4 eV. This particular plot was intended to simulate the absorbance of a 308-nm photon.

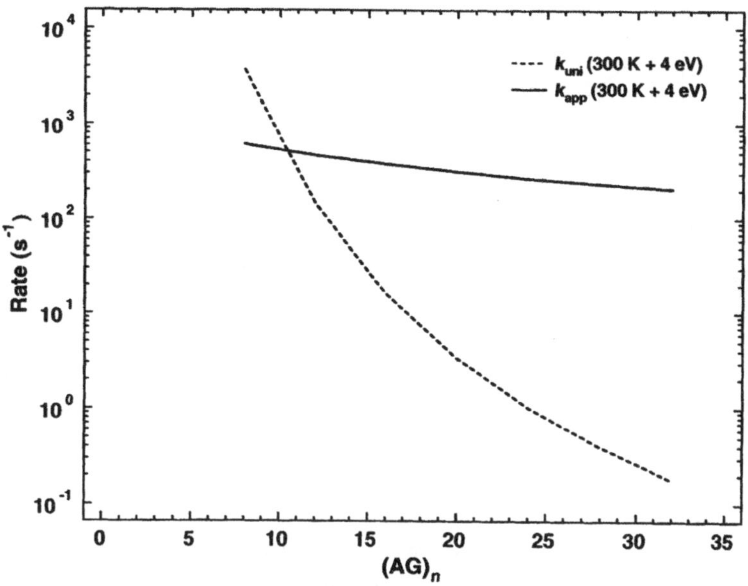

FIGURE 14.10. A comparison of simulated collisional cooling rates k_{app} and dissociation rates k_{diss} for poly$(AG)_n$ as a function of n. [Reprinted from Goeringer and McLuckey (1998) with permission from Elsevier Science.]

The simulation indicates collisional cooling rates of 10^2–$10^3\,s^{-1}$ based on relatively conservative assumptions regarding the efficiency with which single ion–bath gas collisions remove excess energy (Goeringer and McLuckey 1998). The photoactivation simulation is particularly relevant to the ion–ion reaction case in that the input of internal energy into the product ion as a result of the ion–ion reaction effectively deposits energy into the ion relatively quickly.

The profound impact that the presence of the bath gas can have on an experiment in which a discrete and rapid energy input takes place in an bioion is illustrated in Figure 14.11, which indicates the time evolution of the fractional parent ion abundance of $(AG)_8$ ions initially present at 300 K and after the input of 4 eV of excess internal energy.

Note that at $t = 0$, $k_{diss} = 3.7 \times 10^3\,s^{-1}$ whereas at $t = 0.5$ ms, $k_{diss} = 79\,s^{-1}$. By 1 ms, fragmentation has largely been inhibited by collisional cooling such that roughly 55% of the initial parent ion population is not fragmented by the input of 4 eV. The key point here is that, regardless of the type of ion–ion reaction, any energy deposition into a product ion from the ion–ion reaction exothermicity must drive dissociation at rates significantly higher than the collisional cooling rate for fragmentation to be observed.

14.2.4.2. Reaction Exothermicity. The energy to drive dissociation of ion–ion reaction product ions originates from the initial internal energies of the reactants, the ion–ion reaction exothermicity, and any input in energy subsequent to product ion

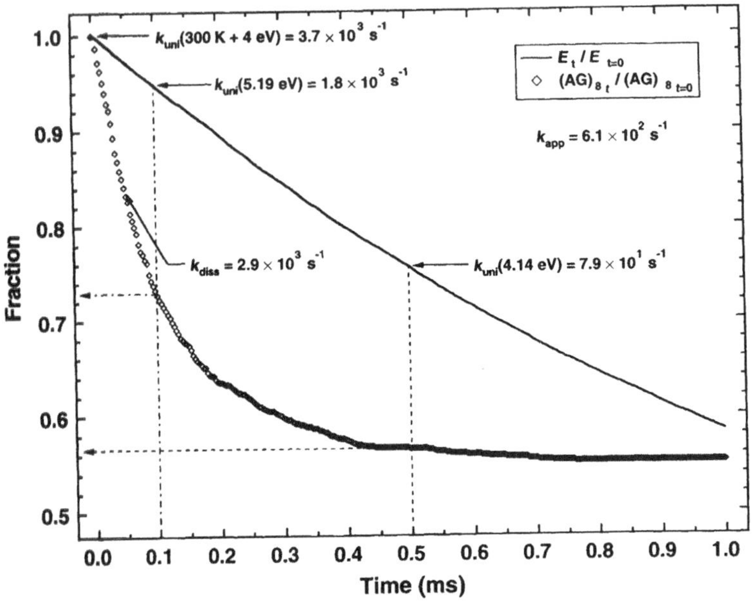

FIGURE 14.11. Time evolution of the fractional abundance of unfragmented parent ions, $(AG)_{8,t}/(AG)_{8,0}$, and the internal energy ratio E_t/E_0 as a function of time after the input of 4 eV into a 300-K ion population. A collisional cooling rate of $610\,s^{-1}$ was used in the simulation and RRKM parameters provided by Griffin and McAdoo (1993) were used to determine the dissociation rate k_{diss} [Reprinted from Goeringer and McLuckey (1998) with permission from Elsevier Science.]

formation. As ion–ion reaction exothermicity increases, the likelihood that sufficient energy is partitioned into internal energy of a product ion to induce dissociation also increases. The effect is likely reflected in Figure 14.12, which compares the product ion spectra obtained from the reaction of the triply charged anion, $(M-3H)^{3-}$, derived from negative electrospray ionization of 5′-*d*(AAAA)-3′ with protonated isobutylene, $C_4H_9^+$ (Figure 14.12a), and with protonated benzoquinoline (Figure 14.12b), BQH^+. Structurally informative fragments known to arise from the unimolecular decomposition of the $(M-3H)^{3-}$ ion (Wu and McLuckey 2004) are observed at relatively low abundance in the data collected with $C_4H_9^+$ whereas no evidence for fragmentation is noted in the reaction with BQH^+.

The enthalpies associated with the reactions are given by relation (14.5). The only difference in enthalpies associated with the experiments of Figure 14.12 arise from the difference in the proton affinities of isobutylene (191.4 kcal/mol) and benzoquinoline (229.6 kcal/mol). Hence, the reaction with $C_4H_9{}^+$ is roughly 38 kcal/mol more exothermic than the reaction with BQH^+. The proton affinity of the $[M-3H]^{3-}$ ion of 5′-*d*(AAAA)-3′ has not been determined but the proton affinity of $(CH_3O)_2OPO^-$, taken as a model of the deprotonated phosphodiester linkage, has a calculated PA of 329.8 kcal/mol (Wu and McLuckey 2003). Multiple charging further increases the PA of the anionic reactant. Therefore, the enthalpies of

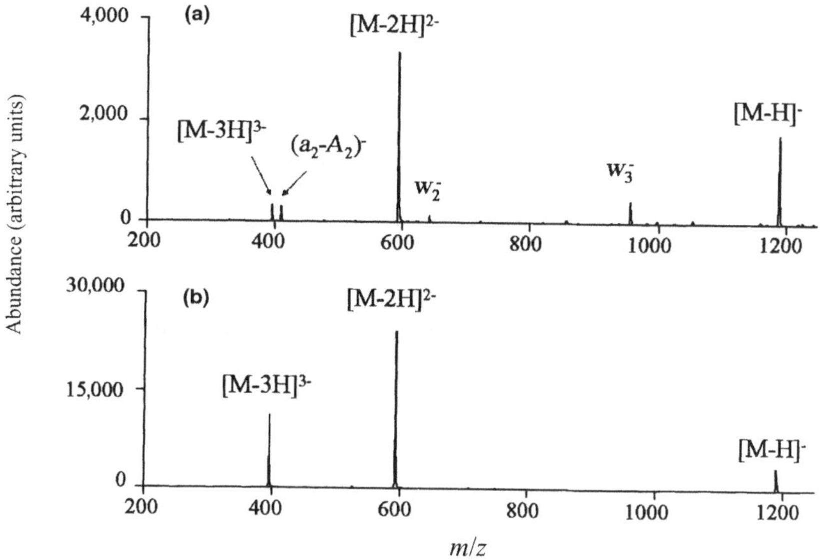

FIGURE 14.12. (a) Product ion spectrum arising from the reaction of the $(M–3H)^{3+}$ ion of 5′-*d*(AAAA)-3′ with protonated isobutylene; (b) product ion spectrum arising from the reaction of the $(M–3H)^{3+}$ ion of 5′-(AAAA)-3′ with protonated benzoquinoline. [Reprinted from Wu and McLuckey (2003) with permission from Elsevier Science.]

the reactions associated with Figure 14.12 are expected to be at least −4.3 eV for BQH^+ and −6 eV for $C_4H_9^+$.

Data for a similar reaction enthalpy comparison have also been collected for electron transfer reactions from the $(M-3H)^{3-}$ ion of 5′-*d*(AAAA)-3′ to $Xe^{+\bullet}$ and to CCl_3^+ (McLuckey et al. 1997) (see Figure 14.13). In this case, the reaction enthalpy is given by relation (14.21) and the difference in the enthalpies of reaction for the two cationic reagents arises from the difference in recombination energies. The ionization energies (IEs) of Xe (IE = 12.1 eV) and $CCl_3^\bullet$ (IE = 7.8 eV) provide estimates of the relevant recombination energies and suggest that electron transfer from $(M-3H)^{3-}$ to ionized xenon is roughly 4.3 eV more exothermic than electron transfer to CCl_3^+. (Note that the use of the IE to estimate the RE is more uncertain with the polyatomic species than with the atomic species. Nevertheless, the exothermicity of the reaction with $Xe^{+\bullet}$ is expected to be significantly greater than that with CCl_3^+.) Examination of the ion–ion reaction data for these two reagents shows that each leads to fragmentation of the oligonucleotide product anion. However, the relative abundance of the intact electron transfer product, $(M-3H)^{2-\bullet}$, is much greater in the data obtained using CCl_3^+ than it is for data obtained using $Xe^{+\bullet}$. In fact, much more signal due to proton transfer to residual neutral species in the vacuum system is observed in the $Xe^{+\bullet}$ data, relative to the signal due to $(M-3H)^{2-\bullet}$, than in the CCl_3^+ data [see also the abundance of the $(M-3H)^{2-\bullet}$ ion relative to the fragment ions]. These results indicate that a significantly greater fraction of the initially formed doubly charged product ions

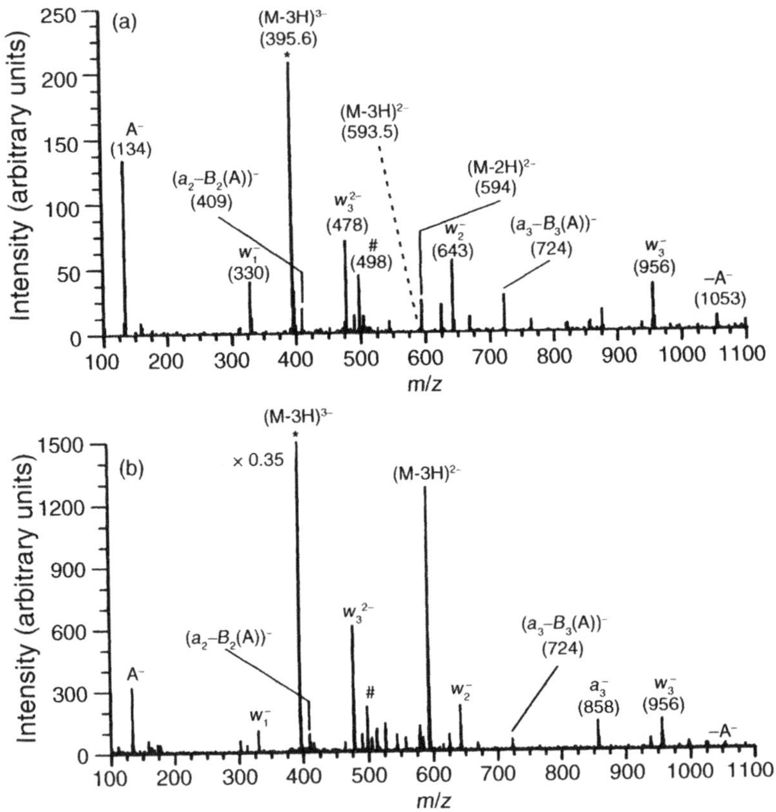

FIGURE 14.13. (a) Product ion spectrum arising from the reaction of the $(M–3H)^{3+}$ ion of 5′-*d*(AAAA)-3′ with $Xe^{+\bullet}$; (b) product ion spectrum arising from the reaction of the $(M–3H)^{3+}$ ion of 5′-(AAAA)-3′ with CCl_3^+. [Reprinted from McLuckey et al. (1997) with permission from Elsevier Science.]

survive in the CCl_3^+ experiment than in the $Xe^{+\bullet}$ experiment, presumably due to the lower overall reaction exothermicity in the former relative to the latter.

14.2.4.3. Energy Partitioning. Reaction exothermicity, of course, provides only an upper limit to the internal energy that can be deposited into a product ion as a result of ion–ion reaction. A key issue is how the energy is partitioned among all possible modes associated with the product ions. Energy partitioning is probably the least well understood aspect of ion–ion reactions of relevance to fragmentation. There is relatively little definitive information on the extent to which energy partitioning may differ for the various types of ion–ion reactions. For example, very little evidence for fragmentation of polypeptide cations has been noted to result from ion–ion proton transfer reactions of the types associated with reaction (14.1), (14.6), or (14.8). On the other hand, fragmentation arising from proton transfer to an oligonucleotide anion has been observed [reaction (14.4); see also Figure 14.12]. It might be argued that the species that accepts the proton will absorb more of the reaction exothermicity than will the species donating the proton because much of the energy will reside initially in the newly formed bond. This might account for the fact that essentially no fragmentation is observed for protein cations, and even noncovalently bound protein complexes, in ion–ion proton transfer reactions (Stephenson and McLuckey 1996a,b; Stephenson et al. 1997), despite the fact that the reaction exothermicities can be relatively high. It might also account for the observation of fragmentation of the oligonucleotide anions. However, the kinetic stabilities of peptide cations and oligonucleotide anions differ.

The dynamics of ion–ion reactions, discussed below, are also expected to play an important role in determination of energy partitioning even for a given reaction type. For example, reactions that proceed via a charged particle transfer at a curve crossing are expected to show different energy partitioning than reactions that proceed via the formation of a longlived chemical complex. Evidence for both reaction mechanisms is evident (see discussion below). The former might be expected to partition some of the reaction exothermicity into relative translation of the products, whereas the latter is expected to deposit much of the reaction exothermicity into the initially formed complex. These mechanisms are expected to compete, and their relative contributions are expected to vary depending on the nature of the reactants.

14.2.4.4. Product Ion Kinetic Stability. For fragmentation to be observed to result from an ion–ion reaction, sufficient internal energy must be present in the product ion to drive fragmentation at rates necessary to avoid cooling. For ions that fragment statistically, the lifetime of the excited product is a function of its internal energy, the number of degrees of freedom over which the energy can be distributed, and the energies and entropies associated with dissociation. The number of degrees of freedom and energies and entropies of activation are fundamental characteristics of the ion, and they determine the internal energy dependent fragmentation rates and, as a consequence, the internal energy dependent lifetimes of the ion. The roles of energy and entropy of activation and numbers of degrees of freedom are discussed separately here. It should be noted that electron capture dissociation (ECD) has been argued to be a nonergodic process (Zubarev et al. 1998); that is, fragmentation does not follow randomization of internal energy. ETD, the ion–ion reaction analog, therefore, may be expected to undergo similar dissociation dynamics. The timeframe of the ion trap experiment precludes determination of dissociation rates greater than roughly $10^3\ s^{-1}$. Therefore, the lifetimes of the ions that undergo dissociation cannot be determined directly. Nevertheless, the kinetic stability of the product ions is key to the observation of fragmentation, regardless of whether it takes place after energy randomization.

Electron transfer from oligonucleotide homopolymeric anions likely provides an example of product ion kinetic stability as the major factor in determining the extent of fragmentation resulting from an ion–ion reaction. Figure 14.14 compares postion–ion reaction spectra from the reaction of the $[M - 7H]^{7-}$ anions derived from 5′-*d*(A)$_{20}$-3′ and 5′-*d*(T)$_{20}$-3′ with $O_2^{+\bullet}$ (IE = 12.07 eV) (Wu and McLuckey 2003). PolyT homopolymer ions are known to be kinetically more stable than either

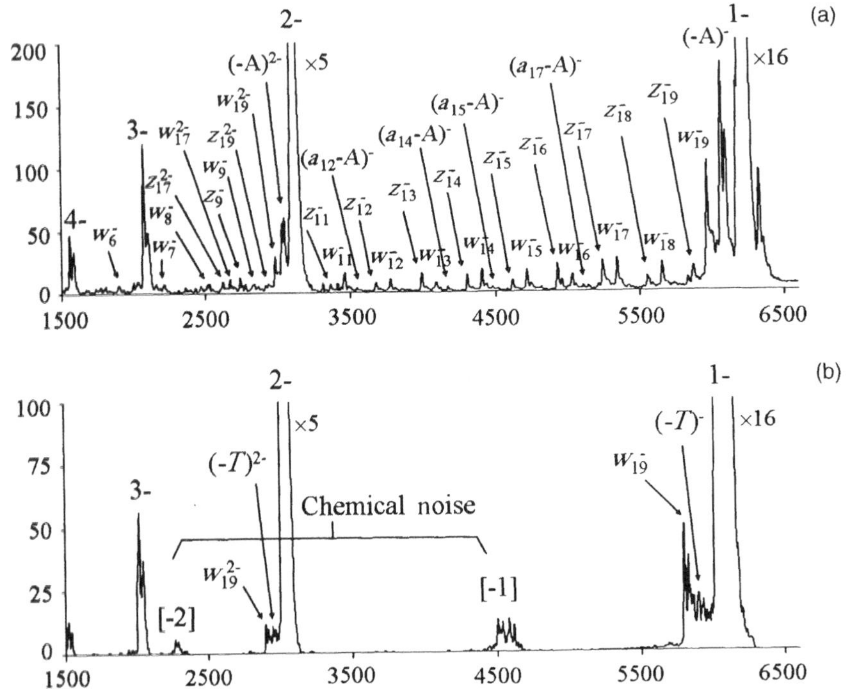

FIGURE 14.14. (a) Postion–ion spectrum derived from the reaction of $[M-7H]^{7-}$ of 5′-$d(A)_{20}$-3′ with $O_2^{+\bullet}$; (b) postion–ion spectrum derived from the reaction of $[M-7H]^{7-}$ of 5′-$d(T)_{20}$-3′ with $O_2^{+\bullet}$. [Reprinted from Wu and McLuckey (2003) with permission from Elsevier Science.]

mixed base or homopolymers of the other nucleotides (Wu and McLuckey 2004). Multiply charged oligonucleotide anions that contain adenosine residues, on the other hand, have shown a tendency to lose adenine and cleave at the 3′ C–O bond of the sugar from which the adenine was lost (McLuckey et al. 1992). Hence, polyA anions are expected to be kinetically less stable than polyT anions. The data of Figure 14.14 show that, as a result of multiple electron transfer reactions, a wide array of low-abundance fragments are observed with the polyA anions, including *w*-type, (*a* − A)-type, and *z*-type ions. Much less fragmentation is observed for the polyT anions. This fragmentation is limited to a relatively small degree of T loss and the formation of the w_{19}^{-} fragment. The different fragmentation behavior is likely to arise largely from the differences in the kinetic stabilities of the homopolymer product ions because differences in reaction exothermicities and numbers of degrees of freedom are expected to be relatively small.

14.2.4.5. Product Ion Degrees of Freedom. The size of a product ion that is formed with excitation arising from an ion–ion reaction is expected to play a major role in the lifetime of the ion, if it behaves statistically. This is clearly reflected in the rapidly falling dissociation rate determined from the RRKM calculation associated with Figure 14.10. Hence, for a given reaction exothermicity, energy partitioning, and entropies and energies of activation, product ion lifetimes are expected to increase with ion size. This phenomenon underlies so-called kinetic shifts associated with the fragmentation of polyatomic ions. Higher internal energies are required in systems with many degrees of freedom to drive dissociation rates at levels comparable to those of smaller ions despite very similar activation energies and entropies. As indicated above, when the dissociation rate becomes comparable to the cooling rate in the ion trap environment, fragmentation is actively inhibited. Partly for this reason, fragmentation of relatively large bioions as a result of an ion–ion reaction has not been observed to be a common phenomenon. A possible exception may be found in electron transfer to protein cations, where nonstatistical behavior has been proposed for the analogous ECD process. It is known that electron capture does not lead to formation of observable fragments for large protein cations (Zubarev 2003). This is consistent with a degree-of-freedom effect. However, it has been posited that dissociation of backbone bonds actually takes place but that intermolecular noncovalent interactions prevent the fragments from separating. Such an interpretation could be applied to any fast activation technique applied to large polyatomic systems, including energy deposition as a result of ion–ion reactions. As some of the mechanistic aspects of ECD and ETD of polypeptide cations have yet to be clarified, the role of the number of degrees of freedom of the product ion remains unclear.

A degree-of-freedom effect has been invoked to account for fragmentation behavior of oligonucleotide product anions formed via electron transfer to rare-gas cations (Stephenson and McLuckey 1997b). Electron transfer to rare-gas cations, such as $Kr^{+\bullet}$ and $Xe^{+\bullet}$, have been shown to lead to extensive fragmentation of relatively small (e.g., 4mers to 6mers) multiply charged oligonucleotide anions (see also Figure 14.13a). As the oligomer length increases, however, the fraction of ion–ion reaction dissociation products decreases. A comparison of Figures 14.13a and 14.14a is probably relevant in this regard. The data were collected with different reagents ($Xe^{+\bullet}$ in the case of Figure 14.13a and $O_2^{+\bullet}$ in the case of Figure 14.14b), but their ionization energies are very similar. Very extensive fragmentation (>90%) was noted for the triply charged anion of 5′-*d*$(A)_4$-3′ in a single-electron transfer step to $Xe^{+\bullet}$ whereas only a small fraction of 5′-*d*$(A)_{20}$-3′ anions were observed to fragment after six or fewer sequential electron transfer reactions with $O_2^{+\bullet}$. Further evidence that oligomer size plays a role in the observation of fragmentation comes from the data of Figure 14.15, which shows preion/ion (Figure 14.15a) and postion–ion (Figure 14.15b) reaction data obtained with multiply charged anions derived from the mixed-base oligonucleotide 16mer 5′-*d*(GTCTTAGCGCTAAGAC)-3′ in reactions with $Xe^{+\bullet}$.

It is apparent that little or no fragmentation of the oligonucleotide anions results from the electron transfer reactions. At least some of the ions of low abundance in the baseline of Figure 14.15b might result from fragmentation, but the preion–ion mass spectrum also shows relatively low levels of background species and these can also contribute to the baseline of Figure 14.15b.

The experiment summarized by Figure 14.15 involves as many as eight consecutive ion–ion electron transfer reactions. The lack of fragmentation suggests

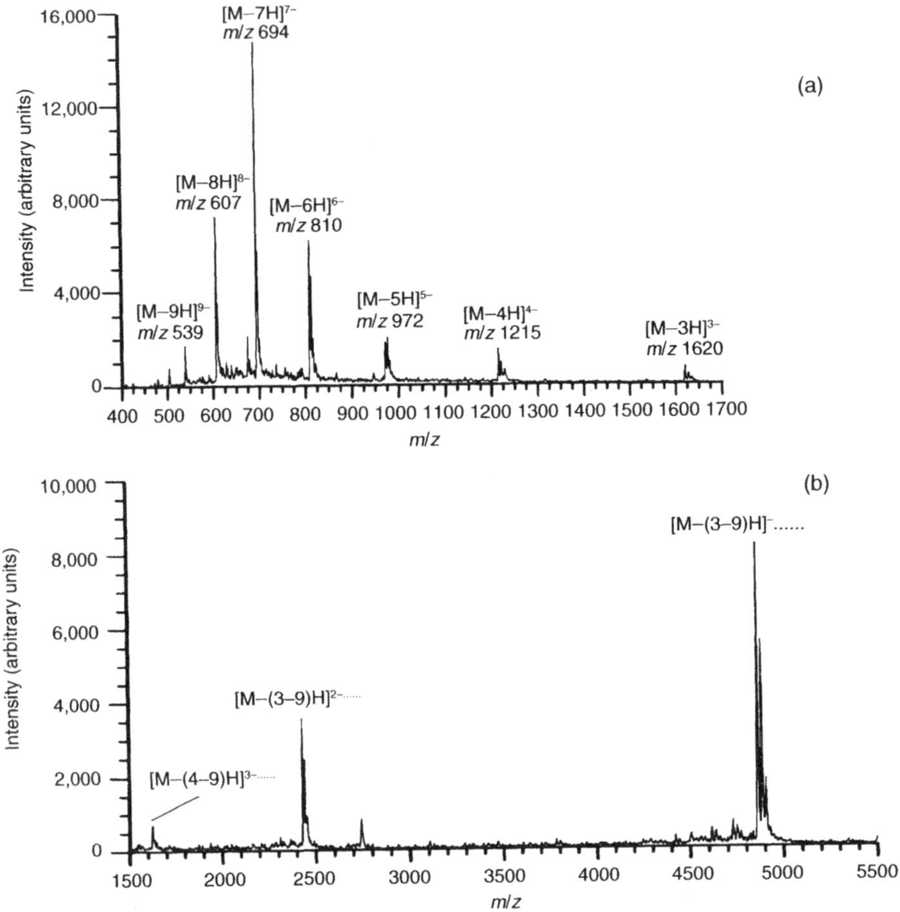

FIGURE 14.15. (a) Negative-ion electrospray mass spectrum of 5′-*d*(GTCTTAGCGCTAA-GAC)-3′; (b) postion–ion reaction data for the anions of (a) after a 250 ms reaction time with $Xe^{+\bullet}$ cations. [Reprinted from Stephenson and McLuckey (1997b) with permission from Wiley.]

that sequential ion–ion reactions are not effective in increasing the net ion internal energies of product ions under the conditions used here. This is consistent with relatively long product ion lifetimes, relatively high cooling rates in the ion trap, and relatively long times between ion–ion reactions. The plot of Figure 14.16 shows the results of a simulation of the ion–ion reaction rates that fit the data of Figure 14.15. The ordinate shows the average time interval between successive ion/ion reactions as a function of charge state. For example, the plot shows that slightly over 100 ms is expected to elapse between formation of a doubly charged anion and its conversion to a singly charged ion.

It is clear that relatively long time intervals between ion–ion collisions are expected at low charge states. However, even for the highest charge states, for which

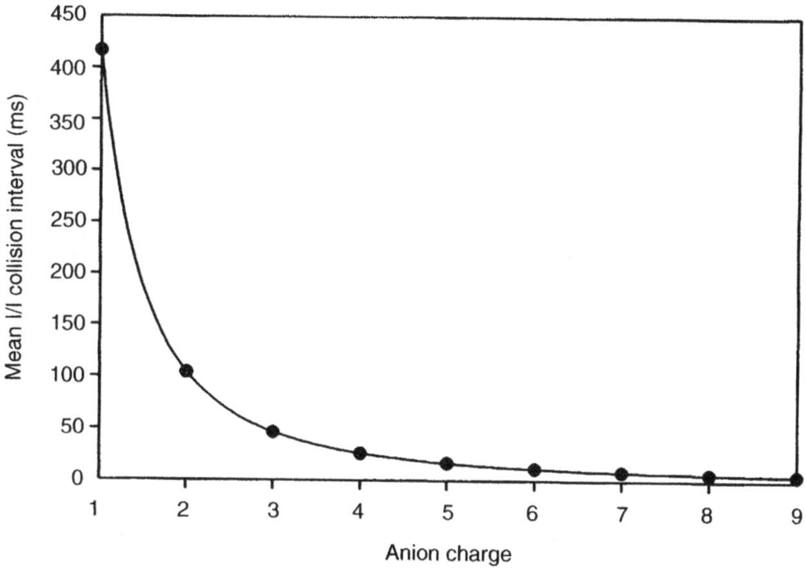

FIGURE 14.16. Mean ion–ion reaction interval as a function of anionic reactant charge state for the conditions used to collect the data of Figure 14.15. [Reprinted from Stephenson and McLuckey (1997b) with permission from Wiley.]

the time between ion–ion collisions is on the order of a millisecond, at least 10^3 ion–helium collisions are expected between ion–ion collisions when helium bath gas is present at roughly 1 mTorr. Hence, there is an important interplay between the cooling rate in the ion trap, the product ion lifetime, and the energy partitioned into the product ion in determining the likelihood for fragmentation.

14.2.4.6. Bath Gas Temperature. The bath gas in an electrodynamic ion trap tends to thermalize the ions trapped therein. A main motivation for its use is to remove translational energy via momentum transfer collisions so that the trapped ion population tends to collapse to the center of the trapping volume. Ion–helium collisions tend to bring the trapped ion population into thermal equilibrium. Therefore, excited ions tend to be cooled, whereas ions that might initially be vibrationally cold are heated by the bath gas. The temperature of the bath gas is not expected to be an important variable in the removal of excess ion kinetic energy because the translational energies of the ions are orders of magnitude higher than the translational energies of the bath gas atoms at generally accessible temperatures (e.g., $\leq$200 °C). However, the bath gas temperature can play a significant role in fragmentation (Asano et al. 1999a). For example, operation at elevated bath gas temperatures has been used to effect thermal dissociation of ions (Asano et al. 1999b, Butcher et al. 1999).

Bath gas temperature has not been extensively studied with respect to its possible role in fragmentation associated with ion–ion reactions. However, more recent

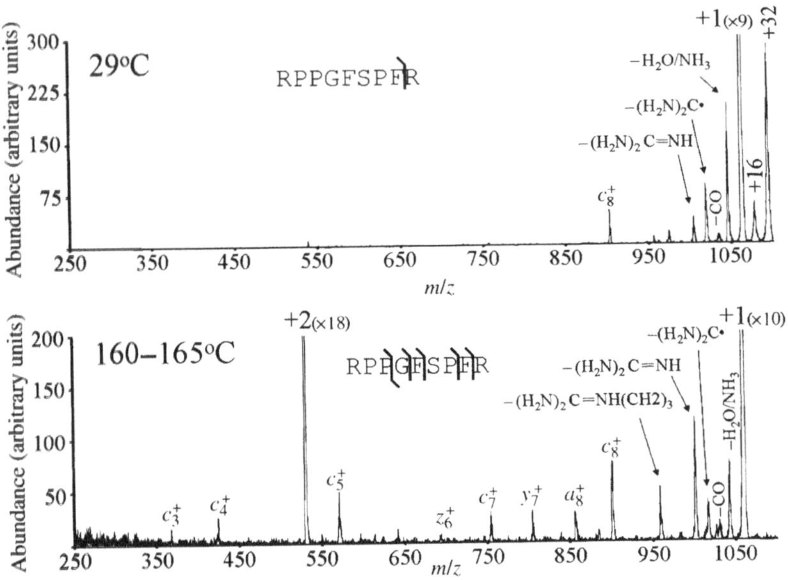

FIGURE 14.17. (a) Postion–ion reaction positive ion spectrum resulting from the reaction of doubly protonated bradykinin with molecular anions of nitrobenzene at a bath gas temperature of 29°C; (b) data collected for the same reactants at a bath gas temperature of 160–165°C. [Reprinted from Pitteri et al. (2005b) with permission from the American Chemical Society.]

results indicate that it plays a significant role in the ETD of peptide cations (Pitteri et al. 2005b). An example is provided in Figure 14.17, which shows the comparison of data derived from the reaction of bradykinin $[M + 2H]^{2+}$ cations with molecular anions of nitrobenzene at two different temperatures. As is commonly observed with many doubly protonated peptides in a room-temperature bath gas, electron transfer does not yield a wide array of fragment ions. At a bath gas temperature of 160–165°C, a wider variety of fragment ions are observed.

Similar data have been observed with other peptides. Specifically, when little fragmentation is observed at room temperature, data collected at elevated bath gas temperatures tend to show a wider array of fragments. Relatively little difference, however, is observed when extensive fragmentation of the peptide ion is already observed at room temperature. It is not possible to determine the mechanism by which the use of elevated bath gas temperature influences ETD from these data alone. Bath gas temperature can affect the structures of the parent ions that undergo the electron transfer reaction, and it can influence the energies of the product ions. Further studies whereby parent ion structures are modeled as a function of temperature are under way to clarify this issue. In any case, these data clearly show that bath gas temperature can be an important factor in fragmentation associated with ion–ion reactions.

14.3. BIOION–ION REACTION DYNAMICS

While the thermodynamics associated with ion–ion reactions are relevant with respect to energy partitioning and fragmentation, the reaction are under kinetic control. Hence, the reaction dynamics play a critical role in determining the relative contributions of proton transfer, electron transfer, and condensation reactions. A discussion of the dynamics of bioion–ion reactions is given here first by addressing small charged particle transfer versus condensation reactions. This discussion is centered, in particular, on proton transfer versus condensation. Having established the phenomena giving rise to charged particle transfer over a distance versus charged particle transfer via a longlived chemical complex and the observation of the complex itself (i.e., condensation), proton transfer versus electron transfer is addressed.

14.3.1. Proton Transfer and Condensation Reactions

Two oppositely charged ions acting under the influence of their mutual attraction can assume either bound or unbound trajectories. In the former case, the trajectories are elliptical in nature with a range of eccentricities ε of $0 < \varepsilon < 1$ (see Figure 14.18). A circular orbit constitutes a special case (i.e., $\varepsilon = 0$). Unbound trajectories are either hyperbolic ($\varepsilon > 1$) or parabolic in shape, the latter constituting the special case of $\varepsilon = 1$.

With no means for removal of some of the initial relative kinetic energy of the ions, all two-body interactions lead to unbound trajectories. If some of the initial relative kinetic energy is removed during the interaction, a bound orbit can result. The means for removal of relative kinetic energy include a collision with a third body or a "tidal effect" (Bates and Morgan 1990, Morgan and Bates 1992), whereby

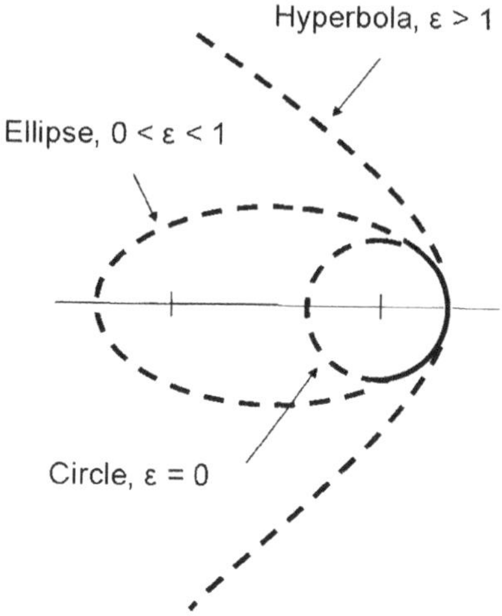

FIGURE 14.18. Orbits of a body moving in a central attractive $1/r$ potential.

the electric field of the oppositely charged ion induces internal motions within an ion.

In discussing the phenomena that can give rise to ion–ion reaction products, it is instructive to examine the cross sections for the various processes,

$$\sigma_{\text{process}} = P_{\text{process}} \pi b^2_{\text{process}} \tag{14.27}$$

where P is the average probability that the phenomenon will occur at classical impact parameter b. The "processes" of interest here include formation of a bound orbit, which eventually will lead to a chemical reaction, proton transfer without formation of a longlived chemical complex, and formation of a chemical complex that may or may not proceed on to proton transfer products. (As discussed further below, electron transfer is another possibility for many combinations oppositely charged reactants.) In the case of the formation of an electrostatically bound orbit, the square of the impact parameter b^2_{orb} is approximated by the classical Thomson model for a three-body interaction (Thomson 1924)

$$b^2_{\text{orb}} \approx \frac{4Z_1^2 Z_2^2 e^4}{4\pi\varepsilon_0(\mu v^2)^2} \tag{14.28}$$

where Z_1 and Z_2 are the unit charges of the ions, e is the electron charge, v is the relative velocity, and μ is the reduced mass, where e is in coulombs, μ in kg, v in m/s, and $1/(4\pi\varepsilon_0)$ in kg $\cdot$ m^3 s^{-2} C^{-2}, and b is in units of meters. The significance of this relation is that it predicts a Z^2 dependence for ion–ion reaction kinetics, which best describes rate measurements made to date in electrodynamic ion traps (Stephenson and McLuckey 1996a, Wells et al. 2003). For example, in the case in which one of the ion populations is in great excess, pseudo-first order kinetics prevail such that the Thomson model predicts

$$\text{Rate} = \pi \text{b}^2_{\text{orb}} \cdot v \cdot [\text{anions}] \tag{14.29}$$

where v represents relative velocity and [anions] represents the effective anion number density. Figure 14.19 shows results from experiments involving multiply protonated ubiquitin ions with singly charged anions derived from a perfluorocarbon (Stephenson and McLuckey 1996a). The plot shows ion–ion reaction rate versus the square of the ubiquitin ion charge. These and other results suggest that formation of a bound orbit is rate-determining.

The distances associated with a chemical reaction are smaller than those associated with formation of an electrostatically bound orbit. However, the trajectories of the ions can lead to distances appropriate for chemical reaction either within an elliptical bound orbit or via hyperbolic trajectory. In discussing chemical reactions, it is useful to refer to a generalized interaction potential, such as that shown in Figure 14.20. This energy diagram is related to that of Figure 14.1 except that it shows potential energy as a function of distance between the ions r

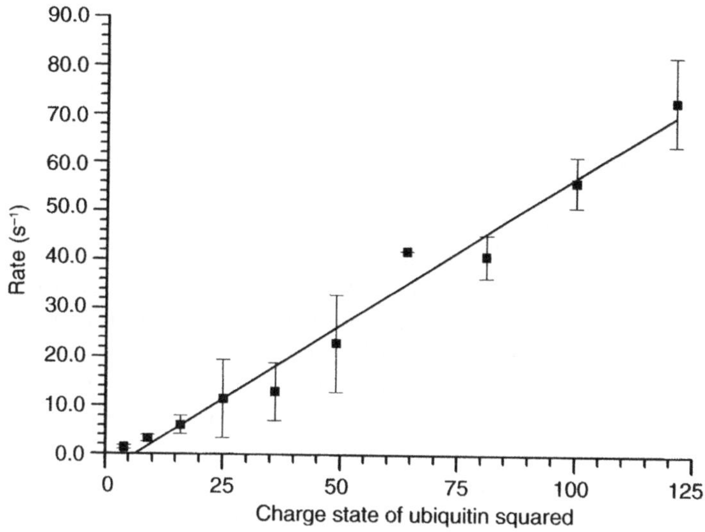

FIGURE 14.19. Proton transfer rate versus the square of the cation charge state for ubiquitin cations and singly charged anions derived from perfluoro-1,3-dimethylcyclohexane. [Reprinted from Stephenson and McLuckey (1996a) with permission from the American Chemical Society.]

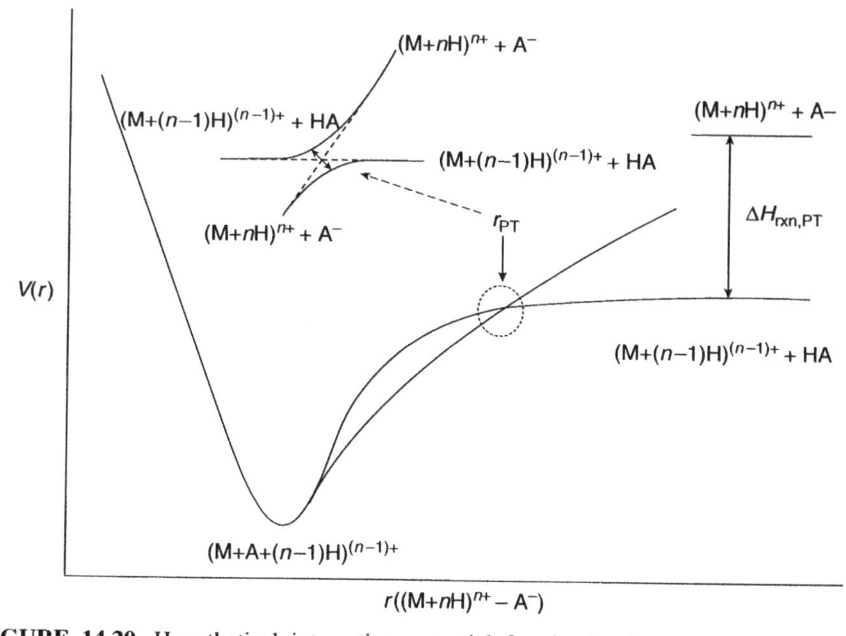

FIGURE 14.20. Hypothetical interaction potential for the ion–ion reaction involving a multiply protonated biomolecule and a singly charged anion.

instead of as a function of reaction coordinate. This diagram shows the short-range repulsive potential associated with collision partners along with the potentials associated with the entrance and exit channels. Of particular interest is the point at which the entrance and exit channel potential curves cross. It is at such an avoided crossing that the probability for a proton transfer without formation of a longlived chemical complex is maximized.

It should be noted that, in some respects, such a diagram is over simplified. For example, it implies that the charge sites are on the surfaces of the ions such that the repulsive part of the potential is found at smaller r values than the crossing point for proton transfer. However, if the charge is buried within an ion, it is possible that a "hard-sphere collision" could occur at larger r than the proton transfer crossing point, r_{PT}. Nevertheless, the diagram of Figure 14.20 is expected to reflect qualitatively the situation for many ion–ion reactions of this type.

Figure 14.20 reflects the possibility for proton transfer at a crossing point, the formation of a complex (i.e., a condensation reaction), and proton transfer via the formation of a complex followed by subsequent dissociation into proton transfer products. For simplicity, a collision in which a longlived complex is formed is referred to as a "hard-sphere collision". The square of the impact parameter for a hard-sphere collision b_{hs}^2 is approximated by (Mahan 1973)

$$b_{hs}^2 \approx r_{hs}^2 \left[1 + \frac{2Z_1Z_2e^2}{4\pi\varepsilon_0 r_{hs}\mu v^2} \right] \tag{14.30}$$

where r_{hs} represents the distance for a physical collision between the ionic partners. An analogous expression applies to the impact parameter for proton transfer and is obtained by replacing r_{hs} with r_{PT} in relation (30) to yield

$$b_{PT}^2 \approx r_{PT}^2 \left[1 + \frac{2Z_1Z_2e^2}{4\pi\varepsilon_0 r_{PT}\mu v^2} \right] \tag{14.31}$$

The proton transfer distance r_{PT} for ground-state reactants and products can be estimated from

$$r_{PT} \approx \frac{Z_1Z_2e^2}{4\pi\varepsilon_0 \Delta H_{PT}} \tag{14.32}$$

where ΔH_{PT}, in the example used here, is given by relation (14.2).

Predictions of the charge state dependence of electrostatically bound orbit formation, proton transfer at the crossing point, and hard-sphere collision for multiply protonated cytochrome c in reaction with singly charged anions of mass 400 Da are summarized in Figure 14.21.

It is clear that neither the charge state dependence associated with proton transfer at a crossing point nor that for hard-sphere collision account well for the observed Z^2 rate dependence. Relatively little rate data have been collected for multiply charged

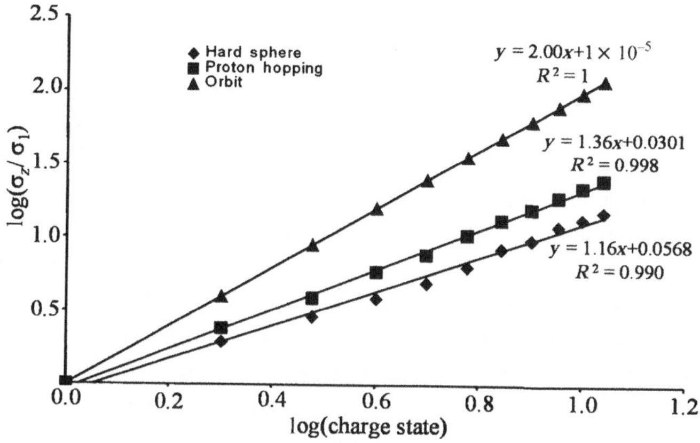

FIGURE 14.21. Plots of the logarithm of the predicted cross section versus the logarithm of the charge of the cytochrome *c* ion for hard-sphere collisions, proton transfer, and stable orbit formation. [Reprinted from Wells et al. (2003) with permission from the American Chemical Society.]

cations in reaction with multiply charged anions. Nevertheless, a Z^2 rate dependence has also been noted. An example is provided in Figure 14.22, which summarizes kinetic data collected for +13 cytochrome *c* cations and +8 cytochrome *c* cations with −5 cytochrome *c* anions.

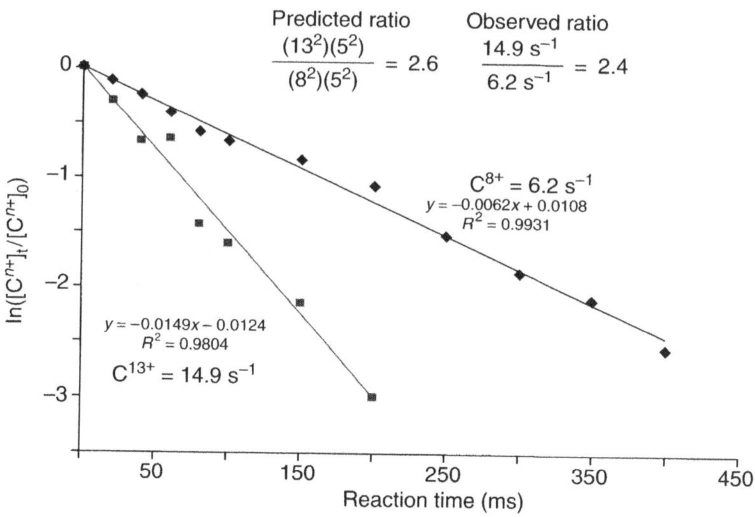

FIGURE 14.22. Reaction rate data (ln([positive reactant]$_t$/[positive reactant]$_{t=0}$) versus reaction time *t*) collected for the reactions of +13 and +8 cytochrome *c* ions with 5-cytochrome *c*. The data were collected with an excess of 5-ions to ensure pseudo-first-order kinetics for the disappearance of the positive reactant. [Reprinted from Wells et al. (2003) with permission from the American Chemical Society.]

The ratio of the rate data for the +13 cation and +8 cation was found to be 2.4 and the predicted value for a Z^2 dependence was 2.6, suggesting that formation of a bound electrostatic orbit plays an important role in establishing the overall ion–ion reaction kinetics for reactions of multiply charged ions of opposite charge.

The issue of competition between proton transfer at crossing points versus formation of a longlived chemical complex was highlighted in the study of multiply charged cations in reaction with multiply charged anions (Wells et al. 2003) (see Figure 14.9). In particular, the formation of products associated with partial proton transfer, represented by the ions labeled $(U)^{5+}$, $(U)^{4+}$, and $(U)^{3+}$ in Figure 14.9a. In the experiment leading to these results, $(M + 8H)^{8+}$ cations derived from ubiquitin were reacted with the $(M - 5H)^{5-}$ anions of cytochrome *c*. An analogous experiment was conducted with $(M + 8H)^{8+}$ cations derived from cytochrome *c* in reaction with $(M - 5H)^{5-}$ derived from cytochrome *c*, the results of which are shown in Figure 14.23.

Three different models were examined for their abilities to simulate the relative abundances of the complex ion, $(2C)^{3+}$, and the proton transfer products, $(C)^{n+}$. One involved all products arising from a chemical complex. Another involved formation of all products via proton transfer at crossing points and hard-sphere collisions without formation of a bound electrostatic orbit. The third involved initial formation of a bound electrostatic orbit within which a competition between proton transfer at crossing points, followed by escape of the products from the orbit due to reduced electrostatic attraction, and condensation takes place. Only the third model could provide even a qualitative agreement with the data using physically realistic parameters. Hence, from the observed kinetics and the relative abundances of proton transfer products and complexes in reactions of protein ions of opposite charge, it

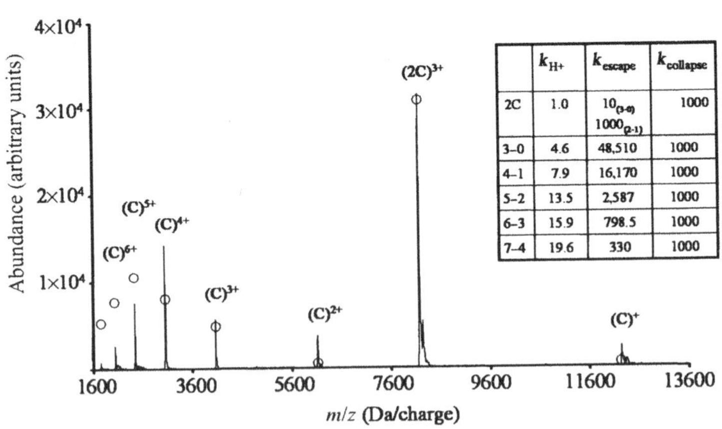

	k_{H+}	k_{escape}	$k_{collapse}$
2C	1.0	$10_{(3\text{-}0)}$ $1000_{(2\text{-}1)}$	1000
3–0	4.6	48,510	1000
4–1	7.9	16,170	1000
5–2	13.5	2,587	1000
6–3	15.9	798.5	1000
7–4	19.6	330	1000

FIGURE 14.23. The postion/ion reaction positive spectrum resulting from the C^{8+}/C^{5-} reaction along with predicted abundances based on a model that involves competition between condensation and proton transfer at crossing points that lead to separation of products that have not undergone complete proton transfer within an electrostatically bound orbit. [Reprinted from Wells et al. (2003) with permission from the American Chemical Society.]

appears that most products from bioion–ion reactions arise from an electrostatically bound orbit.

The extent to which an ion–ion encounter results in a reaction via a longlived chemical complex versus proton or electron transfer (see text below) at a curve crossing, without formation of a longlived chemical complex, has important implications. For example, energy partitioning is likely to depend on which type of interaction predominates. Furthermore, some types of ion–ion reactions can only proceed via a longlived chemical complex. Examples include the metal ion transfer reactions described above, as well as charge inversion via multiple-proton transfer. Therefore, characteristics of a reagent ion, such as hard-sphere cross section, and location of charge sites (i.e., proximity to the surface) are expected to be important in influencing the relative contributions of chemical complex formation and charged particle transfer via flyby encounters.

14.3.2. Electron Transfer

The preceding discussion provides a paradigm for regarding the dynamics of ion–ion reactions that allow for a proton transfer mechanism that occurs at an avoided crossing as well as reactions that occur via the formation of a relatively longlived chemical complex. For many bioion–ion reactions, electron transfer is also a thermodynamically favorable process. Hence, for multiply protonated species, proton transfer and electron transfer can be competitive. For the reactions with singly charged anions, the heats of reaction are given by relations (14.2) (proton transfer) and (14.23) (electron transfer). For most polypeptide cation–reagent anion combinations, the enthalpy of reaction for proton transfer ΔH_{PT} is significantly more negative (i.e., more exothermic) than that for electron transfer, ΔH_{ET} (Gunawardena et al. 2005). This is a significant situation because it implies that electron transfer should be kinetically favored if electron transfer is probable at the relevant crossing point. This follows from the relationship for the square of the impact parameter for electron transfer, which is given by a relationship analogous to that for proton transfer at a crossing point

$$b_{\mathrm{ET}}^2 \approx r_{\mathrm{ET}}^2 \left[1 + \frac{2Z_1Z_2e^2}{4\pi\varepsilon_0 r_{\mathrm{ET}} \mu v^2} \right] \tag{14.33}$$

where the proton transfer distance r_{ET} for ground-state reactants and products can be estimated from

$$r_{\mathrm{ET}} \approx \frac{Z_1Z_2e^2}{4\pi\varepsilon_0 \Delta H_{\mathrm{ET}}} \tag{14.34}$$

Figure 14.24 provides a generalized hypothetical interaction potential that shows a scenario in which both electron transfer and proton transfer as exothermic but the former is less so than the latter.

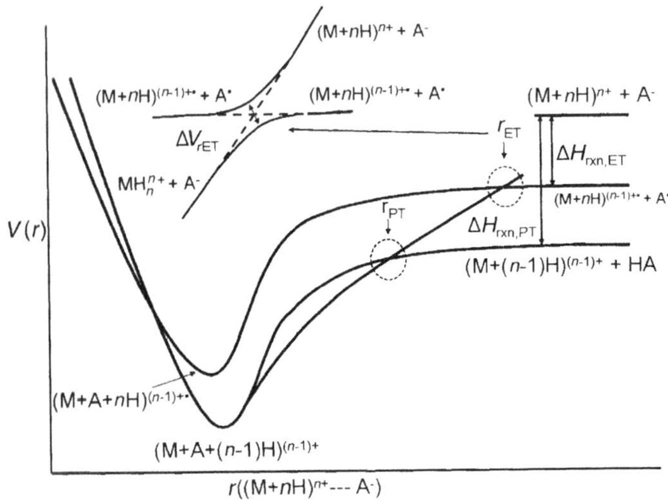

FIGURE 14.24. Hypothetical interaction potential for the ion–ion reaction involving a multiply protonated biomolecule and a singly charged anion in which both ΔH_{PT} and ΔH_{ET} are negative and the absolute value of ΔH_{PT} is greater than that of ΔH_{ET}.

In situations represented by Figure 14.24, it is clear that the crossing point for electron transfer is reached prior to that for proton transfer. Note that the hypothetical interaction potential of Figure 14.24 also shows a crossing for the intermediate longlived chemical complexes $(\mathrm{M}^{\bullet} + \mathrm{A}^{\bullet} + n\mathrm{H})^{(n-1)+}$, the nominal intermediate for electron transfer, and $(\mathrm{M} + \mathrm{HA} + (n-1)\mathrm{H})^{(n-1)+}$, the nominal intermediate for proton transfer. Note that if isomerization between these isomeric species is facile, proton transfer would be expected to dominate from most ion–ion collisions involving protonated polypeptides that lead to a longlived chemical complex. Hence, the probability for electron transfer at the crossing point is the determining factor for the observation of electron transfer.

The usual way to treat the probability for electron transfer at avoided crossings, P_{ET} [see also relation (14.27), where $P_{\mathrm{process}} = P_{\mathrm{ET}}$ in the case of electron transfer], is via Landau–Zener theory (Landau 1932; Zener 1932). Models based on Landau–Zener theory have been applied to ion–ion electron transfer (Olson et al. 1971; Olson 1972) with more recent emphasis on bioion–ion reactions (Anusiewicz et al. 2005; Gunawardena et al. 2005). The dashed lines in the inset of Figure 14.24, representing diabatic states, cross, whereas the solid curves, which represent adiabatic states, do not. For atomic reactants, Landau–Zener theory expresses the probability for transitions between adiabatic surfaces at the crossing where the adiabatic curves are at their closest approach as

$$P_{\mathrm{LZ}} = \exp\left\{ -\left[\frac{\pi(\Delta V_{r_{\mathrm{ET}}})^2}{2\hbar \dfrac{dr}{dt}\left| \dfrac{dV_{\mathrm{I}}}{dr} - \dfrac{dV_{\mathrm{F}}}{dr} \right|} \right] \right\} \tag{14.35}$$

where P_{LZ} represents the so-called Landau–Zener probability for transition between adiabatic states at the avoided crossing, ΔV_{rET} is the shortest distance between the adiabatic curves at the avoided crossing, dr/dt is the radial velocity at this point, and $|dV_I/dr - dV_F/dr|$ is the difference in the slopes of the reactant and product ion channels at the avoided crossing. The energy gap between the adiabatic curves ΔV_{rET} at its minimum point is given by $2H_{12}$, where H_{12} is the coupling matrix element that indicates the strength of electronic coupling between adiabatic states. For molecular systems, the transition probabilities between the vibrational states of the reactants and products also come into play. In the case of molecular systems, Eq. (14.35) is modified by including the relevant Franck–Condon factors (Bauer et al. 1969),

$$P_{LZ,1v';\to 2v''} = \exp\left\{ - \left[\frac{\pi(\Delta V_{r_{ET}})^2 \langle \chi_{1v'} | \chi_{2v''} \rangle^2}{2\hbar \dfrac{dr}{dt} \left| \dfrac{dV_I}{dr} - \dfrac{dV_F}{dr} \right|} \right] \right\} \qquad (14.36)$$

where $+\chi_{1v'}|\chi_{2v''2}$, represents the Franck–Condon overlap between reactant and product vibrational wavefunctions associated with the transition from $1v'$ to $2v''$. As two crossings take place in the course of passage of two ionic reactants in a flyby interaction, a successful electron transfer must involve one diabatic crossing, given by P_{LZ}, and one adiabatic crossing, given by $(1 - P_{LZ})$. This can occur in one of two ways (i.e., an avoided crossing on the incoming trajectory and a curve crossing on the outgoing trajectory or vice versa) so that the total probability for electron transfer, P_{ET} is given by

$$P_{ET} = 2P_{LZ}(1 - P_{LZ}) \qquad (14.37)$$

The maximum likelihood for electron transfer, therefore, is expected to be at P_{LZ} values of roughly 0.5.

While the parameters that determine P_{LZ} in relation (14.36) vary with the nature of the reactants, it is instructive to examine the qualitative behavior that might be expected based on a model and some assumptions made by Olson and coworkers for electron transfer between two singly charged ions. Olson et al. 1971) reported a parameterized coupling matrix element H_{12} given by

$$H_{12} = 1.044\sqrt{\mathrm{EA}}\sqrt{\mathrm{RE}}\left(\frac{\sqrt{2\mathrm{EA}} - \sqrt{2\mathrm{RE}}}{2}\right) r_{ET}\left\{\exp\left[-0.857\left(\frac{\sqrt{2\mathrm{EA}} - \sqrt{2\mathrm{RE}}}{2}\right) r_{ET}\right]\right\} \qquad (14.38)$$

where EA is the electron affinity of A and RE is the recombination energy relevant to the cation (all parameters are in atomic units). Further approximations, which are also expected to be applicable to many bioion–ion reaction scenarios, were also made to simplify evaluation of the exponential term in Eq. (14.36) to give

$$P_{LZ,1v'\to 2v''} = \exp\left\{ -\left[2^{1/2}\pi (r_{ET})^{3/2} \mu^{1/2} H_{12}^2 \langle \chi_{1v'} | \chi_{2v''} \rangle^2 \right] \right\} \qquad (14.39)$$

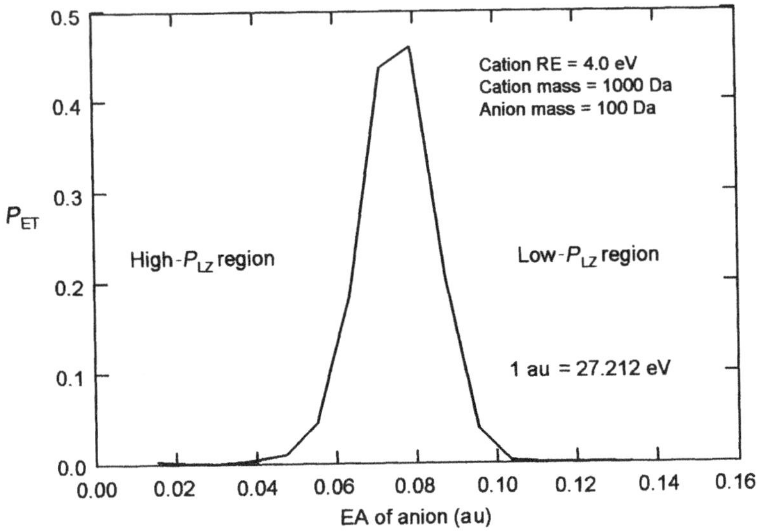

FIGURE 14.25. A plot of P_{ET}, determined from relations (14.37)–(14.39), as a function of the electron affinity of the species used to form the anionic reagent for a hypothetical reaction involving a cation of mass 1000 Da and RE of 4.0 eV and an anion of mass 100 and Franck–Condon overlap of 1.0.

P_{ET} can be estimated using relations (14.38) and (14.39) with the input of the reactant ion masses and charges, RE, EA, and a value for the relevant Franck–Condon factor (atomic units). Figure 14.25 shows the predicted electron transfer probability as a function of the EA of A for a cation mass of 1000 Da, anion mass of 100, cation charge of +1, anion charge of −1, a cation RE of 4.0 eV, and a Franck–Condon overlap of 1.0.

The key result here is that the model suggests that there are two main criteria for high electron transfer probability at a curve crossing: (1) a geometric criterion that requires adequate Franck–Condon overlap and (2) an energy criterion that the EA of the species from which the anion in derived is not too high. Note that relation (14.37) indicates that both high and low values of P_{LZ} lead to low values of P_{ET}. Hence, the model predicts that the electron affinity of A can also be too low. However, since the model accounts only for the ground states, it does not account for the possibility of transitions to excited states of the products. As the EA of A increases, $\Delta H_{rxn,ET}$ decreases and the crossing point occurs at greater distances. Transitions to excited states occur at crossings at greater distances than those associated with the ground states. Hence, for a reaction in which the crossing for the ground states is already too distant for high P_{ET}, transitions to excited states would be even less probable. However, in the case of species with sufficiently low values of EA that the crossing point is too close for high P_{ET}, transitions to excited states may take place at distances favorable for electron transfer. For this reason, reagents with electron affinities that are "too low" in the plot of Figure 14.25 may allow for crossings leading to transitions to excited states that fall in the range for favorable electron transfer. Following this reasoning, it is clear that reagents can have electron affinities that are too high, but it is less clear that reagent electron affinity can be too low when the possibility for transitions to excited states is considered.

A variety of reagent species, A, have been examined with a single polypeptide cation, triply protonated KGAILKGAILR, to evaluate the predictions of the model described above (Gunawardena et al. 2005). To compare the extent to which ETD competes with proton transfer, the percentage of postion–ion reaction ion signals attributable to ETD, indicated as %ETD, were determined for each of the reagents. This value was determined by

$$\%\mathrm{ETD} \equiv \frac{\sum c, z, \text{neutral} \cdot \text{losses}}{\sum \text{postion} - \text{ion products (residual } 3^{+} \text{ excluded)}} \times 100 \qquad (14.40)$$

The results obtained for several dozen reagents are summarized in Table 14.1, which lists the reagent identity, the sum of the Franck–Condon factors for transition from the ground vibrational state of the anionic reagent to the lowest 10 vibrational states of the first electronic excited state of the corresponding neutral, the electron affinity of A, and the %ETD noted for the ion/ion reaction with triply protonated KGAILKGAILR.

The data of Table 14.1 are consistent with the general features of the Landau–Zener-based picture for the probability for electron transfer. None of the reagents species with electron affinities greater than roughly 70 kcal/mol yields observable ETD even if the relevant Franck–Condon overlap is high. This is consistent with the conclusion that the reagent species electron affinity can be "too high." Within the context of the Landau–Zener picture, this is consistent with crossing points that are too distant for a significant P_{ET}. It is less clear from the data in Table 14.1 that the electron affinity of the reagent can be too low. However, the reagents with electron affinities less than ~11 kcal/mol show relatively low %ETD values. The species with the highest %ETD values fall in the range of EA values of 12–47 kcal/mol. The favorable Franck–Condon factor criterion is clearly apparent in the data of Table 14.1. Those reagent anions with very low Franck–Condon overlap show very little ETD even when the electron affinities are within the range of 12–47 kcal/mol (see the data for CS_2, SF_6, and SO_3).

14.4. INSTRUMENTATION FOR THE STUDY OF BIOION–ION REACTIONS

To date, biomolecule ion/ion reaction studies have been conducted either at or near atmospheric pressure prior to sampling products into a mass spectrometer or within an electrodynamic ion trap. The first reports employed a "Y-tube reactor," in which ions formed at different ends of the Y tube were mixed at the center of the tube and drawn into a mass spectrometer (Ogorzalek-Loo et al. 1991, 1992). Several of the reaction phenomenologies, such as proton transfer and charge inversion, were first noted in these studies. A quadrupole mass filter was used to analyze the products of

TABLE 14.1. Summary of Ion–Ion Reaction Data for (KGAILKGAILR + 3H)$^{3+}$ with Various Anionic Reagents[a]

Reagent (A)	Franck–Condon factor $\sum +0\| \leq 10^2$,	EA(A)(kcal/mol)	% ETD
Norbornodiene	1.1×10^{-2}	5.6	7.2
cis-Stilbene	5.2×10^{-3}	10.4	9.8
O_2	9.7×10^{-1}	10.4	4.9
CS_2	4.9×10^{-5}	11.8	<0.01
Azobenzene	1.8×10^{-1}	13.1	48.8
Fluoranthene	3.6×10^{-1}	14.5	37.4
Perylene	4.1×10^{-1}	22.4	20.9
Nitrobenzene	1.4×10^{-1}	23.0	14.7
SF_6	6.7×10^{-11}	24.2	<0.01
SO_2	4.6×10^{-1}	25.5	30.1
m-Dinitrobenzene	2.7×10^{-1}	38.3	26.6
o-Dinitrobenzene	1.2×10^{-4}	38.3	17.2
S_2O	3.5×10^{-1}	43.3	7.3
SO_3	6.9×10^{-8}	43.8	<0.01
p-Dinitrobenzene	1.8×10^{-1}	46.1	16.4
S_3	5.2×10^{-1}	48.3	7.0
O_3	3.8×10^{-1}	48.5	4.8
$NO_2^{\bullet}$	2.3×10^{-1}	52.4	8.5
1,3,5-Trinitrobenzene	6.5×10^{-1}	60.6	7.9
CO_3	9.0×10^{-1}	62.0	<0.01
$I^{\bullet}$	N/A	70.6	<0.01
$CH_3COO^{\bullet}$	5.8×10^{-3}	77.4	<0.01
$NO_3^{\bullet}$	7.6×10^{-1}	90.8	<0.01
$[PDCH\text{-}F]^{\bullet}$	1.3×10^{-3}	96.2	<0.01
H_2PO_4	1.3×10^{-8}	105.4	<0.01
$SF_5^{\bullet}$	4.5×10^{-7}	108.4	<0.01
HSO_4	4.9×10^{-2}	109.5	<0.01
Picric acid	1.3×10^{-5}	113.1	<0.01

[a](For details regarding the sources of the Franck–Condon factors and electron affinities, see Gunawardena et al. (2005).

the reactions, which limited the range over which products could be analyzed to less than a few thousand m/z units. Several variations of ion–ion reactors have more recently been coupled with time-of-flight mass analysis, which provides a wider m/z range and superior mass resolving power and mass measurement accuracy. These have included the passage of ions produced by electrospray through a region exposed to ^{210}Po radioactive decay (α emission) (Scalf et al. 1999, 2000) or through a region exposed to a corona discharge plasma (Ebeling et al. 2000, 2001). In both cases, these atmospheric ionization sources give rise largely to protonated or deprotonated species via a rapid ion–molecule reaction cascade. These species are believed to be the major singly charged reactants for the multiply charged ions of opposite polarity formed via electrospray.

Most biomolecule ion–ion reaction studies published to date have been conducted using an electrodynamic ion trap as a reaction vessel. These devices have the necessary capability to store ions of opposite polarity simultaneously in overlapping regions of space. A variety of ion source configurations have been coupled with three-dimensional quadrupole ion traps. The first studies employed electrospray ionization for the generation of multiply charged anions and in situ ion trap chemical ionization (Herron et al.1995b) or electron ionization (Herron et al. 1995a) for the formation of singly charged cationic reagents. It is more difficult to form high number densities of singly charged anions via in situ electron bombardment in an ion trap than it is to form singly charged cations (McLuckey et al. 1988). For this reason, instruments were adapted to allow for electrospray ionization with ion injection via an endcap electrode and atmospheric sampling glow discharge ionization with ion injection via a hole drilled into the side of the ion trap ring electrode (Stephenson 1997c; and McLuckey Reid et al. 2003). This geometry enabled the study of multiply protonated species in reaction with singly charged anions derived from the glow discharge source. Atmospheric sampling glow discharge ionization (McLuckey et al. 1988) is a relatively bright anion source that can allow for the rapid accumulation of anions with high ionization efficiencies in the ion trap, despite the poor ion injection efficiencies associated with ion injection through the ring electrode. Reactions involving multiply charged ions with singly charged ions have also been studied using electrospray for generation of the multiply charged ion and laser desorption via laser irradiation through a hole in the ring electrode (Payne and Glish 2001). Two quadrupole ion trap instruments have more recently been modified to allow for the study of oppositely charged ions in which both ion polarities are formed via separate electrospray ion sources, each with a distinct atmospheric/vacuum interface (Wells et al. 2002, Badman et al. 2002). A turning quadrupole was incorporated to allow ions of each polarity to be injected in turn via an endcap electrode. These instruments enabled the ion trap studies involving multiply charge cations in reactions with multiply charged anions.

Instruments that employ ion storage with quadrupole arrays, referred to as *linear ion traps*, have become prominent (Schwartz et al. 2002, Hager 2002). These instruments offer advantages over conventional Paul or three-dimensional quadrupole ion traps in terms of ion storage capacity and capture of ions from external ion sources. For trapping of a single-ion polarity, trapping along the axial dimension is accomplished by applying DC potentials to endplates (Hager 2002) or short quadrupole sections at either end of the quadrupole array that are electrically isolated from the central array (Schwartz et al. 2002). Ion–ion reactions have been reported using linear ion traps using several distinct approaches. For example, the first report of electron transfer dissociation of polypeptide ions was carried out in a linear ion trap in which a radiofrequency voltage was applied to the trapping elements for simultaneous axial storage of both ion polarities (Syka et al. 2004). In that work, positive polypeptide ions were formed via electrospray with ion injection into one end of the quadrupole array, whereas the anions were formed via chemical ionization with ion injection into the opposite end of the quadrupole array. Mass-selective ion ejection into the detector was accomplished via radial ion ejection. Ion–ion reactions were also reported in which ions of one polarity were stored in a linear ion trap using a DC voltage applied to the endplates while ions of opposite polarity were transmitted through the quadrupole array (Wu et al. 2004). Ion–ion reaction products of the polarity of the stored ion population were accumulated in the linear ion trap during the period that the oppositely charged ions were transmitted through the quadrupole array and were subsequently analyzed via mass-selective axial ion ejection. Ion–ion reactions via mutual storage of both ion polarities was also reported with the same apparatus (Xia et al. 2005a). Mutual storage was effected by unbalancing the quadrupole rods without the application of a radiofrequency voltage to the endplates. The effect of unbalancing the rods in the quadrupole array is to create an effective radiofrequency barrier for both ion polarities in the axial direction. In the studies describing ion–ion reactions in the transmission mode and via mutual storage using unbalanced quadrupole rods, the multiply charged ions were formed via electrospray and were admitted into the instrument in the axial direction whereas the singly charged ions were injected radially into the side of a quadrupole array.

Until recently (as of 2006), all ion–ion reaction experiments executed with electrodynamic ion traps have employed distinct ionization sources with separate ion paths into the electrodynamic ion trap. Such experiments require specially built or modified instruments to incorporate the distinct ionization sources. Exceptions to the use of multiple distinct ion paths for ion injection into the electrodynamic ion paths have been described. One has involved the use of sonic spray ionization (Hirabayashi et al. 1994), a spray ionization method that forms ions of each polarity simultaneously. Provided that the potentials applied to the ion path can be switched so that each ion polarity can be injected into the electrodynamic ion trap in turn, a single ion source with a single ion path can be used to study ion–ion reactions. Such an approach was demonstrated for a variety of ion–ion reaction types (Xia et al. 2005b). A similar set of experiments were also demonstrated using pulsed dual-electrospray emitters (a nanoelectrospray tip for analyte ions and a conventional electrospray needle for the oppositely charged ions) (Xia et al. 2005c). The ability to use a single optimized ion path for both ion polarities and the ability to use distinct emitters so that solution conditions can be optimized individually provides a high degree of flexibility in studying ion–ion reactions with electrodynamic ion traps.

14.5. CONCLUSIONS

From the foregoing discussion it should be clear that biomolecule–ion/ion reactions are distinct from ion–molecule reactions in several significant ways. Unlike ion–neutral collisions, few of which lead to chemical reactions, virtually all ion–ion collisions result in a chemical reaction of some sort due to the large exothermicities associated with mutual neutralization. Furthermore, due to the importance of the formation of a bound orbit as a rate-determining step, the kinetics of ion–ion reactions are insensitive to the chemical identities of the reactants. This follows from the fact that the distances associated with formation of a bound orbit are generally much larger than those associated with chemical reactions. As a result, bound orbit formation determines the overall rate while the chemical reaction mechanism is determined by interactions that take place at much closer distances. An important consequence of this situation is that kinetics of ion–ion reactions are dependent only on the masses, numbers, charges, and relative velocities of the ions and not on chemical functionalities. The absolute magnitudes of the ion–ion rate constants are several orders of magnitude larger than those of the fastest ion–molecule reactions, which results in useful overall reaction rates ($1–1000\,s^{-1}$ in electrodynamic ion traps) despite the relatively low ion number densities relative to achievable neutral number densities. Furthermore, ionic reactants are readily manipulated by electric fields such that admission, removal, acceleration, and other functions of ionic reactants can be effected with much greater control than with neutral reactants. For this reason, multiple distinct ion–ion reaction steps can readily be incorporated into MS^n experiments.

The predictable kinetics and ease of admission and removal of ionic reactants are attractive characteristics of ion–ion reactions as robust means for manipulating and interrogating biomolecule–ions. The value of such reactions in bioanalysis depends on the utilities of the reactions that can be induced. To date, a variety of applications have been demonstrated. These fall largely into charge state manipulation and structural interrogation categories. Proton transfer reactions, for example, have been demonstrated to be useful for the analysis of whole protein (Stephenson and McLuckey 1996b), oligonucleotide (McLuckey et al. 2002a), and synthetic polymer mixtures (Stephenson and McLuckey 1998a) by reducing charge states so as to minimize overlap from ions of different mass and charge. Product ion spectra have also been simplified by converting product ions largely to the +1 or −1 charge state (Stephenson and McLuckey 1998b). This has permitted tandem mass spectrometry to be applied to whole-protein ions in relatively low resolving power instruments, such as the electrodynamic ion trap (Amunugama et al. 2004). Ion/ion reactions have been shown to be capable of concentrating ions in the gas phase via a technique referred to as "ion parking" (McLuckey et al. 2002b). Sequential stages of ion parking have been shown to be capable of both concentrating and charge-state purifying a protein ion population prior to dissociation of a protein (Reid et al. 2002). Charge inversion reactions promise to be useful in screening mixtures on the basis of the numbers of acidic or basic sites. Two sequential charge inversion reactions have been shown to be able to increase the absolute charge states of either positive ions (He and McLuckey 2003) or negative ions in the gas phase (He and McLuckey 2004). Within the context of structural interrogation, the electron transfer dissociation process offers an approach that is complementary to conventional ion activation techniques (Syka et al. 2004) and appears to be particularly useful in the analysis of posttranslationally modified species. A variation of the ion parking technique, referred to as "parallel parking", has been demonstrated as being capable of achieving single-electron transfer efficiencies on the order of 90% by inhibiting sequential electron transfer reactions (Chrisman et al. 2005). None of the applications just mentioned can be done as well or at all with ion–molecule reactions.

The applications just mentioned are restricted to electron transfer and proton transfer reactions. As noted above, other types of reactions, such as metal ion transfer and complex formation, have also been noted. These, as well as other reaction types not mentioned here, may also prove to have utility in probing biomolecule ions. It seems highly probable that the range of observed reaction phenomenologies will continue to grow as a wider range of instruments are applied to the study of such reactions and as the ability to study such reactions expands to a wider range of research groups. Some of the overall features of ion–ion reactions are already apparent. However, much has yet to be learned about energy partitioning and detailed mechanisms associated with electron transfer dissociation, multiple-proton transfer, metal ion transfer, and similar mechanisms. Bioion–ion reactions, in many ways, remain minimally explored from a fundamental standpoint and largely undeveloped from an application standpoint.

ACKNOWLEDGMENT

Research support for the study of fundamental aspects of biomolecule ion–ion reactions in the author's laboratories has been provided by the Office of Basic Energy Sciences, Division of Chemical Sciences under Award DE-FG02-00ER15105.

REFERENCES

Adams NG, Babcock LM, Molek CD (2003): Ion-ion recombination. In Armentrout RB (ed), *Encyclopedia of Mass Spectrometry: Theory and Ion Chemistry*, Vol. 1, Elsevier, Amsterdam, pp. 555–561.

Amunugama R, Hogan JM, Newton KA, McLuckey SA (2004): Whole protein dissociation in a quadrupole ion trap: Identification of an *a priori* unknown modified protein. *Anal. Chem.* **76**:720–727.

Anusiewicz I, Berdys-Kochanska J, Simons J (2005): Electron attachment step in electron capture dissociation (ECD) and electron transfer dissociation (ETD). *J. Phys. Chem. A* **109**:5801–5813.

Asano KG, Goeringer DE, Butcher DJ, McLuckey SA (1999a): Bath gas temperature and the appearance of ion trap tandem mass spectra of high-mass ions. *Int. J. Mass Spectrom.* **190/191**:281–293.

Asano KG, Goeringer DE, McLuckey SA (1999b): Thermal dissociation in the quadrupole ion trap: Ions derived from leucine enkephalin. *Int. J. Mass Spectrom.* **185/186/187**:207–219.

Badman ER, Chrisman PA, McLuckey SA (2002): A quadrupole ion trap mass spectrometer with three independent ion sources for the study of gas-phase ion/ion reactions. *Anal. Chem.* **74**:6237–6243.

Bates DR (1985): Ion-ion recombination in ambient gas. In Bates DR (ed), *Advances in Atomic and Molecular Physics*, Academic Press, Orlando, FL, pp. 1–37.

Bates DR, Morgan WL (1990): New recombination mechanism—tidal termolecular ionic recombination. *Phys. Rev. Lett.* **64**:2258–2260.

Bauer E, Fisher ER, Gilmore FR (1969): De-excitation of electronically excited sodium by nitrogen. *J. Chem. Phys.* **51**:4173–4181.

Butcher DJ, Asano KG, Goeringer DE, McLuckey SA (1999): Thermal dissociation of bradykinin ions. *J. Phys. Chem. A* **103**:8664–8671.

Chrisman PS, Pitteri SJ, McLuckey SA (2005): Parallel ion parking: Improving conversion of parents to first generation products in electron transfer dissociation. *Anal. Chem.* **77**:3411–3414.

Coon JJ, Syka JEP, Schwartz JC, Shabanowitz J, Hunt DF (2004): Anion dependence in the partitioning between proton and electron transfer in ion–ion reactions. *Int. J. Mass Spectrom.* **236**:33–42.

Coon JJ, Shabanowitz J, Hunt DF, Syka JEP (2005): Electron transfer dissociation of peptide anions. *J. Am. Soc. Mass Spectrom.* **16**:880–882.

Ebeling DD, Westphall MS, Scalf M, Smith LM (2000): Corona discharge in charge reduction electrospray mass spectrometry. *Anal. Chem.* **72**:5158–5161.

Ebeling DD, Westphall MS, Scalf M, Smith LM (2001): A cylindrical capacitor ionization source: Droplet generation and controlled charge reduction for mass spectrometry. *Rapid Commun. Mass Spectrom.* **15**:401–405.

Fenn JB, Mann M, Meng CK, Wong SF, Whitehouse CM (1990): Electrospray Ionization for mass spectrometry of large biomolecules. *Science* **246**:64–71.

Flannery MR (1982): Ion-ion recombination in high pressure plasmas. In McDaniel EW, Nighan WL (eds), *Applied Atomic Collision Physics*, Academic Press, New York, pp. 141–172.

Goeringer DE, McLuckey SA (1998): Relaxation of internally excited high mass ions simulated under typical quadrupole ion trap storage conditions. *Int. J. Mass Spectrom.* **177**:163–174.

Griffin LL, McAdoo DJ (1993): The effect of ion size on rate of dissociation—RRKM calculations on model large polypeptide ions. *J. Am. Soc. Mass Spectrom.* **4**:11–15.

Gunawardena HP, He M, Chrisman PA, Pitteri SJ, Hogan JM, Hodges BDM, McLuckey SA (2005): Electron transfer versus proton transfer in gas-phase ion-ion reactions of polyprotonated peptides. *J. Am. Chem. Soc.* **127**:12627–12639.

Gunawardena HP, McLuckey SA (2004): Synthesis of multi-unit protein hetero-complexes in the gas phase via ion/ion chemistry. *J. Mass Spectrom.* **39**:630–638.

Hager JW (2002): A new linear ion trap mass spectrometer. *Rapid Commun. Mass Spectrom.* **16**:512–526.

He M, McLuckey SA (2003): Two ion/ion charge inversion steps to form a doubly protonated peptide from a singly protonated peptide in the gas phase. *J. Am. Chem. Soc.* **125**:7756–7757.

He M, McLuckey SA. (2004): Increasing the negative charge of a macro-anion in the gas phase via sequential charge inversion reactions. *Anal. Chem.* **76**:4189–4192.

He M, Emory JF, McLuckey, SA (2005): Reagent anions for charge inversion of polypeptide/protein cations in the gas phase. *Anal. Chem.* **77**:3173–3182.

Herron WJ, Goeringer DE, McLuckey SA. (1995a): Gas-phase electron transfer reactions from multiply-charged anions to rare gas cations. *J. Am. Chem. Soc.* **117**: 11555–11562.

Herron WJ, Goeringer DE, McLuckey SA (1995b): Ion-ion reactions in the gas-phase: Proton transfer reactions of protonated pyridine with multiply-charged oligonucleotide anions. *J. Am. Soc. Mass Spectrom.* **6**:529–532.

Hirabayashi A, Sakairi M, Koizumi H (1994): Sonic spray ionization method for atmospheric-pressure ionization mass spectrometry. *Anal. Chem.* **66**:4557–4559.

Landau LD (1932): On the theory of transfer of energy at collisions II. *Phys. Z (USSR)* **2**:46–51.

Mahan BH (1973): Recombination of gaseous ions. In Prigogine I, Rice SA (eds), *Advances in Chemical Physics*, Wiley, New York, pp. 1–40.

McLuckey SA, Glish GL, Kelley, PE (1987): Collision activated decomposition of negative ions in an ion trap mass spectrometer. *Anal. Chem.* **59**:1670–1674.

McLuckey SA, Glish GL, Asano KG, Grant BC (1988): Atmospheric sampling glow discharge ionization source for the analysis of trace organics in ambient air. *Anal. Chem.* **60**:2220–2228.

McLuckey SA, Van Berkel GJ, Glish GL (1992): Mass spectrometry/mass spectrometry of small multiply charged oligonucleotides. *J. Am. Soc. Mass Spectrom.* **3**:60–70.

McLuckey SA, Stephenson, Jr. JL, O'Hair RAJ (1997): Decompositions of odd- and even-electron anions derived from deoxy polyadenylates. *J. Am. Soc. Mass Spectrom.* **8**:148–154.

McLuckey SA, Stephenson JL, Jr. (1998): Ion/ion chemistry of high-mass multiply charged ions. *Mass Spectrom. Rev.* **17**:369–407.

McLuckey SA, Wu J, Bundy JL, Stephenson, Jr. JL, Hurst GB (2002a): Oligonucleotide mixture analysis via electrospray and ion/ion reactions in a quadrupole ion trap. *Anal. Chem.* **74**:976–984.

McLuckey SA, Reid GE, Wells JM (2002b): Ion parking during ion/ion reactions in electrodynamic ion traps. *Anal. Chem.* **74**:336–346.

Morgan WL, Bates DR (1992): Tidal termolecular ionic recombination. *J. Phys. B: Atom. Mol. Opt. Phys.* **25**:5421–5430.

Newton KA, McLuckey SA (2003). Gas-phase peptide/protein cationizing agent switching via ion/ion reactions. *J. Am. Chem. Soc.* **125**:12404–12405.

Newton KA, He M, Amunugama R, McLuckey SA (2004): Selective cation removal from gaseous polypeptide ions: Proton versus sodium ion abstraction via ion/ion reactions. *Phys. Chem. Chem. Phys.* **6**:2710–2717.

Newton KA, Amunugama R, McLuckey SA (2005): Gas-phase ion/ion reactions of multiply protonated polypeptides with metal containing anions. *J. Phys. Chem. A* **109**:3608–3616.

Ogorzalek-Loo RR, Udseth HR, Smith RD (1991): Evidence of charge inversion in the reaction of singly-charged anions with multiply-charged macro-ions. *J. Phys. Chem.* **95**:6412–6415.

Ogorzalek-Loo RR, Udseth HR, Smith RD (1992): A new approach for the study of gas-phase ion-ion reactions using electrospray ionization. *J. Am. Soc. Mass Spectrom.* **3**:695–705.

Olson RE, Smith FT, Bauer E (1971): Estimation of coupling matrix elements for one-electron transfer systems. *Appl. Opt.* **10**:1848–1855.

Olson RE (1972): Absorbing sphere model for calculating ion-ion recombination total cross-sections. *J. Chem. Phys.* **56**:2979–2984.

Payne AH, Glish GL (2001): Gas-phase ion/ion interactions between peptides or proteins and iron ions in a quadrupole ion trap. *Int. J. Mass Spectrom.* **204**:47–54.

Pitteri SJ, McLuckey, SA (2005a): Recent developments in the ion/ion chemistry of high-mass multiply charged ions. *Mass Spectrom. Rev.* **24**:931–958.

Pitteri SJ, Chrisman PA, McLuckey SA (2005b): Electron transfer ion/ion reactions of doubly protonated peptides: The effect of elevated bath gas temperature. *Anal. Chem.* **77**:5662–5669.

Reid GE, Shang H, Hogan JM, Lee GU, McLuckey SA (2002): Gas-phase concentration, purification, and identification of whole proteins from complex mixtures. *J. Am. Chem. Soc.* **124**:7353–7362.

Reid GE, Wells JM, Badman ER, McLuckey SA (2003): Performance of a quadrupole ion trap mass spectrometer adapted for ion/ion reaction studies. *Int. J. Mass Spectrom.* **222**:243–258.

Rutherford E. (1897): The velocity and rate of recombination of the ions of gases exposed to Röntgen radiation. *Philos. Mag.* **44**:422–440.

Scalf M, Westphall MS, Krause J, Kaufman SL, Smith LM (1999): Controlling charge states of large ions. *Science* **283**:194–197.

Scalf M, Westphall MS, Smith LM (2000): Charge reduction electrospray mass spectrometry. *Anal. Chem.* **72**:52–60.

Schlag EW, Grotemeyer J, Levine RD (1992): Do large molecules ionize? *Chem. Phys. Lett.* **190**:521–527.

Schlag EW, Levine RD (1992): Ionization, charge separation, charge recombination, and electron transfer in large systems. *J. Phys. Chem.* **96**:10608–10616.

Schwartz JC, Senko MW, Syka, JEP (2002): A two-dimensional quadrupole ion trap mass spectrometer. *J. Am. Soc. Mass Spectrom.* **13**:659–669.

Stephenson JL Jr, McLuckey SA. (1996a): Ion–ion reactions in the gas phase: Proton transfer reactions involving multiply charged proteins. *J. Am. Chem. Soc.* **118**:7390–7397.

Stephenson JL Jr, McLuckey SA (1996b): Ion/ion proton transfer reactions for protein mixture analysis. *Anal. Chem.* **68**:4026–4032.

Stephenson JL Jr, McLuckey SA (1997a): Gaseous protein cations are amphoteric. *J. Am. Chem. Soc.* **119**:1688–1696.

Stephenson JL Jr, McLuckey SA (1997b): Charge reduction of oligonucleotide anions via gas-phase electron transfer to xenon cations. *Rapid Commun. Mass Spectrom.* **11**:875–880.

Stephenson JL Jr, McLuckey SA (1997c): Adaptation of the Paul trap for study of the reaction of multiply-charged cations with singly-charged anions. *Int. J. Mass Spectrom. Ion Proc.* **162**:89–106.

Stephenson JL Jr, Van Berkel GJ, McLuckey SA (1997): Ion/ion proton transfer reactions of bio-ions involving non-covalent interactions: Homyoglobin. *J. Am. Soc. Mass Spectrom.* **8**:637–644.

Stephenson JL Jr, McLuckey SA (1998a): Reactions of poly(ethylene glycol) cations with iodide and perfluorocarbon anions. *J. Am. Soc. Mass Spectrom.* **9**:957–965.

Stephenson JL Jr, McLuckey SA (1998b): Simplification of product ion spectra derived from multiply-charged parent ions via ion/ion chemistry. *Anal. Chem.* **70**:3533–3544.

Syka JEP, Coon JJ, Schroeder MJ, Shabanowitz J, Hunt DF (2004): Peptide and protein sequence analysis by electron transfer dissociation mass spectrometry. *Proc. Natl. Acad. Sci. USA* **101**:9528–9533.

Thomson JJ, Rutherford E (1896): On the passage of electricity through gases exposed to Röntgen rays. *Philos. Mag.* **42**:392–407.

Thomson JJ (1924): Recombination of gaseous ions, the chemical combination of gases, and monomolecular reactions. *Philos. Mag.* **47**:337–378.

Wells JM, Chrisman PA, McLuckey SA. (2001): Formation of protein-protein complexes in vacuo. *J. Am. Chem. Soc.* **123**:12428–12429.

Wells JM, Chrisman PA, McLuckey SA (2002): "Dueling" ESI: Instrumentation to study ion/ion reactions of electrospray-generated cations and anions. *J. Am. Soc. Mass Spectrom.* **13**:614–622.

Wells JM, Chrisman PA, McLuckey SA (2003): Formation and characterization of protein-protein complexes *in vacuo*. *J. Am. Chem. Soc.* **125**:7238–7249.

Williams ER (1996): Proton transfer reactivity of large multiply charged ions. *J. Mass Spectrom.* **31**:831–842.

Wu J, McLuckey SA (2003): Ion/ion reactions of multiply charged nucleic acid anions: Electron transfer, proton transfer, and ion attachment. *Int. J. Mass Spectrom.* **228**:577–597.

Wu J, McLuckey SA (2004): Gas-phase fragmentation of oligonucleotide ions. *Int. J. Mass Spectrom.* **237**:197–241.

Wu J, Hager JW, Xia Y, Londry FA, McLuckey SA (2004): Positive ion transmission mode ion/ion reactions in a hybrid linear ion trap. *Anal. Chem.* **76**:5006–5015.

Xia Y, Wu J, Hager JW, Londry FA, McLuckey SA (2005a): Mutual storage mode ion/ion reactions in a hybrid linear ion trap. *J. Am. Soc. Mass Spectrom.* **16**:71–81.

Xia Y, Liang X, McLuckey SA (2005b): Sonic spray as a dual polarity ion source for ion/ion reactions. *Anal. Chem.* **77**:3683–3689.

Xia Y, Liang X, McLuckey SA (2005c): Pulsed dual electrospray ionization for ion/ion reactions. *J. Am. Soc. Mass Spectrom.* **16**:1750–1756.

Zener C (1932): Non-adiabatic crossing of energy levels. *Proc. Roy. Soc. Lond. A* **136**: 696–702.

Zubarev RA, Kelleher NL, McLafferty FW (1998): Electron capture dissociation of multiply charged protein cations. a nonergodic process. *J. Am. Chem. Soc.* **120**:3265–3266.

Zubarev RA (2003): Reactions of polypeptide ions with electrons in the gas phase. *Mass Spectrom. Rev.* **22**:57–77.

PART III

THERMOCHEMISTRY AND ENERGETICS

15

THERMOCHEMISTRY STUDIES OF BIOMOLECULES

CHRYS WESDEMIOTIS AND PING WANG

Department of Chemistry
The University of Akron
Akron, OH

Principles of Mass Spectrometry Applied to Biomolecules, edited by Julia Laskin and Chava Lifshitz

15.1. INTRODUCTION AND DEFINITIONS

The thermodynamic stability of a biomolecular system depends on the covalent bonds between its atoms and the intra- and intermolecular interactions between its substituents and subunits. Addition of protons, metal ions, and other Lewis acids to a biomolecule (or removal of protons from it) promotes the development of noncovalent interactions, in particular hydrogen bonding and electrostatic attraction forces that, together with solvation, help stabilize the conformation needed for biological function. Because Lewis acids play an important role in confining the secondary and higher-order structure of biomolecular assemblies, quantitative information about the bond strengths between these species and different types of biomolecules is essential. Mass spectrometry (MS) methods have traditionally been used for the acquisition of such thermochemical data (Lias et al. 1988; Linstrom and Mallard 2005). This chapter briefly reviews the various approaches employed in thermochemistry determinations and discusses the resulting data; discussion is focused on the interactions of biomolecules with protons and metal ions, which are the most widespread Lewis acids in biological systems and have been studied most extensively.

$$B\text{–}X^+ \rightarrow B + X^+ \quad (X^+ = \text{Lewis acid, e.g., proton, metal ion, hydrated or partially ligated metal ion, multiatomic cation}) \tag{15.1}$$

$$\Delta H_{(15.1)} = \Delta H_X \quad [\text{binding energy (affinity) of } X^+ \text{ to B (enthalpy of B–}X^+\text{ bond)}] \tag{15.2}$$

$$\Delta G_{(15.1)} = \Delta G_X \quad (\text{free energy of B–}X^+\text{ bond}) \tag{15.3}$$

$$\Delta G_X = \Delta H_X - T\Delta S_X \tag{15.4}$$

$$\Delta H_H = \text{PA} \quad (\text{proton affinity}) \tag{15.5}$$

$$\Delta G_H = \text{GB} \quad (\text{gas-phase basicity}) \tag{15.6}$$

Since most thermochemistry experiments in the gas phase are conducted under constant pressure, bond enthalpy is the relevant thermodynamic property to correlate thermochemistry to structure. The bonding between protons, metal ions, or other types of Lewis acids and biomolecules may have covalent, electrostatic (i.e., noncovalent), or combined character. In any case, the dissociation enthalpy of the bond formed between a Lewis acid X^+ and a biomolecule B is denoted as the X^+ affinity of B or the binding energy of X^+ to B; the symbol ΔH_X will be used for this variable, as defined in reaction (15.1) and Eq. (15.2). The majority of experimental methods do not furnish enthalpies, but free energies. Equations (15.3) and (15.4) explain the relationship between the enthalpy and free energy of the B–X^+ bond; the latter depends on ΔH_X, the B–X^+ bond entropy (ΔS_X), as well as the temperature (T) of the B–X^+ system [Eq. (15.4)]. With B–H^+ bonds, the acronyms PA (proton affinity) and GB (gas-phase basicity) will be used to describe the enthalpy and free energy, respectively, of the bond formed between a proton and a biomolecule [Eqs. (15.5) and (15.6)].

15.2. METHODS FOR THERMOCHEMICAL DETERMINATIONS

15.2.1. Equilibrium Method

In this method, the equilibrium of X^+ exchange between two biomolecules B_1 and B_2 [Eq. (15.7)] is established in the gas phase using high-pressure mass spectrometry (HPMS) (Kebarle 1988, 1992) or Fourier transform mass spectrometry (FTMS) (Aue and Bowers 1979; Woodin and Beauchamp 1978; Locke and McIver 1980; Hoyau et al. 1999; Gapeev and Dunbar 2001). The intensities of $B_1{-}X^+$ ($I_{B_1-X^+}$) and $B_2{-}X^+$ ($I_{B_2-X^+}$), obtained from the resulting mass spectrum, and the partial pressures of B_1 (P_{B_1}) and B_2 (P_{B_2}), measured by ion gauges, furnish the equilibrium constant of reaction (15.7) according to Eq. (15.8). This equilibrium constant K_{eq} provides the free-energy change of reaction (15.7), as given in Eq. (15.9), which is identical to the free-energy difference of the $B_1{-}X^+$ and $B_2{-}X^+$ bonds and hence provides the *relative* free energy of X^+ binding to B_1 versus B_2:

$$B_1{-}X^+ + B_2 \rightleftarrows B_1 + B_2{-}X^+ \tag{15.7}$$

$$K_{eq} = \frac{P_{B_1} I_{B_2-X^+}}{I_{B_1-X^+} P_{B_2}} \tag{15.8}$$

$$-\mathrm{RT} \ln K_{eq} = \Delta G_{(15.7)} = \Delta G_X(B_2) - \Delta G_X(B_1) = \Delta(\Delta G_X) \tag{15.9}$$

$$\Delta(\Delta H_X) = \Delta(\Delta G_X) + T\,\Delta(\Delta S_X) \tag{15.10}$$

Presently, FTMS is the most widely used method for gas-phase equilibrium studies. Experimentally, $B_1{-}X^+$ and $B_2{-}X^+$ are formed and thermalized in the presence of the corresponding neutral vapors (known P_{B_1} and P_{B_2}) at a set temperature. Then, one of these ions is removed (by ejection), and the evolution of the intensities of $B_1{-}X^+$ and $B_2{-}X^+$ is monitored over time until the intensities level off and equilibrium is reached.

Figure 15.1 illustrates such a measurement for the Na^+ exchange between Phe and AlaOMe (alanine methyl ester) (Gapeev and Dunbar 2001). Repeating such experiments with different pairs of ions leads to a free-energy thermochemical ladder, as displayed in Figure 15.2a (Gapeev and Dunbar 2001), which reveals the *relative* free energies of X^+ binding to the biomolecules investigated in these equilibrium measurements.

In order to convert the ladder of relative free energies to a ladder of *relative* X^+ affinities, the corresponding entropy changes $\Delta(\Delta S_X)$ are needed; see Eq. (15.10). If the equilibria can be measured as a function of temperature, $\Delta(\Delta S_X)$ can be obtained experimentally from the change of $\Delta(\Delta S_X)$ with T (Kebarle 1992; Hoyau et al. 1999). Most experimental setups do not allow, however, for variations in T. In these cases, $\Delta(\Delta S_X)$, which is composed of translational, rotational, and vibrational components, is estimated computationally and the $T\,\Delta(\Delta S_X)$ is added to $\Delta(\Delta G_X)$ to derive the corresponding relative X^+ affinities, $\Delta(\Delta H_X)$. Figure 15.2b illustrates the ladder of *relative* X^+ affinities, $\Delta(\Delta H_X)$, resulting from the free energy ladder in

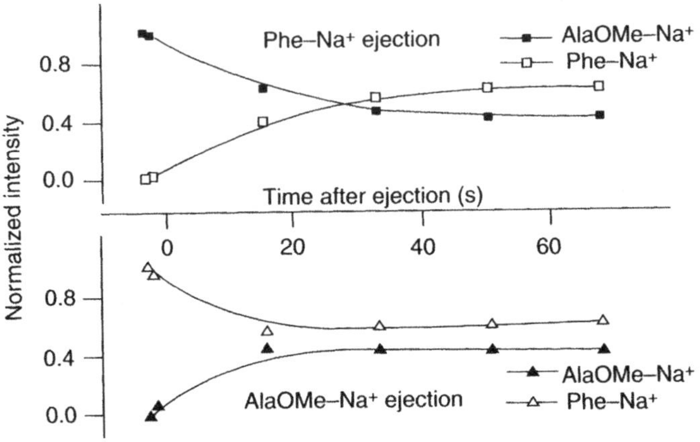

FIGURE 15.1. Approach to equilibrium Phe—Na^+ + AlaOMe ⇄ Phe + AlaOMe—Na^+ from both directions. [Reprinted with permission from Gapeev and Dunbar (2001).]

Figure 15.2a (Gapeev and Dunbar 2001). If the ladder determined in this way contains one or more molecules of known X^+ affinity, established by independent methods (see text that follows), the *relative* affinities can be converted to *absolute* X^+ affinities. Typically, absolute affinities are reported at 0 or 298 K by applying the proper thermal correction to the measured values. The uncertainty of the experimentally derived X^+ affinities depends on how accurately the temperature of the reacting system can be measured and on the correct estimation of the pertinent bond entropies; equilibrium measurements can furnish X^+ binding affinities within $\pm<3$ kJ/mol [Meot-Ner (Mautner) 2003].

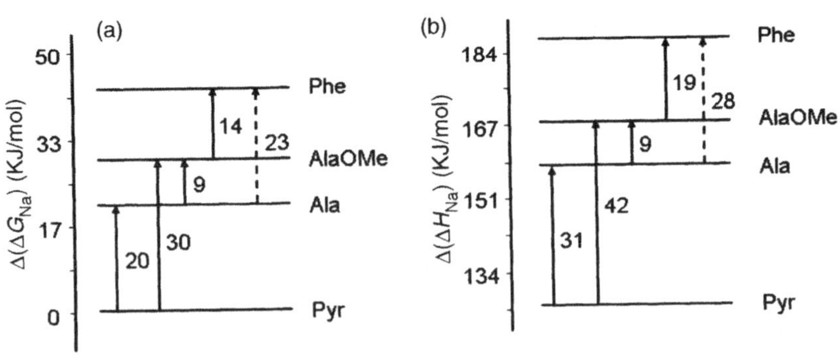

FIGURE 15.2. (a) Relative free energies of Na^+ binding at 370 K. The dashed arrow value was not measured directly but calculated from the other values. (b) Ladder of relative Na^+ affinities obtained from the relative free energies after entropy corrections. The *y* axis gives the corresponding absolute affinities (at 0 K), obtained by anchoring the relative values to the known Na^+ affinity of pyridine (127 kJ/mol) and applying thermal corrections. [Reprinted with permission from Gapeev and Dunbar (2001).]

15.2.2. Bracketing Method

The equilibrium approach presupposes that both biomolecules participating in the ion exchange reaction (15.7), namely, B_1 and B_2, are volatile and thermally stable. This is not true for many biomolecules, such as peptides, saccharides, and nucleotides. In these cases, thermodynamic properties can be measured using the bracketing method in combination with FTMS (Gorman et al. 1992; Wu and Lebrilla 1993; Gorman and Amster 1993a,b; Zhang et al. 1993). The ion $B{-}X^+$, containing the biomolecule of unknown X^+ affinity, is formed by desorption or spray ionization, isolated, and reacted with a series of reference compounds R_i of known X^+ affinity. The transfer of X^+ from B to R_i [reaction (15.11)] is monitored at a constant background pressure of R_i. This reaction is fast when it is exoergic, but does not take place or is slow when is endoergic (Harrison 1992):

$$B{-}X^+ + R_i \rightarrow B + R_i{-}X^+ \tag{15.11}$$

$$\ln \frac{[B{-}X^+]}{[B{-}X^+]_0} = -k_X P_{R_i} t \tag{15.12}$$

From the pseudo-first-order change of the reactant ion intensity [Eq. (15.12)], the rate constant of X^+ transfer from B to R_i, namely, k_X, can be obtained. On the other hand, the collisional rate constant for the encounters between $B{-}X^+$ and R_i, namely k_{col}, can be calculated using the average dipole orientation (ADO) theory (Su and Bowers 1973). The ratio k_X/k_{col}, which is defined as the reaction efficiency of X^+ transfer, is usually ≤ 0.2 for endoergic reactions but much larger for exoergic reactions. The ΔG_X values of the two reference compounds, with which X^+ transfer changes from an endoergic to an exoergic process, "bracket" the free energy of X^+ binding of the biomolecule under study, B. Using this procedure it is possible to determine $\Delta G_X(B)$ within approximately ± 10 kJ/mol.

The bracketing method is exemplified by data summarized in Table 15.1, which concern reaction efficiencies for the deprotonation of $GG{-}H^+$ by a series of reference bases with gas-phase basicities in the range 870–884 kJ/mol. Proton transfer is slow with allylamine (or less basic molecules) and becomes fast with benzylamine (or more basic molecules), bracketing GB(GG) between the GB values of allylamine and benzylamine, at 878 kJ/mol (Zhang et al. 1993; Harrison 1997).

TABLE 15.1. Reaction Efficiencies for Proton Transfer Reaction $GG{-}H^+ + R_i \rightarrow GG + R_i{-}H^+$

Reference Compound R_i	GB_i (kJ/mol)[a]	Reaction Efficiency[b]
3-Fluoropyridine	870.1	0.03
Allylamine	875.5	0.07
Benzylamine	879.4	0.61
n-Propylamine	883.9	0.66

[a]From Hunter and Lias (1998).
[b]From Zhang et al. (1993).

15.2.3. Kinetic Method

Gas-phase equilibria and bracketing reactions concerning X^+ ligand exchange progress through intermediate adducts, in which two ligands are bound to X^+. These intermediates, B_1—X^+—B_2 or B—X^+—R_i, cannot be isolated in experiments conducted at low pressure (e.g., with FTMS, the prevalent method in equilibrium or bracketing studies). The kinetic method involves independent formation of such intermediates in the ionization step and subsequent study of the dissociation kinetics of the intermediates into the individual X^+-bound monomers by tandem mass spectrometry (MS/MS) (Cooks et al. 1994b, 1999; Cooks and Wong 1998):

$$\mathrm{B{-}X^+} + \mathrm{R}_i \xleftarrow{k} \mathrm{B{-}X^+{-}R}_i \xrightarrow{k_i} \mathrm{B} + \mathrm{X^+{-}R}_i \tag{15.13}$$

$$\ln\frac{k}{k_i} = -\frac{\Delta G^{\ddagger} - \Delta G_i^{\ddagger}}{RT_{\mathrm{eff}}} = \frac{\Delta S^{\ddagger} - \Delta S_i^{\ddagger}}{R} - \frac{\Delta H^{\ddagger} - \Delta H_i^{\ddagger}}{RT_{\mathrm{eff}}} \tag{15.14}$$

$$\ln\frac{k}{k_i} = -\frac{\Delta S_{\mathrm{X}}^{\mathrm{app}}(\mathrm{B}) - \Delta S_{\mathrm{X}}^{\mathrm{app}}(\mathrm{R}_i)}{\mathrm{R}} + \frac{\Delta H_{\mathrm{X}}(\mathrm{B}) - \Delta H_{\mathrm{X}}(\mathrm{R}_i)}{RT_{\mathrm{eff}}}$$

$$= \left[-\frac{\Delta(\Delta S_{\mathrm{X}}^{\mathrm{app}})}{R} + \frac{\Delta H_{\mathrm{X}}(\mathrm{B})}{RT_{\mathrm{eff}}}\right] - \frac{\Delta H_{\mathrm{X}}(\mathrm{R}_i)}{RT_{\mathrm{eff}}} \tag{15.15}$$

$$\Delta G_{\mathrm{X}}^{\mathrm{app}}(\mathrm{B}) = \Delta H_{\mathrm{X}}(\mathrm{B}) - T_{\mathrm{eff}}\Delta(\Delta S_{\mathrm{X}}^{\mathrm{app}}) \tag{15.16}$$

$$\ln\frac{k_1}{k_2} \approx \frac{\Delta H_{\mathrm{X}}(\mathrm{B}) - \Delta H_{\mathrm{X}}(\mathrm{R}_i)}{RT_{\mathrm{eff}}} = \frac{\Delta(\Delta H_{\mathrm{X}})}{RT_{\mathrm{eff}}} \tag{15.17}$$

Most commonly, a biomolecule of unknown X^+ affinity (B) is paired with a series of reference molecules of known X^+ affinity (R_i) as shown in Eq. (15.13). The dissociations of B—X^+—R_i into the individual monomers are probed by collisionally activated dissociation (CAD) or other MS/MS methods. Formations of B—X^+ and X^+—R_i are usually the lowest-energy decompositions of an X^+-bound dimer and, most often, either the only or the predominant channels observed on MS/MS (Cooks et al. 1994b; Cooks and Wong 1998); this is exemplified in Figure 15.3 for GlyGly—Na^+—R_i dimers (Kish et al. 2004). Under these circumstances, the abundance ratio of B—X^+ versus X^+—R_i in the MS/MS spectrum is an approximate measure of the rate constant ratio k/k_i of the two competing pathways.

According to transition state theory (Laidler 1987), the natural logarithm of k/k_i depends on the relative free energy of activation of the competing reactions and the temperature of the dissociating dimer ions, as shown in Eq. (15.14), where R is the ideal-gas constant. An effective temperature (T_{eff}) is used in place of a thermodynamic temperature (T), because the dimer ions are not in thermal equilibrium with their surroundings and do not have Boltzmann-shaped internal energy distributions (Cooks et al. 1999). The free energies of activation are composed of entropy and enthalpy components, which are included in Eq. (15.14). Singly charged noncovalent dimers bridged by a central ion generally dissociate

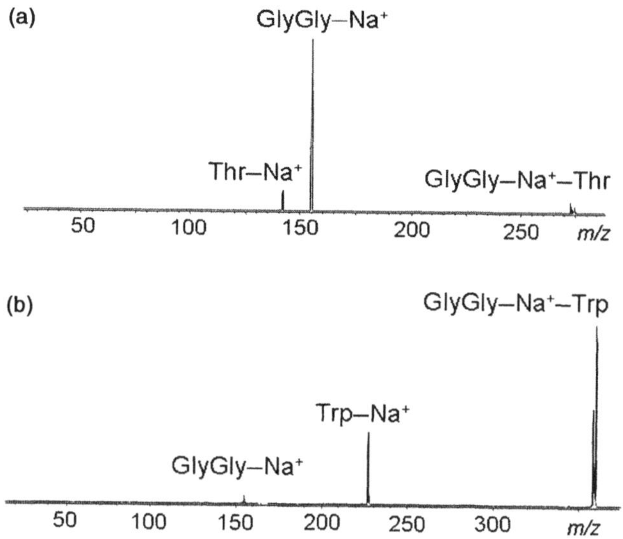

FIGURE 15.3. CAD mass spectra of GlyGly–Na^+–R_i dimers formed by electrospray ionization (ESI) in an ion trap; R_i = (a) Thr and (b) Trp. The spectra were acquired using an excitation amplitude (V_{p-p}) of 0.4 V. [Reprinted with permission from Kish et al. (2004).]

with no reverse activation energies (Cooks et al. 1994b; Cooks and Wong 1998). In such cases, the relative activation enthalpy becomes numerically equal (opposite sign) with the relative X^+ affinity of B versus R_i; see Eq. (15.15). Similarly, the relative activation entropy can be equated with an apparent relative entropy of X^+ binding (opposite sign); an apparent entropy difference $[\Delta(\Delta S_X^{app})]$ is used because of the lack of a true thermal equilibrium, as explained above (Ervin 2002). More recent studies have shown that $\Delta(\Delta S_X^{app})$ depends on the X^+-bound system studied and may vary between ~0 and the actual thermodynamic entropy difference of the B–X^+ and X^+–R_i bonds (Hahn and Wesdemiotis 2003; Wesdemiotis 2004).

The kinetic method experiment renders the k/k_i ratios of the B–X^+–R_i heterodimers. The $\ln(k/k_i)$ values measured under the same MS/MS activation conditions are plotted against $\Delta H_X(R_i)$ to obtain a regression line according to Eq. (15.15); the slope and x intercept of this line provide T_{eff} and an apparent free energy of X^+ binding to B $[\Delta G_X^{app}(B)]$, defined in Eq. (15.16) (Cheng et al. 1993; Kish et al. 2003, 2004). If $\Delta(\Delta S_X^{app})$ is negligible, the apparent free energy of the B–X^+ bond is essentially identical to the corresponding X^+ affinity [Eq. (15.16)] (see also Figure 15.4). In such cases, the dependence of the experimental k/k_i ratios on the corresponding relative X^+ affinities is simplified to Eq. (15.17), known as the "simple" or "classical" kinetic method relationship (Cooks et al. 1994b; Cooks and Wong 1998). In contrast, Eq. (15.15) describes the extended version of the kinetic method (Cooks and Wong 1998; Hahn and Wesdemiotis 2003; Kish et al. 2003, 2004).

The magnitude of $\Delta(\Delta S_X^{app})$ can be appraised by varying T_{eff} with CAD, which is the most widely used MS/MS variant in kinetic method studies; T_{eff} is varied by changing the collision energy and, hence, the internal energy deposited into the

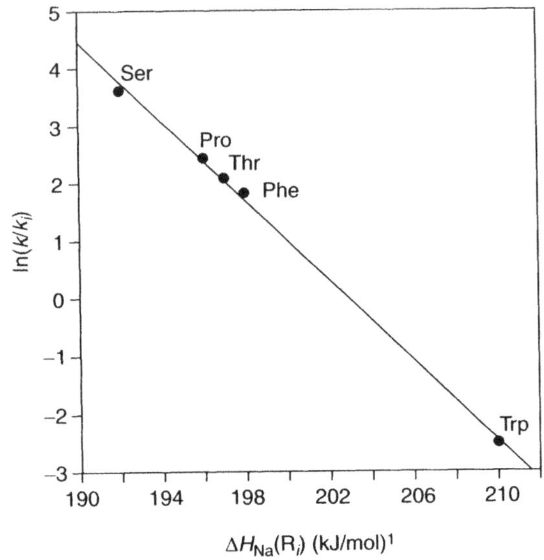

FIGURE 15.4. Plot of $\ln(k/k_i)$ versus $\Delta H_{\mathrm{Na}}(R_i)$ for GlyGly–Na^+–dimers according to Eq. (15.15). The k/k_i ratios were obtained from CAD spectra measured at $V_{\mathrm{p-p}} = 0.4$ V (cf. Figure 15.3). The k/k_i values and, thus, the resulting lines do not change outside experimental error if the excitation amplitude is varied between 0.40 and 0.65 V. The average values of the slopes and x intercepts of the regression lines at all $V_{\mathrm{p-p}}$ voltages yield $T_{\mathrm{eff}} = 356$ K and ΔH_{Na}(GlyGly) = 203 (±8) kJ/mol (error reflects compounded uncertainty in $\Delta H_{\mathrm{Na}}(\mathrm{R}_i)$ and the x intercepts). [Reprinted with permission from Kish et al. (2004).]

decomposing B—X^+—R_i dimers (Cooks et al. 1999). If $\Delta G_X^{\mathrm{app}}(\mathrm{B})$ changes with T_{eff}, a second regression line is constructed by plotting $\Delta G_X^{\mathrm{app}}(\mathrm{B})$ against T_{eff}; according to Eq. (15.16), the latter line provides both $\Delta(\Delta S_X^{\mathrm{app}})$, from the slope, and $\Delta H_X(\mathrm{B})$, from the y intercept.

The procedure outlined necessitates that reference molecules of known X^+ affinity are available. This condition cannot be fulfilled for every biomolecule investigated. An alternative approach is to examine several similarly structured molecules in B_1—X^+—B_2 pairs to determine the order of their X^+ affinities (Cooks et al. 1994b; Cooks and Wong 1998). By varying the B_1 and B_2 molecules paired in the dimer, a ladder of relative affinities is constructed and displayed in staircase format; Figure 15.5 provides an example, referring to the Na^+ affinities of dipeptides (Wang et al. 2006a). In such ladder formats, B_1 is ideally paired with more than one B_2, so that a certain $\ln(k_1/k_2)$ value can be reached via different routes. If Eq. (15.17) applies and $\Delta(\Delta S_X^{\mathrm{app}}) \approx 0$, one-step and cumulative $\ln(k_1/k_2)$ data agree with each other within experimental error (±0.30 on average). This is true for the data in Figure 15.5; for example, the one-step $\ln(k_{\mathrm{GlyLeu}}/k_{\mathrm{GlyAla}})$ value (viz., 1.52) is very similar to the ratio deduced via the stepwise comparison of the GlyAla/AlaAla, AlaAla/Trp, and Trp/GlyLeu pairs, $0.52 + 0.38 + 0.67 = 1.57$. Appreciable apparent entropy differences between the bonds compared in the B_1—X^+—B_2 dimers

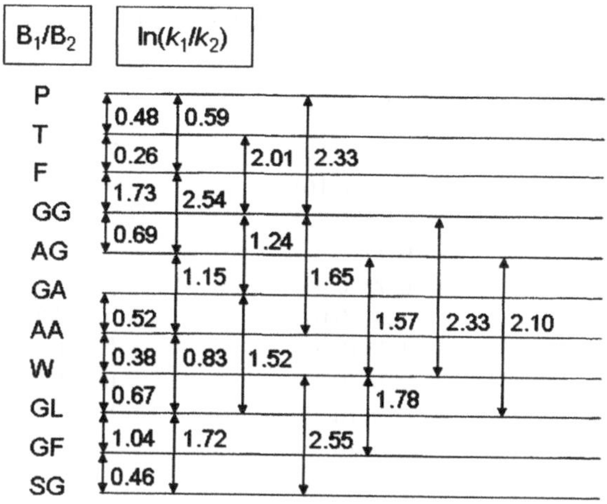

FIGURE 15.5. A $\ln(k_1/k_2)$ ladder constructed from CAD mass spectra of B_1—Na^+—B_2 dimers formed by ESI in an ion trap; ±(0.03–0.51). The components of a dimer are connected by arrows. Na^+ affinities increase from top to bottom (Wang et al. 2006a).

cause significant disagreement between $\ln(k_1/k_2)$ values obtained in one versus several steps because apparent bond entropies have been shown to be nonadditive (Cerda and Wesdemiotis 1995; Lee et al. 1998; Ervin 2002; Hahn and Wesdemiotis 2003; Kish et al. 2003, 2004). If the ladder includes at least three molecules of known X^+ affinity, an effective temperature can be estimated for the decomposing heterodimers according to Eq. (15.17). This T_{eff} can in turn be used to convert a ladder of $\ln(k/k_i)$ values into a ladder of relative X^+ affinities (see Table 15.2). As in equilibrium measurements, relative X^+ affinities can be converted to absolute ΔH_X data if the relative values are anchored to a known X^+ affinity (Table 15.2).

15.2.4. Threshold Collision-Induced Dissociation (TCID)

TCID derives thermochemical information by determining the energy threshold of an endothermic reaction; *threshold* is the minimum energy required to form the products. TCID has extensively been applied to metal ion complexes of biomolecules, B—X^+ (X^+ = metal ion), to determine pertinent metal ion affinities (Rodgers and Armentrout 2000a). B—X^+ is formed in a flowing afterglow or ESI ion source, thermalized, mass-selected, and introduced into an octopole collision cell, where it undergoes collision-induced dissociation (CID) with xenon targets at laboratory-frame collision energies (E_{lab}) varying between 0 and several eV. [The acronyms CAD (defined above) and CID describe the same process.] The internal energy available to drive endothermic reactions, such as the dissociation B—X^+ → B + X^+, is equal to the kinetic energy in the center-of-mass frame E_{cm}, related to E_{lab} according to Eq. (15.18), in which m_{Xe} and $M_{B\text{—}X^+}$ are the masses of the target and mass-selected ion, respectively. The ions exiting the octopole, which include undissociated B—X^+ precursor ions, X^+ fragments, and any other CID

TABLE 15.2. Sodium Ion Affinities of Amino Acids and Dipeptides Determined by Kinetic Method

Amino Acid or Peptide	$\ln(k/k_{GlyGly})^a$	$\Delta(\Delta H_{Na})^b$	ΔH^c_{Na}
Pro	−2.40 (0.17)	−7.2 (1.0)	196 (8)
Thr	−2.03 (0.16)	−6.1 (0.9)	197 (8)
Phe	−1.79 (0.22)	−5.4 (0.9)	198 (8)
GlyGly	0.00	0.0	203 (8)
AlaGly	0.69 (0.06)	2.1 (0.3)	205 (8)
GlyAla	1.24 (0.07)	3.7 (0.5)	207 (8)
AlaAla	1.75 (0.31)	5.3 (1.1)	208 (8)
Trp	2.24 (0.23)	6.7 (1.1)	210 (8)
GlyLeu	2.76 (0.25)	8.3 (1.3)	211 (8)
GlyPhe	3.91 (0.25)	11.8 (1.6)	215 (8)
SerGly	4.55 (0.26)	13.7 (1.9)	217 (8)

[a]Average cumulative $\ln(k/k_{GlyGly})$ values (i.e., Na^+ affinity ranking relative to GlyGly) calculated from the individual $\ln(k_1/k_2)$ data in Figure 15.5. The corresponding pooled standard deviations follow in parentheses.

[b]Calculated using Eq. (15.17) and $T_{eff} = 362$ (± 45) K; T_{eff} was obtained from the correlation of the $\ln(k/k_{GlyGly})$ values of Pro, Thr, Phe, Trp, GlyGly, and Trp with their known ΔH_{Na} (Kish et al. 2003, 2004). The standard deviations (in parentheses) were calculated from the experimental uncertainties in $\ln(k/k_{GG})$ and T_{eff}.

[c]Obtained by anchoring the relative affinities, $\Delta(\Delta H_{Na})$, to ΔH_{Na}(GlyGly) = 203 (± 8) kJ/mol (Kish et al. 2004).

Source: Wang et al. (2006a).

products, are mass-analyzed and detected. The octapole radiofrequency field (RF) collates the ions radially, minimizing scattering losses. Under these conditions, the intensity of the beam entering the octapole, I_0, is essentially equal to the sum of the ion intensities exiting the octopole, $I_0 = I_{B\text{–}X^+} + \Sigma I_{frag}$, where $I_{B\text{–}X^+}$ and I_{frag} correspond to the transmitted precursor and fragment ion intensities, respectively (Ervin and Armentrout 1985). Using $I_{B\text{–}X^+}$ and the individual I_{frag} values, an experimental total reaction cross section σ_{tot} can be calculated via Eq. (15.19), where n is the Xe gas density and l the effective pathlength. The cross section for X^+ (or any other CID fragment) is then calculated via Eq. (15.20) (Ervin and Armentrout 1985):

$$E_{cm} = E_{lab}\frac{m_{Xe}}{m_{Xe} + M_{B\text{–}X^+}} \tag{15.18}$$

$$\frac{I_{B\text{–}X^+}}{I_{B\text{–}X^+} + \Sigma I_{frag}} = \exp(-\sigma_{tot} n l) \tag{15.19}$$

$$\sigma_{X^+} = \sigma_{tot}\frac{I_{X^+}}{\Sigma I_{frag}} \tag{15.20}$$

$$\sigma(\mathrm{E}) = \sigma_0 \sum_i \frac{g_i(E + E_i - E_0)^n}{E} \tag{15.21}$$

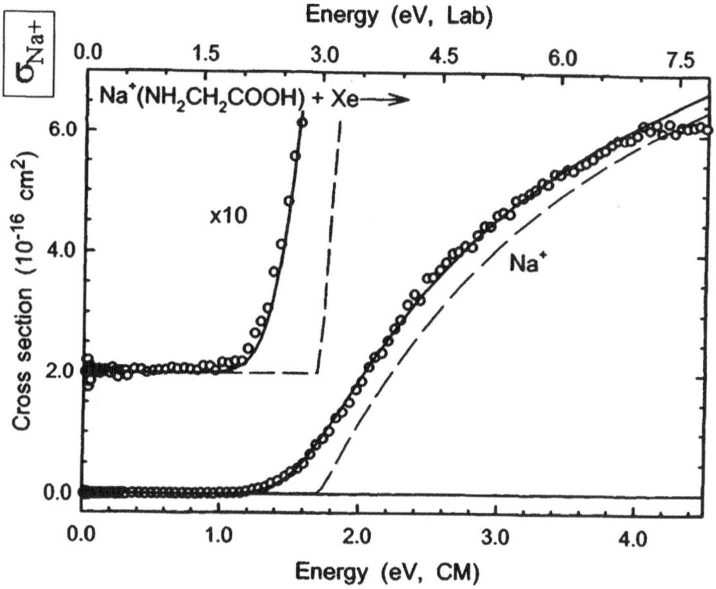

FIGURE 15.6. Zero-pressure extrapolated cross section for the formation of Na^+ from Gly–Na^+ via CID with Xe, as a function of collision energy. Open circles show the experimental data. Solid lines show the best fit using Eq. (15.21) convoluted over the kinetic and internal energy distributions of Gly–Na^+ and Xe. Dashed lines show the model cross-sections for reactants at 0 K and without kinetic energy broadening. [Reprinted with permission from Moision and Armentrout (2002).]

Multiple collisions may cause dissociation below the actual threshold energy. To avoid such problems, the cross section of X^+ is measured at 2–3 Xe pressures and extrapolated to zero pressure (Dalleska et al. 1994). The pressure-corrected cross sections are plotted against E_{cm} and E_{lab}; Figure 15.6 illustrates an example, obtained for the dissociation Gly–Na^+ $\rightarrow$ Gly + Na^+ (Moision and Armentrout 2002). It can be seen that σ_{Na+} decreases exponentially to 0 with decreasing collision energy.

The experimental dissociation threshold can be converted into a bond dissociation energy and enthalpy, if there is no reverse activation energy (i.e., no barrier for the forward reaction in excess of the endothermicity). This is performed by modeling the threshold region of the CID cross sections according to Eq. (15.21), in which σ_0 is an energy-independent scaling factor, n is an adjustable parameter describing the energy distribution transferred in the collision with Xe, E is the relative kinetic energy of B–X^+ and Xe, and E_0 is the threshold for dissociation of the ground electronic and rovibrational (rotational–vibrational) state of B–X^+ (Rodgers and Armentrout 2000a; Armentrout 2003). The model sums over the i vibrational states of B–X^+, where E_i and g_i are the internal energy and population, respectively, of each state ($\Sigma g_i = 1$). The density of rovibrational states is calculated using the Beyer–Swinehart algorithm (Beyer and Swinehart 1973; Stein and Rabinovitch 1977), while the relative populations of these states (i.e., g_i) are calculated at 300 K and for a Maxwell–Boltzmann distribution. The vibrational frequencies and rotational constants needed for this modeling are obtained by semiempirical, density functional, or ab initio methods, depending on the complexity of the system studied.

Since biomolecules can be large and complex, the dissociation of $B{-}X^+$ is usually associated with a kinetic shift, that is, an energy higher than the true threshold necessary for dissociation to occur within the timescale of the experiment ($\sim 10^{-4}$ s). The kinetic shift is accounted for by incorporating Rice–Ramsperger–Kassel–Marcus (RRKM) (Holbrook et al. 1996) or another statistical theory for unimolecular dissociation into Eq. (15.21) in order to predict the dissociation rate constant of $B{-}X^+$ (Rodgers et al. 1997; Armentrout 2003). Application of such a theory requires the rovibrational frequencies of the activated precursor ion and the transition state (TS) leading to the products; these are obtained computationally, assuming that the dissociation of $B{-}X^+$ proceeds over a loose and product-like transition state in the phase space limit (PSL) (Rodgers et al. 1997). If the precursor ion undergoes more than one dissociation, competitive shifts must also be considered for the higher-energy channels. This is accomplished by introducing branching ratios in the statistical model used in the kinetic shift correction (Rodgers and Armentrout 1998; Armentrout 2003).

After the cross sections calculated via Eq. (15.21) are convoluted with the kinetic energy distributions of $B{-}X^+$ and Xe (Chantry 1971; Lifshitz et al. 1978), the parameters σ_0, n, and E_0 are optimized by a nonlinear least-squares analysis to derive the best fit to the experimental data. The threshold values resulting from this procedure correspond to bond dissociation energies at 0 K because all sources of energy are considered in the fitting process (Dalleska et al. 1993; Rodgers and Armentrout 2000a; Armentrout 2003). At 0 K, bond energies and bond enthalpies are identical. The vibrational and rotational frequencies computed for modeling the experimental cross sections allow for calculation of the $B{-}X^+$ bond entropies and for conversion of the experimentally deduced 0 K bond energies to 298 K bond enthalpies and free energies. Table 15.3 summarizes the threshold and thermochemical data obtained through this process for the Gly–Na^+ bond from the measurements in Figure 15.6 (Moision and Armentrout 2002).

TABLE 15.3. Thermochemical Quantities Obtained for the Gly–Na^+ Bond after Modeling TCID Data in Figure 15.6 Using Eq. (15.21) (Uncertainties in Parentheses)

E_0[a] (eV)	E_0(PSL)[b] (eV)	$\Delta S^{\ddagger}_{Na/1000}$[c] (J mol^{-1} K^{-1})	$\Delta H_{Na/0}$[d] (kJ/mol)	$\Delta H_{Na/298}$[d] (kJ/mol)	$T\Delta S_{Na/298}$[e] (kJ/mol)	$\Delta G_{Na/298}$[f] (kJ/mol)
1.74 (0.05)	1.70 (0.05)	40 (5)	164 (6)	166 (6)	34 (5)	132 (6)

[a]Threshold energy without RRKM analysis.
[b]Threshold energy with RRKM analysis. The difference $E_0 - E_0$(PSL) is the kinetic shift.
[c]Transition state entropy at 1000 K; its value is consistent with a simple bond cleavage (Lifshitz 1989).
[d]Na^+ affinities of Gly at 0 and 298 K.
[e]Entropy contribution to free energy at 298 K.
[f]Free energy at 298 K, calculated via Eq. (15.4).
Source: Moision and Armentrout (2002).

15.2.5. Blackbody Infrared Radiative Dissociation (BIRD)

Weakly bound ionic systems that are trapped in the heated cell of a Fourier transform mass spectrometer at low pressures ($<10^{-8}$ Torr); for example, guest–host complexes or metal–ligand clusters involving noncovalent interactions, can dissociate by absorption of infrared photons emitted from the walls of the instrument ("blackbody radiation") (Dunbar 1991, 1994; Dunbar et al. 1995; Dunbar and McMahon 1997). The BIRD method takes advantage of this phenomenon; it measures the rate constant of such a thermally induced dissociation as a function of temperature, to thereby derive the activation energy and entropy associated with this reaction (Dunbar et al. 1995; Price et al. 1996; Schnier et al. 1996, 1998; Dunbar and McMahon 1997; Jockusch et al. 1997; Kitova et al. 2002; Daneshfar and Klassen 2004; Dunbar 2004). From the activation parameters, the enthalpy and entropy of the bond(s) being broken can be extracted, if there is no reverse activation barrier.

BIRD experiments are generally conducted at 25–250°C, which allows for the determination of bond strengths in the ~70–250 kJ/mol range. Dissociation of such weak bonds is usually slow, requiring seconds or larger time intervals of observation. In addition, very low background pressures are necessary to avoid CAD, which would interfere with BIRD and obscure the resulting kinetics. These prerequisites have limited BIRD applications to FTMS instruments, which can provide both long observation times as well as extremely low background pressures.

Because the bonds probed are weak, care must be taken to form the precursor ions with as little internal energy as possible. ESI is ideally suitable for this purpose and has been used predominantly in BIRD studies. Usually, the ions of interest are formed externally and transported into a heated FTMS cell, where they are separated from any other ions coproduced during ESI. Collisions with a pulsed, inert bath gas may follow to remove excess kinetic energy, which could cause unwanted CAD. Subsequently, mass spectra—the abundances of precursor and fragment ion(s)—are monitored as a function of time. For large precursor ions (above ~1600 amu and 500 degrees of freedom), IR photon exchange with the walls is much faster than dissociation, and the ions reach a steady state with a Boltzmann distribution of internal energies (Price and Williams 1997). Under these conditions, termed the *rapid energy exchange* (REX) limit, dissociation kinetics are identical to those obtained in the traditional high-pressure limit, where a Boltzmann distribution is brought on by collisions (Price and Williams 1997; Dunbar 2004).

The relative abundance of the precursor ion, $I_P/(I_P + \Sigma_{frag})$, is calculated from the mass spectra acquired during the BIRD experiment, and its natural logarithm is plotted against time for the various temperatures used. Dissociations with first-order kinetics lead to linear plots, whose slopes yield the corresponding macroscopic rate constants, k_{uni}. Figure 15.7 illustrates an example of such kinetic plots, referring to the dissociation of the noncovalent complex between the single-chain antibody fragment (scFv) from a carbohydrate-binding antibody and the trisaccharide Tal(Abe)Man (*tri*), namely $[\text{scFv} + tri + n\text{H}]^{n+} \rightarrow [\text{scFv} + n\text{H}]^{n+} + tri$ (Kitova et al. 2002).

The dependence of k_{uni} on temperature is given by the Arrhenius relationship, Eq. (15.22). Hence, a new plot of $\ln(k_{uni})$ versus $1/T$ provides the activation energy

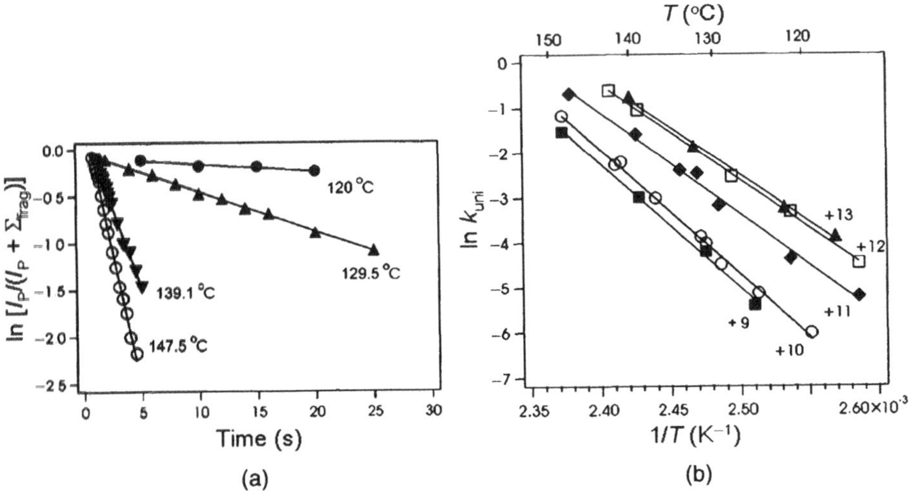

FIGURE 15.7. BIRD experiments with the noncovalent protein/trisaccharide complex $[\mathrm{scFv} + tri + n\mathrm{H}]^{n+}$ formed by ESI; kinetic data (a) for $n = 12$ at the temperatures shown and Arrhenius plots (b) for charge states $n = 9$–13. [Reprinted with permission from Kitova et al. (2002).]

E_a from the slope and the preexponential Arrhenius factor A from the intercept. The Arrhenius plots of several charge states of the noncovalent complex (viz., $n = 9$–12) are included in Figure 15.7 (Kitova et al. 2002). Because ions of the size of this complex (27 kamu) dissociate within the REX limit (Price and Williams 1997), that is, they are in thermal equilibrium with their surroundings, the Arrhenius activation energy becomes a characteristic parameter of the dissociation probed, and structurally diagnostic thermochemical information can be derived from it. In the REX limit, E_a is related to the dissociation threshold (E_0) and the enthalpy of activation ($\Delta H^{\ddagger}$) for the loss of *tri* from the noncovalent complex via Eq. (15.23) (Steinfeld et al. 1999). Similarly, the preexponential Arrhenius factor A corresponds to the entropy of activation (transition state entropy) of this reaction, as indicated in Eq. (15.24), where k_B and h are the Boltzmann and Planck constants, respectively (Steinfeld et al. 1999). The dissociation examined involves the loss of a neutral molecule, which can be assumed to be barrierless; therefore, the value of E_a also is a good estimate for the binding enthalpy of the noncovalent complex (ΔH_{tri}; equivalent to the affinity of scFv for *tri*), i.e. $\Delta H_{tri} \approx E_a$ (Kitova et al. 2004).

$$k_{\mathrm{uni}} = A\ \exp(E_a/RT) \qquad \text{or} \qquad \ln(k_{\mathrm{uni}}) = \ln A - \frac{E_a}{RT} \tag{15.22}$$

$$E_0 \approx \Delta H^{\ddagger} = E_a - RT \tag{15.23}$$

$$A = \frac{k_B T}{h} \exp\left(\frac{\Delta S^{\ddagger}}{R}\right) \tag{15.24}$$

TABLE 15.4. Kinetic Parameters Obtained from BIRD Experiments with $[scFv + tri + nH]^{n+}$ Ions Dissociating to $[scFv + nH]^{+} + tri$ under REX Limit Conditions

n	E_a (kJ/mol)[a]	A (s^{-1})[b]
9	230	$10^{27.9}$
10	231	$10^{28.2}$
11	188	$10^{23.1}$
12	177	$10^{22.0}$
13	174	$10^{21.8}$

[a]Arrhenius activation energy.

[b]Preexponential Arrhenius factor; $A > 17$ characterize loose transition states, while $A < 10$ are consistent with tight transition states.

Source: Kitova et al. (2002).

Table 15.4 lists thermochemical data deduced from the BIRD experiments depicted in Figure 15.7 (Kitova et al. 2002). E_a decreases with the addition of protons 11–13, indicating weakened hydrogen bonding interactions between scFv and *tri*, as the number of ionizing protons exceeds 10, due to increased unfolding of scFv and an inferior binding region in the protein at the higher charge states. The lower binding energy between scFv and *tri* at the higher charge states is also reflected in the respective A values; these decrease with the number of protons, consistent with less hydrogen bonds between scFv and *tri* and the recovery of less rotations after release of *tri*, when the number of protons becomes larger (Kitova et al. 2002).

The interpretation of Arrhenius parameters is not as straightforward if the precursor ion has comparable rates of IR photon exchange and dissociation. In such situations, which apply to systems of smaller size (<500 degrees of freedom), the experimental data must be simulated using the master equation model of Eq. (15.25) in order to determine the actual threshold energy of the reaction (E_0) (Dunbar 1994, 2004; Price et al. 1996; Jockusch and Williams 1998). The master equation is a set of differential equations describing how the precursor ion internal energy evolves with time, as the ion is activated and deactivated by collisions or photon exchange with its surroundings, until it dissociates. The internal energy is divided into bins, with the population of each bin changing with time due to collisions or photon exchange. Equation (15.25) gives the change in ion population for bin i. Ions may enter this bin from another bin (j) by collisional activation (rate constant C_{ij}) or photon absorption (rate constant R_{ij}); conversely, bin i may lose ions by collisional deactivation (C_{ji}), photon emission (R_{ji}), or dissociation (microscopic rate constant D_i) (Dunbar 1994):

$$\frac{dN_i}{dt} = \sum_j (C_{ij}N_j + R_{ij}N_j - C_{ji}N_i - R_{ji}N_i) - D_iN_i \tag{15.25}$$

Initially, a trial value is selected for the threshold dissociation energy E_0 and RRKM (or another kinetic) theory is used to estimate D_i. Reactant vibrational frequencies are obtained by quantum chemical methods. The transition state is not calculated explicitly; rather, the reactant frequencies are used to construct various transition state frequency sets that correspond to a wide range of preexponential A factors and, hence, transition states with different degrees of looseness or tightness. Quantum calculations are also used to obtain the radiative rate constants R_{ij} and R_{ji}, which depend strongly on the respective transition state dipole moments (μ). Collisions, which are unlikely in the low-pressure regime of FTMS, are generally neglected. The master equation is solved over the range of temperatures covered in the BIRD experiment (by matrix algebra) to obtain simulated $\ln(k_{\text{uni}})$–$1/T$ plots. This process is repeated by varying E_0 and the transition state parameters (A and μ), until the agreement between calculated and experimental kinetic plots falls within an accepted error limit (Dunbar 1994; Jockusch and Williams 1998).

Figure 15.8 illustrates a master equation fit for BIRD data concerning the loss of imidazole (Im) from the proton-bound complex of *n*-acetyl-L-alanine methyl ester (AcAlaOMe) and imidazole, specifically, for the dissociation AcAlaOMe–H^+–Im → AcAlaOMe + $^+$H–Im (Jockusch and Williams 1998). BIRD measurements according to Figure 15.7 lead to $E_a = 0.82$ eV (79 kJ/mol) and $\log A = 9.4$. These results are best simulated if Eq. (15.25) is solved using $E_0 = 1.04$–1.19 eV

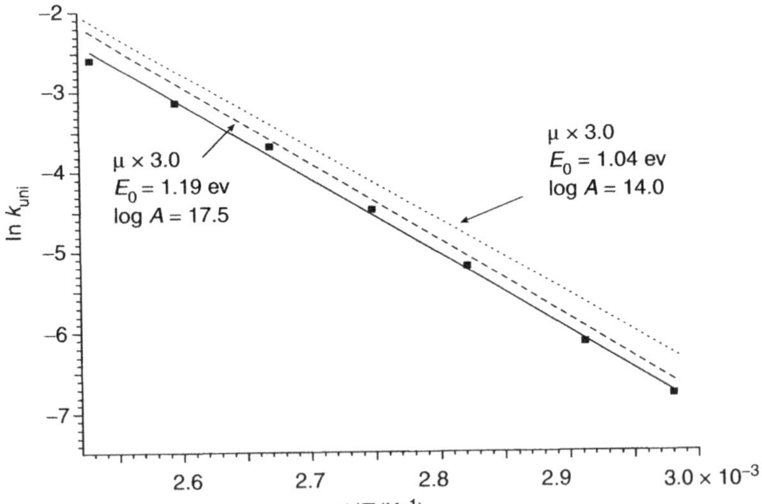

FIGURE 15.8. Experimental and calculated $\ln(k_{\text{uni}})$ versus $1/T$ plots for the dissociation AcAlaOMe-H^+–Im → AcAlaOMe + $^+$H–Im. The solid line represents the experimental data. The dotted and dashed lines result from modeling the data using Eq. (15.25) and the parameters indicated on the figure. The two calculated lines give the highest and lowest E_0 values that fit the experimental data, that is, these lines reproduce the measured E_a and log A within experimental error and provide k_{uni} values that differ from the measured rate constants by a factor of <2. [Reprinted with permission from Jockusch and Williams (1998).]

(108 ± 8 kJ/mol), log $A = 14.0$–17.5, and a transition dipole moment 3 times larger than that calculated at the semiempirical AM1 level of theory (see Figure 15.8). The E_0 obtained this way is in excellent agreement with the bond enthalpy (ΔH) measured by equilibrium experiments at high pressure. Note that the value deduced from the BIRD plots (viz., E_a) is substantially smaller than E_0 and that modeling of the thermal dissociation via Eq. (15.25) is necessary to determine the correct binding energy of the dimer.

It is not necessary to apply the master equation model if the ion under study is small ($<\sim$50 degrees of freedom) and, as a consequence, the rate of dissociation becomes much larger than the rate of thermal photon exchange with the surroundings (Dunbar et al. 1995; Dunbar 2004). Ions with internal energies exceeding E_0 dissociate promptly, thereby completely removing the ion population above E_0. In addition, the population just below E_0 is depleted partially as ions in that region are pumped continually above E_0 by photon absorption and dissociate. The result is a truncated Boltzmann internal energy distribution (Dunbar 2004). In such situations, E_0 can be adequately estimated from the experimental Arrhenius activation energy E_a via Eq. (15.26), where $\langle E_{pop} \rangle$ is the average internal energy of the (truncated Boltzmann) precursor ion population and 3.5 kJ/mol is a correction parameter, accounting for the temperature dependence of the radiation field and the reactive depletion of high-lying energy levels (Dunbar et al. 1995):

$$E_0 \approx E_a + \langle E_{pop} \rangle - 3.5\ (\text{kJ/mol}) \tag{15.26}$$

When applied to the AcAlaOMe–H^+–Im dimer discussed above (87 degrees of freedom), the truncated Boltzmann approximation gives $E_0 = 128$ kJ/mol (Jockusch and Williams 1998), which is significantly higher than the value of 108 kJ/mol determined using the master equation. Hence, for systems of this size, rigorous application of the master equation is required. On the other hand, either the master equation [Eq. (15.25)] or the truncated Boltzmann approximation [Eq. (15.26)] renders very similar E_0 values for the loss of imidazole (Im) from the proton-bound dime Im–H^+–Im, a markedly smaller system (51 degrees of freedom); the corresponding threshold energies are 107 and 105 kJ/mol, respectively (Jockusch and Williams 1998).

15.2.6. Surface-Induced Dissociation (SID)

A moving ion can be activated to fragment by collision with a surface. Surface-induced dissociation (SID) (Cooks et al. 1994a) provides an alternative to the BIRD and TCID methods for studies in fragmentation thermochemistry. SID coupled with FTMS has been employed to determine the energetics and dynamics of the decompositions of protonated peptides (Laskin et al. 2000, 2001, 2002; Laskin 2004). As even small peptides have a considerable number of degrees of freedom, the dissociation of their $[M + H]^+$ ions is affected by kinetic shifts, which reflect the excess internal energy necessary to cause dissociation within the time spent in the mass spectrometer. FTMS permits long observation times (up to several seconds),

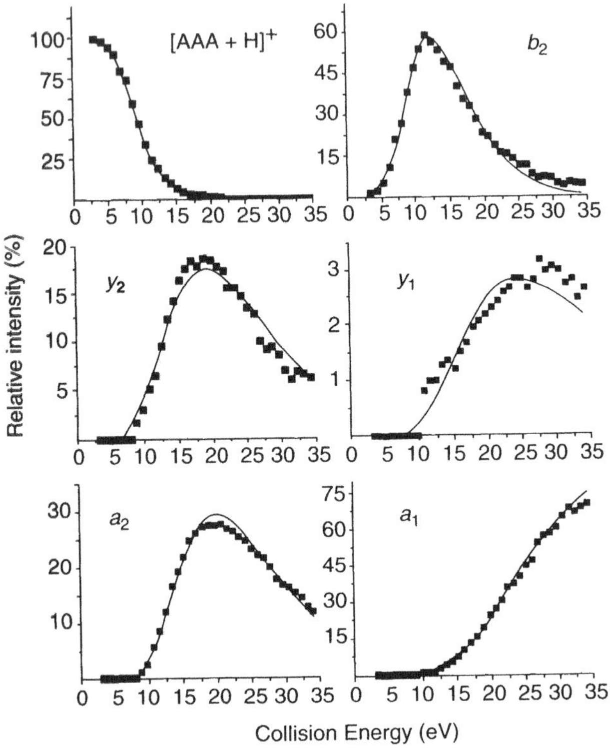

FIGURE 15.9. Experimental (filled squares) and calculated (lines) SID fragmentation efficiency curves (FECs) for the precursor ion and major fragments of $[AAA+H]^+$. [Reprinted with permission from Laskin et al. (2002).]

thereby minimizing kinetic shift effects. Moreover, SID provides fast-ion activation to relatively high levels of internal energy in strictly *one* collision, which essentially eliminates competitive shifts, that is, discrimination against high-energy channels.

SID is performed using gold plates covered with fluorinated self-assembled monolayers (Laskin 2004). The SID spectra of $[M+H]^+$, measured as a function of collision energy for a given time interval (which may vary between microseconds and seconds), are converted to experimental fragmentation efficiency curves (FECs) by plotting relative abundance of ion of interest (precursor ion or fragment) versus laboratory-frame collision energy. Figure 15.9 shows the FECs obtained for $[AAA+H]^+$ and several of its *N*- and *C*-terminal fragments (Laskin et al. 2002). These energy-resolved curves are subsequently modeled using RRKM/quasiequilibrium theory (QET). To facilitate this fitting, the dissociation pathways of the precursor ion are determined independently by CAD and double-resonance techniques. The reaction scheme established this way for $[AAA+H]^+$ is summarized in Scheme 15.1 (Laskin et al. 2002).

For each channel in the experimentally derived reaction scheme, energy-dependent microcanonical rate constants $[k_i(E)]$ are calculated via RRKM/QET; these are used to calculate fragmentation probability (relative abundance of the

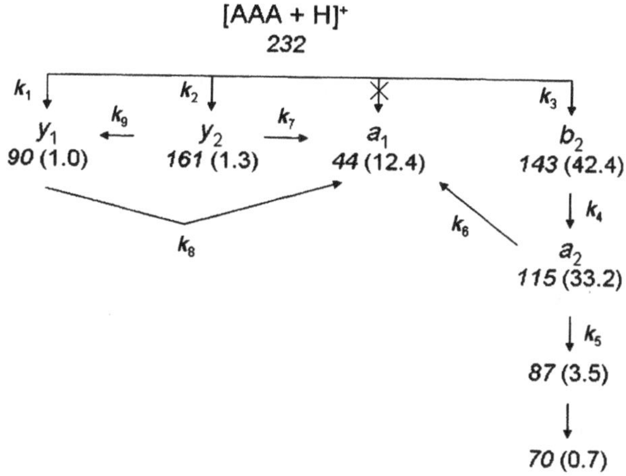

SCHEME 15.1. Dissociation pathways of $[AAA + H]^+$, as revealed by double-resonance CAD experiments via Fourier transform mass spectrometry. Mass-to-charge (*m/z*) ratios are given in italics and relative abundance in parentheses. [Reprinted with permission from Laskin et al. (2002).]

precursor ion or fragment ions, $[F_i(E, t_r)]$ as a function of the internal energy of the precursor ion and the experimental observation time t_r (Laskin et al. 2000; Laskin 2004). $F_i(E, t_r)$ is convoluted with the calculated energy deposition function (EDF) (distribution of internal energies transferred on SID), to obtain the normalized signal intensity for a particular ion in the SID spectrum versus collision energy (i.e. its simulated FEC). In the calculations of $F_i(E, t_r)$ and EDF, the critical energy and activation entropy of the underlying reaction channel and parameters describing the energy deposition function constitute fitting parameters and are varied until the calculated FECs adequately reproduce the corresponding experimental FECs (see Figure 15.9). The thermochemical results of such modeling for the major dissociation pathways of $[AAA + H]^+$ are listed in Table 15.5 (Laskin et al. 2002).

In principle, multiple-collision CAD can be used in place of SID for the acquisition of FECs and the derivation therefrom, after analogous RRKM/QET

TABLE 15.5. Thermochemical Data for Major Surface-Induced Dissociations of $[AAA + H]^+$, Obtained by RRKM/QET Fitting of Experimental Fragmentation Efficiency Curves

Fragment	E_0 (kJ/mol)	$\Delta S^\ddagger$ (J mol^{-1} K^{-1})
$[AAA + H]^+ \rightarrow y_1$	153	10
$[AAA + H]^+ \rightarrow y_2$	161	51
$[AAA + H]^+ \rightarrow b_2$	141	26
$[AAA + H]^+ \rightarrow a_1$	193	—
$b_2 \rightarrow a_2$	125	—
$a_2 \rightarrow a_1$	89	—
$y_1 \rightarrow a_1$	169	—

Source: Laskin et al. (2002).

modeling, of relevant energetics (Laskin 2004). Multiple-collision CAD deposits internal energy in increments and, hence, activates precursor ions more slowly than SID, where all internal energy is deposited in one collision. When the precursor ions reach their lowest dissociation threshold during multiple-collision CAD, they may either dissociate or be further activated. If the rate for ion activation is substantially larger than the dissociation rate, multiple-collision CAD and SID will lead to very similar FECs. If, however, the rate of ion activation is slow compared to the dissociation rate, the precursor ions will be depleted by the lowest-energy fragmentation and competitive, higher-energy channels may either be absent or of very low abundance; as a result, the resulting FECs will be significantly different from those measured via SID. Moreover, the competition between ion activation and dissociation introduces a considerable competitive shift for the higher-energy channels, leading to an erroneously high critical energy for these reactions (Laskin et al. 2002). For example, the E_0 values of y_1 and y_2 from $[AAA + H]^+$ increase by 14 and 17 kJ/mol, respectively, if they are derived via multiple-collision CAD instead of SID (see Table 15.5 for SID values). Because of such problems, SID is the preferable method for reliable energetics (Laskin 2004).

15.2.7. Time-Resolved Photodissociation (TRPD)

TRPD involves measurement of the photodissociation kinetics of slow dissociations and calculation of the respective threshold energies from the observed dissociation rates (Dunbar 2003). TRPD has been applied to peptide radical ions formed by photoionization (PI) of gaseous neutral peptides that were desorbed using laser desorption (Cui et al. 2002; Hu et al. 2003). A photodissociation (PD) pulse is turned on after the precursor ions have been thermalized and any fragments formed in the PI step ("prompt" fragments) have been removed. The PD wavelength can be tuned to deposit a well-defined internal energy and cause fragmentation at a specific chromophore of the peptide; careful choice of the PD laser frequency leads to one fragment. PD mass spectra are acquired at various delay times after the PD pulse and the normalized fragment ion abundance $[I_{frag}/(I_P + I_{frag})]$ is plotted as a function of delay time to obtain a TRPD curve.

An example of TRPD results is depicted in Figure 15.10, which shows curves for the formation of m/z 86 (Leu immonium ion) from the radical ions of LeuTyr and LeuLeuTyr (Hu et al. 2003). The plots of Figure 15.10 unveil the corresponding rate constants k_{uni}, which are $(4.8 \pm 1.8) \times 10^3\ s^{-1}$ for LeuTyr$^{+\bullet}$ and $(2.9 \pm 1.9) \times 10^2\ s^{-1}$ for LeuLeuTyr$^{+\bullet}$ with the PD wavelength used in this experiment. The significantly lower rate constant for the larger peptide agrees well with intramolecular vibrational redistribution (IVR), specifically, randomization, of the deposited energy prior to dissociation. Such behavior is consistent with RRKM theory, according to which fragmentation follows energy randomization.

Thermochemical information can be extracted from TRPD curves by modeling them using RRKM calculations in combination with quantum-chemical theory (Cui et al. 2002; Hu et al. 2003), by a procedure analogous to that outlined in Section 15.2.5. For the experimental data of Figure 15.10, the best fit between experimental

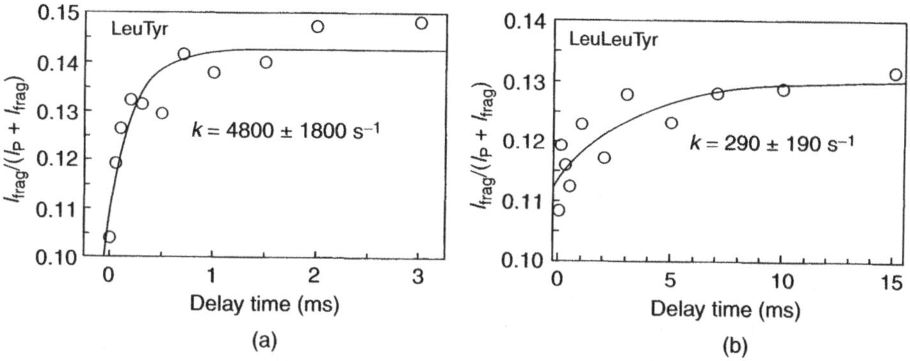

FIGURE 15.10. TRPD curves for the formation of the Leu immonium ion, $^{+}H_2N{=}CHCH_2CH(CH_3)_2$ (m/z 86; frag), from precursors (P) LeuTyr$^{+\bullet}$ (a) and LeuLeuTyr$^{+\bullet}$ (b). Supersonic beams of the peptides were formed by laser desorption using the 1064 nm output of a Nd:YAG laser. Two-photon ionization at 280.5 nm (dye laser) was followed by thermalization of the precursor ions for 1980 ms and PD at 579 nm (dye laser). The PI and PD events took place in an ion trap. After varying trapping delay times, the ions accumulated in the trap were extracted into a time-of-flight mass analyzer for acquisition of their mass spectra. [Reprinted with permission from Hu et al. (2003).]

and calculated TRPD curves is obtained if $E_0 = 1.4$ eV (135 kJ/mol) and $\Delta S^{\ddagger} = 64$–$77\ \mathrm{J\ mol^{-1}\ K^{-1}}$ for the generation of m/z 86 from LeuTyr$^{+\bullet}$ or LeuLeuTyr$^{+\bullet}$. The few biomolecular TRPD studies reported thus far (Cui et al. 2002; Hu et al. 2003) have focused on establishing dissociation rates for information on the ergodicity of the associated fragmentations, not on the underlying thermochemistry.

15.2.8. Kinetic Energy Release (KER) and Kinetic Energy Release Distributions (KERDs)

When precursor ions dissociate, their excess internal energy is partly disposed of as kinetic energy of the fragments (Cooks et al. 1973; Baer and Hase 1996). This kinetic energy release (KER) spreads the translational energies of the fragments, and the associated velocities, leading to broader signals for the fragment ions if mass separation is achieved by a kinetic energy analyzer (as in sector mass and tandem mass spectrometers) or a time-of-flight (TOF) device (as in TOF-based instruments) (Laskin and Lifshitz 2001).

The widths and shapes of fragment ion peaks from decomposition of *metastable* ion beams, that is, precursor ions with keV kinetic energies undergoing spontaneous (unimolecular) dissociation in a field-free region of a mass spectrometer, provide insight about the structures of the decomposing ions as well as the energetics and dynamics of the reaction (Cooks et al. 1973; Holmes and Terlouw 1980; Laskin and Lifshitz 2001). Reactions progressing with no barrier above the dissociation limit (no reverse activation energy) give rise to Gaussian-type fragment peaks, and the corresponding KERs are small. On the other hand, reactions progressing over a transition state that lies higher in energy than the fragments and, thus, having significant reverse activation energies, lead to much wider peaks and larger KERs;

the peak shapes in the latter cases may be broad Gaussian, flat-topped, or dish-topped, depending on the decomposition mechanism (Cooks et al. 1973; Holmes and Terlouw 1980; Holmes 1985).

$$\langle \varepsilon \rangle = 2.16\, \varepsilon_{0.5} \qquad (15.27)$$

Metastable peak shapes are recorded by plotting fragment ion abundance versus E_{lab} or time of flight. The most commonly reported KER ($\varepsilon_{0.5}$) is calculated from peak widths at half-height. With Gaussian-type peaks, $\varepsilon_{0.5}$ also renders the average KER of the dissociation, $\langle \varepsilon \rangle$, via Eq. (15.27) (Holmes and Terlouw 1980). The first derivatives of metastable fragment peaks provide the corresponding kinetic energy release distributions, KERDs, in which the probability for a certain KER value is plotted against ε in center-of-mass coordinates (Holmes and Osborne 1977; Lifshitz and Tzidony 1981; Jarrold et al. 1984). Dissociations with no reverse barrier give rise to KERDs that reach a maximum at very small KER and decrease exponentially thereafter. If there is a reverse barrier, the KERD maximum is shifted toward higher energies and the KERD width is either small (for dish- and flat-topped peaks) or large (for wide Gaussian peaks); the distributions decline nearly exponentially at either side of the energy axis. Binding energies can be derived from KERs and/or KERDs by simulating the experimental data with various theoretical models, for example, phase space theory (PST) (Klots 1971; Chesnavich and Bowers 1979), finite heat bath theory (FHBT; also known as the *evaporative ensemble model*) (Klots 1989, 1991), or the *maximum-entropy method* (Leyh and Lorquet 2003):

$$(NH_3)_nH^+ \rightarrow (NH_3)_{n-1}H^+ + NH_3 \qquad (15.28)$$

$$\Delta E_n = \gamma \langle \varepsilon \rangle \left[1 - \frac{\gamma}{2C(n)} \right]^{-1} \quad \text{and} \quad \Delta H_n = \Delta E_n + RT \qquad (15.29)$$

KERs have not been exploited in thermochemistry studies of biomolecules, but their application to ammonia clusters (Lifshitz and Louage 1989; Wei et al. 1990; Lifshitz 1993) suggests that this approach could be used to quantitate noncovalent interactions between similarly basic biomolecules, such as amino acids and peptides. Proton-bound ammonia clusters dissociate in the metastable time window by elimination of one NH_3 unit [Eq. (15.28)]. The resulting peak shapes are narrow Gaussian and the corresponding KERDs maximize at ~0, indicating that these reactions proceed with no reverse barriers; hence, the activation energy for NH_3 loss and the binding energy of the detaching NH_3 molecule ΔH_n have identical values. According to the FHBT, the average kinetic energy release in such a case correlates with the binding energy as shown in Eq. (15.29) (Lifshitz and Louage 1990), where $C(n)$ is the heat capacity of the cluster ion (size n) and γ is the Gspaan parameter, a dimensionless factor with the value of ~25 for practically all cluster systems and sizes (Klots 1990). The intrinsic heat capacities $C(n)$ can be estimated from ammonia's bulk heat capacity at constant pressure (Lifshitz 1993).

Figure 15.11 shows the binding energies of $(NH_3)_nH^+$ clusters ($n = 4$–17) derived from KERs that were measured in a TOF mass spectrometer (Wei et al.

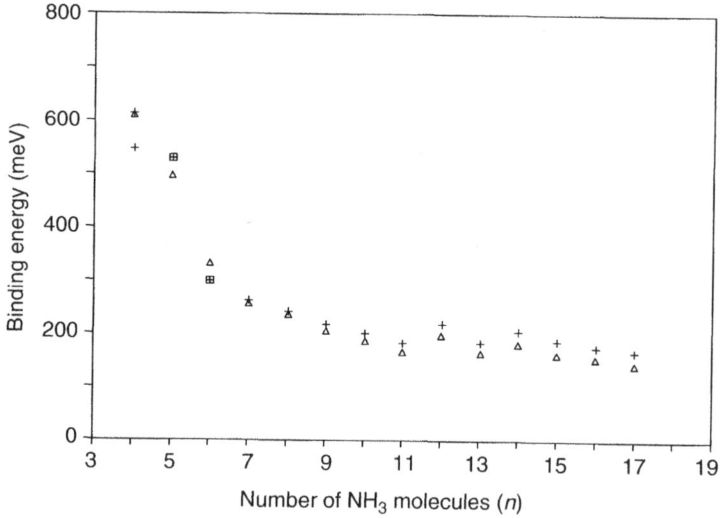

FIGURE 15.11. Binding energies of $(NH_3)_nH^+$ ($n = 4$–17) as a function of cluster size (n); □, literature values from equilibrium measurements [Meot-Ner (Mautner) and Speller 1986]; Δ, derived from KERs using FHBT (Klots 1989, 1991); +, derived using Engelking's QET/RRK theory (Engelking 1987). [Reprinted with permission from Wei et al. (1990).]

1990). There is a sharp drop in the binding energy between $n = 5$ and $n = 6$, when the first solvation shell around a central NH_4^+ ion is completed and additional NH_3 units have to be attached farther away from the charge center (Lifshitz and Louage 1989; Wei et al. 1990; Lifshitz 1993).

15.2.9. Radiative Association Kinetics and Direct Association Equilibria

The majority of methods presented thus far probe *dissociations* in order to determine bond energies. Thermochemical data can also be obtained by examination of the reverse reaction, namely, bond *formation*. In radiative association kinetics, a host–guest complex is generated by collision of an ion (A^+) with a molecule (B) in FTMS instrumentation, and the corresponding bond energy is derived from the measured association kinetics (Ryzhov et al. 1996; Dunbar 1997; Ryzhov and Dunbar 1997; Gapeev and Dunbar 2002). The overall reaction is summarized in Eq. (15.30), where k_{app} is the apparent bimolecular rate constant:

$$A^+ + B \xrightarrow{k_{app}} AB^+ + h\nu \tag{15.30}$$

$$k_{app}(\text{low pressure}) = k_{ra} = \frac{k_f k_r}{k_b + k_r} \tag{15.31}$$

$$\Phi = \frac{k_{ra}}{k_f} \tag{15.32}$$

$$R = \frac{k_r}{k_b} = \frac{\Phi}{1 - \Phi} = \frac{k_{ra}}{k_f - k_{ra}} \tag{15.33}$$

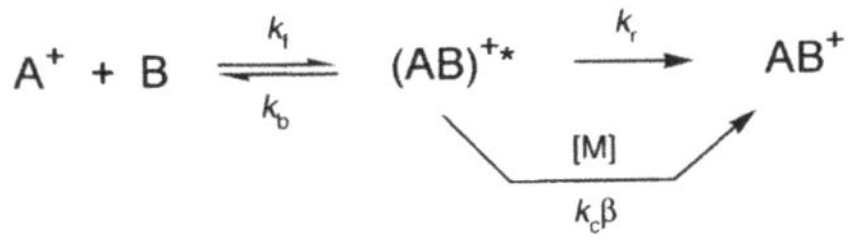

SCHEME 15.2. Radiative association mechanism (Ryzhov et al. 1996).

The detailed pathway traversed during a radiative association is described in Scheme 15.2. The bimolecular encounter of A^+ and B (rate constant k_f) creates a metastable ion–molecule complex, $(AB)^{+*}$, which may redissociate (rate constant k_b) or dispose of some of its internal energy and survive as a stable AB^+ ion. Stabilization can occur via IR photon emission (rate constant k_r) or by collision with a neutral molecule present in the vacuum system of the mass spectrometer [rate constant $k_c\beta$, where β gives the probability of removing enough energy to stabilize $(AB)^{+*}$]. At the low pressures prevalent in FTMS (10^{-7}–10^{-8} Torr), three-body collisional stabilization can be neglected and the overall bimolecular rate constant k_{app} can be equated to a rate constant for radiative association k_{ra}, which is related to k_f, k_r, and k_b, as shown in Eq. (15.31) (Dunbar 1997). The radiative association efficiency Φ is defined as the ratio of k_{ra} and k_f, [Eq. (15.32)]; that is, it gives the fraction of collision between A^+ and B that lead to association product (Dunbar 1997). At the upper limit of $\Phi = 1$, k_{ra} reaches the rate constant for encounters between A^+ and B; this is usually assumed to be identical to the Langevin orbiting rate constant k_{orb}, which can be readily calculated. The competition between radiative association (k_r) and redissociation (k_b) in the metastable $(AB)^{+*}$ complex is described by the ratio R [Eq. (15.33)] (Dunbar 1997). At low efficiencies, $R \approx \Phi$, while at high efficiencies R tends to increase toward infinity; at $R = 1$, exactly one-half of the ion–molecule collisions between A^+ and B result in stabilized and, hence, surviving AB^+ product.

Radiative association experiments monitor the relative intensity of A^+ or AB^+ as a function of time and determine k_{app} from the resulting curves. Measuring k_{app} at different pressures of B and extrapolating to zero pressure yields k_{ra}, from which Φ and R can be obtained if k_f is calculated using the well-known kinetic expression for a point charge colliding with a polarizable molecule or ADO theory (Su and Bowers 1973). The binding energy of AB^+ is subsequently derived by appropriate modeling of the value of $R = k_r/k_b$ (Dunbar 1997). For reactions involving biomolecules, which generally have considerable complexity, the "standard hydrocarbon model" (SHM) allows for the estimation of adequate energetics (Dunbar 1997). SHM presumes that different molecules have similar radiative and dissociative properties, depending mainly on temperature (T), binding energy (E_b), and number of internal vibrational degrees of freedom (S). RRKM theory has been used to generate a series of working graphs depicting the ratio R [Eq. (15.33)], as a function of the parameters T, E_b, and S (Dunbar 1997). These graphs allow one to covert an experimentally determined R ratio into a binding energy E_b for guest–host systems having 30–150 degrees of freedom, temperatures between 100 and 500 K, and binding energies in the range 0.4–5.0 eV. More rigorous modeling, involving ab initio calculations and

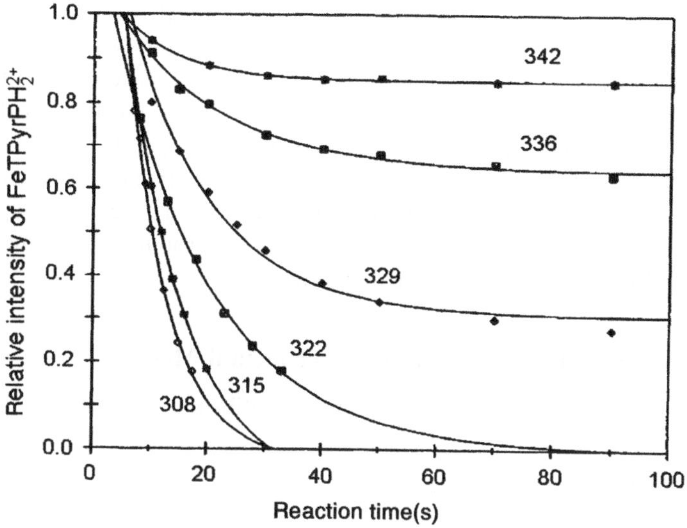

FIGURE 15.12. Change in relative intensity of $FeTPyrPH_2^{2+}$ versus time, resulting from addition of NO at various temperatures (in Kelvins). Lines are exponential fits to the data. The data acquired at 288 K (not included in the figure) are similar with those at 308 K. The abundance of $FeTPyrPH_2^{2+}$ is monitored after an induction period of a few seconds, during which the translationally excited ions are cooled. [Reprinted with permission from Chen et al. (1999).]

variational transition state theory, provides more accurate energetics, but is tractable only with smaller organic or inorganic complexes (Ryzhov et al. 1996).

An application of the radiative association method is illustrated in Figure 15.12 (Chen et al. 1999), which shows the disappearance with time of doubly protonated iron tetrapyridylporphyrin, $FeTPyrPH_2^{2+}$, on its reaction with NO due to the formation of the $FeTPyrPH_2(NO)^{2+}$ complex [NO attached at the iron(II) ion]. The rate of radiative association measured at 288 K at different NO pressures and extrapolated to zero pressure is $k_{ra} = (4.8 \pm 1.5) \times 10^{-11}$ cm^3/s. Using the SHM graphs (Dunbar 1997), the measured quantity corresponds to a binding energy of 111 ± 3 kJ/mol for a system of this size. Repeating the experiments with singly protonated iron tetrapyridylporphyrin, $FeTPyrPH^+$, leads to the same k_{ra} and E_b values, indicating that the binding interaction between the central Fe ion and the NO ligand is not affected by the charge state on the porphyrin ring (Chen et al. 1999).

It is evident from Figure 15.12 that the association rate decreases with temperature. At temperatures >330 K, the relative intensities of $FeTPyrPH_2^{2+}$ and $FeTPyrPH_2(NO)^{2+}$ approach a constant value, as the association of $FeTPyrPH_2^{2+}$ with NO reaches equilibrium. Under these conditions, the *direct association equilibrium* is observable (Chen et al. 1999; Ryzhov and Dunbar 1999). The steady-state intensity ratio of $FeTPyrPH_2^{2+}$ and $FeTPyrPH_2(NO)^{2+}$, combined with the adjusted NO pressure (2×10^{-7} Torr), furnish an equilibrium constant of $K_{eq} = 2.0 \pm 0.6 \times 10^9$ atm^{-1} at 336 K. This K_{eq} can be converted into a free energy of NO binding (ΔG_{NO}) and, after estimation of the respective reaction entropy, into a binding energy; the latter value, $\Delta H_{NO} = 121 \pm 6$ kJ/mol, is in good agreement with that deduced from the association kinetics, as well as with the NO binding energy derived independently via a BIRD experiment on $FeTPyrPH_2(NO)^{2+}$ (108 ± 4 kJ/mol) (Chen et al. 1999). This comparison points out that radiative association kinetics and direct radiative equilibria measurements, which, with the exception of the presented example, have not been used in biomolecular thermochemistry studies, offer additional avenues for probing the thermochemistry of guest–host interactions of biological significance.

15.2.10. Theory and Comparison of Experimental Methods

From the experimental methods presented, equilibrium, bracketing, and kinetic method studies are purely experimental and can yield thermochemical data without the necessity of quantum-chemical or kinetic theory modeling. This characteristic is also true for BIRD studies of large systems (>500 degrees of freedom) that are in thermal equilibrium with their surroundings. Threshold collision-induced dissociation, BIRD experiments of medium-sized biomolecular ions, SID studies, photodissociation studies, kinetic energy release distribution analyses, and radiative association kinetics studies require parallel quantum and/or kinetic theory calculations to fit the experimental data in order to determine thermochemical properties; the quality of the derived data in the latter cases depends strongly on the level and type of theory used and the correct choice of modeling parameters.

Bracketing, TCID, BIRD, SID, photodissociation, KERD, and radiative association studies provide absolute thermochemical quantities, whereas ion exchange equilibrium and kinetic method studies provide only relative thermochemistry and require anchors for converting the relative data measured to absolute values. On the other hand, the last two methods are very sensitive to small differences in binding energy and can sense such differences with high precision. Further, equilibrium studies and photodissociation experiments with supersonic biomolecular beams (such as those described here) require thermally stable and quite volatile samples and thus are limited to small biomolecules that satisfy these criteria.

The best possible characterization of the thermochemical properties of a biomolecule are determined by using more than one of the experimental methods available. Complementary information can also be obtained by quantum-chemical calculations. In fact, computational predictions of binding energies and entropies (or dissociation thresholds) are invaluable for the interpretation of experimental trends and for detecting systematic method/instrument errors or changes of the structure of the biomolecule when it is converted to a solitary gas-phase ion. The choice of theory level depends on the complexity of the system under study. Ab initio methods and density functional theory (DFT) have been the methods of choice for geometry optimization and energy calculations on smaller and medium-sized biomolecules (for example, amino acids and small peptides), while empirical molecular mechanics/dynamics approaches are preferentially used with larger biomolecular systems (e.g., protein complexes). The capabilities and shortcomings of the various methods discussed are exemplified with selected applications in the following sections.

15.3. APPLICATIONS AND EXAMPLES

15.3.1. Proton Transfer Reactions

15.3.1.1. Amino Acids and Peptides. The gas-phase basicities (GBs) and proton affinities (PAs) of all 20 common α-amino acids have been determined by equilibrium, bracketing, and kinetic method experiments (Harrison 1997; Hunter and Lias 1998). Table 15.6 lists the PA ladders determined by select studies using

TABLE 15.6. Proton Affinities of Common α-Amino Acids Measured by Different Methods (kJ/mol) [All Values Anchored to PA(NH_3) = 854 kJ/mol]

Amino Acid	Equilibrium[a]	Bracketing[b]	Kinetic Method[c]	Extended Kinetic Method[e]	Theory		Evaluated[i]
G	882	868	—	888	882[g]	887[h]	886
A	899	911	—	895	890[g]	899[h]	902
C	—	880	—	897	—	—	903
D	—	903	913	909	—	—	909
V	906	911	913	910	—	—	911
E	—	989	930	934	—	—	913
L	909	923	915	911	—	—	915
S	—	903	909	905	—	—	915
I	—	923	917	914	—	—	917
P	925	949	931	937	928[g]	935[h]	921
T	—	938	917	914	—	—	923
F	911	933	920	920	—	924[h]	923
Y	—	933	923	922	—	—	926
N	—	933	929	933	—	—	929
M	—	938	925	927	—	—	935
Q	—	938	949	946	—	—	938
W	—	949	935	936	—	—	949
H	—	938	964	959 979[f]	—	—	988
K	—	997	957	956 984[f]	—	—	996
R	—	—	1026[d]	—	—	—	1051

[a]From equilibrium experiments (Meot-Ner et al. 1979).
[b]Bracketing results (Gorman et al. 1992).
[c]Kinetic method results using Eq. (15.17) (Bojesen and Breindahl (1994).
[d]Wu and Fenselau (1992a).
[e]Extended kinetic method results using Eqs. (15.15) and (15.16) (Afonso et al. 2000).
[f]Wu and Fenselau (1994).
[g]MP2/6-311+G*//B3LYP/6-31G* data from Pepe et al. (2004).
[h]B3P86/6-31G*//B3LYP/6-31G* data from Pepe et al. (2004).
[i]Evaluated mean values from several methods reported in the NIST compilation (Hunter and Lias 1998).

these methods (Meot-Ner et al. 1979; Gorman et al. 1992; Wu and Fenselau 1992a, 1994; Bojesen and Breindahl 1994; Afonso et al. 2000; Pepe et al. 2004). Although general trends are reproduced by the various methods, there are discrepancies in absolute values and in certain orders, originating from the shortcomings of the methods as discussed in Section 15.2. The availability of data from different methods facilitates the detection of outliers and the derivation of "best" average values, which are reported in two compilations (Harrison 1997; Hunter and Lias 1998).

In all studies conducted thus far, there is consensus that (1) arginine is the intrinsically most basic amino acid, (2) histidine and lysine are the next most basic amino acids, and (3) aliphatic amino acids are among the least basic from the 20 commonly occurring L-amino acids. The magnitudes of GB and PA (and of the associated protonation entropy) reflect the structural properties of the amino acids and the capability of their sidechains to stabilize the added proton, either inductively or by intramolecular hydrogen bonding.

Because of the significantly lower volatility and thermal stability of peptides, their protonation thermochemistry has not been studied by equilibrium experiments. Bracketing (Gorman and Amster 1993a; Wu and Lebrilla 1993, 1995; Zhang et al. 1993; Cassady et al. 1995; Ewing et al. 1996; Carr and Cassady 1996) and kinetic method studies (Wu and Fenselau 1992b, 1993; Nold et al. 1999) have revealed that small peptides (consisting of $\leq$7 residues) are more basic than the most basic amino acid residue contained in them. In general, gas-phase basicity increases both with the number and basicity of the amino acid constituents of the oligopeptide. This is attested by the GB values of G, GG, GGG, GGP, and GGH, which have been bracketed at 843, 879, 891, 908, and 948 kJ/mol, respectively (Zhang et al. 1993; Ewing et al. 1996; Carr and Cassady 1996). With isomeric peptides, the position of the most basic residue within the backbone exerts a milder effect on GB, as exemplified by the GBs of GA versus AG (bracketed at 882 vs. 885 kJ/mol, respectively) (Cassady et al. 1995) or those of GKKGG versus GKGKG versus KGGGK (995 vs. 998 vs. 1012 kJ/mol, respectively, measured by the kinetic method) (Wu and Fenselau 1993).

GB and PA refer to the free energy and enthalpy of *protonation*. The *deprotonation* of amino acids, specifically, the heterolytic bond cleavage shown in Eq. (15.34) and the associated ΔG_{acid} and ΔH_{acid} values, as well as the analogous reaction in peptides, have been investigated to a much lesser extent (Locke and McIver 1980; O'Hair et al. 1992). Equilibrium measurements yielded a ΔG_{acid} of 1404 kJ/mol for G and 1396 kJ/mol for A, from which ΔH_{acid} values of 1433 and 1425 kJ/mol, respectively, could be deduced (Locke and McIver 1980). A kinetic method study of 18 out of 20 common α-amino acids revealed the acidity order $G < A < P < V < L < I < L < W < F < T < M < C < S < T < R < N < Q < H$, with the most acidic (H) and least acidic (G) amino acids differing in their ΔG_{acid} and ΔH_{acid} values by only 46 kJ/mol (O'Hair et al. 1992). For comparison, the proton affinities of amino acids span a range of 165 kJ/mol (see Table 15.6) (gas-phase acidity data are not available for peptides):

$$H_2N\text{–}CH(R)\text{–}COOH \rightarrow H_2N\text{–}CH(R)\text{–}COO^- + H^+ \qquad (15.34)$$

The gas-phase acidity of an amino acid is equivalent with the gas-phase basicity of the corresponding conjugate base; thus, ΔG_{acid} and ΔH_{acid} of H_2N—CH(R)—COOH have values numerically identical to those of GB and PA, respectively, of H_2N—CH(R)—COO^-. On the basis of the data given above, the PA of glycine's conjugate base, H_2N—CH_2—COO^-, is 1404 kJ/mol (Locke and McIver 1980). This value decreases dramatically if a permanent positive charge is introduced at the *N* terminus via permethylation. A kinetic method study on the proton affinity of $(CH_3)_3N^+$—CH_2—COO^-, the amino acid betaine, yielded a PA of 1003 kJ/mol (Patrick et al. 1996). This also is the ΔH_{acid} value of the conjugate acid $(CH_3)_3N^+$—CH_2—COOH (protonated betaine). The ~400 kJ/mol decrease in ΔH_{acid} vis à vis glycine reflects the substantially higher acidity of protonated betaine, brought on by the attractive Coulombic forces in the resulting conjugate base (the betaine molecule).

15.3.1.2. Nucleic Acid Constituents. The *protonation* thermochemistry of nucleic acid components, specifically DNA and RNA nucleobases and nucleosides (Figure 15.13), has been examined both by theory (Sponer et al. 2001) as well as bracketing and kinetic method experiments. The experimental results available are summarized in Table 15.7 [Wilson and McCloskey 1975; Meot-Ner (Mautner) 1979; Greco et al. 1990; Liguori et al. 1994, 2000].

Within the nucleobases, proton affinities increase in the order Ura < Thy ≪ Ade < Cyt < Gua. Ab initio theory reveals that the most basic sites and, hence, likeliest protonation sites are O4 for uracil and thymine, O2 or N3 for cytosine (energetically almost equivalent), N1 for adenine, and N7 for guanine (Sponer et al. 2001). The PA order of the corresponding nucleosides changes slightly to Urd ≪ Cyd < Ado < Guo and dThyd ≪ dCyd < dAdo < dGuo, suggesting that the N1 proton in Cyt (which is absent in the nucleosides) stabilizes the protonated site via intramolecular hydrogen bonding. The PA values of ribonucleoside and deoxyribonucleoside with the same nucleobase are very similar, in agreement with protonation at the nucleobase moieties. The H^+ binding energy consistently

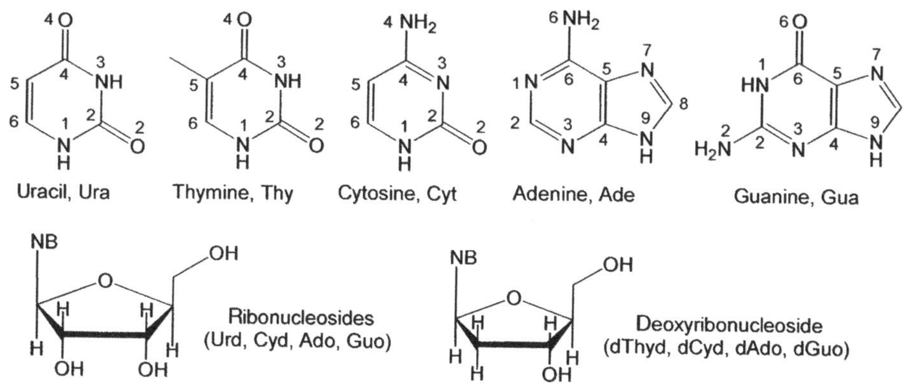

FIGURE 15.13. DNA and RNA nucleobases (NB) and nucleosides. The sugar units are attached at N1 of the pyrimidine and at N9 of the purine nucleobases.

TABLE 15.7. Proton Affinities of RNA and DNA Components (kJ/mol)[a]

Nucleobase	PA		Ribonucleoside	PA	Deoxyribonucleoside	PA
Ura	873[b]		Urd	947	—	—
Thy	881		—	—	dThyd	948
Cyt	953	944[c]	Cyd	983	dCyd	988
Ade	948	939[c]	Ado	989	dAdo	992
Gua	960		Guo	993	dGuo	995

[a]Kinetic method values (Greco et al. 1990; Liguori et al. 1994, 2000) as reevaluated in the NIST compilation (Hunter and Lias 1998), unless noted otherwise.
[b]Bracketing result from Wilson and McCloskey (1975).
[c]Bracketing results from Meot-Ner (Mautner) (1979).

increases on conversion of the nucleobases to nucleosides, presumably because of better intramolecular charge solvation via H bonding interactions in the latter. The increase is markedly higher for Ura and Thy, as compared to Cyt, Ade, and Gua, in which some intramolecular H bonding can already occur in the protonated nucleobases.

The *deprotonation* energetics of nucleobases has been assessed computationally (Chandra et al. 1999; Huang and Kenttämaa 2003, 2004). High-level DFT calculations predict the following ΔH_{acid} values (kJ/mol) for the keto forms, which are the tautomers incorporated in nucleic acids: Ura, 1392 (for deprotonation at N1) and 1444 (N3); Thy, 1400 (N1) and 1446 (N3); Cyt, 1446 (N1) and 1459 (N4); Ade, 1413 (N9) and 1488 (N6); Gua, 1412 (N9) and 1417 (N1) (Huang and Kenttämaa 2003, 2004). The calculations also point out that the most stable guanine isomer in the gas phase is not the 9H tautomer shown in Figure 13, but the 7H tautomer (~2–3 kJ/mol lower in energy); its gas-phase acidities are 1407 and 1414 kJ/mol at N1 and N7, respectively. Further, the enol forms of cytosine and of 9H guanine are found to have very similar stabilities and ΔH_{acid} values with the corresponding keto forms (Huang and Kenttämaa 2003, 2004). The computational data indicate the acidity order Cyt < Ade ≈ Gua < Thy < Ura. Bracketing experiments have provided experimental gas-phase acidities for uracil [1393 and 1452 at N1 and N3, respectively (Kurinovich and Lee 2000; Miller et al. 2004)] and adenine [1393 and 1473 at N9 and N6, respectively (Sharma and Lee 2002)], which are in excellent agreement with the theoretical values. From these data it is evident that the nucleobases are derivatized by ribose or deoxyribose to nucleosides at their most acidic sites.

15.3.2. Metal Ion Thermochemistry

15.3.2.1. Amino Acids and Peptides. The Li^+, Na^+, Cu^+, and Ag^+ affinities of most common α-amino acids and the K^+ affinities of select amino acids have been measured by equilibrium experiments, the kinetic method, and TCID (Bojesen et al. 1993; Cerda and Wesdemiotis 1995, 1999, 2000; Klassen et al. 1996; Hoyau and Ohanessian 1997, 1998; Lee et al. 1998; Hoyau et al. 1999, 2001; Dunbar 2000;

Marino et al. 2000, 2002, 2003; Ryzhov et al. 2000; Gapeev and Dunbar 2001, 2003; Talley et al. 2002; Moision and Armentrout 2002, 2004; Kish et al. 2003; Feng et al. 2003a; Siu et al. 2004; Ruan and Rodgers 2004, 2005; Wang et al. 2006b). The orders of binding energies are given in Table 15.8. The table does not exhaustively cover all data reported about these metal ions, but rather presents representative ladders that illustrate similarities and differences in the binding properties of main group versus transition metal ions.

The most basic amino acids Q, W, H, K, and R (see Table 15.6) also are the most strongly binding metal ion ligands. Metal ion affinities do not follow the same order as proton affinities, however, which reflects the different bonds formed by protons (covalent) and metal ions (electrostatic). PAs increase with the electron density at the protonation site and the ability of the protonated base to intramolecularly solvate the charge site. Conversely, metal ion affinities depend on the number and type of coordination sites that the base (ligand) can offer to the metal ion as well as the geometry of the resulting chelates. These differences are clearly evident with aspartic acid. Its electron-withdrawing sidechain reduces the basicity at the

TABLE 15.8. Relative Metal Ion Affinities of Amino Acids versus Glycine (kJ/mol)

Amino Acid[a]	Li^{+b}	Na^{+c}	K^{+d}	Cu^{+e}	Ag^{+f}
G	0	0	0	0	0
A	6.7	5.7	—	7.1	5.9
C	15.1	14.1	—	36.0	—
D	41.4	41.7	—	20.9	12.6
V	14.2	12.1	—	15.5	10.0
E	47.3	42.8	—	30.1	19.2
L	15.1	13.4	—	17.2	10.5
S	29.3	30.5	—	13.0	13.4
I	15.5	14.8	—	18.0	11.7
P	24.7	35.0	—	20.1	20.9
T	34.7	36.1	—	19.2	19.2
F	28.5	37.3	29.3	33.5	39.7
Y	31.0	39.8	34.2	34.7	40.2
N	—	56.3	—	28.0	34.7
M	36.8	—	—	43.5	54.8
Q	—	60.7	—	41.0	44.8
W	44.8	48.6	43.8	48.1	60.7
H	—	66.7	—	55.6	75.3
K	—	—	—	87.5	82.8
R	—	81.6	—	96.5	>112

[a]Amino acids listed in order of increasing PA (see Table 15.6).
[b]Kinetic method; $\Delta(\Delta G_{Li}) \approx \Delta(\Delta H_{Li})$ from Feng et al. (2003a).
[c]Kinetic method; $\Delta(\Delta G_{Na}) \approx \Delta(\Delta H_{Na})$ from Kish et al. (2003) and Wang et al. (2006b).
[d]$\Delta(\Delta H_K)$ calculated from absolute TCID values (Moision and Armentrout 2004; Ruan and Rodgers 2004).
[e]Kinetic method; $\Delta(\Delta G_{Cu}) \approx \Delta(\Delta H_{Cu})$ from Cerda and Wesdemiotis (1995).
[f]Kinetic method; $\Delta(\Delta G_{Ag}) \approx \Delta(\Delta H_{Ag})$ from Lee et al. (1998).

N terminus, leading to a PA value that is lower than PA(valine). On the other hand, the additional binding site provided to metal ions by the aspartic acid sidechain leads to increased metal ion affinities as compared to all aliphatic amino acids except proline.

The relative affinities in Table 15.8 show that alkali metal ions form particularly strong bonds with amino acids carrying hydroxy or carbonyl groups in their sidechains (S, T, D, E, N, Q). In contrast, the binding energies to transition metal ions are especially enhanced by sulfur-containing sidechains (C, M). These differences are reconciled by the hard/soft-acid/base principle, the alkali metal ions (hard Lewis acids) preferring interactions with hard O sites and the transition metal ions (soft Lewis acids) interactions with soft S sites (Cerda and Wesdemiotis 1995; Lee et al. 1998; Kish et al. 2003; Feng et al. 2003a). Aromatic substituents appear to increase considerably all metal ion affinities, and the corresponding affinity increases are more significant for the larger metal ions (Na^+ and Ag^+).

The most thoroughly studied systems, both by experiment and theory, have been the Na^+ complexes of amino acids. More recent absolute ΔH_{Na} values are summarized in Table 15.9. In general, there is excellent agreement between measured and calculated Na^+ affinities as well as between Na^+ affinities measured by different methods, except for ΔH_{Na}(proline) and ΔH_{Na}(histidine) derived from ligand exchange equilibria, which will be discussed below.

Theory predicts that G and A form bidentate Na^+ complexes, in which the metal ion is bound to the carbonyl oxygen and the amine nitrogen of the neutral (canonical)

TABLE 15.9. Absolute Na^+ Affinities of Amino Acids (ΔH_{Na} at 298 K in kJ/mol)

	Kinetic Method[a]	Equilibrium Experiments[b]	TCID	Theory[e]
G	161	161	166[c]	164[f]
A	167	169	—	167[f]
V	173	173	—	—
C	175	—	—	180[f]
S	192	188	—	200[f]
P	196	175	—	195[f]
F	198	198	208[d]	198[g]
Y	201	—	212[d]	201[g]
W	210	205	220[d]	213[g]
H	227	185	—	232[h]
R	242	—	—	244[h]

[a]Kish et al. (2003), Wang et al. (2006b).
[b]Gapeev and Dunbar (2003).
[c]Moision and Armentrout (2002).
[d]Ruan and Rodgers (2004).
[e]Binding energy in the most stable complex.
[f]MP2/6-311+G(2*d*, 2*p*)//MP2/6-31G* values including zero-point energy (ZPE) and thermal corrections (Kish et al. 2003).
[g]B3LYP/6-311+G(2*d*, 2*p*)//B3LYP/6-31G* values including ZPE, thermal, and basis set superposition error (BSSE) corrections (Gapeev and Dunbar 2003).
[h]MP2/6-311 + G(2*d*, 2*p*)//MP2/6-31G* including ZPE and thermal corrections (Wang et al. 2006b).

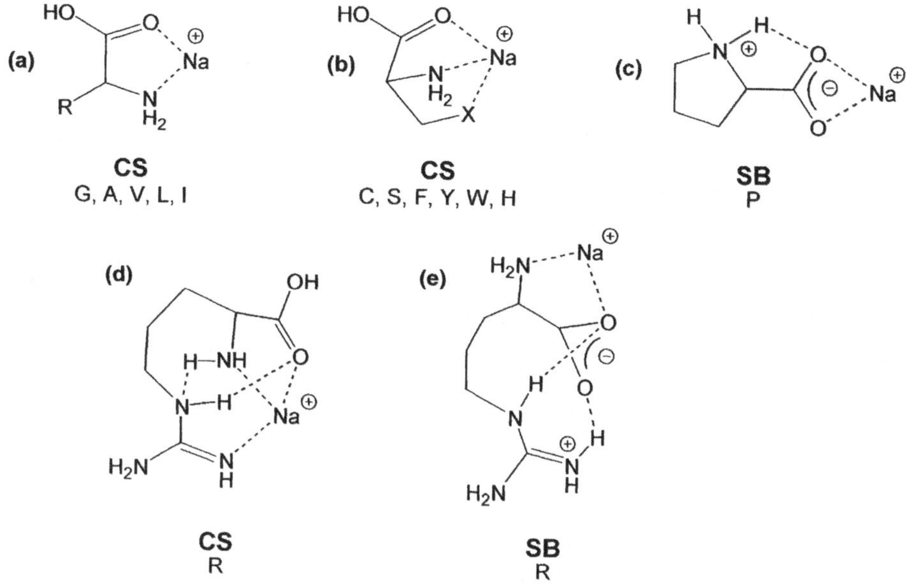

FIGURE 15.14. Geometries predicted computationally for the Na^+ complexes of amino acids: (a) bidentate CS complexes of aliphatic amino acids; (b) tridentate CS complexes of sidechain substituted amino acids; (c) $[P + Na]^+$ complex with SB structure; (d,e) $[R + Na]^+$ complex with (d) CS and (e) SB structure.

form of the amino acid (Figure 15.14) (Hoyau and Ohanessian 1998; Marino et al. 2000; Talley et al. 2002). The same binding motif is expected for the aliphatic amino acids V, L, and I. These structures may be viewed as Na^+ ions solvated by the amino acid ligand; hence, they are termed "charge solvation" (CS) complexes. In sharp contrast, proline, the only secondary protein amino acid, is predicted to form a "saltbridge" (SB) complex, containing the zwitterionic form of proline (Figure 15.14) (Hoyau et al. 1999; Talley et al. 2002; Marino et al. 2003). In the proline zwitterion, the amine group is protonated and the carboxyl group deprotonated. The attractive ion–ion interactions in the $[P + Na]^+$ saltbridge, combined with the higher proton affinity of a secondary versus primary amine lead to a significantly higher Na^+ affinity for P as compared to the other aliphatic amino acids.

The geometries of the Na^+ complexes of the sidechain functionalized amino acids C, S, F, Y, W, H, and R have also been calculated (Hoyau et al. 1999; Dunbar 2000; Gapeev and Dunbar 2003; Siu et al. 2004; Ruan and Rodgers 2004; Wang et al. 2006b). For all expect R, the most stable structures correspond to tridentate CS complexes, with the Na^+ ion bound to the *N*-terminal amine, *C*-terminal carbonyl, and sidechain functional groups (Figure 15.14). The third ligands are the SH and OH groups of C and S, respectively (Hoyau et al. 1999); the π systems of F, Y, and W (Dunbar 2000; Ruan and Rodgers 2004); and the $N^{\delta 1}$ site of the imidazole ring in H ($N^{\delta 1}$ is the site closest to the α-carbon) (Gapeev and Dunbar 2003; Wang et al. 2006b). The histidine unit in the latter complex is the $N^{\varepsilon 2}$ tautomer, in which the $N^{\varepsilon 2}$ site of the imidazole ring is protonated and the $N^{\delta 1}$ site deprotonated. For R, the most basic amino acid (Table 15.6), nearly isoenergetic CS and SB structures exist (Figure 15.14) (Wang et al. 2006b). The CS complex dissociates to yield the NHC(=NH)NH_2 tautomer, which is ~10 kJ/mol less stable than the N=C$(NH_2)_2$ tautomer. The SB complex can dissociate to yield the most stable N=C$(NH_2)_2$ tautomer. The corresponding Na^+ binding energies are 244 and 230 kJ/mol, respectively (Wang et al. 2006b), and suggest that the CS tautomer is probed in the kinetic method experiments (see Table 15.9).

The Na^+ affinities of proline and histidine determined via ligand exchange equilibria (Gapeev and Dunbar 2003) are markedly lower than those from the kinetic method (Kish et al. 2003; Wang et al. 2006b). The differences have been attributed to the involvement of distinct amino acid tautomers in these methods. In the kinetic method investigation, the dimer ions probed were produced in solution and extracted into the gas phase by ESI. This procedure gives rise to the structures given above and whose calculated Na^+ affinities are included in Table 15.9. In the equilibrium studies (see Section 15.2.1), gaseous P and H reagents were generated by thermal desorption and later reacted with Na^+ to form the corresponding complexes in the gas phase. Thermal desorption is believed to produce the neutral (canonical) form of proline and (primarily) the $N^{\delta 1}$ tautomer of histidine in which the $N^{\varepsilon 2}$ site is deprotonated but the $N^{\delta 1}$ site protonated and, hence, unavailable for chelation (Gapeev and Dunbar 2003). The most stable Na^+ complex of canonical proline has CS geometry and a calculated binding energy of 164 kJ/mol, which is fairly similar to the value measured via ligand exchange equilibria (Table 15.9) (Gapeev and Dunbar 2003). Conversely, the lowest-energy Na^+ complex with the $N^{\delta 1}$ tautomer of histidine has saltbridge geometry, with the imidazole ring protonated and the metal ion sequestered at the deprotonated carboxylate terminus (Wang et al. 2006b); the calculated Na^+ binding energy of this structure is 193 kJ/mol and lies fairly close to the Na^+ affinity determined in the equilibrium study (Table 15.9) (Gapeev and Dunbar 2003; Wang et al. 2006b). The $N^{\delta 1}$ and $N^{\varepsilon 2}$ tautomers of H readily interconvert in solution but cannot rearrange to each other in the gas phase. Consequently, formation of the overall more stable Na^+ complex with the $N^{\varepsilon 2}$ tautomer is favored in solution, as observed in the kinetic method experiment (Wang et al. 2006b).

Kinetic method experiments can also lead to underestimated binding energies if there is crowding or severe entropy effects in the heterodimers subjected to dissociation. Such a situation was encountered in an extended kinetic method study of the aromatic amino acids F, Y, and W; pairing them with the nucleobases cytosine, adenine, and guanine (cf. Figure 15.13), resulted in Na^+ affinities of 174, 175, and 180 kJ/mol, respectively (Ryzhov et al. 2000), which are 24–30 kJ/mol lower than the values obtained from F, Y, and W heterodimers with other amino acids (listed in Table 15.9). It was argued that crowding in the dimers with nucleobases obstructed the production of the most strongly bound, tridentate F—Na^+, Y—Na^+, and W—Na^+ conformers, thereby leading to lower ΔH_{Na} values (Ryzhov et al. 2000; Kish et al. 2003).

The examples presented illustrate the power of combining different experimental approaches with theory. For methods entailing reaction of a vapor with a metal ion or metal ion complex (e.g., TCID or equilibrium techniques), the relative volatilities of higher energy tautomers must be considered. This issue also is important in the kinetic method, if the dimer ion precursors of these studies are formed via gas-phase reactions. Additionally, the kinetic method is sensitive to crowding around the metal ion and requires the choice of dimers that can dissociate to yield the lowest-energy structure of the metalated monomer.

Compared to amino acids, the metal ion thermochemistry of peptides is not as well known. TCID studies on the Na^+ and K^+ affinity of GG have been reported (Klassen et al. 1996), and the kinetic method has been applied to examine the Na^+ affinities (ΔH_{Na}) of several di-, tri-, and tetrapeptides (Cerda et al. 1998; Hoyau et al. 1999; Wang et al. 2006a) as well as the free energies of Li^+ and Na^+ binding (ΔG_{Li} and ΔG_{Na}, respectively) of GX dipeptides (Feng et al. 1999). These preliminary investigations have shown that the metal ion binding energy increases with the size of the peptide and the intrinsic affinity of its amino acid constituents. As with amino acids, sidechains with O atoms and aromatic groups bring on considerable enhancements in metal ion affinity. This is illustrated by the following ΔH_{Na} order (kJ/mol relative to GG): GG (0) < AG (2) < GA (4) < AA (5) < GL (8) < GF (12) < SG (14) < FG (19) < AW (25) < WA (28) < GH (37) < HG (48) (Capota and Ohanessian 2005; Wang et al. 2006a). The effect of size is illustrated by the following relative Na^+ affinity scale (kJ/mol relative to GG): GG (0) < GGG (34) < GGGG (58) (Wang et al. 2006a). The corresponding absolute Na^+ affinities can be assessed by adding these increments to ΔH_{Na}(GG). There have been significant discrepancies in the GG–Na^+ bond enthalpies measured by TCID [179 kJ/mol (Klassen et al. 1996)], the kinetic method [181 kJ/mol (Feng et al. 1999)], and the extended kinetic method [181 kJ/mol using nucleobase reference bases and 203 kJ/mol using amino acid reference bases (Kish et al. 2004)]. High-level ab initio calculations predict a 298 K affinity of 200 kJ/mol (Kish et al. 2004), which is in best agreement with the extended kinetic method value of 203 kJ/mol.

The calculations indicate that in the most stable structures of sodiated oligoglycines all carbonyl oxygens are employed in the coordination of the metal ion (Cerda et al. 1998; Wyttenbach et al. 1998; Wang et al. 2006a). The amine terminus may act as an additional ligand or interact via a hydrogen bond with the adjacent NH amide bond; these arrangements are essentially of equal stability (Cerda et al. 1998; Wang et al. 2006a). As the size of the peptide increases, different orientations of the carbonyl groups are possible in the Na^+ complexes, leading to different degrees of backbone folding and hydrogen bonding and to several conformers of very similar energy, all of which are probably present in the ions sampled experimentally (Wang et al. 2006a). It is worth noting that metal ion addition to oligoalanines creates diastereomeric complexes carrying the methyl sidechains in either pseudoaxial or pseudoequatorial position; the equatorial geometries are more stable (Wang et al. 2006a). As with the H–Na^+ complex, the lowest-energy geometries of sodiated GH and HG contain the $N^{\varepsilon 2}$ tautomer of histidine, and in both complexes the metal ion is bound to the $N^{\delta 1}$site of the histidine sidechain as well as to the two carbonyl oxygens of the backbone (Capota and Ohanessian 2005; Wang et al. 2006a).

15.3.2.2. Nucleobases. The extended kinetic method, TCID, and theory have been employed to investigate the Li^+, Na^+, and K^+ affinities of the DNA and RNA

TABLE 15.10. Alkali Metal Ion Affinities of Nucleobases (at 298 K; kJ/mol)

	KM[a]	TCID[b]	Theory[c]
			ΔH_{Li}
Ura	211	214	218 (O4)[d]
Thy	215	212	218 (O4)[d]
Cyt	232	—	291 (**C1** O2/N3) 228 (**C2** O2/N3) 248 (**C3** N1/O2)
Ade	226	230	208 (**A1** N6/N7)[d] 276 (**A2** N3/N9)
Gua	239	—	310 (**G1** 06/N7) 255 (**G2** O6/N7) 231 (**G3** N1/O6) 274 (**G4** N3/N9)
			ΔH_{Na}
Ura	141	136	145 (O4)[d]
Thy	144	136	146 (O4)[d]
Cyt	177	—	213 (**C1** O2/N3) 166 (**C2** O2/N3) 179 (**C3** N1/O2)
Ade	172	143	127 (**A1** N6/N7)[d] 204 (**A2** N3/N9)
Gua	182	—	230 (**G1** 06/N7) 181 (**G2** O6/N7 161 (**G3** N1/O6) 190 (**G4** N3/N9)
			ΔH_{K}
Ura	101	105	108 (O4)[d]
Thy	102	105	107 (O4)[d]
Cyt	110	—	159 (**C1** O2/N3) 111 (**C2** O2/N3) 133 (**C3** N1/O2)
Ade	106	97	79 (**A1** N6/N7)[d] 154 (**A2** N3/N9)
Gua	117	—	179 (**G1** 06/N7) 133 (**G2** O6/N7) 114 (**G3** N1/O6) 149 (**G4** N3/N9)

[a]Extended kinetic method (Cerda and Wesdemiotis 1996).
[b]Rodgers and Armentrout (2000b).
[c]B3LYP/6-3111+G(2*df*, 2*p*) values including ZPE, thermal, and BSSE corrections (Russo et al. 2001a,b). The nucleobase tautomer (Figure 15.15) and binding site of the metal ion are given in parentheses.
[d]MP2/6-311+G(2*d*, 2*p*)//MP2/6-31G* values including ZPE, thermal, and BSSE corrections from Rodgers and Armentrout (2000b): $\Delta H_{Li} = 204$ (Ura), 205 (Thy), 212 (Ade); $\Delta H_{Na} = 144$ (Ura), 144 (Thy), 142 (Ade); $\Delta H_{K} = 109$ (Ura), 108 (Thy), 91 (Ade).

nucleobases listed in Figure 15.13 (Table 15.10) (Cerda and Wesdemiotis 1996; Rodgers and Armentrout 2000b; Russo et al. 2001a,b; Rochut et al. 2004). There is excellent agreement among the experimental and computational results for Ura and Thy (Cerda and Wesdemiotis 1996; Rodgers and Armentrout 2000b; Russo et al. 2001a,b). According to the calculations, the most stable complexes of Ura and Thy have monodentate structures and arise by attachment of the metal ion at O4 of the pyrimidine ring (Russo et al. 2001a,b).

Theory predicts that the most stable form of adenine in the gas phase is the 9H tautomer (**A1** in Figure 15.15); the 7H tautomer (**A2**) lies 34 kJ/mol higher in energy (Russo et al. 2001a,b). This stability order reverses in the alkali metal ion (X^+) adducts, with the X^+ complex of **A2** becoming the most stable tautomer. The $[Ade + X]^+$ complexes of the **A1** and **A2** tautomers are predicted to have bidentate geometries, in which the metal ion interacts with N6/N7 of **A1** and N3/N9 of **A2** (Russo et al. 2001a,b). The computed binding energies in the latter isomers

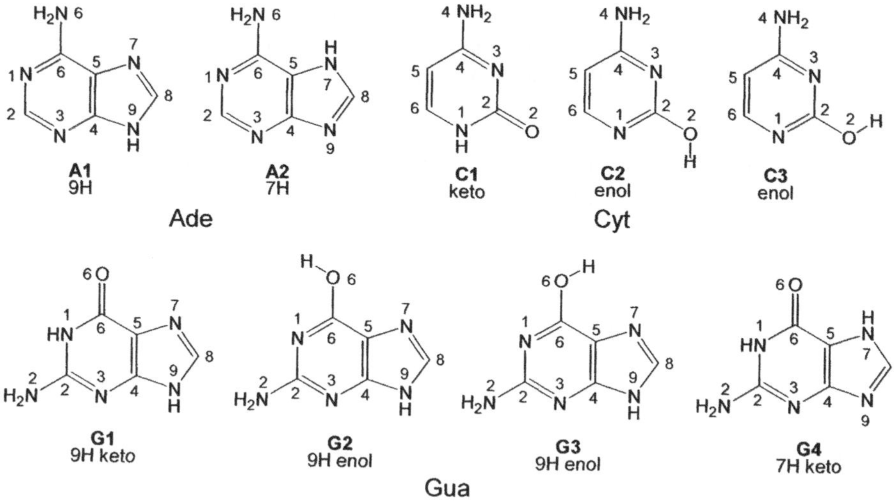

FIGURE 15.15. Tautomers involved in the gas-phase alkali metal ion complexes of adenine, cytosine, and guanine.

(Table 15.10) suggest that the Li^+ complex of **A1** is sampled in both the kinetic method and TCID experiments. The TCID results for the Na^+ and K^+ complexes also are in better agreement with the **A1** complex. On the other hand, the higher Na^+ and K^+ affinities obtained by the kinetic method suggest that the Ade–X^+–B_i dimer ions used in that study (B_i = amino acids G, A, V for Na^+ and pyridine, *n*-propylamine, and aniline for K^+) (Cerda and Wesdemiotis 1996) contained significant amounts of the more strongly bound **A2** tautomer.

The kinetic method data for metalated Cyt and Gua (Table 15.10) are significantly less than the affinities calculated for the most stable conformers of these complexes, which, according to DFT (Russo et al. 2001a,b), involve the keto tautomers **C1** and **G1**, respectively (Figure 15.15). The experimental results can be reconciled by assuming that the complexes generated in the kinetic method experiments contain one or both of the enol tautomers of cytosine (**C2** and/or **C3**) and guanine (**G2** and/or **G3**). The X^+-bound dimers examined by the kinetic method were produced by fast-atom bombardment (FAB) ionization of nucleobase/B_i/metal salt mixtures dissolved in a viscous matrix, such as glycerol (Cerda and Wesdemiotis 1996). The prevailing mode of ion formation in FAB is via ion–molecule reactions of desorbed molecules and ions in the selvedge region (Szekely and Allison 1997). The measured affinities can be explained by assuming that the mixtures subjected to FAB led to a considerable amount of heterodimers containing enol tautomers of Gua and Cyt.

15.3.3. Cluster Ion Thermochemistry: Search for Saltbridges in the Gas Phase

15.3.3.1. Proton-Bound Dimers of Amino Acids. The energetics of dissociation of the proton-bound dimers of several amino acids (AA) and amino acid methyl esters

TABLE 15.11. Threshold Energies, E_0 (kJ/mol), for Dissociation of Proton-Bound Dimers of Amino Acids and Their Methyl Esters According to Eq. (15.35)

AA	E_0	AAOMe	E_0
Gly	111[a]	GlyOMe	106[a]
Ala	108[b]		
Lys	111[b]		
Arg	128[a]	ArgOMe	101[a]
Bet	140[c]		

[a]Price et al. (1997a).
[b]Price et al. (1997b).
[c]Price et al. (1998).

(AAOMe) [Eq. (15.35)] have been investigated by the BIRD approach (Price et al. 1997a,b; 1998). Table 15.11 lists the threshold energies E_0, derived using master equation analysis of the experimental data, as explained in Section 15.2.5.

$$B_2H^+ \rightarrow BH^+ + B \qquad B = \text{AA or AAOMe} \tag{15.35}$$

Amino acids can adopt either canonical (neutral) or zwitterionic structures in the gas phase, whereas only canonical structures are possible for their methyl esters. Hence, the dissociation energies (E_0) of the AAOMe proton-bound dimers provide a measure of the strength of noncovalent interactions in ionic complexes *without* zwitterionic units, that is, complexes held together by charge solvation. Analogously, the dissociation threshold of the proton-bound dimer of betaine (Bet), a permanent zwitterion with the connectivity $CH_3N^+CH_2COO^-$, reveals the minimum energy necessary to break a saltbridge interaction, which is significantly stronger. The E_0 values of $(Gly)_2H^+$, $(Ala)_2H^+$, and $(Lys)_2H^+$ are very similar to that of $(GlyOMe)_2H^+$, in agreement with charge solvation in the proton-bound dimers of Gly, Ala, and Lys (Table 15.11). In sharp contrast, the dissociation energies of the proton-bound dimers of Arg and its methyl ester differ substantially from each other. The higher value for $(Arg)_2H^+$ is consistent with the presence of a saltbridge in this dimer (Price et al. 1997a); this explanation is supported by DFT calculations, which show that the most stable structure of $(Arg)_2H^+$ contains a protonated Arg interacting with an arginine zwitterion (protonated at the basic guanidine group and deprotonated at the *C* terminus).

The higher binding energies in saltbridge structures result from the larger dipole moments of zwitterions versus canonical molecules, which lead to stronger ion–dipole interactions. In the $(Bet)_2H^+$ and $(Arg)_2H^+$ dimers, protonated betaine or arginine are hydrogen-bonded with a zwitterionic Bet or Arg moiety, respectively. The positive charge is delocalized better in the protonated guanidine group of arginine than in the quaternary ammonium group of betaine, which reduces the extent of ion–dipole interactions resulting in a decreased binding energy in the $(Arg)_2H^+$ via à vis the $(Bet)_2H^+$ saltbridge. Such charge delocalization should also be responsible for the slightly decreased binding energy of the charge–solvation complex $(ArgOMe)_2H^+$ vis à vis the charge–solvation complex $(GlyOMe)_2H^+$ (see Table 15.11).

15.3.3.2. Hydrated Amino Acid Complexes. In aqueous solution, amino acids prefer zwitterionic structures, which can develop much stronger hydrogen bonding interactions with the solvent than can neutral, canonical structures. In sharp contrast, all common amino acids favor canonical structures in the solitary environment of the gas phase (Talley et al. 2002; Wang et al. 2006b). Zwitterionic structures become more stable by metal ion addition. For example, the energy difference between canonical and zwitterionic isomers drops from $>+70$ kJ/mol for isolated glycine to $\sim+10$ kJ/mol for its sodiated complex (Hoyau and Ohanessian 1998). Similarly, Na^+ addition to gaseous proline and arginine leads to saltbridge (SB) complexes that are more stable or isoenergetic, respectively, with the corresponding charge–solvation (CS) isomers (see Section 15.3.2.1) (Kish et al. 2003; Wang et al. 2006b). According to DFT calculations, the stability of saltbridges can be further increased by addition of water molecules to these complexes (Jockusch et al. 2001). This expectation has been tested by BIRD experiments on hydrated ions of lithiated valine and the structural analogs alanine ethyl ester (model for nonzwitterionic Val) and betaine (model for zwitterionic Val) (Lemoff and Williams 2004). When energetically excited, these hydrated complexes undergo H_2O loss [Eq. (15.36)]. The dissociation kinetics at temperatures between −60°C and 110°C and subsequent master equation modeling of the experimental data yield the threshold dissociation energies (E_0) given in Table 15.12.

$$\text{B–Li}^+\text{–}(\text{H}_2\text{O})_n \rightarrow \text{B–Li}^+\text{–}(\text{H}_2\text{O})_{n-1} + \text{H}_2\text{O} \qquad (\text{B} = \text{Val, AlaOEt, Bet}) \tag{15.36}$$

With one or two solvent molecule(s), the E_0 values of the Val and AlaOEt complexes match within experimental error and are different from the threshold energy of the betaine complex. This trend changes dramatically when three solvent molecules are added. Now, the valine and betaine complexes show quite similar (although not identical) dissociation energetics; both are substantially higher than that of the alanine ethyl ester complex. Such results provide evidence that the structure of valine changes from canonical (as with AlaOMe) to zwitterionic (as with Bet) with three water molecules. Supporting evidence comes from DFT calculations

TABLE 15.12. Threshold Dissociation Energies for Water Loss from Hydrated Ions of Lithiated Valine, Alanine Ethyl Ester, and Betaine (kJ/mol)[a]

B	$B–Li^+–(H_2O)$	$B–Li^+–(H_2O)_2$	$B–Li^+–(H_2O)_3$
Val	85	59	53
AlaOEt	83	59	36
Bet	79	65	47

[a] ±1 kJ/mol. All data from Lemoff and Williams (2004).

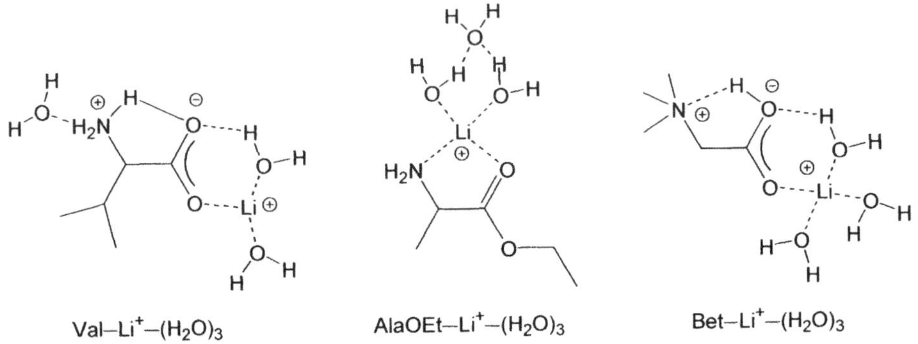

FIGURE 15.16. Most stable structures of B–Li^+–$(H_2O)_3$ complexes based on DFT calculations (Lemoff and Williams 2004).

that predict that the most stable structure of Val–Li^+–(H_2O) involves charge solvation, with Li^+ coordinated between the amine and carbonyl groups of valine (NO coordination) and water attached at the metal ion (Lemoff and Williams 2004). With two water molecules, CS and SB geometries become nearly isoenergetic, and both have the two H_2O molecules attached to the lithium ion (Lemoff and Williams 2004). Finally, with three water molecules, the SB structure of Val–Li^+–$(H_2O)_3$ is the lowest-energy isomer (Figure 15.16). In the latter structure, Li^+ is bound to two water molecules and to zwitterionic Val (OO coordination), while the third water molecule solvates valine's protonated *N* terminus. This hydration motif is uniquely distinct from that predicted for the most stable conformations of the trihydrated AlaOEt and Bet complexes, in which all three water units are attached at the metal ion (Figure 15.16). The saltbridge in Val–Li^+–$(H_2O)_3$ augments the dissociation energy for H_2O loss compared to AlaOMe–Li^+–$(H_2O)_3$; on the other hand, the different arrangement of water molecules in the Val and Bet saltbridges explain the small difference in threshold energies between trihydrated Val and Bet complexes.

Combined BIRD experiments and DFT calculations have also been performed on the monohydrated Na^+ complex of valine (Lemoff et al. 2003) and the monohydrated Li^+ and Na^+ complexes of proline analogs (Lemoff et al. 2005a), glutamine (Lemoff et al. 2005b), and glutamine isomers and analogs (Lemoff et al. 2005b). CS structures prevail in all cases except for the proline systems, which have SB structures; the latter result is not surprising, considering that even in the absence of solvent the alkali metal ion complexes of proline are zwitterionic (Talley et al. 2002; Marino et al. 2003).

Most recently, a TCID and computational study on Gly–Na^+–$(H_2O)_n$, $n = 1$–4, has been reported (Ye et al. 2005). Under TCID conditions, loss of one water unit is the major dissociation pathway, but sequential losses of water and the competitive loss of glycine are also observed at higher energies. The binding energies of water and glycine to the complexes decrease monotonically with increasing number of water molecules. The combined TCID and theoretical data point out that sodiated glycine complexes with up to four water molecules have CS (i.e., nonzwitterionic) structures.

15.4. THERMOCHEMISTRY DATA AS AN AIDE TO MASS SPECTROMETRY ANALYSES

The most widely utilized ionization methods for the mass and tandem mass spectrometry characterization of biomolecules, namely, electrospray ionization (ESI) and matrix-assisted laser desorption/ionization (MALDI), generate quasimolecular ions by protonation, deprotonation, metal ion addition, or a combination thereof (e.g., metal ion addition and deprotonation) (Larsen and McEwen 1998; Dass 2001; Siuzdak 2003; Compton and Siuzdak 2003). The basic and acidic sites within a biomolecule dictate where the resulting positive and/or negative charges are created. Insight into the intrinsic thermochemical properties of biomolecules thus helps predict where these will be charged and what structure(s) they will prefer after ionization.

Positive and negative charges may be viewed as electrophilic or nucleophilic centers, respectively. From their locations, they can promote fragmentation when the quasimolecular ions are energetically activated on MS/MS conditions. Activation is most commonly achieved by collisions with gaseous targets or IR photon absorption, which cause collisionally activated dissociation (CAD) or multiphoton IR photodissociation (MPIRD), respectively. MS/MS conditions may also cause migration (i.e., rearrangement) of the charge to a different location before fragmentation. Charge mobility within a quasimolecular ion again depends on the intrinsic thermochemistry of the charge site. For example, protons attached to arginine residues of a peptide are tightly sequestered because of the high proton affinity of arginine and cannot be mobilized under usual MS/MS conditions (Wang et al. 2005). In such cases, the quasimolecular ions may undergo charge-remote fragmentations.

Unless a biomolecule contains a permanent charge site (e.g., a quaternary ammonium group) or an immobile site as explained above, the fragments generated on its dissociation in an MS/MS experiment will compete for the charge, which will preferably remain with the fragment of higher intrinsic affinity (Wang et al. 2005). Thermochemical data about the protonation, deprotonation, and metal ion thermochemistry of biomolecules and the components of more complex biopolymers are invaluable for deriving decomposition mechanisms from the fragment ion patterns observed in MS/MS spectra and for ultimately constructing accurate algorithms for the computer interpretation of MS/MS data.

How thermochemical data can help us understand ion generation and fragmentation will be briefly illustrated for peptides, which are among the most widely studied biomolecules. On protonation of a peptide, H^+ is preferentially added to the most basic site available, which, on the basis of the PAs and GBs discussed in Section 15.3.1.1, is the sidechain of an Arg, Lys, or His residue, or (in the absence of these) the *N* terminus (Harrison 1997; Wang et al. 2005). With electrospray ionization, the ion having the *N* terminus and all Arg, Lys, and His residues protonated is commonly dominant (Kinter and Sherman 2000). In a protonated peptide with no Arg residue that is energized via CAD or SID, the proton(s) added can migrate along the backbone or to the sidechains, where it can induce fragmentation. This concept of fragment ion production has been termed the mobile proton model (Wysocki et al. 2000). The proton most frequently catalyzes cleavage of the backbone amide bonds, generating ion–molecule complexes that compete for the proton charge. If the *N*-terminal piece is more basic, it keeps the proton and an abundant *N*-terminal *b*-type fragment is generated; conversely, if the *C*-terminal piece is more basic, the complementary *C*-terminal *y*-type ion dominates (Wang et al. 2005). Such a pathway is impossible if the number of protons added to a peptide is equal to or smaller than the number of its arginine residues and no mobile proton is available; in these cases, the MS/MS spectra are usually uninformative, unless the peptides contain acidic residues that enable selective charge-remote fragmentations (Tsaprailis et al. 2000).

Metalated peptides also appear to decompose through ion–molecule complexes, in which there is competition for the metal ion. Since metal ions interact more extensively with amino acid side chains than protons (see Section 15.3.2.1), the fragmentation patterns of protonated and metalated peptides are usually distinct. When bond cleavage takes place, the piece that preserves stabilizing sidechain interactions and has an overall higher metal ion affinity consequently predominates (Feng et al. 2003b; Pingitore and Wesdemiotis 2005; Wang et al. 2006c).

ACKNOWLEDGMENTS

The authors thank the National Science Foundation and the Ohio Board of Regents for generous financial support. The collaboration with Drs. Blas A. Cerda, Michelle M. Kish, Michael J. Polce, and Gilles Ohanessian in the studies discussed in this chapter is gratefully acknowledged.

REFERENCES

Afonso C, Modeste F, Breton P, Fournier F, Tabet J-C (2000): Proton affinities of the commonly occurring L-amino acids by using electrospray ionization-ion trap mass spectrometry. *Eur. J. Mass Spectrom.* **6**:443–449.

Armentrout PB (2003): Threshold collision-induced dissociation for the determination of accurate gas-phase binding energies and reaction barriers. *Top. Curr. Chem.* **225**:233–262.

Aue DH, Bowers, MT (1979): Stabilities of positive ions from equilibrium gas-phase basicity measurements. In Bowers MT (ed), *Gas Phase Ion Chemistry*, Academic Press, New York, Vol. 2, pp. 1–51.

Baer T, Hase WL (1996): *Unimolecular Reaction Dynamics: Theory and Experiments*, Oxford Univ. Press, New York.

Beyer T, Swinehart DF (1973): Number of multiply-restricted partitions [A1] (Algorithm 448): *Commun. Assoc. Comput. Machin.* **16**:379.

Bojesen G, Breindahl T. Andersen U (1993): On the sodium and lithium ion affinities of some α-amino acids. *Org. Mass Spectrom.* **28**:1448–1452.

Bojesen G, Breindahl T (1994): On the proton affinity of some α-amino acids and the theory of the kinetic method. *J. Chem. Soc. Perkins Trans. 2* 1029–1037.

Carr SR, Cassady CJ (1996): Gas-phase basicities of histidine and lysine and their selected di- and tripeptides. *J. Am. Soc. Mass Spectrom.* **7**:1203–1210.

Capota C, Ohanessian G (2005): The low energy tautomers and conformers of the dipeptides HisGly and GlyHis and of their sodium ion complexes in the gas phase. *Phys. Chem. Chem. Phys.* **7**:3744–3755.

Cassady CJ, Carr SR, Zhang K, Chung-Phillips A (1995): Experimental and ab initio studies on protonations of alanine and small peptides of alanine and glycine. *J. Org. Chem.* **60**:1704–1712.

Cerda BA, Wesdemiotis C (1995): The relative Cu(I) ion affinties of amino acids in the gas phase. *J. Am. Chem. Soc.* **117**:9734–9739.

Cerda BA, Wesdemiotis C (1996): Li^+, Na^+, and K^+ binding to the DNA and RNA nucleobases. Bond energies and attachment sites from the dissociation of metal ion-bound heterodimers. *J. Am. Chem. Soc.* **118**:11884–11892.

Cerda BA, Hoyau S, Ohanessian G, Wesdemiotis C (1998): Na^+ binding to cyclic and linear dipeptides. Bond energies, entropies of Na^+ complexation, and attachment sites from the dissociation of Na^+-bound heterodimers and ab initio calculations. *J. Am. Chem. Soc.* **120**:2437–2448.

Cerda BA, Wesdemiotis C (1999): Gas phase copper(I) ion affinities of valine, lysine, and arginine based on the dissociation of Cu^+-bound heterodimers at varying internal energies. *Int. J. Mass Spectrom.* **185/186/187**:107–116.

Cerda BA, Wesdemiotis C (2000): Zwitterionic vs. charge-solvated structures in the binding of arginine to alkali metal ions in the gas phase. *Analyst* **125**:657–660.

Chandra AK, Nguyen MT, Uchimaru T, Zeegers-Huyskens T (1999): Protonation and deprotonation enthalpies of guanine and adenine and implications for the structure and energy of their complexes with water: Comparison with uracil, thymine, and cytosine. *J. Phys. Chem. A* **103**:8853–8860.

Chantry PJ (1971): Doppler broadening in beam experiments. *J. Chem. Phys.* **55**:2746–2759.

Chen O, Groh S, Liechty A, Ridge DP (1999): Binding of nitric oxide to iron(II) porphyrins: Radiative association, blackbody infrared radiative dissociation, and gas-phase association equilibrium. *J. Am. Chem. Soc.* **121**:11910–11911.

Cheng X, Wu Z, Fenselau C (1993): Collision energy dependence of proton-bound dimer dissociations: Entropy effects, proton affinities, and intramolecular hydrogen bonding in protonated peptides. *J. Am. Chem. Soc.* **115**:4844–4848.

Chesnavich WJ, Bowers MT (1979): Statistical methods in reaction dynamics. In Bowers MT (ed), *Gas Phase Ion Chemistry*, Academic Press, New York, Vol. 1, pp. 119–151.

Compton BJ, Siuzdak G (2003): Mass spectrometry in nucleic acid, carbohydrate and steroid analysis. *Spectroscopy* **17**:699–713.

Cooks RG, Beynon JH, Caprioli RM, Lester GR (1973): *Metastable Ions*, Elsevier, Amsterdam.

Cooks RG, Ast T, Pradeep T, Wysocki V (1994a): Reactions of ions with organic surfaces. *Acc. Chem. Res.* **27**:316–323.

Cooks RG, Patrick JS, Kotiaho T, McLuckey SA (1994b): Thermochemical determinations by the kinetic method. *Mass Spectrom. Rev.* **13**:287–339.

Cooks RG, Wong PSH (1998): Kinetic method of making thermochemical determinations: Advances and applications. *Acc. Chem. Res.* **31**:379–386.

Cooks RG, Koskinen JT, Thomas PD (1999): The kinetic method of making thermochemical determinations. *J. Mass Spectrom.* **34**:85–92.

Cui W, Hu Y, Lifshitz C (2002): Time-resolved photodissociation of small peptide ions; combining laser desorption with ion trap/reflectron TOF mass spectrometry. *Eur. Phys. J. D: Atom. Mol. Opt. Phys.* **20**:565–571.

Dalleska NF, Honma K, Armentrout PB (1993): Stepwise solvation enthalpies of protonated water clusters: Collision-induced dissociation as an alternative to equilibrium studies. *J. Am. Chem. Soc.* **115**:12125–12131.

Dalleska NF, Honma K, Sunderlin LS, Armentrout PB (1994): Solvation of transition metal ions by water. Sequential binding energies of $M^+(H_2O)_x$ ($x = 1$–4) for M = Ti to Cu determined by collision-induced dissociation. *J. Am. Chem. Soc.* **116**:3519–3528.

Daneshfar R, Klassen JS (2004): Arrhenius activation parameters for the loss of neutral nucleobases from deprotonated oligonucleotide anions in the gas phase. *J. Am. Soc. Mass Spectrom.* **15**:55–64.

Dass C (2001): *Principles and Practice of Biological Mass Spectrometry*, Wiley-Interscience, New York.

Dunbar RC (1991): Kinetics of low-intensity infrared laser photodissociation. The thermal model and applications of the Tolman theorem. *J. Chem. Phys.* **95**:2537–2548.

Dunbar RC (1994): Kinetics of thermal unimolecular dissociation by ambient infrared radiation. *J. Phys. Chem.* **98**:8705–8712.

Dunbar RC (1997): Modeling radiative association kinetics. *Int. J. Mass Spectrom. Ion Proc.* **160**:1–16.

Dunbar RC (2000): Complexation of Na^+ and K^+ to aromatic amino acids: A density functional computational study of cation-π interactions. *J. Phys. Chem. A* **104**:8067–8074.

Dunbar RC (2003): Photodissociation studies of ion thermochemistry. In Armentrout PB, Gross ML, Caprioli RM (eds), *The Encyclopedia of Mass Spectrometry*, Elsevier, Amsterdam, Vol. 1, pp. 403–417.

Dunbar RC (2004): BIRD (blackbody infrared radiative dissociation): Evolution, principles, and applications. *Mass Spectrom. Rev.* **23**:127–158.

Dunbar RC, McMahon TB, Tholmann D, Tonner DS, Salahub D, Wei D (1995): Zero-pressure thermal radiation induced dissociation (ZTRID) of gas-phase cluster ions: Comparison of theory and experiment for $(H_2O)_2Cl^-$ and $(H_2O)_3Cl^-$. *J. Am. Chem. Soc.* **117**:12819–12825.

Dunbar RC, McMahon TB (1997): Activation of unimolecular reactions by ambient blackbody radiation. *Science* **279**:194–197.

Engelking PC (1987): Determination of cluster binding energy from evaporative lifetime and average kinetic energy release: Application to carbon dioxide and argon ion(1+) ($(CO_2)_n^+$ and Ar_n^+) clusters. *J. Chem. Phys.* **87**:936–940.

Ervin KM (2002): Microcanonical analysis of the kinetic method: The meaning of the "apparent entropy." *J. Am. Soc. Mass Spectrom.* **13**:435–452.

Ervin KM, Armentrout PB (1985): Translational energy dependence of $Ar^+ + XY \rightarrow ArX^+ + Y$ ($XY = H_2$, D_2, HD) from thermal to 30 eV c.m. *J. Chem. Phys.* **83**:166–189.

Ewing NP, Zhang X, Cassady CJ (1996): Determination of the gas-phase basicities of proline and its di- and tripeptides with glycine: The enhanced basicity of prolylproline. *J. Mass Spectrom.* **31**:1345–1350.

Feng WY, Gronert S, Lebrilla CB (1999): Lithium and sodium ion binding energies of N-acetyl and N-glycyl amino acids. *J. Am. Chem. Soc.* **121**:1365–1371.

Feng WY, Gronert S, Lebrilla CB (2003a): The lithium cation binding energies of gaseous amino acids. *J. Phys. Chem. A* **107**:405–410.

Feng WY, Gronert S, Fletcher KA, Warres A, Lebrilla CB (2003b): The mechanism of C-terminal fragmentation in alkali metal ion complexes of peptides. *Int. J. Mass Spectrom.* **222**:117–134.

Gapeev A, Dunbar RC (2001): Cation–π interactions and the gas phase thermochemistry of the Na^+/phenylalanine complex. *J. Am. Chem. Soc.* **123**:8360–8365.

Gapeev A, Dunbar RC (2002): Reactivity and binding energies of transition metal halide ions with benzene. *J. Am. Soc. Mass Spectrom.* **13**:477–484.

Gapeev A, Dunbar RC (2003): Na^+ affinities of gas-phase amino acids by ligand exchange equilibrium. *Int. J. Mass Spectrom.* **228**:825–839.

Gorman GS, Amster IJ (1993a): Gas-phase basicity measurements of dipeptides that contain valine. *J. Am. Chem. Soc.* **115**:5729–5735.

Gorman GS, Amster IJ (1993b): Kinetic and thermodynamic considerations of the bracketing method: Entropy-driven proton-transfer reactions in a Fourier transform mass spectrometer. *Org. Mass Spectrom.* **28**:1602–1607.

Gorman GS, Speir JP, Turner CA, Amster IJ (1992): Proton affinities of the 20 common α-amino acids. *J. Am. Chem. Soc.* **114**:3986–3988.

Greco F, Liguori A, Sindona G, Uccella N (1990): Gas-phase proton affinity of deoxyribonucleosides and related nucleobases by fast atom bombardment tandem mass spectrometry. *J. Am. Chem. Soc.* **112**:9092–9096.

Hahn I-S, Wesdemiotis C (2003): Protonation thermochemistry of β-alanine. An evaluation of proton affinities and protonattion entropies determined by the extended kinetic method. *Int. J. Mass Spectrom.* **222**:465–479.

Harrison AG (1992): *Chemical Ionization Mass Spectrometry*, 2nd ed., CRC Press, Boca Raton, FL.

Harrison AG (1997): The gas-phase basicities and proton affinities of amino acids and peptides. *Mass Spectrom. Rev.* **16**:201–217.

Holbrook KA, Pilling MJ, Robertson SH (1996): *Unimolecular Reactions*, Wiley, New York.

Holmes JL (1985): Assigning structures to ions in the gas phase. *Org. Mass Spectrom.* **20**:169–183.

Holmes JL, Osborne AD (1977): Metastable ion studies, VIII: An analytical method for deriving kinetic energy release distributions from metastable peaks. *Int. J. Mass Spectrom. Ion Phys.* **23**:189–2000.

Holmes JL, Terlouw JK (1980): The scope of metastable peak observations. *Org. Mass Spectrom.* **15**:383–396.

Hoyau S, Ohanessian G (1997): Absolute affinities of α-amino acids for Cu^+ in the gas phase. A theoretical study. *J. Am. Chem. Soc.* **119**:2016–2024.

Hoyau S, Ohanessian G (1998): Interaction of alkali metal cations (Li^+-Cs^+) with glycine in the gas phase: A theoretical study. *Chem. Eur. J.* **4**:1561–1569.

Hoyau S, Norrman K, McMahon TB, Ohanessian, G (1999): A quantitative basis for a scale of Na^+ affinities of organic and small biological molecules in the gas phase. *J. Am. Chem. Soc.* **121**:8864–8875.

Hoyau S, Pelicier J-P, Rogalewicz F, Hoppilliard Y, Ohanessian G (2001): Complexation of glycine by atomic metal cations in the gas phase. *Eur. J. Mass Spectrom.* **7**:303–311.

Hu Y, Hadas B, Davidovitz M, Balta B, Lifshitz C (2003): Does IVR take place prior to peptide ion dissociation? *J. Phys. Chem. A* **107**:6507–6514.

Huang Y, Kenttämaa H (2003): Theoretical estimations of the 298 K gas-phase acidities of the pyrimidine-based nucleobases uracil, thymine, and cytosine. *J. Phys. Chem. A* **107**:4893–4897.

Huang Y, Kenttämaa H (2004): Theoretical estimations of the 298 K gas-phase acidities of the purine-based nucleobases adenine and guanine. *J. Phys. Chem. A* **108**:4485–4490.

Hunter EPL, Lias SG (1998): Evaluated gas phase basicities and proton affinities of molecules: An update. *J. Phys. Chem. Ref. Data* **27**:413–656 (http://webbook.nist.gov/).

Jarrold MF, Wagner-Redeker W, Illies AJ, Kirchner NJ, Bowers MT (1984): Unimolecular and bimolecular reactions in the benzene ion ($C_6H_6^{+\bullet}$) system. Experiment and theory. *Int. J. Mass Spectrom. Ion Phys.* **58**:63–95.

Jockusch RA, Schnier PD, Price WD, Strittmatter EF, Demirev PA, Williams ER (1997): Effects of charge state on fragmentation pathways, dynamics, and activation energies of ubiquitin ions measured by blackbody infrared radiative dissociation. *Anal. Chem.* **69**:1119–1126.

Jockusch RA, Williams ER (1998): Binding energies of proton-bound dimers of imidazole and *n*-acetylalanine methyl ester obtained by blackbody infrared radiative dissociation. *J. Phys. Chem. A* **102**:4543–4550.

Jockusch RA, Lemoff AS, Williams ER (2001): Effect of metal ion and water coordination on the structure of a gas-phase amino acid. *J. Am. Chem. Soc.* **123**:12255–12265.

Kebarle P (1988): Pulsed electron high pressure mass spectrometry. In Farrar JM, Saunders Jr WH (eds): *Techniques for the Study of Ion-Molecule Reactions*, Wiley, New York, pp. 221–286.

Kebarle P (1992): Ion-molecule equilibria, how and why. *J. Am. Soc. Mass Spectrom.* **3**:1–9.

Kinter M, Sherman NE (2000): *Protein Sequencing and Identification Using Tandem Mass Spectrometry*, Wiley-Interscience, New York.

Kish MM, Ohanessian G, Wesdemiotis C (2003): The Na^+ affinities of αamino acids: Side-chain substituent effects. *Int. J. Mass Spectrom.* **227**:509–524.

Kish MM, Wesdemiotis C, Ohanessian G (2004): The sodium ion affinity of glycylglycine. *J. Phys. Chem. B* **108**:3086–3091.

Kitova EN, Bundle DR, Klassen JS (2002): Thermal dissociation of protein-oligosaccharide complexes in the gas phase: Mapping the intrinsic intermolecular interactions. *J. Am. Chem. Soc.* **124**:5902–5913.

Kitova EN, Bundle DR, Klassen JS (2004): Partitioning of solvent effects and intrinsic interactions in biological recognition. *Angew. Chem. Int. Ed.* **43**:4183–4186.

Klassen JS, Anderson SG, Blades AT, Kebarle P (1996): Reaction enthalpies for $M^+L = M^+ + L$, where $M^+ = Na^+$ and K^+ and L = acetamide, N-methylacetamide, N,N-dimethylacetamide, glycine, and glycylglycine, from determination of the collision-induced dissociation thresholds. *J. Phys. Chem.* **100**:14218–14227.

Klots CE (1971): Reformulation of the quasiequilibrium theory of ionic fragmentation. *J. Phys. Chem.* **75**:1526–1532.

Klots CE (1989): Thermal kinetics in small systems. *J. Chem. Phys.* **90**:4470–4472.

Klots CE (1990): Evaporative corrections to high-pressure ionic clustering data. *Int. J. Mass Spectrom. Ion Proc.* **100**:457–463.

Klots CE (1991): Systematics of evaporation. *Z. Phys. D* **20**:105–109.

Kurinovich MA, Lee JK (2000): The acidity of uracil from the gas phase to solution: The coalescence of the N1 and N3 sites and implications for biological glycosylation. *J. Am. Chem. Soc.* **122**:6258–6262.

Laidler KJ (1987): *Chemical Kinetics*, 3rd ed., Harper & Row, New York, p. 112.

Larsen BS, McEwen CN (eds) (1998): *Mass Spectrometry of Biological Materials*, 2nd ed., Marcel Dekker, New York.

Laskin J (2004): Energetics and dynamics of peptide fragmentation from multiple-collision activation and surface-induced dissociation studies. *Eur. J. Mass Spectrom.* **10**:259–267.

Laskin J, Denisov E, Futrell JH (2000): A comparative study of collision-induced and surface-induced dissociation. 1. Fragmentation of protonated dialanine. *J. Am. Chem. Soc.* **122**:9703–9714.

Laskin J, Denisov E, Futrell JH (2001): Comparative study of collision-induced and surface-induced dissociation. 2. Fragmentation of small alanine-containing peptides in FT-ICR MS. *J. Phys. Chem. B* **105**:1895–1900.

Laskin J, Denisov E, Futrell JH (2002): Fragmentation energetics of small peptides from multiple-collision activation and surface-induced dissociation in FT-ICR MS. *Int. J. Mass Spectrom.* **129**:189–201.

Laskin J, Lifshitz C (2001): Kinetic energy release distributions in mass spectrometry. *J. Mass Spectrom.* **36**:459–478.

Lee VW-M, Li H, Lau T-C, Guevremont R, Siu KWM (1998): Relative silver(I) ion binding energies of α-amino acids: A determination by means of the kinetic method. *J. Am. Soc. Mass Spectrom.* **9**:760–766.

Lemoff AS, Bush MF, Williams ER (2003): Binding energies of water to sodiated valine and structural isomers in the gas phase: The effect of proton affinity on zwitterion stability. *J. Am. Chem. Soc.* **125**:13576–13584.

Lemoff AS, Bush MF, Williams ER (2005a): Structures of cationized proline analogues: Evidence for the zwitterionic form. *J. Phys. Chem. A* **109**:1903–1910.

Lemoff AS, Bush MF, Wu C-C, Williams ER (2005b): Structures and hydration enthalpies of cationized glutamine and structural analogues in the gas phase. *J. Am. Chem. Soc.* **127**:10276–10286.

Lemoff AS, Williams ER (2004): Binding energies of water to lithiated valine: Formation of solution-phase structure in vacuo. *J. Am. Soc. Mass Spectrom.* **15**:1014–1024.

Leyh B, Lorquet JC (2003): Kinetic energy release distributions in mass spectrometry. In Armentrout PB, Gross ML, Caprioli RM (eds): *The Encyclopedia of Mass Spectrometry*, Elsevier, Amsterdam, Vol. 1, pp. 17–32.

Lias SG, Bartmess JE, Liebman JF, Holmes JL, Levin RD, Mallard WG (1988): Gas-phase ion and neutral thermochemistry. *J. Phys. Chem. Ref. Data* 17 (Suppl. 1).

Lifshitz C (1989): Recent developments in applications of RRKM-QET. *Adv. Mass Spectrom.* **11**:713–729.

Lifshitz C (1993): Unimolecular and collision induced decomposition of selected cluster ions. In Ng CY, Baer T, Powis I (eds), *Cluster Ions*, Wiley, Chichester, UK, pp. 121–164.

Lifshitz C, Louage F (1989): Magic numbers in kinetic energy releases for unimolecular decompositions of $(NH_3)_nH^+$ ion clusters. *J. Phys. Chem.* **93**:5633–5635.

Lifshitz C, Louage F (1990): Theoretical interpretation for magic numbers in kinetic energy releases. *Int. J. Mass Spectrom. Ion Proc.* **101**:101–110.

Lifshitz C, Tzidony E (1981): Kinetic energy release distributions for $C_3H_6O^{+\bullet}$ ion dissociations: A further test of the applicability of the energy randomization hypothesis to unimolecular fragmentations. *Int. J. Mass Spectrom. Ion Phys.* **39**:181–195.

Lifshitz C, Wu RLC, Tiernan TO, Terwilliger DT (1978): Negative ion–molecule reactions of ozone and their implications on the thermochemistry of O_3^-. *J. Chem. Phys.* **68**:247–260.

Liguori A, Napoli A, Sindona G (1994): Determination of substituent effects on the proton affinities of natural nucleosides by the kinetic method. *Rapid Commun. Mass Spectrom.* **8**:89–93.

Liguori A, Napoli A, Sindona G (2000): Survey of the proton affinities of adenine, cytosine, thymine and uracil dideoxyribonucleosides, deoxyribonucleosides and ribonucleosides. *J. Mass Spectrom.* **35**:139–144.

Linstrom PJ, Mallard WG (eds) (2005): *NIST Chemistry WebBook. NIST Standard Reference Database 69*, June 2005, National Institute of Standards and Technology, Gaithersburg MD 20899 (http://webbook.nist.gov/).

Locke MJ, McIver Jr RT (1980): Effect of solvation on the acid/base properties of glycine. *J. Am. Chem. Soc.* **105**:4226–4232.

Marino T, Russo N, Toscano M (2000): Gas-phase metal ion (Li^+, Na^+, Cu^+) affinities of glycine and alanine. *J. Inorg. Biochem.* **79**:179–185.

Marino T, Russo N, Toscano M (2002): Interaction of Cu^+ and Cu^{2+} ions with α-alanine. A density functional theory. *J. Mass Spectrom.* **37**:786–791.

Marino T, Russo N, Toscano M (2003): Interaction of Li^+, Na^+, and K^+ with the proline amino acid. Complexation modes, potential energy profiles, and metal ion affinities. *J. Phys. Chem. B* **107**:2588–2594.

Meot-Ner (Mautner) M (1979): Ion thermochemistry of low-volatility compounds in the gas phase. 2. Intrinsic basicities and hydrogen-bonded dimers of nitrogen heterocyclics and nucleic bases. *J. Am. Chem. Soc.* **101**:2396–2403.

Meot-Ner (Mautner) M (2003): The proton affinity scale, and effects of ion structure and solvation. *Int. J. Mass Spectrom.* **227**:525–554.

Meot-Ner M, Hunter EP, Field FH (1979): Ion thermochemistry of low-volatility compounds in the gas phase. 1. Intrinsic basicities of α-amino acids. *J. Am. Chem. Soc.* **101**:686–689.

Meot-Ner (Mautner) M, Speller CV (1986): Filling of solvent shells about ions. 1. Thermochemical criteria and the effects of isomeric clusters. *J. Phys. Chem.* **90**:6616–6624.

Miller TM, Arnold ST, Viggiano AA, Stevens Miller AE (2004): Acidity of a nucleotide base: Uracil. *J. Phys. Chem. A* **108**:3439–3446.

Moision RM, Armentrout PB (2002): Experimental and theoretical dissection of sodium cation/glycine interactions. *J. Phys. Chem. A* **106**:10350–10362.

Moision RM, Armentrout PB (2004): An experimental and theoretical dissection of potassium/glycine interactions. *Phys. Chem. Chem. Phys.* **6**:2588–2599.

Nold MJ, Cerda BA, Wesdemiotis C (1999): Proton affinities of the N- and C-terminal segments arising upon the dissociation of the amide bond in protonated peptides. *J. Am. Soc. Mass Spectrom.* **10**:1–8.

O'Hair RAJ, Bowie JH, Gronert S (1992): Gas phase acidities of the α amino acids. *Int. J. Mass Spectrom. Ion Proc.* **117**:23–36.

Patrick JS, Yang SS, Cooks RG (1996): Determination of the gas-phase basicity of betaine and related compounds using the kinetic method. *J. Am. Chem. Soc.* **118**:231–232.

Pepe C, Rochut S, Paumard J-P, Tabet J-C (2004): Ab initio calculations of proton affinities of glycine, proline, cysteine and phenylalanine: Comparison with the experimental values obtained using an electrospray ionization ion trap mass spectrometer. *Rapid Commun. Mass Spectrom.* **18**:307–312.

Pingitore F, Wesdemiotis C (2005): Characterization of dipeptide isomers by tandem mass spectrometry of their mono- versus dilithiated complexes. *Anal. Chem.* **77**:1796–1806.

Price WD, Schnier PD, Jockusch RA, Strittmatter EF, Williams ER (1996): Unimolecular reaction kinetics in the high-pressure limit without collisions. *J. Am. Chem. Soc.* **118**:10640–10644.

Price WD, Jockusch RA, Williams ER (1998): Binding energies of protonated betaine complexes: A probe of zwitterion structure in the gas phase. *J. Am. Chem. Soc.* **120**:3474–3484.

Price WD, Jockusch RA, Williams ER (1997a): Is arginine a zwitterion in the gas phase? *J. Am. Chem. Soc.* **119**:11988–11989.

Price WD, Schnier PD, Williams ER (1997b): Binding energies of the proton-bound amino acid dimers Gly-Gly, Ala-Ala, Gly-Ala, and Lys-Lys measured by blackbody infrared radiative dissociation. *J. Phys. Chem. B* **101**:664–673.

Price WD, Williams ER (1997): Activation of peptide ions by blackbody radiation: Factors that lead to dissociation kinetics in the rapid energy exchange limit. *J. Phys. Chem. A* **101**:8844–8852.

Rochut S, Pepe C, Paumard J-P, Tabet J-C (2004): A computational and experimental study of cation affinity (Na^+) of nucleobases and modified nucleobases by electrospray ionization mass spectrometry. *Rapid Commun. Mass Spectrom.* **18**:1686–1692.

Rodgers MT, Ervin KM, Armentrout PB (1997): Statistical modeling of collision-induced dissociation thresholds. *J. Chem. Phys.* **106**:4499–4508.

Rodgers MT, Armentrout PB (1998): Statistical modeling of competitive threshold collision-induced dissociation. *J. Chem. Phys.* **109**:1787–1800.

Rodgers MT, Armentrout PB (2000a): Noncovalent metal-ligand bond energies as studied by threshold collision-induced dissociation. *Mass Spectrom. Rev.* **19**:215–247.

Rodgers MT, Armentrout PB (2000b): Noncovalent interactions of nucleic acid bases (uracil, thymine, and adenine) with alkali metal ions. Threshold collision-induced dissociation and theoretical studies. *J. Am. Chem. Soc.* **122**:8548–8558.

Ruan C, Rodgers MT (2004): Cation-π interactions: Structures and energetics of complexation of Na^+ and K^+ with the aromatic amino acids, phenylalanine, tyrosine, and tryptophan. *J. Am. Chem. Soc.* **126**:14600–14610.

Ruan C, Rodgers MT (2005): Cation-π interactions with a model for the side chain of tryptophan: Structures and absolute binding energies of alkali metal cation-indole complexes. *J. Phys. Chem. A* **109**:11539–11550.

Russo N, Toscano M, Grand A (2001a): Lithium affinity for DNA and RNA nucleobases. The role of theoretical information in the elucidation of the mass spectrometry data. *J. Phys. Chem. B* **105**:4735–4741.

Russo N, Toscano M, Grand A (2001b): Bond energies and attachment sites of sodium and potassium cations to DNA and RNA nucleic acid bases in the gas phase. *J. Am. Chem. Soc.* **123**:10272–10279.

Ryzhov V, Dunbar RC (1997): Size dependence of radiative association rates. *Int. J. Mass Spectrom. Ion Proc.* **167/168**:627–635.

Ryzhov V, Dunbar RC (1999): Direct association equilibrium in the Fourier-transform ion cyclotron resonance ion trap: The hydration equilibrium of protonated 18-crown-6. *J. Am. Soc. Mass Spectrom.* **10**:862–868.

Ryzhov V, Klippenstein, SJ, Dunbar, RC (1996): Radiative association of NO^+ with 3-pentanone: Rate, binding energy, and temperature dependence. *J. Am. Chem. Soc.* **118**:5462–5468.

Ryzhov V, Dunbar RC, Cerda BA, Wesdemiotis C (2000): Cation-π effects in the complexation of Na^+ and K^+ with Phe, Tyr, and Trp in the gas phase. *J. Am. Soc. Mass Spectrom.* **11**:1037–1046.

Schnier PD, Price WD, Jockusch RA, Williams ER (1996): Blackbody infrared radiative dissociation of bradykinin and its analogs: Energetics, dynamics, and evidence for salt-bridge structures in the gas phase. *J. Am. Chem. Soc.* **118**:7178–7189.

Schnier PD, Klassen JS, Strittmatter EF, Williams ER (1998): Activation energies for dissociation of double strand oligonucleotide anions: Evidence for Watson-Crick base pairing in vacuo. *J. Am. Chem. Soc.* **120**:9605–9613.

Sharma S, Lee JK (2002): Acidity of adenine and adenine derivatives and biological implications. A computational and experimental gas-phase study. *J. Org. Chem.* **67**:8360–8365.

Siu FM, Ma NL, Tsang CW (2004): Competition between π and non-π cation-binding sites in aromatic amino acids: A theoretical study of alkali metal cation (Li^+, Na^+, K^+)-phenylalanine complexes. *Chem. Eur. J.* **10**:1966–1976.

Siuzdak G (2003): *The Expanding Role of Mass Spectrometry in Biotechnology*, MCC Press.

Sponer J, Leszczynski J, Hobza P (2001): Electronic properties, hydrogen bonding, stacking, and cation binding of DNA and RNA bases. *Biopolymers* **61**:3–31 (recent review).

Stein SE, Rabinovitch BS (1977): On the use of exact state counting methods in RRKM rate calculations. *Chem. Phys. Lett.* **49**:183–188.

Steinfeld JF, Francisco JF, Hase WL (1999): *Chemical Kinetics and Dynamics*, 2nd ed., Prentice-Hall, Upper Saddle River, NJ.

Su T, Bowers MT (1973): Ion-polar molecule collisions: The effect of ion size on ion-polar molecule rate constants; the parameterization of the average-dipole-orientation theory. *Int. J. Mass Spectrom. Ion Phys.* **12**:347–356.

Szekely G, Allison J (1997): If the ionization mechanism in fast-atom bombardment involves ion/molecule reactions, what are the reagent ions? The time dependence of fast-atom bombardment mass spectra and parallels to chemical ionization. *J. Am. Soc. Mass Spectrom.* **8**:337–351.

Talley JM, Cerda BA, Ohanessian G, Wesdemiotis C (2002): Alkali metal ion binding to amino acids versus their methyl esters: Affinity trends and structural changes in the gas phase. *Chem. Eur. J.* **8**:1377–1388.

Tsaprailis G, Somogyi Á, Nikolaev EN, Wysocki VH (2000): Refining the model for selective cleavage at acidic residues in arginine-containing protonated peptides. *Int. J. Mass Spectrom.* **195/196**:467–479.

Wang P, Kish MM, Wesdemiotis C (2005): Fragmentation mechanisms of peptide ions. In Caprioli RM, Gross ML (eds), *The Encyclopedia of Mass Spectrometry*, Elsevier, Amsterdam, Vol. 2, Part A, pp. 139–151.

Wang P, Kapota C, Ohanessian G, Wesdemiotis C (2006a): The sodium ion affinities of simple di-, tri-, and tetrapeptides. Submitted for publication.

Wang P, Kapota C, Ohanessian G, Wesdemiotis C (2006b): Amino acids with high intrinsic sodium ion affinities: Binding sites and energetics of Na^+ attachment to asparagine, glutamine, histidine, and arginine. Submitted for publication.

Wang P, Polce MJ, Bleiholder C, Paizs B, Wesdemiotis C (2006c): Structural characterization of peptides via tandem mass spectrometry of their dilithiated monocations. *Int. J. Mass Spectrom.* **249/250**:45–59.

Wei S, Tsang WB, Castleman Jr AW (1990): Dissociation dynamics: Measurements of decay fractions of metastable ammonia clusters. *J. Chem. Phys.* **93**:2506–2512.

Wesdemiotis C (2004): Entropy considerations in kinetic method experiments. *J. Mass Spectrom.* **39**:998–1003.

Wilson MS, McCloskey JA (1975): Chemical ionization mass spectrometry of nucleosides. Mechanisms of ion formation and estimation of proton affinity. *J. Am. Chem. Soc.* **97**:3436–3644.

Woodin RL, Beauchamp JL (1978): Binding of Li^+ to Lewis bases in the gas phase. Reversals in methyl substituent effects for different reference acids. *J. Am. Chem. Soc.* **100**:501–507.

Wu Z, Fenselau C (1992a): Proton affinity of arginine measured by the kinetic approach. *Rapid Commun. Mass Spectrom.* **6**:403–405.

Wu Z, Fenselau C (1992b): Proton affinities of polyglycines assessed by using the kinetic method. *J. Am. Soc. Mass Spectrom.* **3**:863–866.

Wu Z, Fenselau C (1993): Structural determinants of gas phase basicities of peptides. *Tetrahedron* **49**:9197–9206.

Wu Z, Fenselau C (1994): Gas-phase basicities and proton affinities of lysine and histidine measured from the dissociation of proton-bound dimers. *Rapid Commun. Mass Spectrom.* **8**:777–780.

Wu J, Lebrilla CB (1993): Gas-phase basicities and sites of protonation of glycine oligomers (GLY_n; $n = 1$–5). *J. Am. Chem. Soc.* **115**:3720–3725.

Wu J, Lebrilla CB (1995): Intrinsic basicity of oligomeric peptides that contain glycine, alanine, and valine. The effects of alkyl side chain on proton transfer reactions. *J. Am. Soc. Mass Spectrom.* **6**:91–101.

Wysocki VH, Tsaprailis G, Smith LL, Breci LA (2000): Mobile and localized protons: A framework for understanding peptide dissociation. *Org. Mass Spectrom.* **35**:1399–1406.

Wyttenbach T, Bushnell JE, Bowers MT (1998): Salt bridge structures in the absence of solvent? The case of the oligoglycines. *J. Am. Chem. Soc.* **120**:5098–5103.

Ye SJ, Moision RM, Armentrout PB (2005): Sequential bond energies of water to sodium glycine cation. *Int. J. Mass Spectrom.* **240**:233–248.

Zhang K, Zimmerman DM, Chung-Phillips A, Cassady CJ (1993): Experimental and ab initio gas-phase basicities of polyglycines. *J. Am. Chem. Soc.* **115**:10812–10822.

16

ENERGY AND ENTROPY EFFECTS IN GAS-PHASE DISSOCIATION OF PEPTIDES AND PROTEINS

JULIA LASKIN

Fundamental Science Directorate
Pacific Northwest National Laboratory
Richland, WA

Principles of Mass Spectrometry Applied to Biomolecules, edited by Julia Laskin and Chava Lifshitz

16.1. INTRODUCTION

Since the mid-1990s, characterization of complex molecules, particularly biomolecules, became a focus of both fundamental and applied research in mass spectrometry. Most of these studies utilize tandem mass spectrometry (MS/MS) to obtain structural information on complex molecules. MS/MS typically involves the mass selection of a primary ion, its activation by collision or photon excitation, unimolecular decay into fragment ions characteristic of the ion structure and its internal excitation, and mass analysis of the fragment ions. A wide variety of mass filters and ion excitation methods can be employed in these experiments, making tandem mass spectrometry an extremely flexible analytical technique that can be implemented on almost any type of mass spectrometer.

Fragmentation patterns of biomolecules in a mass spectrometer are commonly used as "fingerprints" of their primary and secondary structure. Because the mass spectrum is a snapshot in time of the fragmentation kinetics of the precursor ion, mass spectral "fingerprints" depend on several factors. In particular, the type and amount of fragments observed in the mass spectrum are controlled by the energetics and dynamics of fragmentation, the internal energy distribution of excited ions, and the instrumental time window. Knowledge of fragmentation energetics and mechanisms and of the energy deposition in the ion activation process provides the basis for interpreting and predicting the MS/MS spectra of biomolecules. This chapter discusses the energetics and dynamics of gas-phase dissociation of peptides and proteins, specifically focusing on unimolecular reactions that can be described using statistical theories such as RRKM/QET or phase space theory (PST). Because of the large number of degrees of freedom and the nonresonant nature of most ion activation techniques, dissociation of large molecules commonly satisfies the major assumption of statistical theories, namely, the ergodic assumption that the internal excitation of the ion is randomly redistributed among the vibrational degrees of freedom prior to fragmentation. This implies that the dissociation rate is independent of the mode of preparation of the excited molecule and that dissociation patterns obtained using different ion activation methods can be rationalized within the same model.

The development of statistical theories began in early 1920s and focused mainly on rationalizing experimental rate constants for dissociation of neutral molecules. Similar ideas were later used to establish the theory of mass spectra (Rosenstock et al. 1952) more recently reviewed by Lorquet (1994, 2000). A nice historical overview of the development of the transition state theory was given by Laidler and King (1983). The reader is referred to a number of excellent reviews and books summarizing the basic principles and the current status of the transition state theory (Chesnavich and Bowers 1979; Truhlar et al. 1983, 1996; Baer and Hase 1996; Gilbert and Smith 1990; Holbrook et al. 1996; Forst 2003).

16.2. DISSOCIATION OF POLYATOMIC MOLECULES

16.2.1. Microcanonical Rate Constant

The ergodic assumption ensures that (1) ion activation and dissociation are separated in time and (2) phase space sampled by the excited molecule is determined by the number of vibrational and rotational states of the molecule at a specific internal energy. The next step in the transition state theory is the assumption of the existence of a dividing surface in phase space that irreversibly separates reactants and products. This dividing surface is called the *transition state*, and the reaction is modeled as a unidirectional flux through the transition state. The microcanonical reaction rate constant is given by the RRKM/QET expression

$$k(E) = \sigma \frac{W^{\ddagger}(E - E_0)}{h\rho(E)} \tag{16.1}$$

where E is the internal energy of the molecule, E_0 is the critical energy, $\rho(E)$ is the density of states of the reactant at energy E, $W^{\ddagger}(E - E_0)$ is the sum of states from 0 to $(E - E_0)$ in the transition state, h is Planck's constant, and σ is the reaction path degeneracy.

Both the number and the density of states increase very rapidly with internal energy (E). Because the number of states increases faster with E, the rate constant, $[k(E)]$ is a strong function of the internal energy of the excited molecule. RRKM/QET calculations require knowledge of the critical (or threshold) energy of dissociation and the ensemble of vibrational frequencies of the excited molecule and the transition state. Obviously, the dimensionality of the problem increases dramatically with the size of the molecule. In addition, transition state frequencies are known only for rare cases. Fortunately, RRKM/QET calculations are sensitive not to the details of vibrational frequencies of the molecular ion and transition state but rather to the relative change in frequencies along the reaction coordinate characterized by the activation entropy $\Delta S^{\ddagger}$ (Rosenstock et al. 1979). It follows that the rate–energy dependence of the microcanonical rate constant is determined mainly by two parameters: E_0 and $\Delta S^{\ddagger}$.

It is also interesting to note that uncertainties in vibrational frequencies do not significantly affect the calculated rate constant because they appear in both the numerator and the denominator in Eq. (16.1) in a similar way, resulting in cancellation of errors. Similarly, the rate constant will not be significantly affected by incomplete or nonuniform sampling of the phase space as long as the relative sampling of the reactant and the transition state is the same. On the basis of these considerations, Lorquet suggested that "RRKM/QET formula is a very robust equation. It cannot fail miserably" (Lorquet 2000). The robustness of the RRKM/QET equation and its simplicity account for the remarkable success of statistical theories of unimolecular reactions in general as well as their specific application to mass spectrometry.

16.2.1.1. Threshold Energy and Activation Entropy. Microcanonical rate–energy dependences for two reactions of the same precursor with different threshold energies ($E_{01} < E_{02}$) and activation entropies ($\Delta S_1^{\ddagger} < \Delta S_2^{\ddagger}$) are shown schematically in Figure 16.1. The position of the curve is determined by the threshold energy E_0, while the activation entropy $\Delta S^{\ddagger}$ determines the slope of the curve. The activation entropy characterizes the degree of tightness or looseness of the transition state. Tight transition states are typical for reactions that involve substantial rearrangement in the transition state. These reactions are associated with low or negative activation entropy and the slowly rising rate–energy dependence. Reactions that proceed via a loose transition state are characterized by high activation entropy and steeply rising rate–energy dependence. Figure 16.1 also demonstrates that the competition between two reaction channels is determined by the relative position of the $k(E)$ curves and the

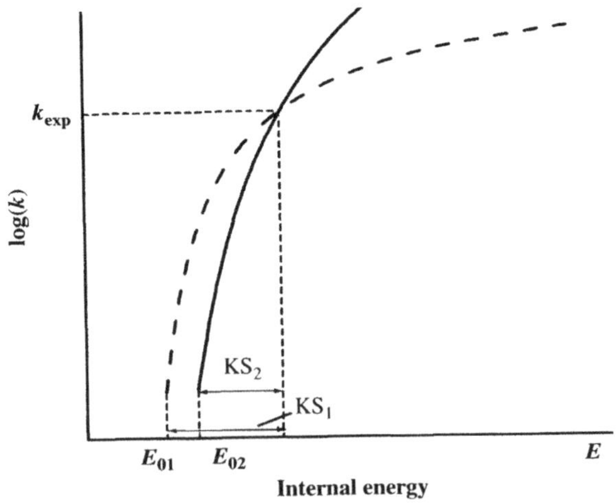

FIGURE 16.1. Schematic drawing of the microcanonical rate–energy dependences on a semilogarithmic scale for two reactions with different threshold energies ($E_{01} < E_{02}$) and activation entropies ($\Delta S_1^{\ddagger} < \Delta S_2^{\ddagger}$). KS$_1$ and KS$_2$ denote kinetic shifts for reactions (1) and (2), respectively.

observation time of the instrument. Long observation times will favor reaction (16.1), while reaction (16.2) will be dominant on the short timescale.

The entropy of the excited molecule (S) and the transition state ($S^{\ddagger}$) can be calculated from statistical thermodynamics as follows

$$S = R \ln Q + RT\frac{d \ln Q}{dT} \tag{16.2}$$

where T is the temperature, Q is the partition function, and R is the molar gas constant. The vibrational partition function is given by

$$Q = \prod_i \left(1 - \exp\left(\frac{-h\nu_i}{k_B T}\right)\right)^{-1} \tag{16.3}$$

where k_B is Boltzmann's constant and ν_i are the vibrational frequencies of the excited molecule or of the transition state. For simplicity, we do not discuss the contribution of rotational degrees of freedom to the activation entropy. However, accounting for rotational contribution in Eq. (16.2) is straightforward [see, e.g., Gilbert and Smith (1990)]. Because the activation entropy is temperature-dependent, it is important to specify the temperature, for which it has been evaluated. Conventionally, activation entropies for gas-phase ionic fragmentation are reported at 1000 K (Lifshitz 1982).

16.2.1.2. Kinetic and Intrinsic Shift. Determination of the dissociation energetics of polyatomic molecules from experimentally measured appearance energies of fragment ions has been pioneered by Chupka and coworkers [for a detailed discussion, see Chupka (1959)]. In a classical photoionization experiment, in which the molecule (M) is ionized and dissociated into a fragment ion (F^+) and a neutral molecule (N) by absorption of one photon:

$$M \xrightarrow{h\nu} M^{+*} \xrightarrow{E_0} F^+ + N \tag{16.4}$$

The dissociation energy E_0 was determined from the difference between the appearance energy (AE) of the fragment ion and the ionization energy (IE) of the precursor molecule (M):

$$E_0 = \mathrm{AE}(F^+) - \mathrm{IE}(M) \tag{16.5}$$

However, it was realized very early that because of the statistical nature of unimolecular dissociation, the molecule may have enough energy but not enough time to dissociate. In this case Eq. (16.5) fails to predict the threshold energy. This is particularly important for larger molecules because rate constants decrease dramatically with increase in the number of vibrational degrees of freedom available for redistribution of the excess internal energy of the molecule $E - E_0$. The internal energy in excess of the threshold energy required to produce detectable dissociation of a polyatomic ion on the timescale of a mass spectrometer is called the *kinetic shift* (KS) Figure 16.1 shows kinetic shifts for hypothetical reactions (16.1) and (16.2) (e.g., KS_1 and KS_2). Originally the kinetic shift was defined for 10 μs reaction time (Lifshitz 1982; 2002)—a typical timescale of sector mass spectrometers that were widely used in early studies on gas-phase ion chemistry. However, a more general definition seems to be appropriate because of the wide variety of mass spectrometric approaches developed more recently for studying dissociation of biomolecules.

The kinetic shift decreases with increase in the observation time t_{obs}. This can be readily rationalized by examining the rate–energy dependence shown in Figure 16.1. The most probable rate constant sampled experimentally k_{mp} is inversely proportional to t_{obs}. As a result, the experimentally measured rate constant is smaller for longer observation time, bringing the dissociation closer to its thermochemical threshold. However, Dunbar and coworkers pointed out that in many cases the system cannot reach its thermochemical threshold because of the competition between the unimolecular decomposition and slow radiative cooling of the excited molecule through the emission of radiation in the infrared (Huang and Dunbar 1990). The "intrinsic" kinetic shift is defined as the amount of energy required to make the dissociation rate equal to one-tenth of the radiative rate (Huang and Dunbar 1990; Lifshitz 1992).

16.2.2. Thermal Dissociation

16.2.2.1. Arrhenius Equation. The famous Arrhenius equation introduced in 1889 and generally accepted only 20 years later summarized in a very simple form the experimental findings on temperature dependence of thermal dissociation rates obtained in the preceding 50 years (Laidler, 1984):

$$k(T) = A\exp\left(-\frac{E_a}{RT}\right) \tag{16.6}$$

where A is the preexponential factor and E_a is the activation energy. It took more than 20 years to establish the physical meaning of the two parameters, A and E_a, in Eq. (16.6).

In 1920 Tolman gave a concise definition of the activation energy as the difference between the average energy of the reacting molecules and the average energy of all the molecules in the heat bath (Tolman 1920):

$$E_a = \langle E_{\text{activated molecules}} \rangle - \langle E_{\text{all molecules}} \rangle \tag{16.7}$$

It was shown later that the activation energy and the threshold energy are connected via the following expression, which is derived from Eq. (16.7) (Klots 1989; Gilbert and Smith 1990; Dunbar 1991; Baer and Hase 1996)

$$E_a = E_0 + \langle E^{\ddagger} \rangle - \langle E \rangle + k_B T \tag{16.8a}$$

or

$$E_{\mathrm{a}} = \Delta H^{\ddagger} + k_{\mathrm{B}}T \tag{16.8b}$$

where k_{B} is Boltzmann's constant; $\langle E^{\ddagger}\rangle$ and $\langle E\rangle$ are the average energy of the transition state and the average energy of all molecules evaluated at temperature T, respectively, and $\Delta H^{\ddagger}$ is the activation enthalpy of reaction.

16.2.2.2. Absolute Reaction Rate Theory. The physical meaning of the preexponential factor was rationalized by the transition state theory in its early formulation—the absolute reaction rate theory—developed by Eyring in 1935 (Eyring 1935). In this formulation the reaction rate in thermal system is given by

$$k(T) = \frac{k_{\mathrm{B}}T}{h}\frac{Q^{\ddagger}}{Q}\exp\left(-\frac{E_{\mathrm{a}}}{RT}\right) \tag{16.9}$$

or

$$k(T) = \frac{k_{\mathrm{B}}Te}{h}\exp\left(\frac{\Delta S^{\ddagger}}{R}\right)\exp\left(-\frac{E_{\mathrm{a}}}{RT}\right) \tag{16.10}$$

The reader is referred to Gilbert and Smith (1990) for detailed derivation of Eq. (16.10). From Eqs. (16.6) and (16.10) the preexponential factor can be expressed as

$$A = \frac{k_{\mathrm{B}}Te}{h}\exp\left(\frac{\Delta S^{\ddagger}}{R}\right) \tag{16.11}$$

showing that the preexponential factor is a measure of the activation entropy $\Delta S^{\ddagger}$.

16.2.2.3. Rapid Exchange Limit. For a system in thermal equilibrium Eq. (16.9) can be derived from the RRKM/QET expression [Eq. (16.1)] by averaging $k(E)$ over the Boltzmann distribution of internal energies $P(E, T)$

$$k(T) = \int_{E_0}^{\infty} k(E)P(E, T)dE \tag{16.12}$$

where

$$P(E, T) = \frac{1}{Q}\rho(E)\exp\left(-\frac{E}{k_{\mathrm{B}}T}\right) \tag{16.13}$$

This corresponds to the so-called high-pressure or rapid energy exchange (REX) limit. In the REX limit the dissociation rate is substantially slower than the rates of activation/deactivation of the molecule. Because the effect of reactive loss of excited molecules on the energy distribution is negligible, the system can be described by an equilibrium (Boltzmann) distribution However, when the system is not in the REX limit, the energy distribution corresponds to a truncated or depleted Boltzmann distribution and the experimentally measured Arrhenius parameters are smaller than the values predicted on the basis of the known reaction thermochemistry. It follows that experimental Arrhenius parameters can be correlated with reaction thermochemistry using Eqs. (16.8b) and (16.11) *only* when dissociation occurs in the REX limit.

Price and Williams (1997) studied the factors that lead to dissociation in the REX limit. They found that thermal kinetics of dissociation of large molecules in the gas phase rapidly approaches the REX limit as the size of the molecule or threshold energy increases or the activation entropy decreases, thereby slowing down the dissociation. They concluded that within the experimental temperature range of 300–520 K and kinetic window of 5–1000 s peptide ions larger than 1.6 kDa dissociate in the REX limit.

An excellent review of thermal kinetics studied using mass spectrometry has been presented by Dunbar (2004). Three different regimes of thermal kinetics comprising large molecules, small molecules, and intermediate-sized molecules and the different approaches to interpreting thermal kinetics in those three regimes have been discussed in detail. Dunbar used the standard hydrocarbon model to calculate the rate of absorption/emission of radiation by molecules of varying size. He then calculated the number of degrees of freedom necessary for the Arrhenius activation energy E_a to come within 10% of the REX limiting value for two different preexponential factors. The results are shown in Figure 16.2. Clearly, the minimum size of the molecule dissociating in the REX limit increases rapidly with increase in activation entropy and with increasing reaction rate. The difference between the prediction of this model and the results obtained by Price and Williams (1997) for peptide ions predicting the transition to the REX limit at considerably smaller sizes was attributed to differences in the rates of energy exchange. Specifically, it has been suggested that exceptional "brightness" of peptides in the IR results in a faster rate of energy exchange, shifting the transition to the REX limit to smaller sizes relative to hydrocarbons.

An important conclusion obtained from such detailed analysis is that "it would be unusual to find a situation where large-molecule behavior was approached for a molecule with much less than 100 degrees of freedom (or molecular weight less than 300 Da). On the other hand, it would not be unusual for a molecule with thousands of degrees of freedom to deviate significantly from the large molecule limit" (Dunbar 2004).

16.2.2.4. Master Equation Modeling. Meaningful energetics and dynamics of dissociation can be obtained from thermal dissociation that does not occur in the REX limit using master equation modeling. Detailed description of this modeling approach can be found in textbooks (Gilbert and Smith 1990; Holbrook et al. 1996). In master equation modeling the entire internal energy range is divided into

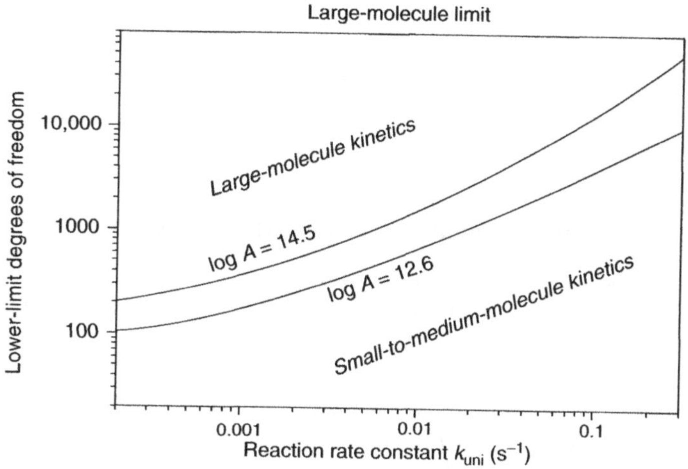

FIGURE 16.2. Model calculation results (standard hydrocarbon model) showing the minimum size necessary for the experimentally determined E_a to lie within 10% of the limiting high-pressure E_{a1} for the same dissociation reaction. Two model transition states are shown (constructed by variation of the vibrational frequencies), one (loose) with $\Delta S(1000\,K) = +14\,J\,K^{-1}\,mol^{-1}$ ($\log A = 14.5$) and the other (tight) with $\Delta S(1000\,K) = -22\,J\,K^{-1}\,mol^{-1}$ ($\log A = 12.6$). Note as a rough conversion for thinking about typical ions, that one degree of freedom corresponds to ~3 Da of mass. For polypeptides (exemplifying biomolecules and other highly polar molecules), Price and Williams (1997) calculated considerably smaller limiting sizes for the large-molecule regime, by a factor of ≥ 3. [Reproduced from Dunbar (2004) with permission from Wiley, copyright 2004.]

small bins. Each bin is populated by exciting collisions or photon absorption and depopulated by collisional deexcitation or photon emission and dissociation. Dissociation rates are calculated using RRKM theory. This approach has been successfully applied for interpreting thermal dissociation of gas-phase ions excited by collisions with the background gas (Goeringer and McLuckey 1997; Asano et al. 1999) or by photon absorption (Dunbar 1991, 1994; Uechi and Dunbar 1992; Lin and Dunbar 1996; Price et al. 1997). However, because master equation modeling requires time-consuming calculations, it has not been applied to thermal kinetics of very large molecules (e.g., proteins).

16.3. HISTORICAL OVERVIEW OF EXPERIMENTAL METHODS FOR MEASURING THRESHOLD ENERGIES AND ACTIVATION ENTROPIES

Knowledge of the internal energy (or the distribution of internal energies) of excited ions is an important prerequisite for studying the energetics and dynamics of gas-phase fragmentation. In the past a number of experimental approaches were established, that studied dissociation of ions with well-defined internal energies. These include photoelectron–photoion coincidence (PEPICO), time-resolved photodissociation (TRPD), time-resolved photoionization mass spectrometry (TPIMS), and threshold collision-induced dissociation (TCID) measurements under single-collision conditions. These methods are briefly described in this section. A comprehensive overview of experimental techniques in gas-phase ion thermochemistry with extensive literature survey can be found elsewhere (Ervin 2001).

16.3.1. Photoelectron–Photoion Coincidence (PEPICO/TPEPICO)

In PEPICO experiments ions are produced and dissociated by single-photon ionization as shown in Eq. (16.4). The internal energy of the excited ion is given by

$$E = h\nu - \mathrm{IE} + E_{\mathrm{th}} - \mathrm{KE}_e \qquad (16.14)$$

where $h\nu$ is the photon energy, IE is the ionization energy, E_{th} is the thermal energy of the molecule, and $\mathrm{KE_e}$ is the kinetic energy of the departing electron. Because $h\nu$ and IE are well known and E_{th} is readily estimated, the internal energy distribution of excited ions can be obtained by measuring the kinetic energy distribution of electrons formed in photoionization. Experimentally, ions are measured in coincidence with photoelectrons and the internal energy distribution of ions is determined using Eq. (16.14). An extension of this technique, in which only zero-energy electrons are detected, resulting in significant improvement in the signal-to-noise ratio (Baer and Hase 1996), is called *threshold PEPICO* (TPEPICO). In TPEPICO experiments the dissociation rate constant of the excited ion is measured as a function of the internal energy of the ion that is given by

$$E = h\nu - \mathrm{IE} + E_{\mathrm{th}} \qquad (16.15)$$

Thus TPEPICO provides a direct measurement of the microcanonical rate constant $k(E)$. Threshold energy and activation entropy are determined by fitting the experimental data with Eq. (16.1).

Because dissociation of small ions is not affected by the kinetic shift, TPEPICO provides a direct measurement of very accurate thermochemical data for these systems [for a more recent review, see Ng (2002)]. Coincidence studies of dissociation of larger polyatomic ions provided a solid basis for rigorous tests of statistical theories and established benchmark data for development of other techniques for studying unimolecular reactions. For specific examples, the reader is referred to a number of excellent reviews (Baer 1986, 2000; Baer and Hase 1996).

16.3.2. Time-Resolved Photodissociation (TRPD)

Coincidence experiments are carried out in a time-of-flight (TOF) mass spectrometer and therefore sample a relatively narrow range of rate constants on a microsecond timescale. This is a significant restriction for larger molecules that undergo slow dissociation. The kinetics of fragmentation of energy-selected ions is more conveniently studied using photodissociation experiments (Dunbar 1984,

2000; Baer and Hase 1996). Time-resolved photodissociation experiments (TRPDs) introduced by Dunbar and coworkers are carried out in trapping devices. Ions are trapped and allowed to undergo collisional and radiative relaxation prior to photoexcitation. Irradiation of ions by a light source produces excited ions with well-defined internal energy:

$$E = h\nu + E_{\text{th}} \tag{16.16}$$

The evolution of a mass spectrum as a function of time enables measurement of dissociation kinetics of excited ions from which the energy-dependent rate constant $[k(E)]$ is extracted. TRPD experiments allow access to a broad range of rate constants from $\sim$1 to $10^6\,\text{s}^{-1}$. The lower limit is determined by the competition between the dissociation and radiative relaxation of excited ions. Access to longer reaction times was crucial for quantitative studies of radiative relaxation of molecules extensively studied by Dunbar and coworkers (Dunbar 1992). However, because photodissociation requires resonant absorption of a photon, the ability to vary the internal energy of the ion in TRPD experiments is fairly limited. For many systems TRPD and TPEPICO experiments complement each other. A beautiful example reported by Dunbar and Lifshitz (1991) is shown in Figure 16.3. TPEPICO

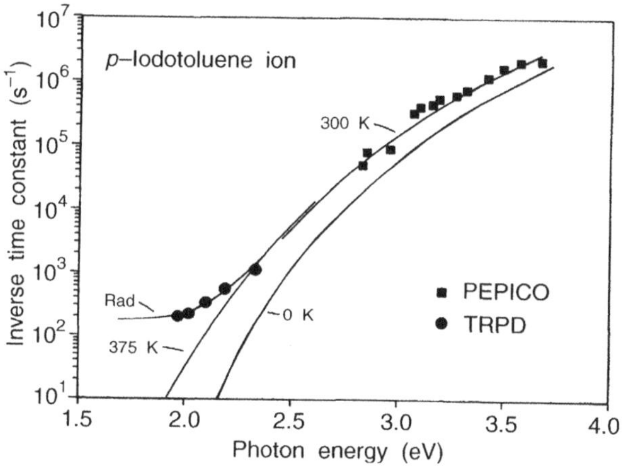

FIGURE 16.3. Dissociation rate constants from TRPD and PEPICO for iodotoluene ion. The curve marked "300 K," which gives a good fit to the PEPICO points, is the predicted RRKM rate–energy curve assuming a dissociation energy of 1.9 eV, corrected for the 300 K thermal internal energy appropriate to the PEPICO experiment. The curve marked "375 K" is the corresponding predicted RRKM curve for a dissociation experiment at 375 K, which is the temperature appropriate to the TRPD experiment. The variant of the 375 K curve marked "rad" is further corrected for IR radiative relaxation of the slowly dissociating ions, giving a curve that fits the observed TRPD points. The curve marked "0 K" is the calculated RRKM rate–energy curve assuming zero thermal internal energy, which is displayed here to show the ideal dissociation kinetics undistorted by thermal energy and radiative relaxation effects. [Reproduced from Dunbar and Lifshitz (1991) with permission from the American Physical Society.]

was used to measure energy-selected rate constants for dissociation of iodotoluene in the range 10^5–10^6 s^{-1}, while TRPD provided rate constants in the range 10^2–10^3 s^{-1} and clearly showed the role of radiative relaxation at longer reaction time on the observed kinetics. The data are in excellent agreement with the RRKM-predicted curve.

First TRPD experiments for peptide radical cations have been carried out by Lifshitz and coworkers (Cui et al. 2002; Hu et al. 2003). These experiments utilized laser desorption of peptides followed by photoionization, trapping and collisional relaxation of ions, and photodissociation. Energy-selected dissociation rate constants showed a strong dependence on the internal energy and the size of the ion. These experiments unambiguously demonstrated that dissociation of gas-phase peptides is statistical and can be described using RRKM/QET. An example of rate–energy dependences determined for peptide ions using TRPD is shown in Figure 16.4. The interpretation of these experiments is at present complicated by the possible contribution of multiphoton dissociation and by a broad thermal distribution that peptide ions have at room temperature. However, they provide a background for further studies of gas-phase dissociation of energy-selected peptide ions.

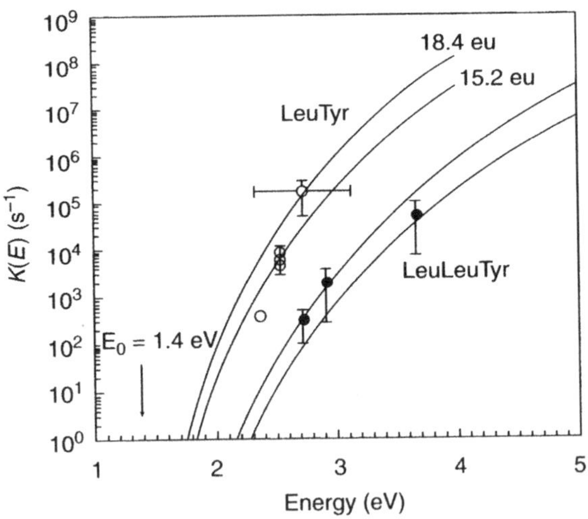

FIGURE 16.4. Microcanonical rate–energy dependencies for the reactions producing the immonium ion from LeuTyr$^{+\cdot}$ and LeuLeuTyr$^{+\cdot}$. The rate constant $k(E)$ is plotted on a logarithmic scale as a function of internal energy E in the reactant ion. The open circles (for LeuTyr), filled circles (for LeuLeuTyr), and error bars are experimental data, and the lines are calculated using RRKM/QET and a model with $E_0 = 1.4$ eV and $\Delta S^{\ddagger} = 15.2$–$18.4$ eu. The horizontal error bar shown for one LeuTyr point demonstrates the width of the thermal energy distribution. [Reproduced from Hu et al. (2003) with permission of the American Chemical Society.]

16.3.3. Time-Resolved Photoionization Mass Spectrometry (TPIMS)

In photoionization mass spectrometry (PIMS) the relative abundance of the parent and fragment ions is measured at different photon energies providing photoionization efficiency curves (PIEs). Numerous examples of early PIMS studies have been summarized by Berkowitz (1979). PIMS experiments are much simpler than PEPICO experiments (described earlier) because they omit the energy selection step. Although ions are produced with a distribution of internal energies, it is assumed that this distribution is given by the photoelectron spectrum of the molecule, which is readily available for a large number of organic molecules. PIE curves are reconstructed using the following steps. Microcanonical rate–energy dependences are calculated using Eq. (16.1). The breakdown curve (BDC) representing the relative abundance of the ion at a particular internal energy is calculated for a given reaction time. The breakdown graph (BDG) representing a collection of BDCs is convoluted with vibrational and rotational thermal distributions at the temperature of the experiment and integrated over the energy deposition function. Threshold energy and activation entropy in step 1 are adjusted to provide the best fit of the PIE curve.

PIMS has been successfully used for a number of relatively small systems characterized by a small kinetic shift. In this case dissociation energy is given by a difference between the appearance energy and the ionization energy, and the reaction entropy can be derived unambiguously from the modeling of PIE. However, it was realized early on that if the fragmentation is associated with a significant kinetic shift, which is characteristic of larger molecules, the values of E_0 and $\Delta S^{\ddagger}$ obtained from the modeling of PIE curves are not unique (Rosenstock et al. 1980; Lifshitz 1982). This ambiguity was resolved by carrying out time-resolved PIMS experiments (TPIMS), in which PIE curves are measured at several reaction times from several microseconds to hundreds of milliseconds (Lifshitz 1982, 1991). It should be noted that these experiments were inspired by time-dependent PEPICO studies carried out by Rosenstock and coworkers (Rosenstock et al. 1979), which sampled much shorter times, between 0.5 and 10 μs.

TPIMS experiments have been carried out for a variety of organic radical cations (Lifshitz 1991; 1997) providing benchmark values of threshold energies and activation entropies. In some cases simultaneous modeling of TPIMS and TRPD data was performed to improve the precision of the dissociation parameters [specific examples can be found in Lifshitz (1997)].

16.3.4. Threshold Collision-Induced Dissociation (TCID)

The methodology of obtaining accurate dissociation energetics of polyatomic molecules from threshold collision-induced dissociation (TCID) experiments has been developed by Armentrout and coworkers and described in a number of reviews (Rodgers and Armentrout 2000; Armentrout 2001, 2003a,b). In these experiments the ion of interest with well-defined internal energy distribution is excited by a single collision with an inert target gas (G), resulting in dissociation:

$$
\begin{aligned}
&M^{+} + G \longrightarrow M^{+*} + G \\
&M^{+*} \xrightarrow{E_0} F^{+} + N
\end{aligned}
\qquad (16.17)
$$

CID cross sections as a function of the center-of-mass (E_{CM}) collision energy determined experimentally are modeled using the following expression

$$
\sigma(E_{CM}) = \sigma_0 \sum_i g_i (E_{CM} + E_i - E_0)^n / E_{CM} \qquad (16.18)
$$

where σ_0 is a scaling factor, n is an adjustable parameter, and E_i and g_i are the internal energy and the population of each rovibrational state of the excited ion, respectively. Detailed description of the modeling of TCID experiments can be found elsewhere (Rodgers et al. 1997). The cross section in Eq. (16.18) is expressed in terms of the energy-dependent microcanonical rate constant, which is calculated using RRKM/QET.

TCID experiments typically sample dissociation rate constants around $10^4\ s^{-1}$. From the previous discussion it is clear that modeling of TCID data alone cannot provide a unique pair of dissociation parameters, E_0 and $\Delta S^{\ddagger}$, for polyatomic molecules. This ambiguity is circumvented by studying reactions characterized by a very loose transition state, for which the activation entropy is given by the phase space limit (PSL). In the PSL the transition state is located at the centrifugal barrier for dissociation and its vibrational frequencies are product-like. Such a loose transition state is commonly associated with cleavage of noncovalent bonds. The validity of the PSL model has been verified by Armentrout and coworkers and Rodgers and coworkers for a number of applications [see Rodgers and Armentrout (2000, 2004) and Armentrout (2001, 2003a,b) and references cited herein].

Alternatively, quantum-chemical calculations are used to determine the structure of the transition state and provide necessary parameters for modeling of TCID data. This procedure has been applied to a number of systems, including protonated triglycine (Shoeib et al. 2001; Muntean and Armentrout 2002; El Aribi et al. 2003).

16.4. WHY IT IS DIFFICULT TO DETERMINE DISSOCIATION PARAMETERS FOR BIOMOLECULES

Early mass spectrometric studies of large molecules encountered two fundamental challenges:

1. Extreme difficulties were associated with production of gas-phase ions of biomolecules. It was found that the ionization efficiency in traditional ionization techniques, such as electron impact, single-photon, or multiphoton ionization,

rapidly decreases with increasing molecular weight. Discussion of the origin of the reduced yield of large radical cations [see Schlag et al. (1992), Becker and Wu (1995), Berkowitz (2000) and references cited therein] is beyond the scope of this chapter. The difficulties in ionization of large molecules have been largely circumvented by the discovery of soft ionization techniques such as electrospray (ESI) (Fenn et al. 1989) and matrix-assisted laser desorption/ionization (MALDI) (Hillenkamp et al. 1991), enabling introduction of ions of biomolecules into the gas phase.

2. Many gas-phase biomolecules did not produce detectable dissociation using single-collision CID—the conventional approach for ion activation in MS/MS experiments at that time. This difficulty in dissociation of large molecules was originally attributed to the inefficiency of the energy transfer in collisions of large molecules with neutral gases [see discussion in Marzluff and Beauchamp (1996)]. However, classical trajectory calculations demonstrated that more than 90% of the center-of-mass collision energy can be transferred to a peptide ion in a single collision (Marzluff et al. 1994). This high energy transfer efficiency is obtained only for head-on collisions. When averaged over all impact parameters, the efficiency drops to 10–15% (Meroueh and Hase 1999, 2000). Our current understanding of low CID efficiency is that two fundamental limits severely constrain CID of large ions in tandem mass spectrometry: (a) the center-of-mass collision energy—the absolute upper limit of energy transfer in a collision process—decreases with increasing mass of the projectile ion for fixed ion kinetic energy and neutral mass, and (b) the dramatic increase in density of states with increasing internal degrees of freedom of the ion decreases the rate of dissociation by many orders of magnitude at a given internal energy [see Eq. (16.1)]. It follows that efficient fragmentation of large molecules requires deposition of a large amount of energy into its internal modes, which in many cases cannot be achieved using single-photon excitation necessary for TRPD or single-collision activation employed in TCID experiments. In addition, PEPICO and TPIMS cannot be applied to study dissociation of nonvolatile molecules. Consequently, conventional approaches developed for accurate determination of dissociation parameters for relatively small polyatomic ions cannot be readily extended to study unimolecular decomposition of large biomolecules (e.g., peptides and proteins). New techniques for studying dissociation of large molecules will be discussed in the next section.

The amount of internal energy required to induce dissociation of large molecules has been estimated in a number of early studies. Schlag and Levine (1989) used a simple RRK expression to estimate the dissociation rate constants of proteins. They found that calculated lifetimes of large molecules were many orders of magnitude longer than typical lifetimes sampled by mass spectrometers and suggested that the dissociation of biomolecules observed experimentally was not consistent with statistical behavior. However, Bernshtein and Oref (1994) used RRKM to show that with very reasonable dissociation parameters large molecules can fragment on a timescale of a mass spectrometer. Griffin and McAdoo (1993) suggested that dissociation of large ions will be detectable only for sufficiently low dissociation thresholds.

TABLE 16.1. Kinetic Shifts (in eV) for Dissociation of Large Ions with 300, 500 and 1000 Vibrational Degrees of Freedom (Average Internal Energy at 300 K in Parentheses) Calculated for Two Reaction Times (10 μs and 1 s) and Different Values of Dissociation Parameters

	$A = 1.2 \times 10^9\ s^{-1}$[a]			$A = 1.2 \times 10^{12}\ s^{-1}$			$A = 1.1 \times 10^{15}\ s^{-1}$		
E_0 (eV)	300 (1.0)	500 (1.8)	1000 (3.6)	300 (1.0)	500 (1.8)	1000 (3.6)	300 (1.0)	500 (1.8)	1000 (3.6)
				$t = 10\ \mu s$					
1.0	18.8	27.4	48.2	5.4	9.1	19.8	3.4	6.0	13.4
1.5	61.8	78.5	116.5	11.6	19.2	39.5	6.8	11.6	24.6
2.0	112.5	137.8	193.1	19.1	31.1	62.3	10.8	18.0	37.4
2.5	165.0	199.0	273.0	27.3	44.3	87.3	15.2	25.1	51.5
				$t = 1\ s$					
1.0	1.8	3.6	8.6	1.6	2.9	7.0	1.2	2.4	6.0
1.5	5.8	10.0	22.0	3.9	6.8	15.3	3.0	5.2	12.0
2.0	11.7	19.5	40.3	6.9	11.9	25.8	5.0	8.8	19.4
2.5	19.0	31.0	62.0	10.5	17.8	37.8	7.5	12.8	27.6

[a]Preexponential factors at 450 K.

The kinetic shift is determined by the dissociation threshold, activation entropy, the size of the ion, and the timescale of the mass spectrometer. The dependence of the kinetic shift on these parameters is illustrated in Table 16.1 along with the average internal energy at room temperature (300 K). Clearly, all four parameters have a very strong effect on the onset for dissociation. The rapid decrease in the kinetic shift as a function of time suggests that it is advantageous to conduct experiments on the long timescale of an ion trap and explains the limited success of some MS/MS experiments conducted on a microsecond timescale. For fast dissociation pathways (with low dissociation energy/high preexponential factor) or long observation time, ion dissociation can be promoted by the thermal energy distribution at room temperature. However, it is difficult to observe dissociation of a large ion in a mass spectrometer when the threshold energy is higher than 2.5 eV. In the following section we will show that threshold energies for dissociation of a large number of peptides and proteins are in the range of 0.8–2 eV, which explains the tremendous success of tandem mass spectrometry in providing sufficient structural information for characterization of biomolecules.

16.5. SURFACE-INDUCED DISSOCIATION OF PEPTIDES

Surface-induced dissociation (SID), introduced by Cooks and coworkers (Mabud et al. 1985; Cooks et al. 1990), provides an efficient means of very fast, single-step excitation of the ion in collision with a surface, in which the internal energy is deposited into the ion in a few picoseconds. Coupling SID with a Fourier transform ion cyclotron resonance mass spectrometer (FTICR MS) provides the distinct advantages of long and variable reaction times (milliseconds to seconds), which both greatly reduces the kinetic shift and allows us to access the lowest-energy dissociation pathways for large molecules. Several studies have demonstrated application of FTICR SID for extracting accurate energetics and dynamics of peptide fragmentation (Laskin et al. 2000b, 2002a; Bailey et al. 2003).

The advantages provided by SID for studying the energetics and dynamics of dissociation of large molecules include very fast ion activation, which eliminates possible discrimination against higher-energy dissociation pathways, and efficient "amplification" of small changes in dissociation parameters (Bailey et al. 2003; Laskin 2004). For example, the difference between threshold energies for dissociation of des-Arg^1–and des-Arg^9–bradykinin is only 0.08 eV, while the values of collision energies required for dissociation differ by 4 eV. Two effects contribute to this "amplification" of subtle variations in threshold energies. First, there is a substantial kinetic shift for dissociation of ions of this size, even on a long timescale of the FTICR experiment. The difference in internal energies required for dissociation of des-Arg^1–and des-Arg^9–bradykinin is 0.8 eV. This corresponds to a tenfold amplification of differences in threshold energies. In addition, the $T \rightarrow V$ transfer efficiency is approximately 20%. It follows that the difference in SID collision energies required to observe fragmentation is about 5 times larger than the difference in the corresponding internal energies. Combining these two factors leads to a 50-fold amplification of the energy difference for slowly fragmenting peptides. Similarly, amplification allows us to accurately measure small differences in reaction entropies using FTICR SID experiments.

16.5.1. Methodology

16.5.1.1. Time- and Energy-Resolved SID Studies. In FTICR SID experiments ions are produced in a high-transmission electrospray source (Laskin et al. 2002b), consisting of an ion funnel interface (Shaffer et al. 1997) followed by three quadrupoles. The quadrupoles provide collisional focusing, mass selection of the ion of interest, and external accumulation of ions. The third (accumulation) quadrupole is held at an elevated pressure (2×10^{-3} Torr) for collisional relaxation of stored ions and their thermalization prior to the SID experiment. After accumulation, the ions are extracted from the third quadrupole and transferred into the ICR cell, where they collide with a self-assembled monolayer (SAM) surface positioned at the rear trapping plate of the cell. Scattered ions are captured by raising the potentials on the front and rear trapping plates of the ICR cell by 10–20 V. The collision energy is defined by the difference in the potential applied to the accumulation quadrupole and the potential applied to the rear trapping plate and the SID target. The ICR cell can be offset above or below ground by as much as ± 150 V. Lowering the ICR cell voltage below ground while keeping the potential on the third quadrupole fixed increases collision energy for positive ions.

Collision-energy-resolved studies provide important information on the appearance energies of fragment ions and the competition between different dissociation pathways. Another dimension is added to SID experiments by varying the delay between the ion–surface collision and the analysis of resulting fragments to conduct kinetic studies. The reaction delay is typically varied from 1 ms to 1 s. Time-resolved fragmentation efficiency curves (TFECs) are constructed by plotting the dependence of the relative abundance of each ion in the spectrum on the kinetic energy of the precursor ion at different reaction delay times.

16.5.1.2. RRKM Modeling of SID Data. Dissociation parameters for peptide ions excited using collisional activation are obtained by fitting experimentally measured fragmentation efficiency curves (FECs) using an RRKM-based model. The FEC is constructed by plotting the relative abundance of the precursor ion and of its fragments as a function of collision energy. Two factors determine the shape and the position of the FEC: fragmentation probability for the particular reaction channel as a function of the internal energy of the precursor ion [the breakdown curve (BDC)] and the distribution of internal energies deposited during collisional activation [energy deposition function (EDF)]. The relative abundance of each ion in the MS/MS spectrum is given by the integrated product of the BDC and EDF [see Eq. (16.21)].

FECs are modeled using the following procedure. Energy-dependent microcanonical rate constants for each reaction channel $k_i(E)$ are calculated using RRKM/QET. The breakdown graph (BDG)—a collection of curves representing relative abundance of the precursor and fragment ions as a function of the internal energy of the excited ion—is constructed by calculating the relative abundance of the precursor ion and fragment ions $[F_i(E, t_r)]$ as a function of the internal energy of the parent ion and the experimental observation time (t_r). For a known fragmentation scheme and reaction time the BDG can be readily calculated using standard equations of formal kinetics.

FECs are obtained by integrating the BDG over the EDF, which is given by the following analytical expression

$$P(E, E_{\text{coll}}) = \frac{1}{C}(E - \Delta)^l \exp\left(-\frac{E - \Delta}{f(E_{\text{coll}})}\right) \tag{16.19}$$

where l and Δ are parameters, $C = \Gamma(l + 1)[f(E_{\text{coll}})]^{l+1}$ is a normalization factor, and $f(E_{\text{coll}})$ has the form

$$f(E_{\text{coll}}) = A_2 E_{\text{coll}}^2 + A_1 E_{\text{coll}} + A_0 \tag{16.20}$$

where A_0, A_1, and A_2 are parameters and E_{coll} is the collision energy. Finally, the normalized signal intensity for a particular fragment (its FEC) is given by the equation

$$I_i(E_{\mathrm{coll}}) = \int_0^{\infty} F(E, t_r)P(E, E_{\mathrm{coll}})dE \tag{16.21}$$

Experimental FECs are compared with simulated curves, and the fitting parameters are varied until the best fit is obtained. The fitting parameters include the critical energy and activation entropy for each reaction channel and parameters characterizing the energy deposition function.

16.5.1.3. Energy Partitioning. Energy partitioning among the ionic and neutral fragments must be taken into account to obtain a good model for the internal energy dependencies for consecutive reaction channels. The internal energy of an ionic fragment formed from a precursor ion with internal energy E will be equal to $E - E_0$ only if a neutral fragment cannot carry away any internal energy. This happens only when the neutral fragment does not have any vibrational degrees of freedom. However, if the neutral fragment is polyatomic, the internal energy of the ionic fragment is less than $E - E_0$ and the fraction of the internal energy carried off by the neutral increases with increasing complexity of the neutral fragment.

Energy partitioning can be estimated by assuming that the excess energy is partitioned statistically among the fragments and by calculating all permutations of the energy partitioning from densities of states of the ionic and neutral fragments. Given the total internal energy in the precursor ion (E), the probability that energy between ε and $\varepsilon + d\varepsilon$ will remain as the internal energy of the ionic fragment is given by (Vestal 1965; Bente et al. 1975)

$$p(E, \varepsilon) = \frac{\rho_1(\varepsilon)\rho_2(E - E_0 - \varepsilon)d\varepsilon}{\int_0^{E-E_0} \rho_1(\varepsilon)\rho_2(E - E_0 - \varepsilon)d\varepsilon} \tag{16.22}$$

where ρ_1 and ρ_2 are the densities of states of ionic and neutral fragments, respectively, and E_0 is the critical energy for reaction.

16.5.2. Examples

The modeling approach outlined earlier was initially tested by studying fragmentation of polyatomic molecules, for which fragmentation energetics are well established by conventional photoionization techniques, such as using bromobenzene and 1-bromonaphthalene cation radicals (Laskin et al. 2000a; Laskin and Futrell 2000). For these two systems the critical energies obtained from the modeling were in excellent agreement with literature values. Furthermore, the uncertainty in critical energies obtained in these studies was comparable to the uncertainties commonly obtained using photoionization or photodissociation techniques—the best-of-class well-characterized methods for studying the energetics of ionic fragmentation. After the initial testing, we have successfully applied this approach to the fragmentation of large gas-phase ions.

16.5.2.1. ***Small Polyalanines.*** Dissociation parameters for different dissociation pathways of small alanine-containing peptides (Laskin et al. 2000b, 2002a) were determined from modeling of FTICR SID and multiple-collision activation (MCA-CID) data. The experiments were performed at fixed reaction delay of 1 s. Major dissociation pathways for protonated trialanine ($AAAH^+$) determined using double-resonance experiments are summarized in Figure 16.5. Three primary reaction channels resulting in formation of the $\mathbf{b_2}$, $\mathbf{y_2}$, and $\mathbf{y_1}$ ions were taken into account in the modeling. The fourth pathway, resulting in formation of the immonium ion ($\mathbf{a_1}$) directly from $(AAA)H^+$, could not be ruled out on the basis of double-resonance ejection. However, RRKM modeling demonstrated that this reaction pathway does not compete with formation of the $\mathbf{b_2}$ and $\mathbf{y_2}$ ions from the protonated precursor ion.

Experimental SID data are compared with modeling results in Figure 16.6, while Table 16.2 summarizes activation parameters for the three primary reaction channels for $(AAA)H^+$ obtained from the modeling of MCA-CID and SID experiments. Although there is good agreement between the critical energies for formation of the

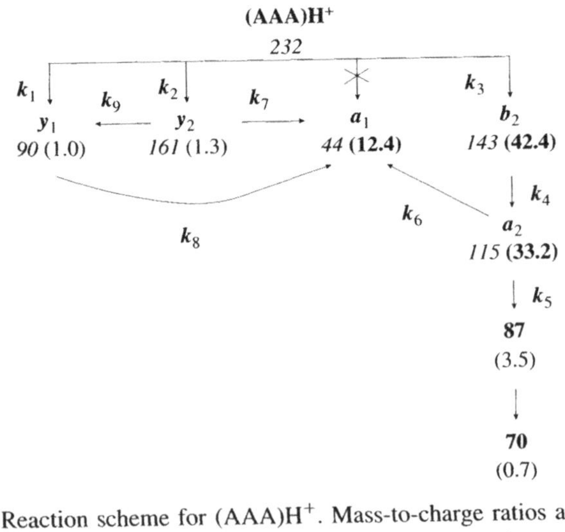

FIGURE 16.5. Reaction scheme for $(AAA)H^+$. Mass-to-charge ratios are shown in italics and relative abundances of fragment ions in parentheses (the abundances are obtained from the high-pressure MCA-CID experiment at 3 eV center-of-mass energy). [Reproduced from Laskin et al. (2002a) with permission from Elsevier Science.]

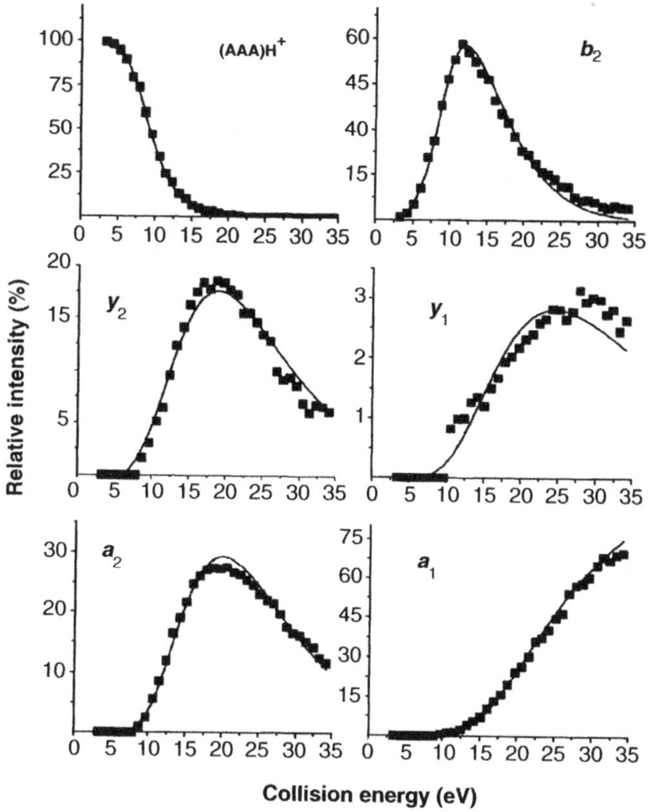

FIGURE 16.6. Experimental and calculated SID fragmentation efficiency curves for the parent and major fragment ions of $(AAA)H^+$. [Reproduced from Laskin et al. (2002a) with permission from Elsevier Science.]

b_2 ion from the protonated peptide [reaction (3)], critical energies for reactions (1) and (2) from MCA-CID experiments are strongly overestimated.

These differences can be readily rationalized using the hypothetical potential energy surface for the formation of two fragments from the parent ion as shown in

TABLE 16.2. RRKM/QET Parameters for Primary Reaction Channels of $(AAA)H^{+}$[a]

Experiment	Reaction	$MH^+ \rightarrow y_1$	$MH^+ \rightarrow y_2$	$MH^+ \rightarrow b_2$
MCA-CID	E_0 (eV); $\Delta S^{\ddagger}$ (eu)	1.72 (0.04); 4.5 (0.5)	1.86 (0.08); 11.9 (1.5)	1.42 (0.04); 4.8 (0.5)
SID	E_0 (eV); $\Delta S^{\ddagger}$ (eu)	1.59 (0.07); 2.5 (1.5)	1.67 (0.05); 12.2 (1.5)	1.46 (0.05); 6.3 (0.8)

[a]Entropy unit (eu) = cal mol^{-1} K^{-1}. Uncertainties are shown in parentheses.

Source: Data adopted from Laskin et al. (2002a).

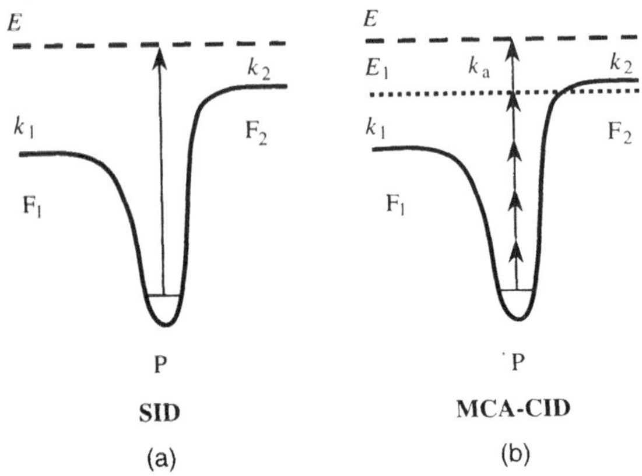

FIGURE 16.7. Schematic representation of the potential energy surface for the formation of F_1 and F_2 from the precursor ion P, activated by ion–surface impact (a) and multiple collisions (b). [Reproduced from Laskin et al. (2001) with permission of the American Chemical Society.]

Figure 16.7. Ion activation by collisions with a surface (a) occurs on a picosecond timescale, which is much faster than dissociation. In contrast, in MCA-CID (b) the internal energy of the ion increases slowly in a stepwise manner. If the molecular ion is activated by collisions above the lowest fragmentation threshold (internal energy E_1), it can either be further activated or decompose to yield F_1. If the rate constant for ion activation k_a is smaller than the dissociation rate constant k_1, then dissociation of ions with internal energy E_1 competes efficiently with ion activation. Consequently, the relative abundance of F_2 ion will be strongly depleted in MCA-CID (or in other slow activation methods such as multiphoton excitation) relative to single-step excitation processes such as SID. The effect is more pronounced when there is a significant difference in threshold energies for the two competing pathways. For example, differences in dissociation thresholds for the primary channels of trialanine shown in Table 16.2 suggest some discrimination in the production of the $\mathbf{y}_1$ ion in MCA-CID and a significant discrimination against the $\mathbf{y}_2$ ion as compared to the single-step activation process.

It is apparent that dissociation parameters obtained for $(AAA)H^+$ from MCA-CID curves using our modeling approach are not reliable. The proposed modeling is based on the assumption that the internal energy is instantly deposited in the parent ion in both multiple-collision activation and surface-induced dissociation. However, this type of modeling is not applicable in the case when ion activation and dissociation are competing. Comparison between the SID and MCA-CID results suggests that the formation of $\mathbf{y}_2$ and $\mathbf{y}_1$ ions from $(AAA)H^+$ competes with the parent ion excitation. From the discussion above it follows that reliable information on the fragmentation energetics of peptide ions can be obtained only from SID data.

Finally, it should be noted that for larger ions, dissociation schemes become increasingly complex, which tremendously complicates the analysis of individual FECs. For this reason, dissociation parameters for larger peptides discussed in the next section have been determined by modeling survival curves (SCs) of the precursor ion.

16.5.2.2. Larger Peptides. In Section 16.3.3 we mentioned that modeling photoionization experiments conducted at different reaction times provides a unique pair of dissociation parameters (E_0 and $\Delta S^\ddagger$). It follows that conducting SID experiments at different reaction times is a reliable method for reducing the degeneracy of the model toward E_0 and $\Delta S^\ddagger$. Figure 16.8a shows the evolution of SCs of singly protonated RVYIHAF as a function of the reaction delay. Time dependence of the value of $E_{50\%}$, namely, the collision energy, at which fragmentation efficiency equals 50%, is shown in Figure 16.8b. The dramatic decrease in $E_{50\%}$ from short to long times is followed by the gradual leveling off, which results from competition between dissociation and radiative cooling of excited precursor ions. For radiative cooling to compete efficiently with dissociation on this timescale, the radiative rate should be of the order of 10–20 s^{-1}, which is confirmed by the modeling (Laskin et al. 2003; 2004; Bailey et al. 2003).

Table 16.3 summarizes the dissociation parameters obtained for three groups of peptides studied in our laboratory. Listed in the table are the number of vibrational degrees of freedom (DOF), threshold energies (E_0), reaction entropies ($\Delta S^\ddagger$), and preexponential factors at 450 K (A). All threshold energies shown in Table 16.3 are in the range from 1 to 1.62 eV, while the preexponential factors vary by six orders of magnitude (Laskin 2004). Comparison between the dissociation energetics of peptides demonstrates that lower threshold energies and smaller preexponential factors are obtained when a peptide ion fragments selectively. For example,

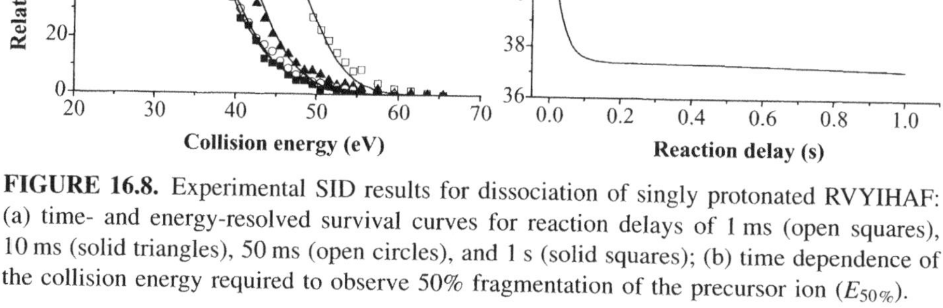

FIGURE 16.8. Experimental SID results for dissociation of singly protonated RVYIHAF: (a) time- and energy-resolved survival curves for reaction delays of 1 ms (open squares), 10 ms (solid triangles), 50 ms (open circles), and 1 s (solid squares); (b) time dependence of the collision energy required to observe 50% fragmentation of the precursor ion ($E_{50\%}$).

TABLE 16.3. Mass-to-Charge Ratio of Ion (m/z), Number of Degrees of Freedom (DOF), Dissociation Parameters ($E_0, \Delta S^{\ddagger}$), Kinetic Shifts (KS), and Values of $E_{50\%}$ at 1 s Reaction Time and Average Energy Transfer Efficiencies ($\%T\text{–}V$)

	m/z	DOF	E_0, (eV)	$\Delta S^{\ddagger}$ (eu)[a]	A (s^{-1})[b]	KS (eV)	$E_{50\%}$ (eV)	$\%T\text{–}V$
RPPGFSPF	904	378	1.09	−23.3	2×10^{8}	3.6	23.5	19.5
PPGFSPFR	904	378	1.17	−22.2	4×10^{8}	4.3	27.5	18.8
LDIFSDF	856	351	1.20	−0.4	2×10^{13}	2.6	16.1	21.5
LDIFSDFR	1012	420	1.24	−7.9	5×10^{11}	4.2	24.6	19.8
RLDIFSDF	1012	420	1.31	−8.6	3×10^{11}	4.8	27.3	20.1
LEIFSEFR	1040	438	1.33	−12.5	5×10^{10}	5.3	32.6	20.2
RVYIHAF	905	384	1.26	−6.8	8×10^{11}	5.9	36.8	19.8
RVYIHPF	931	396	1.19	−11.5	8×10^{10}	6.3	35.5	21.9
DRVYIHPF	1046	435	1.00	−28.8	1×10^{7}	5.0	36.5	20.5
RVYIHDF	950	393	1.01	−23.2	2×10^{8}	4.2	30.2	21.2

[a]Entropy unit (eu) = cal mol^{-1} K^{-1}.
[b]A is the preexponential factor at 450 K calculated using Eq. (16.11).

des-Arg[9]– and des-Arg[1]–bradykinin have only a few primary reaction fragments, and peptide ions containing both arginine and aspartic acid residues undergo specific fragmentation C-terminal to aspartic acid (Yu et al. 1993; Qin and Chait 1995; Summerfield et al. 1997; Tsaprailis et al. 1999).

For the series of peptides with the LDIFSDF motif, we found that addition of arginine residue to peptides containing aspartic acid results in a very small increase in dissociation threshold while the dynamics of dissociation is affected dramatically. The A factors for nonselective fragmentation are two orders of magnitude higher than the A factors characteristic of selective cleavages. A similar result is obtained by adding an acidic residue (D) to an arginine-containing peptide (RVYIHPF or RVYIHAF). This indicates that selective cleavages either are associated with substantial rearrangements or require a very specific conformation in the transition state in order to undergo dissociation.

Interestingly, the position of the basic residue (R) in a peptide sequence has a measurable effect on the dissociation threshold. The dissociation threshold for LDIFSDF**R** is 0.07 eV lower than the dissociation threshold **R**LDIFSDF. Replacing aspartic acid (D) with a glutamic acid residue (E) results in a 0.09 eV increase in dissociation threshold. This is correlated to the difference in acidities of the corresponding sidechains. However, the position of the aspartic acid residue in DRVYIHPF and RVYIHDF when R remains close to the N-terminus affects only the reaction entropy. These observations provide important insights on mechanisms of peptide fragmentation that cannot be deduced from fragmentation spectra alone.

Finally, we note that entropy plays a significant role in peptide fragmentation. For example, transition from selective to nonselective fragmentation in peptides with the LDIFSDF motif is associated with minor changes in the dissociation threshold but dramatic changes in activation entropy. Comparison of dissociation thresholds obtained in our studies with thermal data also revealed that activation entropy strongly influences the difference between the Arrhenius energy and threshold energy.

16.6. BLACKBODY INFRARED RADIATIVE DISSOCIATION (BIRD)

16.6.1. General Background

In Section 16.2.2 we presented basic equations describing thermal kinetics in the rapid exchange (REX) limit. We mentioned that the REX limit is achieved when the rate of energy exchange with the surroundings is much greater than the dissociation rate. Traditionally, thermal kinetics of both ions and neutral molecules has been studied under high-pressure conditions. It was found that the high-pressure rate constant does not depend on pressure. In contrast, in the low-pressure limit, when activation becomes the rate-limiting step, the dissociation rate constant is proportional to pressure. This behavior was rationalized using the Lindemann–Hinshelwood mechanism (Gilbert and Smith 1990; Baer and Hase 1996), which assumes that in thermal systems molecules are excited by nonreactive collisions with the background gas and that reaction and collisional activation are well separated in time.

The Lindemann–Hinshelwood mechanism predicts that the rate constant extrapolates to zero at zero pressure. However, in 1994 McMahon and coworkers showed that at pressures below 10^{-8} Torr dissociation rates of weakly bound cluster ions stored in a FTICR mass spectrometer were independent of pressure (Tholmann et al. 1994; Tonner et al. 1995). It was suggested that stored ions continuously exchange energy with surroundings by absorption and emission of infrared photons from the photon flux generated by the vacuum chamber walls. Interaction with blackbody radiation heats the ions to the temperature of the surrounding walls and results in spontaneous dissociation of ions in the absence of activating collisions (Dunbar and McMahon 1997).

Figure 16.9 shows a schematic view of the dependence of the dissociation rate constant on pressure. This dependence can be rationalized by assuming that ions are excited/deexcited both by collisions with background gas (M) and by photon exchange with the surroundings [Eq. (16.23)]. The resulting unimolecular rate constant (k_{uni}) is given by Eq. (16.24)

$$AB^+ + M/h\nu \underset{k_{-1,c}/k_{-1,\text{rad}}}{\overset{k_{1,c}/k_{1,\text{rad}}}{\rightleftarrows}} AB^{+\cdot} \overset{k_d}{\Rightarrow} A^+ + B \tag{16.23}$$

$$k_{\text{uni}} = k_d\left(\frac{k_{1,c}[M] + k_{1,\text{rad}}}{k_{-1,c}[M] + k_{-1,\text{rad}} + k_d}\right) \tag{16.24}$$

where $k_{1,c}$ and $k_{-1,c}$ represent rate constants for excitation and deexcitation by collisions, $k_{1,\text{rad}}$ and $k_{-1,\text{rad}}$ are rate constants for excitation and deexcitation by absorption and emission of IR photons, and k_d is the dissociation rate constant. At

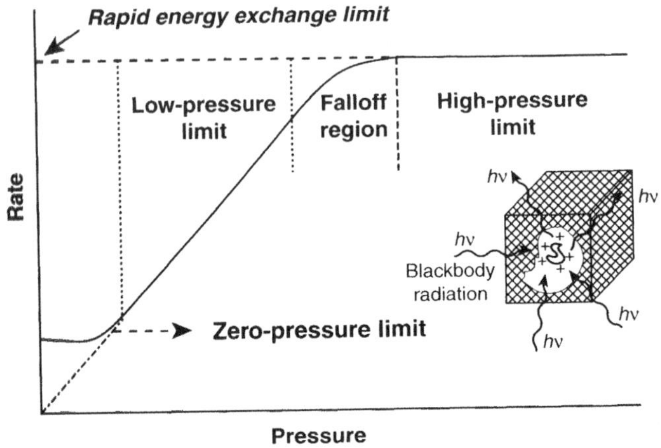

FIGURE 16.9. Modified Lindemann–Hinshelwood plot showing effects of pressure on unimolecular dissociation rates of small molecules (solid line). Classically, the rate constant in the low-pressure limit extrapolates linearly to zero in a collisionless environment. Recent experiments show that at low pressure ($<10^{-8}$ Torr), the unimolecular rate constant is nonzero and independent of pressure, and that molecular activation occurs by absorption of blackbody radiation. The rapid energy exchange (REX) limit for larger molecules is indicated by the dashed line. [Reproduced from Price et al. (1996b) with permission of the American Chemical Society.]

sufficiently high pressures when $k_{1,c}[M] \gg k_{1,rad}$ and $k_{-1,c}[M] \gg k_{-1,rad}$, Eq. (16.24) can be simplified to give the standard Lindemann–Hinshelwood expression (Gilbert and Smith 1990; Baer and Hase 1996). However, at low pressures activation/deactivation by collisions becomes negligibly slow and can be ignored. Equation (16.24) clearly shows that in this regime the experimentally measured rate constant k_{uni} is independent of pressure and hence cannot be extrapolated to zero.

These findings provided a foundation for a new technique called *blackbody infrared radiative dissociation* (BIRD), which has been successfully used for studying the energetics of fragmentation of both small and large molecules. For additional information, readers are referred to more detailed reviews on this topic (Dunbar 1996, 2003, 2004; Price et al. 1996a; Dunbar and McMahon 1997; Lifshitz 2001).

16.6.2. Advantages for Large Molecules

In 1996 Williams and coworkers showed that under readily achievable experimental conditions, activation of large ions by blackbody photons is significantly faster than the dissociation processes (Price et al. 1996b). As a result, ions equilibrate with the blackbody radiation field inside the vacuum chamber, and the internal energy distribution of a population of ions is given by a Boltzmann distribution corresponding to the temperature of the vacuum chamber. Because ions are in the REX limit Arrhenius parameters for dissociation of large ion can be readily determined from the

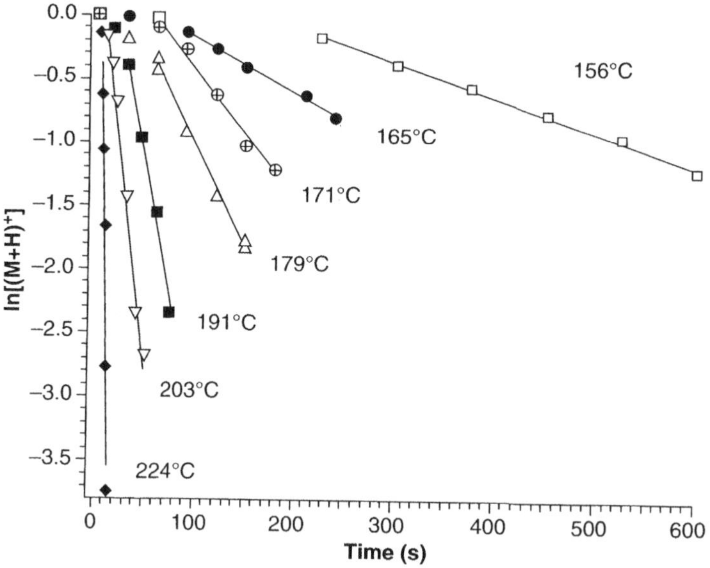

FIGURE 16.10. Data for the dissociation of bradykinin fit to unimolecular kinetics at the temperatures indicated. The rate constants in order of increasing temperature are 0.0027, 0.0042, 0.013, 0.016, 0.043, 0.068, and 0.32 s^{-1}. [Reproduced from Schnier et al. (1996) with permission of the American Chemical Society.]

temperature dependence of the dissociation rate constant. This allows quantitative studies of thermal kinetics of dissociation of large ions in the absence of collisions.

In BIRD experiments large ions produced using an electrospray source are transferred into the ICR cell that is equilibrated to the temperature of the experiment. The relative abundance of the precursor ion is recorded at different reaction times. This experiment is repeated at several temperatures. An example of experimental data obtained in BIRD experiments is shown in Figure 16.10. Linear plots obtained at different temperatures indicate that the disappearance of the precursor ion follows the first-order kinetics. Rate constants extracted from the slopes of kinetic plots are used to construct Arrhenius plots [$\log k(T)$ vs. $1/T$] as shown in Figure 16.11, from which Arrhenius parameters (E_a and A) are obtained. In this example a family of a bradykinin peptide and its analogs was studied using BIRD. Differences in slopes and intercepts of Arrhenius plots shown in Figure 16.11 indicate differences in the energetics and dynamics of dissociation of these ions.

16.6.3. Examples: Peptides and Proteins

BIRD has been extensively used to study the energetics and dynamics of dissociation of large molecules in the REX limit. Williams and coworkers studied dissociation of a number of peptides (Schnier et al. 1996, 1997), proteins (Jockusch et al. 1997; Gross et al. 1997) and oligonucleotides (Schnier et al. 1998). Table 16.4 lists

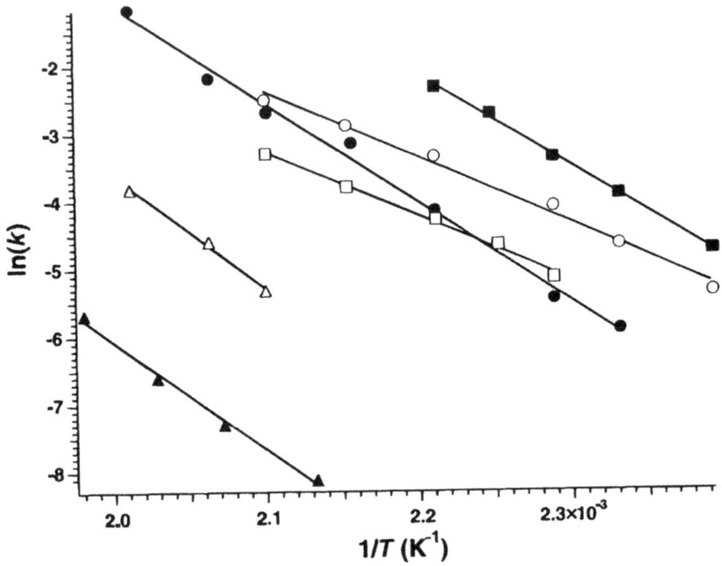

FIGURE 16.11. Arrhenius plot for the dissociation of singly protonated bradykinin (•), des-Arg[1]–bradykinin (□), des-Arg[9]–bradykinin (■), methyl ester of des-Arg[9]–bradykinin (▲), Lys-bradykinin (△), and doubly protonated bradykinin (○). [Reproduced from Schnier et al. (1996) with permission of the American Chemical Society.]

TABLE 16.4. Measured Zero-Pressure Arrhenius Activation Parameters for Dissociation of Singly and Doubly Protonated Bradykinin and Its Variants[a]

	Singly Protonated		Doubly Protonated	
Peptide	E_a (eV)	A (s^{-1})	E_a (eV)	A (s^{-1})
[Ala[6]]–bradykinin	1.1	10^{11}	0.74	10^{7}
Bradykinin	1.3	10^{12}	0.84	10^{8}
des-Arg[1]–bradykinin	0.82	10^{7}	—	—
des-Arg[9]–bradykinin*	1.2	10^{12}	—	—
des-Pro[2]–bradykinin	1.1	10^{11}	0.77	10^{7}
[Lys[1]]–bradykinin	1.0	10^{9}	—	—
Lys–bradykinin	1.4	10^{12}	1.2	10^{10}
[Thr[6]]–bradykinin*	1.2	10^{12}	—	—
Methyl ester of [Ala[6]]–bradykinin	0.76	10^{7}	—	—
Methyl ester of bradykinin	0.61	10^{5}	0.82	10^{6}
Methyl ester of des-Arg[1]–bradykinin	0.94	10^{7}	—	—
Methyl ester of des-Arg[9]–bradykinin	1.3	10^{9}	—	—

[a]Activation energies have standard deviations between 0.03 and 0.1 eV. Those processes for which calculations indicate the measured Arrhenius parameters are lower than the "high-pressure" limit values, indicated by an asterisk.

Source: Data reproduced from Schnier et al. (1996) with permission of the American Chemical Society.

Arrhenius parameters reported for bradykinin and its analogs. Activation energies are in the range from 0.6 to 1.3 eV, while preexponential factors range from 10^5 to $10^{12}\,s^{-1}$. Very low preexponential factors obtained for some peptide ions are quite unusual and have not been reported for smaller systems, for which typical preexponential factors are in the range 10^{10}–$10^{16}\,s^{-1}$. Low preexponential factors reported for peptide ions indicate that peptide fragmentation requires significant rearrangement of the precursor ion.

Dissociation parameters of variants of bradykinin along with fragmentation pathways were used to evaluate the effect of the structure and the charge state of the ion on the energetics and dynamics of dissociation (Schnier et al. 1996). It was found that modification of the residues between the two terminal arginines had only a minor effect on the dissociation parameters with average dissociation energy for the singly protonated ions of 1.2 and average preexponential factor of $10^{12}\,s^{-1}$. In contrast, methylation of the carboxylic acid group of the *C*-terminus reduces the E_a of bradykinin from 1.3 to 0.6 eV and the *A* factor from 10^{12} to $10^5\,s^{-1}$. In addition, this modification quenches the loss of NH_3 from bradykinin—one of the major dissociation channels. The small preexponential factor indicates that the methyl ester dissociates through a very "tight" transition state. Similar results are observed for [Ala^6]-bradykinin. It was proposed that the most stable form of singly protonated bradykinin is a salt bridge structure, in which both arginine residues are protonated and interact with a negatively charged carboxylate of the *C*-terminus. Methylation eliminates the saltbridge form of the parent ion and destabilizes it relative to the transition state for dissociation.

Interestingly, methylation of the *C*-terminus has no effect on dissociation parameters of doubly protonated bradykinin. The doubly protonated methyl ester of bradykinin has an E_a of 0.82 eV, comparable to the value of 0.84 eV for bradykinin itself. However, this value is 0.2 eV higher that the activation energy for dissociation of singly protonated methyl ester of bradykinin, indicating that the Coulomb repulsion is not the most significant factor in the activation energy of this ion. The similar activation parameters for doubly protonated bradykinin and its methylated counterpart indicate that methylation does not perturb the structure of the doubly protonated ion as significantly as it does for the singly protonated ion.

Similarly to singly protonated precursors, replacement of a single amino acid residue between the terminal arginines had only a minor effect on dissociation parameters. However, it was found that for these ions the presence of a second charge reduces the activation energy by 0.4 eV and decreases the *A* value by a factor of 10^4. Lower activation energy could be attributed to destabilization of the peptide as a result of Coulomb repulsion between the charges, while the substantial reduction in *A* factors for doubly protonated precursors is more difficult to rationalize. Interestingly, the addition of a basic residue to the *N*-terminus of the peptide (Lys-bradykinin) stabilizes both the singly and the doubly protonated ion with respect to dissociation. Williams and co-workers suggested that addition of the *N*-terminal lysine residue helps solvate the charge on the *N*-terminus more effectively, thereby reducing its influence on the ion dissociation.

Klassen and coworkers studied thermal dissociation of a number of large proteins and protein–ligand complexes (Felitsyn et al. 2001; Kitova et al. 2002a,b; Wang et al. 2003). Table 16.5 shows experimental Arrhenius parameters for dissociation of different charge states of the homopentamer (B_5) of the lethal Shigalike toxin I, B_5 bound to a decavalent P*k*- based oligosaccharide ligand (STARFISH, S)—the analog of the natural trisaccharide receptor globotriaoside. In this example, Arrhenius activation energies range from 32.4 to 80.4 kcal/mol (1.4–3.5 eV) and preexponential factors range from $10^{16.2}$ to $10^{39.2}$ s^{-1}. Observation of unusually high preexponential factors has become a subject of active discussion. It has been suggested that softening of numerous vibrational modes during dissociation of protein–ligand complexes is responsible for large preexponential factors (Felitsyn et al. 2001). It has been also noticed that experimental Arrhenius parameters

TABLE 16.5. Experimentally Measured Arrhenius Parameters (E_a and log A) for Loss of a Subunit From B_5^{n+} and B_5S^{14+} [a]

Parent ion	Products	log(A)	E_a	T(K)	ΔE_{corr}	E_0	$\Delta S^‡$
B_5^{14+}	B^{6+}/B_4^{8+}, B^{5+}/B_4^{9+}, B^{4+}/B_4^{10+}	18.5 ± 0.4	35.4 ± 0.9	413	7.1	28.3 ± 1.3	23.5 ± 0.5
	B^{6+}/B_4^{8+}	19.7 ± 0.6	38.4 ± 1.1	413	8.8	29.6 ± 1.8	29.0 ± 0.9
	B^{5+}/B_4^{9+}	17.3 ± 0.5	34.0 ± 0.9	413	5.5	28.5±1.6	18.0 ± 0.5
	B^{4+}/B_4^{10+}	16.2 ± 0.6	32.4 ± 0.9	413	4.0	28.4 ± 1.8	12.9 ± 0.5
B_5^{13+}	B^{6+}/B_4^{7+}, B^{5+}/B_4^{8+}, B^{4+}/B_4^{9+}	23.7 ± 0.6	44.9 ± 2.0	423	14.9	30.0 ± 2.1	47.3 ± 1.2
B_5^{13+}	B^{6+}/B_4^{7+}, B_5^{+}/B_4^{8+}, B^{4+}/B_4^{9+}	23.0 ± 1.1	46.2 ± 1.1	428	14.3	31.9 ± 2.3	44.1 ± 2.1
	B^{5+}/B_4^{8+}	24.8 ± 0.9	48.6 ± 1.8	428	16.8	31.8 ± 2.4	52.3 ± 1.9
	B^{4+}/B_4^{9+}	23.0 ± 0.8	45.6 ± 1.6	428	14.3	31.3 ± 2.2	44.1 ± 1.5
B_5^{12+}	B^{5+}/B_4^{7+}, B^{4+}/B_4^{8+}	26.2 ± 0.6	51.8 ± 1.1	433	19.3	32.5 ± 1.5	58.7 ± 1.3
	B^{5+}/B_4^{7+}	25.7 ± 0.5	51.6 ± 1.0	433	18.6	33.0 ± 1.3	56.4 ± 1.1
	B^{4+}/B_4^{8+}	26.1 ± 0.6	52.0 ± 1.2	433	19.1	32.9 ± 1.5	58.2 ± 1.3
B_5^{11+a}	B^{4+}/B_4^{7+}	39.2 ± 1.2	80.4 ± 2.3	463	40.7	39.7 ± 2.6	118.2 ± 3.6
B_5S^{14+}	B^{6+}/B_4S^{8+}	30.2 ± 1.0	58.6 ± 1.8	423	23.5	35.1 ± 2.3	77.0 ± 2.5
	B^{5+}/B_4S^{9+}, B^{4+}/B_4S^{10+}	28.8 ± 1.2	56.9 ± 2.3	423	21.7	35.2 ± 2.9	70.6 ± 2.9
	B^{6+}/B_4S^{8+}	30.4 ± 1.1	59.3 ± 2.0	423	23.8	35.5 ± 2.5	77.9 ± 2.8

[a] E_a, ΔE_{corr}, E_0 are in kcal/mol; $\Delta S^‡$ is in eu; A is in s^{-1}.

Sources: Table reproduced from Felitsyn et al. (2001) with permission of the American Chemical Society; calculated Tolman corrections (ΔE_{corr}), threshold energies (E_0) and activation entropy ($\Delta S^‡$) at 415 K reproduced from Laskin and Futrell (2003) with permission of the American Chemical Society.

obtained for the B_5 and its complex are highly correlated; this has been attributed to the limited kinetic window (1–600 s) characteristic of BIRD experiments, which naturally limits the range of measurable rate constants. Arrhenius activation energies for dissociation of proteins and their complexes with various ligands were used to discuss the effect of the local protein environment and the structure of the ligand on the protein stability and the strength of protein–ligand interactions.

16.6.4. Interpretation of Arrhenius Parameters Using Tolman's Theorem

In Section 16.2.2 we discussed the meaning of Arrhenius parameters. According to the Tolman's theorem, the Arrhenius activation energy is given by the difference between the average energy of the TS and the average energy of the ensemble of excited molecules, $P(E)$. Because the latter is not affected by the details of the potential energy surface in the TS, it is interesting to examine the variation in the internal energy distribution of the TS of varying degree of tightness or looseness. This functional form of the energy distribution of the TS is given by $k(E)P(E)$ and can be calculated in a straightforward manner. Figure 16.12 shows the result of such a calculation for three hypothetical dissociation reactions of a peptide ion with 400 vibrational degrees of freedom with activation entropies of −37, 0, and 76 eu respectively. Because the spacing between the vibrational levels in the tight TS ($\Delta S^{\ddagger} = -37$ eu) is smaller than in the ground state, the corresponding distribution is shifted significantly (by close to 1 eV) toward lower energies as compared to the TS distribution with zero entropy effect. The increased spacing between the vibrational levels in the system that fragments via loose TS accounts for the significant shift of the corresponding distribution toward higher internal energies. Clearly, the difference between the average energy of the TS and the ensemble of excited molecules is a strong function of the degree of tightness or looseness of the TS.

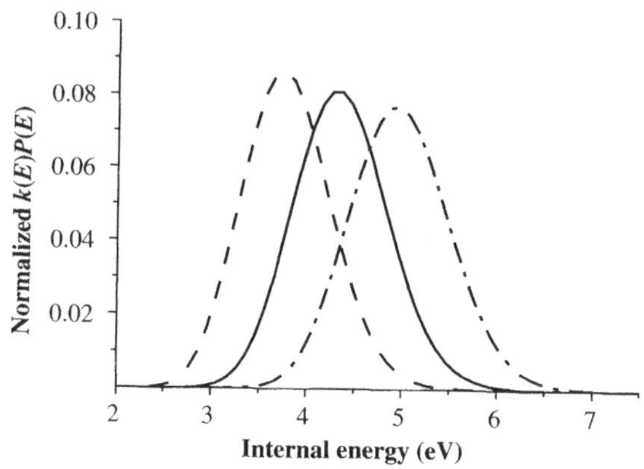

FIGURE 16.12. Normalized $k(E)P(E)$ distributions at 450 K as a function of the internal energy of the precursor ion E. Solid line corresponds to $\Delta S^{\ddagger} = 0$; dashed line corresponds to $\Delta S^{\ddagger} = -37$ eu ($A = 10^5\,\text{s}^{-1}$); dashed–dotted line corresponds to $\Delta S^{\ddagger} = 76$ eu ($A = 10^{30}\,\text{s}^{-1}$).

For thermal kinetics in the REX limit the modified Tolman equation [Eq. (16.8a)] can be rewritten in the following form

$$E_a = E_0 + \Delta E_{corr} + k_B T \tag{16.25a}$$

where ΔE_{corr} is the Tolman correction factor, given by

$$\Delta E_{corr} = k_B T^2 \frac{\partial \ln Q^\ddagger/Q}{\partial T} \tag{16.25b}$$

and $Q^\ddagger$ and Q are partition functions of the transition state and the excited molecule, respectively. The preexponential factor can be calculated using Eq. (16.11) from the entropy of activation given by Eq. (16.26):

$$\Delta S^\ddagger = k_B \ln \frac{Q^\ddagger}{Q} + k_B T \frac{d \ln Q^\ddagger/Q}{dT} \tag{16.26}$$

Clearly, the Tolman correction factor [Eq (16.25b)] is correlated with the activation entropy through the second term in Eq. (16.26).

Figure 16.13 shows the dependence of the two terms in Eq. (16.26) on log(A). Because the ratio of the partition functions of the ground and the transition states depends on the degree of tightness or looseness of the transition state, both terms increase with increase in log(A). Further, the derivative of the ratio of the partition functions shows a stronger—almost linear—dependence on log(A). It follows that

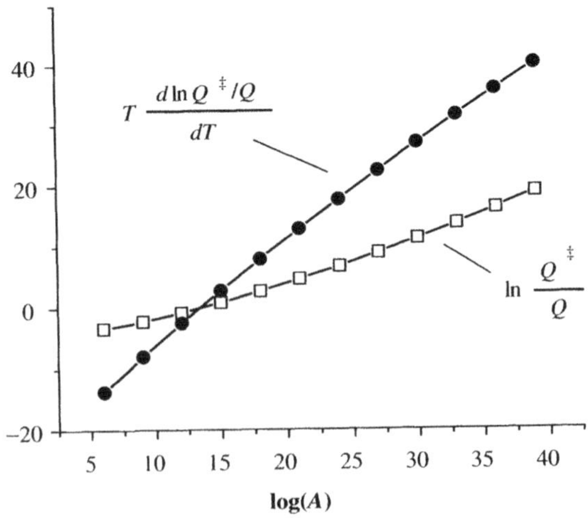

FIGURE 16.13. The dependence of the two contributions to the activation entropy [Eq. (16.26)] on the value of log(A).

Tolman's correction factor also increases almost linearly with $\log(A)$. If the spacing between vibrational levels in the TS is decreased relative to the reactant molecule (loose TS), $Q^\ddagger$ is larger than Q and ΔE_{corr} is positive. However, for reactions that proceed via tight TS (low preexponential factor, negative activation entropy), $Q^\ddagger$ is smaller than Q and ΔE_{corr} is negative.

These findings have an important implication on the relationship between the Arrhenius activation energy and the energy threshold for reaction summarized as follows:

$$\begin{array}{ll} \Delta S^\ddagger \ll 0 & E_a \ll E_0 \\ \Delta S^\ddagger = 0 & E_a \cong E_0 \\ \Delta S^\ddagger \gg 0 & E_a \gg E_0 \end{array} \qquad (16.27)$$

For reactions proceeding via loose TS, the Arrhenius activation energy is higher than the threshold energy for the reaction, while for reactions proceeding via a very tight transition state (low preexponential factors), the Arrhenius activation energy is lower than the threshold energy.

For reactions with $\Delta S^\ddagger = 0$, the difference between E_a and E_0 is on the order of $k_B T$. This difference is usually smaller than the experimental uncertainty and can be ignored in most cases. The correction factor is quite small for unimolecular reactions, for which preexponential factors are in the range 10^{10}–$10^{16}\ \text{s}^{-1}$. However, when dissociation is characterized by very low or very high preexponential factors, Tolman's correction becomes quite significant with the result that the Arrhenius activation energy is strongly correlated with the preexponential factor (Laskin et al. 2002c; Laskin and Futrell 2003). Figure 16.14a shows a plot of the Arrhenius activation energies for close to 60 dissociation reactions of proteins and protein–ligand complexes reported in the literature (Felitsyn et al. 2001; Kitova et al.

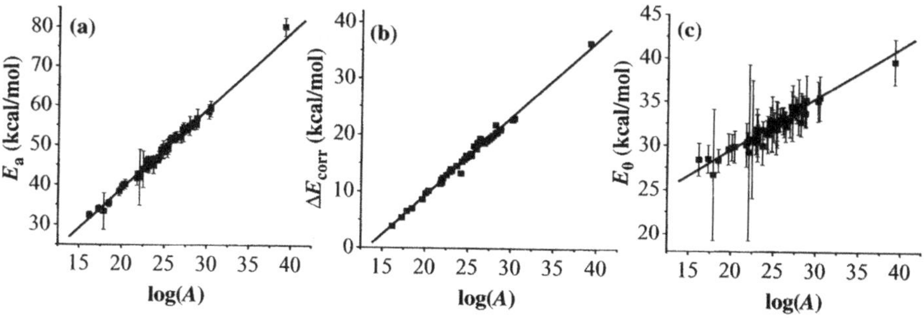

FIGURE 16.14. Arrhenius activation energies for dissociation of proteins and protein–ligand complexes (a), Tolman's correction factor (b), and threshold energy (c) as a function of $\log(A)$ for reaction. Lines are linear fits through data points. [Reproduced from Laskin and Futrell (2003) with permission of the American Chemical Society.]

2002a,b; Wang et al. 2003) as a function of $\log(A)$. There is a clear linear dependence of E_a on $\log(A)$ regardless of the identity of the proteins and protein–ligand complexes. The correlation between E_a and $\log(A)$ implies that the Arrhenius activation energy contains a significant entropic contribution and therefore does not necessarily reflect the thermochemical stability of the ion. Differences in thermochemical stabilities of ions can be accessed only from the differences in threshold energies for reactions.

The dependence of Tolman's correction factor on $\log(A)$ is plotted in Figure 16.14b, while Table 16.5 lists the calculated values of ΔE_{corr} at an average experimental temperature along with estimated threshold energies and activation entropies for a number of selected systems. As expected, the correction factor increases almost linearly with $\log(A)$. ΔE_{corr} ranges from 4 to 24 kcal/mol for most of the systems and reaches a value of 40.7 kcal/mol for the largest $\log(A)$ of 39.2. This demonstrates that the correction factor becomes very large for large values of the preexponential factor. Figure 16.14c shows the dependence of threshold energies on $\log(A)$. Although some correlation between $\log(A)$ and E_0 remains, most of the threshold energies are the same within the experimental error bars. It can therefore be concluded that dissociation of these proteins and protein–ligand complexes is characterized by the same threshold energy of 32.5 ± 2.5 kcal/mol. However, activation entropies for these reactions deduced from the corresponding preexponential factors are quite different, ranging from 13 eu for the lowest $\log(A)$ of 16.2 to 118 eu for the highest $\log(A)$ of 39.2. This implies that entropy is the major driving force for these reactions.

These results demonstrate that even when Arrhenius activation energies are correctly extracted from the data, they do not necessarily represent actual dissociation energies even in a relative sense. On fundamental grounds comparing threshold energies is the only way to address the relative stability of different gas-phase ions. Equations (16.11), (16.25b), and (16.26) can be used to deduce threshold energies and activation entropies from the experimental values of E_a and A.

16.6.5. Entropy–Enthalpy Compensation

The correlation between E_a and $\log(A)$ shown in Figure 16.14a is a direct indication of the well-known entropy–enthalpy compensation. This effect has been reported for many types of host–guest complexes and has been extensively discussed in the literature (Grunwald and Steel 1995; Exner 2000; Sharp 2001; Liu and Guo 2001; Cooper et al. 2001; Cornish-Bowden 2002; Houk et al. 2003). The term "enthalpy–entropy compensation" implies that highly endothermic reactions are kinetically favored and are characterized by a large positive entropy effect, while exothermic reactions are associated with a large adverse entropy effect. It was found that plots of reaction enthalpies versus entropies often give straight lines with excellent correlation coefficients. Moreover, in many cases the slope of such plots was close to the average temperature of the experiment. It can be argued that formation of a more stable complex (more negative ΔH) is associated with substantial rearrangement of the interacting molecules or the solvent, resulting in a more negative entropy effect. This explains the origin of the extrathermodynamic relationship between ΔH and ΔS. However, it is not clear why the relationship between ΔH and ΔS should be linear. Several explanations have been given in the literature for the origin of the linear dependence between the entropy and enthalpy of reaction. A detailed review on this topic and an exhaustive literature survey can be found elsewhere (Liu and Guo 2001). The present discussion is limited to a few simple examples that are more relevant to the gas-phase studies.

The simplest form of the compensation is obtained if for some reason the range of the experimentally measured values of ΔG is small. In this case, linear correlation of ΔH and ΔS with the slope close to the average experimental temperature follows immediately from the relationship $\Delta G = \Delta H - T\Delta S \approx$ Constant. Sharp (2001) analyzed a number of examples of such compensation using solution data for protein reactions. He concluded that in many cases linear plots of ΔH vs. ΔS simply manifest the similarity between the free energies of reactions. Analysis of gas-phase data shown in Figure 16.14a shows that the free energy of all protein dissociation reactions reported so far is 25.9 ± 0.8 kcal/mol. This explains the excellent correlation between E_a and $\log(A)$ found for these systems.

Krug et al. (1976a,b) presented a detailed analysis of propagation of errors and suggested that in many cases compensation plots arise from experimental errors. This analysis specifically focused on the thermochemical values derived from Van't-Hoff and Arrhenius plots. They found that for data points taken at n temperatures the correlation coefficient between ΔH and ΔS (or E_a and $\ln A$) is given by the following equation:

$$\rho = \frac{\sum 1/T}{\sqrt{n \sum (1/T)^2}} \tag{16.28}$$

Clearly, as the temperature range decreases, the correlation coefficient approaches unity. The statistical correlation can be circumvented by applying a simple transformation that makes the determination of the slope and the intercept of the Arrhenius or Van't-Hoff plots independent. This is achieved by centering the experimental data about the mean experimental temperature, T_C. Krug et al. demonstrated that the slope and the intercept of the following plot are not correlated ($\rho = 0$):

$$\ln k = \{\ln A - E_a 1/T_C/R\} - E_a/R\{1/T - 1/T_C\} \tag{16.29}$$

Sharp (2001) used the statistical analysis test suggested by Krug and coworkers to evaluate the significance of entropy–enthalpy compensation plots obtained for a number of proteins. He found that no statistical significance at the 95% confidence level could be attributed to the experimental entropy–enthalpy plots if their slopes are within the $\pm 2\sigma$ interval of the average experimental temperature T_C.

Cornish-Bowden (2002) presented an interesting numerical example of the statistical entropy–enthalpy compensation. He generated a random set of points of

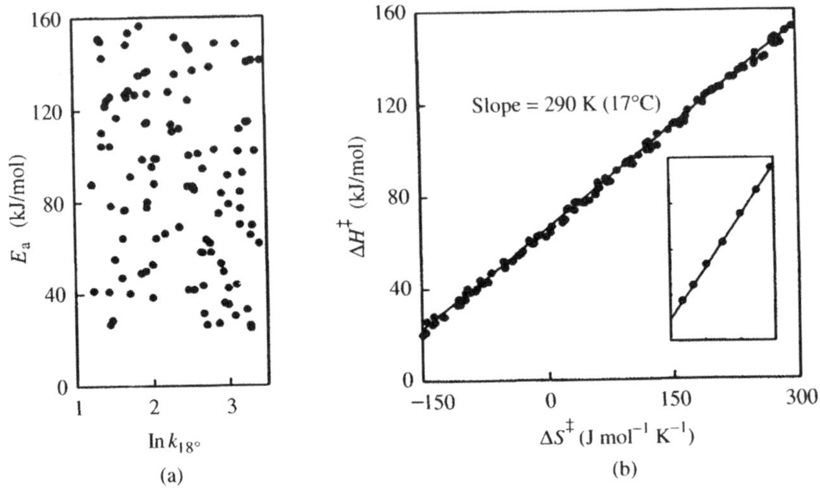

FIGURE 16.15. (a) A random set of 100 points ($k_{18°C} = 2.3 \pm 1.15$; $E_a = 30$–160 kJ/mol); (b) enthalpy–entropy compensation plot; the $\Delta H^{\ddagger}$ and $\Delta S^{\ddagger}$ values for each of the points are estimated according to the Arrhenius plot analysis of a corresponding point in (a). [Reproduced from Cornish-Bowden (2002) with permission from the Indian Academy of Science.]

E_a and the natural logarithm of the rate constant at 18°C (Figure 16.15a) and derived the corresponding values of ΔH and ΔS using Eqs. (16.8b) and (16.11). The entropy–enthalpy plot generated using this approach is shown in Figure 16.15b. Clearly, there is an excellent correlation between the values of ΔH and ΔS, and the slope of the plot is very close to the assumed experimental temperature. The origin of the correlation is illustrated in Figure 16.16. A small error in the slope of the Arrhenius plot (ε) is translated to an error of ε/T_C in the intercept. Assuming that

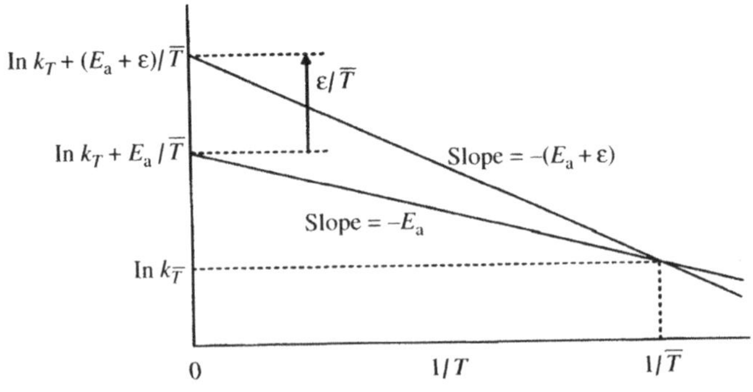

FIGURE 16.16. Origin of the correlation between entropy and enthalpy of activation. If the slope of an Arrhenius plot is decreased by an error ε while maintaining the ordinate value constant at a temperature $T = T_C$, then the rotation of the line will produce an error ε/T_C in the intercept. [Reproduced from Cornish-Bowden (2002) with permission from the Indian Academy of Science.]

variations in ΔS originate primarily from experimental errors, the slope of the plot of ΔH versus ΔS is close to T_C.

Another cause of the compensation is the temperature dependence of the thermodynamic functions:

$$\Delta H(T) = \Delta H^{\circ}_{0\,K} + \int_0^T \Delta C_P dT \tag{16.30a}$$

$$\Delta S(T) = \Delta S^{\circ}_{0\,K} + \int_0^T \frac{\Delta C_P}{T} dT \tag{16.30b}$$

where ΔC_P is the difference in heat capacities of the products and the reactants. Cooper et al. (2001) showed that if ΔC_P is independent of temperature and the range of experimental temperatures δT is small enough, Eqs. (16.30a) and (16.30b) can be simplified to give

$$\Delta H(T) = \Delta H(T_C) + T_C[\Delta S - \Delta S(T_C)] \tag{16.31}$$

which corresponds to a straight line with the slope of T_C.

Finally, it can be shown that a nearly linear increase in ΔH versus ΔS is predicted from Eqs. (16.30a) and (16.30b) when the temperature dependence of the heat capacity is taken into account. Indeed, for each vibrational degree of freedom enthalpy and entropy are given by the following expressions (Davidson 1962)

$$H(T) - H^{\circ}_{0\,K} = RT\frac{u}{e^u - 1} \tag{16.32a}$$

$$S(T) - S^{\circ}_{0\,K} = R\left[\frac{u}{e^u - 1} - \ln(1 - e^{-u})\right] \tag{16.32b}$$

where $u = h\nu/k_B T$. It can be shown that the two terms on the right-hand side of Eq. (16.32b) depend on temperature in a similar way. As a result, the plot of *H*/*RT* versus *S*/*R* follows an almost linear dependence. However, the slope of this plot is smaller than unity because of the contribution of the $\ln(1 - e^{-u})$ term. Figure 16.17a shows a compensation plot obtained for one harmonic oscillator with a characteristic frequency of 2000 cm^{-1}, while Figure 16.17b shows the corresponding plot for a hypothetical protein with 6300 vibrational degrees of freedom. Although in general the dependence of ΔH on ΔS is not linear, the deviation from linearity, which results from a contribution of low-frequency vibrational modes, is quite small and is not significant in a narrow range of temperatures sampled in a typical kinetic experiment.

16.7. COMPARISON WITH ENTROPY EFFECTS IN PROTEIN ASSOCIATION REACTIONS

Gas-phase studies on the energetics and dynamics of dissociation of peptides and proteins clearly demonstrated that reaction entropies vary over a very broad range

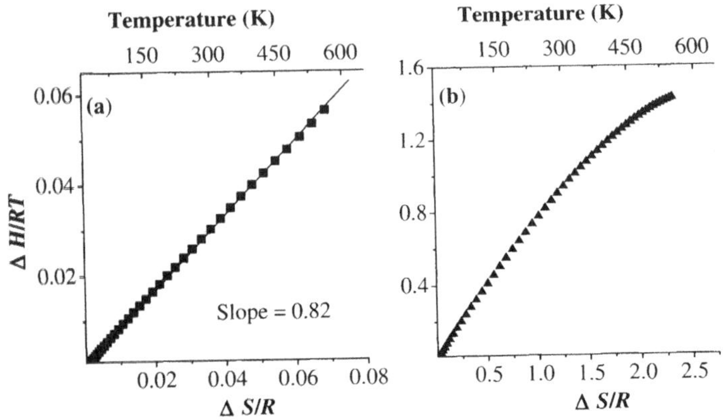

FIGURE 16.17. Correlation between vibrational entropy and enthalpy: (a) for one oscillator ($\nu = 2000\ \text{cm}^{-1}$); (b) for a hypothetical protein with 6300 vibrational degrees of freedom and the entropy of the reaction $\Delta S^{\ddagger} = 40$ eu.

from −40 to 120 eu (eu = cal/mol · K), while the threshold energies for dissociation typically fall into a relatively narrow range of 0.8–2 eV. The lower range of entropies is characteristic of peptide fragmentation, while dissociation of proteins and protein–ligand complexes in the gas phase is associated with a large positive entropy effect in excess of 10 eu [$\log(A) > 16$].

These entropy effects can be compared to the entropy cost of protein association—the opposite of the entropy effect for dissociation—that has been extensively discussed in the literature [see, e.g., Doty and Myers (1953), Finkelstein and Janin (1989), Searle and Williams (1992), Tidor and Karplus (1994), Tamura and Privalov (1997), Amzel (1997), Janin (1997), D'Aquino et al. (2000), Yu et al. (2001), and Siebert and Amzel (2004)]. Several contributions to entropy loss in protein association reactions have been considered. It has been shown that translational entropy loss is an important factor in these reactions. Indeed, loss of three translational degrees reduces the entropy of the system by 40–50 eu depending on the molecular weight of the system. Tidor and Karplus (1994) calculated 48 eu decrease in entropy for dimerization of insulin. Similar values were reported by Finkelstein and Janin (1989). Doty and Myers (1953) estimated the translational (77 eu) and rotational entropy loss (47 eu) for two rigid spheres of size and mass of insulin. However, much smaller entropy effects were obtained by these authors experimentally, suggesting that the dissociation of insulin dimer is very different from the simple breaking apart of two monomers. Although translation and overall rotation are important in association/dissociation reactions, translational and rotational entropy gain or loss varies very slowly with the molecular weight of the complex. It follows that similar translational and rotational entropy effects are expected for association/dissociation of small and large molecules.

Vibrational degrees of freedom of large molecules can contribute significantly to the net entropy effect. For example, loss of translational and rotational entropy in dimerization of insulin is partly compensated by increase in vibrational entropy of about 24 eu. (This increase is a result of the presence of six new vibrational degrees of freedom in the complex). As a result, the overall change of the external entropy (entropy not associated with hydration and conformational changes) is close to −67 eu. (Tidor and Karplus 1994). Because this estimation relies on the ideal-gas equations and does not take into account the effect of solvent, it should be relevant for comparison with the gas-phase dissociation studies discussed in this chapter. It should be noted that much smaller entropy effects have been reported for binding and folding of proteins in solution [see, e.g., Searle and Williams (1992), Tamura and Privalov (1997), and Neumann et al. (2002)].

Conformational entropy changes are by far the most difficult to evaluate because it requires calculation of the multidimensional energy profile of a molecule as a function of the angles of all rotatable bonds. D'Aquino et al. (2000) developed a toolbox for evaluation of conformational entropy changes for binding small organic molecules to macromolecular targets. Cheluvaraja and Meirovitch (2004) used a so-called complete hypothetical scanning Monte Carlo approach to calculate entropy and free-energy changes for transitions between the helix, extended and hairpin conformations of $(Gly)_{10}$ and $(Gly)_{16}$ in the absence of solvent. The largest entropy effect of 39 and 71 eu at 100 K for $(Gly)_{10}$ and $(Gly)_{16}$, respectively, was obtained for the extended-helix transition. This demonstrates that conformational entropy may contribute significantly to the total entropy effect in gas-phase fragmentation reactions.

Finally, it should be emphasized that the discussion above applies to the thermodynamic entropy of reaction, namely, the difference between the entropy of the products and reactants ΔS. However, entropy effects derived from Arrhenius plots reflect the difference between the entropy of the transition state (or activated complex) and the reactant $\Delta S^{\ddagger}$. Grabowy and Mayer (2004) demonstrated that these entropy effects can be quite different. They used variational transition state theory to determine activation entropies $\Delta S^{\ddagger}$ for the dissociation of a series of acetonitrile–alcohol dimers in the gas phase and compared them to thermodynamic entropies of reactions. The latter were calculated by accounting for the difference in translational, rotational and vibrational entropies of the dimer and of the two monomers. For all proton-bound dimers the thermodynamic entropy was higher than the activation entropy and changes in $\Delta S^{\ddagger}$ were not mirrored by the corresponding changes in the thermodynamic entropy. It follows that the activation entropies derived from gas-phase kinetic studies are determined by the interplay between the vibrational and conformational entropy effects, while the sum of translational and rotational entropy can be used as an upper limit for the activation entropy.

16.8. CONCLUDING REMARKS

Since the early 1990s tandem mass spectrometry has been successfully used to determine structures of peptides and proteins. The peptide sequence can be reconstructed from the MS/MS mass spectrum based on the fragments that are formed from the molecular ion. The type and amount of fragmentation in MS/MS spectra are determined by several factors, including the method of ion activation, the internal energy distribution of excited ions, the energetics and dynamics of fragmentation, and the instrumental time window. Understanding the energetics and mechanisms of peptide fragmentation is essential for the development of MS/MS approaches for sequencing and characterization of biomolecules. However, studying dissociation energetics of peptides and proteins is challenging because most of the well-developed experimental approaches that have been successfully employed in the studies of small and medium-size ions are simply not applicable to the fragmentation of large molecules.

Several approaches have been developed for studying the energetics and entropy effects in peptide and protein dissociation reactions. Experiments and simulations demonstrated that fragmentation of protonated peptides and proteins is characterized by relatively low dissociation thresholds in the range from 0.5 to 2 eV, with most values lying below 1.5 eV. In contrast, entropy effects for dissociation of large ions vary significantly. Unexpectedly high ($\leq 10^{39}\ s^{-1}$) and low ($10^4\ s^{-1}$) preexponential factors were obtained from Arrhenius plots for dissociation of peptides and proteins. Moreover, it was found that Arrhenius parameters for these systems are strongly correlated. This correlation is a direct consequence of the relative change in the spacing between vibrational levels of the reactant and the transition state for reaction. Converting E_a into the threshold energy E_0 using Tolman's theorem reveals the true magnitude of the correlation between molecular complexity and stability. However, the absolute value of the Tolman correction factor depends on the choice of transition modes that change in the course of the reaction. Because gas-phase unimolecular dissociation of molecules of this size is largely unexplored and nothing is known about the number of modes involved in the dissociation process, it is difficult to determine with confidence the absolute values of the threshold energies for dissociation from thermal kinetics data.

How do we describe the transition state for such complex dissociation processes? Can we obtain an accurate absolute value for the dissociation threshold using Tolman's correction? What can we learn about the reaction dynamics from experimentally measured entropy effects? These are some of the open questions for future studies.

ACKNOWLEDGMENTS

The author acknowledges with gratitude very helpful discussions with Jean Futrell on various topics presented in this chapter. The support from the Office of Basic Energy Sciences of the U.S. Department of Energy (project 40457) during the preparation of this chapter and during the course of research on activation and dissociation of large ions conducted by our group is gratefully acknowledged. Surface-induced dissociation studies described in this chapter were conducted at the W. R. Wiley Environmental Molecular Sciences Laboratory (EMSL), a national scientific user facility sponsored by the U.S. Department of Energy and located at Pacific Northwest National Laboratory. PNNL is operated by Battelle for the U.S. Department of Energy.

REFERENCES

Amzel LM (1997): Loss of translational entropy in binding, folding, and catalysis. *Proteins* **28**:144–149.

Armentrout PB (2001): Reactions and thermochemistry of small transition metal cluster ions. *Annu. Rev. Phys. Chem.* **52**:423–461.

Armentrout PB (2003a): Threshold collision-induced dissociations for the determination of accurate gas-phase binding energies and reaction barriers. *Top. Curr. Chem.* **225**:233–262.

Armentrout PB (2003b): Guided ion beam studies of transition metal-ligand thermochemistry. *Int. J. Mass Spectrom.* **227**:289–302.

Asano KG, Goeringer DE, McLuckey SA (1999): Thermal dissociation in the quadrupole ion trap: Ions derived from leucine enkephalino. *Int. J. Mass Spectrom.* **187**:207–219.

Baer T (1986): The dissociation dynamics of energy-selected ions. In Prigogine I, Rice SA (eds), *Advances in Chemical Physics*, Wiley, New York, pp. 111–202.

Baer T (2000): Ion dissociation dynamics and thermochemistry by photoelectron photoion coincidence (PEPICO) spectroscopy. *Int. J. Mass Spectrom.* **200**:443–457.

Baer T, Hase WL (1996): *Unimolecular Reaction Dynamics: Theory and Experiments*, Oxford Univ. Press, New York.

Bailey TH, Laskin J, Futrell JH (2003): Energetics of selective cleavage at acidic residues studied by time-and energy-resolved surface-induced dissociation in FT-ICR-MS. *Int. J. Mass Spectrom.* **222**:313–327.

Becker CH, Wu KJ (1995): On the photoionization of large molecules. *J. Am. Soc. Mass Spectrom.* **6**:883–888.

Bente PF, McLafferty FW, McAdoo DJ, Lifshitz C. (1975): Internal energy of product ions formed in mass-spectral reactions—degrees of freedom effect. *J. Phys. Chem.* **79**:713–721.

Berkowitz J (1979): *Photoabsorption, Photoionization and Photoelectron Spectroscopy*, Academic Press, New York.

Berkowitz J (2000): The quantum yield of ionization. *Phys. Essays* **13**:248–255.

Bernshtein V, Oref I (1994): Unimolecular dissociation of very large polyatomic molecules. *J. Phys. Chem.* **98**:136–140.

Cheluvaraja S, Meirovitch H (2004): Simulation method for calculating the entropy and free energy of peptides and proteins. *Proc. Natl. Acad. Sci.* **101**:9241–9246.

Chesnavich WJ, Bowers MT (1979): Statistical methods in reaction dynamics. In *USA* Bowers MT (ed.), *Gas Phase Ion Chemistry*, Vol. 1, Academic Press, New York, pp. 119–151.

Chupka WA (1959): Effect of unimolecular decay kinetics on the interpretation of appearance potentials. *J. Chem. Phys.* **30**:191–211.

Cooks RG, Ast T, Mabud A (1990): Collisions of polyatomic ions with surfaces. *Int. J. Mass Spectrom. Ion Proc.* **100**:209–265.

Cooper A, Johnson CM, Lakey JH, Nollmann M (2001): Heat does not come in different colours: Entropy-enthalpy compensation, free energy windows, quantum confinement, pressure perturbation calorimetry, solvation and the multiple causes of heat capacity effects in biomolecular interactions. *Biophys. Chem.* **93**:215–230.

Cornish-Bowden A (2002): Enthalpy-entropy compensation: A phantom phenomenon. *J. Biosci.* **27**:121–126.

Cui W, Hu Y, Lifshitz C (2002): Time resolved photo dissociation of small peptideions—combining laser desorptionwith ion trap/reflectron TOF mass spectrometry. *Eur. Phys. J. D* **20**:565–571.

D'Aquino JA, Freire E, Amzel LM (2000): Binding of small organic molecules to macromolecular targets: Evaluation of conformational entropy changes. *Proteins Struct. Funct. Genet.* 93–107.

Davidson N (1962): *Statistical Mechanics*, McGraw-Hill, New York.

Doty P, Myers GE (1953): II. Low molecular weight proteins: Thermodynamics of the association of insulin molecules. *Discuss. Faraday Soc.* **13**:51–58.

Dunbar RC (1984): Photodissociation in the ICR ion trap. In Bowers MT (ed), *Gas Phase Ion Chemistry*, Vol. 3, Academic Press, New York, pp. 129–166.

Dunbar RC (1991): Kinetics of low-intensity infrared laser photodissociation. The thermal model and applications of the Tolman theorem. *J. Chem. Phys.* **95**:2537–2548.

Dunbar RC (1992): Infrared radiative cooling of gas-phase ions. *Mass Spectrom. Rev.* **11**: 309–339.

Dunbar RC (1994): Kinetics of thermal unimolecular dissociation by ambient infrared radiation. *J. Phys. Chem.* **98**:8705–8712.

Dunbar RC (1996): New approaches to ion thermochemistry via dissociation and association. In Babcock LM, Adams NG (eds), *Advances in Gas Phase Ion Chemistry*, Vol. 2, JAI Press, Greenwich, CT.

Dunbar RC (2000): Photodissociation of trapped ions. *Int. J. Mass Spectrom.* **200**:571–589.

Dunbar RC (2003): Blackody infrared radiative dissociation: BIRD. In Armentrout PB (ed), *Encyclopedia of Mass Spectrometry*, Vol. I, Elsevier, Oxford, pp. 371–381.

Dunbar RC (2004): BIRD (blackbody infrared radiative dissociation: Evolution, principles, and applications. *Mass Spectrom. Rev.* **23**:127–158.

Dunbar RC, Lifshitz C (1991): Slow time-resolved photodissociation of p-iodotoluene ion. *J. Chem. Phys.* **94**:3542–3547.

Dunbar RC, McMahon TB (1997): Activation of unimolecular reactions by ambient blackbody radiation. *Science* **279**:194–197.

El Aribi H, Rodriquez CF, Almeida DRP, Ling Y, Mak WWN, Hopkinson AC, Siu KWM (2003): Elucidation of fragmentation mechanisms of protonated peptide ions and their products: A case study on glycylglycylglycine using density functional theory and threshold collision-induced dissociation. *J. Am. Chem. Soc.* **125**:9229–9236.

Ervin KM (2001): Experimental techniques in gas-phase ion thermochemistry. *Chem. Rev.* **101**:391–444.

Exner O (2000): Entropy-enthalpy compensation and anticompensation: Solvation and ligand binding. *Chem. Commun.* 1655–1656.

Eyring H (1935): The activated complex in chemical reactions. *J. Chem. Phys.* **3**:107–115.

Felitsyn N, Kitova EN, Klassen JS (2001): Thermal decomposition of a gaseous multiprotein complex studied by blackbody infrared radiative dissociation. Investigating the origin of the asymmetric dissociation behavior. *Anal. Chem.* **73**:4647–4661.

Fenn JB, Mann M, Meng CK, Wong SF, Whitehouse CM (1989): Electrospray ionization for mass-spectrometry of large biomolecules. *Science* **246**:64–71.

Finkelstein AV, Janin J. (1989): The price of lost freedom—entropy of bimolecular complex-formation. *Protein Eng.* **3**:1–3.

Forst W (2003): *Unimolecular Reactions*, Cambridge Univ. Press, Cambridge, UK.

Gilbert RG, Smith SC (1990): *Theory of Unimolecular and Recombination Reactions*, Blackwell Scientific Publications, Oxford, UK.

Goeringer DE, McLuckey SA (1996): Evolution of ion internal energy during collisional excitation in the Paul ion trap: A stochastic approach. *J. Chem. Phys.* **104**:2214–2221.

Grabowy JAD, Mayer PM (2004): Entropy changes in the dissociation of proton-bound complexes: A variational transition state study. *J. Phys. Chem. A* **108**:9726–9732.

Griffin LL, McAdoo DJ (1993): The effect of ion size on rate of dissociation—RRKM calculations on model large polypeptide ions. *J. Am. Soc. Mass Spectrom.* **4**:11–15.

Gross DS, Zhao YX, Williams ER (1997): Dissociation of heme-globin complexes by blackbody infrared radiative dissociation: Molecular specificity in the gas phase? *J. Am. Soc. Mass Spectrom.* **8**:519–524.

Grunwald E, Steel C (1995): Solvent reorganization and thermodynamic enthalpy-entropy compensation. *J. Am. Chem. Soc.* **117**:5687–5692.

Hillenkamp F, Karas M, Beavis RC, Chait BT (1991): Matrix-assisted laser desorption ionization mass-spectrometry of biopolymers. *Anal. Chem.* **63**: A1193–A1202

Holbrook KA, Pilling MJ, Robertson SH (1996): *Unimolecular Reactions*, Wiley, New York.

Houk KN, Leach AG, Kim SP, Zhang XY (2003): Binding affinities of host-guest, protein-ligand, and protein-transition-state complexes. *Angew. Chem. Int. Ed.* **42**:4872–4897.

Hu YJ, Hadas B, Davidovitz M, Balta B, Lifshitz C (2003): Does IVR take place prior to peptide ion dissociation? *J. Phys. Chem. A* **107**:6507–6514.

Huang FS, Dunbar RC (1990): Time-resolved photodissociation of methylnaphthalene ion—an illustration of kinetic shifts in large-ion dissociations. *J. Am. Chem. Soc.* **112**: 8167–8169.

Janin J (1997): The kinetics of protein-protein recognition. *Proteins* **28**:153–161.

Jockusch RA, Schnier PD, Price WD, Strittmatter EF, Demirev PA, Williams ER (1997): Effects of charge state on fragmentation pathways, dynamics, and activation energies of ubiquitin ions measured by blackbody infrared radiative dissociation. *Anal. Chem.* **69**:1119–1126.

Kitova EN, Bundle DR, Klassen JS (2002a): Thermal dissociation of protein-oligosaccharide complexes in the gas phase: Mapping the intrinsic intermolecular interactions. *J. Am. Chem. Soc.* **124**:5902–5913.

Kitova EN, Bundle DR, Klassen JS (2002b): Evidence for the preservation of specific intermolecular interactions in gaseous protein-oligosaccharide complexes. *J. Am. Chem. Soc.* **124**:9340–9341.

Klots CE (1989): The Arrhenius equation—a centennial revisit. *J. Chim. Phys. PCB* **86**: 2119–2124.

Krug RR, Hunter WG, Grieger RA (1976a): Statistical interpretation of enthalpy-entropy compensation. *Nature* **261**:566–567.

Krug RR, Hunter WG, Grieger RA (1976b): Enthalpy-entropy compensation. 1. Some fundamental statistical problems associated with analysis of Vant Hoff and Arrhenius data. *J. Phys. Chem.* **80**:2335–2341.

Laidler KJ (1984): The development of the Arrhenius equation. *J. Chem. Ed.* **61**:494–498.

Laidler KJ, King MC (1983): The development of transition-state theory. *J. Phys. Chem.* **87**:2657–2664.

Laskin J (2004): Energetics and dynamics of peptide fragmentation from multiple-collision activation and surface-induced dissociation studies. *Eur. J. Mass Spectrom.* **10**:259–267.

Laskin J, Bailey TH, Futrell JH (2003): Shattering of peptide ions on self-assembled monolayer surfaces. *J. Am. Chem. Soc.* **125**:1625–1632.

Laskin J, Bailey TH, Futrell JH (2004): Fragmentation energetics for angiotensin II and its analogs from time-and energy-resolved surface-induced dissociation studies. *Int. J. Mass Spectrom.* **234**:89–99.

Laskin J, Byrd M, Futrell J (2000a): Internal energy distributions resulting from sustained off-resonance excitation in FTMS: I. Fragmentation of the bromobenzene radical cation. *Int. J. Mass Spectrom.* **196**:285–302.

Laskin J, Denisov E, Futrell JH (2000b): A comparative study of collision-induced and surface-induced dissociation. 1. Fragmentation of protonated dialanine. *J. Am. Chem. Soc.* **122**:9703–9714.

Laskin J, Denisov E, Futrell JH (2001): Comparative study of collision-induced and surface-induced dissociation. 2. Fragmentation of small alanine-containing peptides in FT-ICR MS. *J. Phys. Chem. B.* **105**:1895–1900.

Laskin J, Denisov E, Futrell JH (2002a): Fragmentation energetics of small peptides from multiple-collision activation and surface-induced dissociation in FT-ICR MS. *Int. J. Mass Spectrom.* **219**:189–201.

Laskin J, Denisov EV, Shukla AK, Barlow SE, Futrell JH (2002b): Surface-induced dissociation in a Fourier transform ion cyclotron resonance mass spectrometer: New instrument design and evaluation. *Anal. Chem.* **74**:3255–3261.

Laskin J, Bailey TH, Denisov EV, Futrell JH (2002c): On the relative stability of singly protonated des-Arg1 and des-Arg9 bradykinins. *J. Phys. Chem. A* **106**:9832–9836.

Laskin J, Futrell J (2000): Internal energy distributions resulting from sustained off-resonance excitation in FT-ICR MS: II. Fragmentation of the 1-bromonaphthalene radical cation. *J. Phys. Chem. A* **104**:5484–5494.

Laskin J, Futrell JH (2003): Entropy is the major driving force for fragmentation of proteins and protein-ligand complexes in the gas-phase. *J. Phys. Chem. A* **107**:5836–5839.

Lifshitz C (1982): Time-resolved appearance energies, breakdown graphs, and mass spectra: the elusive "kinetic shift." *Mass Spectrom. Rev.* **1**:309–348.

Lifshitz C (1991): Dissociative photoionization in the vacuum UV region with ion trapping. *Int. J. Mass Spectrom. Ion Proc.* **106**:159–173.

Lifshitz C (1992): Recent developments in applications of RRKM-QET. *Int. J. Mass Spectrom. Ion Proc.* **118**:315–337.

Lifshitz C (1997): Energetics and dynamics through time-resolved measurements in mass spectrometry: Aromatic hydrocarbons, polycyclic aromatic hydrocarbons and fullerenes. *Int. Rev. Phys. Chem.* **16**:113–139.

Lifshitz C (2001): Some recent aspects of unimolecular gas phase ion chemistry. *Chem. Soc. Rev.* **30**:186–192.

Lifshitz C (2002): Kinetic shifts. *Eur. J. Mass Spectrom.* **8**:85–98.

Lin C-Y, Dunbar RC (1996): Zero-pressure thermal-radiation induced dissociation of tetraethylsilane cation. *J. Phys. Chem.* **100**:655–659.

Liu L, Guo QX (2001): Isokinetic relationship, isoequilibrium relationship, and enthalpy-entropy compensation. *Chem. Rev.* **101**:673–695.

Lorquet JC (1994): Whither the statistical theory of mass spectra? *Mass Spectrom. Rev.* **13**:233–257.

Lorquet JC (2000): Landmarks in the theory of mass spectra. *Int. J. Mass Spectrom.* **200**: 43–56.

Mabud MDA, Dekrey MJ, Cooks RG (1985): Surface-induced dissociation of molecular-ions. *Int. J. Mass Spectrom. Ion Proc.* **67**:285–294.

Marzluff EM, Beauchamp JL (1996): Collisional activation studies of large molecules. In Baer T (ed), *Large Ions: Their Vaporization, Detection and Structural Analysis*, Wiley, New York, pp. 115–143.

Marzluff EM, Campbell S, Rodgers MT, Beauchamp JL (1994): Collisional activation of large molecules is an efficient process. *J. Am. Chem. Soc.* **116**:6847–6948.

Meroueh O, Hase WL (1999): Collisional activation of small peptides. *J. Phys. Chem. A* **103**:3981–3990.

Meroueh O, Hase WL (2000): Energy transfer pathways in the collisional activation of peptides. *Int. J. Mass Spectrom.* **201**:233–244.

Muntean F, Armentrout PB (2002): Modeling kinetic shifts for tight transition states in threshold collision–induced dissociation. Case study: Phenol Cation. *J. Phys. Chem B* **106**:8117–8124.

Neumann L, Abele R, Tampe R (2002): Thermodynamics of peptide binding to the transporter associated with antigen processing (TAP). *J. Mol. Biol.* **324**:965–973.

Ng CY (2002): Vacuum ultraviolet spectroscopy and chemistry by photoionization and photoelectron methods. *Annu. Rev. Phys. Chem.* **53**:101–140.

Price WD, Schnier PD, Williams ER (1996a): Tandem mass spectrometry of large biomolecule ions by blackbody infrared radiative dissociation. *Anal. Chem.* **68**:859–866.

Price WD, Schnier PD, Jockusch RA, Strittmatter EF, Williams ER (1996b): Unimolecular reaction kinetics in the high-pressure limit without collisions. *J. Am. Chem. Soc.* **118**: 10640–10644.

Price WD, Schnier PD, Williams ER (1997): Binding energies of the proton-bound amino acid dimers Gly.Gly, Ala.Ala, Gly.Ala, and Lys.Lys measured by blackbody infrared radiative dissociation. *J. Phys. Chem. B* **101**:664–673.

Price WD, Williams ER (1997): Activation of peptide ions by blackbody radiation: Factors that lead to dissociation kinetics in the rapid energy exchange limit. *J. Phys. Chem.* **101**:8844–8852.

Qin J, Chait BT (1995): Preferential fragmentation of protonated gas-phase peptide ions adjacent to acidic amino-acid-residues. *J. Am. Chem. Soc.* **117**:5411–5412.

Rodgers MT, Armentrout PB (2000): Noncovalent metal-ligand bond energies as studied by threshold collision-induced dissociation. *Mass Spectrom. Rev.* **19**:215–247.

Rodgers MT, Armentrout PB (2004): A thermodynamic "vocabulary" for metal ion interactions in biological systems. *Acc. Chem. Res.* **37**:989–998.

Rodgers MT, Ervin KM, Armentrout PB (1997): Statistical modeling of collision-induced dissociation thresholds. *J. Chem. Phys.* **106**:4499–4508.

Rosenstock HM, Stockbauer R, Parr AC (1979): Kinetic shift in chlorobenzene ion fragmentation and the heat of formation of the phenyl ion. *J. Chem. Phys.* **71**:3708–3714.

Rosenstock HM, Stockbauer R, Parr AC (1980): Photoelectron-photoion coincidence study of the bromobenzene ion. *J. Chem. Phys.* **73**:773–777.

Rosenstock HM, Wallenstein MB, Wahrhaftig AL, Eyring H (1952): Absolute rate theory for isolated systems and the mass spectra of polyatomic molecules. *Proc. Natl. Acad. Sci. USA* **38**:667–678.

Schlag EW, Grotemeyer J, Levine RD (1992): Do large molecules ionize? *Chem. Phys. Lett.* **190**:521–527.

Schlag EW, Levine RD (1989): On the unimolecular dissociation of large molecules. *Chem. Phys. Lett.* **163**:523–530.

Schnier PD, Price WD, Jockusch RA, Williams ER (1996): Blackbody infrared radiative dissociation of bradykinin and its analogues: Energetics, dynamics, and evidence for salt-bridge structures in the gas phase. *J. Am. Chem. Soc.* **118**:7178–7189.

Schnier PD, Price WD, Strittmatter EF, Williams ER (1997): Dissociation energetics and mechanisms of leucine enkephalin $(M + H)^+$ and $(2M + X)^+$ ions (X = H, Li, Na, K, and Rb) measured by blackbody infrared radiative dissociation. *J. Am. Soc. Mass Spectrom.* **8**:771–780.

Schnier PD, Klassen JS, Strittmatter EE, Williams ER (1998): Activation energies for dissociation of double strand oligonucleotide anions: Evidence for Watson-Crick base pairing in vacuo. *J. Am. Chem. Soc.* **120**:9605–9613.

Searle MS, Williams DH (1992): The cost of conformational order—entropy changes in molecular associations. *J. Am. Chem. Soc.* **114**:10690–10697.

Shaffer SA, Tang KQ, Anderson GA, Prior DC, Udseth HR, Smith RD (1997): A novel ion funnel for focusing ions at elevated pressure using electrospray ionization mass spectrometry. *Rapid Commun. Mass Spectrom.* **11**:1813–1817.

Sharp K (2001): Entropy-enthalpy compensation: Fact or artifact? *Protein Sci.* **10**:661–667.

Shoeib T, El Aribi H, Siu KWM, Hopkinson AC (2001): A study of silver (I) ion-organonitrile complexes: Ion structures, binding energies, and substituent effects. *J. Phys. Chem. A* **105**:710–719.

Siebert X, Amzel LM (2004): Loss of translational entropy in molecular associations. *Proteins* **54**:104–115.

Summerfield SG, Whiting A, Gaskell SJ (1997): Intra-ionic interactions in electrosprayed peptide ions. *Int. J. Mass Spectrom. Ion. Proc.* **162**:149–161.

Tamura A, Privalov PL (1997): The entropy cost of protein association. *J. Mol. Biol.* **273**:1048–1060.

Tholmann D, Tonner DS, McMahon TB (1994): Spontaneous unimolecular dissociation of small cluster ions, $(H_3O^+)L_N$ and $Cl^-(H_2O)_N$ ($N = 2$–4), under Fourier-transform ion-cyclotron resonance conditions. *J. Phys. Chem.* **98**:2002–2004.

Tidor B, Karplus M (1994): The contribution of vibrational entropy to molecular association—the dimerization of insulin. *J. Mol. Biol.* **238**:405–414.

Tolman RC (1920): Statistical mechanics applied to chemical kinetics. *J. Am. Chem. Soc.* **42**:2506–2528.

Tonner DS, Tholmann D, McMahon TB (1995): Deuterium-isotope effect on the radiatively induced unimolecular dissociation of small cluster ions. *Chem. Phys. Lett.* **233**:324–330.

Truhlar DG, Garrett BC, Klippenstein SJ (1996): Current status of transition-state theory. *J. Phys. Chem.* **100**:12771–12800.

Truhlar DG, Hase WL, Hynes JT (1983): Current status of transition-state theory. *J. Phys. Chem.* **87**:2664–2682.

Tsaprailis G, Nair H, Somogyi A, Wysocki VH, Zhong WQ, Futrell JH, Summerfield SG, Gaskell SJ (1999): Influence of secondary structure on the fragmentation of protonated peptides. *J. Am. Chem. Soc.* **121**:5142–5154.

Uechi GT, Dunbar RC (1992): The kinetics of infrared-laser photodissociation of normal-butylbenzene ions at low-pressure. *J. Chem. Phys.* **96**:8897–8905.

Vestal ML (1965): Theoretical studies on the unimolecular reactions of polyatomic molecule ions. I. Propane. *J. Chem. Phys.* **43**:1356–1369.

Wang WJ, Kitova EN, Klassen JS (2003): Bioactive recognition sites may not be energetically preferred in protein-carbohydrate complexes in the gas phase. *J. Am. Chem. Soc.* **125**:13630–13631.

Yu W, Vath JE, Huberty MC, Martin SA (1993): Identification of the facile gas-phase cleavage of the Asp Pro and Asp Xxx peptide-bonds in matrix-assisted laser-desorption time-of-flight mass-spectrometry. *Anal. Chem.* **65**:3015–3023.

Yu YB, Privalov PL, Hodges RS (2001): Contribution of translational and rotational motions to molecular association in aqueous solution. *Biophys. J.* **81**:1632–1642.

INDEX

Principles of Mass Spectrometry Applied to Biomolecules, edited by Julia Laskin and Chava Lifshitz

Printed and bound by CPI Group (UK) Ltd, Croydon, CR0 4YY
18/09/2026
14964975-0002